Microstructure and Properties of High-Temperature Superconductors

Ivan A. Parinov

Microstructure and Properties of High-Temperature Superconductors

With 309 Figures and 44 Tables

Ivan A. Parinov
South Federal University
Vorovich Mechanics & Applied
Mathematics Research Institute
Stachki Avenue 200/1
Rostov-on-Don Russia 344090
ppr@math.rsu.ru

Library of Congress Control Number: 2007923179

ISBN 978-3-540-70976-3 Springer Berlin Heidelberg New York

Springer is a part of Springer Science+Business Media
springer.com

Typesetting: by the author and Integra, India using a Springer LaTeX macro package
Cover design: eStudio Clamar S.L., F. Steinen-Broo, Pau/Girona, Spain

Printed on acid-free paper SPIN: 11802174 5 4 3 2 1 0

To My Wife, NINA

Preface to the English Edition

In 2006, the scientific society is celebrating the twentieth anniversary of the discovery of high-temperature superconductivity by George Bednorz and Alex Müller. Dynamically developing researches in this field give new scientific results. This caused a significant modernization of the English edition compared to the Russian one [808], which has been written in 2003. Considerable changes have been introduced in Chaps. 1–3 and Appendix A, in particular, new Sect. 3.1.2 is devoted to acoustic emission study of BSCCO/Ag tapes under bending. New Chapter 4 is devoted to carbon problem in HTSC and includes "old" text from Sects. 2.6 and 7.7 of the Russian edition, and "new" text (Sections 4.3 and Appendix B), presenting mathematical modeling of the brittle carbonate formation and following fracture during interaction of YBCO with CO_2. The main aims of the monograph have been retained and connected with Material Science of HTSC and their mathematical modeling. Comparatively, less attention has been devoted to Physics of HTSC. The main results as before have been related to the YBCO and BSCCO families, while the main trends in R&D of other superconductors have also been marked. The author would be grateful for reports of typographical and other errors to be sent via the following web page http://www.math.rsu.ru/niimpm/strl/welcome.en.html, where an up-to-date errata list will be maintained.

I would like to thank all those who have contributed to the preparation of this manuscript at final stage, especially Jacqueline Lenz and Dhivya Balarajan.

Ivan Parinov
Rostov-on-Don, December 2006

Preface to the Russian Edition

The discovery in 1986 of high-temperature superconductors (HTSC) on the basis of copper oxides with the temperature of superconducting transition, that is greater than the temperature of low-cost, non-toxic and accessible liquid nitrogen (77 K), marked qualitative jump in the development and application of new technical conductors, devices for energy transmission, transformation and storage. Together with enough high critical temperatures T_c, an intrinsic brittleness of oxide cuprates, the layered anisotropic structure and the super-short ($\sim$ 1 nm) coherence length, ξ, presenting itself a spatial characteristic of superconducting electrons, are other main features defining HTSC microstructure and properties. Due to the above-mentioned peculiarities, even the existence of an intergranular boundary could be enough to suppress superconductivity, but the structure-sensitive properties of HTSC systems depend very much on the weak links of intergranular boundaries by manufacturing them in the polycrystalline form, demonstrating coexistence of inter- and transgranular currents. Also, superconductivity can be destroyed at the attainment of the critical value of the external magnetic field H_{cm}. The interfaces of the "superconductor–normal metal", "superconductor–insulator" and other types based on them are the localization places of different defects. The microstructure features, connected with phase composition, domain structure, crystallographic properties, an existence of structure defects, pores, microcracks, inclusions, etc., define directly useful properties of HTSC materials and composites. The main goal of the present monograph is to study microstructure, strength, electromagnetic and superconducting properties. Another aim includes discussion of the optimization directions for the fabrication techniques, superconducting compositions, external loading and thermal treatments to obtain HTSC, possessing improved and more controlled physical and mechanical properties. The link "composition–technique–experiment–theory-model" investigated in the book, assuming considerable HTSC defectiveness and structure heterogeneity, forms the whole picture of modern representations on the microstructure, strength and connected with them the structure-sensitive properties of the materials considered. Special attention in the book is devoted to Bi–Sr–Ca–Cu–O and Y–Ba–Cu–O families that today are most prospective for applications among HTSC.

The monograph is addressed to students, postgraduate students and specialists taking part in the development, preparation and research of new materials. The author thanks the Russian Foundation for Basic Research, Russian Department of Education and Science, Soros Foundation and the American International Program COBASE (Collaboration in Basic Science and Engineering) grants which during the last decade have rendered considerable financial support and promoted to publish this book.

I am also grateful to colleagues and close scientific workers, who have directly or indirectly contributed to the book. In particular, I wish to thank V.P. Zatsarinny, D.N. Karpinsky, E.A. Dul'kin, E.M. Kaydashev, E.V. Rozhkov, A.A. Polyanskii and D.C. Larbalestier.

Corrections and proposals the book will be considered with thanks. They could be to presented by E-mail: ppr@math.rsu.ru.

I. A. Parinov
September 2003

Contents

1

Superconductors and Superconductivity: General Issues

1.1 Superconductivity Discovery

The discovery of superconductivity, that is, the phenomenon by which current flow in material occurs without noticeable energy dissipation, has been recognized as one of the greatest scientific achievements of twentieth century. This phenomenon is accompanied by a sudden drop of the electrical resistance to zero[1] by cooling below the *critical temperature* (T_c), which is the temperature of superconducting transition that is defined for every specific material.

In 1911, Heike Kamerlingh-Onnes when researching the properties of some metals in the vicinity of liquid helium (4.2 K) had found that mercury cooled to ~ 4.25 K losing its electrical resistance, that is, transformed into superconductor [782]. In the next years, the superconductivity of some other metals, several alloys and intermetallic compounds had also been discovered (Fig. 1.1)[2]; however, they demonstrated very low critical temperatures (maximal value $T_c = 23.2$ K for Nb_3Ge), some increasing the liquid helium temperature. This circumstance impeded the practical applications of superconductors in tremendous degree due to the high cost of liquid helium ($\sim$ \$25 per liter) and difficulties in its preparation. The long absence of noticeable successes in the increasing of critical temperature (last record was achieved in 1973 for compound Nb_3Ge) raised the highly restrained moods of scientists, who worked in this field in the middle of 1980. So, experimentalists discussed the issue of the perspectives of the T_c increasing for Nb_3Sn upto a value of 30 K with the application of very exotic techniques [193], and theorists predicted a 40 K T_c ceiling [416].

The situation changed dramatically in 1986 with the discovery of the so-called high-temperature superconductivity (HTSC) in non-traditional

---

[1] On the basis of the sensitivity of modern equipment, it may be argued that the resistivity of superconductors is no more 8×10^{-25} Ω cm. For comparison, we note that the resistivity of high-purity copper is of the order 10^{-9} Ω cm at 4.2 K.

[2] The total classification of known superconductors is presented in Appendix A.

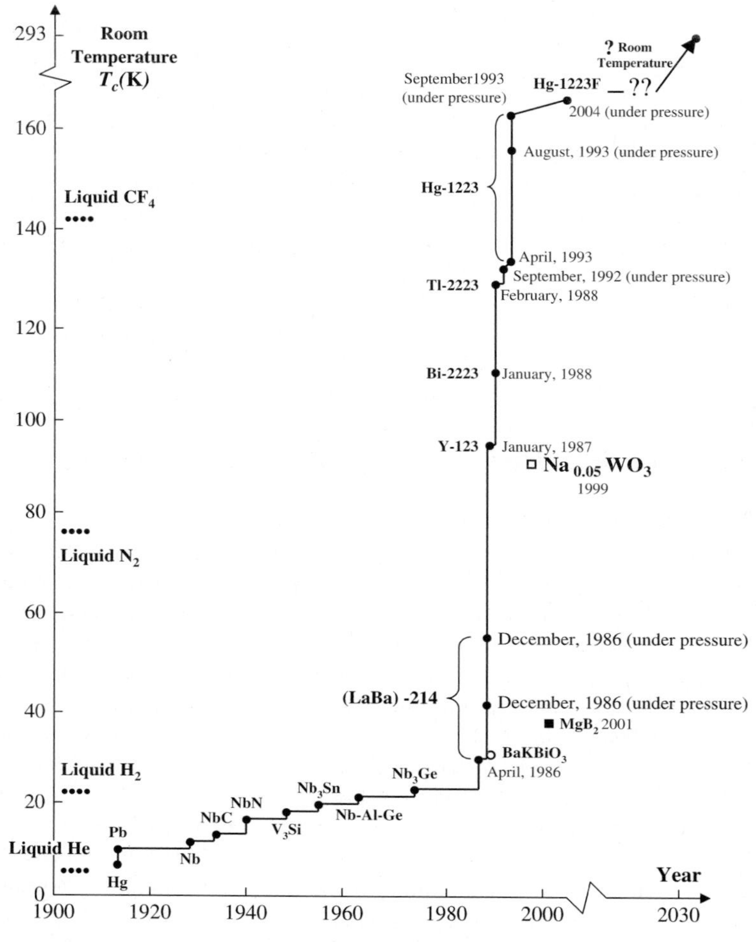

Fig. 1.1. History of superconductors' discovery and of T_c-increasing

compounds, namely cuprates. In spite of the long time and tremendous forces of world scientific society, the mechanism of superconductivity for these materials is not yet stated. The distinctive feature of HTSC is that all these compounds have atomic CuO_2 plane, playing key role as the origin of superconductivity. The first from HTSC was $La_{2-x}Ba_xCuO_4$ with $T_c = 30$ K (George Bednorz and Alex Müller, April 1986) [58]. The main peak of discoveries took place between December 1986 and March 1987, when the existence of the critical temperature $T_c > 30$ K was confirmed by Chu's group and Kitazawa's group, independently [152], and then the superconductivity in $YBa_2Cu_3O_7$ at 93 K, attained jointly by Chu's group and Wu's group [1150], was published. These achievements opened a new epoch in the superconductivity investigations as they brought down the liquid nitrogen temperature barrier of 77 K. Low cost ($\sim$ \$0.5 per liter), simple conditions for its preparation and utilization led to a considerable progress, in the next years, in the development, manufacture and initial application of high-temperature superconductors. The discovery of HTSC initiated real ecstasy in scientific society (the number of researchers in this field had increased more than order in 1987), and also the tremendous interest of press and public. In one year, the critical temperature increased on 70 K, whereas, for the previous 75 years of superconductivity researches the growth of T_c was 20 K only! Scientists, as simple people, called 1987 "the year of progress in physics." Public interest caused a sharp increase in financing of HTSC studies in the world. In the following years, other compounds were discovered (see Fig. 1.1 and Appendix A) with critical temperature also above the liquid nitrogen temperature, namely $Bi_2Sr_2Ca_2Cu_3O_{10+x}$ (T_c = 110 K) [653], $Tl_2Ba_2Ca_2Cu_3O_{10}$ (T_c = 125 K) [972] and $HgBa_2Ca_2Cu_3O_8$ (T_c = 134 K) [980]. The record critical temperature, 164 K, was achieved at high pressure (30 GPa) in $HgBa_2Ca_2Cu_3O_{8+x}$ family in September 1993 [153].

In 1988, Sheng et al. presented results of measurements of the electrical resistance in $Tl_2Ba_2Ca_{x-1}Cu_xO_{2x+4}$ samples [971, 972]. The results and data obtained by Hazen et al. [400] demonstrated that at increasing number x the critical temperature $T_c(x)$ grows by the next way: $T_c(1) = 90$ K, $T_c(2) = 110$ K, $T_c(3) = 125$ K. By the linear dependence on x, the temperature $T_c = 300$ K is attained at $x = 10$. However, the values of $T_c > 125$ K for thallium oxides were not defined. Analogous predictions for the behavior of T_c on increasing x had been fulfilled for $Bi_2Sr_2Ca_{x-1}Cu_xO_{2x+4}$ family [438]. Regretfully, the dependence also destroys at $x > 3$.

Thus, numerous studies of HTSC have stated that maximal critical temperature, T_c, is reached in compounds with three CuO_2 layers per elementary cell. Moreover, it is necessary to carry out two conditions: (i) to reach the small distance, $d_{Cu-O} = a/2$, between atoms of copper and oxygen (where a is the period of two-dimensional lattice in the CuO_2 plane) and (ii) the hole concentration in CuO_2 layers should be near the optimal value of $p = 0.16$ (in account per copper atom). All these conditions have been realized in mercury HTSC Hg-1223 with fluorine additives (Hg-1223F) in 2004. The maximal $T_c = 138$ K (at $P = 0$) has been attained in samples with $a = 0.38496$ nm, and a record $T_c = (166 \pm 1.5)$ K at $P = 23$ GPa [721]. In mercurial HTSC, a linear dependence of T_c on the lattice constant, a, is observed, namely: T_c increases with decreasing of a. The critical temperature $T_c \sim 100$ K in Hg-1201 at $a = 0.388$ nm, and $T_c = 138$ K in Hg-1223F at $a = 0.38496$ nm. Then, in order to reach $T_c = T_{room} = 293$ K, it is necessary that $a \approx 0.374$. However, a decreasing of T_c is observed at increasing number of the CuO_2 layers, $x > 3$, that obviously occurs due to a *buckling* of the CuO_2 layers at diminishing a. All the above-mentioned cuprates are hole-doped. In 1989, single cuprate family was discovered to be electron-doped: (Nd,Pr,Sm)CeCuO with critical temperature $T_c = 24$ K.

Periodically appeared information on the sightings of superconductivity at temperatures even above 300 K today are doubtful because they do not satisfy the four criteria to determine the superconductivity existence, stated in 1987 by the researchers who discovered the HTSC [152]: (i) zero resistivity, (ii) Meissner's effect marked (when on the decrease of temperature and magnetic field below the critical values, the total displacement of magnetic flux from conductor, transforming into superconductor, is observed), (iii) high reproducibility of results and (iv) high stability of effect. Today, the mistakes of the superconductivity effect statement at temperatures above 130 K are linked usually with two general problems, namely:

(1) with experimental mistakes of non-practiced experimentalists having insufficient knowledge of the modern measuring methods or with *what* should be demonstrated to prove the superconductivity of material (the last cause is met very seldom by development of this field);
(2) with discovery of "new" superconductors really being known superconducting chemical compounds involving some additional components.

Note that the last problem is very difficult even for experienced researchers.

The main cause that decreased the interest of scientific society considerably to HTSC in the middle of 1990 and decreased considerably financial supporting (in particular, in USA), directed to these aims, concluded that the high-temperature superconductors could not replace low-temperature superconductors and attain sufficiently wide applications. The main obstacle is the material brittleness, which is intrinsic for HTSC compositions. It forms microstructure defects (voids, microcracks, damage, etc.) during sample preparation and loading that can be rapidly transformed to macrodefects and degrade the superconductivity properties. Moreover, the critical current density, J_c, was found to be the key parameter for engineering application than the value of T_c. Its value is the limit magnitude of direct undamped electric current in superconducting sample, above which the sample transforms to normal (i.e., non-superconducting) state. In particular, the $(Bi, Pb)_2Sr_2Ca_2Cu_3O_{10+x}/Ag$ tapes possess critical current density up to 80 kA/cm^2 (at 77 K and 0 T) that changes in dependence on the conductor length and shape [287]. At a temperature of 4 K and in the absence of the magnetic field, their value of J_c increases three times. The volume critical current density, $j_c = (2-5) \times 10^6$ A/cm^2 in $Bi_2Sr_2CaCu_2O_{8+x}$ monocrystal, found on the basis of the measurement of the critical current in Cu_2O_4 surface layer (i.e., pair of parallel CuO_2 layers) at $T = 4.5$ K [1052], evidently, is also the limit for Bi-2212 thin films and tapes.

Melt-processed Y(*RE*)BCO bulks (where *RE* is the rare-earth elements), the most prospective for applications, demonstrate values of $J_c > 100$ kA/cm^2 (at 77 K and 0 T) that decrease rapidly with increase of temperature and magnetic field [908]. For comparison, the low-temperature superconductors of Nb–Ti family possess $J_c = 300$ kA/cm^2 (at 4 K and 5 T), but Nb_3Sn samples demonstrate $J_c = 100$–200 kA/cm^2 (at 4 K and 10 T) [598].

Remarkable discoveries were also made in another (no cuprate) superconducting systems. In particular, in 2001 in broadly accessible and very cheap MgB_2 system, $T_c = 39\,K$ [756] was obtained. These achievements again increased sharply an interest to superconductivity in the world.

1.2 Progress and Prognosis of Superconductivity Applications

At the beginning of 1990, the epoch of engineering applications of HTSC (Fig. 1.2) was started marked by considerable technical achievements that stated the use of HTSC in specific products and devices. Growth of HTSC

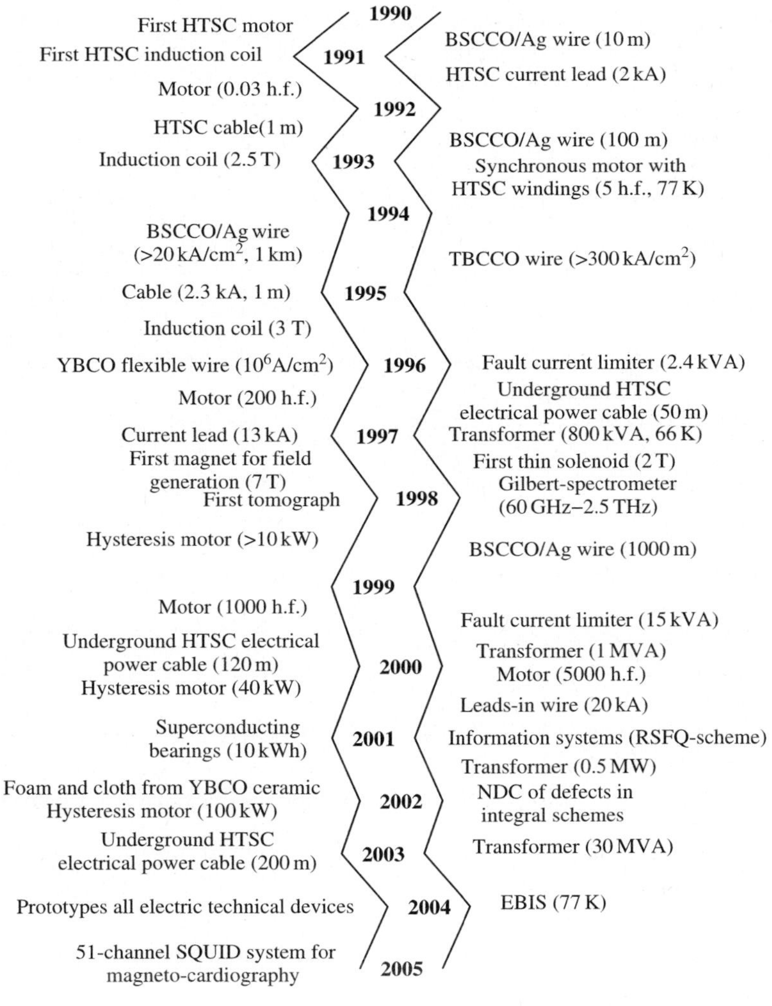

Fig. 1.2. Progress attained in the development and manufacture of HTSC products

applications attained and predicted in perspective to comparison with low-temperature superconductivity (LTSC) is present in Fig. 1.3. The areas of existing and potential applications of superconductivity (both HTSC and LTSC) could be depicted in the form of symbolic "tree" (Fig. 1.4):

I. *Technical directions.* These are present at the "roots" of the tree. However, the set of the disciplines pointed does not cover all possible applications. The experience shows that practically every scientific and technical direction – and also most social ones – can impact upon and be impacted positively by the future of superconductivity. Then we proceed clockwise, considering the branches of the tree.

II. *Electronics.* In this area, thin-film and SQUID-based systems,[3] as well as cellular and satellite communicating gear, are the most close to achieving commercial success. The next development of superconductivity applications in the area of electronic and information systems promises very fruitful competition between semiconductor and superconductor technologies, a competition that involves the state of development of cost-effective refrigeration and packaging of circuits and devices. It may be assumed that existing advantages of

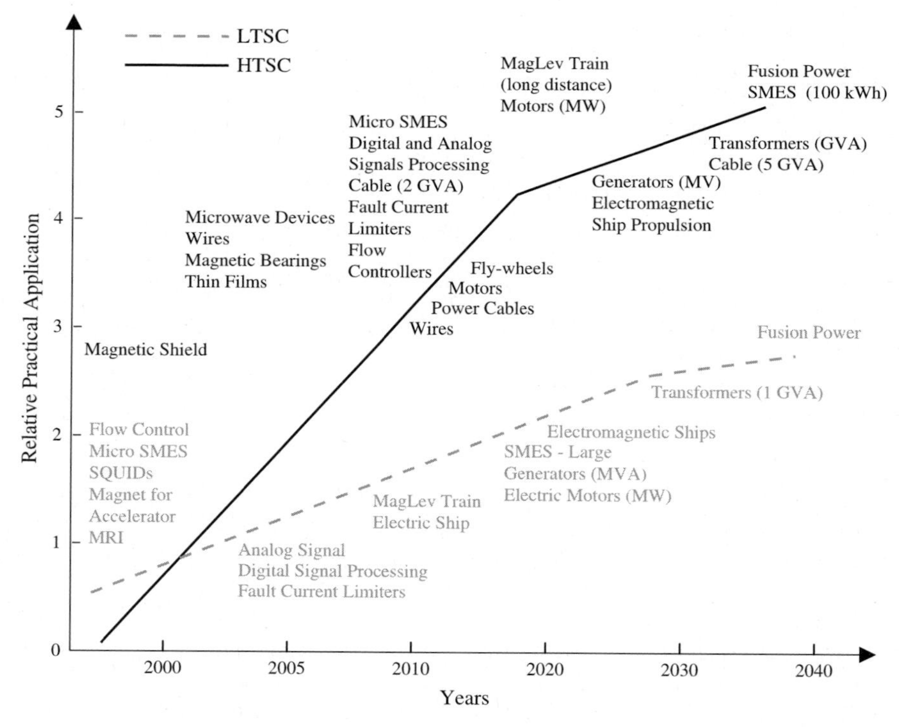

Fig. 1.3. HTSC and LTSC applications reached and predicted [898]

[3] SQUID is an acronym of superconducting quantum interference device.

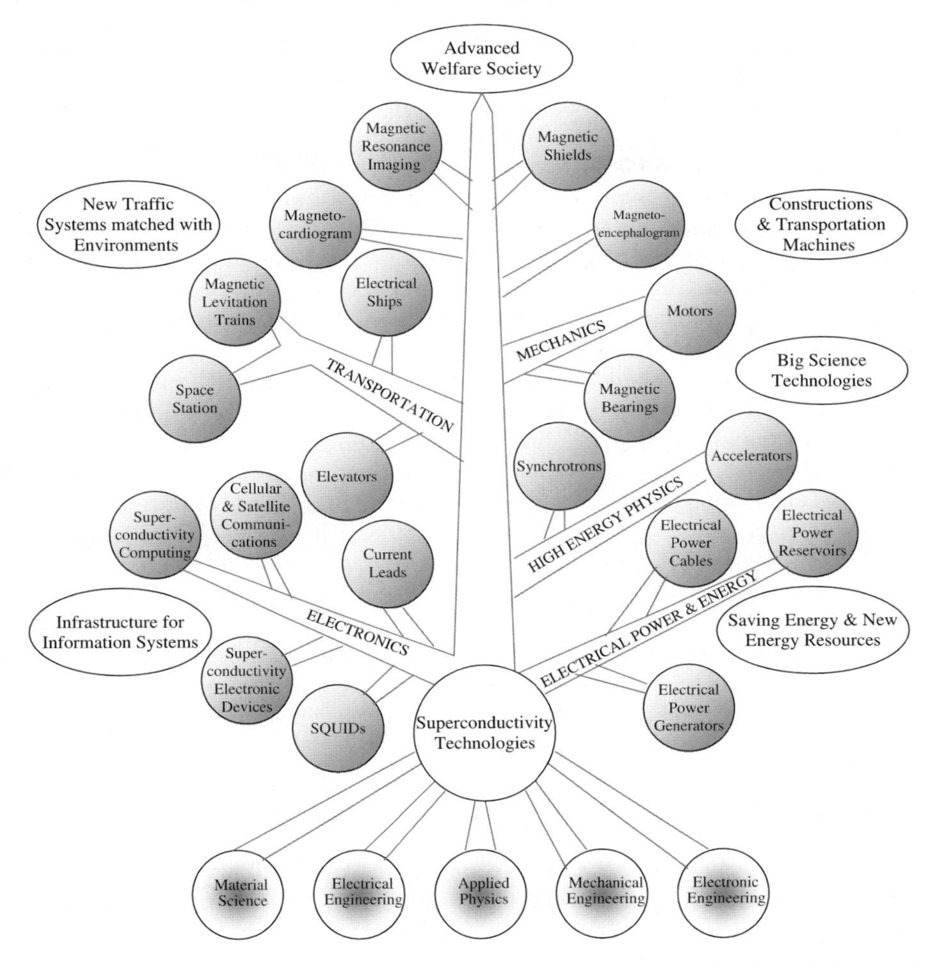

Fig. 1.4. Symbolic "tree" representing wide perspectives of already existing and potential applications of superconductivity [1045]

semiconductors will decrease more and more with solution of problems intrinsic for HTSC and taking into account unique properties of superconducting devices, namely, high rate of switching, small consumed power, weakened noise and high sensitivity to external electromagnetic radiation.

Today, the application of superconducting electronics, in particular, includes passive super-high-frequency (SHF) filters for system of communications, Josephson standards of voltage, hot electron bolometers (HEB). One of the main goals is R&D of superconducting computer. Superconducting Josephson qubits are two-level quantum systems, in which electric charge or magnetic flux is the degree of freedom. Josephson contacts, that is, two superconducting layers with dielectric buffer layer, present their basis. In

1999, the solid realization of the qubit in SQUID loop was proposed [471]. This idea is based on using "exotic" superconductors with symmetry of parameter of the superconducting order, D, which is lower than symmetry of crystalline lattice (e.g., HTSC $YBa_2Cu_3O_7$, possessing d-wave symmetry of D in superconducting CuO_2 layers). In 2000, a new computer architecture, namely, *hybrid technology multi-threaded architecture* (HTMT), was started. The schemes of *rapid single flux quantum* (RSFQ) are the basis of the computer, and it is assumed that the superconducting supercomputer will carry out up to 10^{15} floating-point operations per second. The RSFQ conception is based on using shunted Josephson junctions. In 2003, a formation of the *entangled state* between two solid superconducting qubits [834] has been demonstrated. In 2004, a coherent coupling of Josephson qubit with harmonic modes of transmitting line (for "charge" qubit [1124]) and with plasma oscillations (for "flux" qubit [148]) has been carried out. By disposing Josephson qubits on one chip, it is possible in principle to organize their coupling, by means of electromagnetic waves. In the case of actual HTSC applications, it is necessary to reach 5% current spread per plate with 10^3 Josephson junctions.

The areas of application of HTSC products are divided into passive and magnetic ones based on the state (diamagnetic or ferromagnetic) of the sample in working regime (see Table 1.1). The capability of superconductor pushes off external magnetic field is used in the diamagnetic state (in this case, an existence of origin of the external direct field is assumed). On the other hand, superconductor itself becomes the origin of magnetic field in the ferromagnetic state (state with "freezed" magnetic flux).

III. *Transportation.* The next main directions of superconductivity applications in the aviation and astronautics have been formed, namely:

(1) Systems of electromechanical start.

In the construction of the newest single-pass motor of air- spacecraft, the necessary air inflow is attained at super-sound velocities of apparatus; it only demands initial acceleration of the craft. For this aim, it is possible to use the take-off superconducting platform does not leaving the airport limits, but

Table 1.1. Passive and magnetic applications of HTSC

Passive applications	Magnetic applications
Magnetic bearings	Quasi-continuous magnets
Fly-wheels	Magnetic separation
Cryogenic pumps	MagLevs
Filling systems	Magnetic captures
Hysteresis motors (rotor)	Motors (stator)
Linear carrying capacitors	Magnetic dampers
Inertial transformers	Control of bundles of the charge particles
Electric drives	Magnetic apparatus
Magnetic screens	

racing aircraft up to sound velocity. The electromagnetic systems of start from Earth to its orbit, using HTSC magnet or linear synchronous motor, present themselves the high-speed electromagnetic guns for start to space of the heavy useful loads. Analogous starting systems are applied for replacement of the first step of the multi-step rockets.

(2) Levitation systems.

The passive superconducting magnetic suspenders for contactless suspension of shafts may be used to stabilize satellites, but electromechanical storages of energy to feed HTSC bearings.

(3) Electric motors, working on the principles of hysteresis and "freezed" magnetic flux.

The hysteresis motors developed consisting of cylindrical or disc HTSC rotors are based on the use of two types of bulk superconducting ceramics, namely (i) monocrystalline (YBCO) or bulk (BSCCO) elements and (ii) melt-textured YBCO samples. The constructive elements of motors of the synchronous–asynchronous type are the HTSC bulks, but ac-synchronous and dc-unipolar electric machines use HTSC wires (from BSCCO).

The main directions of superconductivity applications in various transport systems cover the following:

(1) High-speed trains and other kinds of ground transport on magnetic suspension.

The superconducting (SC) electromagnetic systems (using linear electric motors to speed up the truck) are applied in the development of high-speed ground transportation based on the SC windings placed in the route and SC magnets located in the vehicle. By this, in order to economically, accumulate and use energy effectively, the SC storages of electric energy that feed HTSC bearings are applied. Magnetic levitation (MagLev) systems based on the use of magnetic forces to "suspend" one object above another have been developed in Japan, the USA, Germany, France, Switzerland, China and Canada. HTSC systems have been demonstrated in China during realization of Germanic China project directed to create the levitated transportation mean. During the last 25 years, great efforts have been taken in Japan to develop high-speed trains on magnetic suspension by the application of superconducting systems. In particular, for the passenger train on magnetic suspension, caused by superconductivity, a speed of over 1000 km/h [898] has been demonstrated! Today, an analogous train has been developed, called the Japanese Linear Chuo Shinkansen Project (MLX) with nominal speed of 500 km/h. At the same time, trains with superconducting magnetic suspension compete with high-speed trains (e.g., France train TGV and German Transrapid System with speed of 420 km/h). The Inductrack MagLev System is developed in the USA. The Swissmetro Project (speed of 320 km/h) should be noted. The German Transrapid System is based on "classical" electromechanical technologies, requiring small air gap (< 12 mm) between the vehicle and the truck. The Japanese MagLev is based on active superconductivity magnetic systems, permitting large air gap (> 80 mm). The USA MagLev is based on passive superconductive magnets, also permitting large air gap ($>$ 80 mm). Swissmetro presents a unique example of MagLev systems: it is designed to work under partial vacuum ($<$ 10 kPa). The vehicles run in tunnels of small inside diameter (5 m), which require partial vacuum in order to reduce the aerodynamic resistance. Independent magnetic systems permit to have medium air gap of 20 mm. Thus, based on the melted HTSC ceramic, the superconducting magnets are able to capture high magnetic fields (tens of Tesla) compared with the classical magnets, and allow to create ecologically pure high-speed ground transport systems with broadened air gap (the thickness of magnetic suspension). This defines the possibility to design railroads with lesser financial and operation expenses.

(2) Electromobiles and automobiles.

HTSC electric motors for electromobiles and automobiles act on the principles of hysteresis and "freezed" magnetic flow. The superconducting wires, bulks and thick films are used in construction. The next electric devices and machines, using superconductivity, that are development under now are:

- ac-synchronous and dc-unipolar electric motors, using HTSC wires (from BSCCO) and acting at the temperature of liquid hydrogen;
- hysteresis motors with cylindrical and disc YBCO rotors that are developed on the basis of monocrystalline YBCO- or BSCCO-ceramic, and also melt-textured YBCO bulks;
- motors of synchronous–asynchronous type, consisting of HTSC bulks;
- motors with rotor from layered composites, namely YBCO bulks (thick films) with intermediate ferromagnetic inserts (steel plates);
- HTSC levitated systems, realized in electromechanical storages of energy and levitated suspenders that use high-speed electromobiles.

(3) High-speed ships and Navy.

The linear synchronous electric motors are applied in starters of aircraft carriers. The superconducting magneto-hydrodynamic systems are used in Navy with the aim of decreasing the noise detection of motor in torpedoes and in the development of high-effective motors for fast-moving ships. There are researches directed to the use of superconductivity in the ship impulse source of energy, for example, for aircraft carrier. The Navy has intention of creating a warship with unique superconducting equipment and apparatus [356], including: (i) engines, (ii) magnetic mine-sweepings for fishing out of sea mines, (iii) board systems of feeding on the basis of inductive storage and (iv) radio-locator, using superconducting magnet in the protection system of the ship.

Superconducting electric engines have small sizes and weight, consume small capacity. Their important advantage consists in the absence of mechanical moving details that exclude acoustic noise at the ship movement. Fast discharge is the advantage of superconducting storage of the electric energy, which could be important for applications in guns, catapults, torpedoes with electromagnetic launching. The clearance of the coastal waters from sea mines is another important problem of the Navy. The mechanism of sea mine launching is connected with their sensitivity to the magnetic image of ship (i.e., with perturbations of Earth's magnetic field under influence of magnetic mass of the ship). Small ship with superconducting magnet can imitate the magnetic image of moving ship, thus collecting the mines. Under large magnetic field (for this, the superconducting magnet is used) the mines can be blown up sufficiently far from the magnet. Today, HTSC magnet, which may be placed in helicopter and models fields, imitating moving ship, has been developed. With the aim to find and counteract small, low-flying rockets, the Navy develops powerful high-frequency radio-locator on the base of superconducting magnet. In this case, interaction of electronic ray with magnetic field creates electromagnetic radiation with frequency proportional to the magnetic field.

IV. *Medicine.* At the top of our technological tree are seen benefits, connected with increase of human welfare and, especially, with development of medicine. The main problem is to use the uniquely useful, ecologically safe and energy-saving potential of superconducting materials and magnets. In particular, it relates to use of *magnetic resonance imaging* (MRI). This method rapidly established itself as a new and virtually indispensable medical diagnostic tool. The perspectives of the method are caused by progress in the technique of electronic image and cryogenic refrigeration. The base application of MRI involves visualization of concentrations of "hydrogen molecules" or "liquid" content of the various organs in the body. The use of even higher field superconducting magnets (so-called *functional MRI*) allows to state very accurately the distribution of other chemical elements in the human body that exist in much more limited concentrations than hydrogen. Another perspective direction is the use of SQUIDs that are the most sensitive detectors of flux or magnetic fields. SQUIDs are used to measure brain waves and brain functions. Similar excitement surrounds the use of SQUIDs in cardiology. Their other applications cover microbiology, biomedicine, high-energy physics, nanoparticle magnetism, non-destructive control, archeology, geology and SQUID microscopy. Medical HTSC tomographs can be used to control the quality of goods, in particular of food products.

V. *Mechanical systems.* The application of superconducting motors shows that such motors could be at least 20% more efficient than typical present products. The total energy saving on use of this technology could be enormous. In particular, 5000 horsepower motors have been demonstrated [219]. Other examples of superconductivity applications are the devices of energy accumulation and storage based on the levitation phenomenon, high-field magnets for laboratory research, superconducting magnets for specialized fabrication processes of the chemical and pharmaceutical products, industrial growing of silicon crystals, material separation and so on.

VI. *Scientific research.* The application of superconducting magnets has probably been the most radical event for high-energy physics. Note, in particular, most powerful in the world an accelerator of high-energetic charged particles (Large Hadron Collider), constructed in Europe by CERN[4] consortium in 2007. In total, approximately 1200 tons of NiTi superconducting wires and cables were required, which were supplied by several companies of Europe, Japan and the USA.

Scanning HTSC SQUID microscopes have highest sensitivity to weak magnetic fields in broad frequency diapason. They are able to register magnetic fields generated by vortex currents, magneto-tactile bacteria and currents of leakage in integral schemes. Electron-ray origins of multi-charged ions are used widely in laboratories. The parameters of HTSC origin [758] are sufficient to create the multi-charged ions, namely, helium-like xenon and neon-like uranium. It is used to study a surface modification of different materials by the multi-charged ions, using the scanning probe microscope.

Record parameters of magnetic field transducers (2006) [1192]:

(1) Sensitivity of LTSC and HTSC *SQUIDs* is equal to $10^{-15}\,\mathrm{T/Hz}^{1/2}$ (1 Hz, 4 K, screened room) and $5 \times 10^{-14}\,\mathrm{T/Hz}^{1/2}$ (1 Hz, 77 K, screened room), respectively.
(2) Sensitivity of *giant magnetic resistance (GMR) spin valves* is equal to $4 \times 10^{-10}\,\mathrm{T/Hz}^{1/2}$ (1 Hz, 300 K, without screening) and $4 \times 10^{-11}\,\mathrm{T/Hz}^{1/2}$ (1 Hz, 4.2 K, without screening).
(3) Sensitivity of *GMR spin valves with superconducting transformer of flux* is equal to $10^{-12}\,\mathrm{T/Hz}^{1/2}$ (1 Hz, 77 K, without screening) and $3 \times 10^{-13}\,\mathrm{T/Hz}^{1/2}$ (1 Hz, 4.2 K, without screening).
(4) Sensitivity of *atomic magnetometer with sizes of microelectronic chip* (at minimal consumed capacity) is equal to $5 \times 10^{-11}\,\mathrm{T/Hz}^{1/2}$ (10 Hz)

VII. *Electric power.* The superconductivity could render potentially enormous influence on the electric power industry, in particular, the power generation, power storage, power transformation and distribution and also the improvement and assurance of power quality. Even though the efficiency of power transmission lines, motor generators and especially electric transformers is already very impressible, energy losses due to use of ordinary *resistive* copper and aluminum conductors are enormous. Taking into account the basic influence that energy renders on all economic sectors the use of superconductivity in this field is connected with even more enormous perspectives. LTSC superconducting windings for electric power generator equipment to reduce energy losses have been carried out in a number of countries. While some very challenging engineering problems were successfully overcome, the economics, defined by using liquid helium and existence of high magnetic fields, pose extra challenges. In this case, the application of HTSC, which do not lose their properties at liquid nitrogen temperature, provides, in perspective, considerable advantages. For example, while the earlier approach involved the use of homopolar machine design – to have the windings exposed to dc-magnetic

[4] CERN is European Laboratory for Particle Physics.

fields only – the present HTSC design involves ac-synchronous approaches, which also maintain dc-fields around the superconducting windings [220]. In addition to lower losses, through the utilization of superconducting windings instead of copper, the decrease of size and weight should also lead to lower overall cost of product. Finally, greater reliability and potentially longer equipment lifetimes can be expected due to a constant and much cooler operating temperature environment.

The problems of thermonuclear energetic are directly connected with the use of high-field superconducting magnets (the proposals of TOKAMAK, LHC, ITER[5], etc.). While low-temperature superconductors will continue to be employed initially, the potential applications of HTSC in this field have great perspectives.

A very attractive feature of superconducting magnets is the fact that they can be considered to be "electromagnetic batteries." They can store enormous amounts of energy for long time sufficiently. Compared to electrochemical batteries, SMES[6] systems – although initially more costly – are considered to be more efficient, environmentally clean and less expensive for long term [100].

Successful feasibility demonstrations of superconducting transformers with HTSC windings have been carried out in Japan, Germany and the USA [682]. The economic savings, operational reliability and environmental benefits are the main advantages of superconductors, used in this case. Because of HTSC capabilities, they can readily and rapidly replace existing transformers, while providing higher ratings, overload protection capability and smaller footprints for equivalent ratings of conventional transformers. Superconducting cables are of considerable interest to the electric industry because they offer: (i) efficiency gains in transmission lines, reducing resistive losses, and (ii) opportunities to replace existing underground cables with much higher capacity cables, than in existing pipelines, that also lead to their quantitative decrease. The requirements to HTSC wires for different electric technical devices are presented in Table 1.2, and properties of main superconductors are shown in Table 1.3.

Another unique application of superconductivity is the development approaches for single-phase and multi-phase systems having significant near-term commercial potential. In particular, fault current limiters are devices utilizing the ability of superconductors to act like a resistance "switch," capable of exhibiting zero resistance, when superconducting, but capable of returning to a higher resistance state, when critical temperature, current or magnetic field limits are exceeded.

SuperFoam, synthesized from $YBa_2Cu_3O_z$ ceramic [878], may be ideal material for the fault current limiters. Advantages of superconducting foam under "tape" and "bulk" devices from BSCCO include the following ones [772]: (i) it endures critical currents at $T = 77\,K$, which are significantly higher,

---

[5] ITER is International Thermonuclear Energy Research.

[6] SMES is superconducting magnetic energy storage.

Table 1.2. The requirements of HTSC wires for different electric technical devices [761]

Device	J_c (A/cm^2)	Field (T)	Working temperature (K)	I_c (A)	Wire length (m)	Strain (%)	Twisting radius (m)	Cost ($/kA m)
Fault current limiters	$10^4 - 10^5$	0.1 – 3	20–77	$10^3 - 10^4$	1000	0.2	0.1	10–100
Big motors	10^5	4.5	20–77	500	1000	0.2–0.3	0.05	10
Generators	10^5	4–5	20–50	> 1000	1000	0.2	0.1	10
SMES	10^5	5–10	20–77	10^4	1000	0.2	1	10
Cables	10^4–10^5	< 0.2	65–77	100 per strand	100	0.4	2 (in cable)	10–100
Transformers	10^5	0.1–0.5	65–77	10^2–10^3	1000	0.2	1	10

Table 1.3. Main superconductors and their properties [761]

Material	Crystalline structure	Anisotropy	T_c (K)	H_{c2} (T)	H^*(T)	ξ(nm)	$\lambda(0)$(nm)	De-pairing current density, at 4.2 K (A/cm^2)	Critical current density (A/cm^2)	ρT_c (µΩ cm)
NbTi (Nb, 47 wt%)	b.c.c.	Neglected	9	12 (4 K)	10.5 (4 K)	4	240	3.6×10^7	4×10^5 (5 T)	60
Nb_3Sn	A-15, cubic	Neglected	18	27 (4 K)	24 (4 K)	3	65	7.7×10^8	$\sim 10^6$	5
MgB_2	P6/mmm, hexagonal	2–2.7	39	15 (4 K)	8 (4 K)	6.5	140	7.7×10^7	$\sim 10^6$	0.4
YBCO	Orthorhombic, layered perovskite	7	92	$>$ 100 (4 K)	5–7 (77 K)	1.5	150	3×10^8	$\sim 10^7$	$\sim$ 40–60
Bi-2223	Tetragonal, layered perovskite	50–100	108	$>$ 100 (4 K)	0.2 (77 K)	1.5	150	3×10^8	$\sim 10^6$	$\sim$ 150–800

compared to Bi-2212 phase, having problems with pinning centers at this "high" temperatures, (ii) it has sufficiently high electric resistance at room temperature in order to dissipate in heat an energy of supercurrent and (iii) it switches rapidly from and in superconducting state. The last two properties are forced in significant degree due to the foam includes open pores, in which there is cryogenic cooler. This provides continuously a directed contact with liquid nitrogen, compensating bad heat capacity of ceramics in the bulks.

1.3 Superconductivity Phenomena

1.3.1 Critical Field

Not only thermal fluctuations destroy superconductivity effect, but a sufficiently intense magnetic field can restore resistance and return the superconductor to its normal, that is, resistive state. This field is called *the critical field of bulk material*, H_{cm}, or the thermodynamic critical field. The temperature dependence of H_{cm} is well described by the empirical formula:

$$H_{\mathrm{cm}}(T) = H_{\mathrm{cm}}(0)[1 - (T/T_{\mathrm{c}})^2] \; . \tag{1.1}$$

This dependence is shown in Fig. 1.5a, which essentially represents the H–T phase diagram of the superconducting state. Within area of S, any point in H–T plane corresponds to the superconducting state. The critical field can be related to the critical temperature T_{c}. Since the critical field is zero at T_{c}, then in this case there is a second order phase transition. The entropy is continuous at T_{c} in zero magnetic field, but the specific heat is discontinuous. When $H \neq 0$, transition takes place at $T < T_{\mathrm{c}}$ and becomes of first order.

The zero resistance state can also be destroyed by large enough electrical currents, stated by the Silsbee criterion [989]. It defines an existence of *critical current*, the application of which through bulk superconductor causes initiation on its surface of the critical magnetic field, destroying superconductivity.

1.3.2 Josephson Effects

The quantum nature of superconductivity is provided by the so-called *weak superconductivity*, or Josephson effects [500]. They were predicted in 1962 and soon verified experimentally. The term "weak superconductivity" refers to situation in which two superconductors are coupled together by a weak link.

There are two Josephson effects, namely stationary (dc) and non-stationary (ac).

dc-Josephson effect. Let us apply a current through a weak link (or so-called Josephson junction). If the current is small enough, it passes through the weak link without resistance, even if the material of the weak link itself is not superconducting (e.g., it is an insulator in a tunnel junction). Through the weak link, the electrons of the two superconductors merge into a single

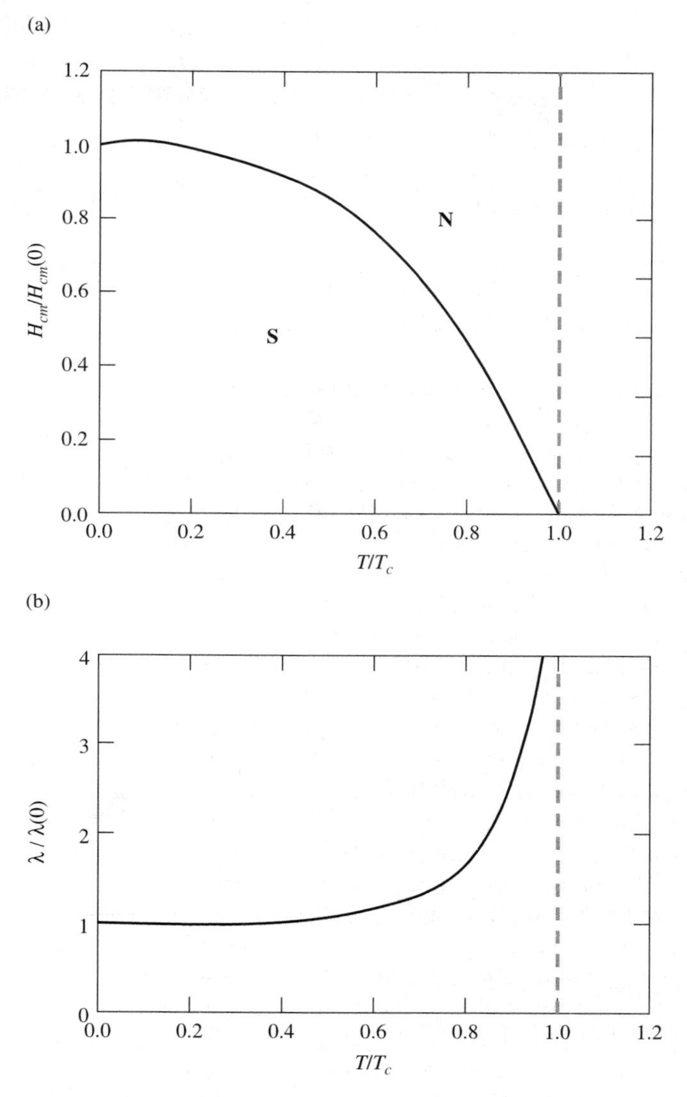

Fig. 1.5. Temperature dependence of the critical field, H_{cm} (**a**); temperature dependence of penetration depth, λ (**b**)

quantum body. In other words, all superconducting electrons on both sides of the weak link are described by the same wave function. The presence of the weak link should not change significantly the wave functions on the two sides, compared to what they had been before the link was established.

ac-Josephson effect. Let us increase the dc- current through the weak link until a finite voltage appears across the junction. Then, in addition to a dc-component V, the voltage will also have an ac-component of angular frequency ω, so that

$$\hbar\omega = 2eV \,, \tag{1.2}$$

where $\hbar$ is Planck's constant and e is the electron charge.

1.3.3 The Meissner Effect

Not only is the superconductivity phenomenon a state of zero resistance, but a superconductor is not simply an ideal conductor to being a piece of metal with zero resistance. Consider the behavior of an ideal conductor in an external magnetic field that is weak enough so as not to destroy the specimen's ideal conductivity. Suppose that initially the ideal conductor is cooled down below the critical temperature in zero external magnetic field. After that an external field is applied. From general considerations, it is easy to show that the field does not penetrate the interior of the sample (Fig. 1.6). Let us prove this with the help of Maxwell's equations. As the induction $\boldsymbol{B}$ changes, an electric field $\boldsymbol{E}$ must be induced in the specimen due to the equation

$$\mathrm{curl}\boldsymbol{E} = -c^{-1}\partial\boldsymbol{B}/\partial t \,, \tag{1.3}$$

where c is the speed of light in vacuum.

In the ideal conductor $\boldsymbol{E} = 0$, since

$$\boldsymbol{E} = j\rho \,, \tag{1.4}$$

where ρ is the resistivity (which in our case is zero) and $\boldsymbol{j}$ is the density of the induced current.

It follows that $\boldsymbol{B}$ = const and taking into account that $\boldsymbol{B} = 0$ before applying the external field, we arrive at $\boldsymbol{B} = 0$ in any point of the ideal

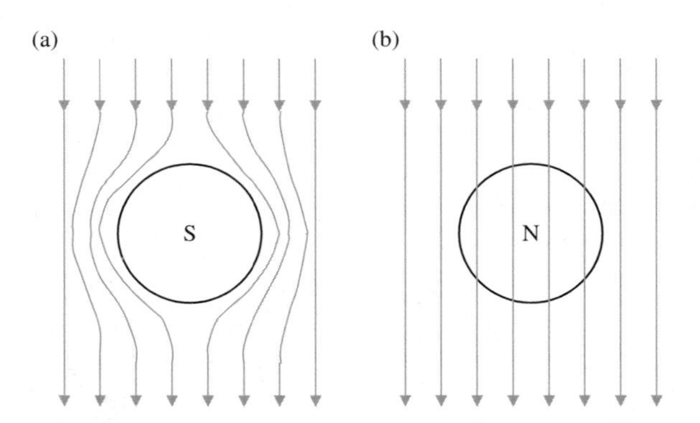

Fig. 1.6. Dependence of magnetic state of an ideal conductor on its history at $T < T_c$ and $H > 0$: (**a**) magnetic field applied to an ideal conductor at $T < T_c$; (**b**) field applied at $T > T_c$

conductor also after the field is applied. However, the same situation (an ideal conductor at $T < T_c$ in an external magnetic field) can be reached through another sequence of events, that is, by first applying the external field to a "warm" sample and then cooling it down to $T < T_c$. In that case, electrodynamics predicts an entirely different result. At $T > T_c$, the resistivity of the sample is finite and, therefore, the magnetic field penetrates into it. After cooling the specimen down through the superconducting transition, the field remains in it (see Fig. 1.6b).

The experiment by Meissner and Ochsenfeld [693] showed that the superconductor is not only an ideal conductor. They revealed that at $T < T_c$ the field inside a superconducting specimen was always zero ($\boldsymbol{B} = 0$) in the presence of an external field, independent of which a procedure has been chosen to cool the superconductor below T_c. Therefore, the superconductor is the perfect diamagnetic having negative magnetic susceptibility (i.e., ratio of magnetization to magnetic field) and stating zero field inside itself.

So, if condition $\boldsymbol{B} = 0$ is independent of the specimen's history, the zero induction can be treated as an intrinsic property of the superconducting state at $H < H_{cm}$. Furthermore, it implies that we can treat a transition to the superconducting state as a phase transition, but the superconducting state obeys the equations

$$\rho = 0 \,, \tag{1.5}$$

$$\boldsymbol{B} = 0 \,. \tag{1.6}$$

1.3.4 The Isotope Effect

The isotope effect provided a crucial key to the development of the superconductivity theory of Bardeen–Cooper–Schrieffer (BCS) discussed in Sect. 1.5. It was found that for a given element, T_c was proportional to $M^{-1/2}$, where M is the isotope mass. The vibration frequency of a mass M on a spring is proportional to $M^{-1/2}$. The same relation holds for the characteristic vibration frequencies of the atoms in a crystal lattice. So, the existence of the isotope effect indicated that although superconductivity is an electronic phenomenon, it is nevertheless connected with the vibrations of the crystal lattice in which the electrons move. However, after the development of the BCS theory, for some conventional (i.e., obeying the BCS theory) superconductors, the exponent of M is not $-1/2$, but near zero (see Table 1.4).

Table 1.4. Isotope effect ($T_c \propto M^{-\alpha}$) [581]

Element	Mg	Sn	Re	Mo	Os	Ru	Zr
α	0.5	0.46	0.4	0.33	0.21	0 (±0.05)	0 (±0.05)

In high-temperature superconductors, the exponent is also near zero. This fact caused research of HTSC mechanisms that are distinct on the electron–phonon interactions.

1.3.5 Penetration Depth and Coherence Length

Detailed studies have shown that provided the magnetic field remains weak enough, it gradually diminishes at the superconductor surface over a depth λ of some hundreds of angstroms order. This is called the *penetration depth.* When an external magnetic field is applied, a direct thermodynamic current appears at the superconductor surface in such a way as to screen the bulk from the applied field. The phenomenological theory of superconductivity, proposed by Gorter and Casimir and based on the assumption that there are two components of conducting electronic "fluid" in superconducting state, namely "normal" and "superconducting," gave the name *two-fluid model* [326]. The properties of the "normal" component are identical to system of electrons in the normal metal, but the "superconducting" component is responsible for the anomalous properties. In the framework of the two-fluid model of superconductor, the temperature dependence of the penetration depth λ is approximated well by [325]

$$\lambda(T) = \frac{\lambda(0)}{[1 - (T\,/\,T_c)^4]^{1/2}} \; . \tag{1.7}$$

The divergence of λ at transition ($T \to T_c$) shows that we pass continuously from the normal metal, in which $\lambda = \infty$, to the superconducting state (see Fig. 1.5b). In non-zero field, λ varies discontinuously from $\lambda = \infty$ to a finite value, and the transition is then first order.

Taking into account sharpness of the superconducting transition at the absence of magnetic field and also significant dependence of the observed depth λ on the admixture concentration, Pippard concluded that superconducting state should be characterized by a finite coherency length of electron impulse (Δp) [847], but no infinite one as in the theory of the brothers F. and H. London [644]. Therefore, the order parameter changes smoothly at the distance ξ, called *the coherence length.* Based on the estimation made by Pippard for dependence λ on the field, the length of $\xi \sim 1\,\mu\text{m}$. The coherence length is the distance between two electrons of Cooper pair, to be in essence space characteristic of superconducting electrons, on which value it could be judged by using the relation of uncertainties

$$\xi \approx \frac{\hbar}{\Delta p} \approx \frac{\hbar \nu_F}{kT_c} \; , \tag{1.8}$$

where ν_F is Fermi velocity and k is Kelvin constant.

To explain the dependence of depth λ on the length of the free run of electrons l, Pippard proposed that an effective coherence length $\xi(l)$ is connected with corresponding value of ξ_0 for pure metal by the equation

$$\frac{1}{\xi(l)} = \frac{1}{\xi_0} + \frac{1}{Al} , \tag{1.9}$$

where A is the constant near unit.

In the two limit cases, the evident equations for λ may be obtained in the form:

$$\lambda = \lambda_{\mathrm{L}} \left[\frac{\xi_0}{\xi(l)} \right]^{1/2} \quad \text{at } \xi << \lambda (\text{London limit}) , \tag{1.10}$$

where

$$\lambda_{\mathrm{L}} = \left(\frac{mc^2}{4\pi n_s e^2} \right)^{1/2}$$

is the London penetration depth; m and e are the mass and charge of electron, respectively; n_s is the quantitative density of superconducting electrons.

The condition $\xi << \lambda$ is carried out for pure metals near T_{c} (where $\lambda \to \infty$) and also for alloys and thin admixture films, where l and ξ are decreased or limited by the electronic scattering at defects, admixtures or at film boundaries, so that $\xi \to l$ at $l \to 0$. The contrast limit case ($\xi >> l$), corresponding to most part of bulk superconductors ($l \to \infty$), at the temperatures leads to far from T_{c},

$$\lambda_{l \to \infty} = \left(\frac{\sqrt{3}}{2\pi} \xi_0 \lambda_{\mathrm{L}}^2 \right)^{1/2} \quad \text{at } \xi >> \lambda (\text{Pippard limit}) . \tag{1.11}$$

1.4 Magnetic Properties of Superconductors

According to their magnetic properties, superconductors are divided into *type-I superconductors and type-II superconductors.* Type-I superconductors include all superconducting elements except niobium. Niobium, superconducting alloys and chemical compounds are type-II superconductors. The high-temperature superconductors also belong to this group. The main difference between the superconductors of both types lies in their different response to an external magnetic field. The Meissner effect is observed in type-I superconductors, only.

1.4.1 Magnetic Properties of Type-I Superconductors

Let us consider the magnetization curve of a superconductor representing a long cylinder in a longitudinal external magnetic field $\boldsymbol{H}$. When the field $\boldsymbol{H}$ increases, the induction inside the sample does not change at first; it remains at $\boldsymbol{B} = 0$. As soon as H reaches the value of H_{cm}, the superconductivity is destroyed, the field penetrates into the superconductor and $\boldsymbol{B} = \boldsymbol{H}$. Therefore, the magnetization curve $B = B(H)$ has the form shown in Fig. 1.7a. The magnetic induction $\boldsymbol{B}$ and magnetic field $\boldsymbol{H}$ are related to each other by the expression:

$$\boldsymbol{B} = \boldsymbol{H} + 4\pi \boldsymbol{M}\,, \tag{1.12}$$

where $\boldsymbol{M}$ is the magnetic moment per unit volume (Fig. 1.7b).

In type-I superconductors, the diamagnetism (i.e., negative value of magnetic moment and its contrast direction to magnetic induction vector) remains up to the value of field H_{cm}. This supercooling of normal phase is caused by the difficulty of local nucleating a superconducting region within a normal region in type-I superconductors. If there are defects (e.g., dislocations) in the crystal lattice, these can modify such theoretical behavior by smoothing out transitions. Main magnetic properties of type-I superconductors are led from the (1.5) and (1.6) [594]:

(1) *Magnetic field lines outside a superconductor are always tangential to its surface.* Indeed, it is known from electrodynamics that magnetic field lines (i.e., lines of the magnetic induction) are continuous and closed. This can be written as the equation div $\boldsymbol{B} = 0$. Then, the components of $\boldsymbol{B}$ that are normal to the surface must be equal on both sides of the surface, that is, inside and outside of the sample. However, the field in the interior of a superconductor is absent and, consequently, $\boldsymbol{B}_{\text{n}}^{(\text{i})} = 0$. It follows that the normal component $\boldsymbol{B}_{\text{n}}^{(\text{e})}$ at the outside of the superconductor's surface is zero, too: $\boldsymbol{B}_{\text{n}}^{(\text{e})} = 0$, that is, the magnetic field lines are tangential to the surface of the superconductor.

(2) *A superconductor in an external magnetic field always carries an electric current near its surface.* This is one of the consequences of the first property, because from Maxwell's equation curl $\boldsymbol{B} = (4\pi/c)\boldsymbol{j}$ and the requirement $\boldsymbol{B} = 0$, it follows that the volume current in the interior of the superconductor is zero ($\boldsymbol{j} = 0$) and only a surface current is possible. Then, from relation between the surface current $\boldsymbol{j}_s$ and the magnetic field at the surface of the superconductor $\boldsymbol{H}$

$$\boldsymbol{j}_{\text{s}} = (c/4\pi)(\boldsymbol{n} \times \boldsymbol{H})\,, \tag{1.13}$$

where $\boldsymbol{n}$ is the unit vector along the normal to the surface, it is obvious that the surface current is completely defined by the magnetic field at the surface of a superconductor. In other words, the surface current assumes a value such that the magnetic field generated by it inside a superconductor is exactly equal in value and opposite in direction to the external field. This assures zero total field in the interior: $\boldsymbol{B} = 0$.

(3) *In a simply connected superconductor (i.e., body inside of which an arbitrary closed path can be reduced to a point without crossing the boundaries of the body), surface currents can exist only when the superconductor is placed in an external magnetic field.* Indeed, if the surface current remains after switching off the external field, it would create its own field in the superconductor, which is impossible.

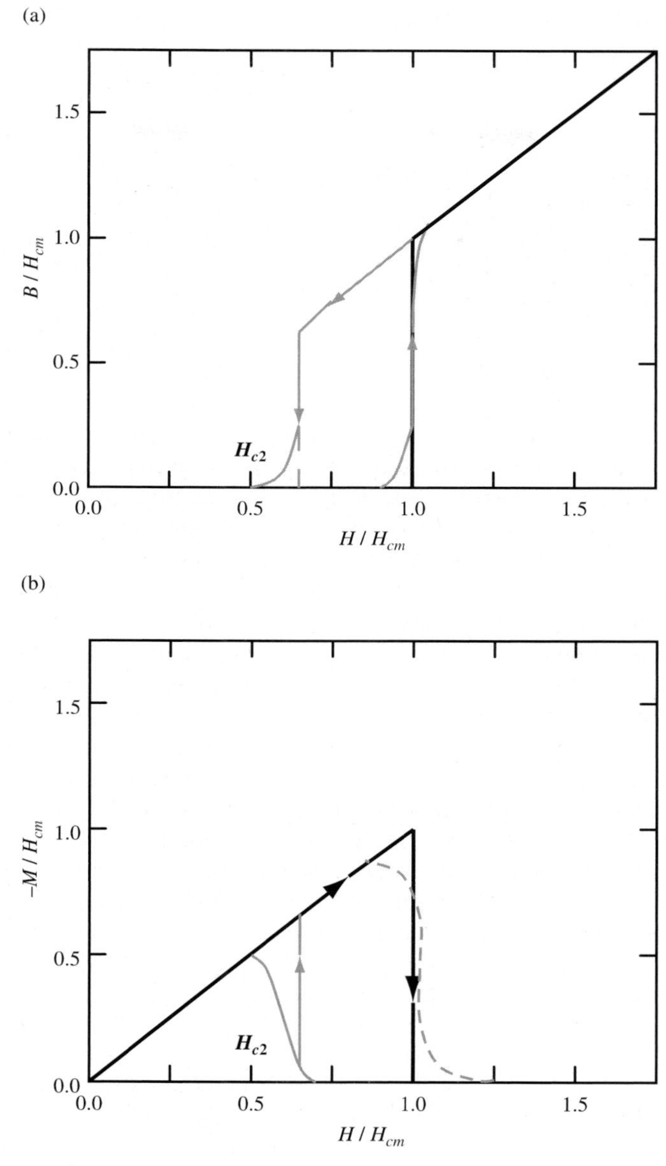

Fig. 1.7. Magnetic induction (**a**) and magnetization (**b**), as a function of applied field in type-I superconductors. When the magnetic field is reduced, superconductivity reappears at a field $H_{c2} < H_{cm}$. *Dotted lines* correspond to impure samples [617]

1.4.2 The Intermediate State

Consider the behavior of a superconducting sphere, placed in an external magnetic field (Fig. 1.8). Since the magnetic field lines are always tangential to the surface of a superconductor, then in this case, it is obvious that the field lines have a higher density at the equator thereby producing a local increase of the magnetic field. At the same time, the field at the poles is absent. Far away from the sphere, where any perturbations are averaged, the uniform external field H is lower than that at the equator. Therefore, an attainment of the field at the equator of the critical value of H_{cm} the magnitude of $H < H_{\mathrm{cm}}$ and it is not permitted for the whole sphere to revert to the normal state. On the other hand, it is not permitted for the whole sphere to be superconducting either, because the field at the equator has already reached the critical value. This contradiction involves a co-existence of alternating superconducting and normal regions within the sphere, which is called *the intermediate state*. The interfaces between these regions are always parallel to the external field, while in the cross-section perpendicular to the field they may assume various intricate configurations.

Assume that before a superconducting body goes into the intermediate state, the maximum field at its surface (in the case of a sphere, at the equator) is H_{m}, and the external field far away from the body is H. Then, it is obvious that, on the one hand, $H_{\mathrm{m}} > H$ and, on the other hand, H_{m} is proportional to H with the proportionality factor dependent on the exact shape of the body. It can be written in the form $H_{\mathrm{m}} = H/(1-n)$. The values of the demagnetizing factor n for different forms are present in Table 1.5.

With the help of Table 1.5, it is possible to calculate the field H corresponding to the transition into the intermediate state for a body of a certain shape. The transition takes place when the field H_{m} reaches the value of H_{cm}. In particular, for a sphere it occurs at the external field $H = H_{\mathrm{cm}}(1-1/3) = 2/3H_{\mathrm{cm}}$.

Let us now consider the conditions for thermodynamic equilibrium in the intermediate state. Assume that the field in one of the normal regions exceeds H_{cm}. Then it must destroy the superconductivity of the adjacent superconducting regions. Conversely, if the field in a normal region is less than H_{cm},

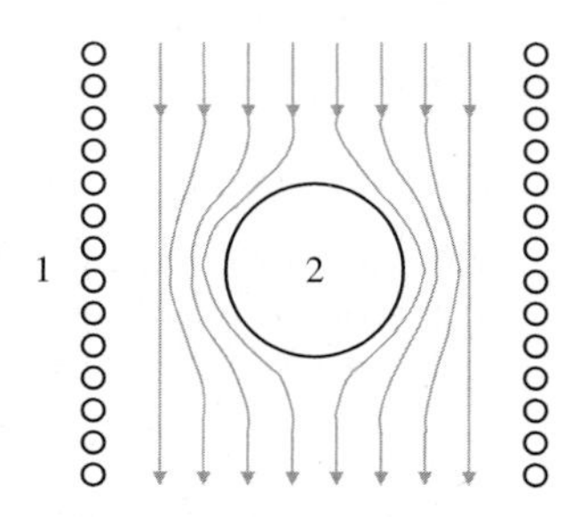

Fig. 1.8. Superconducting sphere in the homogeneous field of a solenoid. *Numbers* denote winding of the solenoid (1) and superconducting sphere (2)

Table 1.5. The demagnetizing factor for different forms of superconductor

Sample geometry	Cylinder in parallel field	Cylinder in transverse field	Sphere	Thin plate in perpendicular field
n	0	1/2	1/3	1

this region must be superconducting. Therefore, a stable co-existence of the normal and superconducting regions is only possible, if the field in the normal regions equals to H_{cm}.

1.4.3 Magnetic Properties of Type-II Superconductors

In contrast to type-I superconductors, type-II superconductors do not show Meissner effect and are characterized by the *mixed state*, but no intermediate state. In this case, electric resistance is absent, but a magnetic field penetrates into superconductor in a quite extraordinary way. Let us consider a type-II superconductor in the form of a long cylinder placed in a longitudinal magnetic field, which is increased from zero. First, the cylinder pushes out all field, causing the zero magnetic induction in the interior of the cylinder. This means that the Meissner effect is observed at this stage. However, beginning from a certain value of the field, there is a non-zero (i.e., finite) induction in the cylinder. This field is called *the lower critical field* and denoted by H_{c1}. With next increase of the external field H, the induction will build up until the average field in the cylinder becomes equal to the external field H, but the cylinder itself goes to the normal state. This will happen at the so-called *upper critical field* H_{c2}. In 1963, Saint-James and de Gennes have shown that the superconductivity will remain even at $H > H_{\mathrm{c2}}$ in a thin surface layer, until $H \leq 2.392kH_{\mathrm{c}}$, where k is Ginzburg–Landau parameter, H_{c} is the thermodynamic critical field, that is, $H \leq 1.695H_{\mathrm{c2}}$ [907]. This field $H = 1.695H_{\mathrm{c2}}$, also destroying superconductivity in the surface layer has been named by *the third critical field* and denoted by H_{c3}. However, based on the measurements of magnetic susceptibility of the Nb cylinders under fields above H_{c2} [129, 570], it has been stated that the ratio of $H_{\mathrm{c3}}/H_{\mathrm{c2}}$ depends on processing method of surface, but was above the theoretical value of 1.695 [907] in all cases. The near-surface critical current was absent in the fields above $H_{\mathrm{c3}}^{\mathrm{c}}(T) = 0.81H_{\mathrm{c3}}(T)$, where factor 0.81 was independent of the temperature, surface quality and admixtures. The last two parameters led to alteration of the ratio $H_{\mathrm{c3}}/H_{\mathrm{c2}}$ from 1.86 to 2.57. It could be assumed [129, 570] that below H_{c3} down to $H_{\mathrm{c3}}^{\mathrm{c}}$ there is only local near-surface superconductivity in the form of separate sites, which contribute to the magnetic susceptibility, but do not provide surface critical current. They couple together only in the field below $H_{\mathrm{c3}}^{\mathrm{c}}$ and a single coherent superconducting state forms.

1.5 Theories of Superconductivity

The first theory which had success in describing the electrodynamics of superconductors, was the phenomenological theory of brothers F. and H. London (1935). It had introduced two equations, in addition to Maxwell's ones, governing the electromagnetic field in a superconductor [644]. These equations provided a correct description of the two basic properties of superconductors: absolute diamagnetism and zero resistance to a dc-current. The London theory did not resolve the microscopic mechanism of superconductivity on the level of electrons. Therefore, the question: "*Why* does a superconductor behave according to the London equations?" remained beyond its scope.

According to the London theory, electrons in a superconductor may be considered as a mixture of superconducting and normal electrons. The quantitative density of the superconducting electrons, n_{s}, decreases with increasing of temperature and attains zero at $T = T_{\mathrm{c}}$. On other hand, at $T = 0$ the value of n_{s} is equal to the total density of conductivity electrons. This postulates two-fluid model of superconductor proposed by Gorter and Casimir. A flux of superconducting electrons meets no resistance. Obviously, such a current cannot generate a constant electric field in a superconductor because, if it did, it would cause the superconducting electrons to accelerate infinitely. Therefore, under stationary conditions, corresponding to absence of an electric field, the normal electrons are at rest. In contrast, in the presence of an ac-electric field, both the normal and the superconducting components of the current are finite and the normal current obeys Ohm's law.

The London equations provided a description for the behavior of the superconducting component of the electronic fluid in both dc- and ac-electromagnetic field. They also helped to understand a number of aspects of the superconductors' behavior in total. However, by the end of the 1940, it was clear that one question at least was not answered in the framework of the London theory. For the interfaces between adjacent normal and superconducting regions, the theory predicted a negative surface energy: $\sigma_{\mathrm{ns}} < 0$. This implied that a superconductor in an external magnetic field could decrease its total energy by turning into a mixture of alternating normal and superconducting regions. In order to make the total area of the interface within the superconductor as large as possible, the size of the regions must be as small as possible. This was supposed to be the case even for a long cylinder in a longitudinal magnetic field, in contradiction to experimental evidence existed at that time. Experiments showed that such a separation of the normal and superconducting regions took place only for samples with a non-zero demagnetizing factor (the intermediate state). Moreover, the layers were rather thick ($\sim$1 mm), which could only be the case if $\sigma_{\mathrm{ns}} > 0$, which also contradicted the London theory.

This contradiction was reconciled by a theory proposed by Ginzburg and Landau, which was also phenomenological but took into account quantum effects [313]. In that moment, it became clear why it is so important to include quantum effects in considerations. Assume that there is a wave function (or order parameter) Ψ, describing the electrons quantum-mechanically. Then, the squared amplitude of this function (which is proportional to n_s) must be zero in a normal region, increase continuously through the normal-superconducting interface and finally reach a certain equilibrium value in a superconducting region. Therefore, a gradient of Ψ must appear at the interface. At the same time, as is known from quantum mechanics, $|\nabla\Psi|^2$ is proportional to the density of the kinetic energy. So, quantum effects taken into account lead to an additional positive energy stored at the interface, which creates the opportunity to obtain $\sigma_{ns} > 0$.

Tremendous importance of the Ginzburg–Landau theory consisted in that it introduced quantum mechanics into the description of superconductors. It assumes the description of the total number of superconducting electrons by a wave function, depending on spatial coordinates (or equivalently, a wave function of n electrons is a function of n coordinates, $\Psi(\boldsymbol{r_1}, \boldsymbol{r_2}, \ldots, \boldsymbol{r_n})$). Based on this, the theory established *the coherent (coupled) behavior* of all superconducting electrons. Indeed, in quantum mechanics, a single electron in the superconducting state is described by a function $\Psi(\boldsymbol{r})$. If we now have n_s absolutely identical electrons (where n_s, the superconducting electron number density, is a macroscopically large number), and all these electrons behave coherently, it is obvious that the same wave function of a single parameter is sufficient to describe each of them. This idea permitted to involve in the description the superconductivity quantum effects, at the same time, retaining macroscopic features of the material.

It is interesting that analysis of experimental data on the basis of the Ginzburg–Landau theory permitted to estimate effective charge, e^*, which rendered approximately two times greater than electron charge. However, then nobody had any idea about coupling of electrons, while the Ogg's paper (1946) that was published some years ago contained the idea about coupling of the electrons with their further Bose–Einstein condensation.

The Ginzburg–Landau theory was built on the basis of the theory of second-order phase transitions (the Landau theory) [595] and, therefore, it is valid only in the vicinity of the critical temperature. By applying the Ginzburg–Landau theory to superconducting alloys, Abrikosov in 1957 developed a theory of the so-called type-II superconductors [3]. It turned out that superconductors need not necessarily have $\sigma_{ns} > 0$. Materials that provide this condition are type-I superconductors and transition in external magnetic field from superconducting state to normal one for them is the phase transition of I type. However, the majority of superconducting alloys and chemical compounds demonstrate $\sigma_{ns} < 0$, and they are type-II superconductors. For type-II superconductors, there is no Meissner effect; magnetic field penetrates inside the material but in a very unusual way, that is, in the form of *quantized vortex lines* (quantum effect on the macroscopic scale!). Superconductivity in these materials can survive up to very high magnetic fields, and transition in external magnetic field from superconducting state to normal one for them is the phase transition of II type. Abrikosov found theoretically vortex structures in superconductors, and thus explained experiments of Shubnikov, who together with co-authors discovered them as long ago as 1937 by observing the unusual behavior of some superconductors in external magnetic field [948]. Abrikosov assumed that mixed state of superconductor (or Shubnikov phase) is the vortex state, in which superconducting vortices form periodic lattice [3].

Neither the London nor the Ginzburg–Landau theory could answer the question: "What are those 'superconducting electrons,' whose behavior they were intended to describe?" It was 46 years since the discovery of superconductivity, but at the microscopic level a superconductor remained a mystery. This issue was finally resolved in 1957 by the work of Bardeen, Cooper and Schrieffer [50], in which the so-called BCS theory was presented. In 1958, an important contribution was also made by Bogolyubov [73], who developed mathematical methods which were widely used in studies of superconductivity. The BCS theory described the process of coupling of the conductivity electrons with formation of pairs and regrouping into one quantum state. In contrast to ideal electrons, all pairs could be in the same quantum state, forming macroscopic quantum wave. In these conditions, electric current exists for account of a motion the all aggregate of *paired electrons* without scattering of energy of the single electrons, that is, electric resistance. In the BCS theory, the formation of pairs is explained by the existence of a certain type of indirect interactions between electrons. This interaction has a character of attraction and, therefore, is a contrast to Coulomb force that pushes away one electron on other. Displacing the conductivity electron in metal during its motion causes a local deformation of crystalline lattice on account of attraction forces, with which it acts on positive ions (cations of crystal). This deformation (one from the types of the lattice excitations) can interact with the second electron. Due to this process, there is an attraction of two electrons that render to be "coupled" for account of crystalline lattice. This interaction exists on account of link of the electrons with quantums of excitations, that is, with lattice phonons. The formed pairs of electrons are called *Cooper pairs*. An existence of interaction "electron–lattice–electron" leads to change of energetic spectrum of electrons. It is known that in the solid the energetic levels accessible for electrons form consequent zones. Conductor is characterized by the conductivity zone occupied partially up to the energy of E_{F}, called *the Fermi level.* In superconductor the BCS theory predicts at this energy E_{F} the beginning of a restricted zone (*energy gap*) that separates base state of system of the coupled electrons from excited states occupied by usual (uncoupled) electrons. Width, Δ, of this zone is equal to an energy that it is necessary to apply in order to break the pair and consequently for the failure of superconductivity. One depends on a temperature, namely has maximum value at the zero temperature and is equal to zero at the $T = T_{\mathrm{c}}$. In several cases, superconductivity can exist even in the absence of a restricted zone.

The BCS theory explained directly the isotope effect, discovered in 1950, and, it was very important to find the critical temperature, T_{c}, through phonon and electron characteristics. This opened a possibility to search sensibly new superconductors with higher T_c from materials with great values of Debye temperature and constant electron–phonon coupling.

In 1959, the microscopic theory of superconductivity was elaborated further by Gor'kov, who developed a method to solve the BCS problem, using Green's functions [327–329]. He applied this method, in particular, to find microscopic interpretations for all phenomenological parameters of the Ginzburg–Landau theory, as well as to define the theory's range of validity. The works of Gor'kov completed the development of the Ginzburg–Landau–Abrikosov–Gor'kov theory.

In 1964, Little and Ginzburg expressed independently an idea about probable non-phonon mechanism of superconductivity in low-dimensional (quasi-1D or quasi-2D) systems. It was shown that substitution of phonons by excitons (excitations of sub-system of the coupled electrons) should permit, in principle, to increase T_c up to 50–500 K. However, search of these superconductors was unsuccessful.

The soliton (or bisoliton) model of superconductivity was considered for the first time by Brizhik and Davydov in 1984 [92] in order to explain the superconductivity in organic quasi-1D conductors discovered by Jérome and co-authors in 1979 [484].

In 1986, trying to explain the superconductivity in heavy fermions discovered in 1979, Miyake and co-authors considered the mechanism of superconductivity based on the exchange of antiferromagnetic spin fluctuations [708]. The calculations showed that the anisotropic even-parity couplings are assisted, and the odd-parity as well as the isotropic even-parity are impeded by antiferromagnetic spin fluctuations.

Interest in the research of superconductivity obtained powerful impulse in 1986 due to the discovery of oxide high-temperature superconductors,[7] made by Bednorz and Müller. The classical BCS theory was unable to explain many of their properties. The electron–phonon mechanism became questionable.

By analyzing the layered structure of cuprates, Krezin and Wolf proposed, in 1987, a model of high-temperature superconductivity based on the existence

[7] In fact, in the early 1970s, a compound of lanthanum and copper oxides was synthesized in Moscow (USSR). This research was not connected with superconductivity, but the researchers were looking for good and cheap conductors. At low temperatures, the conductivity of this new material showed an abnormal behavior. The scientists understood the significance of this abnormality, but, nevertheless, they could not continue the next experiments because in these years there were great difficulties with liquid helium. Moreover, they did not reveal persistence in defense their questionable results. The mysterious compound was put away in a cupboard and forgotten. Thus, in 1986, Bednorz and Müller discovered high-temperature superconductivity practically in the same compound that had been synthesized before in the USSR [846]. So, the possibility of the superconductivity discovery in cuprates a few years earlier, than in heavy fermions and organic superconductors, was missed.

of two energy gaps, namely superconducting and induced [579, 580]. Indeed, different experiments performed after 1987 have demonstrated the existence of two gaps; however, they both have the same superconducting origin.

In 1987, Gor'kov and Sokol proposed existence of a new type of microscopic and dynamical phase separation [330] that was later discovered in other theoretical models

In this year, Anderson proposed a model of superconductivity in cuprates, separating the coupling mechanism and the mechanism for the establishment of phase coherence [23].

In 1988, Davydov suggested that high-temperature superconductivity occurs due to the formation of bisolitons, as it takes place in organic superconductors [178]. In 1990, he presented a HTSC theory based on the concept of a moderately strong electron–phonon coupling [179, 180]. The theory utilizes the concept of bisolitons or electron (hole) pairs coupled in a single state due to local deformation of the –O–Cu–O–Cu– chain in CuO_2 planes.

In the early 1990s, based on the Anderson's assumption, namely, that in cuprates the coupling mechanism and the mechanism for the establishment of phase coherence are different, some theorists autonomously proposed that independently of the origin of coupling mechanism, spin fluctuations cause the long-range phase coherence in cuprates.

In 1994, Alexandrov and Mott showed that, in cuprates, it was necessary to distinguish the "internal" wave function of a Cooper pair and the order parameter of the Bose-Einstein condensate, which may have various symmetries [13].

In 1995, Emery and Kivelson emphasized that superconductivity requires coupling and long-range phase coherence [249]. They demonstrated that in cuprates, the coupling can occur above T_c without the phase coherence.

In the same year, Tranquada et al. found the presence of coupled, dynamical modulations of charges (holes) and spins in Nd-doped $La_{2-x}Sr_xCuO_4$ (LSCO) by using neutron diffraction [1076].

In 1997, Emery, Kivelson and Zachar presented a theoretical model of HTSC based on the presence of charge stripes in CuO_2 planes [250].

In 1998, Chakraverty et al. attempted to prove that the theory of bi-polaron superconductivity of HTSC contradicted with experiments and was theoretically discrepant [134]. In answer of Alexandrov [11], was been stated contrary opinion, namely: the negation of the bi-polaron superconductivity of HTSC is the result erroneous approximation for energetic spectrum of bi-polarons and erroneous application of the bi-polaron theory carried out by Chakraverty et al. Based on *two-zone* model, he obtained a formula for T_c, which was free from adjusted parameters and included besides basic constants the concentration of carriers, n, and penetration depths of magnetic flux λ_{ab} and λ_c along two mutually perpendicular crystallographic directions. The substitution of test values of n, λ_{ab} and λ_c (for Y-123) estimated $T_c \sim 100\,\mathrm{K}$. It proves self-consistency of the bi-polaron approach and testifies HTSC to be in the regime of Bose–Einstein condensation.

In 1999, analysis of tunneling and neutron scattering measurements, carried out by Mourachkine, showed that in $Bi_2Sr_2CaCu_2O_{8+x}$ (Bi-2212) and $YBa_2Cu_3O_{6+x}$ (YBCO), the phase coherence is established due to spin excitations [732, 733], which cause the appearance of the so-called magnetic resonance peak in inelastic neutron scattering spectra [899].

In this year, Cronstrom and Noga determined a new solution of BCS equations in approximation of mean field, which pointed to the existence in thin superconducting films (or in superconducting bulks with layered structure) of type-III phase transitions [167]. The critical temperature of this transition increases at decreasing of the layer thickness and is independent of isotope mass. The electronic heat capacity is a continuous function of temperature, but has discontinuity of derivative.

In 1999, Leggett defined a very simple dependence, $T_c(n)$ for "calcium" HTSC (where n is the number of CuO_2 layers per elementary cell): $T_c(n) = T_c(1)+T_0(1-1/n)$, where T_0 is the own constant for each Bi, Hg and Tl family, in which the CuO_2 layers are separated by calcium layers [615]. In particular, it is followed from this formula that $[T_c(3) - T_c(2)]/[T_c(2) - T_c(1)] = 1/3$, which agrees with test data 0.25–0.28 and 0.25–0.34 for HTSC on the basis of Hg and Tl, respectively.

In 2000, Tang measured the critical temperature, T_c, of ultra-thin HTSC films $YBa_2Cu_3O_7$ depending on their thickness d. It has been observed approximately linearly on d a diminishing of T_c with decreasing of the film thickness at $d < 10\,\text{nm}$ [1046]. The dependence $T_c(d)$ is well described by the empirical formula: $T_c = T_{c0}(1 - d_m/d)$, where $T_{c0} = 90\,\text{K}$ and $d_m = 1.56\,\text{nm}$. The critical thickness, d_m, is near to the thickness of one elementary cell along c-axis that supports quasi-2D nature of HTSC superconductivity.

In 2001, Kivelson proposed the following way for increasing of T_c: it is necessary to create multi-layer systems with different concentrations of carriers in various layers, so the layers with low concentration of carriers provide their coupling, but the layers with high concentration of carriers guarantee phase rigidity [553]. Maximal critical temperature of HTSC is found by competition of two effects: (i) the coupling interaction weakens at increasing of the charge carrier concentration, x, that is connected with properties of doped Mott's dielectric, but (ii) the density of super-fluid component, which controls the system rigidity in relation to phase fluctuations, increases with growth of x. Thus, optimal T_c is reached at the boundary of the region with prevalence of the phase alignment and region with prevalence of the coupling interaction.

In the same year, based on tunneling measurements, Mourachkine provided evidence that the quasiparticle peaks in tunneling spectra of Bi-2212 crystals are caused by condensed soliton-like excitations, which form the Cooper pairs [734–736].

In 2002, Cui proposed a possible responsibility for superconductivity of relativistic attraction of electrons. At least, there are two types of collective movement in superconductors, which can suppress usual Coulomb repulsion of electrons as the attraction component becomes predominant. This movement is caused by combination of electron gas and phonons in conventional superconductors and by itself electron gas (or electron liquid) – in HTSC. The repulsion and attraction between electrons balance approximately each other in majority of matters; therefore, the theory of electron gas (i.e., non-interacting particles) works well. However, the repulsion predominates over attraction in some matters, and then electron sub-system demonstrates properties directly contrary to superconductivity [168].

In this year, Laughlin showed that cuprate HTSC in undoped state should be considered as no dielectrics, but superconductors with very great gap and extremely small super-fluid density [605]. Laughlin named these superconductors "gossamer superconductors." In practice, a brittleness of coupled state of this superconductor creates obstacles for stating superconductivity in all volumes. However, he assumed that the wave function of this coupled state can serve as a good starting point for understanding of correlations between Mott's dielectric and HTSC.

In 2003, Hussey et al. proved an existence of 3D Fermi surface, based on investigations of angle oscillations of magnetic resistance in HTSC $Tl_2Ba_2CuO_{6+\delta}$ [459]. Thus, almost a 20-year-old argument about coherency or incoherency of electronic states along the c-axis was solved in favor of coherency.

In 2004, Homes et al. obtained the next universal scaling relationship between physical values, characterizing normal and superconducting states of HTSC [435]: $\rho_s = A s_{dc} T_c$ (where ρ_s is the density of super-fluid component, s_{dc} is the static specific conductivity and T_c is the critical temperature), which should be carried out for *all* HTSC, without dependence on the value of T_c, type of carriers (holes or electrons), doping level, crystalline structure and current direction (parallel or perpendicular to CuO_2 planes). In this case, the proportionality factor, $A = 120 \pm 25$, if ρ_s is measured in s^{-2}, s_{dc} in $(\Omega\,cm)^{-1}$ and T_c in K. The straight dependence, $\rho_s(s_{dc}T_c)$, includes even the points for low-temperature superconductors Pb and Nb. It covers diapason above of five degree of magnitude on each of the coordinate axes. This empirical dependence is surprising, if taking into account the principally different character of current transfer in various crystallographic directions (i.e., coherent in ab-plane and incoherent along c-axis).

In this year, Alexandrov again explained HTSC physics on the basis of bi-polarons mechanism. HTSC properties weakly differ from usual metals, namely: there is standard BCS phenomenon, only Bogolyubov's quasi-particles (Cooper's pairs with d-symmetry) transfer current. The magnetic mechanism of superconductivity exists, but as subsequent of electron–phonon coupling. Because constants of electron–phonon coupling are greater than 1, polarons and super-light bi-polarons must arise. In this case, the critical temperature is found as [12]

$$T_c = 1.64[eR_H/(\lambda_{ab})^4(\lambda_c)^2]^{1/3}\,, \tag{1.14}$$

where e is electron charge, R_H is Hall constant, λ_{ab} and λ_c are the penetration depth of magnetic field in ab-plane and along c-axis, respectively.

In 2006, Honma and Hor proposed to distinguish 2D and 3D hole concentration [436]. It is well known that the critical temperature of HTSC is a universal function of this parameter (the hole concentration is usually found by number of holes per Cu atom in CuO_2 layer). Their analysis shows that T_c is defined by 3D *concentration*, n_h, and universal bell-like dependence takes place just for $T_c(n_h)$.

In this year, Terashima et al. demonstrated experimentally in HTSC a change of character of the anomaly for the law of electron dispersion (so-called "kink") in the vicinity of Fermi level at partial substitution of Cu atoms by Zn and Ni atoms, not much differing in mass, but in other spin states (that may be considered as "magnetic isotope-effect") [1052].

Obviously, the above list of mechanisms and superconductivity theories is not complete. Finally note that beginning from 1987, theorists proposed more than 100 models of high-temperature superconductivity, in particular, based on the representations about polarons, plasmons, excitons, solitons, super-exchange and direct interaction between electrons. Today, the intensive discussion of HTSC mechanisms and high critical temperatures, T_c, to being proper for these materials is continued. Some review results of different HTSC theories can be found in the overviews and monographs [99,172,299,452,666,737,986].

1.6 High-Temperature Superconductors

1.6.1 General Remarks on Type-II Superconductors

High-temperature superconductors placed in the center of our research are the type-II superconductors. Therefore, their properties and superconductivity mechanisms are considered in more detail. The term type-II superconductors was first introduced by Abrikosov in his classical paper [3], where he assumed a detailed phenomenological theory of these materials' behavior, based on the Ginzburg–Landau theory, and explained their magnetic properties. Initially, Abrikosov's theory was greeted with certain skepticism: so much out of the ordinary was in its predictions. However, at the next development of physics of superconductors this theory obtained numerous experimental supports. Finally, several years later it was accepted in total, when it consequently explained the complex behavior of superconducting alloys and compounds, in particular the very high critical fields of some materials. As has been noted for type-II superconductors, the energy of an interface between a normal and a superconducting region $\sigma_{ns} < 0$. Total displacement of external field from superconductor does not lead to a state with the least energy, if the contribution of surface energy of the interface between two phases is significant. Therefore, in this case, the energetically favorable state is that at which superconductor of corresponding shape (any one, besides an infinitely long cylinder placed in a parallel magnetic field) is divided into great number of alternating superconducting and normal regions.

The magnetization curve of a type-II superconductor in the form of a long cylinder, placed in a parallel magnetic field, is shown schematically in Fig. 1.9. As long as the external field is $H < H_{c1}$, the field in the interior of the sample is absent ($B = 0$). However, at $H_{c1} < H < H_{c2}$, a steadily increasing field penetrates the superconductor in the form of flux lines (vortices). It remains below the external field H, and superconductivity of the sample is not destroyed. At a certain field $H = H_{c2}$, the field in the interior becomes equal to external field H, and the bulk superconductivity disappears. In contrast to the type-I superconductors, the superconductivity may easily originate in a heterogeneous way in these materials. In this case, the field of the superconductivity initiation H_{c2} may be well above the critical thermodynamic field H_{cm}. Between H_{c1} and H_{c2}, the material has no electric resistance and contains a lattice of flux lines, which can be simplistically treated as normal regions. This is the mixed state, also known as the Shubnikov phase [987]. In dirty materials (with defects) vortices remain anchored onto impurities, leading to significant hysteresis and even paramagnetization curves, as shown in Fig. 1.9.

Thus, the above H_{c1} type-II superconductors do not show the Meissner effect. Magnetic field penetrates into these materials in the form of quantized flux lines (vortices), each of which has a normal core, which can be approximated by a long thin cylinder with its axis parallel to the external magnetic field. Inside the cylinder, the order parameter, $\Psi = 0$. The radius of the cylinder is of the order ξ, the coherence length. The direction of the super-current, circulating around the normal core, is such that the direction of the magnetic field, generated by it, coincides with that of the external field and is parallel to the normal core. By this, the vortex current circulates into region with radius of the order λ. The size of this region is by far above a value ξ, because $\lambda >> \xi$ for type-II superconductors [935].

Each vortex carries one magnetic flux quantum. Penetration of vortices in the interior of a superconductor becomes thermodynamically favorable at $H > H_{c1}$. Inside the superconductor, the vortices arrange themselves at distances $\sim\lambda$, from each other, so that in the cross-section, they form a regular triangular or square lattice (see Fig. 1.10). This state of superconductor (at $H_{c1} < H < H_{c2}$) is the mixed state, because it is characterized by a partial penetration of the magnetic field in the interior of the sample. Once formed at H_{c1}, the vortex lattice persists at much higher fields. As the external field increases, the lattice period steadily decreases and the density of the vortices rises. Finally, at a field $H = H_{c2}$, the vortex lattice becomes so dense that the distance between the neighboring vortices, that is, the lattice period attains the order ξ. This means that the normal cores of the vortices come into contact with each other and the order parameter Ψ becomes zero over the total volume of the superconductor, that is, a second-order phase transition occurs.

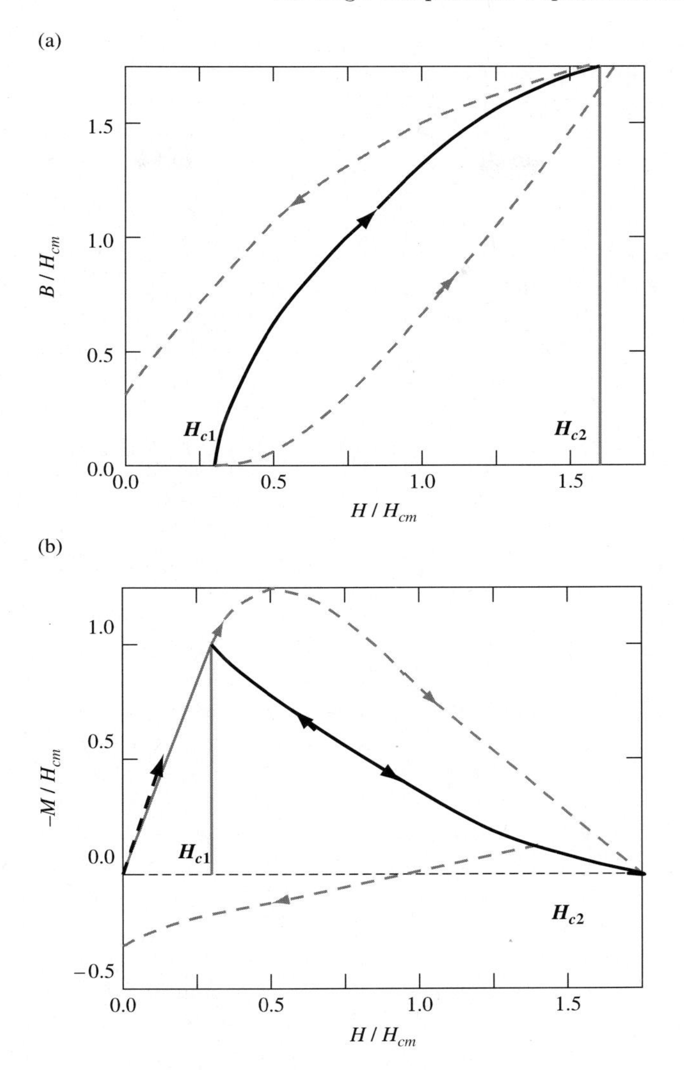

Fig. 1.9. Magnetic induction (**a**) and magnetization (**b**) as a function of applied field in type-II superconductors. *Dotted lines* correspond to impure samples [617]

1.6.2 Doping of Cuprates

The simplest copper oxide perovskites are insulators. In order to become superconducting, they should be doped by charge carriers. There are two ways to increase the number of charge carriers in cuprates chemically: (i) to substitute metallic atoms in the intermediate planes by higher-valence atoms and/or (ii) to change the number of oxygen atoms. Doping increases the number electrons or holes at the Fermi level. The concentration of charge carriers

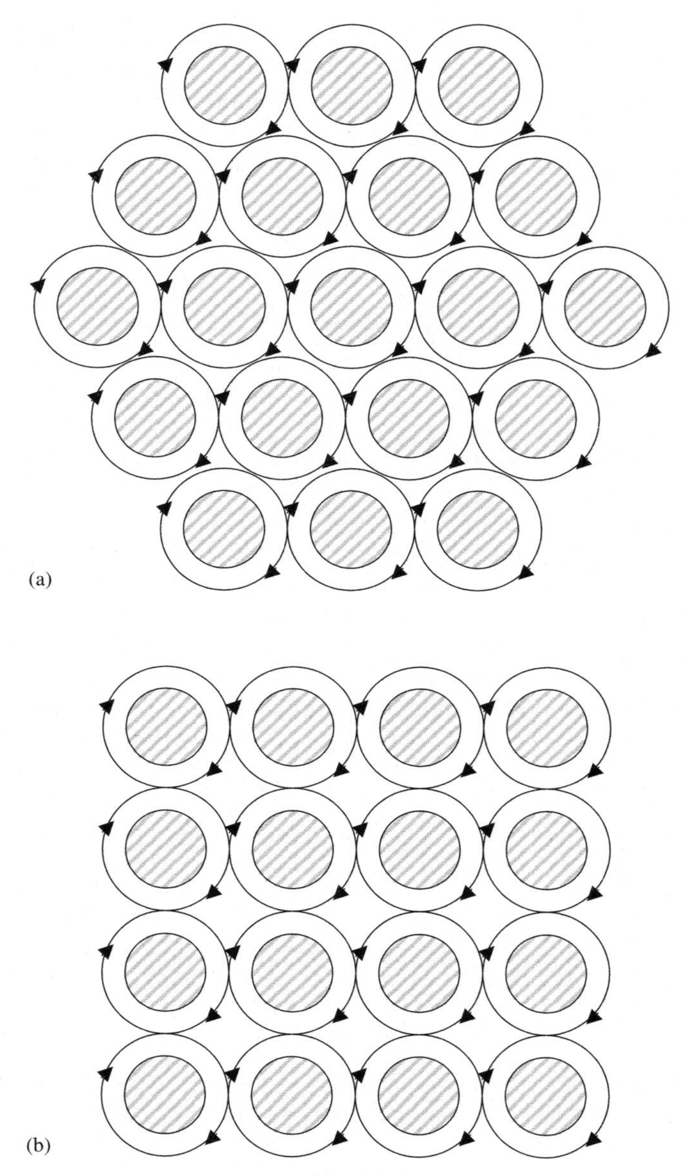

Fig. 1.10. Mixed state of a type-II superconductor. Superconducting vortices form a regular triangular (hexagonal) (**a**) or square (**b**) lattice. Vortex cores (*dashed regions*) are normal

in HTSC is low ($\sim 5 \times 10^{21}$), in comparison with conventional superconductors ($\sim 5 \times 10^{22} - 10^{23}$). However, due to the large coherence length in conventional superconductors, only a 10^{-4} part of the electrons, located near the Fermi surface, participate in coupling. At the same time, in cuprates, $\sim$ 10% of all conduction electrons (holes) form the Cooper pairs.

In conventional superconductors, the critical temperature increases monotonically with growth of charge carriers: $T_c(p) \propto p$. In cuprates, this dependence is non-monotonic. In most of hole-doped cuprates, (but not in all) the $T_c(p)$ dependence has the bell-like shape and can be approximated as [866]

$$T_c(p) \cong T_{c,\max}[1 - 82.6(p - 0.16)^2] , \tag{1.15}$$

where $T_{c,\max}$ is the maximum critical temperature for a given compound.

Superconductivity occurs within the limits, $0.05 \leq p \leq 0.27$, which vary slightly in various cuprates. Thus, the different doping regions of the superconducting phase may be chosen such as *the underdoped, optimally doped and overdoped* regions (Fig. 1.11). The insulating phase at $p < 0.05$ is called *the undoped* region, but above 0.27, cuprates become *metallic*.

The ratio between the maximum critical temperatures in hole-doped and electron-doped cuprates is 135 K/24 K = 5.6. Obviously, it may be assumed that the electron–hole asymmetry has fundamental character: superconducting hole-doped compounds will always have the critical temperature a few times higher than the same electron-doped superconductor.

In cuprates and in many other compounds with low dimensionality, the distribution of charge carriers is inhomogeneous. Moreover, in cuprates, this distribution is inhomogeneous in a micro-, as in a macroscopic scale (Fig. 1.12). In the undoped region ($p < 0.05$), doping holes are preferably distributed inhomogeneously into CuO_2 planes, and they form dynamical 1D charge strips (so-called *charge-strip phase*). In the undoped region, these strips have diagonal shapes and locate not along–O–Cu–O–Cu–bonds, but along the diagonal–Cu–Cu–Cu–direction, as shown in Fig. 1.12. In undoped cuprates, the concentration of holes is low, but the distance between charge strips, separated by 2D insulating antiferromagnetic domains, is large. The charge-strip

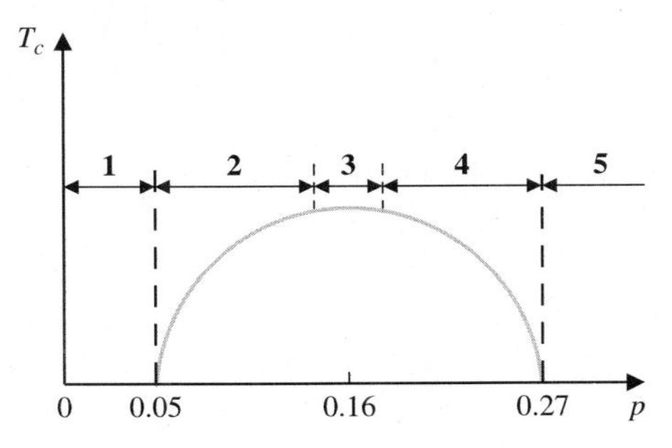

Fig. 1.11. Critical temperature as a function of doping [737]. *Numbers* mark the next regions: 1, undoped ($p < 0.05$); 2, underdoped ($0.05 \leq p \leq 0.14$); 3, optimally doped ($0.14 < p < 0.18$); 4, overdoped ($0.18 \leq p \leq 0.27$); and 5, metallic phase ($p > 0.27$)

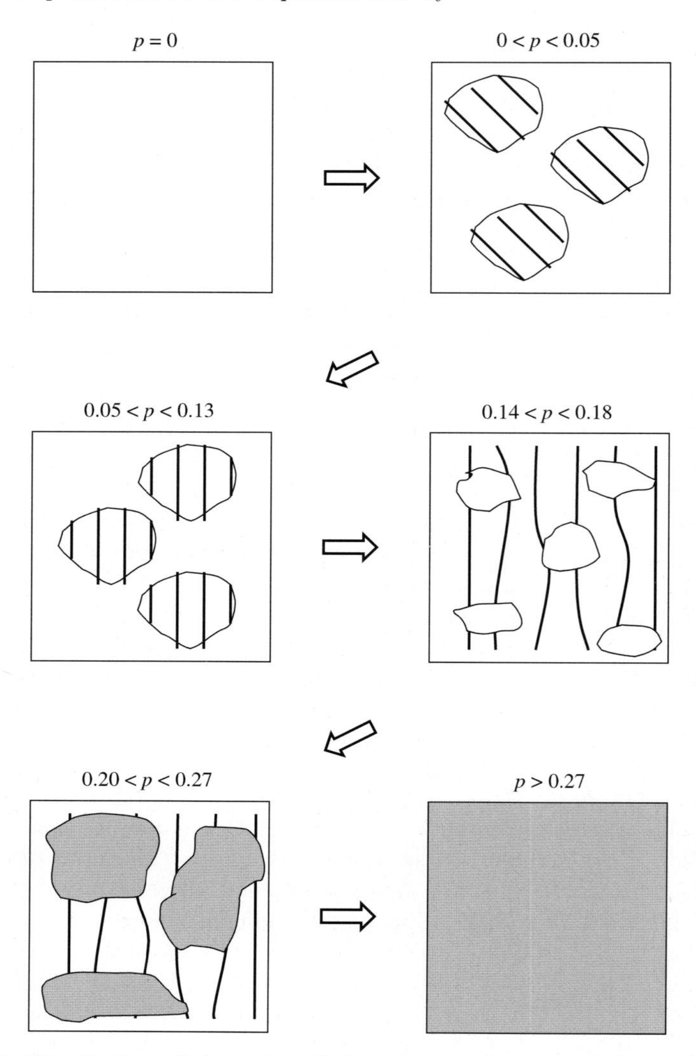

Fig. 1.12. Distribution of charges in CuO_2 planes as function of doping [737]. The antiferromagnetic and metallic phases are shown in *white* and *gray*, respectively. The *lines* depict charge strips

phase is distributed inhomogeneously: there are two types of small islands, containing either the antiferromagnetic or the charge-strip phase.

In the underdoped region ($0.05 < p < 0.13$), charge strips are vertical (or horizontal) and located closer to each other. In this region, the average distance between charge strips d_s is approximately proportional to $1/p$ and saturates at $p = 1/8$ (Fig. 1.13). Above $p = 1/8$, the distance between strips is practically constant. As p increases, the concentration of antiferromagnetic regions decreases, but the two types of islands, containing the antiferromagnetic

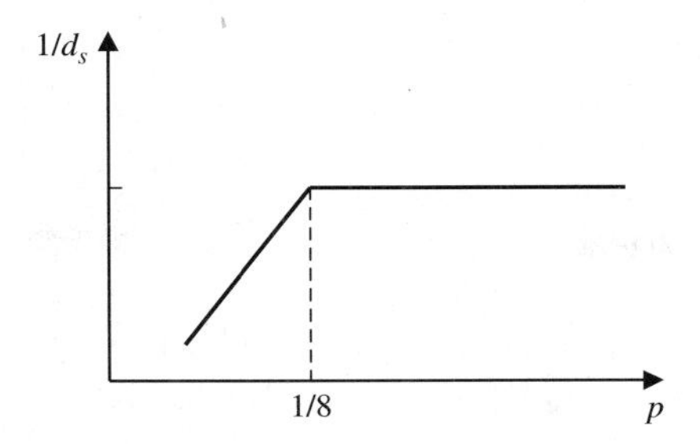

Fig. 1.13. Dependence of incommensurability $\delta(\propto 1/d_s)$ of spin fluctuations on doping level [1154]

phase and the vertical charge-strip phase, still co-exist. The dynamical charge strips can move in the transverse direction, and they are *quasi-1D*.

In the near optimally doped region ($p \sim 0.16$) and in the overdoped region ($0.2 < p < 0.27$), the average distance between charge strips remains almost constant (see Figs. 1.12 and 1.13). Therefore, as the doping level increases, new doped holes cover antiferromagnetic islands, which completely vanish at $p = 0.19$. Above this value, small metallic islands start appearing. Above $p = 0.27$, the charge-strip distribution becomes homogeneous in 2D CuO_2 planes, and cuprates transform in non-superconducting (normal) metals.

1.6.3 Coherence Length and HTSC Anisotropy

Despite the fact that there is no definite theory to explain high critical temperatures of HTSC, their magnetic and superconducting properties can be well described in the framework of the classical BCS/Ginzburg–Landau theory. They demonstrate a set of properties that are similar to conventional low-temperature superconductors. In particular, superconductivity in cuprates occurs due to coupling of electrons. Moreover, there is energy gap in a spectrum of electron excitations that is caused by electron coupling. Non-monotonous dependence $T_c(p)$ (see Fig. 1.11) is similar to non-monotonous behavior $T_c(p)$ of superconducting semiconductors. Finally, isotope effect also exists in cuprates, while it is directly found by the concentration of holes [737].

The main difference from conventional superconductors is caused by intrinsic material properties, for example, the extremely short coherence length ξ (in conventional superconductors $\xi = 400 - 10^4$ Å). Short coherence length is a consequence of the big energy gap and the small Fermi velocity. Due to the extremely short coherence length, even a grain boundary can be sufficient to suppress superconductivity in cuprates. In particular, the grain boundaries can be used to fabricate devices of Josephson type (in the form of epitaxial films on bicrystalline substrates), based on the existence of weak links.

The second important property of high-temperature superconductors is their huge anisotropy caused by the layered crystalline perovskite structure.[8] For example, $Bi_2Sr_2CaCu_2O_8$ (Bi-2212) crystal, presented in Fig. 1.14a, consists of the sequence of CuO_2 planes, alternating with other oxide layers. The basic block is the CuO_2 double layer (intercalated by Ca). These blocks are separated by four oxide layers namely two SrO and two BiO ones. In Fig. 1.14b, a process of intercalation of the additional Ca/CuO_2 plane is present to form $Bi_2Sr_2Ca_2Cu_3O_{10}$ (Bi-2223) crystal.

Due to 2D structure of cuprates, the coherence length depends on the crystallographic direction, namely along c-axis the value ξ_c is far lesser than in ab-plane (ξ_{ab}). In the different hole-doped cuprates, $\xi_{ab} = 10$–35 Å, at the same time, $\xi_c = 1$–5 Å. As a rule, the coherence length of cuprates with low critical temperature is longer than in cuprates with high T_c (see Table 1.6). In electron-doped NCCO, the coherence length is sometimes longer than in other hole-doped cuprates. Small values of ξ_c mean that transport along the c-axis is not coherent, even in the superconducting state. For example, $\xi_c \sim 1$ Å in Bi-2212 that is sometimes shorter, compared to the distance between layers.

Two critical fields $H_{c2||}$ and $H_{c2\perp}$, directed parallel and perpendicular to the basic ab-plane, respectively, correspond to two principle axes (in ab-plane and along c-axis). The notation is as follows. The upper critical field perpendicular to the ab-plane, $H_{c2\perp}$, is determined by vortices (with magnetic flux Φ_0), whose screening currents flow parallel to this plane. Then, for the dependence between critical field and coherence length, we have the following equation [935]:

$$H_{c2\perp} = \frac{\Phi_0}{2\pi\xi_{ab}^2} \,. \tag{1.16}$$

The indices of ab or c of parameters λ and ξ show the directions of the screening currents.

Due to the high-temperature superconductors possess the layered crystal structure superconductivity in HTSC is confined to the CuO_2 planes. They are separated from neighboring planes by weakly superconducting, normal or even insulating regions of the crystal. Three-dimensional phase coherence is provided by the Josephson currents, which flow between above planes.

If we assume a homogeneous order parameter and use description of anisotropy in the framework of the Ginzburg–Landau theory, then for parameter $H_{c2||}$ we have

$$H_{c2||} = \frac{\Phi_0}{2\pi\xi_{ab}\xi_c} \,; \tag{1.17}$$

then the anisotropy ratio is

$$\frac{H_{c2||}}{H_{c2\perp}} = \frac{\xi_{ab}}{\xi_c} \,. \tag{1.18}$$

[8] Crystalline structures of some HTSC are present in Appendix A.

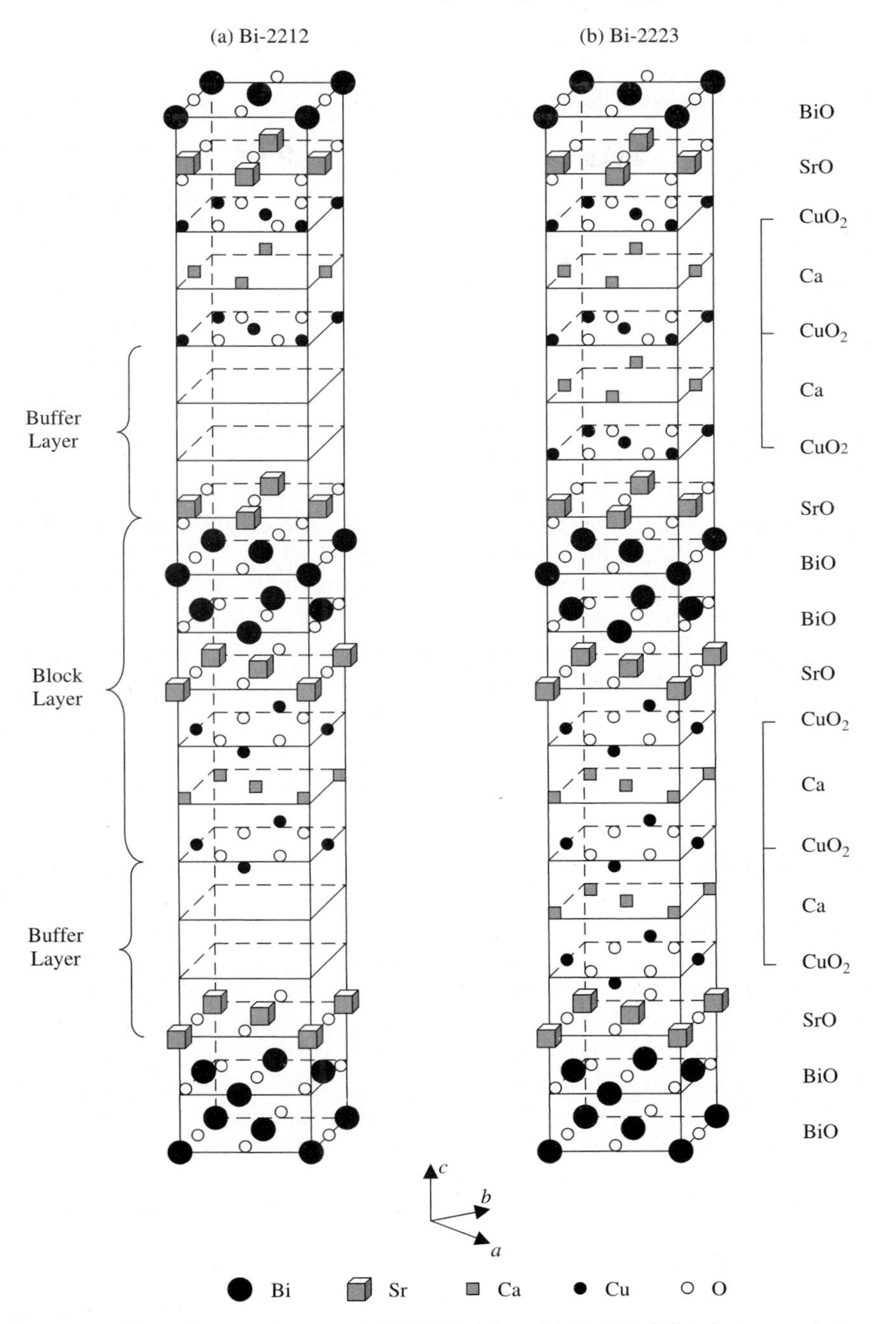

Fig. 1.14. Crystalline structures of Bi-2212 (**a**) and Bi-2223 (**b**). The layered structure of Bi-2212 can be divided into block layers and buffer layers for intercalation of additional Ca/CuO_2 plane, forming Bi-2223. The height shown of buffer layers in the structure of Bi-2212 is expanded to compare with real one for clearness of comparison of both structures [119]. Below, the principal axes a, b and c are shown

Table 1.6. Characteristics of optimally doped cuprates [737]

Material	T_c(K)	ξ_{ab} (Å)	ξ_c (Å)	λ_{ab} (Å)	λ_c (Å)	$B_{c2\|\|}$ (T)	$B_{c2\perp}$ (T)
NCCO	24	70–80	∼ 15	1200	260000	7	–
LSCO	38	33	2.5	2000	20000	80	15
YBCO	93	13	2	1450	6000	150	40
Bi-2212	95	15	1	1800	7000	120	30
Bi-2223	110	13	1	2000	10000	250	30
Tl-1224	128	14	1	1500	–	160	–
Hg-1223	135	13	2	1770	30000	190	–

In the case of weakly anisotropic material, such as $YBa_2Cu_3O_{7-x}$, this representation is sufficient. However, for anisotropic material, such as $Bi_2Sr_2CaCu_2O_8$, the value of ξ_c then would be of the order of 0.1 nm, that is, approaching atomic scales. In any case, this contradicts the assumption of a homogeneous order parameter. In cuprates, low critical fields $B_{c1||}$ and $B_{c1\perp}$ are very small. For example, in YBCO $B_{c1||} \sim 2 \times 10^{-2}$ T and $B_{c1\perp} \sim 5 \times 10^{-2}$ T. It is interesting that anisotropy of values B_{c1} has different sign than of B_{c2}, namely $B_{c2\perp} < B_{c2||}$ and $B_{c1\perp} > B_{c1||}$. In conventional superconductors, $B_{c2} \propto T_c^2$; at the same time, $B_{c2} \propto T_c^{\sqrt{2}}$ [737] in cuprates with low T_c. The extreme anisotropy is also responsible for many particular effects associated with the flux line lattice in high-temperature superconductors.

1.6.4 Vortex Structure of HTSC and Magnetic Flux Pinning

The layered structure of the cuprate superconductors, with the superconductivity arising within the CuO_2 planes, causes the properties of a single vortex. The orientation of the CuO_2 planes is defined by the crystallographic a- and b-axes. The CuO_2 planes are coupled to each other by Josephson junctions. Lawrence and Doniach proposed the phenomenological model for this layered structure [607]. The Lawrence–Doniach theory contains the anisotropic Ginzburg–Landau and London theories as limiting cases, when the coherence length ξ_c in c-direction exceeds the distance between layers s. In this limit, the anisotropy may be considered in terms of the reciprocal mass tensor with the principal values $1/m_{ab}$, $1/m_{ab}$ and $1/m_c$. Here m_{ab} and m_c are the effective masses of Cooper pairs, moving into ab plane and along the c-axis, respectively. If the interlayer coupling is weak, then we have $m_{ab} \ll m_c$. In the framework of the anisotropic Ginzburg–Landau limit, the extended relations (1.18) may be obtained:

$$\left(\frac{m_c}{m_{ab}}\right)^{1/2} = \frac{\lambda_c}{\lambda_{ab}} = \frac{\xi_{ab}}{\xi_c} = \frac{H_{c2||}}{H_{c2\perp}} = \frac{H_{c1\perp}}{H_{c1||}} . \tag{1.19}$$

If magnetic field is oriented along the c-axis, the flux lines reduce to stacks of 2D point vortices or pancake vortices. A detailed modeling of the layered cuprate superconductor in terms of a stack of thin superconducting films in the framework of the Lawrence–Doniach theory has been carried out [158,283]. Energetically, the perfect stacking of the pancake vortices along c-axis is favorable, than a more disordered structure. At the same time, compared to a continuous flux line, as it exists in the conventional superconductors, a stack of the pancake vortices has additional degrees of freedom for thermal excitations As an example, we consider the displacement of a single pancake vortex, presented in Fig. 1.15. This displacement is equivalent to the excitation of a vortex–antivortex pair (Kosterlitz–Thouless transition), possessing the interaction energy [452]: $U(r) = \varphi_0^2/\mu_0 r$, where φ_0 is the quantum of magnetic flux, μ_0 is the vacuum permeability and r the distance between the vortex and antivortex. For 2D screening length, Λ, we have the binding energy $U(r) \approx \varphi_0^2/\mu_0\Lambda$. Interpreting the displacement of a single pancake vortex as an evaporation process, the evaporation temperature, T_D, has form [158]:

$$T_\mathrm{D} \approx \frac{1}{k_\mathrm{B}} \frac{\varphi_0^2}{\mu_0 \Lambda} \,. \tag{1.20}$$

When an external field is nearly parallel to the ab-plane, the vortex core preferably runs between the CuO_2 layers. When the coupling between layers is weak, vortex lines along the ab-plane are referred to as Josephson vortices or strings. For any magnetic field direction not parallel to the ab-plane, the

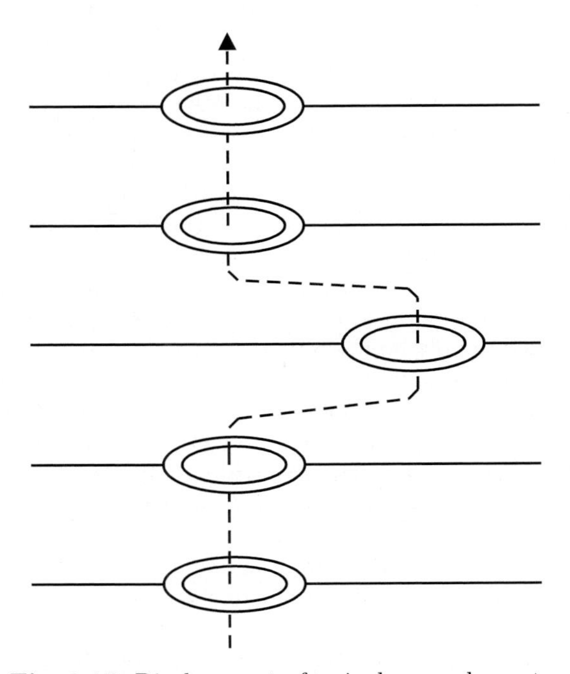

Fig. 1.15. Displacement of a single pancake vortex

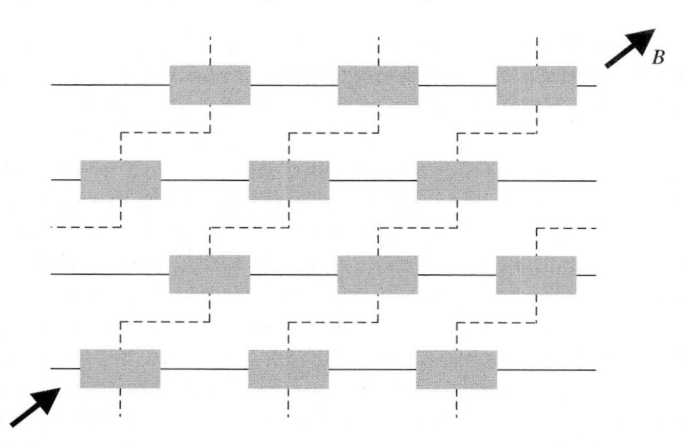

Fig. 1.16. Pancake vortices (*gray rectangles*) coupled by Josephson strings (*horizontal dashed lines*)

pancake vortices existing in the CuO_2 planes are coupled by such Josephson strings, as shown in Fig. 1.16.

Investigation of the temperature and magnetic field dependence of the magnetization in the powder samples of Ba–La–Cu–O [740] discovered *an irreversibility line* in the H–T phase space (H is the magnetic field, T is the temperature). Above this line, the magnetization is perfectly reversible with no detectable magnetic flux pinning. However, below this line the hysteresis of *magnetization* arises, and the equilibrium vortex distribution can no longer be established due to *the magnetic flux pinning.* Soon after this discovery, a similar line was found in YBCO single crystal [1172]. Due to these and similar observations, the concept of "vortex matter" may be stated, taking *a liquid, glassy or crystalline* state in the phase diagram. These features have important influence on the transport processes associated with vortex motion. The vortex lattice, originally proposed by Abrikosov [03], consisted of a regular configuration of the magnetic flux lines in the form of a triangle (hexagonal) or square lattice that minimized their interaction energy (see Fig. 1.10). In HTSC, thermal energies are large enough to melt the Abrikosov vortex lattice (at $H = H_{\mathrm{m}} < H_{\mathrm{c2}}$), forming a vortex liquid over a large part of the phase diagram. In addition to the high temperatures, there is the structure of magnetic flux lines, consisting of individual, more or less strongly coupled pancake vortices, which promote this *melting transition.* In order to avoid energy dissipation at existence of transport current, each vortex should be fixed at pinning center. In this case, linear and plane defects are most effective. Increasing the number of defects is capable of moving the line $H_{\mathrm{m}}(T)$ at phase diagram into a region of greater values of H and T. There is a universal field, H_1, such that $H_1(T) < H_{\mathrm{m}}(T) < H_{\mathrm{c2}}(T)$, at which thermodynamic fluctuations of order parameter lead to tearing of vortices from lengthy pinning centers [281]. This field presents upper boundary of the irreversibility field, $H_{\rm irr}(T)$, at which the dissipation begins.

The simplest theoretical description of a melting transition is based on the Lindemann criterion, according to which a crystal melts if the thermal fluctuations $\langle u^2 \rangle^{1/2} = c_{\rm L} a$ of the atomic positions are of the order of the lattice constant a. The Lindemann parameter $c_{\rm L} \approx 0.1$–0.2 depends only slightly on the specific material. This approach has been used to determine the melting transition of vortex lattices. In particular, expressions for the melting temperature in the 2D and 3D cases have been derived [71,82,84,284,1064]. A schematic phase diagram for melting of the solid vortex lattice for 3D material such as YBCO with field applied parallel to the c-axis is shown in Fig. 1.17.

The various parts of the vortex phase diagram are caused by competition of four energies, namely thermal, vortex interaction, vortex coupling between layers and pinning. The thermal energy pushes the vortex structure towards the liquid state, the interaction energy favors lattice state, the coupling energy tends to align the pancake vortices in the form of linear stacks and the pinning energy generates disorder. An interaction of these energies, whose relative contributions vary strongly with magnetic field and temperature, results in the complex phase behavior defined by the vortex matter. Then, the irreversibility line may be interpreted as the melting line, above which the state of the vortex liquid is attained, and below which there is the state of the vortex glass or vortex lattice. The vortex glass state is connected with magnetic flux pinning in the sample, disrupting any vortex motion. The first clear evidence for vortex lattice melting has been obtained from transport measurements (electrical

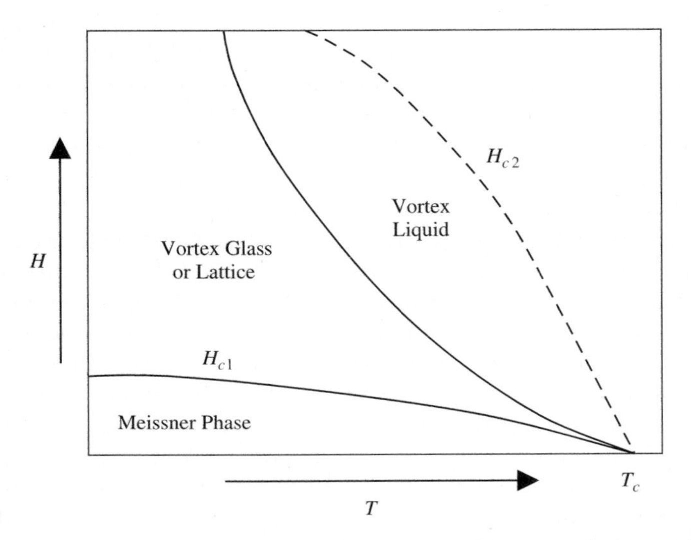

Fig. 1.17. Schematic phase diagram for a three-dimensional material such as YBCO [452]

resistance) for de-twinned single crystals of YBCO with $\boldsymbol{H}$ parallel to the c-axis. At a well-defined freezing temperature of the magnetic flux T_{m}, which depends on the magnetic field, a sudden drop to zero of the resistivity was observed, defining the onset of strong pinning in the vortex solid. The sharp drop of the resistivity at T_{m} demonstrates *a first-order freezing transition* [165]. The first order vortex-lattice melting transition has been observed in thermodynamic measurements, using a high-quality single crystal of BSCCO with $\boldsymbol{H}$, again parallel to the c-axis [1191]. Review devoted to vortex matter and its melting transition has been presented in [166].

Early investigations of transport processes in HTSC demonstrated the power dependence of volt–ampere I–V characteristic (I is the current, V is the voltage) [561,1189] that in the following has been selected as a criterion for the freezing transition into limits of the superconducting vortex glass structure. In another interpretation, a distribution of the activation energy is used for this [347].

Weakening of flux pinning by melting of the vortex lattice is expected only when there are many more flux lines to compare with existing pinning centers. At the same time, in the opposite case, softening of the vortex lattice often leads to stronger pinning than in a rigid vortex lattice. This is explained by the concept that the atomic-scale defects (also as oxygen vacancies) can act as pinning centers for HTSC (the case is often realized in practice). Therefore, melting of vortex matter does not necessarily result in a reduction of pinning. Because of the complexity of this question, there is no simple answer (I see reviews [83,84]). Flux pinning is caused by spatial inhomogeneity of the superconducting material, leading to local depression in the Gibbs free energy density of the magnetic flux structure. Due to the short coherence length in HTSC, inhomogeneities, even on an atomic scale, can act as pinning centers. As these important examples, we note deviations from stoichiometry, oxygen vacancies in the CuO_2 planes, and twin boundaries. The separation of a flux line into individual pancake vortices also promotes pinning caused by atomic size defects.

An original discussion of magnetic flux pinning caused by atomic defects in the superconducting CuO_2 planes (in this case by oxygen vacancies) has been carried out in [528,1102]. By this, the elementary pinning interaction of vortices with the oxygen vacancies was calculated, and the vacancy concentration was related to the critical current density. The various structure defects in HTSC, acting as *pinning centers*, were considered in review [1148], but a detailed research of pinning effect on magnetic relaxation has been carried out in paper [1173]. An advance in solution of the problem of the statistical summation of pinning forces has been attained in the framework of the Larkin–Ovchinnikov theory of collective pinning [603]. In this theory, the elastic deformation of the vortex lattice in the presence of a random spatial distribution of pinning centers plays a central role, but the increase of the elastic energy is balanced by the energy gained by passing the flux lines through favorable pinning sites. A discussion of corresponding physical basis is represented in monograph [1064]. We shall return to the discussion of the pinning problem in other chapters of our book.

1.6.5 Interactions of Vortices with Pinning Centers

As has been noted, for the attainment of high density of the critical current, it is necessary that microstructure of superconductor retained vortex lines of magnetic flux on the moving, caused by the Lorentz forces. It is reached only by pinning of vortices on the microstructure heterogeneities (or defects). However, no any defect can effectively interact with vortex lines. For example, in conventional superconductors, vacancies, individual atoms of secondary phases or other similar tiny defects are not effective pinning centers due to obvious causes: as a rule, a specific size of vortex (coherence length) is far greater than atomic size, that is, proper size of this defect. Therefore, vortex line simply "does not notice" them. On the contrary, the structure defects with size of $\sim \xi$ and greater become effective ones in this sense, and they can cause high density of critical current.

However, in the case of HTSC there is another situation. Here, the coherence length is extremely short, and point defects have sizes commensurable with ξ. Therefore, consider in more detail a situation arisen on example of vortex interaction with a cavity in superconductor.

Consider an infinite superconductor, containing a defect in the form of a cylindrical cavity. How will a single vortex parallel to the cavity interact with it? Assume that the diameter of the cavity, d, satisfies the inequality $d > \xi(T)$. If the vortex is far away from the cavity, its normal core (of diameter $\sim 2\xi$) stores a positive energy (relative to the energy of the superconductor without the vortex), because the free energy of the normal state exceeds that of the superconducting state by $H_{\mathrm{cm}}^2/8\pi$ (per unit volume). Then, the energy of the normal core (per unit length) is

$$\frac{H_{\mathrm{cm}}^2}{8\pi}\pi\xi^2 . \tag{1.21}$$

On the other hand, if the vortex is trapped by the cavity, that is, passes through its interior, then it does not have a normal core and, accordingly, the energy of the system is reduced by the amount of (1.21). This means that the vortex is attracted to the cavity. The interaction force per unit length, f_{p}, can be found easily, if we recall that the energy changes by the value of (1.21), when the vortex changes its position near the edge of the cavity by $\sim \xi$

$$f_{\mathrm{p}} \approx H_{\mathrm{cm}}^2\xi^2/8 . \tag{1.22}$$

For a spherical cavity of diameter d, the interaction force f_{pd} caused by the vortex can be obtained from (1.22) in the form

$$f_{\mathrm{pd}} \approx H_{\mathrm{cm}}^2\xi d/8 . \tag{1.23}$$

To get an idea of how large this force is, let us find the current (j) that must be applied in the direction perpendicular to the vortex in order to produce the Lorentz force $f_{\mathrm{L}} > f_{\mathrm{pd}}$. It is known that the Lorentz force per unit length of a vortex is $j\Phi_0/c$ (Φ_0 is the magnetic flux of the vortex, c is the speed of light in vacuum) [935]. Then, the force applied to the part of the vortex which actually interacts with the defect is $j\Phi_0 d/c$. Equating this to f_{pd}, from (1.23), we obtain

$$j = \frac{cH_{\mathrm{cm}}^2}{8\Phi_0}\xi \,. \tag{1.24}$$

Furthermore, since $H_{\mathrm{cm}} = \Phi_0/2\sqrt{2}\pi\lambda\xi$ [935], then from (1.24) we obtain

$$j = \frac{cH_{\mathrm{cm}}}{16\sqrt{2}\pi\lambda} \,. \tag{1.25}$$

It may be shown that j in the last expression is of the same order of magnitude as the Cooper pair-breaking current [935]. Thus, in order to tear the vortex off a spherical void, it is necessary to apply the maximum possible current for a given superconductor.

The above discussion is also applied to a superconductor, containing tiny dielectric inclusions. It also remains valid (at least by the order of magnitude) for normal metal inclusions, provided the size of the inclusions is larger than ξ. The restriction is caused by the *proximity effect*,[9] which is essential only at distances of the order of the coherence length ξ from the interface. Therefore, the various types of inclusions present effective pinning centers in superconductors. This property is widely used in technical applications, when large critical currents and magnetic fields act.

As effective pining centers in superconductors we also note dislocations, dislocation walls, grain boundaries and interfaces between different superconductors. The above analysis of vortex-defect interactions forms the basis for interpretation of effects caused by the so-called *columnar defects* [713], produced in high-temperature superconductors by irradiation. In this case, the normal cylindrical amorphous regions are formed in the superconductor. The nuclear tracks represent themselves as cylindrical regions with diameter being only 5–10 nm, that is, of the order of the coherence length ξ. As we have just seen, cavities of such sizes are effective pinning centers. This is one of the very few possibilities for producing artificial defects of sizes comparable with the coherence length in HTSC. In given materials, the coherence length is so short that any metallurgical methods used to produce pinning centers, such as precipitates or grain boundaries, have small effect, due to the disproportional size of defects, or not applicable at all, due to the brittleness of the material.

[9] If thin layer of a superconductor is brought in contact with layer of a normal metal, then pairs of electrons with summary zero impulse, formed in the superconductor will stretch on both layers. Proximity effect depends on the interface nature and the relative thickness of both layers.

Each defect is capable of trapping approximately one vortex. Hence the optimum pinning efficiency can be expected at magnetic fields, for which the vortex lattice period is less than the average distance between the amorphous tracks. When the Lorentz force[10] caused by the applied field becomes greater than an interaction force with defect or inhomogeneity, then a displacement of vortex flux leads to energy dissipation and initiation of finite electrical resistance. This state of vortex structure is called the *resistive state.*

As may be expected, vortex moving caused by the Lorentz force and corresponding resistance to magnetic flux strongly depend on three unique properties of HTSC, namely high critical temperature, small coherence length and layered anisotropic structure. Combining the above features strongly facilitates vortex moving and destroys superconductivity. Due to this, the *resistive transition* essentially expands in magnetic field. This is one from causes, stimulating tremendous efforts of material scientists directed to change of the HTSC manufacture technology with aim to decrease the vortex moving, making more active the magnetic flux pinning. Obviously, this problem will be complicated with the discovery of superconductors, possessing even greater temperatures of transition into superconducting state.[11] Composition features and modern manufacture techniques for basic HTSC systems will be considered in Chap. 2.

1.7 Weak Links of Josephson Type

In 1962, Josephson published a theoretical paper [500] predicting the existence of two remarkable effects. One was supposed to find them in superconducting tunnel junctions. The basic idea of the *first effect* was that a tunnel transition should be able to sustain a superconducting (i.e., zero-voltage) current. The critical value of this current was predicted to depend on the external magnetic field in a very unusual way. If the current exceeds the critical value, which is a characteristic of a particular junction, the junction begins to generate high-frequency electromagnetic waves. This phenomenon was called *second Josephson effect.*

Soon after their discovery, both effects obtained experimental confirmation [968, 1167]. Moreover, it soon became clear that the Josephson effects exist not only in tunnel junctions, but also in other kinds of the so-called weak links, in particular, short sections of superconducting circuits, where the critical current is substantially suppressed [630,631,1004]. So, initiating weak superconductivity is based on the quantum nature of the superconducting state that assumes the existence of condensate of the Cooper pairs. This means that all electron pairs in the superconducting state occupy the same

[10] In general case, the Lorentz forces can be initiated as transport current, as screening currents caused by sample magnetization.

[11] Nevertheless, it is not obligatory that superconductors with transition temperatures near to room temperature will possess oxide structure.

quantum level and are described by a single wave function, common to all of them. Their behavior is mutually conditioned and they are coherent.

Consider two bulk superconductors having the same temperature and completely isolated from each other. The behavior of the superconducting electrons in each of them is subjected to own wave function, when they are in the superconducting state. Then, because the temperatures and the materials of the superconductors are identical, the amplitudes of the wave functions must also be the same. However, the phases, in this case, are arbitrary. This situation remains as long as the superconductors are isolated from each other. Let us establish a weak contact between them, that is, the contact that is weak enough so as not to change radically the electron states of the two superconductors, but sufficient to initiate a perturbation. In this case, a new wave function is formed that is general for the joint superconductor, which can be considered as a result of interference between the wave functions of both its pieces. Therefore, phase coherence is a direct result of establishing the weak link.

The weak links (Josephson junctions, JJs) can be classified in the following way:

(1) *Devices without concentration of current such as tunnel junctions of the "superconductor–insulator–superconductor" type* (S–I–S) (Fig. 1.18a). The thickness of an insulating layer, as a rule, is about 1–2 nm, and the critical current density is in the order of 10^4 A/cm^2, that is, much less than the critical current density of the bulk superconductor, in particular, exceeding 10^5 A/cm^2 at 77 K and 0 T in Y(*RE*)BCO melt-processed samples.[12]

(2) *Layered structure of the "superconductor–normal metal–superconductor" type* (S–N–S). It involves the normal layer with thickness of $\sim 1\,\mu$m (Fig. 1.18b). The wave functions of the superconducting electrons penetrate the normal metal due to the proximity effect. In the region of their overlap, the wave functions interfere, establishing phase coherence between bulk superconductors. If the amplitude of the superconducting wave function in the weak link is small, then the critical current is also small.

(3) *Layered structures where a normal layer between two superconductors is replaced by a doped semiconductor or another superconductor with a small critical current density.* For example, if a narrow superconducting film is covered by thin film of a normal metal (Fig. 1.18c), then the amplitude of the superconducting electron wave function in the film is reduced, where the film contacts with normal metal, due to proximity effect. This causes local decrease of the critical current density, that is, weak link formation.

(4) *Devices with concentration of current.* The critical current density in the weak link is the same as in the bulk, but the absolute value of the critical current is much less. A superconducting film with a short narrow constriction (Dayem bridge) falls into this category, provided the size of the

[12] High-quality tunnel Josephson junctions, made of niobium with a barrier layer of aluminum oxide, attain the critical current density of $\sim 10^4 - 10^5$ A/cm^2.

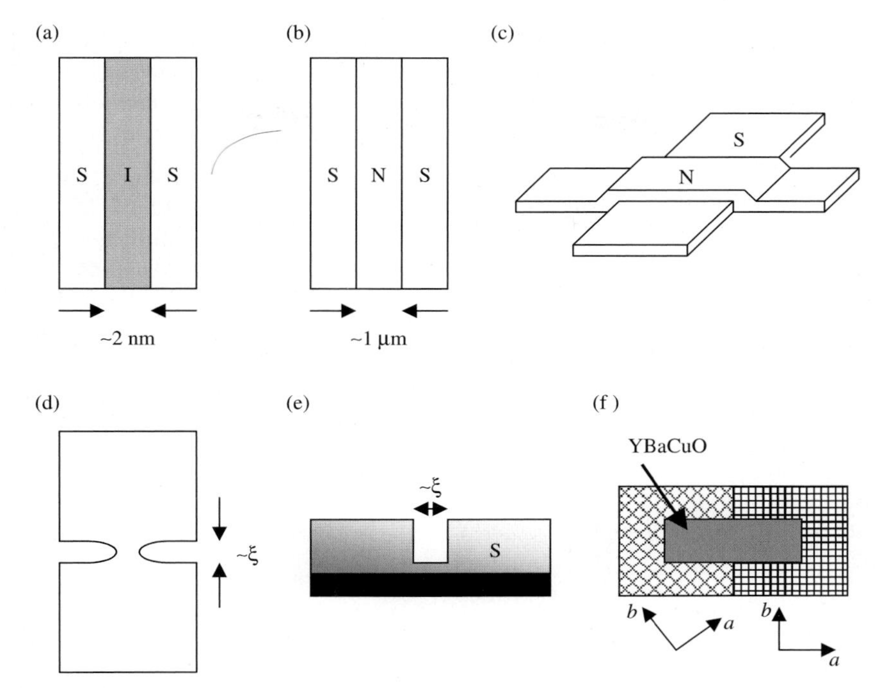

Fig. 1.18. Different types of weak links: (**a**) tunnel junction (S–I–S); (**b**) sandwich structure (S–N–S); (**c**) normal film (N) causes local suppression of the order parameter of superconducting film (S); (**d**) Dayem bridge; (**e**) bridge of variable thickness (longitudinal cross-section); (**f**) grain-boundary junction

constriction is of the order of the coherence length, ξ (see Fig. 1.18d). Another example is a bridge of variable thickness, such that the thickness of the basic film is hundreds of nanometers, at the same time, the thickness of the bridge itself is only several tens of nanometers (Fig. 1.18e).

(5) *Grain boundary (or bicrystal) transition.* It is a typical one for high-temperature superconductors (Fig. 1.18f). Due to the extremely short coherence length in HTSC ($\xi \sim 1\,\text{nm}$), defects in their crystalline structure can act as weak links. The best-controlled defects can be produced between two regions of an epitaxial high- temperature film with various crystal orientations (grain boundaries). The critical current density of such a weak link can be varied, changing the misorientation angle between two crystallites.

HTSC Josephson junctions can also be classified into the following three classes [355]:

– junctions without interfaces, in which the weak coupling is achieved by locally degrading the superconducting properties of a HTSC thin-film microbridges by focused electron or ion beam irradiation;

- junctions with intrinsic barriers or interfaces formed, for example, using the intercrystalline boundaries of different crystallographic orientations;
- junctions with extrinsic interfaces, in fabrication of which artificial barriers from normal metals and insulators are used.

From the standpoint of the damage and material strength problem, the last two classes of the Josephson junctions display most interest. HTSC JJs with extrinsic interfaces are prepared by using the film technology, and they themselves present multi-layered structures (hetero-structures) with rectilinear and inclined interfaces. There are different technological possibilities, which are discussed in detail in the following two chapters, for increasing superconducting properties of these systems. For example, an overdoping by calcium of intergranular boundaries in multi-layered structures and YBCO super-lattices permits to significantly increase J_c at all temperatures up to T_c, but also at magnetic fields up to 3 T [174].

2

Composition Features and HTSC Preparation Techniques

Microstructure, strength and other HTSC properties are defined by the existence of numerous components. This circumstance supposes different chemical, physical and mechanical influences during numerous technological operations to prepare final sample from initial powders. Super-sensitiveness of HTSC final properties to the technical conditions, their manufacture and to composition features, and also numerous effects have determined various ways of the oxide superconductor preparation. HTSC samples in the forms of films at the mono- and polycrystalline substrates, coated conductors, tapes and superconducting bulks are the most interesting for applications. Their preparation techniques are considered in this chapter.

2.1 YBCO Films and Coated Conductors

In general, in order to synthesize HTSC films (as YBCO family as another ones) in situ and ex situ methods are used. In the first case, the film crystallization takes place directly during their deposition and an epitaxial growth occurs under corresponding conditions. In the second case, the films are deposited initially under low temperature, that is insufficient to form necessary crystalline structure, and then the films are sintered in O_2 atmosphere that leads to the crystallization of the necessary phase (e.g., this sintering temperature is equal to 900–950°C for YBCO films). Most one-stage methods are realized under thermal treatment that is much more lower compared with two-stage methods. The high-temperature annealing forms large crystallites and rough surfaces, defining small critical current density. Therefore, the in situ methods have advantages initially. According to the preparation methods and the deposition of HTSC components on a substrate, there are physical methods of deposition, including all possible evaporations and scatterings, and also the chemical methods of precipitation.

Methods of vacuum co-evaporation. These assume simultaneous or successive (layer-by-layer) co-precipitation of HTSC components evaporated from different sources by using, for example, electron beam guns or resistive evaporators. The films, prepared by this technique, yield their superconductive properties to the samples, manufactured by methods of laser evaporation or magnetron scattering. Methods of vacuum co-evaporation are used in two-stage synthesis, when the structure of films, scattered in first stage has no principle consideration, as also the oxygen contents in them.

Laser evaporation. This is highly effective in the HTSC thin-film deposition. This method is simple in realization, demonstrates high rate of deposition and permits contact with small targets. Its main advantage is the evaporation, equally well, of all chemical elements contained in the target [521]. The films, having the same quality as the targets, may be prepared in the target evaporation under concrete conditions. The distance between target and substrate and also the oxygen pressure are the important technical parameters. Their right selection allows, on one hand, non- overheating of the growing film by energy of plasma, evaporated by laser, and accompanying formation of very big grains, but, on the other hand, to state an energetic regime that is necessary for film growth at perhaps very low temperatures of substrate. The high energy of the components deposited and existence in the laser flame of monatomic and ionized oxygen permit the preparation of HTSC thin films in one stage. In this case, the films are mono-crystalline or possess high texture with c-axis orientation (the c-axis is perpendicular to the substrate plane). The main disadvantages of the laser evaporation are the following: (i) small region in which stoichiometric films could be deposited, (ii) heterogeneity of their thickness and (iii) surface roughness. By using the radiation methods in film preparation, the interesting dependences may be stated, for example, between the degree of a-axis orientation of YBCO film,[1] the substrate temperature and the material, and also the deposition rate. These dependences for the ion beam sputtering method in deposition of CeO_2 buffer layer are presented in Figs. 2.1 and 2.2. Obviously, by using the sapphire substrate in the preparation of the a-axis-oriented YBCO thin film higher rates of deposition are necessary. Moreover, film surfaces with smaller roughness are obtained on sapphire compared with $SrTiO_3$, defining its better superconducting properties.

Magnetron scattering. This permits to obtain in one stage YBCO films, not yielding their superconducting properties to the samples, deposited by the laser evaporation method. Moreover, they have more homogeneous thickness and higher smoothness of surface. As at the laser evaporation, the plasma formation at magnetron scattering creates high-energetic atoms and ions that

[1] Due to the high anisotropy of HTSC, films with c-axis orientation only have good transport and screening properties. At the same time, films with a-axis orientation, possessing greater coherence length in direction that is perpendicular to the surface and distinguished by high smoothness, could be convenient to prepare qualitative HTSC Josephson junctions, consisting of successively deposited "HTSC-normal metal" (or "dielectric–HTSC") layers. Films demonstrating mixed orientation are not desirable in all cases.

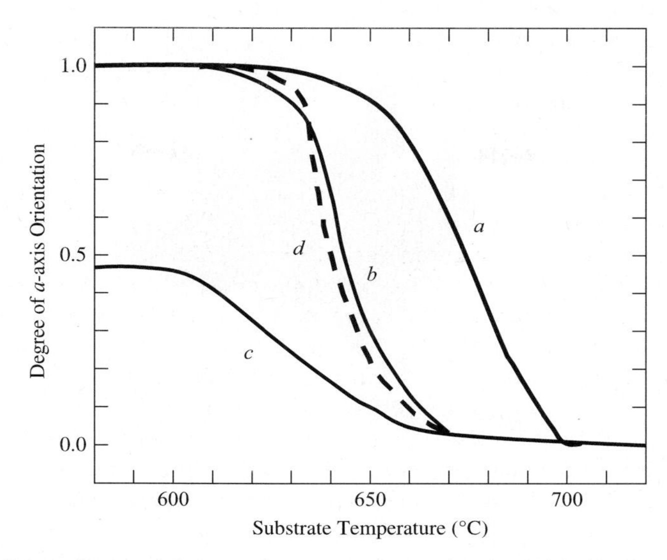

Fig. 2.1. Dependence of degree of a-axis orientation of YBCO film upon the substrate temperature. *Solid line* (**a**) indicates deposition rate of 20 Å/min on sapphire; (**b**) for 10 Å/min; (**c**) for 5 Å/min. *Dashed line* (**d**) is for 5 Å/min on $SrTiO_3$ [486]

permit to obtain HTSC films in one stage at not high temperatures. Here, the distance of "target–substrate" is also important. At small distance and insufficient pressure of environment, the substrate is subjected to intensive bombardment by the negative ions of oxygen that fracture the structure of growing film and its stoichiometry. In order to solve this problem, the set approaches are used [175], including the protection of the substrate from the bombardment of energetic ions and its location at optimum distance from gas-discharge plasma to ensure high rate of deposition and successive growth of the film at the maximum low temperatures.

In situ YBCO thin films prepared by out-axis magnetron scattering and possessing optimum electric properties have demonstrated $T_c = 92$ K and $J_c = 7 \times 10^6$ A/cm^2; and $T_c = 89$ K and $J_c = 2 \times 10^6$ A/cm^2 at optimum smoothness of surface [222,223].[2]

Chemical precipitation from vaporous phase of metal-organic compounds. The core of the method is the transportation of metallic components in the form of steams of the volatile metal-organic compounds in a reactor, a mixing with gaseous oxidizer, decomposition of the steams and condensation of the oxide film on substrate. This method permits to obtain HTSC thin films possessing parameters, which are compared with samples prepared by the physical

[2] The varieties of impulse laser deposition, used to obtain YBCO films and coated conductors with high texture on the different mono- and polycrystalline substrates with and without buffer layers, permit to reach the critical current density $J_c = 2.4$ MA/cm^2 at 77 K and zero magnetic field [271].

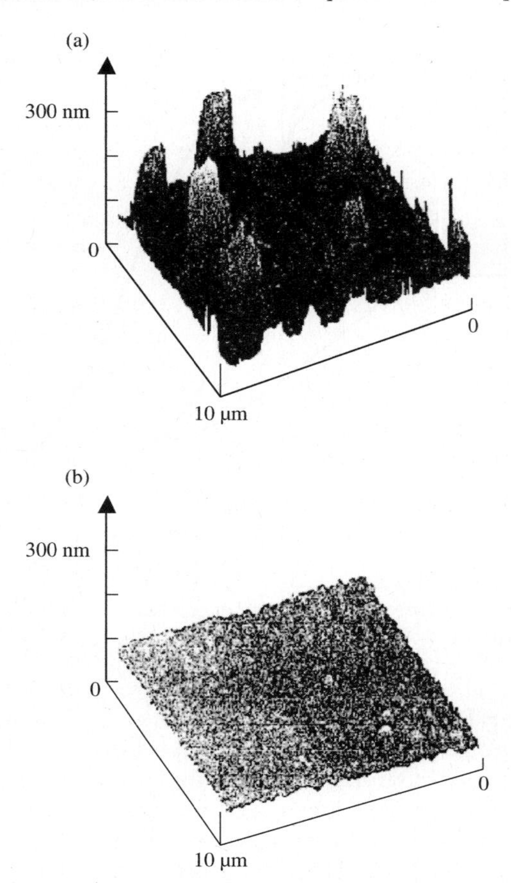

Fig. 2.2. Atomic force microscopy images of YBCO films deposited on (**a**) $SrTiO_3$ and (**b**) sapphire substrate with CeO_2 buffer layers. Sputter conditions for the films are identical [331]

methods of deposition. The comparative advantages of the method considered to last ones are the following: (i) a possibility to obtain homogeneous films on details with non-plane configuration and possessing a great area; (ii) more high rates of precipitation under conditions of high quality; (iii) a flexibility of process at the preliminary stage of technological regime due to the smooth change of the vaporous phase composition.

The following techniques are as the most widespread ones:

- preparation of films and coated conductors (CC) with two-axes texture by using the *ion-beam-assisted-deposition (IBAD)* [464,1151];
- preparation of films and coated conductors by using the *rolling-assisted biaxially textured substrates* (*RABITS*) [336,337,769,1181];

- preparation of two-axes ordered buffer layers by using the *inclined substrate-pulsed laser deposition* (*ISPLD*) [395];
- preparation of thin films by the *magnetron sputtering* [1042];
- preparation of buffer layers by the *electron beam evaporation* [680], *laser ablation* [646,893], *ion beam sputtering* [331,486] and *rf-sputtering* [147, 632,701];
- preparation of thick epitaxially grown films by the *liquid phase epitaxy* (*LPE*) [1155];
- preparation of films by the *electrophoretic deposition* [704];
- preparation of buffer layers by the *surface oxidation epitaxy* (*SOE*) [678];
- preparation of coated conductors by using the *metal-organics decomposition* (*MOD*) [985];
- preparation of thick films on the base of precursor films by the *metal-organic Chemical vapour deposition* (*MOCVD*) [253].

As substrates in various techniques, the Ni-based alloys (Inconel, Hastelloy, Ni-Cr, Ni-V, etc.), metals (Ni, Ag, Zr), oxides (Al_2O_3, $SrTiO_3$, $NdGaO_3$, $LaAlO_3$, MgO, PrO_2) and non-metals (Si, glass) have been used. As buffer layers YSZ, CeO_2, MgO, $SrTiO_3$, ZrO_2, $BaZrO_3$ and NiO have been used.

As it has been noted, the critical current could be enhanced due to the formation of artificial defects in the sample, playing role of the magnetic flux pinning centers. With this aim, the defects with linear sizes, l, which are near to the superconducting coherence length (e.g., $l = 2$–4 nm in Y-123 at $T < 77$ K), and density of order of $(H/2) \times 10^{11}\,\mathrm{cm}^{-2}$, where H is in Tesla, are the best of all. This high density of defects with nanometer sizes is reached with difficulty at the stage of sample fabrication; therefore the ionic irradiation is usually used for this goal. In [399], a method has been proposed for fabrication of Y-123 films, consisting of non-superconducting Y-211 particles with $l \sim 8$ nm, density of which was larger than $10^{11}\,\mathrm{cm}^{-2}$. This method consists of alternative deposition of the Y-123 and Y-211 layers, and further formation of the Y-211 "islands" (because of mismatches of the lattice periods of (2–7)% for Y-123 and Y-211 phases). In this case, the thickness of Y-123 layers and Y-211 layers was equal to ~ 10 and ~ 1 nm, respectively, but Y-211 nanoparticles were distributed uniformly into Y-123 matrix. These films have much more larger current density, compared to Y-123 films without defects, moreover, one decreases much more slower at the enhancement of magnetic field.

From the view of material strength, the main aim is to select neighboring layers with like thermal and crystallographic properties, and also to prevent chemical reaction between them. It is necessary to note, that the critical current density of the film on polycrystalline substrate diminishes considerably at the existence of high-angle intergranular boundaries [206], causing the problem of weak links. Therefore, for tape applications, it is very important to prepare films with highly in-plane aligned crystalline structure, that is, to form the ordered structure not only along the c-axis, but along the a-direction. As a rule, thick YBCO-coated conductors are obtained layer-by-layer (metal substrate–buffer layer–superconducting layer), see Fig. 2.3. Every layer including metallic substrate plays a special role in the long length conductors. The candidate materials and the thickness of each layer are also presented in Fig. 2.3.[3]

In general, thick YBCO-coated conductors consist of epitaxially grown films on appropriate substrate with or without few buffer layers. Therefore, the substrate or the buffer layer must have a suitable textured surface structure to provide the required in-plane alignment growth of the HTSC crystalline films, resulting in the avoidance of weak links due to the existence of only low-angle grain boundaries. Typical problems, decided now at design and manufacture of coated conductors, are presented in the Table 2.1.

The general problem of films, prepared by the liquid phase epitaxy and electrophoretic deposition, demonstrating high rates of precipitation and sufficiently low cost, is the crack formation due to stress relaxation during sample cooling and different thermal properties of buffer layer and film. If we assume that fixed cracks are extended through the whole film thickness, we may follow a calculation for cracks in brittle films on elastic substrates [1057,1060]. According to the model, the total energy is the sum of the strain energy in the

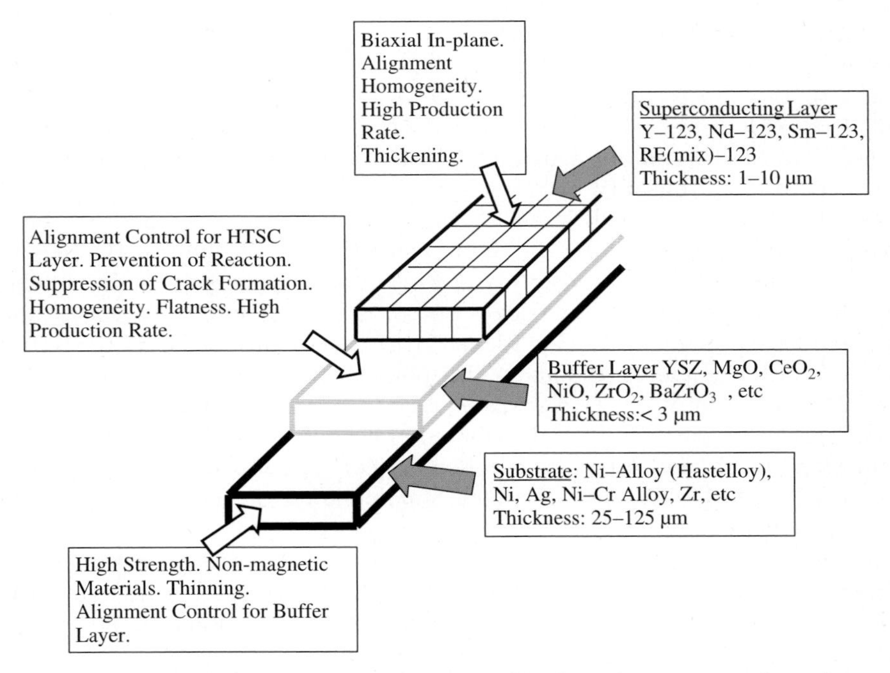

Fig. 2.3. The problems, solved in the design of high-qualitative-coated conductors and including the development of basic structure, estimation of key factors and use of candidate materials for each layer [985]

[3] For simplicity and clearness, a thin passivation layer for stabilization, insulation and encapsulation is excluded.

Table 2.1. Nearest goals of R&D for three different types of YBCO-coated conductors [985]

Type	Structure or technique features	Substrate	Buffer layer	HTSC layer	Key processes	Target
Textured substrate	Non-reactive, high strength	Ni, Ag clad materials: Ni–Cr, Ni–V, Ni-based alloys, etc.	None or NiO, ZrO_2, $BaZrO_3$, MgO, YSZ, Y_2O_3, CeO_2 etc.	Y-123, *RE*-123, etc.	Rolling/annealing; surface polishing; surface oxidation epitaxy; buffer layer; HTSC layer; evaluation (J_c, etc.)	Length: 10–100 m; substrate thickness: $\leq 100\,\mu m$; $J_c \geq 10^5 - 10^6$ A/cm^2 (77 K)
Aligned buffer layer	ISPLD, IBAD	Polycrys-talline Hastelloy, Ni-based alloys, etc.	YSZ, MgO, CeO_2, etc.	Y-123, *RE*-123, etc.	Substrate polishing; IBAD process; ISPLD process; HTSC layer; evaluation (J_c, etc.).	Length: 100–1000 m; substrate thickness: $\leq 100\,\mu m$; $J_c \geq 10^4$–10^5 A/cm^2 (77 K); production rate: > 1 m/h
Rapidly grown HTSC layer	MOD, LPE	Ni, Ag clad materials: Ni-based alloys, etc.	None or MgO, YSZ, NiO, $BaZrO_3$, CeO_2, etc.	Y-123, *RE*-123, etc.	Substrate polishing; buffer layer; homogeneous seed film; MOD process; LPE process; evaluation (J_c, etc.)	Length: 1–10 m; substrate thickness: $\leq 100\mu m$; HTSC thickness: $\geq 5\mu m$; $J_c \geq 10^5$–10^6 A/cm^2 (77 K); production rate: > 1 m/h

cracked region and the energy necessary to create the cracks, that is, the new surface energy. The crack spacing, l, will be adjusted in order to minimize the total energy of the system to [1060]

$$l \approx 5.6\sqrt{K_{Ic}^2 h/(E\varepsilon)^2}\ , \tag{2.1}$$

where K_{Ic} is the fracture toughness of the mode I, h the film thickness, E the Young's modulus and ε the strain. The total energy of the cracked film cannot exceed the strain energy of the uncracked film. This results in minimum crack spacing and is consistent with the requirement that the energy release by cracking should be higher than the energy necessary to create the cracks. This means that below a critical thickness

$$h_c = 0.5K_{Ic}^2/(E\varepsilon)^2\ , \tag{2.2}$$

no cracks will appear.

Based on this approach, the predicted dependence of the average crack spacing l on the thickness h of a-axis-oriented YBCO films on (100)$SrPrGaO_4$ substrates is shown in Fig. 2.4. At the same time, the decreasing micron-sized particles to sub-micron-sized particles in colloids may diminish the film cracking at its electrophoretic deposition consideraldy (see Fig. 2.5).

In order to obtain YBCO samples with high properties, a preparation of qualitative precursor powders is also very important. A number of techniques for processing YBCO powders have been reported, including conventional solid-state reaction, precipitation [106], plasma spray [415], freeze drying [495], spray drying, combustion synthesis [571], sol–gel method [722], acetate method [686] and flame synthesis [889].

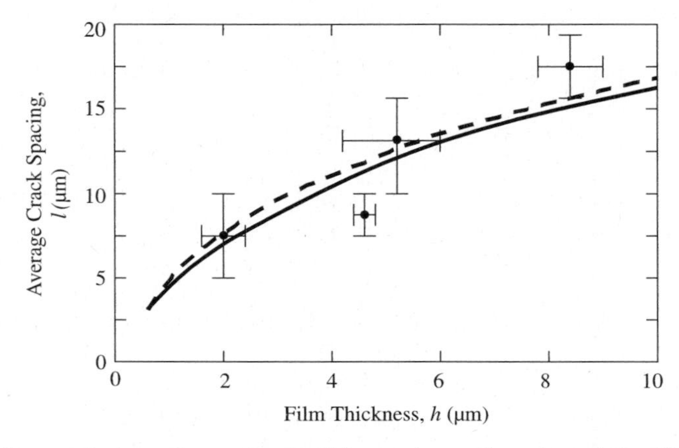

Fig. 2.4. Theoretical crack spacing (*solid curve*) as a function of the film thickness compared with the experimental data and a fit (*dashed curve*) to the data obtained from a-axis-oriented YBCO LPE films grown on (100) $SrPrGaO_4$ [05]

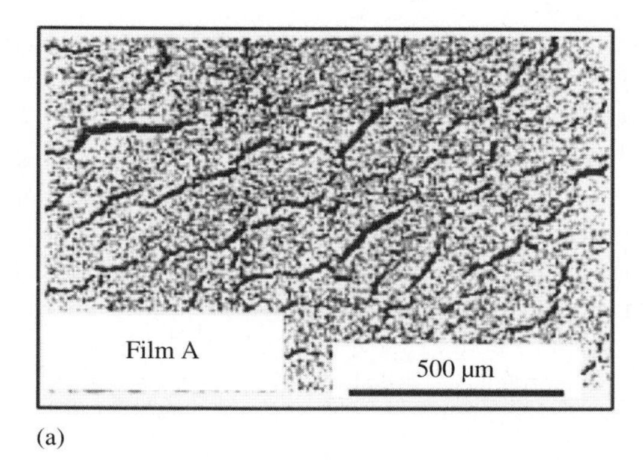

(a)

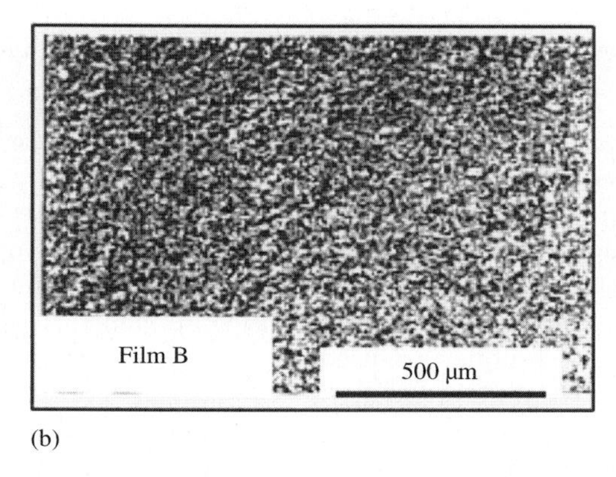

(b)

Fig. 2.5. Scanning electron microscopy images of sintered films. The images of (**a**) and (**b**) show films fabricated from colloids, consisting of micron-sized particles and sub-micron-sized particles, respectively. The heat treatment conditions for both are the same [926]

The fabrication of a ceramic superconducting powder is presented in the flow chart of Fig. 2.6. First, the raw materials are weighed in a desired molar ratio and then mixed by conventional mixing/milling or liquid-solution mixing. The homogeneity, obtained by conventional mixing, is limited by the particle size of the powders, but the best mixing is generally obtained from powders with a particle size less than 1 μm. For ultra-fine powders (particle size much smaller than 1 μm), the particles tend to segregate, thus resulting in poor mixing. This problem may be minimized by liquid-solution mixing, ensuring the precise compositional control and molecular-level chemical homogeneity. Moreover, this technique eliminates contamination from grinding and milling media that occurs during a conventional mix/milling process. For a multi-component system like HTSC, the mixing plays a key role in obtaining high phase purity. Better mixing also translates into faster reaction kinetics.

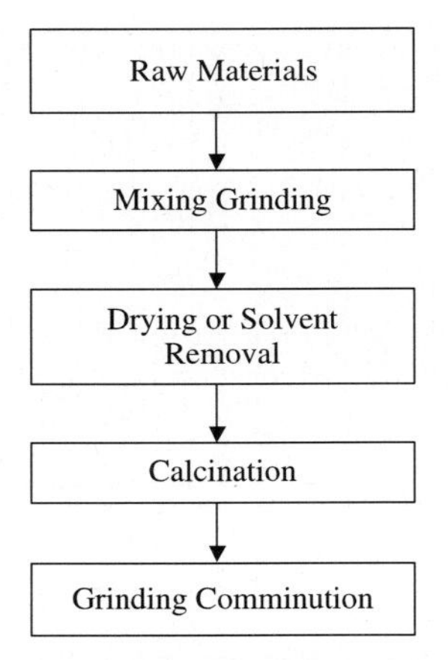

Fig. 2.6. A flow chart for the synthesis of ceramic superconducting powders

These powders can be calcined at lower temperatures and/or shorter time to achieve the desired phase purity.

The next step is drying or solvent removal, that is necessary to preserve the chemical homogeneity obtained by mixing. For multi-component systems, solvent removal by slow evaporation can lead to a very inhomogeneous residue due to the difference in solubilities of various components. In order to minimize this effect, various techniques are used, including, in particular, the processes of filtration, sublimation, etc. [960].

After drying, the powders are calcined in a controlled atmosphere for reaction until reaching a final composition and phase assemblage. The reaction

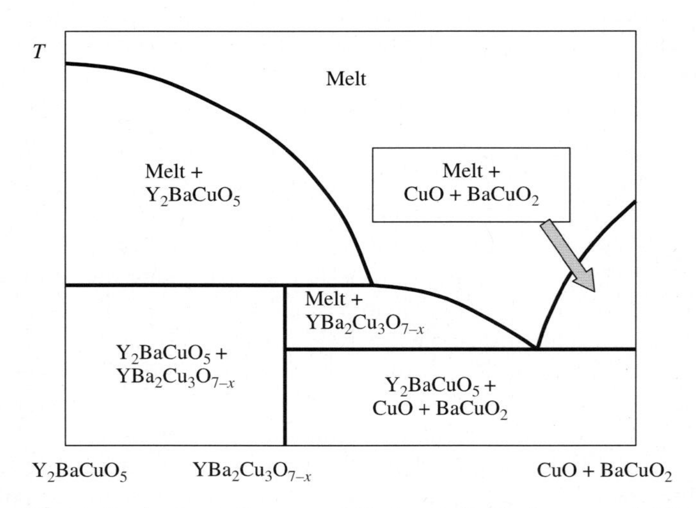

Fig. 2.7. A phase diagram of Y_2O_3–CuO–BaO system [318]

path for HTSC systems depends on various processing parameters, such as calcination temperature and time, heating rate, atmosphere (oxygen partial pressure) and starting phases. The powders can also be synthesized directly from solution using pyrolysis techniques [889] or an electro-deposition technique [65].

The phase diagram, presented for $YBa_2Cu_3O_x$ in Fig. 2.7, shows that even small fluctuations of compositions can lead to the formation of normal (non-superconducting) phases, namely Y_2BaCuO_5, CuO and $BaCuO_2$. An application of precursors with carbon content also complicates the formation of the $YBa_2Cu_3O_x$ phase and may lead to diminishing superconducting properties. In detail, the carbon problem will be considered in Chap. 4.

2.2 BSCCO Films, Tapes and Wires

Now for the preparation of Bi-2212 thick films on Ag and MgO substrates, the set of techniques are applied, namely melt-processing [405,407], electrophoretic deposition [49], doctor-bladed [520], dip-coated [1067] and organic precursor [397] films. Last three procedures are presented in Fig. 2.8. It should be noted that the fabrication of Bi-2223, possessing higher superconducting properties, by using melt processing is impossible because of PbO loss in high-temperature process [253]. The powder for doctor-bladed and dip-coated films can be made using any of the processes listed in Table 2.2.

In the *doctor-bladed process*, one makes a "green" film (i.e., film before heat treatment) from the organic/powder mixture by pouring a pool of the slurry on a flat surface (e.g., a piece of glass), then leveling the slurry with a straight-edged blade, located above the flat surface at the desired film thickness (this allows to carry out corresponding control during its preparation). The film is dried, cut into strips that are placed on the silver foil and finally melt processed.

For the *dip-coated films*, the silver foil is passed through the organic/powder mixture, which adheres to the foil. The thickness of the film is controlled, changing the organic compounds, modifying their proportions in the mixture and adjusting the solids loading in the mixture. After passing through the organic/powder mixture, the coating is dried, the organics are burned out and the film is melt processed.

For the *organic precursor films*, a solution of organometallic compounds of Bi, Sr, Ca and Cu is deposited on the Ag foil, the solvent is burned out and the process is repeated until the desired layer thickness is built up. Finally, the film is melt processed. A Bi-2212 film can also be made, painting Bi-2212 powder onto a silver foil. Here, the powder is mixed with an organic liquid having a high vapor pressure (e.g., butanol). This slurry is brushed onto the foil and finally melt processed.

Since the pioneering work of Heine et al. [405] on Bi-2212 conductors, almost all Bi-2212 conductors have been melt processed. This method is also

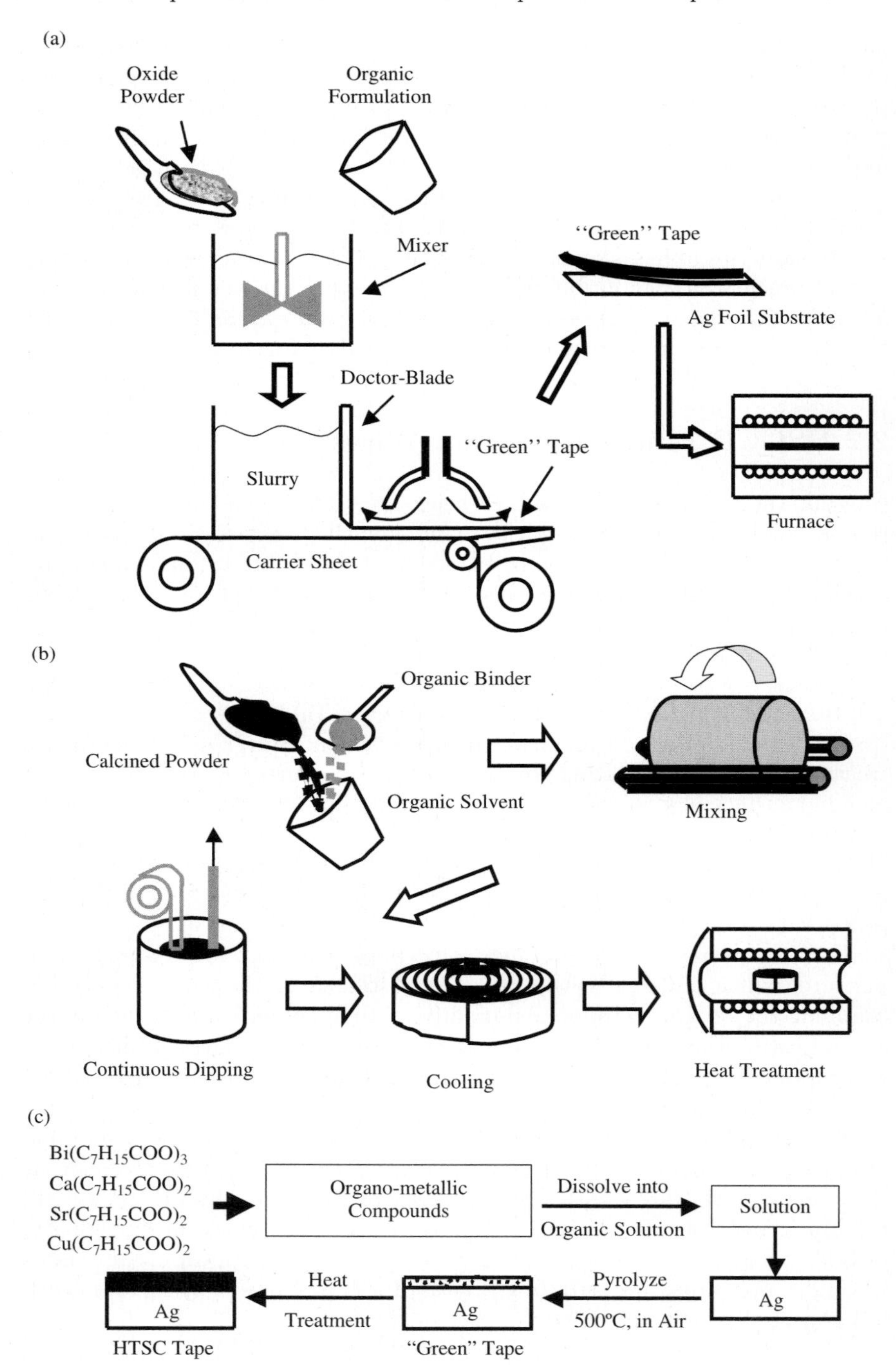

Fig. 2.8. Schematic diagram to make (**a**) doctor-bladed, (**b**) dip-coated and (**c**) organic precursor films [397]

Table 2.2. Synthesis techniques to make Bi-2212 and Bi-2223 powders

Method	Description	Advantages	Disadvantages
1	2	3	4
Solid-state reaction [409]	Mix oxides, peroxides, carbonates or nitrates of Bi, (Pb), Sr, Ca and Cu. React at elevated temperature where no melting occurs. Grind sample and refire. Repeat until reaction is complete	Simple, inexpensive technique	Large grain size of reactants can cause slow reactions. Product can have large grain size. Impurities can introduce during grinding
Co-precipitation [409]	Dissolve Bi, (Pb), Sr, Ca and Cu compounds in acid. Add base to precipitate cations. Fire precipitate to yield the desired phase	Intimate mixing of cations	Not all cations may precipitate out at the same rate, causing segregation. Initial composition and precipitate composition may be different
Aerosol spray pyrolysis [409]	Make solution, containing cations. Produce fine mist of the solution and pass it through a hot furnace to form a powder of mixed oxides. Fire mixed powder to yield the desired phase	Intimate mixing of cations. Product has very fine grain size (1–2 μm). Product can have low carbon content (using nitrates)	Species can lose, partially Pb, during pyrolysis. Powder, formed in pyrolysis is not fully reacted to the desired phase
Burn technique [07]	Form nitrate solution of cations. Add organic species, such as sugar, to solution. Heat solution to remove water then heat powder at elevated temperature. The sugar (fuel) and nitrate ion oxidant react (i.e., burn) at elevated temperature, yielding a high temperature that forms mixed oxides. Fire this powder to yield the desired phase	Intimate mixing of cations. Powder can have fine grain size	Species can lose, partially Pb, during burn process. Powder formed in burn process is not fully reacted to the desired phase. Product may contain carbon

(continued)

Table 2.2. (continued)

Method 1	Description 2	Advantages 3	Disadvantages 4
Freeze drying [409]	Spray aqueous nitrate solution of Bi, (Pb), Sr, Ca and Cu into liquid nitrogen. Collect frozen droplets and freeze dry them to remove water. Fire dried powder to yield the desired phase	Intimate mixing of cations. Product can have low carbon content	Cations may demix during freeze drying, if the temperature is not carefully controlled. Nitrates, presented after freeze drying, may melt during firing, leading to large grains of non-superconducting phases
Liquid mix method [838]	Form nitrate solution of cations then add glycol or citric acid. Heat to remove water and form polymerized gel, then heat to elevated temperature to yield the desired phase	Intimate mixing of cations. Powder can have fine grain size	Product may contain carbon
Micro-emulsion [587]	Form suspension of micro-droplets of aqueous nitrate solution of Bi, (Pb), Sr, Ca and Cu in oil. Add base to form precipitates. Separate precipitate from oil by washing in solvent. Fire precipitate to yield the desired phase	Intimate mixing of cations. Powder can have fine grain size	Product may contain carbon
Sol–gel [91]	Form alkoxide solution of cations. Add water or alcohol to cross-link molecules, forming gel through polymerization and condensation reactions. Heat to elevated temperature to burn the organics and yield the desired phase	Intimate mixing of cations. Powder can have fine grain size	Method is better suited to making films, than bulk powders

named by *partial melt processing*, that reflects the fact that Bi-2212 melts incongruently forming liquid and crystalline phases (i.e., partial melt). [4] Numerous heating schedules that are currently used for the melt process of Bi-2212 can be generalized for tapes and wires (Fig. 2.9a), and also for films (Fig. 2.9a). Both schedules could be divided into four stages [409].

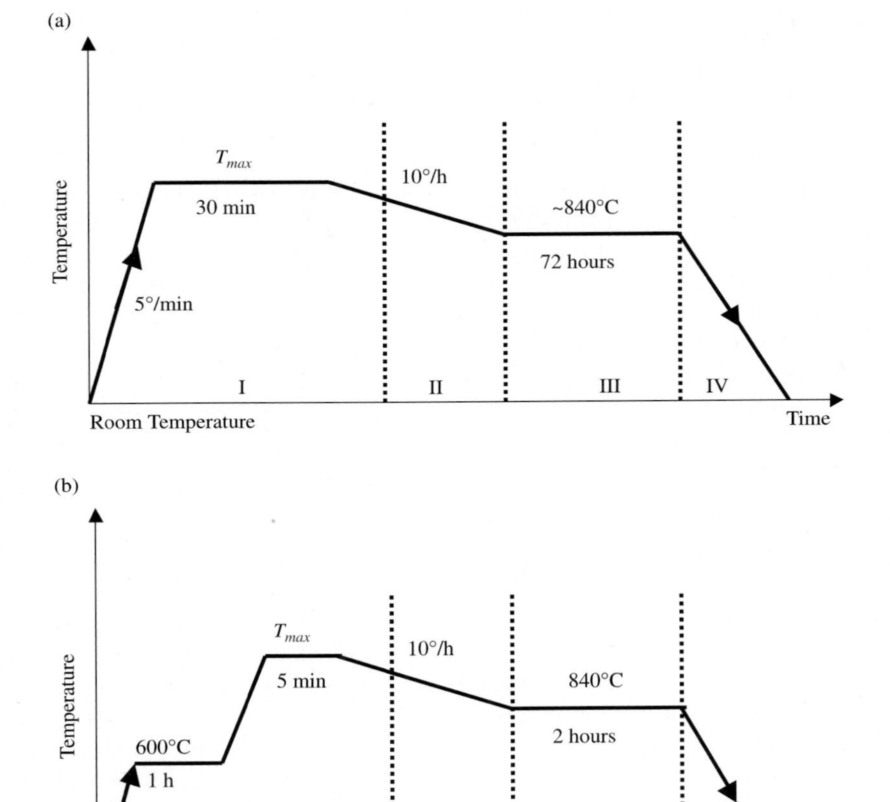

Fig. 2.9. Generic Bi-2212 melt-processing schedules used for (**a**) tapes and wires and (**b**) films. The schedules have been divided into four regions that are described in the text [409]

[4] It should be noted that the term melt processing clearly distinguishes from a lower-temperature processing, where only a portion of the Bi-2212 is melted, which is called *liquid-assisted processing* [410].

In Region I, the conductor is heated above the melting point of Bi-2212, held at this temperature for a short time and then cooled. In this stage, Bi-2212 powder melts incongruently, forming liquid and non-superconducting crystalline phase. At the end of this stage, Bi-2212 formation begins. At melting, the Bi-2212 powder releases oxygen, which is not a problem for films that are open to the environment, but can be a problem for tapes and wires, where it can cause the silver sheath to bubble. On the other hand, at elevated temperatures, used during melting, Bi evaporation from the melt is possible [924]. The silver sheath of tapes and wires prevents this process, but this is actual danger for films. Also, it should be noted that critical condition to prepare homogeneous highly aligned microstructure is the minimum fraction of crystalline phases in the melt at the beginning of Bi-2212 phase formation.

In Region II, Bi-2212 forms from the melt, where growth and alignment of Bi-2212 grains occur. Since Bi-2212 melts incongruently, at cooling it should be formed by a reaction between liquid and crystalline phases. In this case, in order to form a homogeneous highly aligned microstructure, the melt must contain small grain of non-superconducting phases when Bi-2212 begins to form. The cooling rates, used for the melt process of Bi-2212 conductors should be sufficiently fast to intensify Bi-2212 phase formation and minimize fraction of normal phases, which are present in the final product always. The problems with having non-superconducting phases in the fully processed conductor are that the phases are too large to pin flux, they block the supercurrent path and diminish useful properties of the conductor. In order to obtain high superconducting properties, a highly aligned grain structure is necessary. Plate-like Bi-2212 grains grow from the melt because growth of Bi-2212 is faster in the *ab*-plane than in the *c*-direction. This two-dimensional growth is critical for the alignment that develops during cooling. The experiments have shown [409] that the cooling rate and thickness of the oxide melt affect the alignment: slower cooling and thinner oxide yield higher alignment. The misorientation angle for a given grain size decreases with decreasing oxide thickness. The alignment mechanism requires that the large, properly oriented grains grow at the expense of the smaller misoriented grains. Films that are < 20–$25\,\mu m$ thick align easily. In thicker films, the alignment is usually not uniform throughout, being higher close to the free surface than near the silver interface. It has been suggested that in the films the Bi-2212 growth and alignment begin at the free surface and proceed into the oxide layer. Since films are two-dimensional, one would expect a higher alignment in thinner films because of the smaller misorientation angle a grain could have and still grow to a given length. Moreover, as the aligned grains grow in films, they may rotate misaligned grains into alignment near the free surface more easily compared with the nearest neighborhood of the superconductor/metal interface. The free surface in films is also important from the point of general alignment of the film structure. Thick films (50–100 μm) can have a 20–25 μm thick layer of aligned Bi-2212 grains at the free surface with poorly aligned Bi-2212 below this layer [409]. In general, the grain alignment and J_c are higher in films than tapes.

In Region III, the formation and alignment of Bi-2212 is maximized. At the same time, in Region IV, where the sample cools to room temperature, the critical alteration of microstructure occurs. High J_c requires fast cooling rate for films (of order 1200°C/h [984]) and analogous rates for tapes and wires [977]. However, at very high cooling rates the conductors may crack, leading to low J_c [984]. At the same time, the Bi-2212 can decompose with slow cooling rates (at < 300°C/h) forming Bi-2201, possessing lower superconducting properties.

In order to minimize the number of intergranular weak links, that are proper for oxide superconductors, a high degree of crystallographic texture must be obtained. One possible route by which a strong crystallographic texture can be produced is to melt process the material under the effect of an elevated magnetic field [275]. In this case, the driving force for grain alignment is provided by the anisotropic paramagnetic susceptibility, exhibited by the superconductor grains. When a superconductor grain is placed in a magnetic field, the axis of maximum susceptibility aligns with the magnetic field direction. As a result, in the case of superconductors, such as BSCCO and YBCO the grains should align with the c-axis parallel to the external magnetic field [53,1100]. Scanning electron microscopy (SEM) backscattered images of polished cross-sections of Bi-2212 thick films with two different thicknesses, processed in zero field and at 10 T magnetic field, are shown in Figs. 2.10

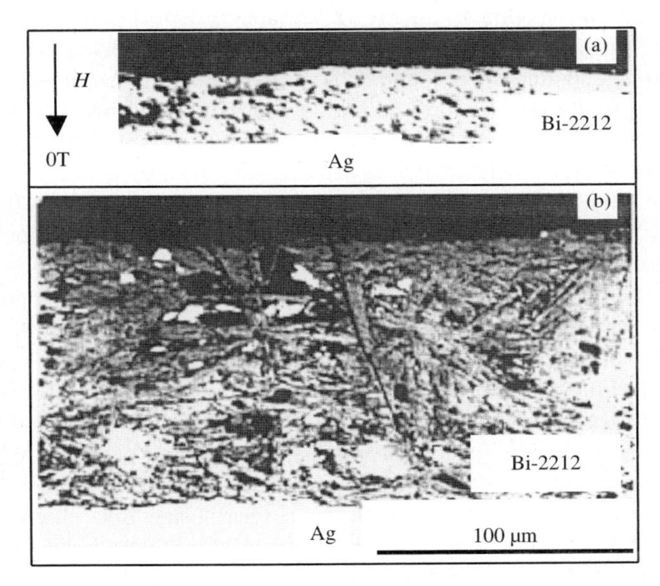

Fig. 2.10. SEM images of cross-sections of Bi-2212 films melt processed in the absence of magnetic field [275]

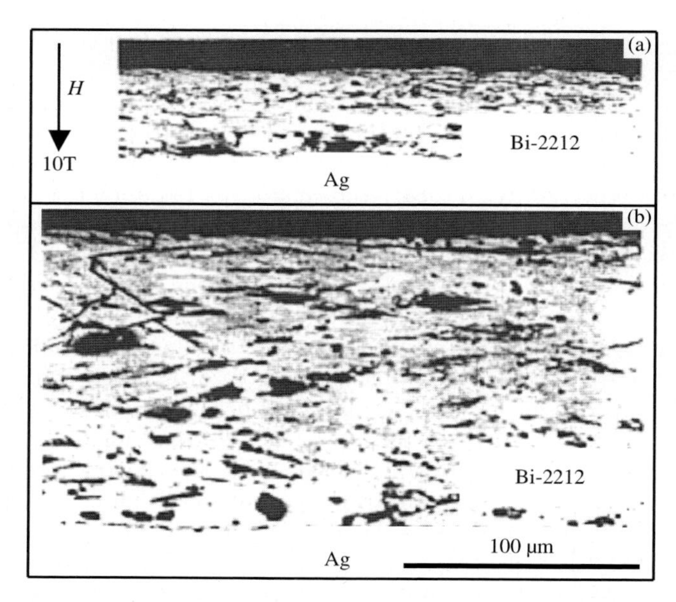

Fig. 2.11. SEM images of cross-sections of Bi-2212 films melt processed under 10 T magnetic field [275]

and 2.11, respectively. In the first case, the degree of texture decreases with the increasing film thickness. In Fig. 2.12, the transport critical current densities of the films measured at 4.2 K are plotted as function of their thickness for films processed under zero and 10 T magnetic field. Analogous results of a higher degree of alignment at the higher magnetic field are also observed in Bi-2212 tapes (see Fig. 2.13).

The superconductors Bi-2212/Ag are also prepared by using *melt-solidification method* [396]. In its modification, *pre-annealing* and *intermediate rolling* are applied to improved the alignment of superconducting grains [699]. At the same time, monocore and multifilament superconducting tapes and wires in silver sheath (Bi-2212/Ag and Bi-2223/Ag) are prepared most successively by using the *oxide-powder-in-tube (OPIT) method* [216,1091,1159]. Short Bi-2223/Ag multifilament tapes have demonstrated $J_c > 80\,\mathrm{kA/cm^2}$ at 77 K using this technique [287]. However, this multistage method is characterized by numerous technical parameters and procedures which define (together with initial composition) useful properties of the final sample. In this case, it is very important to prepare superconducting Bi-2212 and Bi-2223 powders, possessing better phase composition, grain size and shape, and also chemical purity. Corresponding technical operations have been presented in Table 2.2. Today, the standard approaches to obtain Bi-2223 precursor powders include the so-called methods of *one-powder synthesis* and *two-powder*

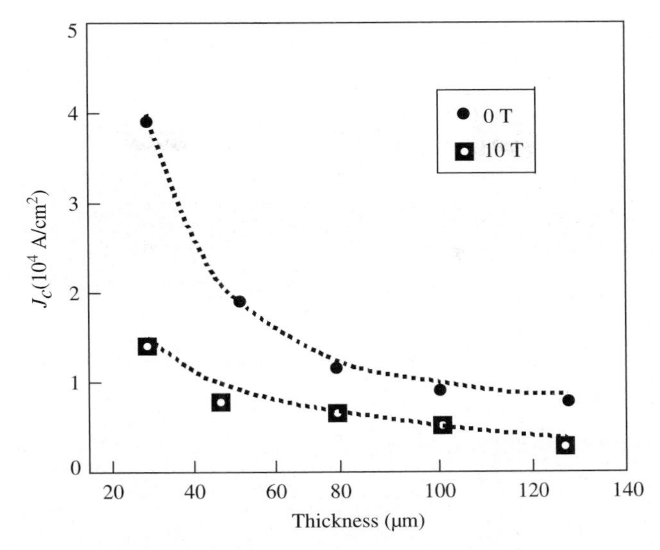

Fig. 2.12. Transport current densities at 4.2 K, zero field, as a function of thickness for films processed under 0 and 10 T magnetic fields [275]

synthesis [213]. In the first case, the precursor is prepared as the result of calcination of the oxides and carbonates mixture. In the second case, sintering of the mixture of two cuprate compositions is carried out. The OPIT method for making conductors is shown schematically in Fig. 2.14 and includes the following procedures. After the preparation of precursor powder, it is packed in a tube from Ag or its alloy. Here, silver is used, in particular, because of its property to diffuse oxygen at high temperatures. This permits to control the oxygen pressure during superconductor fabrication [409]. The tube is sealed and mechanically worked into desired conductor form. Usually the Ag tube is drawn to a small diameter (~ 2 mm), using set of cone holes. Then, this wire is rolled into a flat tape with thickness ~ 0.1 mm. Multifilament superconductors are prepared, placing separate wires in a tube with a large diameter and carrying out analogous technical operations. At final stage, the samples are subjected to set of thermal treatments. During chemical reaction, necessary superconducting phases form at this stage. Mechanical deforming helps to align and construct the texture of crystalline structure in this technological link. Thus, the brittle superconducting oxide is encircled by the sheath from silver or its alloy which protects the superconducting core from chemical and thermodynamic influences. Figure 2.15 shows different OPIT conductor forms that have been made.

The use of silver as sheath material of BSCCO tapes and wires is caused by the compatibility of Ag with superconducting powders at high temperatures of sintering. However, for tapes with silver sheath, there is the problem of

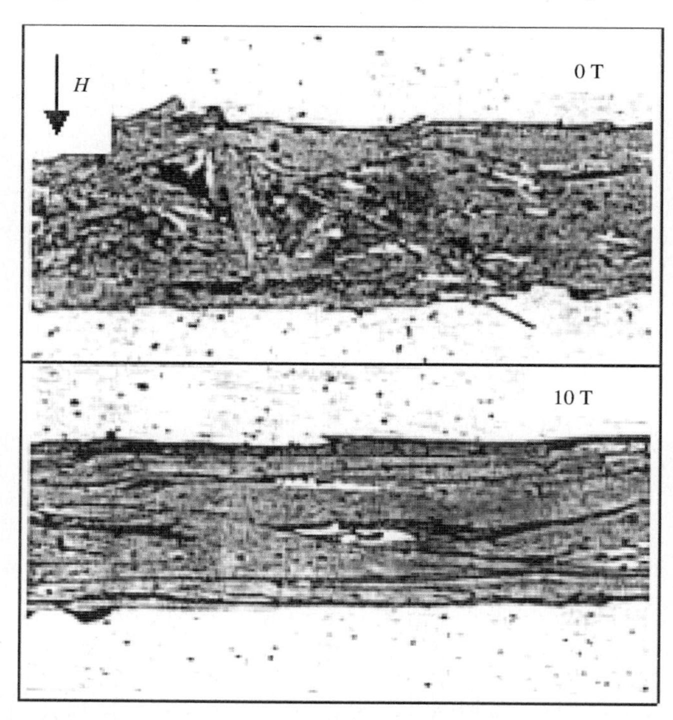

Fig. 2.13. SEM images of cross-sections of Bi-2212 tapes melt processed under 0 and 10 T magnetic fields [275]

formation of an undulating oxide/Ag interface, which is known as *sausaging*. This occurs along the tape after its drawing, rolling and pressing (Fig. 2.16) [385,665], and also silver creeps at high temperatures [320]. Moreover, HTSC tape sheath, used in concrete devices or samples, must satisfy definite demands, namely (i) they must possess sufficient mechanical strength in order to bear strains, stated by high electric and magnetic fields, that may crack brittle superconducting core; (ii) they must possess low heat conduction in current conductor and high electric resistance to decrease current losses in cables [1177]. In order to solve these problems, the silver alloys with some metals, namely Au, Cu, Mg, Mn, Pd, Zr, (Ni, Y), (Zr, Al) and (Mg, Ni) are used [06,16,954,991,1063]. Thus, in total, there is a problem to obtain optimum microstructures of superconducting composite, taking into account different behaviours of the silver (alloy) sheath and oxide core at thermal treatment [385,1130].

Now, OPIT method is generally applied to prepare BSCCO superconductors. The advantage of this HTSC system is that the texturing and alignment of crystallites, influencing significantly superconducting properties of tapes, may be simply achieved at the final stage of the sample fabrication. In this

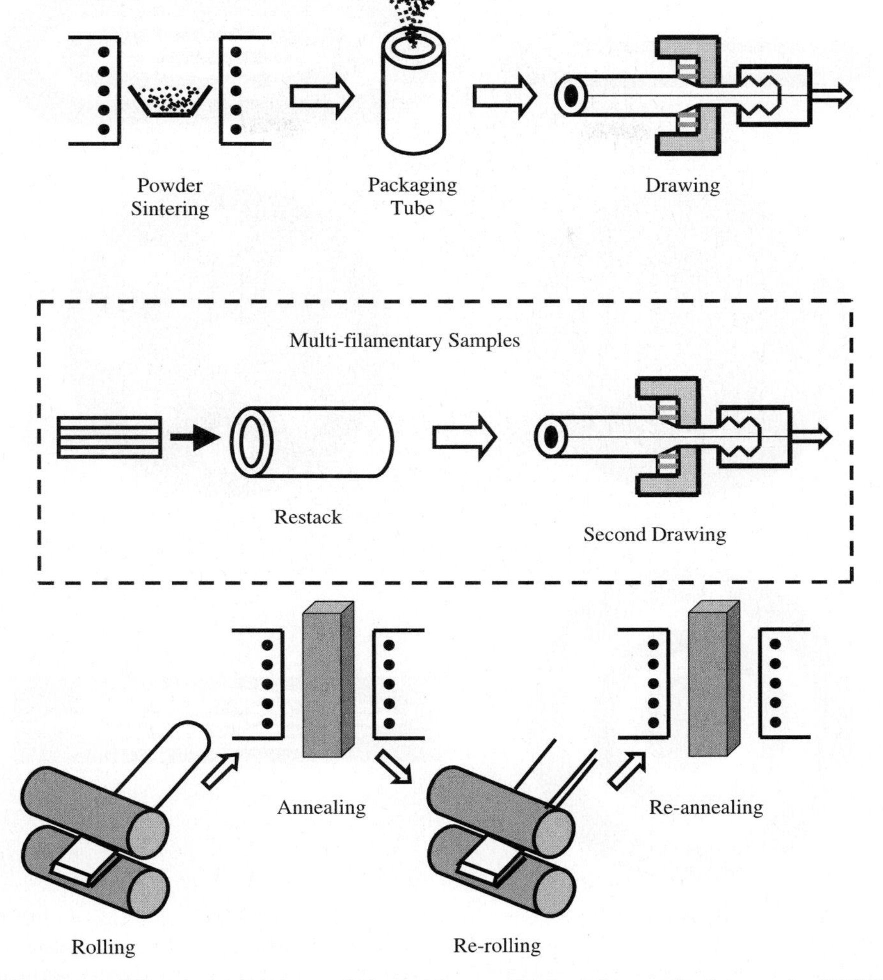

Fig. 2.14. Schematic diagram of the oxide-powder-in-tube method to make HTSC wires and tapes [409]

case, the tape deformation leads to shear in BSCCO structure along the double layer Bi–O, acting as glass phase, due to the weak joints of the adjacent layers to each other. As a result, the grains form an aligned structure at the next heating. The YBCO family has a worse structure of grains and demonstrates higher isotropy compared with BSCCO and also possesses small intergranular transport currents that are caused by weak links. Therefore, all attempts to use OPIT method for this family were not successful [649,1021].

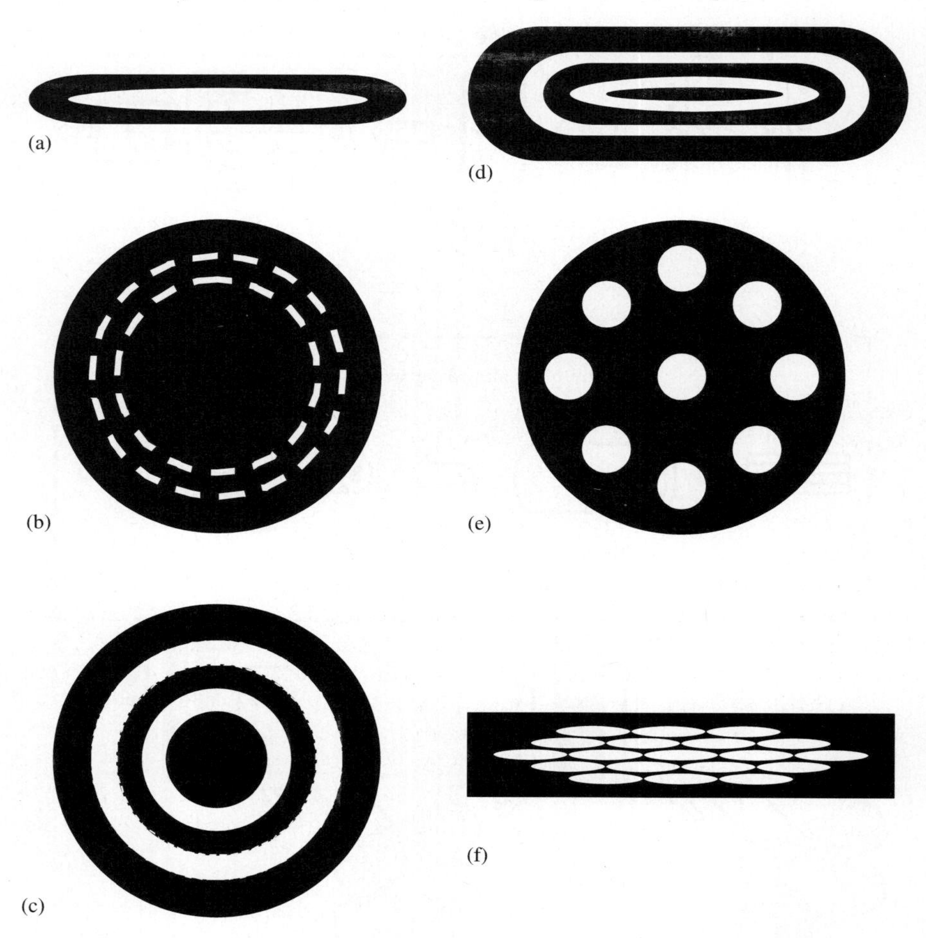

Fig. 2.15. Representative configurations for Ag-sheathed wires and tapes. *Black regions* are Ag and *white regions* are BSCCO: (**a**) monocore tape, (**b**) coaxial multifilament wire, (**c**) wire with two BSCCO cores, (**d**) tape rolled from the wire in (**c**), (**e**) filament wire and (**f**) filament tape

At long sintering of Bi-2223/Ag monocore tapes, a diminishing of magnetic flux pinning with corresponding decrease in critical current, I_c (Fig. 2.17[5]), which could be caused by decreasing Pb content in the superconducting core at increasing calcination time [670,671], is observed. However, we propose the more probable causes decreasing I_c to be inevitable processes of pore transformation at intergranular boundaries with pore displacement into

[5] Obviously, the initial increase in critical current is connected with improvement in quality of intercrystalline boundaries. However, this factor has secondary value at longer sintering.

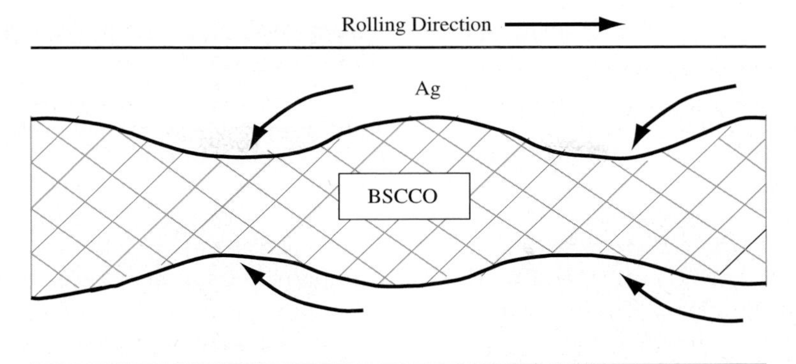

Fig. 2.16. Longitudinal cross-section of a BSCCO tape, showing undulations in the Ag/oxide interface. This is known as sausaging

growing grains, forming close porosity at prolonged sintering of the sample [801,1185].

In total, an increasing of low pinning, which is intrinsic to BSCCO family (compared to YBCO family), is one of the main goals of thermal treatment, applied to prepare BSCCO/Ag samples by OPIT technique [87,282,493]. There are other problems demanding their solution during tape processing by using OPIT technique, namely (i) an initiation of the local heterogeneities into powders during mechanical deformation (e.g., pressing) [673]; (ii) the bubble formation into silver sheath due to isolation of gases [384,568]; (iii) a possibility of the microstructure perturbations because of violation in crystallite alignment, initiated near secondary phases due to the formation of crack-like

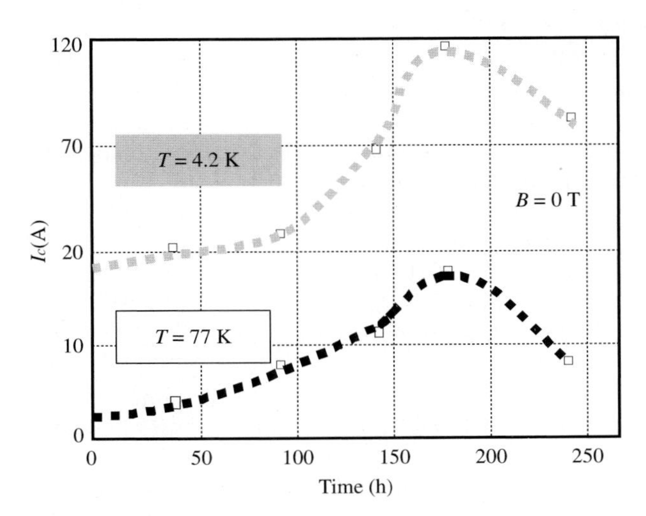

Fig. 2.17. The critical current vs sintering time ($B = 0\,\mathrm{T}$, $T = 4.2$ and $77\,\mathrm{K}$)

defects during inelastic strain of pressed powder [831], and also due to various deviations from optimum-phase composition [783]; (iv) difficulties in the texture formation of HTSC connected with the existence of secondary phases, heterogeneity and redistribution of material during thermal treatment [385,624]; (v) impossibility to totally heal the microcracks formed during mechanical deformation using high-temperature treatment [854]. Finally, an increased length of tape initiates an action of scaling factor, which influences considerable decrease in the critical current density. Due to this, the doping additions of Ag (Ag_2O, $AgNO_3$) are added to HTSC for improvement of structure-sensitive properties [915,1001]. The complexity and multistage character of the OPIT technique, and also smaller more than order of magnitude the critical current density of tapes of compare with films, define modifications of this standard technique of the BSCCO/Ag tape processing. As examples, we point the following:

- using the hot extrusion (Fig. 2.18a) to obtain long-scale ($l > 150\,\mathrm{m}$) Bi-2223/Ag tapes [673];
- intermediate mechanical deforming (one-axis pressing [341] (Fig. 2.18b) or cold rolling [831]) of tapes after first sintering, that form Bi-2223 phase in order to increase the density of superconducting core and corresponding J_c,
- application of deformation methods which are alternative to rolling, namely semi-continuous pressing [592], consecutive pressing [32] and periodic pressing [672];
- modifications of the standard rolling: (i) out-center rolling with eccentricity, using two concentric rollers [565], (ii) sandwich rolling, in which a tape locates between two thick steel plates fastened with springs [1133], (iii) transversal (differing from standard or longitudinal) rolling [359], (iv) two-axis rolling, that replaces the wire-drawing [450], (v) groove rolling (Fig. 2.18c) [457];
- pressing Bi-2223/Ag tapes at cryogenic temperatures, based on relative increase in silver hardness compared with superconducting oxide at liquid nitrogen temperatures [164];
- application of hot isostatic pressing (Fig. 2.18d) to solve the problem of porosity and microcracks in Bi-2223/Ag tapes [77];
- using an expensive pressure to pack a superconducting core at the temperatures of Bi-2223/Ag phase formation [888];
- using the *wind-and-react (W&R), react-and-wind (R&W)* and *react-wind-sinter (RWS)* techniques of continuous winding at the final stage of Bi-2212/Ag tape processing, which decrease the damage of long-scale samples during their winding [78];
- using the *wind-react-and-tighten (WRAT)* technique, that combines the advantages of the W&R and R&W techniques and consists of free winding and reaction in the winding with final introduction of isolation after completion of thermal processes and tightness of the winding to its final size [765].

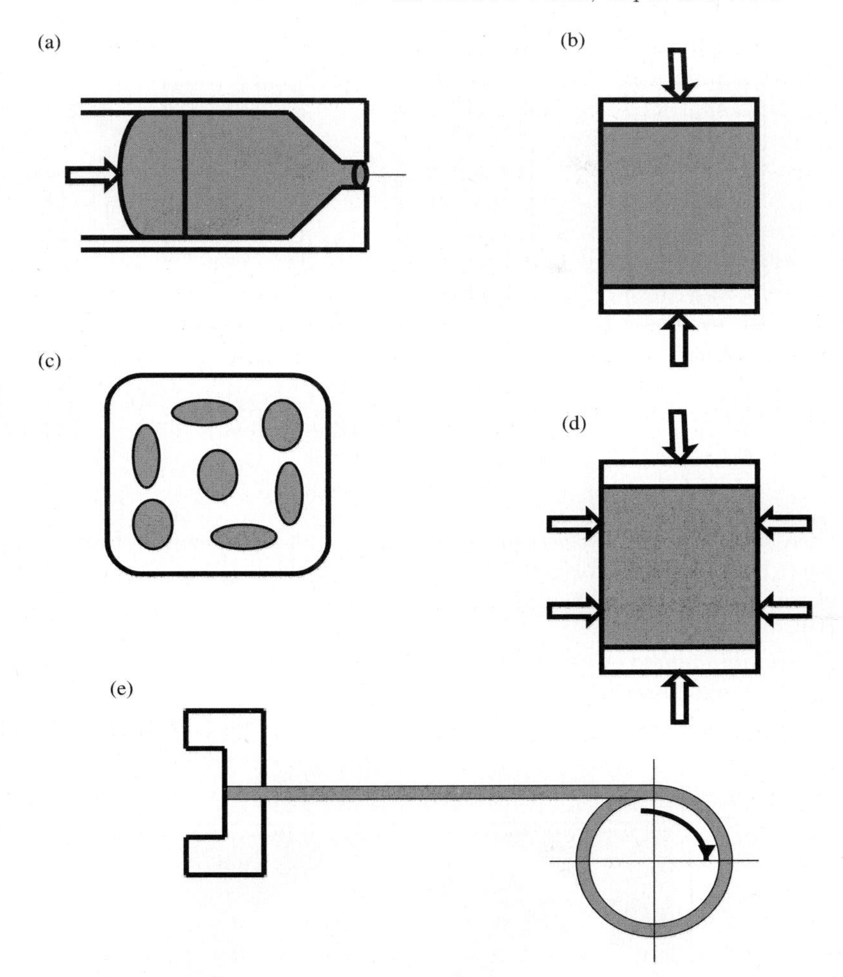

Fig. 2.18. Some additional technical operations used to improve quality of HTSC prepared by OPIT technique: (**a**) extrusion, (**b**) one-axis pressing (**c**) groove rolling (a cross-section of seven-filament wire is shown), (**d**) isostatic pressing and (**e**) continuous winding

Finally, new techniques to prepare the composite tapes and wires are designed, that use optimum geometry and placement of components or advantages of superconducting and ceramic materials, namely:

- *Tape-in-rectangular-tube (TIRT)* technique to prepare superconducting filaments with *c*-axis oriented in the direction that is not perpendicular to the tape surface (Fig. 2.19) [850].
- Processing round Bi-2212/Ag wires by using *stranded-and-formed method (SAFM)*, in which first, filaments are bundled to consist of a segment, and

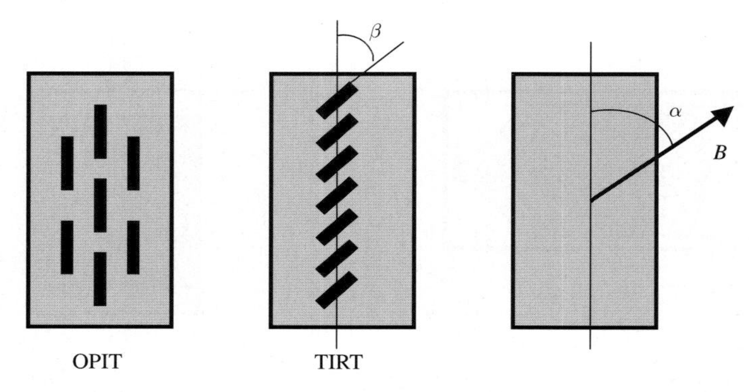

Fig. 2.19. Schematic view of the cross-sections of the OPIT and TIRT tapes. The mutual orientation of the filaments and that of the field with respect to the tape broad face are expressed by the angles β and α [850]

then three segments are stranded, drawn and formed to a final round-shape wire. Round wires fabricated by SAFM are more tolerant against bending strain than the usual round wires and permit to reach high J_c (Fig. 2.20) [18].

- *Wrapping method* of BSCCO/Ag tapes around core to decrease the transport current losses (Fig. 2.21) [1027].
- *Rotation-symmetric-arranged tape-in-tube wire (ROSATwire)* technique with Bi-2212/Ag tape-shaped multifilaments, possessing triple rotation

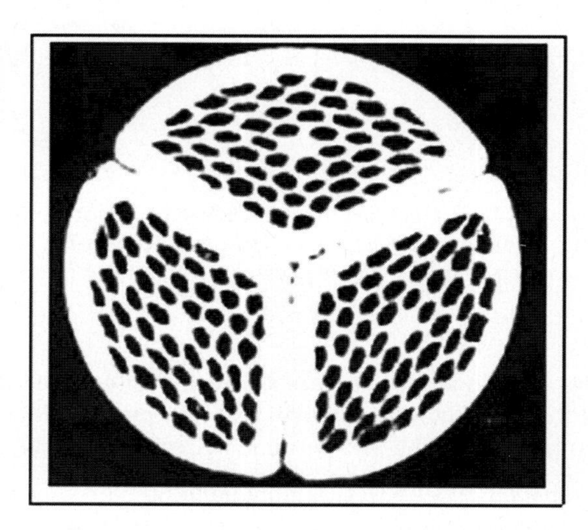

Fig. 2.20. Cross-section of a round wire fabricated by stranded-and-formed method [18]

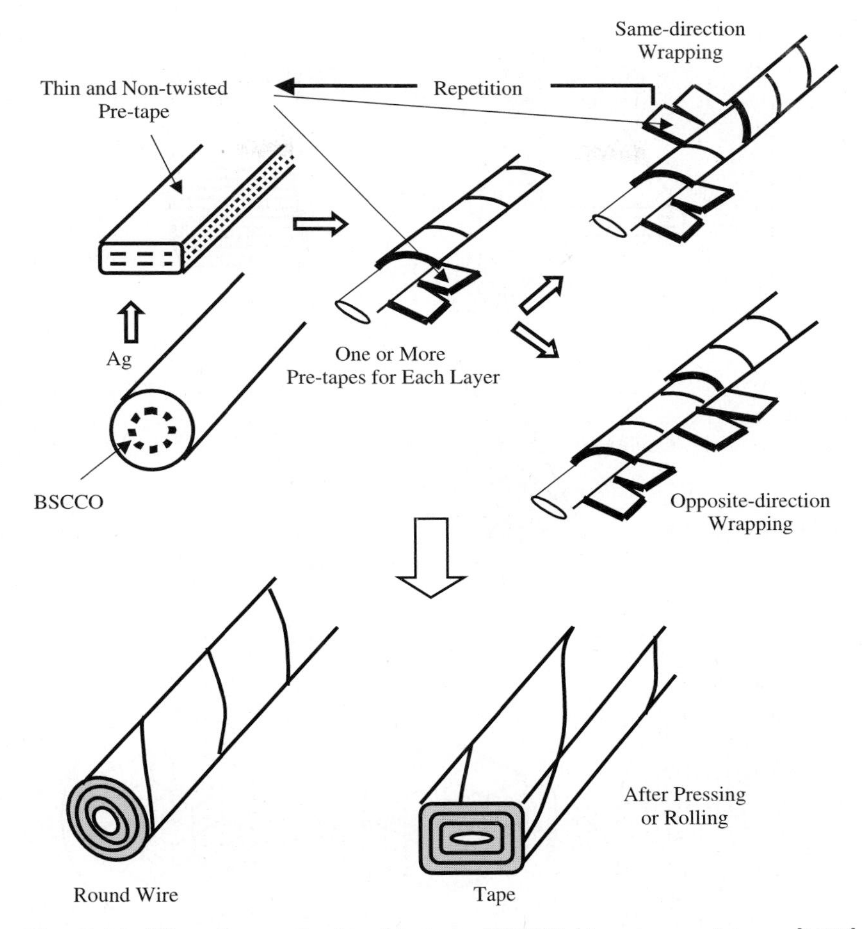

Fig. 2.21. Wrapping method to fabricate BSCCO/Ag wires and tapes [1027]

symmetry and having good crystal alignment in each filament. Since the present wire structure yields complete symmetrical arrangement of the tape-shaped filaments, it is no longer necessary to use a rolling, but permits to use a drawing (Fig. 2.22) [780].

- Manufacture of Bi-2223/Ag one- and multifilament composite tapes with oxide barrier layers between filaments to diminish electric and magnetic losses (Fig. 2.23) [315,449,574].
- Development of multifilament tapes with central part, that consists of Bi-2223 filaments, ensuring transport current flow, and surrounded by the barrier ceramic layer and YBCO thin film to screen external magnetic field and also to protect Bi-2223 filaments (Fig. 2.24) [45].

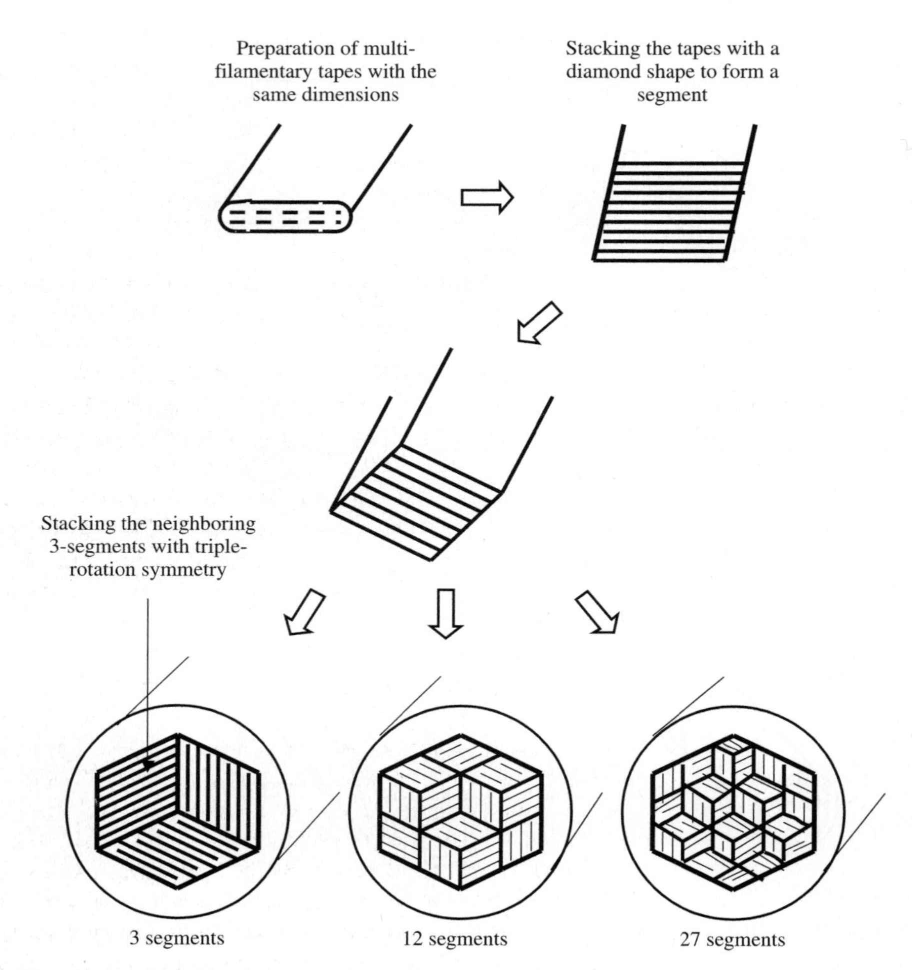

Fig. 2.22. Fabrication process of rotation-symmetric-arranged tape-in-tube wire with various Bi-2212/Ag segments [780]

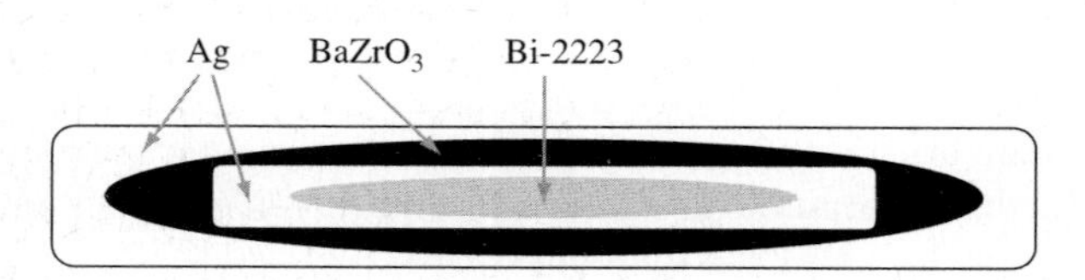

Fig. 2.23. Typical cross-section of Bi – 2223/Ag/$BaZrO_3$/Ag monocore tape [574]

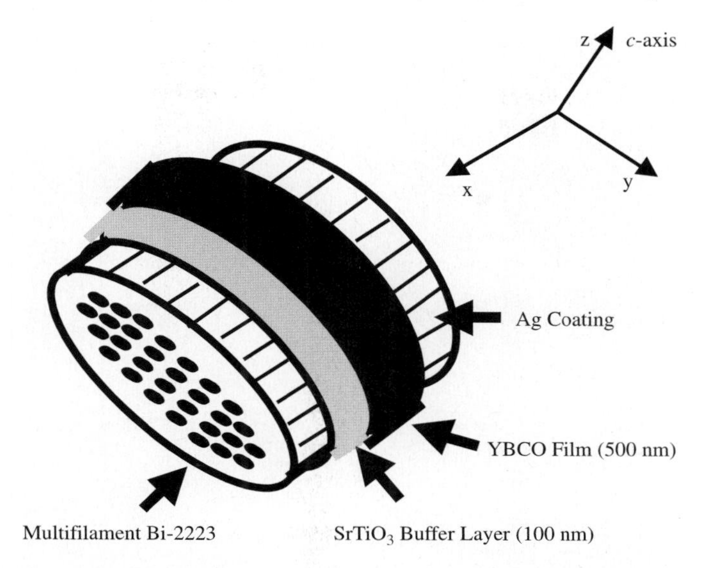

Fig. 2.24. Schematic diagram of Bi-2223 multi-filament tape with Y-123 screening layer [45]

2.3 Tapes and Wires, Based on Thallium and Mercurial Cuprates

Processing techniques of the thallium thin films (Tl-1212, Tl-1223, Tl-2012, Tl-2212 and Tl-2223) coincide in many details with corresponding fabrication techniques of YBCO thin films. Best results have been obtained by using the following substrates: $LaAlO_3$ and $NdGaO_3$ [432,604,753,754], and also buffer layers: YSZ and CeO_2 [881] due to good agreement of their crystallographic properties. Moreover, MgO, $LaGaO_3$, $SrTiO_3$ have been used as substrates [604]. Tl-1223 thick films and wires with high superconducting properties have been prepared on substrates from Ag and ZrO_2 [724,725,1025]. An expensive doping by fluorine has permitted to improve considerably the structure-sensitive properties of Tl-1223 samples [380].

Hg-1212 and Hg-1223 thick films demonstrating maximum critical temperatures, T_c, have been prepared in two stages [723]: (1) a precursor layer, that consists of $Ba_2Ca_2Cu_3O_7$ and doping oxides, that is, PbO, Bi_2O_3 and ReO_2 has been deposted (2) superconducting phase has been formed by heating the precursor film at the partial pressure of mercury. The precursor layers could be obtained by using the precipitation processes including: (i) spray pyrolysis [990]; (ii) pulverization of powder with solvent [695]; (iii) sol–gel method [1078]; and (iv) application of powder-polymeric suspension [1178]. The majority of researches use mono-crystalline ceramic substrates (e.g., YSZ) [990,1078,1178]. Moreover, Ni-based substrates with Cr/Ag buffer layer have been used [695]. The increased superconducting properties and

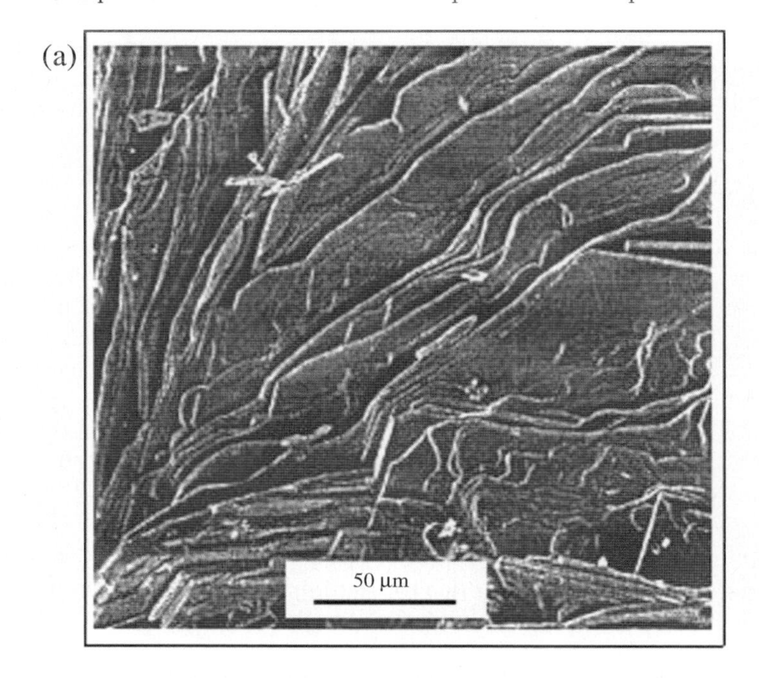

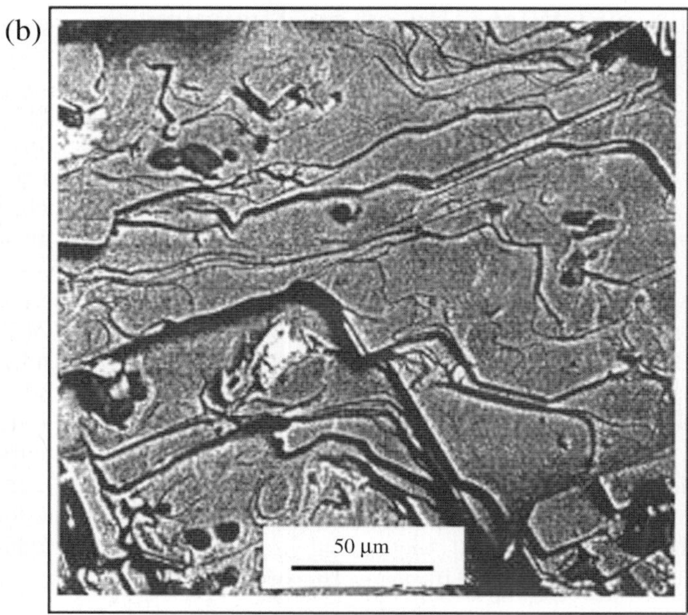

Fig. 2.25. (**a**) SEM micrograph of (Hg, RE)-1212 sample synthesized at 820°C, (**b**) backscattered electron micrograph of the same sample [923]

improved formation of the mercurial conductor are reached by Re, Pb and F doping [269,378,379]. Optimum synthesis conditions permit to obtain good aligned (Fig. 2.25a) and phase purity (Fig. 2.25b) structures of thick films.

2.4 BSCCO Bulks

Bi-2212 superconducting coverings for tubes and rods are synthesized by the diffusion reaction between Sr-Ca-Cu oxide substrate, possessing high melting temperature, and Bi-Cu oxide coating layer [1032] (or mixing Bi–Cu and AgO_2 [1156,1157,1160]) with low melting temperature (Fig. 2.26). This technique permits to fabricate diffusive superconducting layer with thickness 150 μm (for roads of diameter 3 mm), demonstrating dense homogeneous structure of plate-like grains that are aligned almost perpendicular to the substrate surface (Fig. 2.27).

The weakening of pinning properties at above 20 K is intrinsic for Bi-2212 system. The practical requirement is to enhance this threshold that causes Pb doping of bulks in the *partial-melting technique* [08]. In order to obtain high-dense-textured Bi-2212 bulks, the *hot-forging technique* is used [324] and also the *solid-state reaction method* [876]. The addition of PbO increases fraction, formation rate and stability of Bi-2223 phase [1037]. However, fabrication of Bi-2223 necessary bulks with for applications properties is the very difficult problem. *Hot-pressing technique*, used during processing of Bi-2223 bulks, promotes void elimination under pressure and high temperature. Moreover, it improves the crystallite alignment in the sample, causing an increasing of the critical current density [448]. At the same time, J_c can decrease at prolonged sintering of superconducting bulks due to corresponding decrease of the ceramic density. Hot pressing causes high density near the theoretical level for Bi-2223 ($\sim$ 6.31 g/cm^3). The sufficiently long sintering before hot pressing rises Bi-2223 phase content in the sintered samples. At the same time, in order to reach necessary properties of final superconducting ceramics, special control of processing parameters is required.[6] Room temperature pressing is not effective because of high resistance to strains. By increasing the pressing temperature, this problem becomes less sharp. In this case, closed porosity is formed and the sample density increases. This increases the contact area between Bi-2212 and secondary phases. The formation of Bi-2223 phase is based on the epitaxial growth of the Bi-2223 crystallites into Bi-2212 matrix in accordance with the chemical reaction:

$$\mathrm{Bi}-2212 + \text{secondary phases} \rightarrow \mathrm{Bi}-2223. \tag{2.3}$$

Therefore, the elevated contact area causes an acceleration of chemical reaction and Bi-2223 formation phase. The sample densification also increases

[6] This, in the first place, is explained by that Bi-2223 phase is stable only in very small temperature intervals (in difference from Bi-2212 phase) and kinetics of its formation are very slow to obtain mono-phase material.

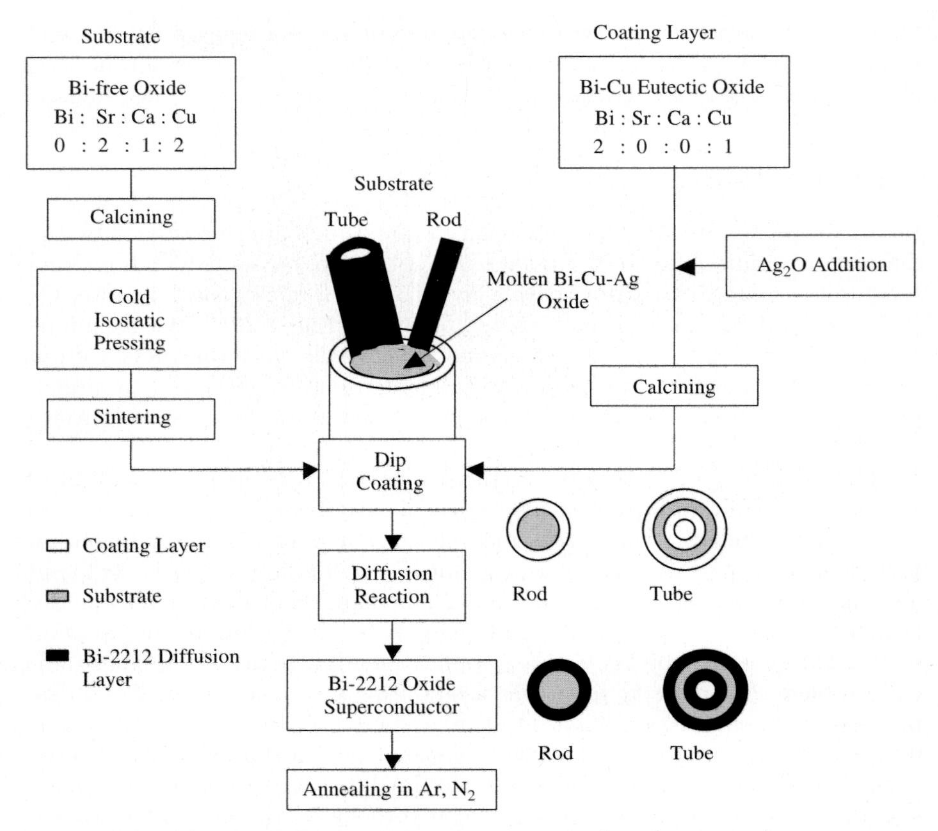

Fig. 2.26. Specimen preparation procedure [1156,1160]

a connectivity of superconducting grains (Bi-2223 or Bi-2212), which is found by a quality of intercrystalline boundaries. The increase of the grain connectivity causes a rise of the critical current in the sample. However, during the hot pressing, an excessive elimination of liquid phase, that is rich by Cu and Pb, is possible. This leads to considerable phase segregation in the sample, but a deficiency of the secondary phases prevents total transformation of Bi-2212 phase into Bi-2223 phase during hot pressing, decreasing T_c and J_c. Preliminary sintering causes a frame formation from Bi-2212 crystallites, preventing the secondary phase losses. After that, the hot pressing redistributes secondary phases around Bi-2212 grains, accelerating the phase transformation considerably. Displacement of amorphous Bi-2212 phase and secondary phases at intergranular boundaries heals intercrystalline defects. At the same time, preliminary prolonged sintering can lead to the expense of larger part of secondary phases on the growth of arbitrary oriented Bi-2223 crystallites.

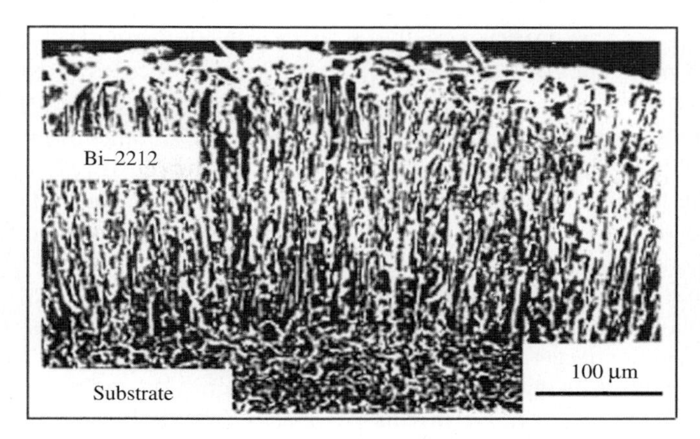

Fig. 2.27. SEM micrograph taken on the fractured cross-section of the Bi-2212 rod (Bi-2212 diffusion layer and substrate). The sample is obtained at 850°C for 20 h on the (0213) substrate [1157]

In this case, amorphous Bi-2212 phase has sufficient time to re-crystallize. Therefore, one could not be used to increase the grain connectivity during hot pressing. Thus, preliminary super-long sintering before the hot pressing, do not permit to fabricate bulks with high J_c. Moreover, expensive liquid phase, not dissolved into Bi-2223, forms non-superconducting phases and diminishes J_c [783].

One-axis cold pressing can lead to partial orientation of BSCCO plate-late grains in the pressing direction, taking into account great microstructure anisotropy [959]. Therefore, the sample re-crystallization, presenting an orientated grain growth, provides maximum texture during next sintering, corresponding to external loading. There is a good correlation between the final angle distribution of BSCCO plates and sample *squashing*, namely a lower misorientation corresponds to higher loading [770,900]. Then, the hot forging followed by squashing of sample leads to improvement of alignment of Bi-2223 plates [901]. These samples show $J_c = 8\,\mathrm{kA/cm^2}$ (zero magnetic field). This presents interest for current conductors and current limiters. Another multistage processing of Bi-2223 bulks includes one-axis cold pressing of superconducting powders, then annealing without pressure and finally the hot forging [1043].

Taking into account the plate-late morphology of BSCCO grains, *magnetic-melt-processing-texturing* (*MMPT*) technique has been developed, combining at high temperature two different parameters, namely an application of in situ magnetic field (8 T) and one-axis loading (60 MPa) [771]. This technique has permitted to obtain materials with a good texture. At the same time, *hot plastic deformation* technique permits the formation of sharp crystallographic texture and high density of defects in the crystalline lattice, which are effective pinning centers, in the BSCCO bulks. As the result, the critical current density $J_c > 10^5\,\mathrm{A/cm^2}$ has been reached in Bi-2212 ceramic [173,467,468].

In order to form microstructure features – pinning centers – in BSCCO systems, following methods are used: (i) irradiation of samples after processing by protons [584], heavy ions [1055] or neutrons at doping HTSC by uranium [392], that forms super-thin amorphous cylindrical inclusions, the so-called "columnar defects" (Fig. 2.28); (ii) introducing defects, connected with doping additions, disperse particles and dislocations [04,292,669]; and (iii) technique change, supposing an introduction of the mixed particles (e.g., Pb) into Bi-2212 phase [658]. The enhanced pinning due to irradiation of superconductor by high-energetic ions, leads to significant increasing of the critical current density, J_c, sometimes in the order of magnitude and more. However, the critical temperature, T_c, in this case, as a rule decreases, according to the common law of the superconductivity suppression by structure. At small irradiation doses, small ($\sim 1\,\mathrm{K}$) increasing of T_c (so-called "effect of small doses") is sometimes observed. Usually, the irradiation dose is selected so as to reach the maximal value of J_c at small decreasing of T_c. First, in [04], very significant (on 15 K) growth of T_c in Bi-Pb-Sr-Ca-Cu-O thick films after their irradiation by argon ions has been found. The origin films consisted of mixing of Bi-2212 and Bi-2223 phases with the domination of the Bi-2212 phase (they has $T_c = 85\,\mathrm{K}$). The ionic irradiation stimulated the processes of local melting of the film and diffusion of atoms, resulting in the formation of high-T_c Bi-2223. Thus, the ionic irradiation can be an alternative technique to fabricate Bi-2223 phase, which is known for its great sensitivity to conditions of sample synthesis and annealing.

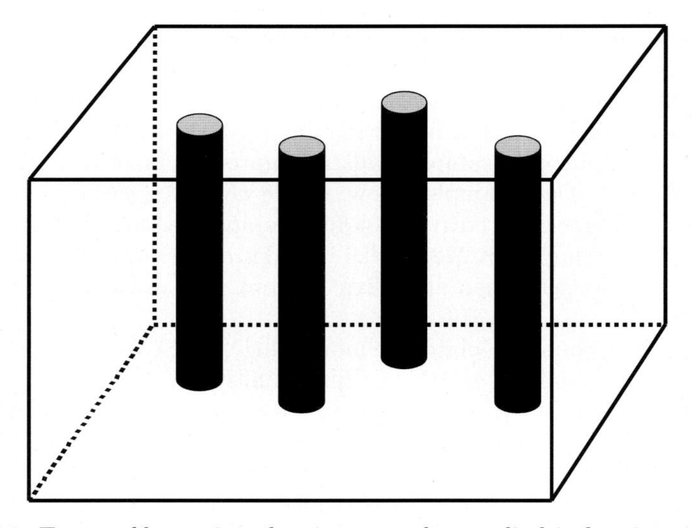

Fig. 2.28. Traces of heavy ions forming amorphous cylindrical regions in HTSC

An intensive effect in improvement of superconducting properties is decreasing of cooling rate of the sample that increases J_c and T_c and forms more perfect microstructure of superconductor [829]. In order to improve mechanical and superconducting properties of BSCCO during processing, the next additions are introduced into the samples: inclusions of Ag (Ag_2O, $AgNO_3$ [999], Al_2O_3 strengthening filaments of Al_2O_3 [679,1147] and $ZrO_2 \bullet Y_2O_3$ [679], whiskers of MgO [1185,1186,1187], particle dispersion of MgO and additions of high-dense polyethylene [97]. A schematic description of the solid-state processing method of the $(MgO)_w$/BPSCCO composite is shown in Fig. 2.29, and SEM micrographs taken from etched surfaces of polished monolithic and composite specimens are presented in Fig. 2.30. Moreover, in order to strengthen BSCCO bulks, the glass-epoxy tapes are used [1044]. At the same time, an application of doping additions of TiO_2 and ZrO_2, which leads to toughening of superconductor, can simultaneously cause considerable decomposition of Bi-2223 phase and diminish T_c [351,702].

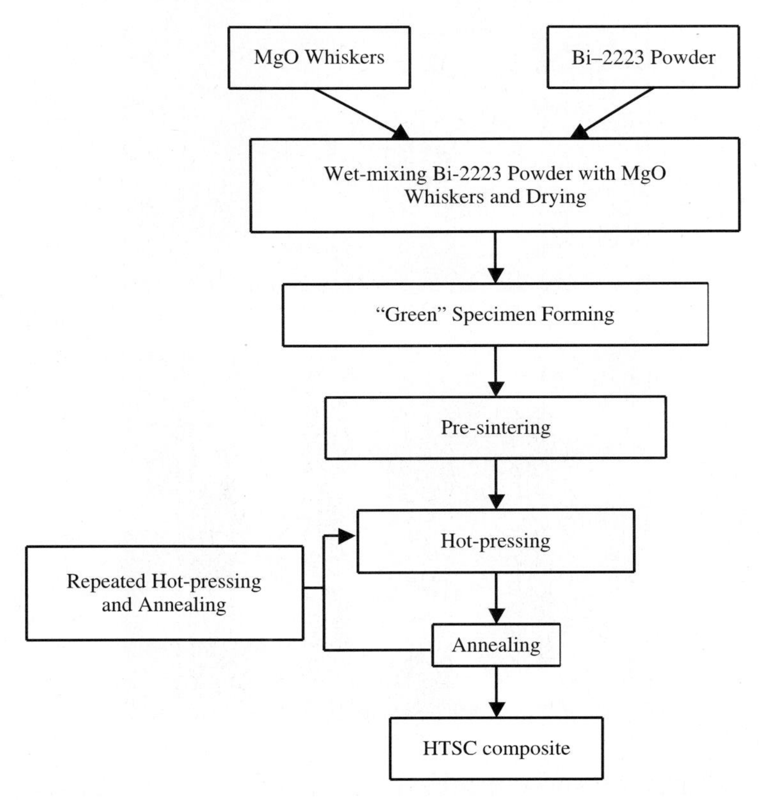

Fig. 2.29. Solid-state processing of $(MgO)_w$/BPSCCO composite [1185]

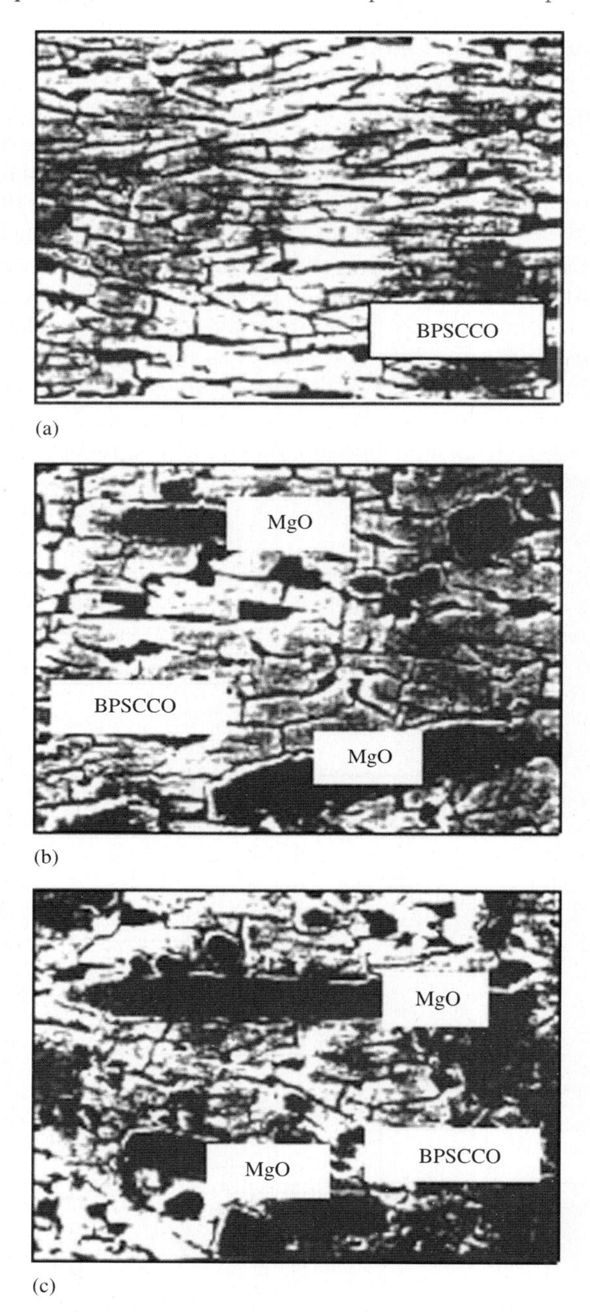

Fig. 2.30. Microstructure of BPSCCO phase in polished and etched cross-sections parallel to the hot-pressing direction of (**a**) a monolithic BPSCO, (**b**) and (**c**) a $(MgO)_w$/ BPSCCO composite with 10 and 20% concentration of whiskers, respectively (all specimens have undergone a three-cycle hot-pressing and annealing) [1186]

2.5 Y(*RE*)BCO Bulks

Intercrystalline boundaries can diminish considerably the transport properties of YBCO bulks and increase sensitivity to applied magnetic fields. Therefore, it is very expedient to create large grain specimens with maximum number of strongly connected intercrystalline boundaries. As a rule, $Y(RE)Ba_2Cu_3O_{7-x}$ crystallites (123 phase), prepared by the melt-processing technique, include $Y(RE)_2BaCuO_5$ particle dispersion (211 phase). The elevated critical current density in these samples can correlate with the concentration of the 211 inclusions [961] and assumes that fine size of these particles causes directly elevated pinning. At the same time, the sufficiently large particles of 211 normal (non-superconducting) phase obviously diminish the superconducting properties. Therefore, in order to optimize the ceramic properties, it is necessary to control the concentration and the size of 211 particles in precursor powder used in YBCO processing [643]. After fabrication of the precursor powder, it is pressed into pellets, which is then calcined and melted. The pellets, subjected to insignificant pressing, demonstrate a lost density compared with pellets subjected to cold isostatic pressing. Further sintering causes their densification next. However, the hard aggregates of grains, formed during the cold isostatic pressing, can prevent alignment of some crystallites, and, therefore limit an effective densification during the following sintering. Then, it could be assumed that the sintering is the optimum process to create high-dense HTSC samples. It should be noted that 211 particles grow into precursor bulk during sintering. It is not desired with the view of preservation of fine granularity of the 211 phase. Therefore, it is necessary to exclude this effect. Thus, the cold isostatic pressing in some cases may be more practical to prepare optimum precursor samples, which are then subjected to melting [642].

All techniques, based on melt processing and used to prepare YBCO large-grain ceramics, are characterized by the peritectic reaction at $T_{\mathrm{p}} = 1015°\mathrm{C}$ owing to the formation of 123 phase from 211 phase and liquid component [642]:

$$\underset{(211\ \text{phase})}{Y_2BaCuO_5} + \underset{(\text{liquid phase})}{3BaCuO_2 + 2CuO} \rightarrow \underset{(123\ \text{phase})}{2YBa_2Cu_3O_{6.5}}\,, \tag{2.4}$$

or in another form [640]:

$$\underset{(211\ \text{phase})}{Y_2BaCuO_5} + \underset{(\text{liquid})}{Ba_3Cu_5O_{6.72}} + \underset{(\text{gas})}{0.42\,O_2} \rightarrow \underset{(123\ \text{phase})}{2YBa_2Cu_3O_{6.28}}\,. \tag{2.5}$$

Pseudo-binary diagram, showing two peritectic transformations, is presented in Fig. 2.31. The 211 phase and liquid could be formed by fast heating a pre-sintered precursor sample to a temperature that is considerably above T_{p}. Then 123 phase forms very slowly during the cooling of YBCO partial-melting through point T_{p}. In this case, 123 phase are added up to 30 wt% of the 211 particles before the melting process with the aim to create local pining centers and to prevent liquid loss during melting [641]. Solidification processes, depicted by the formulas (2.4) and (2.5), put definite requirements

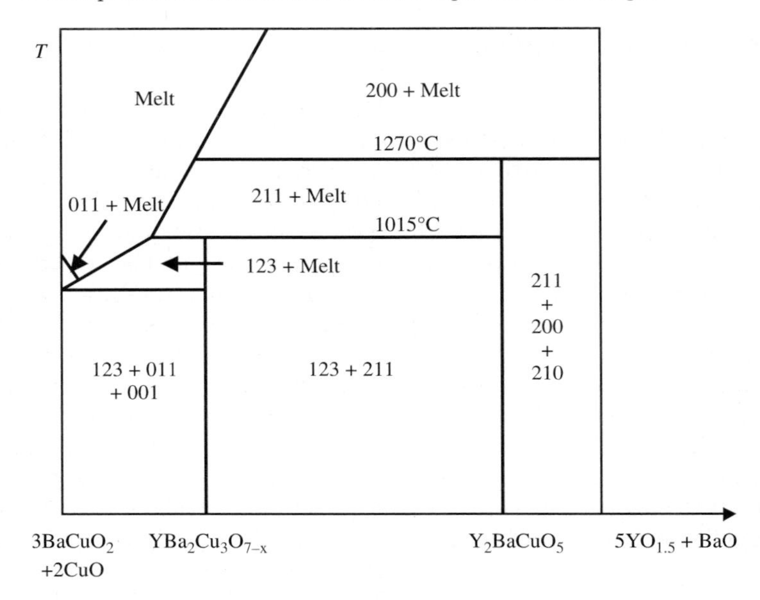

Fig. 2.31. Pseudo-binary diagram along the Y_2BaCuO_5–$YBa_2Cu_3O_{7-x}$ line showing two peritectic transformations [29]

on YBCO precursor samples [642], namely (i) 211 particles must be sufficiently fine in the initial sample precursor in order to form fine-grain dispersion in the sample fabricated. (ii) The material should be able to retain the liquid phase, appeared in result of the peritectic reaction into samples at the temperatures, which are considerably above T_p. It is necessary in order to form 123 phase at cooling. Successive process is dependent on the precursor homogeneity and density and also on the size distribution of the 211 particles. (iii) The precursor sample should not contain foreign compositional and surface admixtures, which form the heterogeneous nucleation sites of grains, and hence limit the grain sizes, that may be reached during the grain growth.

Obviously, it is very possible an enlargement of 211 fine-size inclusions during melting. As rule, the 211 enlargement occurs above peritectic temperature, solely. Therefore, the 211 inclusions with diameter up to 50 μm can form in the final ceramic sample. There are other problems, namely a very slow rate of solidification, a necessity to control thermal gradients during high-temperature treatment, a limited size of domains obtained, that is accompanied by their misorientation, a microcracking formation and heterogeneous composition. The progress could be reached by changing processing parameters and using doping additives.

Oxygen release increases sharply during melting. This forms voids, causing the corresponding rise of the final specimen volume. At the same time, a surface tension, connected with the melted state and decreased rate of the oxygen release, leads to specimen densification at increasing temperature. As it is shown in tests, the sample densification, controlled by the surface tension, is the dominating factor [642].

Figure 2.32 demonstrates the YBCO cross-sections obtained at various heating rates. Even, using the same (in sizes) pellets does not preserve from considerable differences in geometry, distribution and structure of porosity in the melt-processed superconductors. Considerable changes of the sample geometry and homogeneity have been observed at the heating rate of 30°C/h because of void formation with sizes above 1 mm (Fig. 2.32a). However, the void sizes diminished together with the heating rate (Fig. 2.32b and c) that caused their maximum homogeneity in the case of heating rate of 10°C/h [640]. At the same time, a void diminishing, caused by the following decrease of the heating rate, can lead to unfavorable consequences, in particular to elevated loss of liquid phase. This is not desirable to a high degree in view of stoichiometry and can lead to loss of control for composition and grain size during superconductor processing. Therefore, improved physical properties should be reached, using other methods, in particular by changing the specimen density and processing parameters of the peritectic solidification [642].

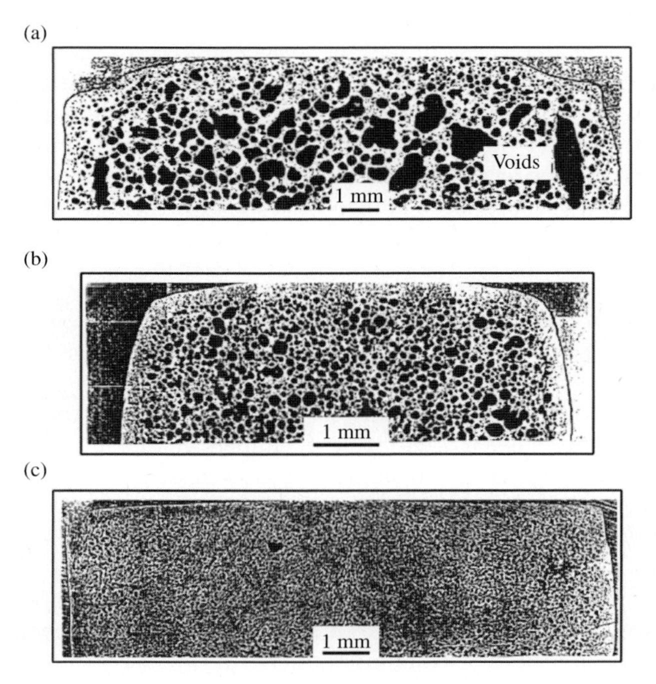

Fig. 2.32. Optical micrographs of the polished YBCO cross-sections obtained at various heating rates: (**a**) 30, (**b**) 20 and (**c**) 10°C/h [642]

HTSC applications require high-aligned grains and specimens with perfect texture that is reached more easily in smaller specimens in the presence of thermal gradients. The use of large thermal gradient, leading to better alignment, could cause microcracking of the sample. The latter spreads out as a consequence of the anisotropic thermal expansion of 123 phase and secondary phases ($BaCuO_2$, CuO), which precipitate at grain boundaries. The microcracks perpendicular to the *ab*-plane are the most detrimental for HTSC and can develop during cooling or re-oxygenation of the sample. Then, purification, associated with the crossing of the hot zone, can lead to chemical heterogeneities in long samples (which are always obtained using a melt-zone-type technique) owing to the different diffusion coefficients of the species. One observes in such a case a progressive decrease of the barium and copper amount, correlated with an increase of the amount of 211 phase during the process. Additives, such as Y_2O_3, which increase the viscosity of the melt, can drastically limit these phenomena. The more rapid heating rate of 123 phase causes the finer 211 precipitates. This means that the more rapid is the heating above the peritectic temperature, the higher is the temperature where the decomposition of 123 phase begins. The lower the stability of the 123 phase, the more rapid is the precipitation of the 211 phase, that is, the higher the number of nucleation sites. In rapid heating, defining the fine-sized 211 phase a long plateau allows the particle coarsening by Ostwald ripening, due to the bigger grains consuming the smaller ones. In fact, as for the growth of the 123 phase, two mechanisms can be supposed to limit the growth of the 211 particles dispersed in a liquid, namely (i) diffusion of a solute in the liquid; and (b) reaction at the interface between 211 phase and liquid. Finally, the maximum temperature, reached above the peritectic temperature, controls the amount of liquid and its viscosity. So one expects a large influence on the growth of the 211 phase. Taking into account these considerations, different *methods of melt crystallization* have been proposed during the development of melting techniques, namely:

- *Melt-textured growth* (*MTG*) [491]. The textured growth of 123 superconducting phase from melting, in which the 123 ceramic is used as initial precursor.
- *Liquid-phase processing* (*LPP*) [913]. The liquid-phase technique based on decreasing of prolonged treatment of samples at maximum temperatures with the aim to prevent an undesired growth of the 211 particles, that occurs intensively above the peritectic temperature.
- *Zone melting* (*ZM*) [684]. This technique applies zone melting to obtain long samples using 123 ceramic as initial precursor.
- *Quench-melt-growth* (*QMG*) [745]. The growth of 123 superconducting phase from melting, based on the super-fast cooling, in which the 211 phase forms in results of rapid interaction Y_2O_3 phase with melt into high-temperature region, where the 211 normal phase to be thermodynamically stable.
- *Melt-powder-melt-growth* (*MPMG*) [297]. This process, using the same thermal schedule as the QMG process, introduces a drastic crushing of

the quenched mixture after quenching in order to obtain a fine distribution of the Y_2O_3 phase, leading further to a finer 211 phase, which is uniformly distributed.

- *Powder-melt process* (*PMP*) [1199]. The same as QMG process, starting from a $Y_2BaCuO_5 + 3BaCuO_2 + CuO$ mixture instead of pre-synthesized 123 phase. The result is equivalent to the QMG process, but without any overheating of the sample.
- *Solid-liquid-melt-growth* (*SLMG*) [976]. Using the same thermal profile as the PMP process, the SLMG process differs by the starting material: a mixture of $Y_2O_5 + BaCuO_2 + 2CuO$ is used here. These two techniques differ essentially in the morphology of the 211 phase: needle shape in the case of SLMG instead of quasi-spheres for PMP process.
- *Microwave-melt-texture-growth* (*MMTG*) [146]. This technique was developed to take into account the high thermal gradient for both increasing the solidification rate, and directing this process.
- *Magnetic-melt-texturing* (*MMT*) [190]. This technique uses a static magnetic field in order to modify the microstructure of a material and to obtain samples with a good texture. This process exploits magnetic anisotropy of elementary cell that may be elevated by replacing Y ion with ion of rare-earth element (*RE*).
- Doping YBCO by the *RE* ions in order to form local stresses caused by the deformation mismatch of the YBCO and *RE*BCO crystalline lattices, that is an important origin of the magnetic flux pinning and corresponding increase of J_c [627,628].

The thermal cycle, used in a classical MTG process, is shown in Fig. 2.33. Here, the partial melting of a pre-sintered 123 sample is induced between 1100 and 1200°C, preserving practically the geometry of the precursor sample. Finally, the sample is very slowly cooled (1–2°C/h) under an oxygen atmosphere in a thermal gradient [1126].

Introduction of silver [992], resin impregnation [647,1069,1070] and also using other additives improve the structure-sensitive properties of Y(*RE*)BCO. In general case, these additives are used to decrease the size and morphology modification of the 211 phase. They act in three ways, namely: (i) changing the 123/211 surface energy, (ii) changing kinetics of diffusion process, and (iii) forming nucleation sites for the 211 phase. In particular, an additional large dispersion of the 211 particles (up to 40 wt%) to the stoichiometric 123 phase is able to diminish the 211 particle size, formed during decomposition of the 123 precursor sample [745]. To improve superconducting microstructure and properties, the doping additives used are: Ag (Ag_2O, Au/Ag) [998,1171], Pt (PtO_2) [509,777], Sn ($SnO_2, BaSnO_3$) [186,685,717], Zr ($ZrO_2, BaZrO_3$) [132,322], Ce (CeO_2, $BaCeO_3$) [185,187,664], Ca [391] and SnO_2/CeO_2 [664].

The formation of preferable orientation of grains is necessary for maximum use of anisotropic properties of HTSC in the specific products. The specimen texture could be formed and controlled by using the *top-seeded-melt-growth*

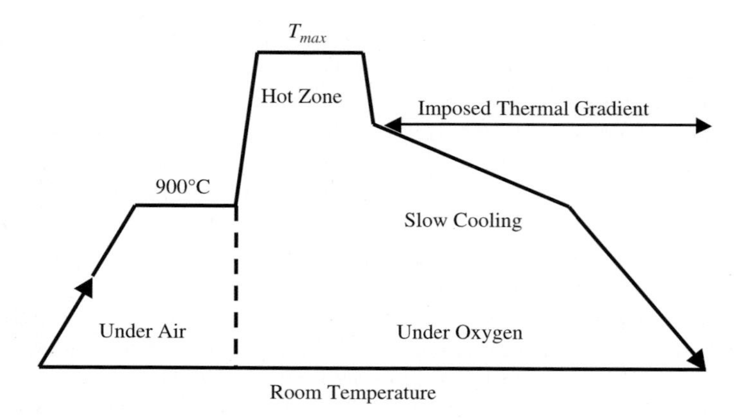

Fig. 2.33. Typical thermal cycle used in the MTG process [1126]

(*TSMG*) technique [729]. In this process, the crystallites-seeds from *RE* equivalents of the 123 phase are introduced.[7] They have the higher temperature of the peritectic decomposition. In this case, the 123 phase nucleates and grows in the specific direction (Fig. 2.34). Moreover, the weak links decrease considerably accompanied by proper increase of the grains. The peritectic solidification of YBCO by using seeds leads to growth morphology of faceted grains with symmetry, depending on nucleus crystallite, for example, SmBCO (Sm-123), NdBCO (Nd-123) or other *RE* barium cuprate. However, standard procedure by using seeding (e.g., SmBCO) proposes high overheating of the sample, that is necessary to (i) avoid numerous formation of nuclei, (ii) increase liquid in the sample and (iii) eliminate an effect of precursor microstructure [1126]. In order to get over the numerous nucleation and arbitrary growth of 123 crystals, beginning from substrate material at prolonged process of TSMG, as observed in test, the increasing of the formation temperatures for different *RE*BCO phases together with their ionic radiuses could be used [728]. Then, a selection of corresponding *RE* composition permits to diminish an undercooling area $\Delta T = T_\mathrm{p} - T_\mathrm{g}$ (where T_p is the peritectic temperature and T_g is the temperature of grain growth in the sample under constant conditions) at slow cooling to homogeneous temperature (Fig. 2.34c). In this case, is the grain growth rate decreased and unstable solidification eliminated. The kind of cooling process, that is, slow continuous cooling or jump-like decreasing of temperature during one or some stages renders considerable effect on quality of the final crystallographic structure [479].

Carrying out the TSMG process in an air atmosphere with seeds from Nd, Sm, Eu and Gd forms *RE*BCO samples with decreased critical temperature T_c accompanied by broad superconducting transition. This is caused by

[7] There are *hot seeding* and *cold seeding* [480]. In the first case, the seed is placed on YBCO sample at room temperature, in the second case, the seed is placed at temperature T_max above the peritectic temperature (Fig. 2.34c).

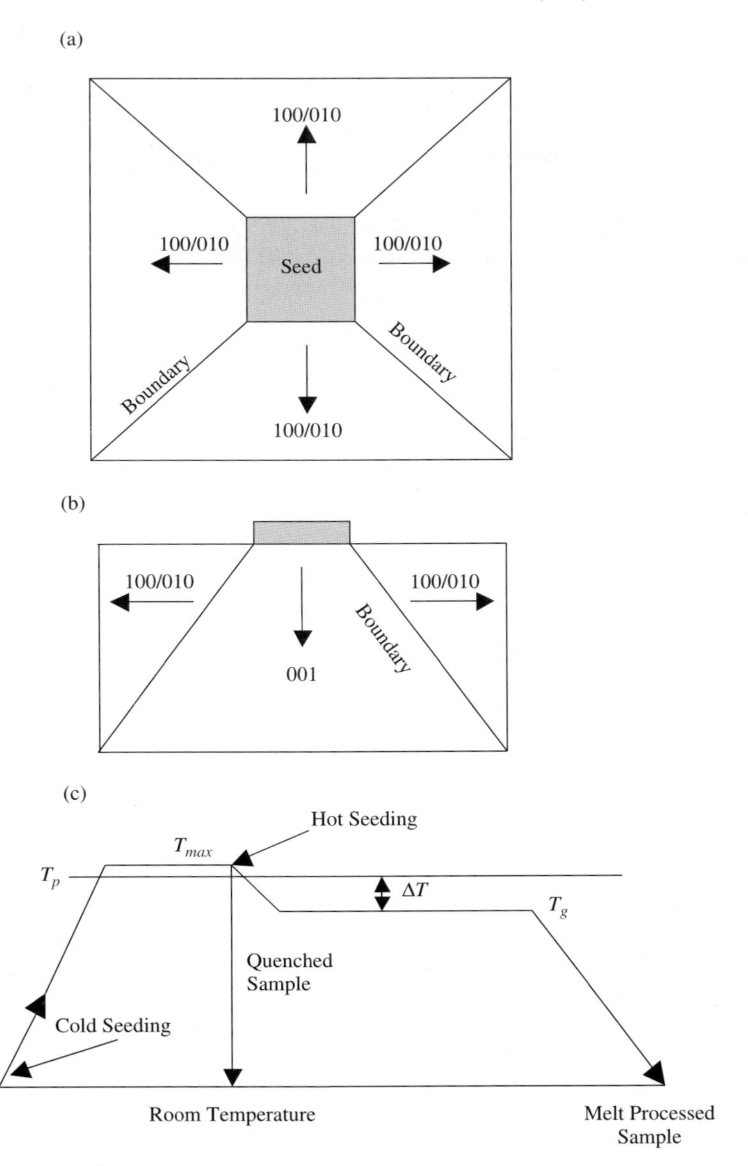

Fig. 2.34. Schematic diagram of the top-seeded melt growth process applied to form and control specimen texture (**a**) view from above, (**b**) view at the side and (**c**) applied heating schedule of the TSMG processing

the formation of hard precipitation of $RE_{1-x}Ba_{2-x}Cu_3O_{7-x}$ because of near cation radius between pointed *RE* and Ba. Carrying out the TSMG process in the controlled atmosphere with low partial pressure of oxygen (1 or 0.1%

of oxygen in Ar) (*oxygen-controlled-melt-growth-process* (*OCMG*)) [749], it is possible to except formation this precipitation. In order to reduce the duration of TSMG process, the *multi-seeded-melt-growth* (*MSMG*) technique [932] is used with corresponding thermal treatment. This process consists of some seeds on the YBCO sample. Now, *RE*BCO samples demonstrate the critical current density that is above $100\,\mathrm{kA/cm^2}$ ($77\,\mathrm{K}$ and $0\,\mathrm{T}$) with corresponding control of composition and microstructure [908]. Considerable magnetic fields can be trapped in large-grain melt-processed HTSC (e.g., $10\,\mathrm{T}$ at $45\,\mathrm{K}$ [880]), that is much more than in conventional magnets, and it is very important for various applications.

3

Experimental Investigations of HTSC

3.1 Experimental Methods of HTSC Investigations

3.1.1 Special Techniques

The complexity of HTSC structures and properties has resulted in a small number of directed observations and test dependencies of the type "structure–property." Among methods of three-dimensional observation the magnetic vortex structures in HTSC, the *small angle neutron scattering* and *spin precession of polarized muons* have an important application as the means of microscopic research of the local magnetic fields, but the *Bitter decoration* method is treated as the most often used method of spatial resolution [84]. In particular, the Bitter decoration method utilizes small ferromagnetic particles for decorating the magnetic domain structure. When these particles are sprinkled on a material, displaying at its surface an inhomogeneous distribution of magnetic flux density, the particles are attracted to the regions with the largest value of the local magnetic field. This method has demonstrated high effectiveness in visualization of the vortex structures localized at defects [208, 1116].

The great achievement in the magneto-optical characterization of the vortex structures consists of using the Bi-doped iron garnet thin films [470], which operate from below 4.2 K to above 500 K, that is, they very well fit the temperature window used in HTSC. These films come in two distinct varieties, one with magnetization vector perpendicular to the surface of the film (first investigations) and another with magnetization vector in-plane. As a result, a magnetic resolution of 10 μT has been reached. The spatial resolution can reach values less than the garnet film thickness, if the magnetic field gradients are sufficiently strong. The resolution of 0.4 μm has been reached for imaging obtained using films with thickness of 2 μm. Some achievements in using the *magnet-optical imaging (MOI)* technique for visualization of the vortex structures in different HTSC systems have been presented in the overviews [559, 858, 1119]. In combination with a digital camera and an image-processing system, time-resolved observations are possible. This method ensures the directed visualization of the superconducting currents and Abrikosov vortices, penetrating the sample. In this case, the total picture of defects is identified from voids and microcracks in polycrystal down to smallest heterogeneities in single crystallites. MOI technique has been used to research the structure of single crystals [209, 469, 559], bi-crystals [833, 860, 1088] and polycrystalline materials [469, 832, 1141]. It should be noted that for the best understanding of HTSC intrinsic properties, it is very important to carry out comparative investigations of the "perfect" (i.e., monocrystalline) material and intergranular boundaries [523, 1128, 1129]. In particular, the discovery of the lamellar-like defects, which are weak links in Bi-2212 crystallites obtained by the directed solidification technique [1081, 1129], can be used to interpret comparative data on superconducting properties of Bi-2212 polycrystals and related materials.

The *microscopic Hall probe array* method is a promising novel technique to measure the component of the local magnetic field perpendicular to the sample surface [1190, 1191]. With this aim, two-dimensional Hall sensors for research of electronic gas are etched in a GaAs/Al GaAs hetero-structure. The active area of the Hall elements was initially $\geq 10\times10\,\mu\mathrm{m}^2$ and was gradually reduced to $3 \times 3\,\mu\mathrm{m}^2$. In the last case, the arrays have been fabricated, including up to 20 Hall elements with the magnetic field sensitivity better than $10\,\mu$T. In addition to a research of the vortex-lattice melting transition [1191], this method has been utilized for measurements of local magnetic relaxation in crystals of $YCa_2Cu_3O_{7-\delta}$ [1190] and $Nd_{1.85}Ce_{0.15}CuO_{4-\delta}$ [312].

The rapid development of the *scanning probe microscopy*, following the pioneering work [68], has also contributed to vortex imagine research. The achievements in the investigation of HTSC systems by using this method are presented in the overview [183].

The *scanning SQUID microscopy* provides highest sensitivity of vortex structures [548]. The acting microscope consists of a mechanical xy-scanning system combined with an integrated miniature SQUID-magnetometer. The SQUID is fabricated with $1\,\mu$m Nb–AlO_x–Nb junctions and octagonal pick-up loop with a diameter of $10\,\mu$m integrated on a silicon substrate and has a sharpened tip. The tip is brought in immediate contact with the sample during the scanning and serves for distance regulation. The spatial resolution (about $10\,\mu$m) is limited by the size of the pick-up loop.

By using *scanning tunneling microscopy* at low temperatures, the spatially varying quasiparticle density of states may be measured, yielding important information about electronic properties of vortices and vortex structures [412, 413]. *Scanning tunneling spectroscopy* in the mixed state of HTSC is difficult because of the problem of the sample surface quality and the strict requirements for highly reproducible tunneling conditions. The first successive experiments have been fulfilled on YBCO single crystals with vector $\boldsymbol{B}$ oriented along c-axis [656]. The experiments on Bi-2212 single crystals brought an unexpected result; they showed that the spectra inside the vortices differ weakly from the zero field spectra, in contrast to the situation in YBCO [882].

Therefore, the detection of the vortices in BSCCO is much more difficult than in the YBCO system.

Magnetic force microscopy has been used to imagine vortex systems and research their interaction with defects in the HTSC thin films [1183]. The imagines of Abrikosov- and Josephson-vortices trapped in thin-film YBCO washer dc-SQUIDs have been obtained, using low-temperature scanning electron microscopy at liquid nitrogen temperature [525]. During scanning of the sample surface by the electron beam, the signal is generated due to the beam-induced local displacement of the vortices, resulting in a change of the flux coupled from the trapped vortices into the SQUID loop. In this way, $1/f$-noise sources in the device have been identified. The spatial resolution of this method is about 1 μm.

Lorentz microscopy for vortex imaging uses a high-voltage field emission electron microscope, generating a coherent electron beam. The electron phase shift due to vortices in a tilted specimen is detected in an appropriately defocused image [389]. This method has been used to investigate the vortex lattice in single crystals Bi-2212 [390] and include the imaging of vortex motion [1074].

Other applications of the experimental methods for researching of HTSC structures are as follows:

- methods for measuring electromagnetic properties ($J_c - H$, $E - H$, magnetic sensitivity, etc.), used to separate components of the magnetic flux pinning and grain connectivity, causing J_c [243, 440, 830];
- optical and scanning electron microscopy methods, used to study intercrystalline boundary structure, surface morphology and phase composition [306, 723, 724];
- atomic-force microscopy, used to research surface morphology [754];
- X-ray diffraction methods, used to research texture and phase composition [724, 754];
- spectroscopy methods to study microstructure and phase composition [6, 626];
- electron probe microanalysis to investigate chemical composition of surfaces and interfaces [1186];
- using of the transparent organic analogs to study peritectic reaction in YBCO [938];
- using of picture of the YBCO volume magnetization to identify internal defects [707];
- research of individual filaments escaped from BSCCO tapes [116].

It should be noted that the broad spectrum of methods of the surface and interface reconstructions and also experimental approaches developed for semiconductor structures and nanomaterials [716] can be used to study HTSC structures at nanosize scale.

The measuring methods of mechanical and strength properties of HTSC repeat in many aspects the tests, which have been applied broadly for various ceramics and composites. In particular, the so-called R-curves are used to study fracture toughness of volume samples, changing with crack propagation [1171]. The fracture toughness has also been estimated by using the method of Vickers indentation [614,868]. In this case, there are specific features connected, for example, with the cryogenic temperature effects on superconductors or with very small characteristic sizes of superconducting components. In order to estimate elastic and acoustic properties, ultra-sound methods are used [323, 876]; anisotropic stresses at sample surface are computed by application of X-ray diffraction [381]; mechanical behavior, strength and fracture toughness of sample bulks are computed during tests fulfilled at room and cryogenic (77 K) temperatures, using tension/compression [1187], schemes of the three-point bending [712] and four-point bending [324], rapture [712] and bending [322] of notched specimens, creep at compression [902] and also investigations of thermal cycling with cycles between room and cryogenic temperatures [1070]. The concept of effective volume and Weibull's distribution function are used for test estimations of strength characteristics [908]. In order to study bulk sample density the methods of immersion [970], picnometry [789] and Archimedes [770] are applied. The methods of system analysis are developed, for example, for design of material properties and critical behavior of HTSC materials and composites [804,487,1152]. In order to estimate stress–strain state of superconducting bulks, special devices are designed, in particular with integrated gauges, which are able to measure strains with required precision under different thermal and magnetic fields, taking into account anisotropy of properties in ab-plane and in the c-axis direction [710, 711]. Several methods for the mechanical tests of BPSCCO bulks are shown in Fig. 3.1. Figure 3.2 shows two schematic illustrations of how the strain gauges and Hall probe for the measurements of the thermal and magnetic properties are positioned on the YBCO surface.

BSCCO/Ag tape superconductors are tested at room and cryogenic temperatures: on tension [915], longitudinal compression (stability loss) [1117], transversal compression [246], compression and tension by using U-shaped spring dais [1051], bend under three-point bending [776], by hand [573] or by using special devices [575], around cylindrical surfaces (one side [575] and two side [854]) and helical surfaces (an investigation of winding on tension and bending) [768]. They are subjected to thermal cycling including change of temperature from room to cryogenic value and back [1118], to fatigue tests, including cycles of the tape bending and straightening [659], to cyclic loading and unloading [552] or cyclic changing of external tension [1023], low-cyclic [1050] and multi-cyclic (up to 10^7 cycles) [434] fatigue under longitudinal tension, to tension in the c-axis direction and multi-cyclic (up to 10^5 cycles) fatigue [433]. Several test methods for HTSC tapes are shown in Fig. 3.3.

In order to evaluate supercondunting core density of tapes, the measuring methods of microhardness, using Vickers diamond pyramid [1159] and Knoop

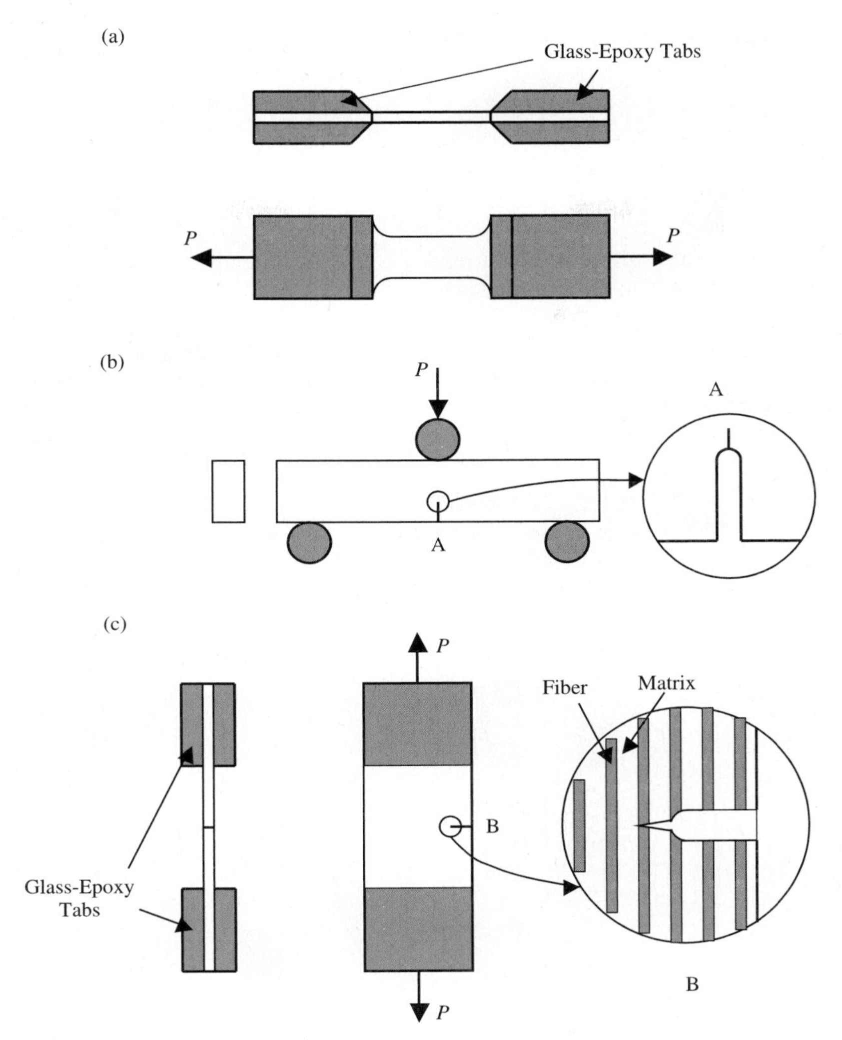

Fig. 3.1. Specimens for evaluation of the mechanical behavior of monolithic BPSCCO and unidirectional Al_2O_3/BPSCCO composite in: (**a**) tensile test, (**b**) three-point bending notched specimen and (**c**) single-edge notched fracture test [712]

non-symmetric indenter [827], are applied. Relative changing of density is estimated for wire-drawing by using registration of its lengthening in dependence on decreasing of the cross-section diameter [384, 568]. The microhardness profiles (i.e., monitoring of transversal cross-sections in different points of their diameters or main axes) are applied to estimate superconducting core density of tapes and wires after deformation (drawing, extrusion, rolling and pressing; see Fig. 3.4) [458] or single filaments of multifilament samples [457]. The

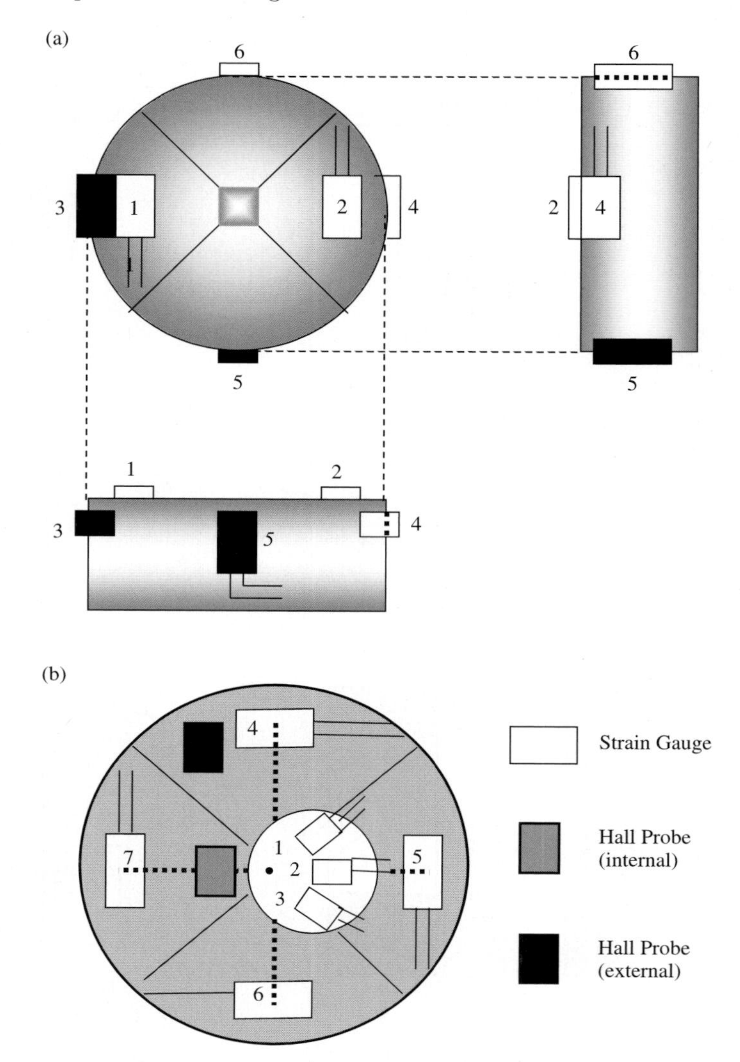

Fig. 3.2. (**a**) Schematic illustration showing how the strain gauges are mounted on the YBCO surface (gauges 1 and 2, positioned on the top surface, are used for the measurements of the strain along the *ab*-plane, and gauges 3–6, fixed on the side, are used to measure the strain along the *ab*-axis on the *ac*-plane (gauges 3 and 4) and along the *c*-axis (gauges 5 and 6) [711]; (**b**) schematic illustration of how the strain gauges and Hall probe are attached to the YBCO sample for the measurement of the mechanical properties and trapped magnetic field (Nos. 1–3 are internal gauges of strain, Nos. 4–7 are external gauges of strain) [710]

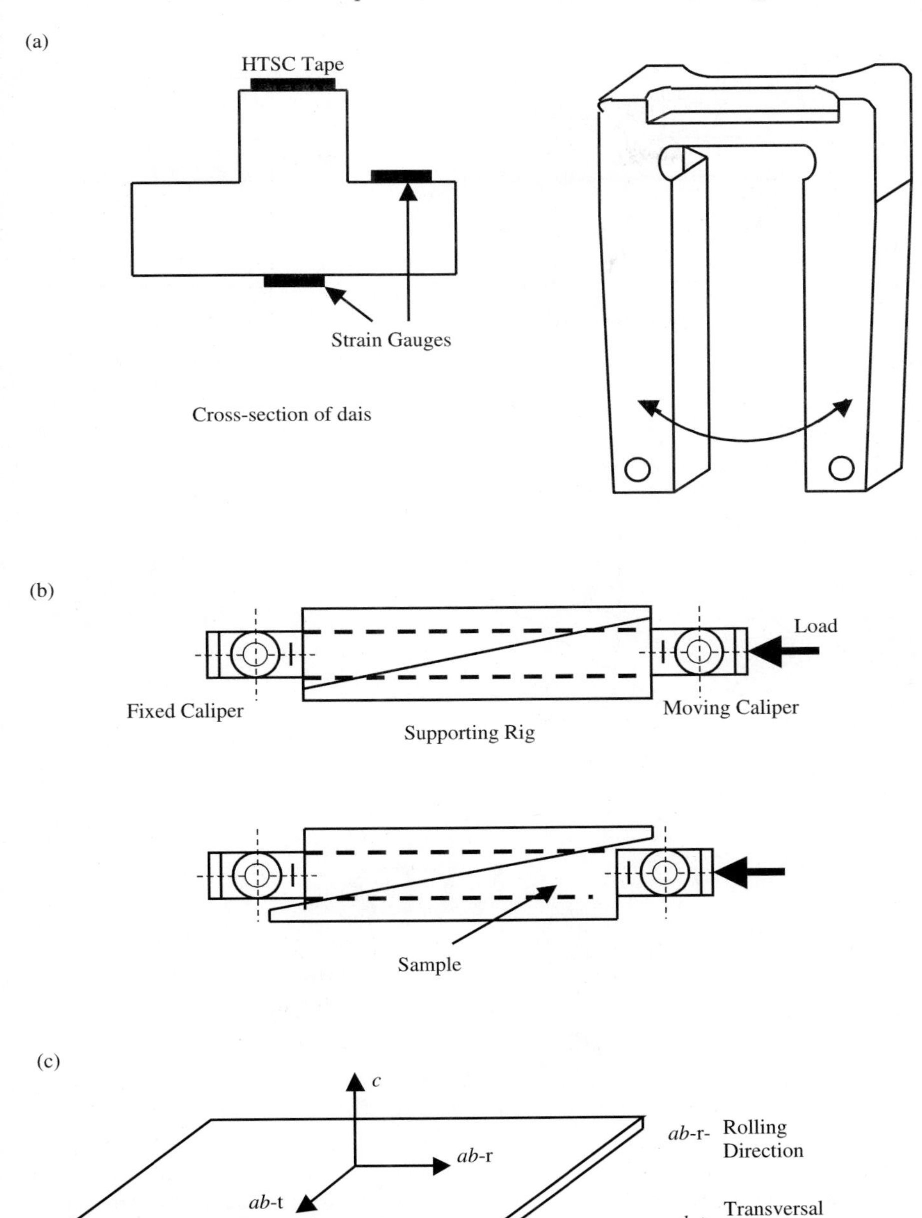

Fig. 3.3. Test methods for investigation of HTSC tapes: (**a**) on compression and tension by using U-shaped spring dais with soldered sample [550], (**b**) on stability loss at longitudinal compression [1117], (**c**) on tension or cyclic fatigue along c-axis [433]

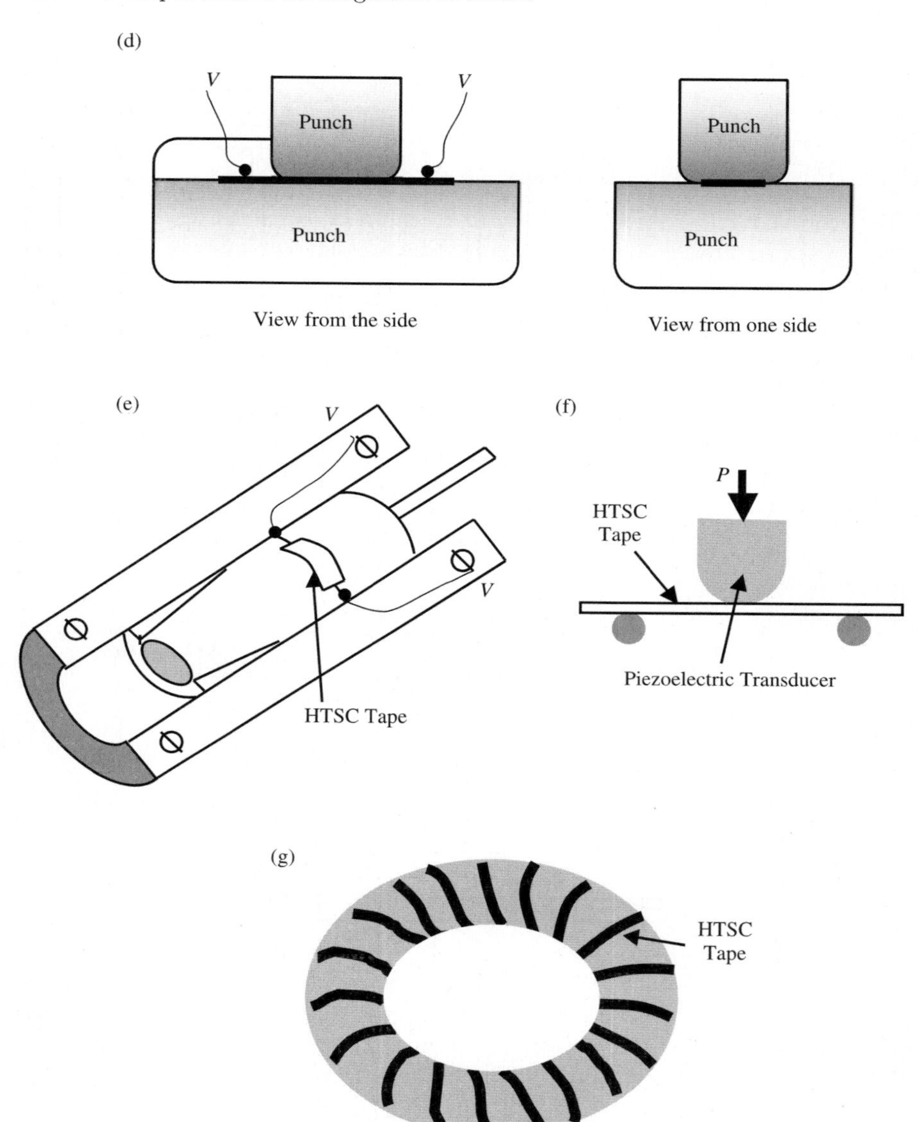

Fig. 3.3. (**d**) on transversal compression by using two punches with evaluation volt–ampere characteristic [246], (**e**) on bending by using measuring devices of critical current depending on continuously decreasing (or increasing) diameter of bending [1051], (**f**) on three-point bending by using a device in which loading is carried out by the piezoelectric transducer of acoustic emission [820], (**g**) on bending around helical surface [768]

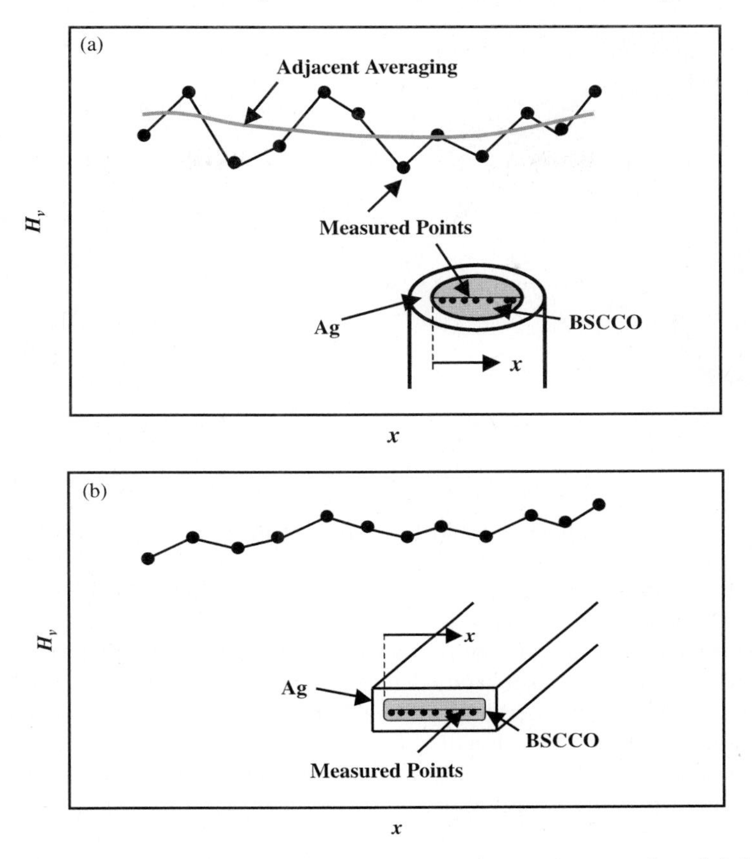

Fig. 3.4. Proper microhardness profiles of HTSC wires and tapes after: (**a**) drawing and/or extrusion, (**b**) rolling and/or pressing [458]

ultrasound vibration method is used to evaluate the microcracking density [21, 996]. The method for investigation of mechanical aging that model mechanical effects, developing during a conductor stranding, was proposed in [676]. This test models bending, twisting and tensile forces, experienced by the HTSC wires during conductor stranding (Fig. 3.5). Moreover, to predict thermal cracking, the wire aging is studied under conditions of elevated temperatures [676].

3.1.2 Acoustic Emission Method

Acoustic emission (AE) is the physical phenomenon connected with irradiation of elastic waves by a solid under loading due to dynamic local reformation of the material's internal structure. This method has been applied to research different physical and mechanical properties of HTSC, accompanying

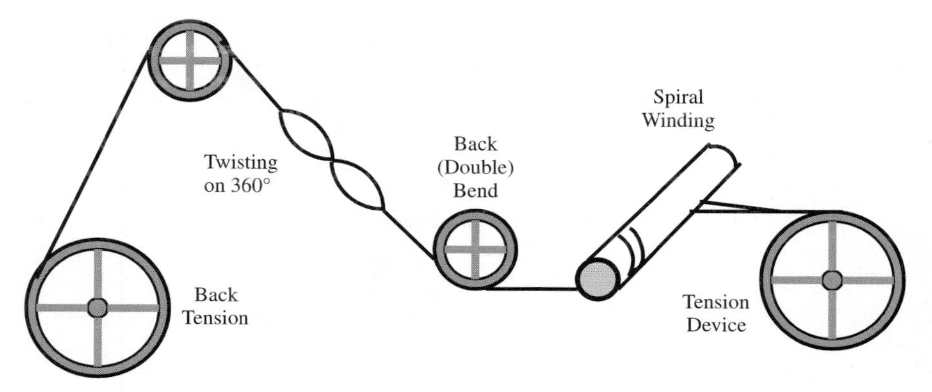

Fig. 3.5. Schematic of mechanical aging test, designed to simulate the conductor stranding process [676]

microstructure and phase transformations during fabrication and loading of the superconductors at various internal and external loading. Intensive AE has been displayed at $YBa_2Cu_3O_{7-x}$ sintering [230] and thermal treatment [1011]. It is observed due to microcracking, caused by anisotropic compression of crystalline lattice in narrow temperature range at cooling after sintering. Acoustic emission has been studied in YBCO ceramic at the sample heating from liquid nitrogen temperature [1017]. The observed AE has been correlated with the microstress relaxation at the grain boundaries due to the thermal expansion anisotropy of grains. As possible mechanisms of stress relaxation, the dislocation displacement and microcracking initiation at most unfavorably oriented grain boundaries have been assumed. The oxidation kinetics of YBCO ceramic has also been studied by using AE method [231]. The oxidation has submitted to exponential law. The results showed that a narrow surface layer oxidized initially, but the oxygen solubility into material was controlled by the rate of the volume diffusion. YBCO ceramic has been investigated in the 25–700°C range [229]; AE irradiation has been registered in the 260–300°C; and it has been shown that this was caused by the structure phase transition of 90° phase into 30° one during redistribution of oxygen into material. AE method has been used to register the structure transformations in HTSC during thermal cycling [963]. The effect of the cyclic change of current in YBCO ceramic has been studied by the AE method at 77 K [233]. In this case, a penetration of magnetic flux lines into sample under the action of own magnetic field has been stated. A pick of acoustic emission in $Tl_2Ba_2CuO_{6+x}$ sample at critical temperature T_c [788] has been observed. The relaxation anomalies stated have been correlated with a change of charge state of linear defects and parametric changes of the cuprate layer structure.

The processes of microcracking and microplasticity of $GdBa_2Cu_3O_y$ and $Bi_2Ca_2Sr_2Cu_3O_y$ monocrystals and also of YBCO ceramic have been studied by the local indentation method accompanied by registration of AE signals [80]. Comparison of AE data and fracture pictures permitted to estimate a character of inelastic strain of HTSC. Then, it may be concluded that AE method can operatively control mechanical properties of HTSC. In the case of $GdBa_2Cu_3O_y$ monocrystals, the AE signals have been caused by formation of characteristic microcracks at the angles of the indenter imprint. In the process of local loading of $Bi_2Ca_2Sr_2Cu_3O_y$ crystals, formation of wedge-shaped defects, accompanied by initiation of low-frequency component of AE signals, was observed. It may be assumed that these defects are result of joint processes of microcracking and de-lamination in BSCCO. The AE method was also applied to research inelastic strains of HTSC ceramics during macroscopic mechanical tests [79] and microindentation [340]. AE was used to investigate secondary sintering of Bi-2223/Ag tape during heating after primary sintering and rolling [232]. In this case, it has been assumed that the process of liquid-phase healing of cracks, formed during rolling, has appeared as a possible source of AE. This result leads to the conclusion on the possibility of application of the AE method as the non-destructive control method in the processing of Bi-2223/Ag tapes.

In order to answer the issue of the possibility to apply the AE method for estimation of microdamage initiation and propagation during bending of Bi-2223/Ag tapes, a test device has been designed, which is presented in Fig. 3.6 [820]. In the tests carried out at room temperature, the following parameters were estimated: (i) sample displacement under indenter (Δl), (ii) the applied loading ($P_{\rm m}$), (iii) the number of AE impulses (ΣN), (iv) a sum of AE signal amplitudes (ΣA), (v) AE signal activity ($\dot{N}$), (vi) the loading time (t).

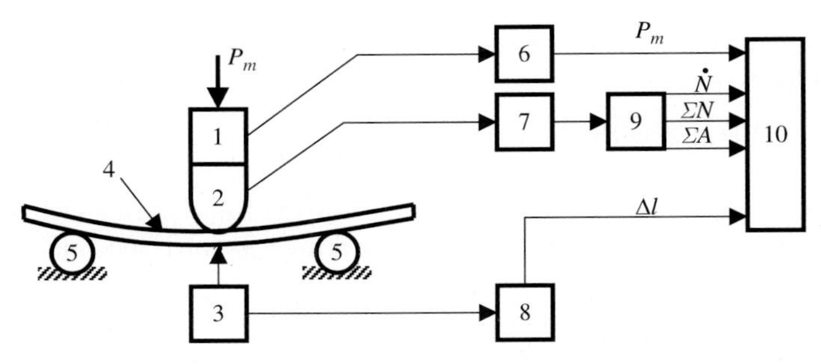

Fig. 3.6. Schematic illustration of test device. Dinamometer (*1*), cylindrical indenter of 20 mm diameter with AE transducer (*2*), measurer of displacements (*3*), Bi-2223/Ag tape (*4*), supports (*5*) with 10 mm diameters and 35 mm spacing, tensometric devices of agreement (*6,8*), preliminary amplifier (*7*), system block of AE signal treatment (*9*) and registering device (*10*)

Deformation was estimated as [776] $\varepsilon = 6d\Delta l/L^2$ and deformation rate as $\dot{\varepsilon} = 6dV_n/L^2$, where $V_n = \Delta l/t$ is the loading speed, d is the thickness tape, L is the distance between supports. A test specimen is loaded consistently in several sites. The AE signal activity was discontinuous in the form of single bundles, characterizing non-uniformity of the fracture process or defect initiation under loading, which was proper for brittle fracture.

First, a contribution in the AE activity from deformation of a silver sheath was stated. For this, the tests of pure silver tapes with the same sizes and conditions of loading, as for the tested Bi-2223/Ag tapes, have been carried out. Moreover, it has been proved that loading device did not cause acoustic noises. In order to confirm a character of fracture and correlation of the AE activity with the sample damage during loading, a comparison was carried out between the obtained test data and the experimental results (obtained by A. A. Polyanskii [856] by MOI method [859]) of critical current I_c and critical current density J_c in magnetic fields directed along c-axis of the tapes. Magneto-optical images demonstrated pictures of the magnetic flux arrangements directly in regions loaded for two regimes: zero field cooled (ZFC) and field cooled (FC). In ZFC, the sample was cooled below T_c in the absence of a field, and then a field was applied. In the resulting image (Figs. 3.7 and 3.8) the sample was shielding the magnetic field. For the FC regime, the sample was cooled below T_c in the presence of a magnetic field, and then the field was turned off, so the sample was trapping magnetic flux (Fig. 3.9).

The test results for monocore tapes #1 and #3 and also for multifilamentary tape #2 by using the AE and MOI methods are presented in Tables 3.1 and 3.2 and Figs. 3.7–3.9. As it follows from Table 3.1 and Figs. 3.7 and 3.8, in total, there is a good correlation between results obtained by the above two methods. Moreover, for sample #2, a tendency of the damage increasing with increasing strain and deformation rate is obvious. The correlation of the results in the case of sample #3 (see Tables 3.2 and Fig. 3.9) is less obvious. Based on the AE tests of this sample, the following conclusions are possible:

(1) At sites 2 and 3 with the smallest critical current, AE signals were also observed. The site 3 (with the smallest critical current) was loaded in the tests twice. It may be assumed that during first loading (when electric hindrances were observed), a sample cracking was initiated that was forced at repeated loading.
(2) An absence of AE signals at the sites 1 and 4 (at the existence of cracks, revealed by the MOI method) is related with the possible absence of a good acoustic contact of the sample with the transducer or/and due to the cracks that could be formed at the following accompaniment of the sample (i.e., at marking of the sample, its straightening or/and in the following measurements).

Additionally, the AE tests of the Bi-2223/Ag monocore tapes were carried out with the following measurement of I_c (at 77 K). At the testing of three

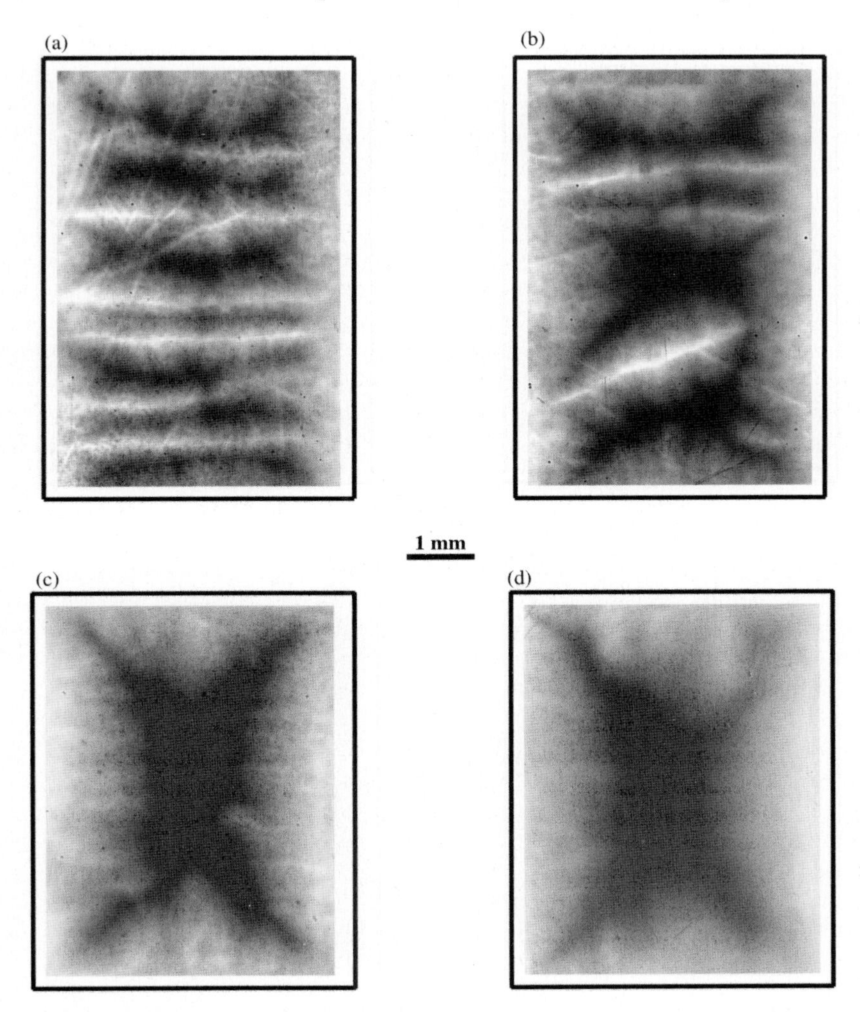

Fig. 3.7. Magneto-optical images of the monocore tape #1, ZFC regime (T = 13 K, H = 600 Oe): (**a**) site 1 (two-side bending), six cracks are seen; (**b**) site 2, 3 cracks are seen; (**c**) site 3, no cracks; and (**d**) site 4, weak defects [820]

samples with different strain rates (one test for every sample) there is correlation between changes of the strain rate and critical current. We obtained the critical currents $I_c = 31$, 24 and 18 A for deformable samples (for initial unstrained sample $I_c = 39$ A). They corresponded to strain rates: $\dot{\varepsilon}/d = 0.00030$, 0.00049 and $0.00054 (\mathrm{s\,mm})^{-1}$, normalized in the thickness, at visible absence of the dependence on strain. Corresponding correlation was also observed for acoustic emission.

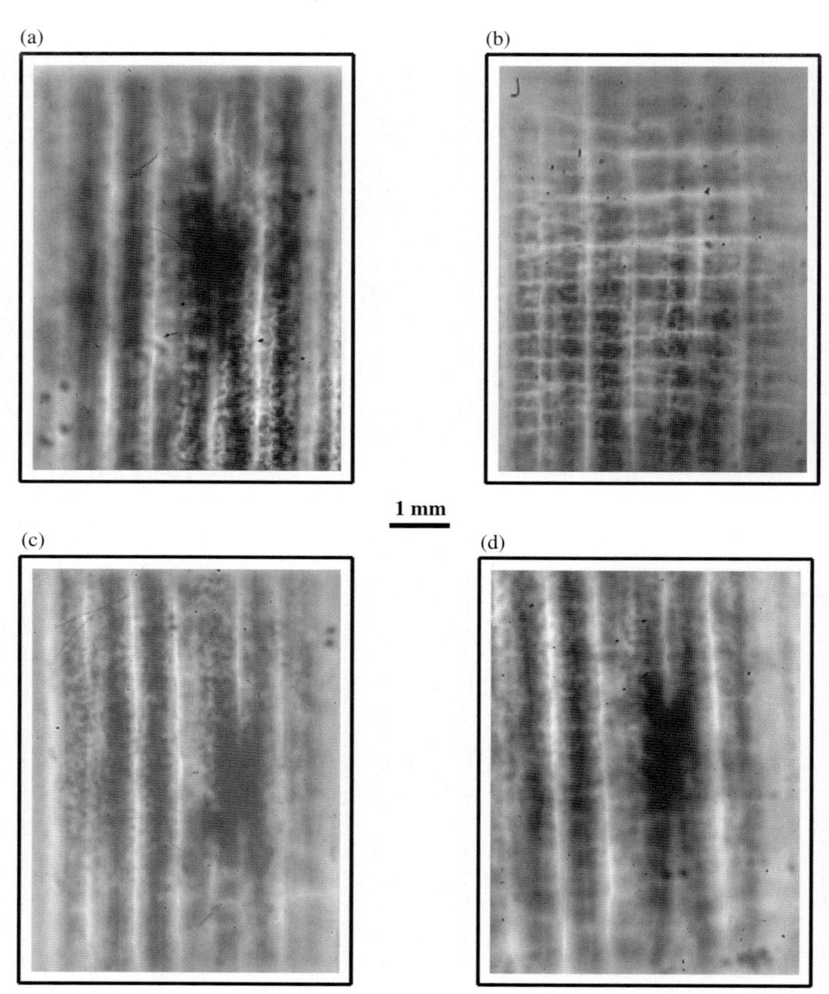

Fig. 3.8. Magneto-optical images of the multi-filamentary tape #2, ZFC regime (T = 13 K, H = 1200 Oe): (**a**) initial unstrained tape; (**b**) site 1, 10–12 great cracks are seen; (**c**) site 2, no cracks; (**d**) site 3, some cracks are possible [820]

Thus, an increase of the strain rate caused a forcing of the AE activity, accompanying greater microdamage and corresponding diminishes of critical current. Hence, it is necessary to take into account the sample strain rate at evaluation of the superconducting tape damage, additionally to the sample deformation.[1]

[1] The sample strain is only taken into account at mechanical treatment of the deformational behavior of structure-sensitive properties, usually.

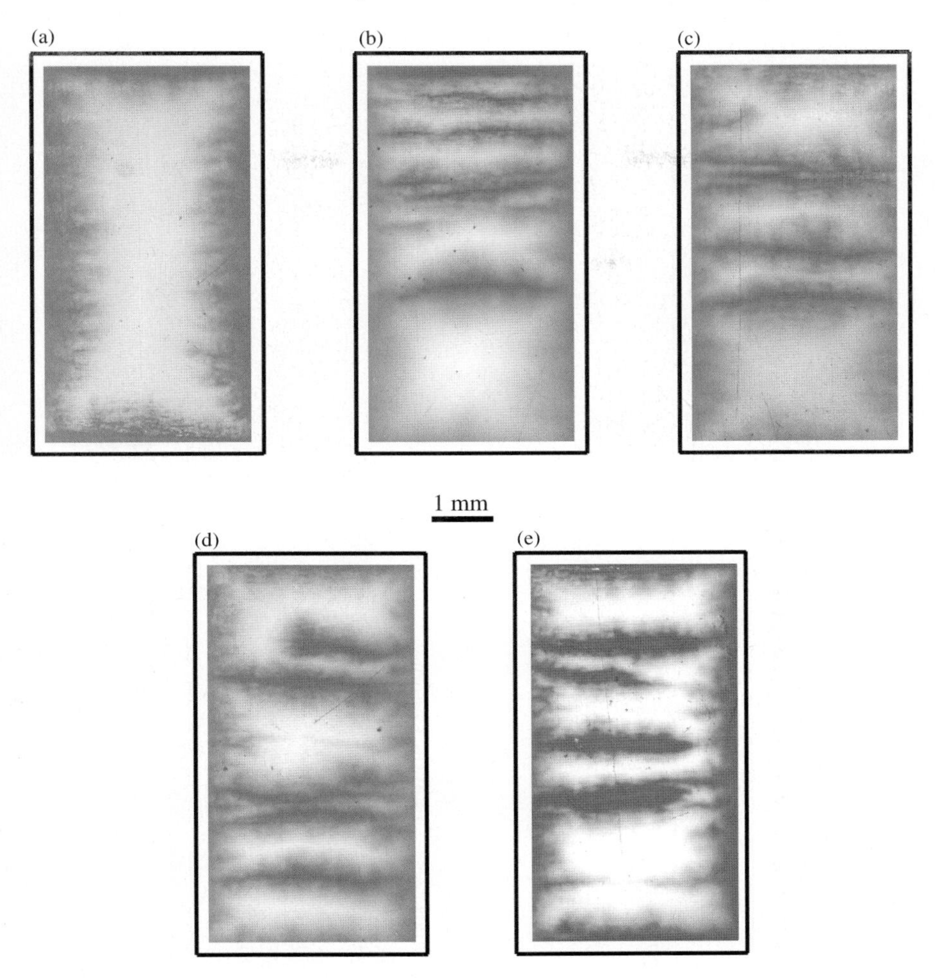

Fig. 3.9. Magneto-optical images of the monocore tape #3, FC regime at H = 600 Oe (T = 12 K, H = 0 Oe): (**a**) initial unstrained tape; (**b**) site 1, 4–5 great cracks are seen; (**c**) site 2, 3–4 great cracks are seen; (**d**) site 3, 5 great cracks are seen; (**e**) site 4, 4–5 great cracks are seen [820]

Based on the above tests, it could be concluded that AE method is suitable to estimate the microdamage formation and propagation during Bi-2223/Ag tape bending. An improvement of the test results may be related to using the designed interference method [817] and the created optic-holographic device for estimation of small displacements [705, 818].

Table 3.1. Test results for the samples #1 and #2 obtained by the AE and MOI methods

No. of site	Δl (mm)	V_{n} (mm/s)	ε/d (mm^{-1})	$\dot{\varepsilon}/d$ $(\mathrm{s\,mm})^{-1}$	$\dot{N}$ (imp/s)	ΣN (imp)	ΣA ($\times 10^{-7}$ m)	P_{m} (g)	MOI data
#1, site 1	3	0.1	0.0147	0.00049	–	–	–	5	–
#1, site 1[a]	1.5	0.06	0.0073	0.00029	80 (in 6 bundles)	800	0.2	5	6 cracks
#1, site 2	2.4	0.15	0.0118	0.00074	130 (in 3 bundles)	400	–	5	3 cracks
#1, site 3	2.9	0.2	0.0142	0.00098	–	–	–	5	No cracks
#1, site 4	2.9	0.26	0.0142	0.00127	1500 (in 1 bundle)	3000	0.6	5	Weak defects
#2, site 1	3	0.2	0.0147	0.00098	$\approx$ 1300 (in 1 bundle)	4000	3	80	10–12 great cracks
#2, site 2	1.4	0.06	0.0069	0.00029	–	–	–	75	No cracks
#2, site 3	2.9	0.08	0.0142	0.00039	150 (in 4 bundles)	600	0.2	80	some cracks are possible

[a] At second loading of considered site from the opposite side

Table 3.2. Test results of AE and critical current for sample #3

Property	Site 1	Site 2	Site 3	Site3[a]	Site 4
Δl (mm)	6	6.5	4.3	6.5	6.5
V_n (mm/s)	0.33	0.23	0.3	0.3	0.26
ε/d (mm^{-1})	0.0294	0.0319	0.0211	0.0319	0.0319
$\dot{\varepsilon}/d$ (s mm)$^{-1}$	0.00162	0.00113	0.00148	0.00148	0.00127
$\dot{N}$ (imp/s)	–	300 (in 9 bundles)	AE with electric hindrances	400 (in 1 bundle)	–
ΣN (imp)	–	1000		400	–
$\Sigma A(\times 10^{-7}$ m)	–	1		0.2	–
P_m (g)	14	18	18	20	22
I_c (A)	23	18	–	14	21
J_c (kA/cm^2)	15.6	12.2	–	9.5	14,2
AE data	No AE	There are active defects	AE is not distinguished	Weak AE	No AE

[a] At second loading of considered site from the same side. For original no-deformed edge of the sample placed at the outer side of supports at the first loading, the measured critical current had the following values: $I_c = 45$ A and $J_c = 30.6$ kA/cm^2 (at $T = 77$ K). Maximum thickness of Bi-2223 core is equal to 60 μm, and $S = 1.47 \times 10^{-3}$ cm^2 is the cross-section area

3.2 Intergranular Boundaries in HTSC

The properties of high-angle grain boundaries are believed to control the macroscopic $J_c(H)$ characteristics of all polycrystalline HTSC. This control occurs because most high-angle grain boundaries act like barriers to the current and have electromagnetic properties, such as Josephson junction-like properties [205]. On the other hand, in the melt-processed polycrystalline YBCO, actual currents can penetrate through intergranular boundaries misoriented up to 30° and more. So, HTSC properties are connected closely with distribution of the intergranular boundary misorientations [205, 649].

From both the high-field flux-pinning viewpoint and low-field, Josephson junction-based electronics viewpoint, there is strong motivation to develop a detailed picture of the grain boundary structure and microstructure and to describe their effects on the electromagnetic properties of the grain boundaries. The superconducting coherence length, ξ, defines the grain boundary thickness that may be penetrated by the supercurrent. In this case, barriers with thickness up to a few coherence lengths can still show superconducting coupling, albeit of reduced strength. At the same time, the grain boundaries in HTSC are defects with thickness in the 0.5–1.0 nm range, indeed, approaching ξ [40, 582]. Super-short coherence length and great values of n in the power dependence of E–J (where E is the electric field intensity and J is the current density) proper for HTSC [370, 372], sign, that the defects with size of some nanometers can prevent supercurrent and create obstacles with effective size, that is considerably higher than nominal size of defect. There are other planar defects, causing magnetic flux pinning and supercurrent percolation in HTSC, namely (i) twinning in YBCO [288, 593, 622], (ii) stacking faults [272, 677], (iii) colonies of low-angle c-axis intercrystalline boundaries [272, 677, 1096], (iv) twist intergranular boundaries of the "brick wall" type [104] and low-angle a-, b-axis boundaries of the "railway switch" type [1186] in BSCCO, (v) overgrowth of Bi-2223 phase into Bi-2212 phase [559], (vi) overgrowth of superconducting phase into silver sheath [887], (vii) amorphous and normal (non-superconducting) phases [289], (viii) voids, microcracks and other crack-like defects, which are proper for all oxide superconductors [403, 743], (ix) macrodefects into Josephson junctions [1014], and also various dislocation networks [634], discussed in detail below. Different types of planar defects causing the structure-sensitive properties of HTSC, are shown in Fig. 3.10.

Polycrystalline YBCO samples can be divided into two broad classes, according to microstructure features, namely (i) specimens heated above peritectic temperature with the aim of obtaining oblong well-orientated grains [14] and (ii) samples sintered with grain structure that is near to equal-axes one [975]. The experimental studies relate to investigation of both individual isolated grain boundaries and polycrystalline samples with averaged effects of great number of the intercrystalline boundaries. Both test types often supplement each other. So, the study of high-angle boundaries in YBCO thin

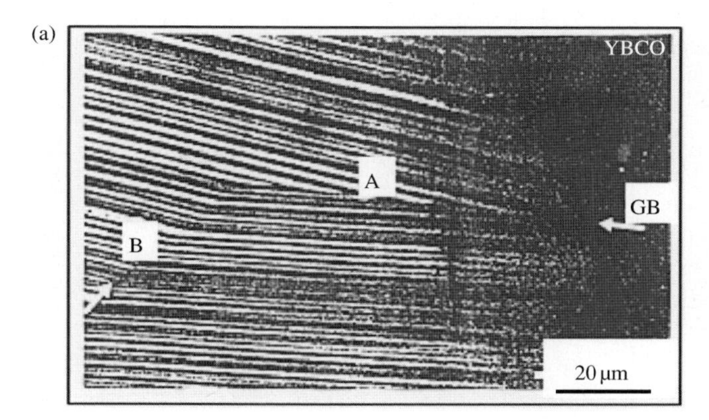

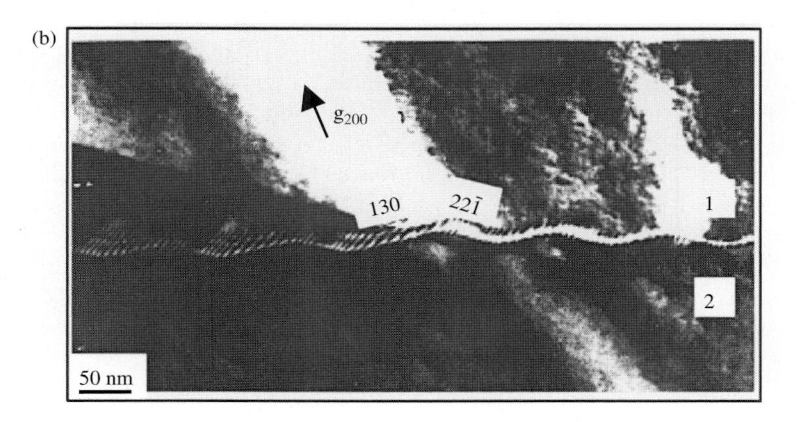

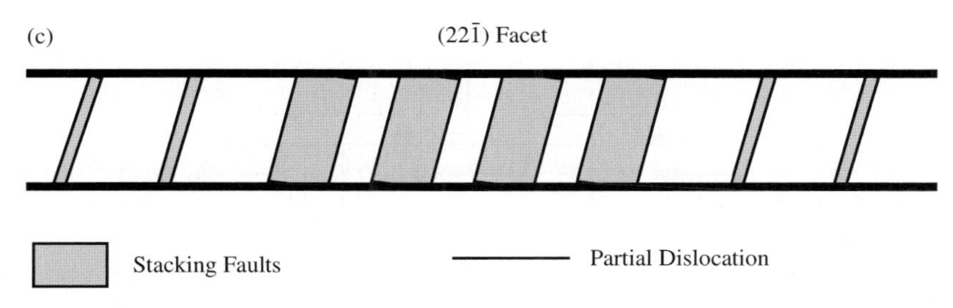

Fig. 3.10. Different types of planar defects in HTSC: (**a**) {110} twinning boundaries in the 10°-bicrystal of YBCO (A and B are the macroscopic facets) [1082]; (**b**) diffraction contrast image of the spatial dislocation network configuration in grain boundary ($g = [200]$ is the diffraction vector), the *dark* and *bright* areas are associated with long-range strain contrast in facet junctions [1079]; (**c**) schematic diagram, showing the arrangement of dislocations in the $(22\bar{1})$ facet [1079]

(d)

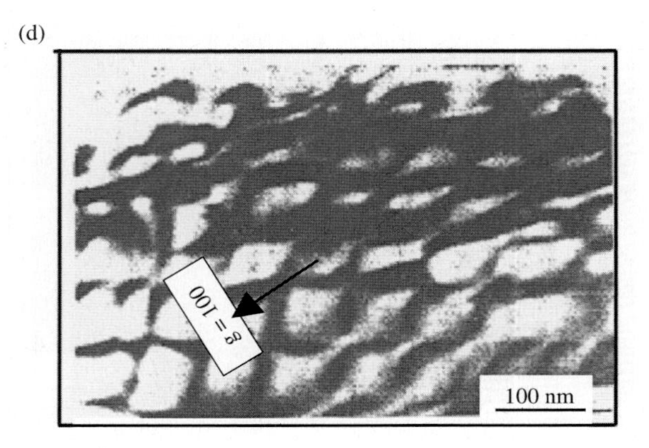

(e)

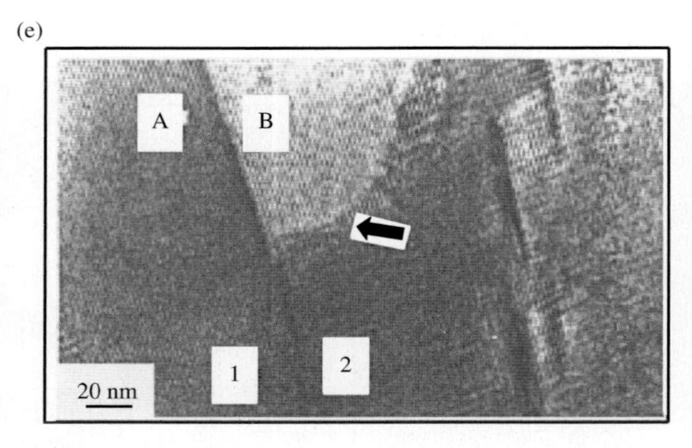

(f)

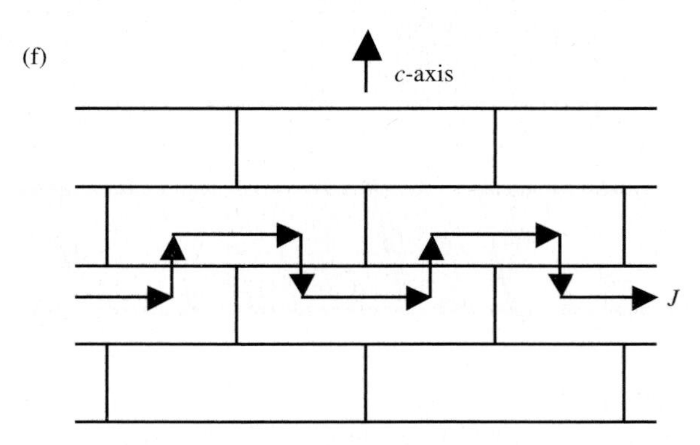

Fig. 3.10. (**d**) screw dislocation in twist intergranular boundary [1131]; (**e**) lattice fringe image, showing examples of a twist grain boundary (grains A and B), a low-angle c-axis tilt colony boundary (colonies 1 and 2) and a low-angle ab-axis tilt colony boundary (*arrow*) [242]; (**f**) twist intergranular boundaries of the "brick wall" type [1186]

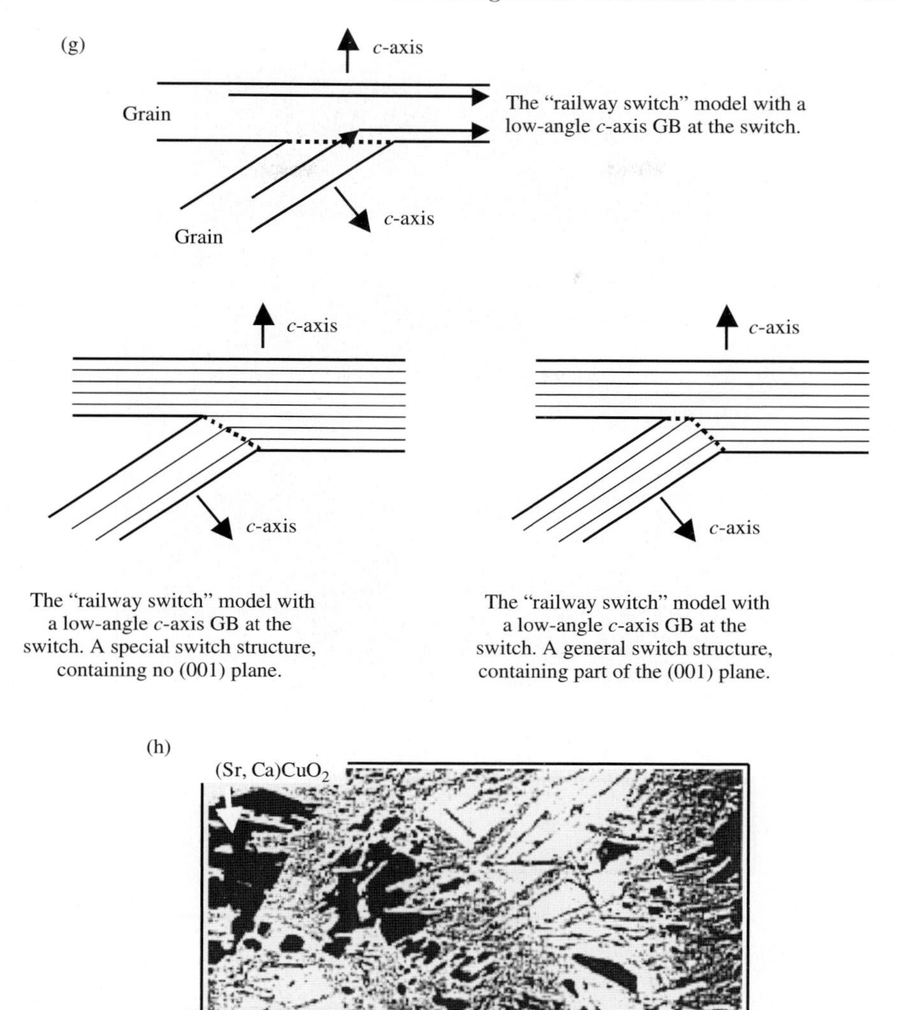

Fig. 3.10. (**g**) different models of the grain boundaries of the "railway switch" type [1186]; (**h**) overgrowth of Bi-2223 phase into Bi-2212 phase [289]

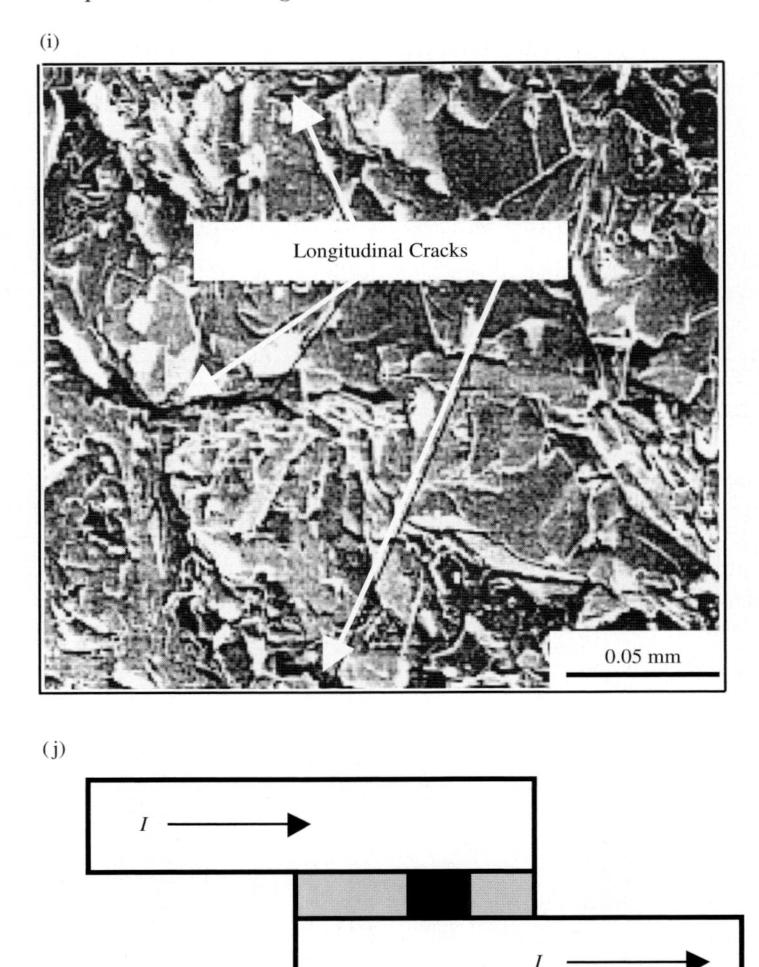

Fig. 3.10. (**i**) longitudinal cracks in BPSCCO superconducting core [790], (**j**) scheme of macrodefect (*black area* in *gray* intermediate layer) in Josephson junction [1014]

epitaxial films onto bicrystalline $SrTiO_3$ substrates [205] has confirmed the behavior of intercrystalline boundaries of Josephson type. Earlier, this result has been predicted by using tests on bulks [841]. On the other hand, the tests on oriented YBCO bicrystals [38] showed that some grain boundaries, including high-angle ones, demonstrate under high magnetic fields much more high transport properties compared with behavior that is proper for weak links subjected to Josephson effects. Analogous phenomena take place in bulks, in which there is often small intergranular current component that does not depend on field [248].

The general results are usually based on the bicrystal investigation of three main types, namely (i) epitaxial thin films that are grown on bicrystalline [205, 353, 474] or otherwise specially prepared [324, 619, 1122] substrates with predetermined misorientation relationships and boundary planes (on a macroscopic scale); (ii) very large grain (millimeters) melt-processed bicrystals (MPB) that contain considerable volume fractions of secondary phases and have random misorientation relationships and macroscopically meandering boundary planes [279, 795]; and (iii) flux-grown bicrystals (FGB) with dimensions of the order of a few hundred microns, misorientation relationships of the θ [001] type and comparatively straight, but uncontrolled boundary planes with large tilt sites [38, 39, 599].

The study of electric resistance of the [001] symmetrical tilt boundaries at the normal state of sample revealed common dependence of electromagnetic properties on the misorientation angle θ. The value of $J_c(gb)/J_c(g)$ (where $J_c(gb)$ and $J_c(g)$ are the critical current densities of the boundaries and neighboring grain interiors, respectively) decreased rapidly with a $1/\theta$-like dependence for θ values up to about 20°. For $\theta > 20°$, an approximately θ-independent, uniformly low value of $J_c(gb)/J_c(g) \approx 0.01$ was observed [205, 206]. The same behavior of grain boundaries with low/high angles of misorientation is observed in all HTSC. In particular, epitaxial bicrystal experiments using thin films of Bi-2212 [20, 681, 1129], Tl-1223 [755], Tl-2212 [122] and Tl-2223 [921, 1129], all showed a qualitatively similar dependence of $J_c(gb)$ on θ. The $1/\theta$ dependence/plateau behavior of $J_c(gb)/J_c(g)$ vs θ led researchers [122, 205, 754] to postulate that the dislocations, covering the structure of intercrystalline boundaries, determine their electromagnetic properties. As the misorientation angle of the boundary increases, barriers with weak link properties progressively pinch off the supercurrent. At some critical transition angle, the barrier regions completely control the boundary properties and a weak link forms. The dependence on θ suggests that the primary grain boundary dislocations (PGDBs), constituting the boundary structure, cause the weak link behavior [205, 353]. In this model, the boundary is frequently described as a uniform array of like edge dislocations with spacing, which are characteristics of symmetrical tilt boundaries as specified by Frank's formula [877].

There are many attributes of dislocation network, including the weak link behavior, namely (i) overlapping of dislocation cores with cation stoichiometry [754]; (ii) overlapping of elastic strain fields, caused by PGDBs [887]; (iii) agreement of lattice parameters through stoichiometry on oxygen for elimination of lattice mismatch at intercrystalline boundary [1202, 1203]; (iv) local alterations of T_c and chemical potentials due to elastic strains [373]. The test data show the transition from strong to weak coupling at $\theta \approx 10°$,[2] which is most evident in thin-film YBCO bicrystals [205, 353]. This transition could be explained by narrowing of superconducting connecting dislocation cores [205] or overlapping of their strain fields [149]. This model assumes absence of superconductivity of dislocation cores and limitation of the supercurrent paths by channels, connecting these cores, and also by neighboring boundary regions. The regions of perfect crystalline structure into space between dislocations or symmetric facets in the absence of the void-like defects can be potential sources of current-carrying paths defined by strong links [94]. The orthorhombic crystalline structures of cuprate superconductor could considerably complicate investigation of their intercrystalline boundaries. In addition to different lattice parameters a and b, YBCO family possesses a good developed twinning structure [344], but BSCCO family demonstrates non-regulated modulation along b-axis [1117]. Both the lattice parameters and YBCO superconducting properties change with oxygen concentration. In particular, the lattice parameters a and c increase with decreasing oxygen in the structure, causing an expansion of unit cell. Therefore, the deficiency of oxygen and corresponding weakening of superconducting properties may be intrinsic to intercrystalline boundaries in YBCO.

The electromagnetic properties of weakly coupled boundaries vary substantially, and perhaps systematically, with position along the grain boundary. The patches of "better" material are separated by weak or non-superconducting regions, as shown schematically in Fig. 3.11. One of the main causes of these structure and composition alterations could be the local oxygen depletion and oxygen disorder in YBCO structure [94, 218, 1201]. Other possible microstructure sources of heterogeneity include: (i) cation composition modulation, existing within the boundary plane [372, 1107]; (ii) "wavy" boundaries, facets with arbitrary configurations and facet junctions, observed in epitaxial thin-film bicrystals of YBCO and Bi-2212 [09, 420, 1077]; (iii) strain fields, caused by intrinsic and extrinsic intercrystalline dislocations and also by regular distribution of facets [1079, 1080]; (iv) the intersections of twin planes with the intercrystalline boundary plane in the case of YBCO [43]; and (v) oscillatory changes in misorientation of the neighboring grains due to twinning in YBCO [43, 1079, 1080]. Different faceted structures are shown in Fig. 3.12.

The superconductor crystals that nucleate on one substrate crystal can often grow past the substrate boundary and over the "other" crystal for appreciable distances before impinging on a crystal growing in the "correct" orientation. This overgrowth, which may be more prevalent for certain misorientation angles, results in "wavy" boundary topography, causing boundary morphology and properties [505, 606, 1080]. On the other hand, the twinning in each YBCO single crystal alone results in discrete, systematic changes in the local misorientation relationship along the boundary, which should produce

[2] Strong link behavior can be observed in melt-processed YBCO bicrystals [280, 795] and in flux-grown YBCO bicrystals [43] at angle θ up to 20°.

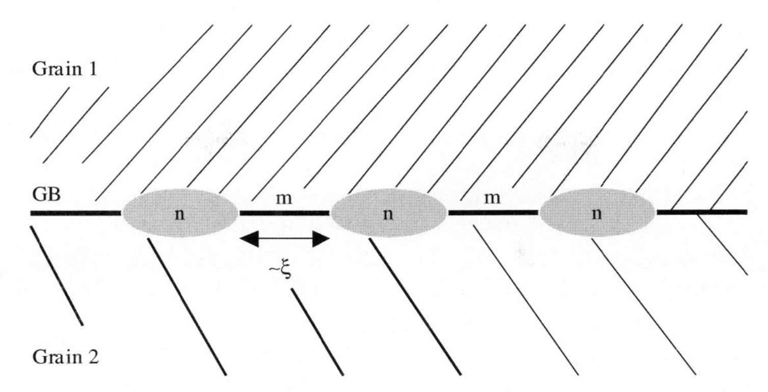

Fig. 3.11. Schematic illustration of the heterogeneity, proposed for weak-linked grain boundaries in YBCO. The current flows through microbridges (labeled by m) that exist between non-superconducting regions (labeled n). The electrical character of the microbridges is assumed to be normal, superconducting or insulating depending on the model used

concomitant oscillations of PGDBs, as observed for a high-angle boundary in sintered material [305].

There are numerous examples of PGDBs at the low- and high-angle YBCO intercrystalline boundaries. Some high-angle boundaries include regularly positioned amorphous regions [697, 887]. Nevertheless, usual observations show high degree of the intercrystalline structure localization, ranging only on 1–2 elementary cells to neighboring grains [149, 305, 606]. These observations agreed with the width of dislocation cores, which are observed on representative intercrystalline boundaries in ceramics [697].

For the framework of description of the high-angle boundary structure in terms of PGDB [41], concentrated picture of the "good/bad" regions may be used, according to defect distribution. Intercrystalline dislocations are localized sufficiently in order to create periodic field of the elastic physically signed strains. In this case, the accommodation dislocations of the lattice mismatch at [100] boundary are partial dislocations and can dispose sufficiently far from the boundary into one from grains [125, 305, 754]. These partial dislocations are joined with boundary by stacking faults, which rather introduce copper excess in a boundary region. Nevertheless, this boundary can act as strong link despite its extended structure and possible stoichiometry [255, 586, 619].

Nanofacets are observed at different intercrystalline boundaries [128, 697]. In this case, the facet planes are often dictated by a crystalline structure. Facets are observed on boundaries, possessing relatively simple crystallography and low-coincidence index (Σ) values, where Σ is the fraction of lattice sites of one crystal that are coincident with the other [754], and also on boundaries with complex geometry [606]. Facets aspire to align parallel to planes with lowest index of one from crystals and are observed, as rule, in (100), (010) and (001) planes. The facet formation introduces considerable

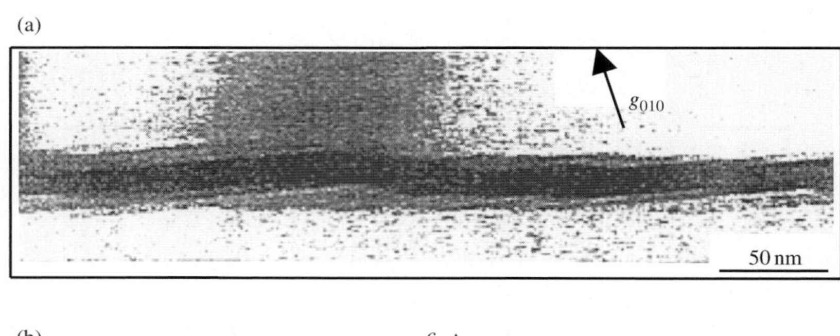

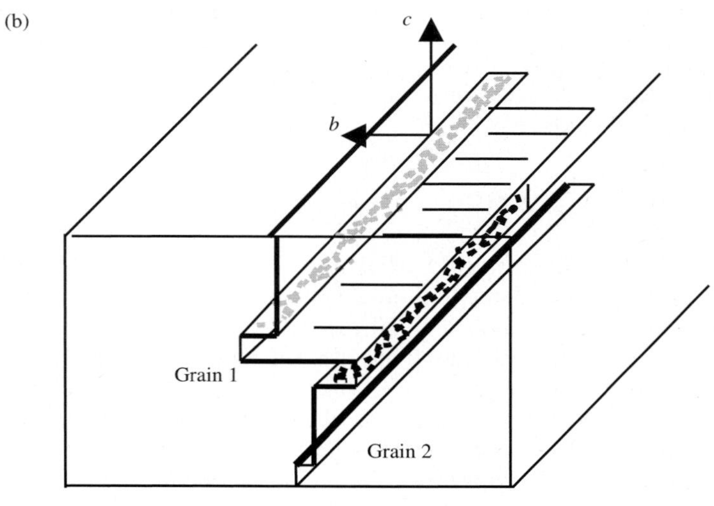

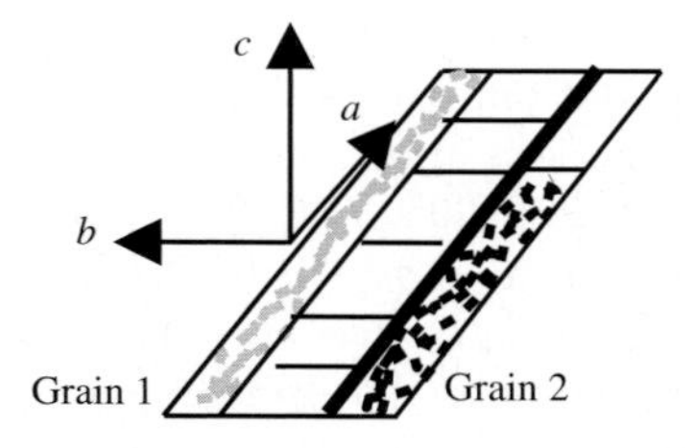

Fig. 3.12. Faceted structures in HTSC: (**a**) a $g||$ [010] diffraction-contrast TEM image of nominally pure [001] tilt boundary in 10° [001] bicrystal of Bi-2212 [1082]; (**b**) perspective and [001] plan view schematics of the "stepped" boundary topography, composed of pure symmetric tilt facets and pure twist facets, that was observed in many sections of the [001] tilt 2212 bicrystals [1082]

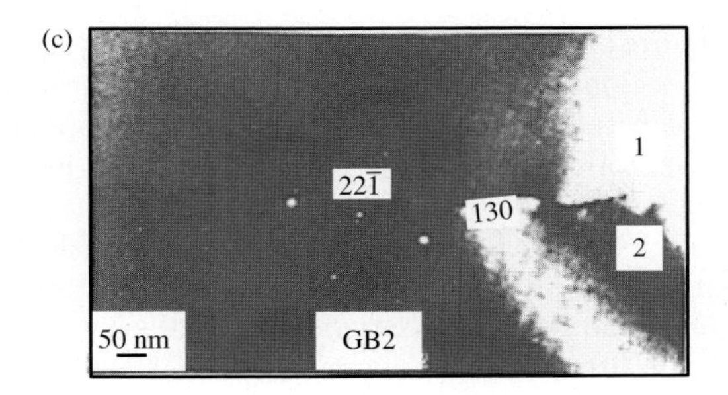

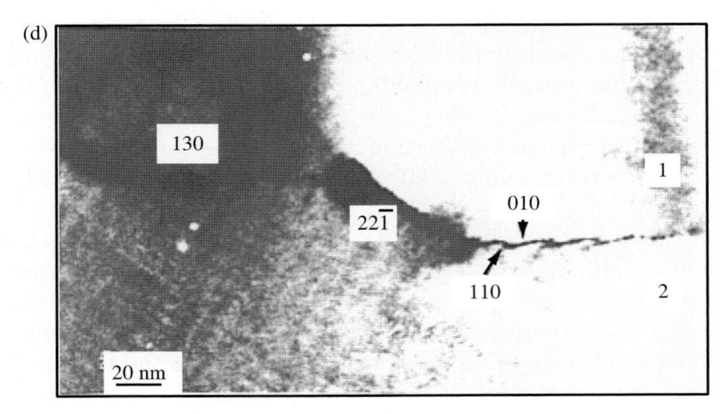

Fig. 3.12. (**c**) diffraction-contrast TEM image showing saw-tooth faceting onto (130) and $(22\bar{1})$ plane in GB2 [1079];(**d**) enlargement of a portion of (**c**), showing that the (130) facets further facet onto (010) and (110) subfacets; the dot-like strain contrast along the subfacets is produced by the grain boundary dislocations [1079]

structure heterogeneity in the boundary. Due to limited size of facets, macroscopic boundaries with different planes may be considered as mixture of some types of the boundary facets of fixed structure. The intersection lines of facets possess their own atomic structure, and secondary dislocations are often associated with them [43]. In this structure level, heterogeneities, which are important for superconducting properties, include saw-tooth structure of facets, additional faceting of facets, initially appearing plane, intensive strains at facet junctions, inhomogeneously distributed dislocations at boundaries in the range of individual facets and favorable conditions for dislocation-division with Burgers vector $\boldsymbol{b} = \langle 100 \rangle$ in partial dislocations at the center of some facets (see Fig. 3.12c and d) [1079].

The facet junctions can influence stoichiometry of intercrystalline boundaries; in particular, it is possible for the concentrated excess of copper to be connected with facet distribution in volume bicrystals with a low misorientation. Some more oscillations of stoichiometry on copper with period of about 100–200 nm are observed at high-angle boundaries [42]. Contrast strains take place in the regions with copper excess, but the facet morphology and the strain origin have not exactly been stated. These copper redistributions together with facet topography and/or strains in facet junctions can cause intercrystalline electromagnetic properties.

Dissociation of PGDBs, forming pairs of partial dislocations, can considerably influence the value of the critical angle at transition from weak links to strong ones due to the accompanying decreasing of dislocation spacing. During PGDBs dissociation, superconducting channels narrow and can lead analogously to microbridges, demonstrating Josephson effect, that is , weak link behavior [630]. This transition could be expected by decreasing below the coherence length (about 1 nm in *ab*-plane). These narrow superconducting channels, caused by the dislocation dissociation in partial ones, form weak links at intercrystalline boundary (one could be strongly connected, if consists of PGDBs only). Thus, a transition to lower energy structure of boundaries (at dislocation dissociation in partial ones) will suppress an intercrystalline supercurrent owing to mechanism of the superconducting channel narrowing.

Strain fields near tilt boundaries must considerably differ from fields near twist boundaries due to differences between parallel rows of edge dislocations and cross-shaped lattices of screw dislocations, covering their structures, respectively. While the boundaries along [001] direction have, as a rule, one row of dislocation, in the case of mixed orientation of grains, there are at least two intersecting rows of dislocations. This alters a form of channel with strong coupling owing to inclusion of both rectangular sites, located between parallel rows of dislocations, and almost point contacts of high-angle boundaries with mixed orientation (Fig. 3.13). These boundaries retain a good connected

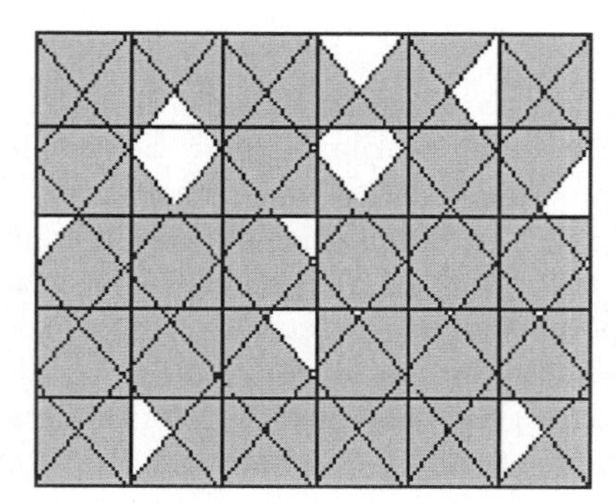

Fig. 3.13. Schematic presentation of possible structures for intercrystalline boundaries in general case. The *horizontals, verticals* and *diagonals* show dislocation network, *grey* regions present barriers for supercurrent, *white sites* are the channels of strong links. Contrast to the model of boundaries directed along [001] direction (see Fig. 3.11), the strongly connected channels at these boundaries to rather point, than linear contact

system of flaw pattern of the supercurrent (see test results [280]), despite far smaller number of "strong" channels predicted by the PGDBs model [205, 206].

Faceted structures in YBCO and Bi-2212 differ in the type of facets, as well as in regularity of their distribution. At nanoscale in Bi-2212, there is non-regular distribution of symmetric tilt and twist [001] facets. At the same time, very regular saw-tooth distributions of tilt facets dominate in YBCO [1079, 1107]. In this case, the strains observed in YBCO bicrystals are always in accordance with saw-tooth facet intercrystalline structure [1079, 1080]. Elevated strains could be caused by incomplete elimination of long-range component because of small number of dislocations that are contained at the facets with limited size.

Despite differences in the dislocation topography and distribution at the boundaries in Bi-2212 and YBCO, there is one common feature concerning the Burgers vectors of these dislocations, which may be correlative with superconductivity of boundaries. It is known that existence of the partial dislocations is common for intercrystalline boundaries in Bi-2212 and YBCO [1079, 1080, 1107]. Moreover, in both materials, Burgers vector value and direction correspond to distance between two nearest atoms of oxygen in the CuO_2 planes of unit cells. Then, the strain caused by these dislocations can initiate local superconductivity. In particular, the shear strain in $\{110\}$ planes is associated with the beginning of superconductivity in YBCO on the basis of analysis of the anomaly in thermal expansion factors in a-axis and b-axis directions [110, 111, 439].

It is known that a layer of material with different superconducting properties exists at YBCO grain boundaries. This layer appears to be hole-deficient YBCO [39, 94, 1202]. However, key questions about the nature of this layer remain unanswered: (i) What is the average width of the hole-depleted zone, and does it depend on the form of the material (bulk or thin film) or processing method? (ii) Are the magnitude of depletion and width of the depleted zone uniform along the boundary? (iii) What is the microstructural origin of the holes' depletion?

There is correlation between the value of hole depletion and initiation of effects, demonstrating the weak link behavior [39]. The hole-concentration profile studies are based on the concept that the hole depletion zone width exceeds both the superconducting coherence length for YBCO and the apparent structural width of the boundary. The widths reported vary from one material form to another: $\sim 8\,\mathrm{nm}$ wide affected zone at polycrystalline thin-film boundary [94]; the width of the depletion zone in FGBS is $\sim 60\,\mathrm{nm}$ [39] and, in sintered materials, a width the order of 10–20 nm [1202]. All of these values exceed the width of the structurally disordered region ($\sim 1\,\mathrm{nm}$) and the metal cation non-stoichiometry ($\sim 5\,\mathrm{nm}$) [42], and also the coherence length ξ. Unfortunately, it is not possible to associate the reduced density of holes with a specific atomic defect such as oxygen non-stoichiometry or oxygen disorder, these being the most likely candidates. In this case, there may be important differences in the structure, composition and electronic properties of grain boundaries in thin films as compared to their bulk counterparts. Therefore, it is expected that hole density variations might occur on a number of different length scales in both strongly and weakly coupled boundaries, and also in superconductors with different forms and sizes.

Thus, the main features of grain boundaries in the considered HTSC systems (Fig. 3.14 illustrates some of them) include:

- the cores of the grain boundary dislocations, where the nearest neighbor configurations are disrupted and the strains are very large;
- the ribbons of stacking fault, disposing between dislocations that dissociate to form pairs of partial dislocations;
- the variety of changes in boundary plane that occur as facets form, for example, due to twinning;
- the different dislocation distributions in the various facets and the uneven arrangement of dislocations that occur in smaller facets;
- the compositional and structural changes, including heterogeneous but periodic copper excess (XS Cu), that are caused by local diminution and disorder of oxygen, and also cation modulations;
- the electron–hole depletion that occurs at boundaries and also appears to be non-uniform;
- stress–strain states of different length scales that arise as a result of the dislocation network, dislocation dissociation, facet junctions, phase transitions, twinning, initiation and growth of pores and microcracks.

These features comprise a boundary that is heterogeneous on a variety of length scales with respect to atomic distributions, electronic structure, composition and stress–strain state. The above heterogeneities influence all structure-sensitive properties of bicrystals and polycrystalline materials.

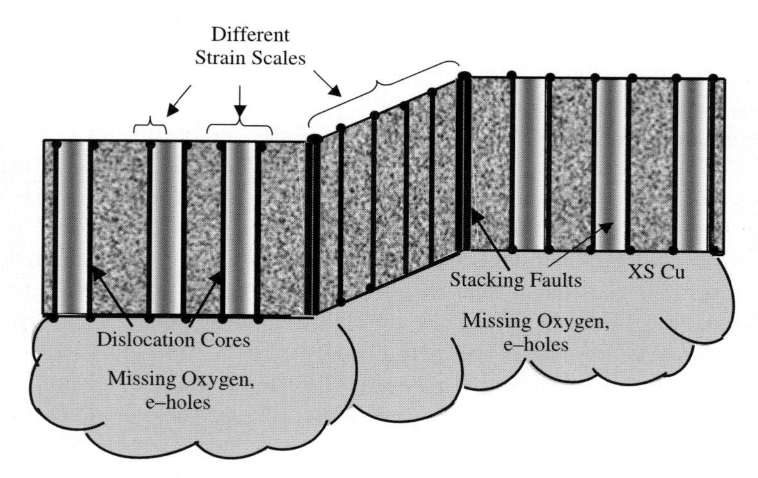

Fig. 3.14. Schematic illustration of various types and scales of structural and compositional heterogeneities at [001] tilt boundaries in YBCO and in other HTSC

3.3 Superconducting Composites, Based on BSCCO

3.3.1 BSCCO/Ag Tapes

Investigations of BSCCO/Ag tapes show that the *c*-axis alignment reaches maximum near the "superconductor–silver sheath" interface, but is often observed in the tape center [273, 600]. This microstructure assumes that an oxide core of Bi-2223 tapes consists of the plate-late grain colonies, in which the grains are divided by twist boundaries. The boundaries of colonies could be divided into four types, namely [14, 411, 638] (i) the boundary in the *c*-axis direction approximately parallel to the *c*-axis of colony; (ii) the colony boundary adjacent to basis plane at one side of the boundary; (iii) the colony boundary, neighboring to basis planes at both sides of the boundary; and (iv) three-point junctions of colonies. Generally, in the above cases, the twist boundaries are low-angle ones existing at the screw dislocations. Despite high texture, high-angle boundaries are often observed in the tapes. Majority of colony boundaries in the *c*-axis direction are the mixed type, including tilt and twist components and also dislocation network. The value of J_c could be controlled by residual layers of Bi-2212 phase at the twist [001] boundaries [1096, 1097] and also by Bi-2201 inclusions [1132], if their content is sufficiently great.

Amorphous buffer layers can exist at all four types of boundaries and considerably influence conductive properties, because their thickness usually exceeds considerably the superconducting coherence length [14, 1186]. The triple point junctions of the colonies in the form of triangles (Fig. 3.15a) are usually formed in tapes owing to plate-late growth of the colonies and misorientation between them. In these triple point junctions, the amorphous phase and admixtures are often observed in the tapes even with high J_c. Obviously, the triple point junctions of colonies, exceeding on some orders of magnitude the coherence length, render negative effect on the transport current in the tape. Secondary phases, consisting of amorphous structure and admixtures, form into Bi-2223 colonies separated particles of $(Ca, Sr)_2CuO_3$ and other oxides (see Fig. 3.15b) [828]. Large particles of secondary phase lead to arbitrary orientation of Bi-2223 colonies and serious distortion of crystalline lattice, creating numerous defects in these regions. Mutual intergrowths of Bi-2212 and Bi-2223 phases that can be usually observed in colonies in the oxide core of the tape eliminate near silver sheath [1097]. At specific thermal treatment, the intergrowth of superconducting phase is possible in silver [887]. The propagation of these intergrowths increases Ag/BSCCO interface and assumes an energetic advantage of their growth along concrete crystallographic planes of Ag. Then, the silver texture is able to effectively influence superconducting grain alignment.

There are other microstructure features of Bi-2223/Ag tapes, namely [001] edge dislocations, high density of stacking faults, bending and interrupting of (001) planes in local small regions, which can destroy perfection and continuity

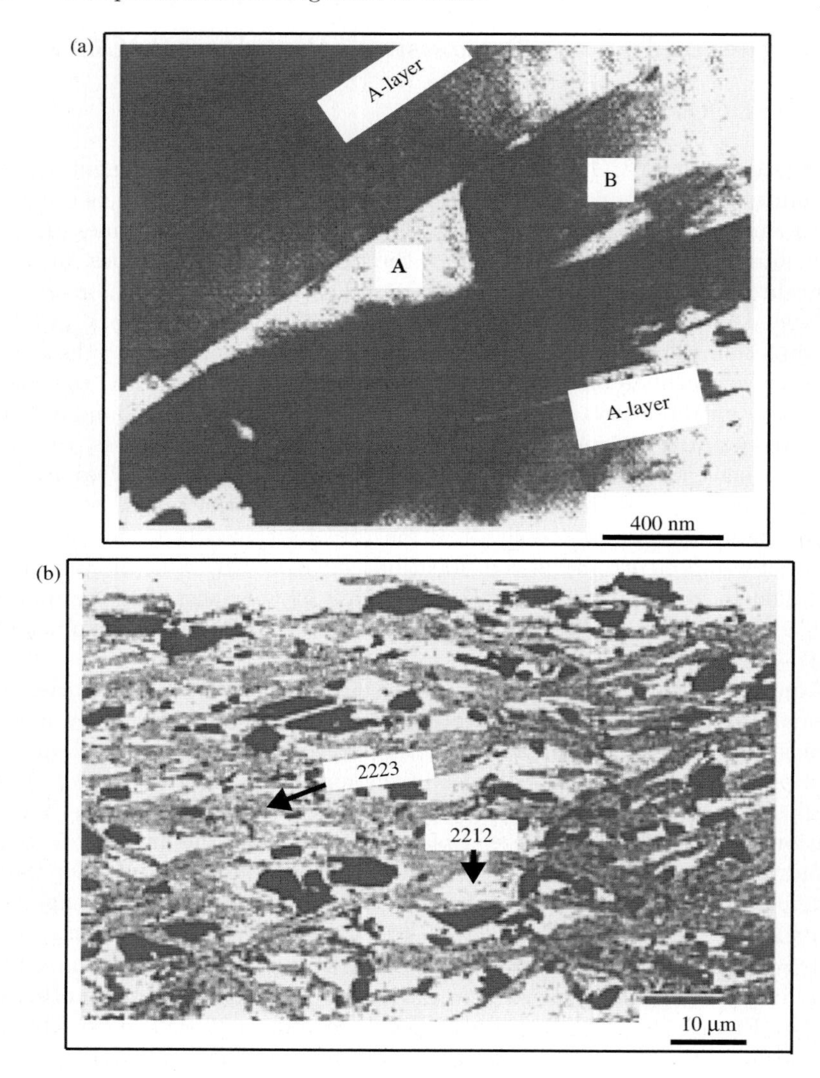

Fig. 3.15. (**a**) Thin amorphous layer at the boundaries of colonies with two neighboring basis planes (B) and amorphous structure (A) in triple-point junction of colonies [14]; (**b**) inclusion of admixed non-superconducting phase of $(Ca, Sr)_2CuO_3$, shown by using more *dark color* [828]

of CuO_2 plane, decreasing J_c. At the same time, strain around these defects can act as effective magnetic flux-pinning centers, improving superconducting properties [793]. The dislocation density in monolithic tapes near Ag/Bi-2223 interface can exceed on one order of magnitude the corresponding value in the core center [1117]. A more high plastic deformation near the interface in the tape processing is explained. An addition of silver dispersion permits to form relatively uniform density of core. In this case, an increased dislocation density leads to higher and homogeneous residual stresses in the core, allowing the preparation of superconductor with better properties. Moreover, an increasing of the irreversible strain ε_{irr} (the strain level, above which J_c decreases irreversibly) is caused by fact that the silver inclusions deflect crack and pin their surfaces in the mechanism of crack bridging, preventing the growth of crack into superconducting core [1117]. In this case, $AgNO_3$ additives render better effect on the formation of Bi-2223 phase and the value of J_c compared with Ag or Ag_2O dispersion [1000]. Fabrication of layered tape compositions of Bi-2212/Ag with intermediate buffer layer of silver provides the improved superconducting properties at bending compared with single tapes, having the same cross-sections of superconducting component [466]. The level of irreversible strain also changes in dependence on the number of superconducting filaments. So, the bending strain above 0.1% at the surface of monocore tape leads to decreasing of critical current (I_c). At the same time, the tape with 1296 filaments can sustain strains of 0.7% without actual diminution of I_c [925].

The test data (Fig. 3.16a) of tension of the 61 filamentary Bi-2223/Ag tapes under $T = 20\,K$ and $B = 0$–$8\,\mathrm{T}$ show small changes (5% decreasing of I_{c0} at $\varepsilon = 0$) in the range of small strain (upon to $\varepsilon_{irr} = 0.4\%$) [873]. Above this level, there is intensive fracture of superconducting filaments, destroying current paths and leading to sharp diminution of critical current. This behavior is observed for all measured magnetic fields. In this case, corresponding normalized values of I_c/I_{c0} (see Fig. 3.16b) state generalized curve for all considered values of the magnetic field, causing independence of mechanical properties from field. Moreover, the moment of irreversible decreasing of I_c by straining does not depend on external magnetic field, implying ε_{irr} to be an intrinsic parameter of superconductor, found by its material properties. This result is confirmed by comparison of corresponding curves $I_c = I_c(\varepsilon)$, obtained at various temperatures (see Fig. 3.16c). Small differences between values of ε_{irr}, observed in broad temperature range, are explained by material differences of test samples, rather than by some other factors. It should also be noted that a strain increasing much more that ε_{irr} can demonstrate saturation at decreasing of I_c. In many cases, the critical current makes up 20% from initial value even after a strain of 0.8% [246].

The most important effects on superconducting properties are rendered by voids and microcracks, forming in the BSCCO core during multi-stage thermal treatments [831, 1196]. The distribution of microcracks in Bi-2223/Ag tapes is introduced by mechanical deformation during superconductor processing. Comparative studies of longitudinal-rolled and transversal-rolled "green" (i.e., without thermal treatment) tapes, and also of tapes prepared by using one-axis pressing, give the following results [359]. The longitudinal-rolled and transversal-rolled tapes demonstrate smooth interfaces with a small number of short cracks, respectively, in the direction of the tape width and along its length. At the same time, the one-axis pressed tapes show wavy interfaces

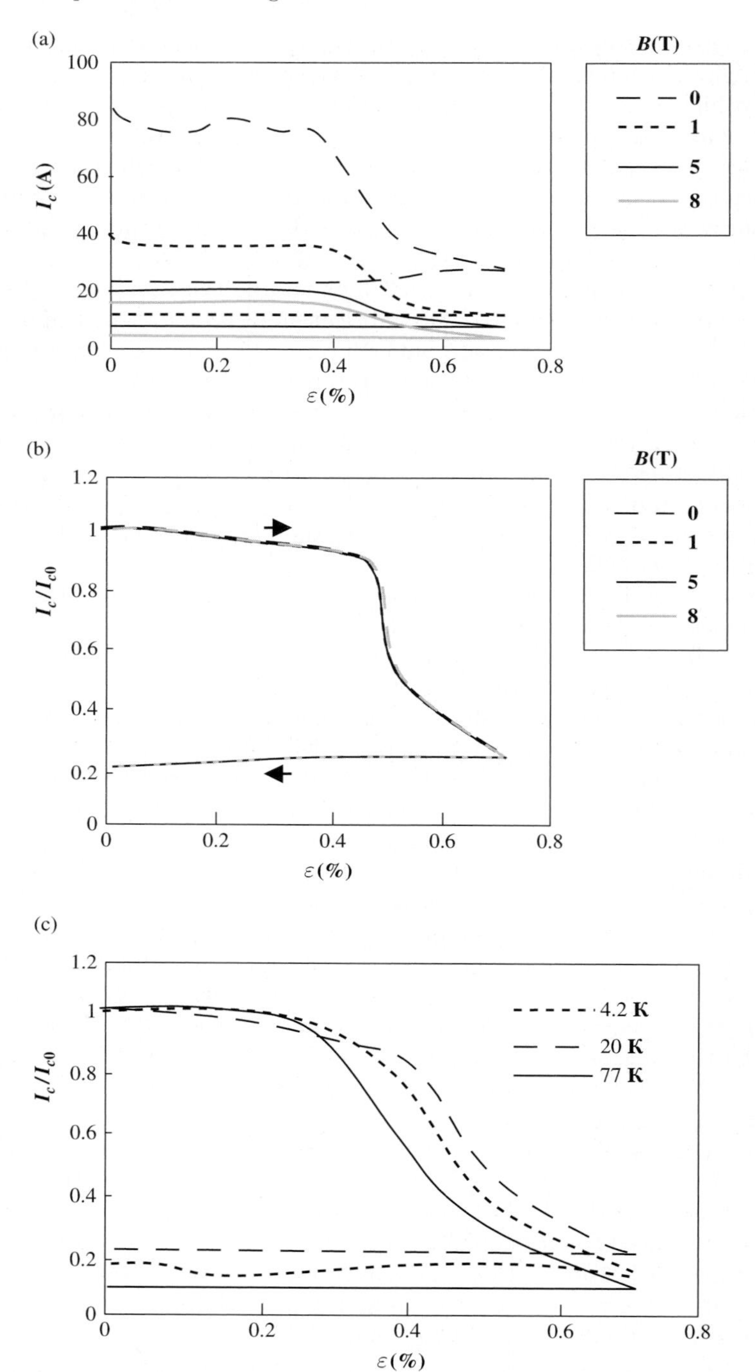

Fig. 3.16. The test data of the Bi-2223/Ag tape tension: (**a**) critical current and (**b**) its normalized value, obtained at T = 20 K and B = 0–8 T; (**c**) comparison results for different temperatures [873]

and long cracks directed to the sample length. These results are explained by the concept that in the case of rolling, maximum shear stress will develop along plane with normal vector, having component coinciding with the rolling direction, promoting the formation of transversal cracks that block transport currents. In the case of pressing, the stress state twists at 90° about the normal to the tape plane, causing cracking along current direction (Fig. 3.17). It should be noted that the contact length (of roller with material deformed) during rolling in the rolling direction is far lower than in perpendicular direction (along the axis of rollers) and agrees with acting friction force. Then, the strains $\varepsilon_x >> \varepsilon_z$ develop in the central part of oxide core under conditions of

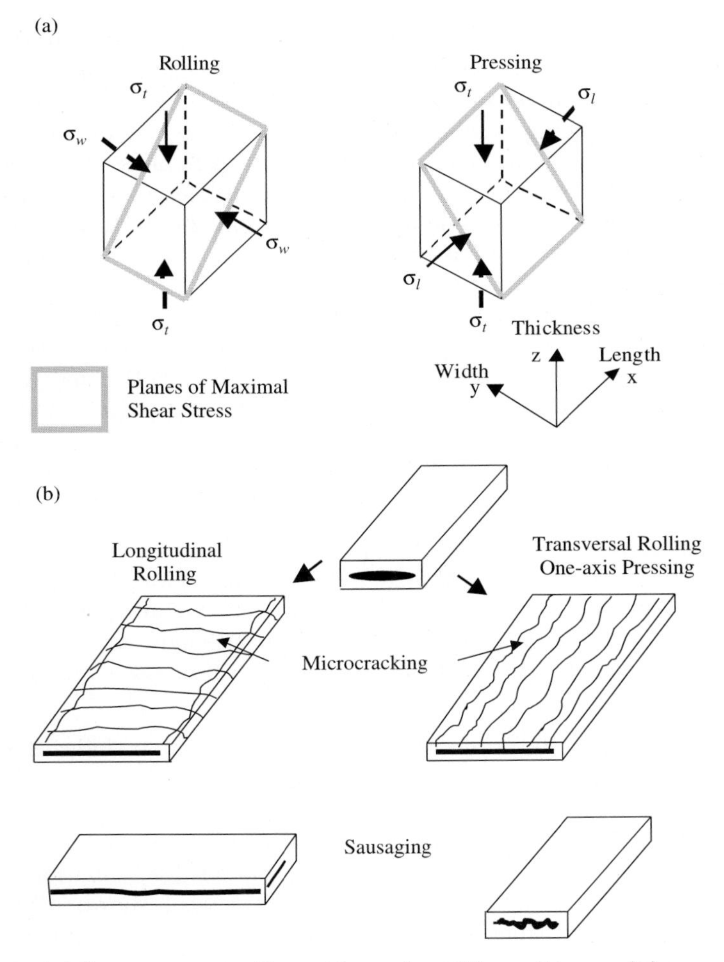

Fig. 3.17. (**a**) Stress state at rolling and pressing of "green" tapes; (**b**) corresponding behavior of microcracking and sausaging

restraining of the surrounding material. However, the situation is different at the tape edges because of absence of the restraining in the direction outside of the free side surface. It leads to side extension of the tape, strains $\varepsilon_z > \varepsilon_x$ and possible longitudinal cracking of the sample (Fig. 3.17b).

In the case of one-axis pressing, the situation is complicated because of high friction between tape and punches, causing considerable heterogeneity of stress–strain state, which grows with decreasing of the ratio $h_{\text{Ag}}/d_{\text{Ag}}$, where h_{Ag} and d_{Ag} are the thickness and width of silver layer in the tape, respectively. Therefore, the pressing effect on the core density will be directly found by the pressing parameters (i.e., pressure and friction) and also by the tape composition and sizes.

Decreasing of microcracks and porosity may be reached by using modification of standard rolling [450, 565, 1133], alternative methods of pressing [32, 592, 672], intermediate deforming [341, 831], regulating the rate of the tape thickness decreasing under mechanical loading [637], changing radii of rollers in rolling [995], applying excessive pressure at temperature of Bi-2223 phase formation [888], controlling the process of the Bi-2223 phase decomposition–restoration at high temperatures [828], sintering samples after their deforming [59, 857] and optimizing the cooling rate [440, 586, 829]. In these cases, a hot deformation in BSCCO systems improves weak links, but a cold deformation increases a magnetic flux pinning [636].

Pressing of Bi-2223/Ag tapes under cryogenic temperatures permits to transfer more mechanical deformation energy to the oxide core, due to a relative increasing in the hardness of silver to oxide in this thermal treatment. Then, the fracturing of grains during deformation can produce multiple new surfaces, promoting better final alignment of superconducting grains [164]. In this sense, it may be assumed that the mechanism of the grain alignment under pressing would be expected to involve at least two steps:

(1) Initial fracturing of the grains to produce multiple surfaces. The degree of fracture depends on the deformation introduced into the grains.
(2) Rotation and/or movement of the fractured grains into $(00l)$ alignment. It is suggested that powder flow is the main mechanism for the rotation and/or movement of the granular bulk. Powder flow also exposes the multiple surfaces created by the fracturing step.

The room temperature process (RTP) of deformation includes the following steps. A small amount of force is applied and a small degree of fracture occurs. Further force is applied and the silver sheath begins to deform plastically. As a result, there are both powder flow and fracturing of the grains. At this step only few new surfaces are created and a ridge formation starts to occur (the ridge formation mechanism in the RTP tapes is shown in Fig. 3.18). The tape density increasing in the final step reduces the amount of powder flow, and the fractured grains are further broken into small pieces.

The cryogenic process (CP) of deformation (77 K) includes the following steps. First, a force is applied that fractures the platelets. The silver sheath is

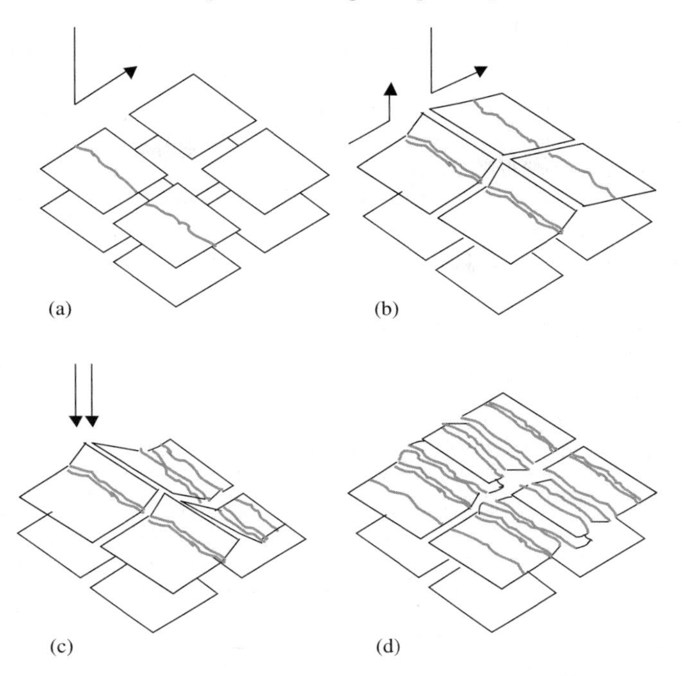

Fig. 3.18. A schematic representation of the steps, involved in the ridge formation in the RTP tapes [164]

harder and powder flow does not occur at the same amount of applied force as for the RTP tapes. This causes the platelets to fracture, but with very small powder flow. This step creates many new surfaces in the superconductor, but the surface density of the tape increases, as there is no lateral powder flow. As the force increases, the silver sheath becomes significantly deformed and both powder flow and fracturing occur. As the force continues to increase, the silver continues to plastically deform, and the tape density increases, reducing the amount of powder flow. The fractured grains are further broken into small pieces. The RTP and CP are demonstrated in Fig. 3.19. In conclusion, note that the increased silver sheath hardness at cryogenic deformation increases rate of deformation in the Bi-2223/Ag processing compared to usual conditions. As a result, an optimum process of tape fabrication improves connectivity and texture of grains, and also J_c [558]. Nevertheless, a densification of superconducting core may be accompanied by aggravation of Bi-2223 texture and diminishing of J_c [1061]. Then, the texture of Bi-2223 superconducting grains could be improved as a result of directed growth of the Bi-2212 grains due to residual stresses during tape [1198] or during reaction of Bi-2212 with secondary phases [696].

In the case of substitution of silver sheath by a silver alloy, all alloying additions form in the sheath particles of oxides as a result of the internal

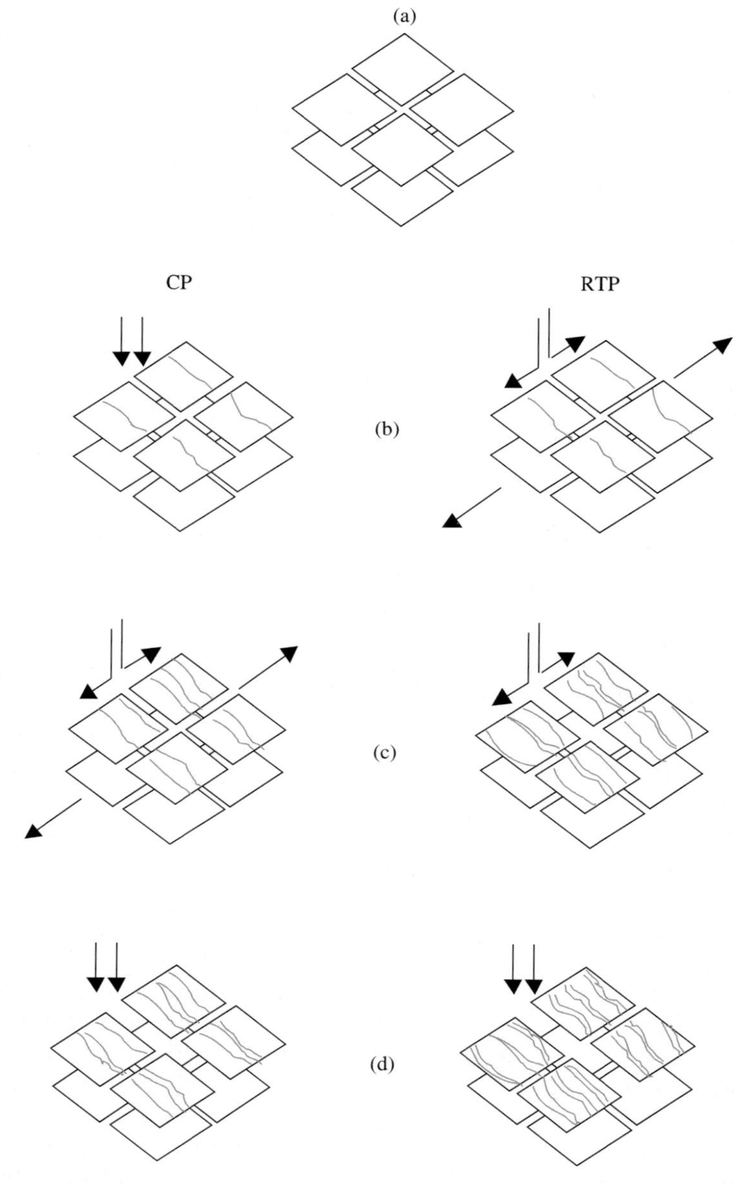

Fig. 3.19. A schematic representation of the steps, involved in the grain alignment mechanisms for both cryogenically pressed (CP) and room temperature pressed (RTP) tapes [164]

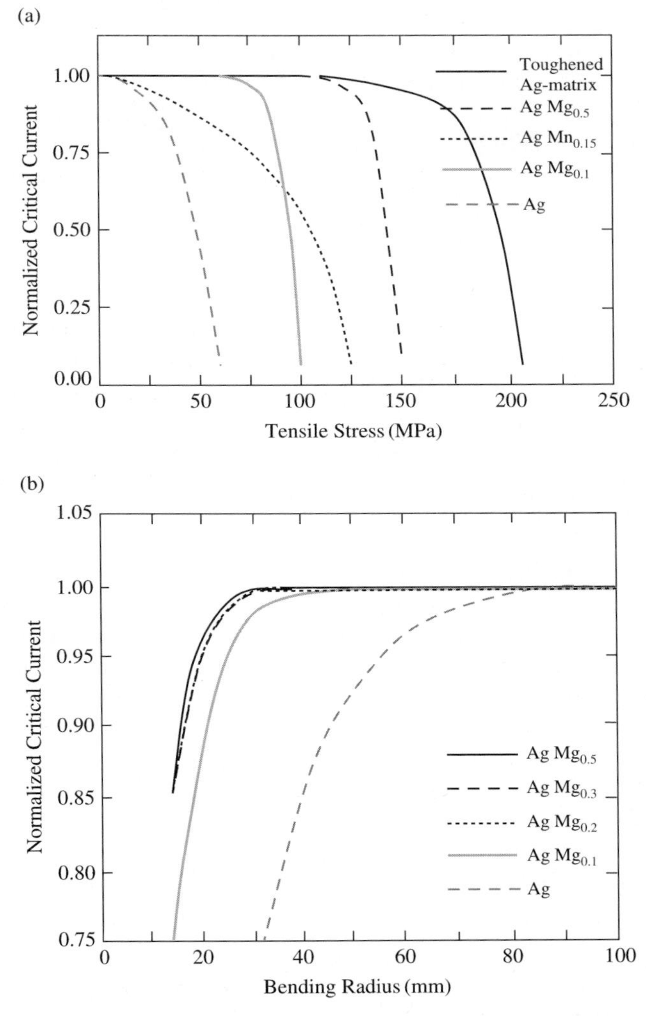

Fig. 3.20. Improved mechanical properties of the alloyed Ag-sheath Bi-2223 tapes with 55 filaments in comparison with pure Ag-sheath Bi-2223 tapes: (**a**) tension and (**b**) bending [282]

oxidation and promote formation of Ag_2O particles, serving as seeds. Moreover, they reduce grain sizes of the sheath and also decrease pre-stress of oxide core, arising from differences in thermal expansion coefficients of sheath and ceramics that is revealed at tape cooling from sintering temperature to cryogenic one of HTSC application [16]. All these factors result in microhardness enhancement of the alloyed silver sheaths, as well as an increase of the ultimate strength and the yield strength of a composite as a whole. It is accompanied by the growth of J_c stability relative to thermal cycling. Proper test data (tension and bending) demonstrating improved mechanical properties of tapes with sheath from silver alloy in comparison to tapes possessing silver sheath are shown in Fig. 3.20. At the same time, the doped Ag-sheath leads to some decreasing of critical current density, amount of Bi-2223 phase and T_c, and also to growth of secondary phase and the core thickness variability (sausaging) [826]. Besides, there is little silver in the superconducting core in the case of the alloyed Ag-sheath. At the same time, Ag intensively penetrates from the sheath into ceramics of an unalloyed sheath composite during annealing and forms particles of pure silver there. These particles provide an enhancement of J_c. The above-mentioned decreasing of Bi-2223 phase amount and T_c is the result of the internal oxidation of the alloyed Ag-sheath and accompanying decrease of oxygen flow to the ceramics [626]. It should be noted that the critical current density in multi-filamentary tapes is very sensitive to technical parameters and microstructure properties. In the case of silver sheath, the value of J_c in the central part of the tape is found to be considerably higher relative to the side sites [290] owing to heterogeneous strain caused by rolling. The silver alloys with improved mechanical properties decrease heterogeneous strains during rolling and increase J_c in the side sites [1193]. Further decreasing of sausaging and differences in the values of critical current density in the central and side sites causes fabrication of the alloyed Ag-sheath multi-filamentary tapes, possessing higher J_c in comparison with corresponding counterparts, having pure silver sheath (Fig. 3.21).

Investigations of the after-heat-treatment BSCCO-core Vickers microhardness, assuming directed conformity between Vickers microhardness (H_V) and volume density of the material [1159], state a correlation between high density of superconducting phase and high values of J_c [458, 826]. These measurements, carried out at different stages of the oxide-powder-in-tube process, estimate an effectiveness of different thermal treatments. In particular, a pressing

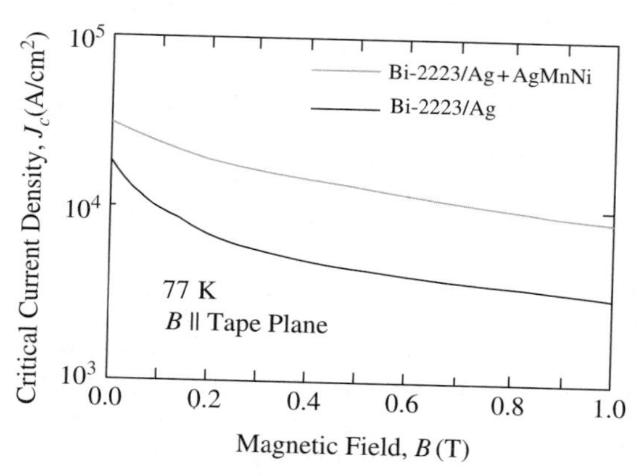

Fig. 3.21. Critical current density vs magnetic field at 77 K [1193]

leads to a greater increasing of J_c in comparison with rolling due to a greater oxide core density after first processing, than after the second one. Moreover, the degree of transformation from Bi-2212 to Bi-2223 phase is always found to be greater in the pressed tapes [458, 826]. Corresponding results for different thermal treatment (i.e., cycles of "heat–strain") are shown for rolled and pressed Bi-2223/Ag monocore tapes in Fig. 3.22. It is also noted

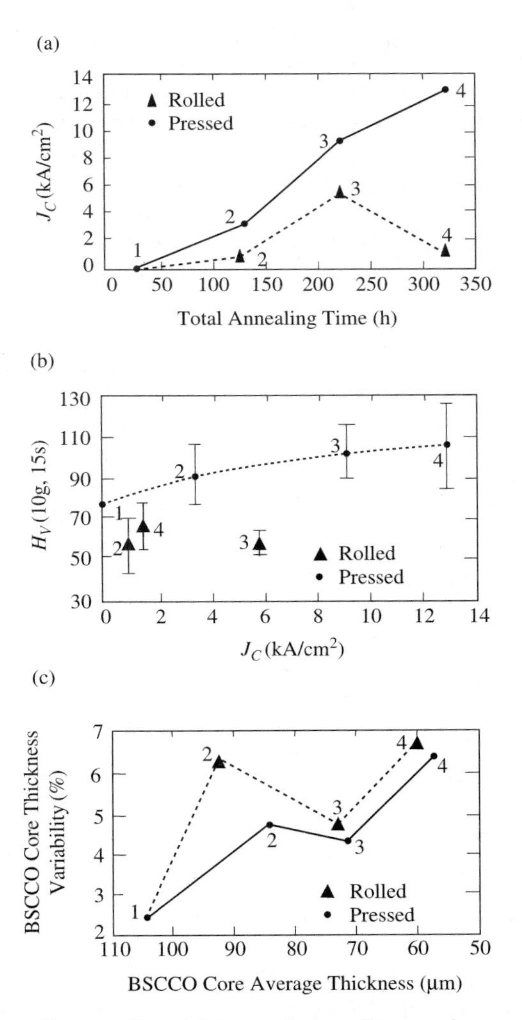

Fig. 3.22. Comparative results of intermediate rolling and pressing between heat cycles for Bi-2223/Ag tapes: (**a**) J_c (0 T, 77 K) vs total heat treatment time; (**b**) Vickers hardness vs critical current density; (**c**) BSCCO core thickness variability as a function of average core thickness. The *numbers* point to the number of heat treatments, which the tape is subjected to [826]

that the sausaging increases with the increasing of superconductor thickness in the process of intermediate mechanical deformation between heat cycles. At this thermal mechanical regime, the rolling leads to greater sausaging in comparison with the pressing (see Fig. 3.22c). Moreover, an increase of microhardness in the case of prolonged heating correlates almost linearly with the growth of the critical current density (see Fig. 3.22a) [826]. The measurement of the Knoop microhardness (H_K) estimates difference of the J_c behavior in *ab*-plane (that is parallel to the rolling plane) and along *c*-axis direction (that is perpendicular to the rolling plane). So, in the first case, the microhardness H_K remains approximately constant, at the same time; in the second case, it increases with the annealing time (Fig. 3.23) [568]. Destroying of the microhardness growth with the density increasing permitted to state heterogeneities, linked with formation of more dense blocks of grains, separated by cracks [519] that was caused by dominating mechanisms of the core deformation, namely sliding and fracture of grains.

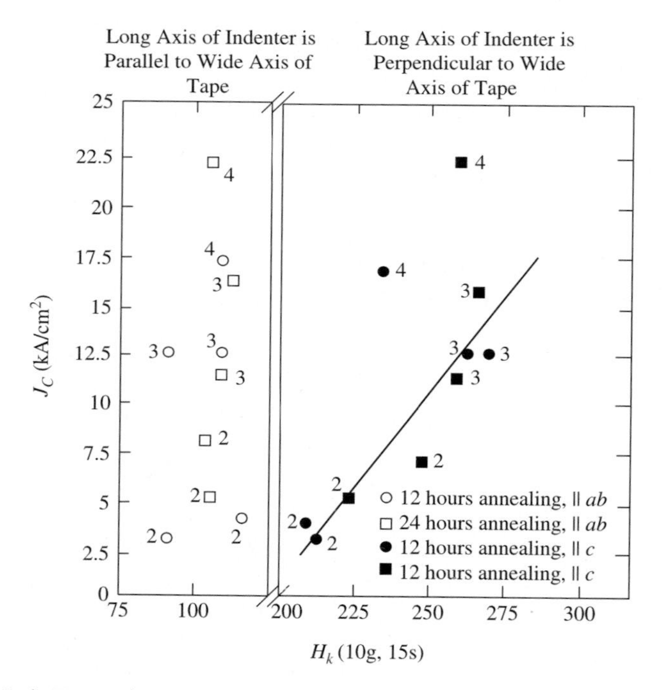

Fig. 3.23. J_c (0 T, 77 K) as a function of Knoop microhardness of BSCCO transverse cross-sections. The data to the left of the axis break is for H_K parallel to the rolling plane (i.e., approximately || *ab*-plane), and the data to the right is H_K perpendicular to the rolling plane (i.e., approximately || *c*-axis). The *numbers* point to the number of heat treatments, which the tape is subjected to [568]

Key factors, stating critical strains and stresses for superconducting tapes, are the following: (i) connectivity and alignment of grains [273, 600], (ii) uniformity and high texture of ceramic core [385, 519], (iii) interface regions Ag/BSCCO and sheath material [626], (iv) temperature during loading [433]. For multi-filamentary tapes, number and distribution uniformity of filaments play considerable role in the estimation of mechanical properties [854, 873]. Critical mechanical characteristics can be considerably improved, adding Ag ($Ag_2O, AgNO_3$) dispersion in the superconducting core. For example, the value of ε_{irr} increases more than two times compared to the monolithic tape in the case of addition into monocore tape of 7 wt% $AgNO_3$. In the case of multi-filamentary tape, this value increases more than three times in comparison with the monolithic monocore tape [1001]. This growth of mechanical properties compensates well a small decreasing of J_c, observed in the case of $AgNO_3$ additives [32].

Based on MOI method, intensive researches of BSCCO/Ag samples under bending and at different values of strain permitted to observe the stage of microcrack formation and also to understand how magnetic flux penetrated superconductor [558, 855]. It is confirmed that the defects of the void type or non-superconducting phase type are the sites of crack initiation [247, 836]. The bending, also as another strain, changes the microstructure and critical current in two directions: (i) as a result, the intergranular contacts worsen, decreasing critical current density, and (ii) preliminary existing defects are developed that diminish local critical current due to decreasing of effective superconducting cross-section [451, 855]. Intergranular contact, destroyed at bending, could be restored after straightening of tape and the tape cooling down to cryogenic temperature [573]. The observed effect is explained by the core compression transferred by silver sheath on cooling. The compression causes a sliding of BSCCO grains, restoring the broken contacts. Obviously, this effect will depend on the core density and texture. In this case, the lattice of transversal microcracks, formed during the longitudinal rolling, possesses elevated sensitivity to the tape bending [854]. In bending, most tensile and compressing loading exist at external surfaces of the tape and render effect mainly on the metal (alloy)/ceramic interfaces. Fatigue tests of the tape bending/straightening type identify the mechanism of microcracking formation in these interfaces and the microcrack growth in ceramic. It is found that tensile loading favors formation of intercrystalline cracks. At the same time, the compressing stresses form transcrystalline cracks, depending on the orientation of the *ab*-plane [659]. When tape is subjected to cyclic strains, which are lower than ε_{irr}, the cracks are transcrystalline type irrespective of stress kind.

The tensile tests of Bi-2223/Ag tapes demonstrate three typical stages, namely (i) very narrow region of elastic behavior, (ii) sufficiently broad stage of microcrack initiation and growth and (iii) a macroscopic flow, accompanied by multiple cracking and macrocrack formation [786, 787]. Fatigue tests of monocore and multi-filamentary tapes show that the microcracks do not reach the macrofracture threshold in the second stage [552, 1023]. An effect of tension on the sharpness of transition, depicted in curves of the "transport electric field–current density" (E–J) dependence (that is very important for devices, based on HTSC), may also be estimated experimentally [550]. Typical V–I curves dependent on strain for multi-filamentary Bi-2223/Ag tapes are shown in Fig. 3.24. As shown in the figure, before sharp decreasing of I_c its degradation takes place initially in the region of lower electric field. The test curves also estimate change of value n, defining the dependence V–I^n.

Joint effect of strain and external magnetic field defines considerable decreasing of I_c even in the smallest growth of the field [451]. Two current-carrying paths co-exist in the tapes: one is through the Josephson junction network, consisting of weakly linked grains, and the other is through the strongly linked grains. Investigation of these path show that in addition to material cracking, strain deteriorates the grain connectivity, leading to the easier suppression of I_c by magnetic field in the low field region. For the well-connected current-carrying path, strain on one hand damages well-connected grain boundaries through cracking or sliding. On the other hand, strain can introduce new defects in grains, increasing intercrystalline components of the magnetic flux pinning. Typical dependencies of transport currents, flowing through weak-linked boundaries (i.e., the Josephson junction network) I_{cw} and through strong-linked grains I_{cs} (the resulting critical current $I_c = I_{cw} + I_{cs}$), on applied magnetic field and strain are shown in Fig. 3.25 [451].

3.3.2 Irreversibility Lines for BSCCO

As has been noted in Sect. 1.6.4, the current-carrying capability of superconductor was found by the irreversibility line, presented by the "magnetic field–temperature" non-linear dependence. Above this line, there is a region with the condition that direct current decreases rapidly, making superconductor to be useless for many applications. This property is connected with the material anisotropy. Due to the greatest anisotropy of Bi family in comparison

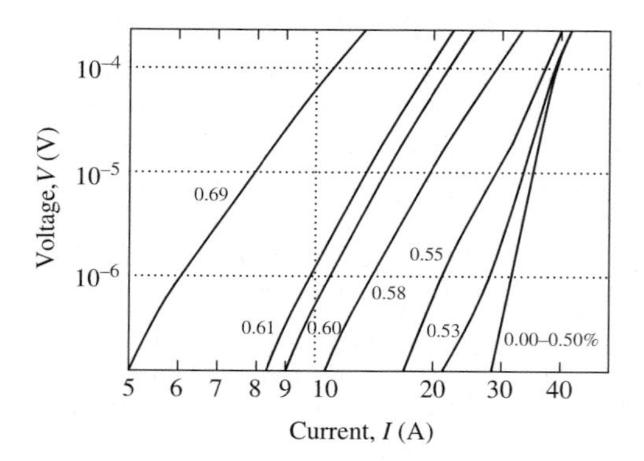

Fig. 3.24. Characteristics of transport current V–I vs strain in multi-filamentary Bi-2223/Ag [550]

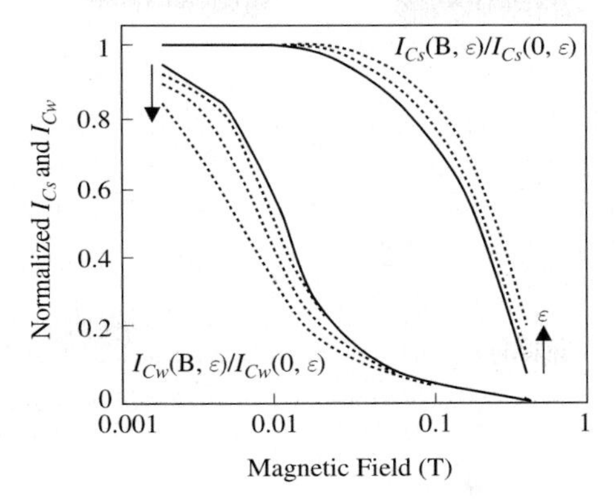

Fig. 3.25. Normalized I_{cw} and I_{cs} as a function of magnetic field, showing the strain effects on both the strong-link and weak-link current-carrying paths [451]

with other HTSC, its irreversibility line is the lowest among high-temperature superconducting families (Fig. 3.26). Therefore, the irreversibility line is one of the main problems of BSCCO techniques. Comparative test data, obtained for single Bi-2212 crystals, Bi-2212/Ag and Bi-2223/Ag tapes [214], show that the irreversibility line for Bi-2223/Ag tapes occupies a region of higher temperatures than the irreversibility line for Bi-2212/Ag tapes, which is higher than the one for Bi-2212 crystals (Fig. 3.27). Defects introduced, in particular by using mechanical loading, act as pinning centers and are responsible for trapping of vortex lines. The irreversibility lines are a measure of connectivity of the CuO_2 superconducting planes and the vortices trapping (i.e., pinning strength) [214, 472]. All three samples (Bi-2212 crystals, Bi-2212/Ag and Bi-2223/Ag tapes) are superconductors of the same family and possess identical spacing between CuO_2 planes. Moreover, the samples investigated in [214] had comparative sizes. Hence, it may be assumed that shear of the irreversibility line location characterizes directly the pinning strength and is thus maximum for Bi-2223/Ag tapes, followed by Bi-2212/Ag tapes and smallest for Bi-2212

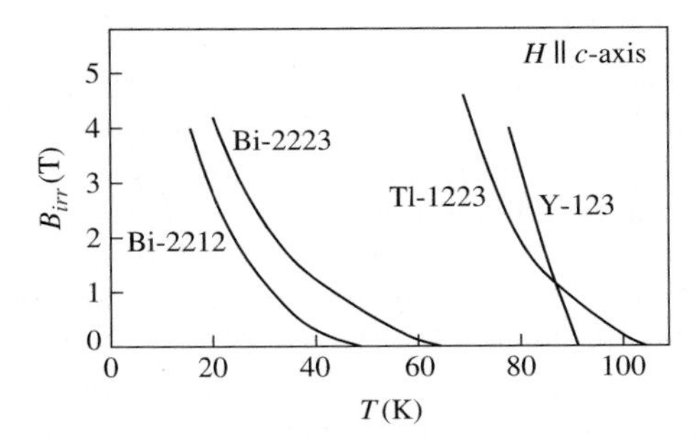

Fig. 3.26. Irreversibility lines for Bi-2212, Bi-2223, Y-123 [151] and Tl-1223 [1065]

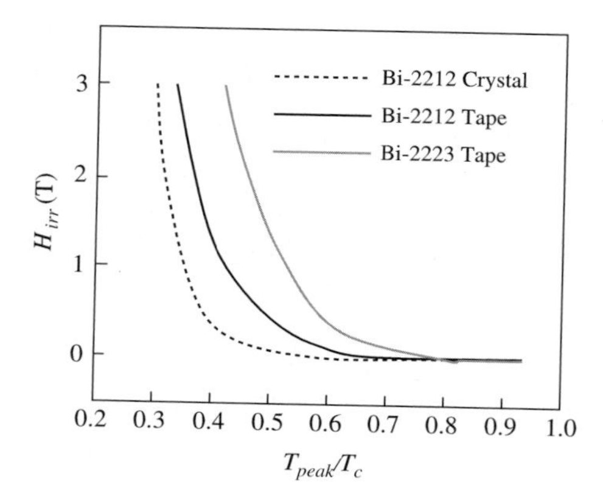

Fig. 3.27. Irreversibility lines for single Bi-2212 crystals, Bi-2212/Ag and Bi-2223/Ag tapes, possessing comparative sizes [214]

crystals. Elevated magnetic flux pinning could be due to higher dislocation density in tape in comparison with crystal. At the same time, the difference between Bi-2212/Ag and Bi-2223/Ag tapes, in this case, is rather caused by different processing conditions. So, in order to obtain the Bi-2212/Ag tapes, the melt-texturing technique has been used, different from Bi-2223/Ag tapes prepared by using the solid-phase reaction techniques and partial application of liquid-phase sintering. In this case, the defects, introduced during mechanical strain in Bi-2212/Ag tapes, are healed partially on melting. At the same time, the solid-phase reaction in the Bi-2223/Ag tapes does not lead to considerable decreasing of material damage [214]. The experiments show [472] that the dislocation line density in Bi-2223/Ag tapes is approximately greater by an order of magnitude, than in Bi-2212/Ag tapes and single Bi-2212 crystals.

The comparison of irreversibility lines for 27 filamentary tapes, processed by using different techniques with application of hot-pressing, rolling and cold-pressing, has been carried out in [385]. All three tapes had the same sizes and have been subjected to the same thermal and mechanical treatments before final sintering. Therefore, in this case also, shear of irreversibility line must be attributed only to the characteristic of strength pinning. The cold strain must create more defects than hot deformation because in the latter case, the grains possess elevated mobility. Then, it may be expected that dislocation density in the tapes, obtained by using cold-pressing technique, is more, than in the case of hot-pressing technique. This states a higher location of the irreversibility line in the tapes, obtained by cold-pressing method, in comparison with the rolled tapes. The latter, in its turn, exceeds corresponding data for tapes, fabricated by hot-pressing technique (Fig. 3.28).

Then, the dependence J_c–H for the hot-pressed sample (Fig. 3.29) shows improved behavior in the region of low magnetic field in comparison with other two samples. The fast diminution of J_c at low external field is proper

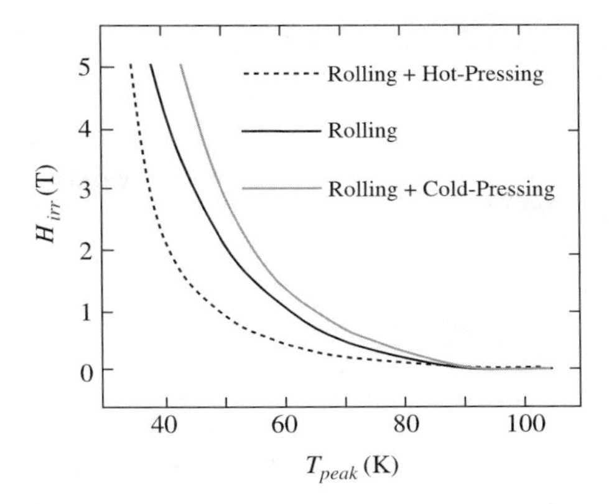

Fig. 3.28. Irreversibility lines for the cold-pressed, rolled and hot-pressed 27-filamentary Bi-2223/Ag tapes, possessing the same sizes [385]

for Josephson weak links at intercrystalline boundaries. In this case, the rolled tapes demonstrate maximum diminution of critical current under the low field. Under high magnetic field, there is a plateau due to magnetic flux pinning in single grains. The dependence J_c–H for the hot-pressed and cold-pressed tapes is almost parallel to one another, when applied field is parallel to the tape surface. In the case of perpendicular field to the tape surface, prepared by using cold pressing, the critical current density decreases slowly in external field, exceeding 100 mT, in comparison with the hot-pressed and rolled tapes. This result proves that hot-pressing increases grain connectivity and

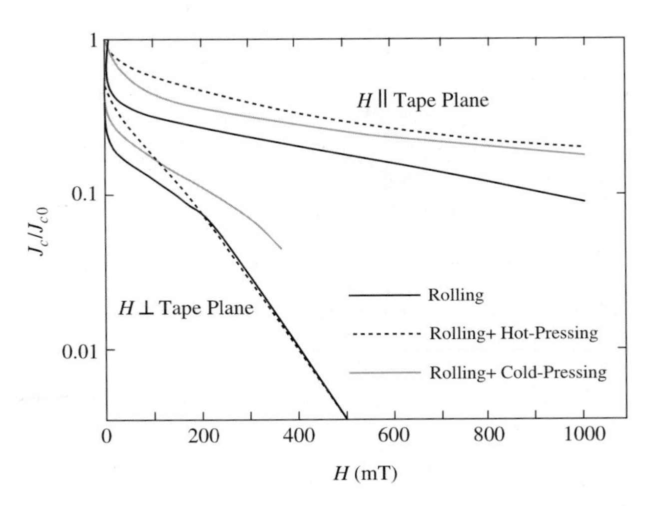

Fig. 3.29. Dependencies $J_c - H$ for the cold-pressed, rolled and hot-pressed 27 filamentary Bi-2223/Ag tapes at 77 K in magnetic fields that are parallel and perpendicular to the surface tape [385]

therefore improves weak link behavior in low fields. At the same time, cold isostatic pressing forms a great number of defects, forcing the magnetic flux pinning [385].

3.3.3 BSCCO Bulks

An addition in BPSCCO (Bi-2223) bulks of thermo-mechanical and chemical compatible reinforcement improves the HTSC microstructure, superconducting and mechanical properties. In particular, an addition of MgO whiskers, oriented parallel to *ab*-plane of superconducting grains [1186], together with carrying out cyclic procedures of hot-pressing and annealing, increase the average grain size both in the *ab*-plane and on the sample thickness. At the same time, when a whisker concentration (V_w) is 20% and more, there is considerable constraint of the BPSCCO phase grain growth, especially in the *ab*-plane, which is most favorable for improvement of superconducting properties. By introducing whiskers, the current-carrying capability of superconductor increases in comparison with analogous monolithic sample at the same grain size (Fig. 3.30).[3] Additional pinning centers, for example, microdamages, defects of crystalline lattice or low chemical substitutions, can form at the MgO/BPSCCO interfaces. These interfaces remain clear without chemical reaction zone even after prolonged annealing (Fig. 3.31), which is an additional factor of J_c increasing. Long thermal treatment can also contribute to obtain denser BPSCCO microstructure with lower content of admixtures, causing better superconducting properties [1184].

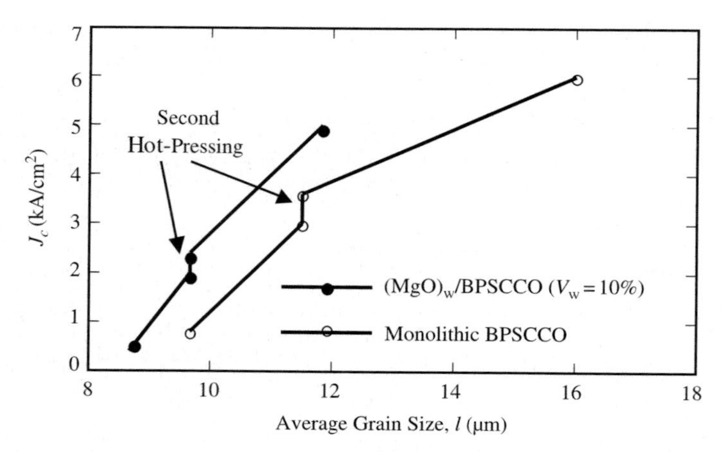

Fig. 3.30. Effects of average grain size (l) on critical current density (J_c) in monolithic BPSCCO and $(MgO)_w$/BPSCCO composite with $V_w = 10\%$ [1186]

[3] An addition of Al_2O_3 filaments is less advantageous due to physical and chemical incompatibility of BPSCCO with Al_2O_3 leading to chemical reactions on interfaces at elevated temperatures of processing, decreasing J_c [1147].

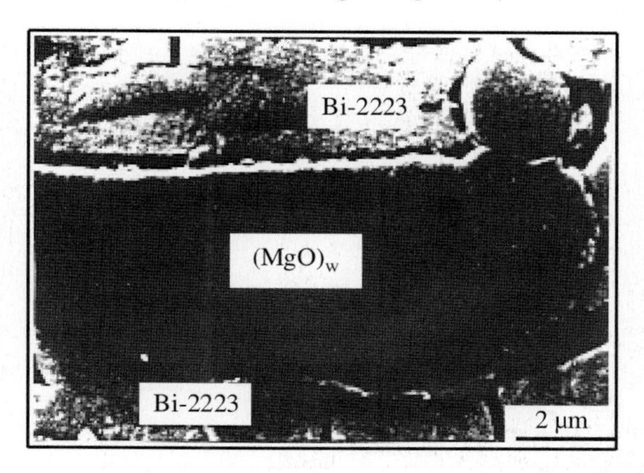

Fig. 3.31. High-magnification SEM micrograph of the interface region in a polished and etched cross-section of a $(MgO)_w$/BPSCCO composite with $V_w = 10\%$ [1186]

In order to reach desired phases, microstructures and structure-sensitive properties of superconductor, it is important to control technological parameters. In the case of hot-pressing technique, these parameters include temperature, pressure and process duration. Based on the tests, the optimum temperature of hot-pressing for BPSCCO bulk is equal to 800–855°C [1019]. An initial hot-pressing leads to J_c increasing in HTSC directly after this procedure, as well as following annealing (Fig. 3.32). However, as has been discussed in Sect. 2.4 single hot-pressing alone cannot create a dense and alignment microstructure with large grains. Moreover, it is impossible to reach high purity of the Bi-2223 phase. Sometimes, a diminution of its content is observed because of partial decomposition, caused by combination of high temperature and high pressure [1185]. Due to the above causes, it is necessary to carry out additive annealing or cyclic procedures of "annealing–pressing" to increase J_c (Fig. 3.33). Next, annealing improves superconductor properties, but it should be noted that exceeding its definite duration (this mark is equal to 160 in Fig. 3.32) causes a diminution of J_c. There are other causes for decreasing of the critical current density: (i) stopping of densification in the absence of additional pressure [801, 1185] and (ii) definite alterations of the superconductor chemical composition because of a possible evaporation of Bi and Pb during prolonged annealing [670, 671].

High texture along c-axis due to one-axis cold pressing could be formed in the samples with MgO dispersion and in the superconductors with polymeric matrix (BPSCCO/polyethylene) [97]. An addition of silver in different forms into the bulk leads to improved superconducting and mechanical properties. In this case, $AgNO_3$ additives show greater effect on formation of superconducting phase and J_c in comparison with Ag and Ag_2O particles [999]. Introducing Al_2O_3 filaments into BPSCCO matrix does not lead to considerable decrease in superconducting properties [1147], but provokes substantial toughening

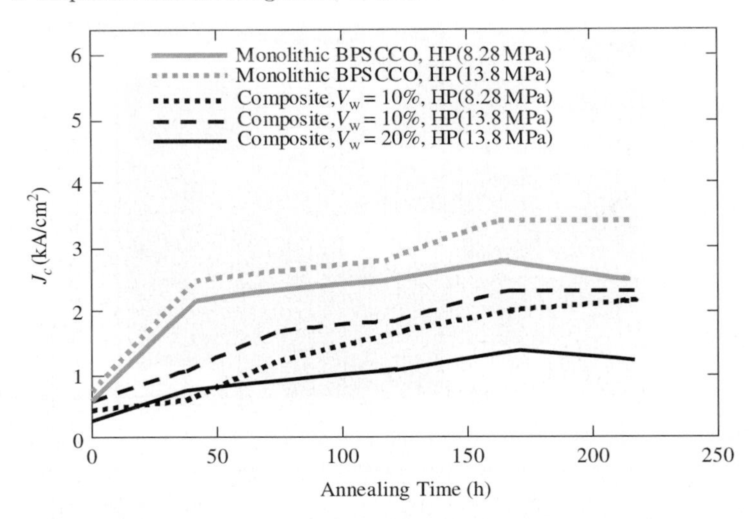

Fig. 3.32. Effect of annealing time on J_c (77 K and zero field) for monolithic BPSCCO and $(MgO)_w$/BPSCCO composite (first, hot-pressed at 820°C with different pressures and then continuously annealed at 832°C in 8% O_2) [1185]

of superconductor simultaneously, caused by non-regular and rough surfaces of fracture [712]. In the case of room temperature (293 K), macroscopically plane surfaces of crack are observed corresponding to brittle fracture with small de-laminations (Fig. 3.34a) and microscopically short pulled-out fibers (see Fig. 3.35a). At the same time, at cryogenic temperature (77 K), there

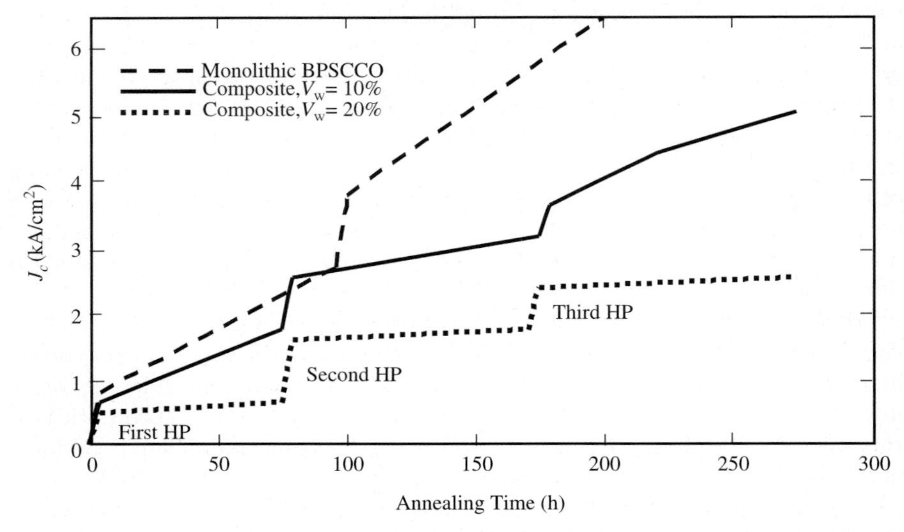

Fig. 3.33. J_c (77 K and zero field) of monolithic BPSCCO and $(MgO)_w$/BPSCCO composite with V_w = 10% and V_w = 20% in 2-cycle and 3-cycle hot-pressing and annealing processes [1185]

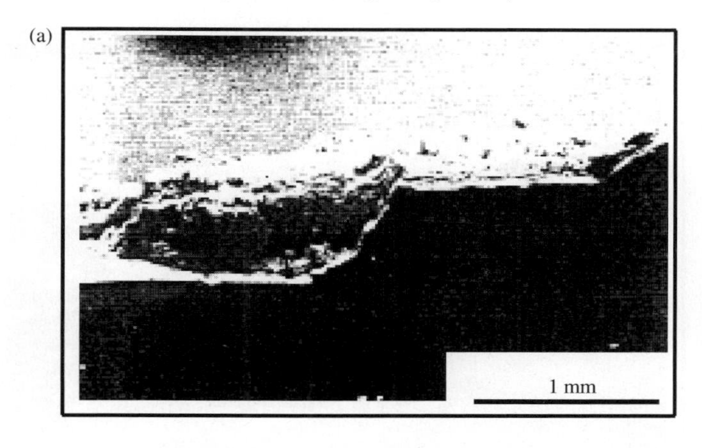

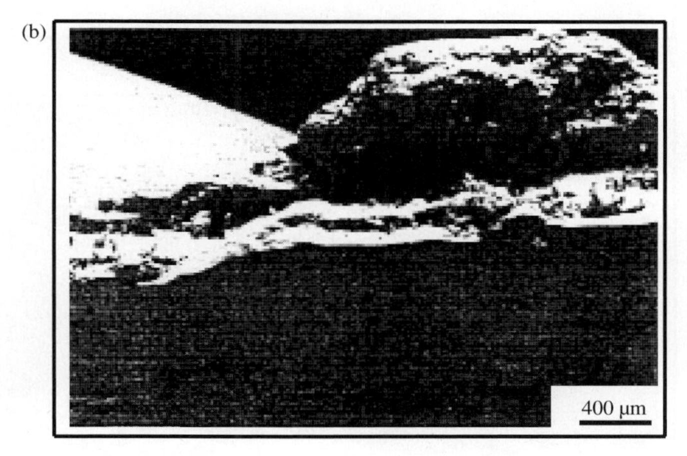

Fig. 3.34. Macroscopic failure modes in Al_2O_3/BPSCCO composite at: (**a**) 293 K and (**b**) 77 K [712]

are intensive de-laminations at the fracture surface (Fig. 3.34b) and pulled-out long fibers (Fig. 3.35b). The pulled-out fibers are covered in both cases by BPSCCO grains oriented preferably along fibers (Fig. 3.36), demonstrating strong interfaces between fibers and matrix. Due to brittleness of the BPSCCO monolithic sample at 77 K and as its fracture is controlled by microstructure defects, a considerable weakening of Al_2O_3/BPSCCO composite takes place at 77 K because of intensive cracking of matrix, caused by thermal stresses. Corresponding weakening of interface region between matrix and fibers causes greater increasing of fracture toughness at 77 K in comparison with room temperature. The toughening mechanisms, acting at cryogenic temperature, include the macrocrack deflection, branching and blunting and also crack propagation along the fiber direction. Thus, the differences of the fracture

(a)

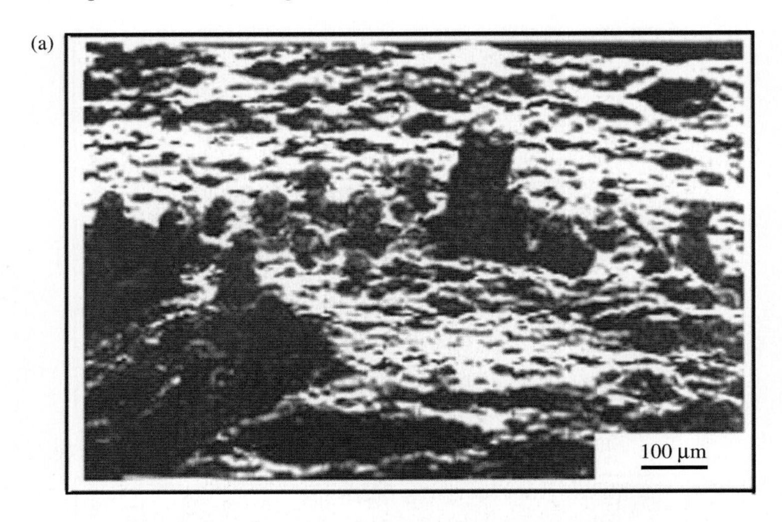

(b)

Fig. 3.35. SEM micrographs of fracture surfaces of Al_2O_3/BPSCCO composite at: (**a**) 293 K and (**b**) 77 K [712]

toughness and fracture modes at room and cryogenic temperatures may be explained by thermal stresses, forming in the composite during cooling down to cryogenic temperature. At the same time, strength of the Al_2O_3/BPSCCO composite may decrease in comparison with monolithic superconductor, if the interfaces between matrix and fibers are weak [679]. In this case, the fibers act as pores and also as stress concentrations.

Best thermodynamic compatibility of MgO [215] assumes application as toughening elements of high-strength MgO whiskers that ensure significant increasing stiffness and fracture toughness of the monolithic superconductor

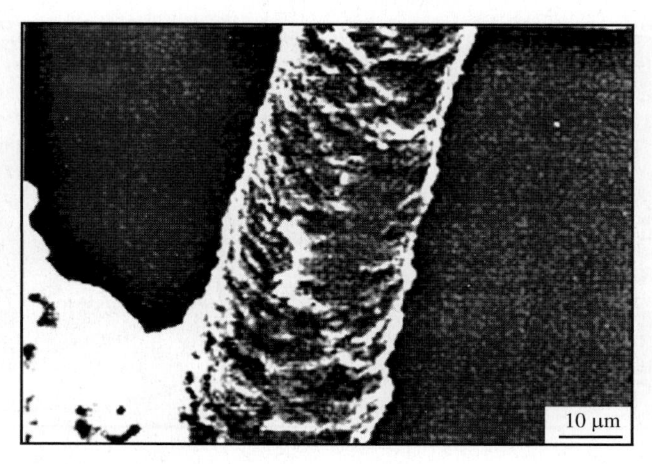

Fig. 3.36. Surface morphology of a pulled-out fiber observed in fracture surface of Al_2O_3/BPSCCO composite at 77 K [712]

[1187]. The toughening mechanisms of this composite are related to whisker fracture, whisker/matrix interface de-bonding, pushing of whiskers by crack surfaces and also the fracture modes caused by the crack deflection (Fig. 3.37). An appreciable increasing of plane-strain fracture toughness (K_{Ic}) is observed both for monolithic BPSCCO specimens and $(\mathrm{MgO})_{\mathrm{w}}$/BPSCCO composite after cooling from room temperature (293 K) to cryogenic (77 K) (Fig. 3.38). For monolithic ceramic, it can be related to greater surface energy (γ) at lower temperature by using Griffith's formula [348] as

$$K_{\mathrm{Ic}}(T) = \left[\frac{2E_{\gamma}(T)}{1-\nu^2}\right]^{1/2} , \tag{3.1}$$

where E and ν are Young's modulus and Poisson's ratio, respectively.

A formation of small residual thermal stresses may be expected in the $(\mathrm{MgO})_{\mathrm{w}}$/BPSCCO composite that introduce minimum number of defects at interfaces during cooling to cryogenic temperature because of proximity of the thermal expansion factors for whiskers and superconducting matrix and also because of their small sizes. Hence, an elevated fracture toughness of composite is caused in total by the surface energies of BPSCCO matrix and MgO whiskers [1187]. At the same time, an initial increasing of strength with volume fraction of whiskers may be replaced by its decrease of growth of the whisker fraction up to 20% and more. It could be explained by microdefect formation, including grain boundaries and BPSCCO/MgO interfaces, that creates significant stress concentrations. An increasing of the whisker content leads to the same effect on change of elastic module. By introducing MgO particles in BPSCCO matrix, the bending strength increases. In this case, a decreasing of compliance in comparison with monolithic sample is possible. However, an application of polymeric addition increases the composite compliance.

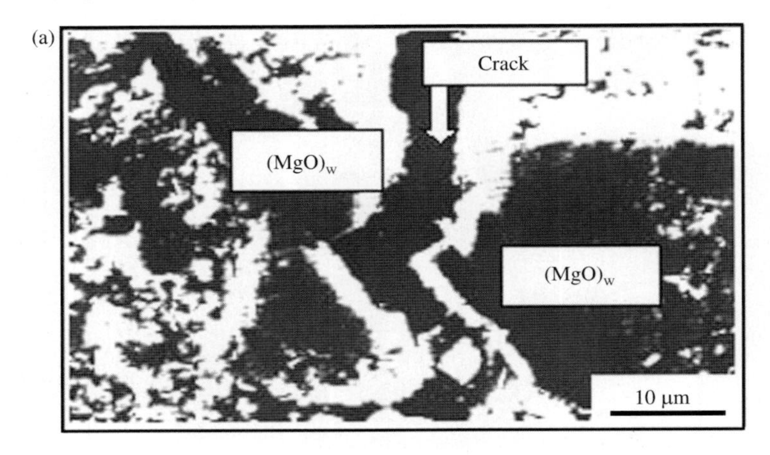

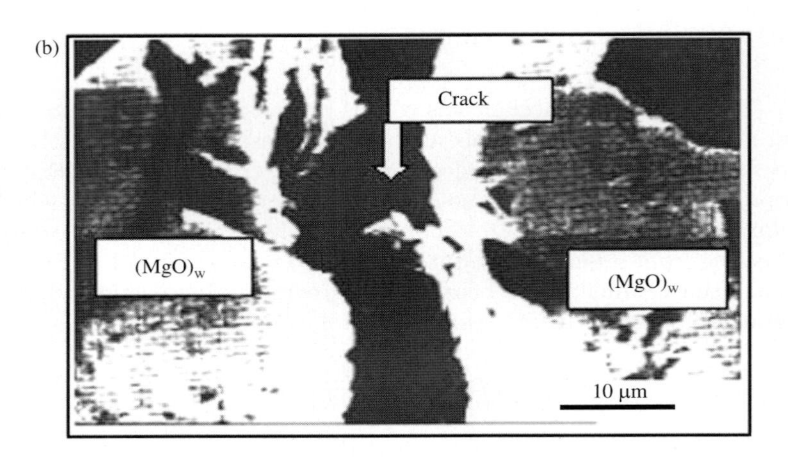

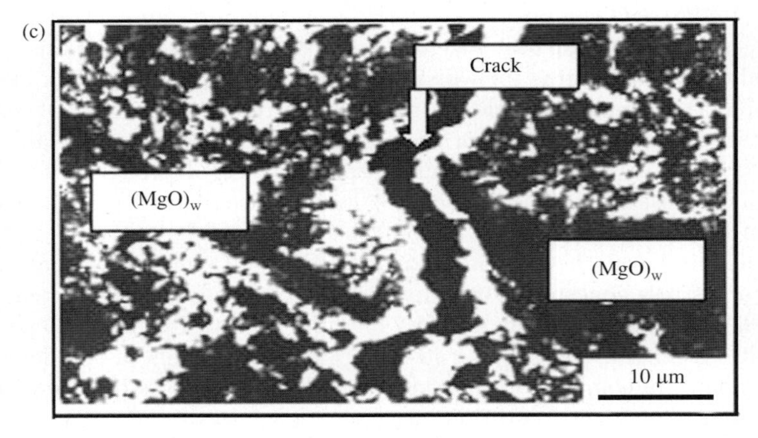

Fig. 3.37. SEM micrographs, showing multiple toughening mechanisms in fractured specimens (three-point notched bending) of $(MgO)_w$/BPSCCO composite, tested at room temperature: (**a**) whisker breakage and whisker/matrix interface de-bonding, (**b**) whisker pulled out and (**c**) crack deflection [1187]

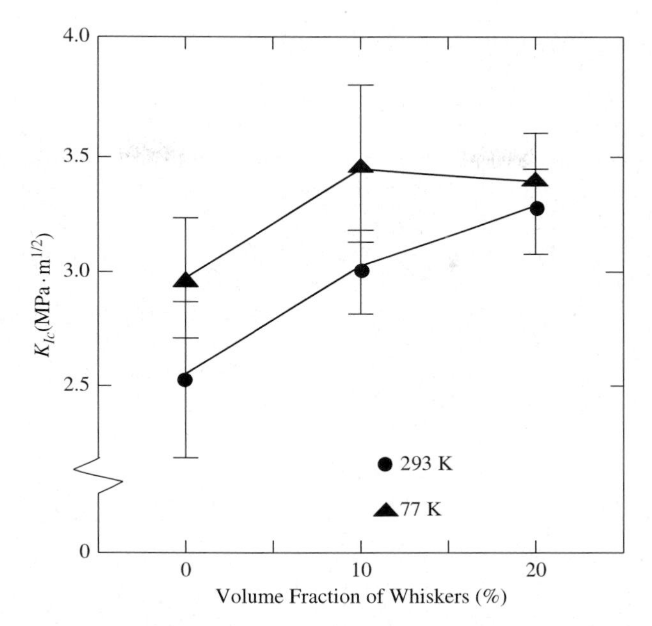

Fig. 3.38. Fracture toughness of monolithic BPSCCO and $(MgO)_w$/BPSCCO composite at room temperature and 77 K [1187]

3.4 Melt-Processed Y(*RE*)BCO

3.4.1 Microstructure Features

Now, the melt-processing techniques, presented in Sect. 2.5, are considered the main perspective for preparation of large-grain YBCO samples with high values of J_c. Therefore, the main attention will be directed to investigation of the melt-processed Y(*RE*)BCO samples. The normal phases, chemical and physical heterogeneities at boundaries of a Y-123 (or 123) phase cause weak links, leading to decreasing of J_c in YBCO. In the case of melt-processed multi-domain samples, a liquid is often observed at boundaries of these regions as a result of incomplete peritectic reaction (so-called defects caused by grain growth) and is one of the causes of small J_c [196]. Moreover, a non-symmetric HTSC structure, leading to anisotropy of transport current, is also proper for small critical current density. So, J_c in *ab*-plane is much more greater than the corresponding value along *c*-axis [376]. The value of J_c in the melt-processed YBCO strongly depends on the microstructure of superconducting grains [535, 744, 1134]. Compared to the conventionally sintered samples, many more defects are included in the interior of melt-processed YBCO grains. The defect density is a function of fabrication conditions including peritectic heat treatment and subsequent oxygenation treatment [535] and also a special concentration of the *ab*-planar defects (microcracks or lamellae boundaries) around trapped Y-211 (or 211) particles [196, 1134]. In this case, these defects do not considerably influence the transport current parallel to the *c*-axis [1134]. The Y-211/Y-123 interface [744] and the structure defects [1134] are proposed as flux-pinning centers of a Y-123 phase. Other advantages, associated with the Y-211 additives, are the following: the improved fracture toughness [298], homogeneous microstructure [536] and suppression of the microcrack formation [744].

The microstructure of melt-processed YBCO, as a rule, consists of Y-123 lamellae (pseudo-grains), oriented almost parallel to the *ab*-planes and possessing common *c*-axes (Fig. 3.39). However, the formation process of this structure is yet to be understood. For example, it is argued in [639] that lamellae formation occurs during peritectic solidification and is associated with the irregular growth morphology of a Y-123 growth front in the presence of a build-up of Y-211 particles. On the other hand, it is suggested that the lamellae structure forms during the tetragonal to orthorhombic phase transition (TOPT), which occurs at 400–500°C, in subsequent sample oxidization rather than at high temperature during peritectic solidification [1134]. The latter point of view can be supported by the absence of lamellae in a tetragonal, unoxygenated Y-123 grain, at simultaneous existence of a number of lamellae in the oxygenated Y-123 grain [530]. Therefore, it may be suggested that the driving force for lamellae formation is the stress, induced during the TOPT, and the subsequent instability of the Y-123 phase.

Besides twins and microcracks, another defect frequently observed in the melt-textured microstructure is a CuO stacking fault. The stacking fault presents itself as an additional CuO layer between two BaO layers [1134]. These additional CuO layers exist together with partial dislocations both in a Y-123 matrix and around trapped Y-211 particles (Fig. 3.40). In [10], it was proposed that CuO stacking faults formed during high-temperature peritectic

Fig. 3.39. Scanning electron micrograph, showing $BaCuO_2$ lamellae along the *ab*-plane of a Y-123 domain [530]

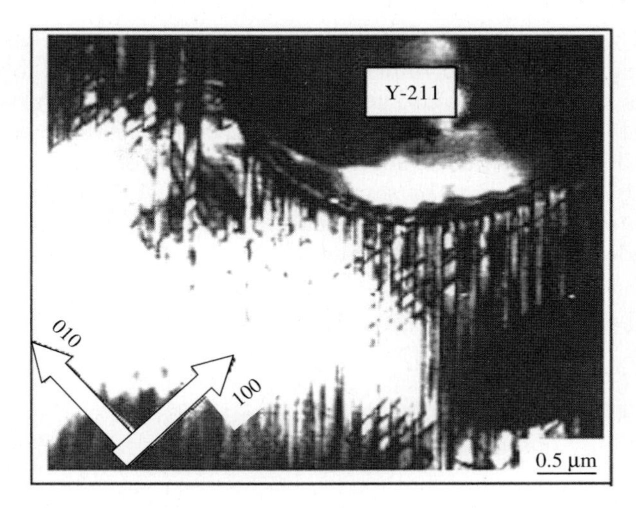

Fig. 3.40. Stacking faults and partial dislocations, developed around a trapped Y-211 particle [530]

reaction. However, in [541], it was found that the stacking faults were formed during oxygen annealing in the same manner as that of the lamellae, because Y-123 was not stable at oxygen annealing and decomposed into other stable secondary phases, following the reaction [1145]:

$$4YBa_2Cu_3O_{7-x}+(1/2-3/2\delta+2x)O_2 \rightarrow 2Y_2BaCuO_5+3Ba_2Cu_3O_{6-\delta}+CuO \,. \tag{3.2}$$

The Y-123 decomposition seems not to be associated with the formation of a Y-211 phase, because yttrium-enriched phase regions are observed around the trapped Y-211 particles [1134]. Similar to the $BaCuO_2$ lamellae, a driving force for the formation of CuO stacking faults may be stress induced by the TOPT and also stress caused by the difference in thermal expansion between Y-211 and Y-123 phases. The stresses caused by the TOPT in the melt-processed YBCO are many more than corresponding values of sintering ceramic due to comparatively greater Y-123 grain size. It also can provoke decomposition of Y-123 phase around trapped Y-211 particles. In the case of application of the melting processing (PMP), stacking faults form in the sample because of incomplete diffusion during peritectic reaction between Y-211 particles and the liquid ($BaCuO_2$ and CuO). The amount of defects could be decreased using an additional annealing of the sample under elevated temperature. The considerable decreasing of J_c, observed after this annealing, proves that stacking faults are proper magnetic flux-pinning centers [1194].

There are two main sources of void formation in Y(*RE*)BCO. (i) Inert gas in technological atmosphere remains in the form of voids in sample after its fabrication. In this case, the void number increases with the partial pressure of inert gas, which cannot diffuse from the sample [824]. (ii) A great amount of oxygen realises as a result of decomposition of Y-123 phase, forming voids [909]. In this case, there are two ways for oxygen evaporation: usual diffusion and O_2 bubbles, moving through liquid. In low-temperature region, oxygen diffusion is dominated, but at high temperatures, bubble formation occurs, which may be accompanied by sample deformation [197, 909]. A melting in pure oxygen renders effective the decreasing of Sm-123 porosity; however, in this case, a decreasing of superconducting properties is possible [465].

3.4.2 Growth Processes in Seeded Sample

The growth of Y-123 crystal occurs in the form of a parallelepiped with (100), (010) and (001) habit planes [475]. Basically, three modes of crystal growth from the seed are observed (Fig. 3.41), namely:

(1) Epitaxial growth of single Y-123 crystal from the seed with its *c*-axis parallel to the sample axis (the case of using Nd-123 plate-shaped seed and direct seeding at the beginning of solidification) [197]. In this case, there are five growth sectors (GSs), namely: four *a*-GSs with habits perpendicular to the [100], [$\bar{1}$00], [010] and [0$\bar{1}$0] directions and a *c*-GS with habit perpendicular to the [00$\bar{1}$] direction. The top angle of the *c*-GS depends on the ratio of the growth rates in the *c*- and *a*-directions.
(2) A cubic Nd-123 seed leads to five domain samples with *c*-GSc dominating in each grain [1066]. In this case, four narrow *a*-GSs develop along both sides of the 90° boundaries in each grain. These 90° high-angle grain boundaries between grains are strongly coupled and *ab*-microcracks do not disturb current in *ab*-planes.
(3) Seed from MgO single crystal does not dissolve in the partially melted sample and it can be placed on a cooled green sample before melt processing. Due to the misfit of the MgO and 123 lattices, it often happens that samples with the *ab*-plane parallel to the sample axis are produced. They consist of two *c*-GSc and three *a*-GSs [429].

In Y-123 crystals, grown from partially melted bulks, subgrains (SGs) can form, which present crystal regions, divided by low-angle grain boundaries [197]. In general, these boundaries are not to act as weak links; however, some low-angle grain boundaries can act as weak links in high-field regions [200]. Therefore, for high-field applications, it is necessary to control the low-angle grain boundaries, that is, it is important to study the formation mechanism of the subgrains. Five different types of subgrain can be classified, according to growth directions active in their formation [202, 203, 204], namely: (i) *a*-subgrains (*a*-SGs) have subgrain boundaries (SGBs) parallel to the *a*-axis; (ii) *a*-*a* subgrains (*a*-*a*-SGs) have SGBs tilted from the *a*-direction (Fig. 3.42a); (iii) *a*-*c* subgrains (*a*-*c*-SGs) have SGBs parallel to the *c*-axis and develop at the stepped planar *a* growth front; (iv) *c*-subgrains (*c*-SGs) have SGBs parallel to the *c*-axis; and (v) *c*-*a* subgrains (*c*-*a*-SGs) have SGBs parallel to the *a*-axis and develop at the stepped planar *c* growth front. The first

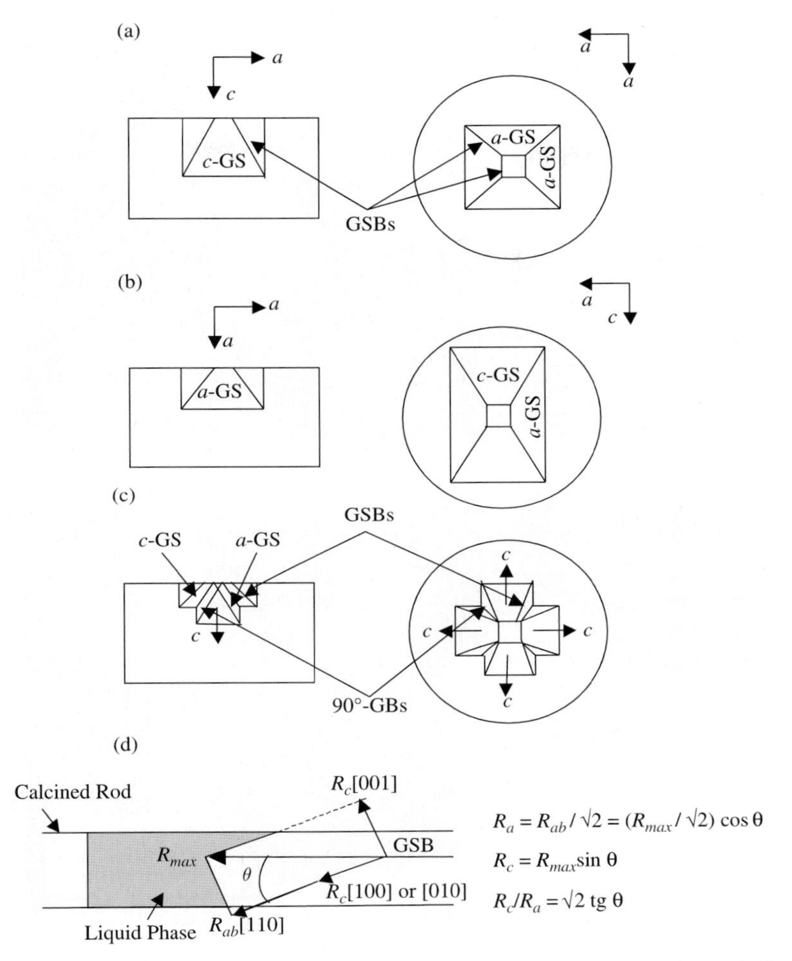

Fig. 3.41. Schematic illustrations of GSs, divided by GS boundaries (GSBs), formed in the melt-grown 123 bulk: TSMG-processed single-grain bulks with the *c*-axis parallel (**a**) or perpendicular (**b**) to the sample axes and in the five-grain bulk (**c**). Schematic illustration of longitudinal interface for typical sample, prepared by directional solidification (**d**) [197]

three SGs form in the *a*-GS and the last two SGs form in the *c*-GS. It is interesting that the *c*-SGs are always larger than the *a*-SGs in cross-section. The arrangement of subgrains in the single-grain melt-grown bulk is demonstrated in Fig. 3.43.

In particular, subgrains are manifested by the changes in the direction of microcracks parallel to the *ab*-plane (Fig. 3.44) [202]. In this case, the SGBs are clean and not cracked.. The growth-related subgrains, observed in melt-grown Y-123, is supposed to be formed by the dislocation arrangement into

(a)

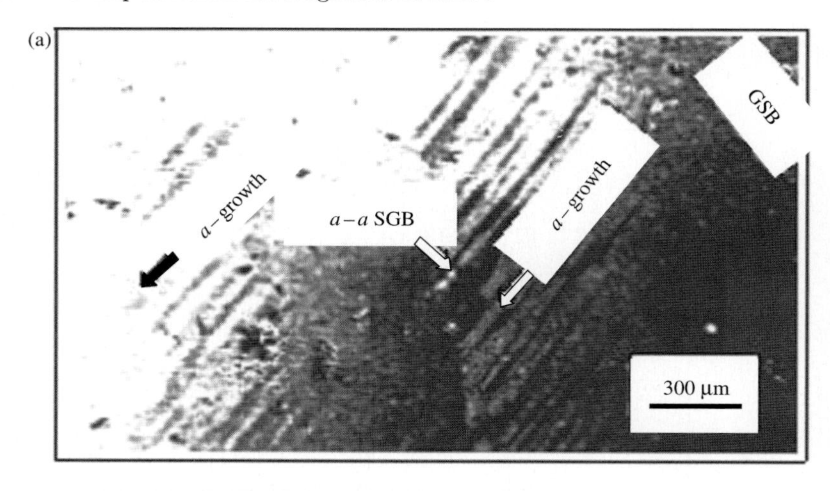

(b)

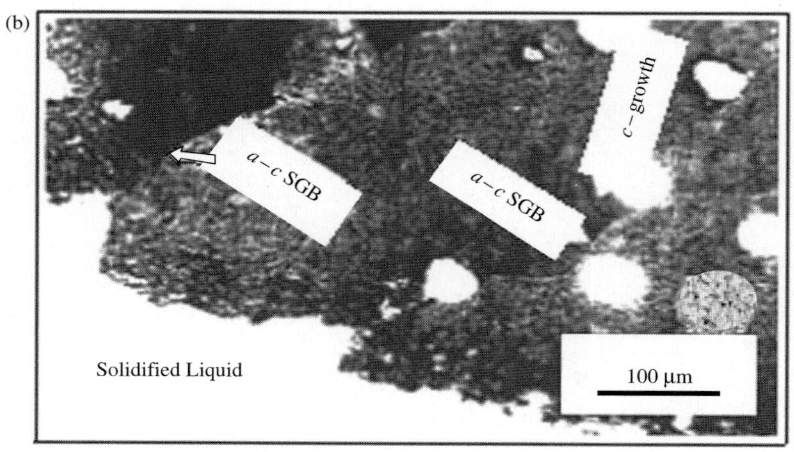

Fig. 3.42. Polarized optical micrographs of TSMG-processed Y-123/Y-211: (**a**) a higher macrostep at the growth front along *a*-axis, forming *a*-*a*-SGB; (**b**) the cross-section both parallel to the *c*-axis and the growth direction, *c*-*a*-SGBs, which are connected with the inner corners of the steps, form on the stepped growth front [197]

dislocation walls during the crystal growth [478]. The subgrain formation is assisted by the edge dislocations with Burgers vectors parallel to the growth front. In the case of melt-grown 123 grains, the dislocation formation due to Y-211 particle seems to be the most plausible mechanism because Y-123/Y-211 interface is incoherent [203]. The density of dislocations should be proportional to the density of 211 particles, which is supported by the fact that the subgrain size increases with an increase in the average spacing between 211 particles [199]. Moreover, it is observed that the subgrain size is reduced with an increased cooling rate [1022]. The fact that subgrains are not observed for

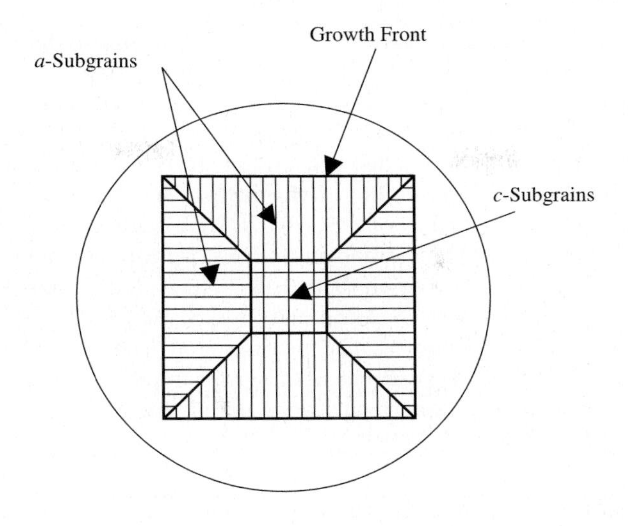

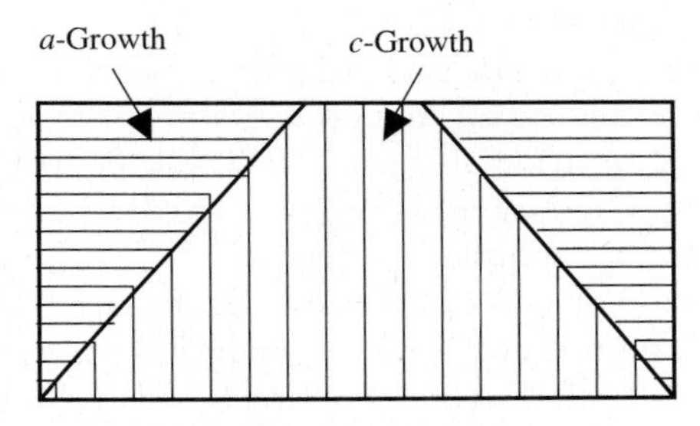

Fig. 3.43. The subgrain arrangement in the single-grain melt-grown bulk [197]

Fig. 3.44. The tilting of *ab*-microcracks at SGBs in a section parallel to the (100) plane [197]

single crystals supports the idea that the incorporation of the 211 particles into a 123 matrix is responsible for the subgrain formation.

The consequence of the proposed model of subgrain formation is the existence of an incubation period free of subgrains (e.g., subgrain-free region) at the beginning of growth, when dislocation density is not high enough. Such regions are observed at the seed and at the GSBs (Fig. 3.45). The regions free from subgrains can also form along SGBs. In such regions, the grain growth occurs perpendicularly to the main growth front, forming a step at the growth front (Fig. 3.46 [197]). The region B in Fig. 3.46 is the barrier for GSBs to continue from part A to part C of the [010] a-GS. Hence, the incubation period free of subgrains appears at the beginning of the C part of the [010] a-GS. At the same time, the presence of subgrain-free regions along the GSBs and SGBs shows that cellular growth does not define subgrain formation in melt-grown Y(*RE*)BCO bulks [197].

3.4.3 Behavior of 211(422) Disperse Phase

A uniformity of Y-211 phase distribution within a Y-123 matrix is necessary to increase superconducting and mechanical properties of a Y-123 phase. For this, two problems should be solved, namely (i) the spherical Y-211-free regions [532] and (ii) 211 segregation along specific crystallographic orientations of a Y-123 phase[4] [1106]. The formation of spherical Y-211-free regions is attributed to the formation of spherical pores due to gas (oxygen) evolution during incongruent melting of a Y-123 phase [532]. Figure 3.47 demonstrates spherical pores in liquid, pores filled by liquid and Y-211-free regions. When a Y-123 powder compact is heated above a peritectic temperature, oxygen gas is released, forming spherical pores in the liquid. If oxygen gas diffuses out of the pores, they will disappear by the liquid filling process. Compared to the liquid motion to pores, the mobility of solid 211 particles is relatively slow. It makes a non-uniform Y-211 distribution around the liquid pockets (these regions are shown by circles in Fig. 3.47b). During peritectic reaction, liquid pockets turn into spherical Y-211-free regions. At the same time, due to the lower Y-211 density around the liquid pockets, the reaction to form a Y-123 phase is not easy. Therefore, unreacted liquid phase ($BaCuO_2$ and CuO) is often observed in the center of Y-211-free regions [530]. The size and amount of pores are dependent on the heating rate to a peritectic temperature [642]: larger pores are developed at higher heating rate. Prolonged holding at the partial melting state can eliminate spherical pores by providing enough time for diffusion of oxygen gas out of the sample [1008]. However, the prolongation of this process leads to significant coarsening of Y-211 particles and decreasing of J_c [642].

[4] Heterogeneity of 211 particle distribution can also be classified in the following forms [197]: (i) an increasing of the 211 phase density along c-axis direction, (ii) a development of particle heterogeneity at 123 grain and subgrain boundaries and (iii) oscillations of 211 concentration perpendicular to the c-axis.

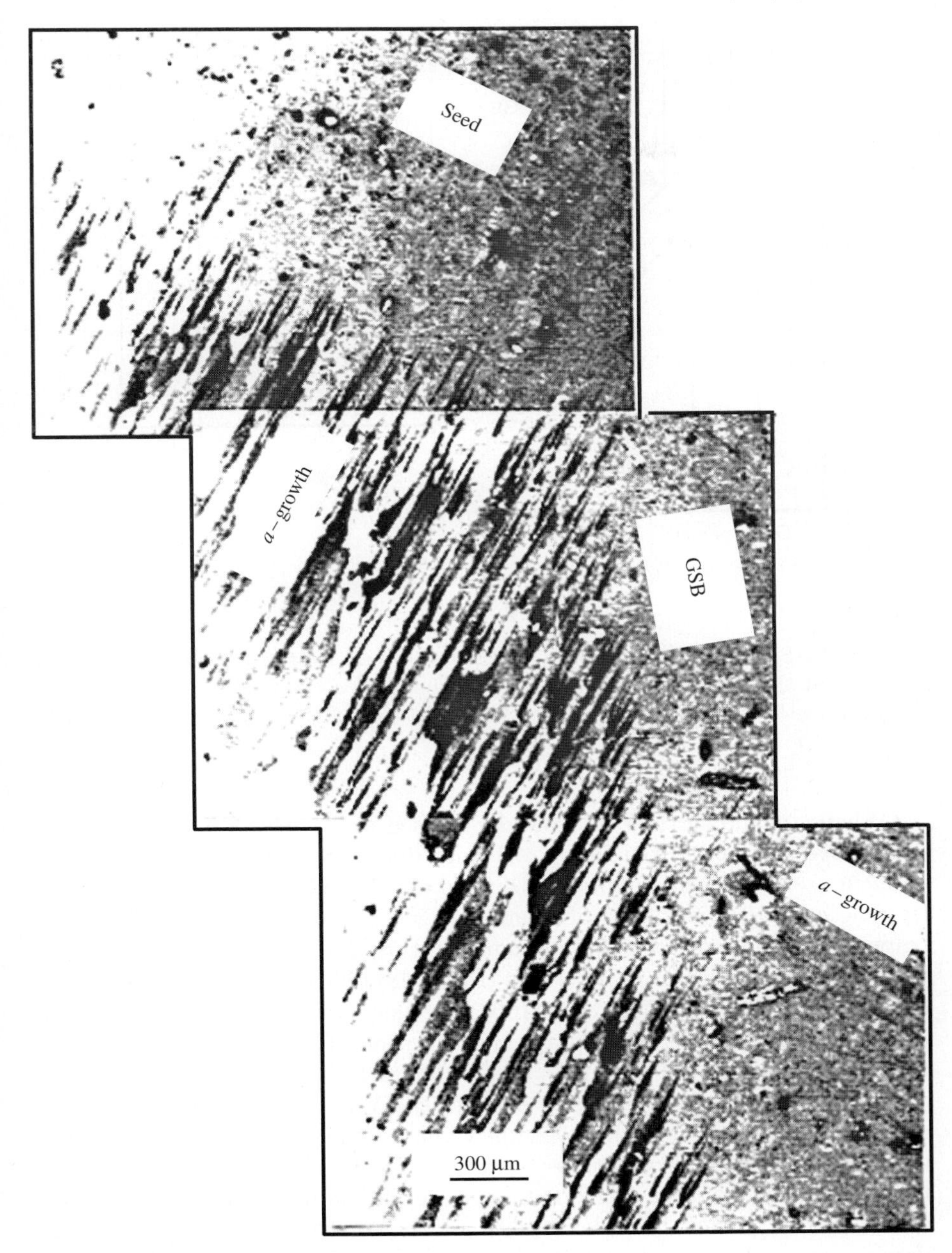

Fig. 3.45. Polarized optical micrograph of TSMG-processed Y-123/Y-211 [197]

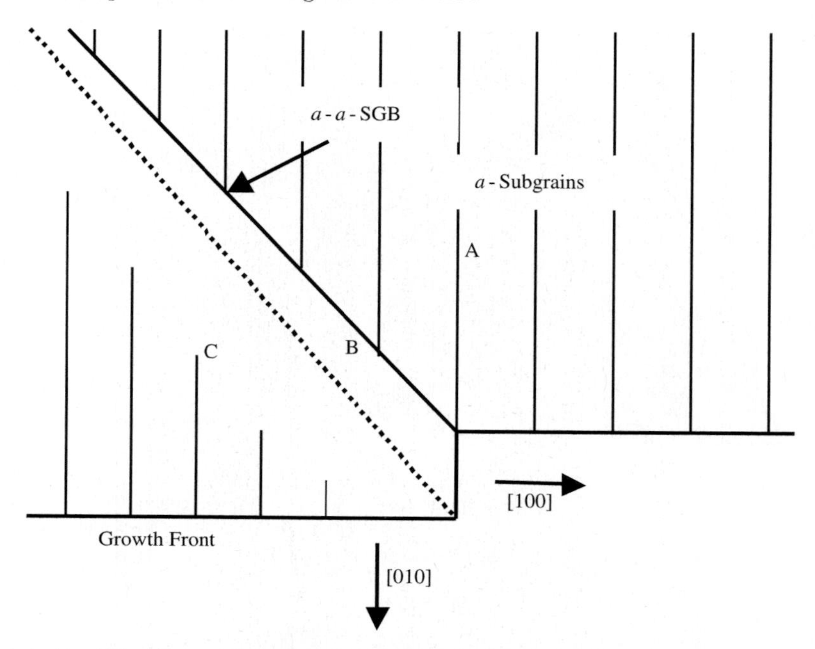

Fig. 3.46. Schematic view of the *a*-*a*-SGB, formed by the growth in the [100] direction at the step on the growth front of the [010] *a*-GS. The layer B, grown by [100] *a*-growth, is a barrier for dislocation walls (*a*-SGBs) to travel from part A to part C of the [010] *a*-GS

When stoichiometric Y-123 powder is used as a starting material for the melt processing, Y-211 particles are formed due to incongruent melting. Most of them are consumed completely to form a Y-123 phase during the peritectic reaction. However, some of them often remain unreacted in the form of trapped particles within Y-123 domains. In this case, the trapped Y-211 particles form X-like linear Y-211 tracks (Fig. 3.48) along the diagonal directions of Y-123 domains [532, 534, 707, 1106]. The shape of the Y-211 track pattern is dependent on the crystallographic orientation of the polished surfaces of the Y-123 domains, but generally, the Y-211 tracks meet the corners of Y-123 domains. The boundary planes to produce the Y-211 tracks are {110} planes of a Y-123 matrix [530]. Other secondary phases ($BaCeO_3$, $BaSnO_3$, etc.) that are formed by impurity additions do not make such a pattern. When the particle density is low, most of them are segregated in liquid at the Y-123 domain boundary [532]. When their density is high, they are trapped within the Y-123 domains normal to the growth fronts in the form of agglomerates [534].

In the case of using Y-211 excess powder, instead of linear Y-211 tracks, a planar segregation mode is developed. Specific crystallographic parts of Y-123 domains are filled with Y-211 particles, but other parts are free from Y-211 particles. The planar segregation of Y-211 particles appears as a symmetrical

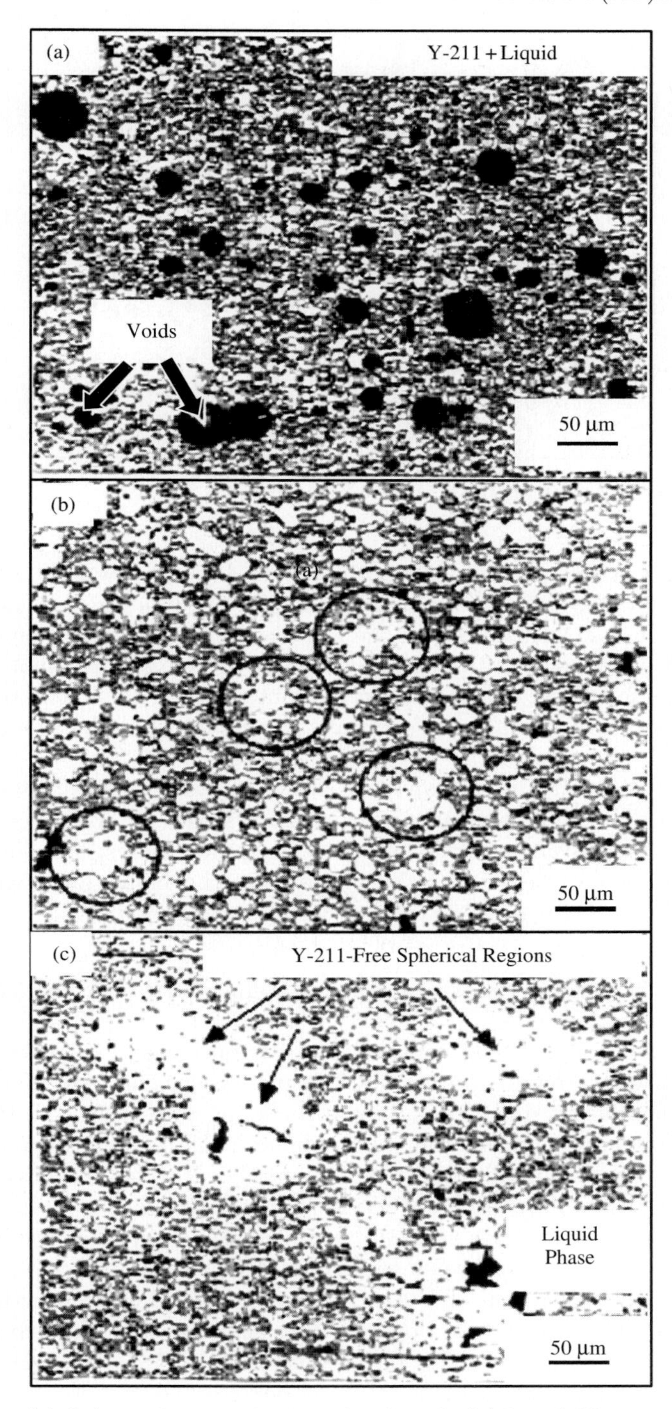

Fig. 3.47. (**a**) Spherical pores, developed in liquid, (**b**) liquid filling into pores and (**c**) spherical Y-211-free regions in a Y-123 domain [530]

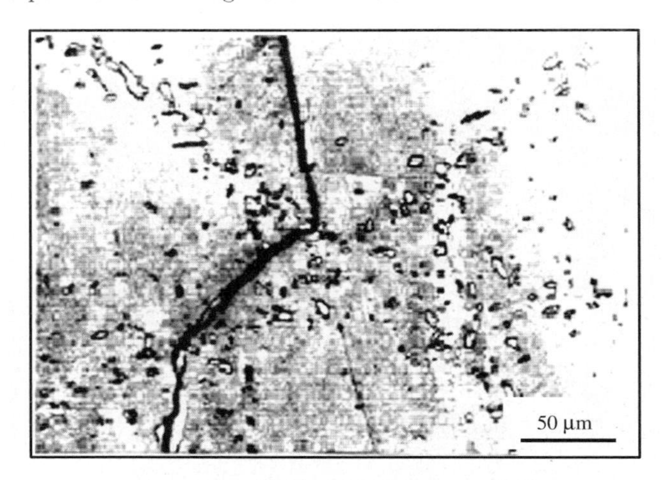

Fig. 3.48. X-like Y-211 pattern [530]

butterfly-like pattern presented by both (100) and (010) growth interfaces on a polished surface [534]. The formation of planar Y-211-free regions is dependent on the size of Y-211 particles and the growth rate of Y-123 domains. Large-sized Y-211 particles are trapped randomly; at the same time, small-sized Y-211 particles form a segregation pattern, containing Y-211-free regions. The Y-211 segregation pattern is preferentially developed when the cooling rate is low [530].

The Y-211 density in the liquid of the Y-211 excess system is relatively higher than that of the stoichiometric Y-123 system. The presence of many particles at the advancing solid interface decreases the critical velocity for particle trapping by increasing the viscosity of the liquid (η) given in the form [1015]:

$$\eta^* = \eta(1 + 2.5 f_\mathrm{p}) , \tag{3.3}$$

where η^* is the effective viscosity and f_p is a volume fraction of particles ahead of the growing interface.

Even at a constant Y-211 content, the distribution of Y-211 particles can be changed with growth rate of Y-123 fronts. As can be seen in Fig. 3.49, Y-211 particles are trapped randomly at a cooling rate of 20°C/h through a peritectic temperature, while they make planar Y-211 segregation patterns at a rate of 5°C/h. This implies that the formation of Y-211 segregation pattern is a function of the growth rate of Y-123 interfaces [530].

The spatial distributions of Y-211 particles have been studied in the samples obtained by using *RE*-seeds with application of the undercooling technique. A macrosegregation of Y-211 particles occurs during growth process from seeded liquid. It depends on the interface direction and growth rate, and also introduced undercooling of the sample [251, 252]. The Y-211 particle density into Y-123 matrix is higher at lower degree of the undercooling sample

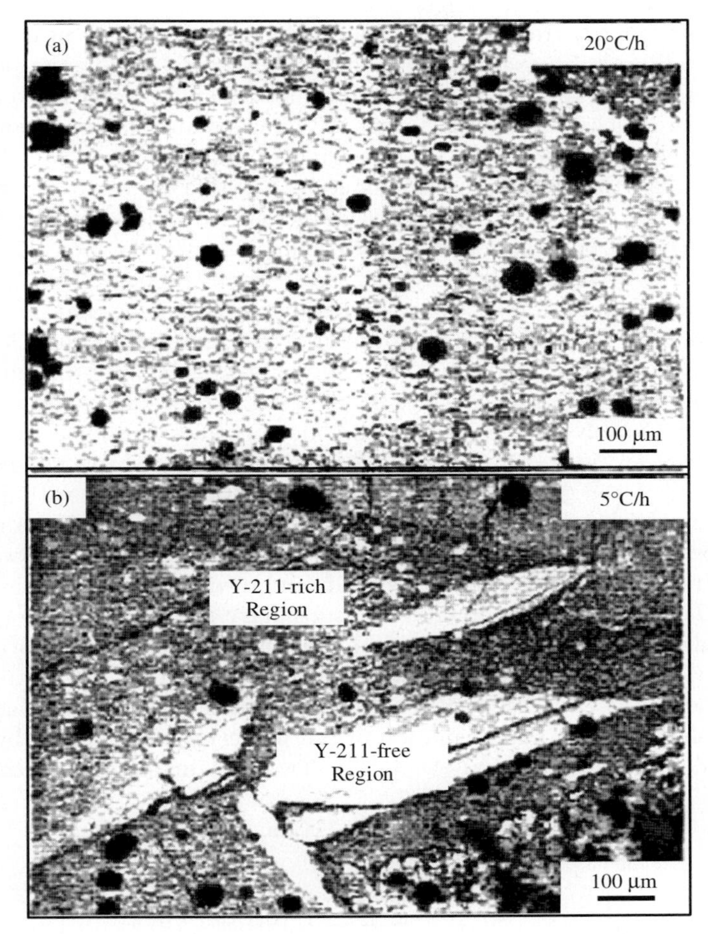

Fig. 3.49. Y-211 distribution within Y-123 domains of samples melt-processed at cooling rates of (**a**) 20°C/h and (**b**) 5°C/h [530]

below peritectic temperature ($\Delta T = T_\mathrm{p} - T_\mathrm{g}$) [251], showing that the particle trapping is caused by the growth rate of Y-123 interface. However, the single relationship between the growth rate and undercooling (ΔT) is not sufficient to explain the macrosegregation. The interaction of the advancing solid–liquid interfaces and the particles in the melt must be taken into account. By considering the particle radius, the growth rate and the number of particles per unit volume in the vicinity of the growth front, two critical growth ranges may be stated [188]. When $15°\mathrm{C} \le \Delta T \le 25°\mathrm{C}$ both growth rate and number of particles per unit volume exhibit low values. Therefore, only the largest particles can be trapped by Y-123/Y-211 interface. When $\Delta T \ge 25°\mathrm{C}$, the undercooling is increased, leading to an increase of the growth rate that causes the smallest particle trapping also. The Y-211 particle density near seed can be smaller than at the sample edge. It may be explained by the processes of their pushing, trapping, ripening and coalescence, occurring in liquid, and also by microcracks, observed between 123 lamellae and spherical pores near seed, which cause lower density of this sample part.

When using the hot-seeding technique by Nd-123 crystals at lowered oxygen gas pressure, the Nd-422 particle sizes increase considerably with distance from seed (Fig. 3.50) that correlate with a time of crystallization [123]. However, in this case, an increase of density of the small trapped Nd-422 particles is not observed, that is, processes of small particle pushing and large particle trapping are realized at the definite growth conditions of NdBCO grain. Generally, the growth rate of Nd-123 grain is greater in the *c*-axis direction, than in the *ab*-plane at any values of ΔT (Fig. 3.51). The growth rate increases considerably with growth of ΔT and becomes non-linear in Nd-123 samples even at $\Delta T = 5°C$, causing the processes of unstable solidification [123]. The observed gradual decreasing of growth rate together with duration of the sample fabrication time at lowered content of oxygen gas can be explained by increased volume fraction of Nd-422 particles into liquid, which, in this case restrains diffusive flux directed from growing front to melt.

As ΔT increases, many undesirable subsidiary nuclei begin to form. The growth mode of Y-123 crystal can be classified into four groups of morphology (Fig. 3.52), namely: planar facet mode ($\Delta T \leq 30°C$), cellular growth mode ($30°C < \Delta T \leq 40°C$), cellular growth with undesirable nuclei ($40°C < \Delta T \leq 45°C$) and formation of the polycrystalline sample by nuclei with random orientation ($\Delta T > 45°C$). Two-step undercooling technique,

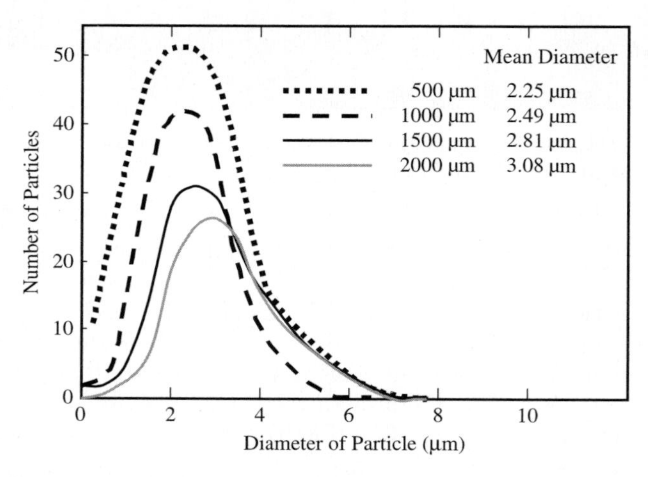

Fig. 3.50. Size distribution for Nd-422 particles into Nd-123 matrix as function of distance from NdBCO seed. The sample is grown at decreased pressure of O_2, 1030°C, during 50 h [123]

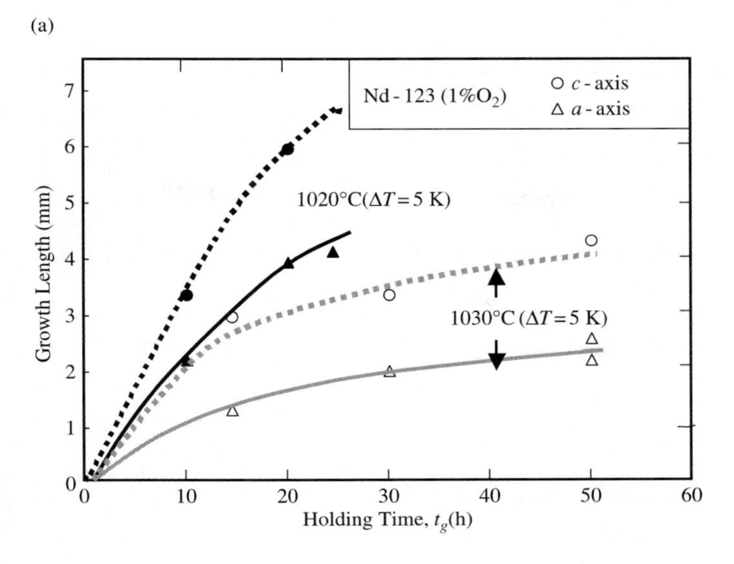

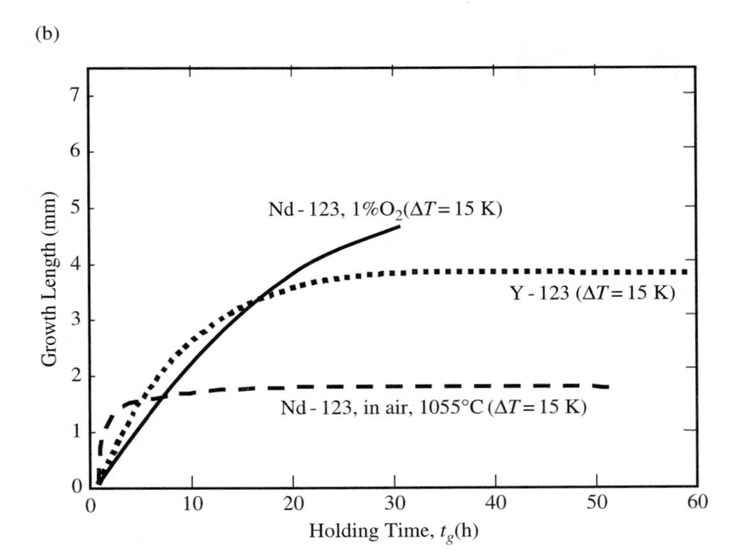

Fig. 3.51. Growth rate of NdBCO grain: (**a**) as function of undercooling ΔT and (**b**) at different processing conditions [123]

in which the Y-123 nucleus is stabilized at the first undercooling step, where the planar growth mode is maintained, and then crystal growth is accelerated at a second undercooling step, permits to reduce considerably the required processing time for single crystal fabrication compared to the conventional technique without degradation of the sample magnetic properties and texture

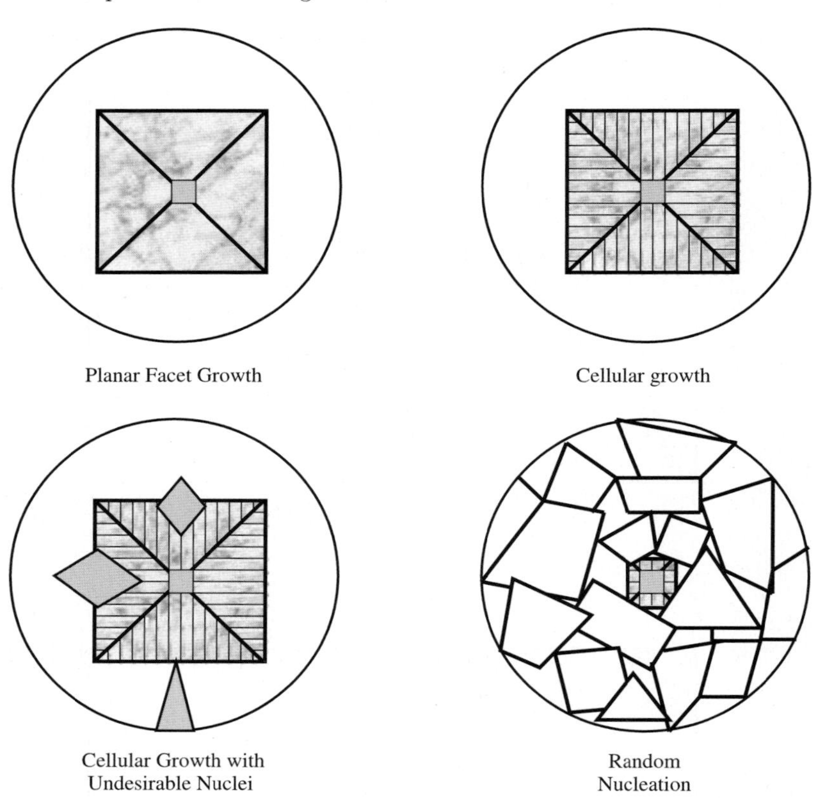

Fig. 3.52. Schematic illustrations of the growth morphology of Y-123 depending on the degree of undercooling, ΔT [479]

[479]. The dependencies of the Y-123 phase growth rate at {100} and {001} facets on undercooling degree are subjected to parabolic law. Apparently, it changes a direction of crystalline growth and combines orientation of the *ab*-plane with direction of sample solidification [251].

By using some seeds (MSMG technique), accelerating processing of melted Y(*RE*)BCO samples, a decreasing of levitation force due to residual products of melt at intergranular boundaries [932] is possible. In order to study this problem samples were prepared by using two seeds with different crystallography [531]. The obtained superconductors with junctions (100)/(100) and (110)/(110) demonstrated better properties (i.e., levitation force and trapped magnetic field) than the samples with junctions (100)/(001) and (110)/(001).

In melted YBCO samples with and without PtO_2 additives, 211 particles are disposed so that 211 longer axes are parallel or perpendicular to the *ab*-direction or the *c*-axis of crystalline lattice of 123 phase [124]. The orientation relations between 123 matrix and trapped 211 particles ($[001]_{123} \parallel [001]_{211}$

and $[110]_{123} \parallel [111]_{211}$) are stated in [37]. In contrast to these results, investigations with application scanning electron microscopy [1053] show absence of definite relationships between crystallographic orientations of 123 matrix and 211 inclusions. Due to own magnetic properties, the 211 particles align along an applied magnetic field, when sample is prepared by the melt-processing technique in the presence of the magnetic field [237]. However, it may be observed that Nd-422 particles [1168] and Sm-211 particles [1026] are arranged parallel to growth direction in one-direction solidified YBCO system even without magnetic field. The 211 particles have polygonal form with faceted surfaces, possessing interface anisotropy. The particle anisotropy increases with addition of PtO_2 [1015] and CeO_2 [535] that are used to control the 211 particle sizes. Additives form 211 needle-shaped particles in liquid. In order to diminish a free energy of interface, 211 particles should be reoriented in liquid to dispose with the least surface energy before the moment of their trapping by the growing 123-front. An alignment of 211 particles also depends on the growth rate of 123 interfaces, size and shape of 211 particles [530].

3.4.4 Effects of Doping Additives

To improve the structure-sensitive properties of HTSC, metal oxide additions are used. Among the impurity elements, PtO_2 [777] and CeO_2 [534] are known to be effective materials to increase J_c through the refinement of Y-211 particles and/or possible substitution in Y-123 lattices. The impurity elements not only reduce the Y-211 size effectively [777], but also change its morphology [1106]. During heating of Y-123 powder with doping additives to a peritectic temperature, the additives react with a Y-123 phase to form $BaCeO_3$ [537] and $Ba_4CuPt_2O_9$ [1180]:

$$YBa_2Cu_3O_{7-x} + CeO_2 \rightarrow BaCeO_3; \tag{3.4}$$

$$YBa_2Cu_3O_{7-x} + PtO_2 \rightarrow Ba_4CuPt_2O_9. \tag{3.5}$$

This causes Ba deficiency in the composition of the remainder and formation of the free CuO phase. The reaction of the latter with Y-123 phase leads to the pseudo-peritectic reaction, forming a Y-211 phase prior to peritectic melting [530]:

$$YBa_2Cu_3O_{7-x} + CuO \rightarrow Y_2BaCuO_5 + \text{liquid}\,. \tag{3.6}$$

These Y-211 particles, formed before melting, serve as seeds for the Y-211 phase during peritectic reaction, increasing the number of peritectic Y-211 particles [349].

One of the ways to reduce the size of Y-211 particles is to provide heterogeneous nucleation sites for peritectic Y-211 particles. The $Ba_4CuPt_2O_9$ compound, which is formed as result of PtO_2 addition in accordance with reaction (3.5), can act in this role due to the similar lattice parameters between

$Ba_4CuPt_2O_9$ and a Y-211 phase [777]. A similar heterogeneous nucleation mechanism acts in the case of CeO_2 additives. Y_2O_3 particles, forming as a result of the reaction between CeO_2 and a Y-211 phase:

$$Y_2BaCuO_5 + CeO_2 \rightarrow BaCeO_3 + Y_2O_3 + CuO\ , \tag{3.7}$$

can be similar nucleation sites for the peritectic Y-211 phase [1115].

Y-211 particles coarsen easily in liquid by the Ostwald ripening process [64]. However, it is difficult to estimate the degree of contribution of Y-211 coarsening to the Y-211 refinement because of the nucleation and growth of Y-211 particles, taking place simultaneously. The attempt to separate one process from the other was concluded in a melt infiltration into Y-211 compacts, where no Y-211 nucleation was involved [533]. The technique was applied to Y-211 powder compacts with and without CeO_2 and PtO_2 additions [533, 539]. The results show that the addition of either CeO_2 or PtO_2 significantly suppresses the Y-211 growth in liquid (Fig. 3.53).

Shape control of Y-211 particles is required to yield fine Y-211 particles because the dissolution behavior of Y-211 particles in liquid during the peritectic reaction depends on the Y-211 shape [538]. It is a function not only of the initial powder composition, but also of the type of precursor powder [718]. In general, the shape of peritectic Y-211 particles of a stoichiometric Y-123 system is a prism-like one, while the shape of a Y-211 excess system is an almost equiaxed granular with facets [530]. During a peritectic reaction the faceted Y-211 particles are dissolved in liquid and their shape is changed to an irregular form with round surfaces. Sometimes, a longer Y-211 particle is divided into several parts, demonstrating that Y-211 particles with a high aspect ratio are better in reducing size than block-like or equiaxed Y-211 particles. Introducing CeO_2 and PtO_2 additions changes the Y-211 shape in liquid

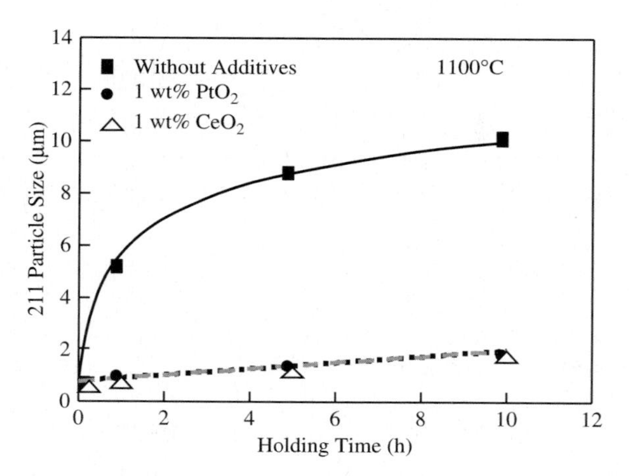

Fig. 3.53. Variation of Y-211 particle size against holding time at 1100° C in samples with and without PtO_2/CeO_2 addition [530]

to more anisotropic, that is, the needle-like one (Fig. 3.54). These transformations can be a result of change by the additives of the Y-211/liquid interfacial energy.

The melt-processed YBCO samples with $BaSnO_3$ addition could be grown with higher rate than in the case of addition absence. This implies that SnO_2 doping improves effectively yttrium diffusive factor [186]. However, in this case, ripening of 211 particles is often observed. CeO_2/SnO_2 additives (at greater amount of CeO_2) permit to increase growth rate of 123 oxide and to reach submicron size of secondary phase particles, demonstrating square shape and distributed uniformity in the matrix [664]. Partial substitution of yttrium by calcium in large-grain YBCO introduces additional hole-carriers

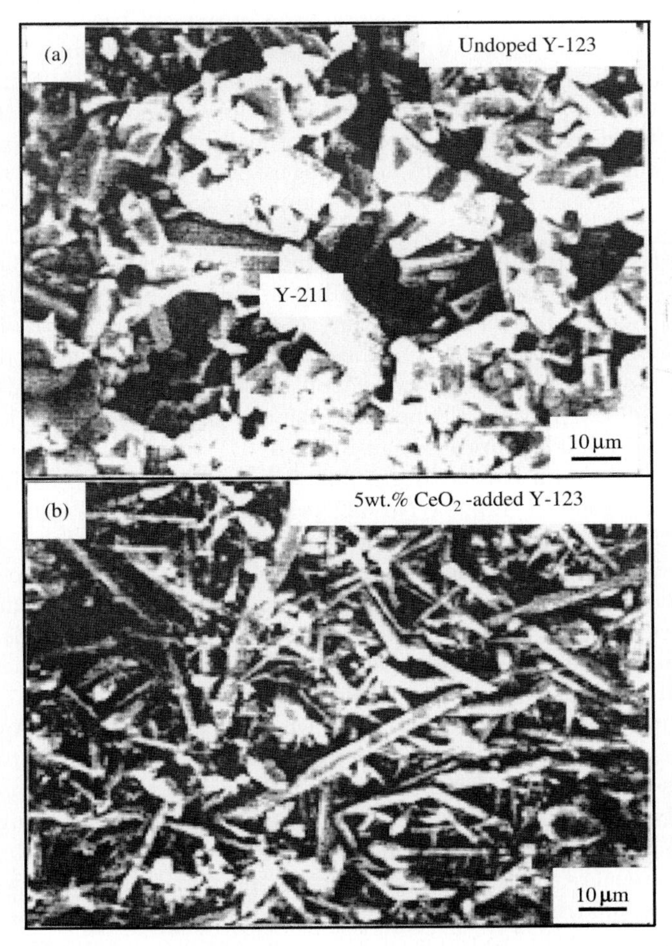

Fig. 3.54. Y-211 particles in liquid of (**a**) undoped Y-123 sample and (**b**) 5 wt% CeO_2-added Y-123 sample [530]

in CuO_2 planes, ensuring a regime of excessive doping [382, 383]. These additives permit to improve the structure and to reach sub-micron size of 211 particles without using Pt [391]. Then, it has been shown on the test data for monocrystals [555] that the substitution Y/Ca occurred without strain. However, the substitution Ba/Ca or Cu/Ca, which competes with the substitution Y/Ca, becomes predominant at the compression or tension strain above $\sim 6\%$. This is found due to Y and Ca ionic radii being approximately the same. At the same time, Ba and Cu ionic radii on 20% are greater and smaller than Ca ionic radius, respectively. Because the regions are highly deformed near intergranular boundaries in polycrystallite samples, Ca substitutes mainly Ba and Cu, but not Y, decreasing the local strain. In this case, the formation energy of oxygen vacancies in CuO_2 planes and Cu–O links decreases at the deformation. Therefore, there is segregation of oxygen vacancies in the undoped, by Ca of the $YBa_2Cu_3O_{7-\delta}$ samples, near intergranular boundaries, and superconducting properties of the YBCO diminish because of the decreasing of hole carriers. The Ca additives decrease strain and make the formation of oxygen vacancies disadvantageous energetically. The chemical composition of intergranular boundaries and adjoining regions remains to be relative to stoichiometrical one, and J_c increases compared to the value in the samples without Ca.

Higher temperature incongruent melting of Nd-123 and Sm-123 compared to Y-123 [729, 1179] and greater solubility of Nd and Sm in liquid compared to Y [1169] increase both a ripening Sm-211/Nd-422 and a growth of Sm-123/Nd-123 regions. Addition of PtO_2 does not significantly influence decreasing of particle size compared to the effect on Y-211 particles [730]. A precursor with small Nd-422 particles defines dispersion of small Nd-422 particles in the final sample [746]. The improved Sm-211/Nd-422 structure can be reached by combining CeO_2 additives with grinding of precursor powder [542]. In this case, an anisotropy of Sm-211/Nd-422 particles into melt due to change of energy of the *RE*-211(422)/liquid interface is increased. Doping of precursor by Au/Ag additives diminishes Nd-422 particles down to sub-micron level that is approximately three times smaller than that in the absence of additives and decreases ripening of Nd-422 particles in peritectic region. In this case the doped grains consist of spherical Nd-422 inclusions compared to needle ones that are observed usually in monolithic sample [479].

An increase of critical current together with the amount of Ag has discrepant character [316,779]. Nevertheless, the presence of silver (up to 20 wt%) in YBCO samples, which demonstrate initially weak links, can improve them significantly [914]. This is manifested in the form of corresponding increasing of sample density, orientation and growth of superconducting grains. As result, an increasing of superconducting properties together with silver amount (Fig. 3.55) is possible. The increase of J_c for small amounts of Ag (≤ 20 wt%) may be due to the acceleration of densification and grain growth. At the same time, the decrease of J_c for large amounts of Ag (> 20 wt%) may be due to the excessive increasing of the normal (non-superconducting) phase; 20 wt% of Ag_2O can increase critical current density in the *ab*-plane (J_c^{ab}) by ~ 200% and critical current density along *c*-axis (J_c^c) by ~ 150% at 5 K. These results are correlated with a reduction of the microcrack density parallel to the *ab*-planes in YBCO/Ag_2O composite. The proper increase of superconducting properties is depicted in Figs. 3.56 and 3.57. Silver additives can additionally improve the flux pinning that also leads to increasing of superconducting properties [694].

3.4.5 Mechanical and Strength Properties

Microcracks form in the *ab*-plane of Nd-123 orthorhombic phase during cooling even at small numbers of Nd-422 particles due to the difference of thermal expansion between matrix and inclusions. These planar defects are absent both in the Y-123 tetragonal phase [203, 535] and in the Nd-123 tetragonal phase with small Nd-422 particles (Fig. 3.58) [201], that is, they are not defined by crystallization process [376]. In the samples with much larger Nd-422 particles, the *ab*-microcracks also develop in the tetragonal parts of the sample (Fig. 3.59). The crack spacing in the orthorhombic Nd-123 is much smaller than in the tetragonal part. This suggests that for both the orthorhombic and tetragonal 123 phases there is a critical size of 422 (211) particles, below which *ab*-microcracking associated with 422 (211) particles does not occur. The higher critical particle size for the tetragonal 123 phase is associated with the absence of additional shortening of the *c*-axis due to oxygen uptake or/and influenced by the higher fracture toughness of the tetragonal 123. Besides the *ab*-microcracks, the *c*-microcracks are also observed in melt-grown YBCO. These *c*-cracks stop at the boundary between orthorhombic

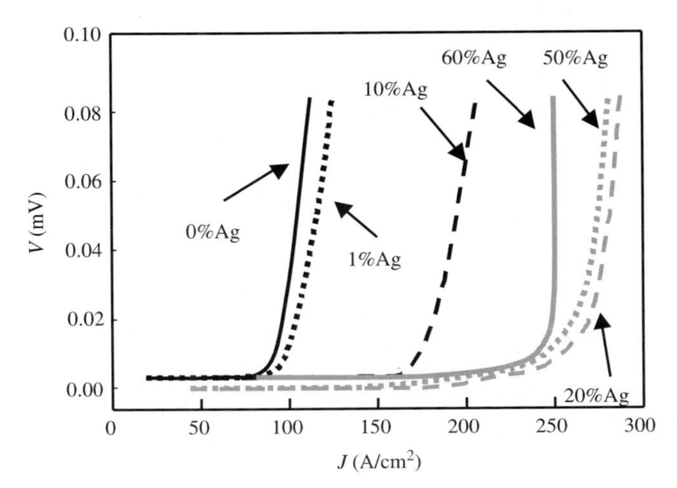

Fig. 3.55. *V*–*J* diagrams for different Ag-doping of YBCO [914]

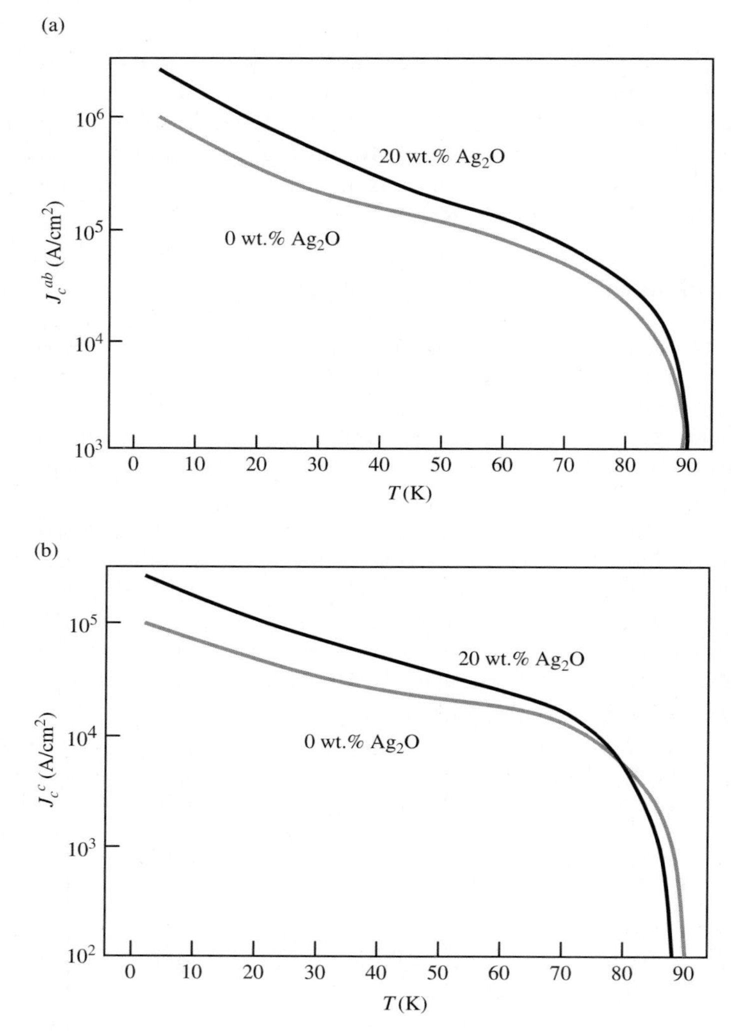

Fig. 3.56. Temperature dependencies of the self-field critical current density: (**a**) at $H \| c$ and (**b**) at $H \| ab$ for 0 wt% and 20 wt% Ag_2O [694]

and tetragonal phase (see Fig. 3.60), demonstrating formation of the c-cracking during tetragonal–orthorhombic phase transition [198].

Figure 3.61 shows interesting change of the five-domain sample shape. An originally circular-shaped sample can deform during solidification. The c-growth parts of the sample are bulged and the a-growth parts have shortened radii when the sample is observed from above.[5] Due to this, there is specific macrocracking of the sample.

---

[5] Higher 211 concentration at 90° boundaries confirms existence of the a-GSs along 90° boundaries [197].

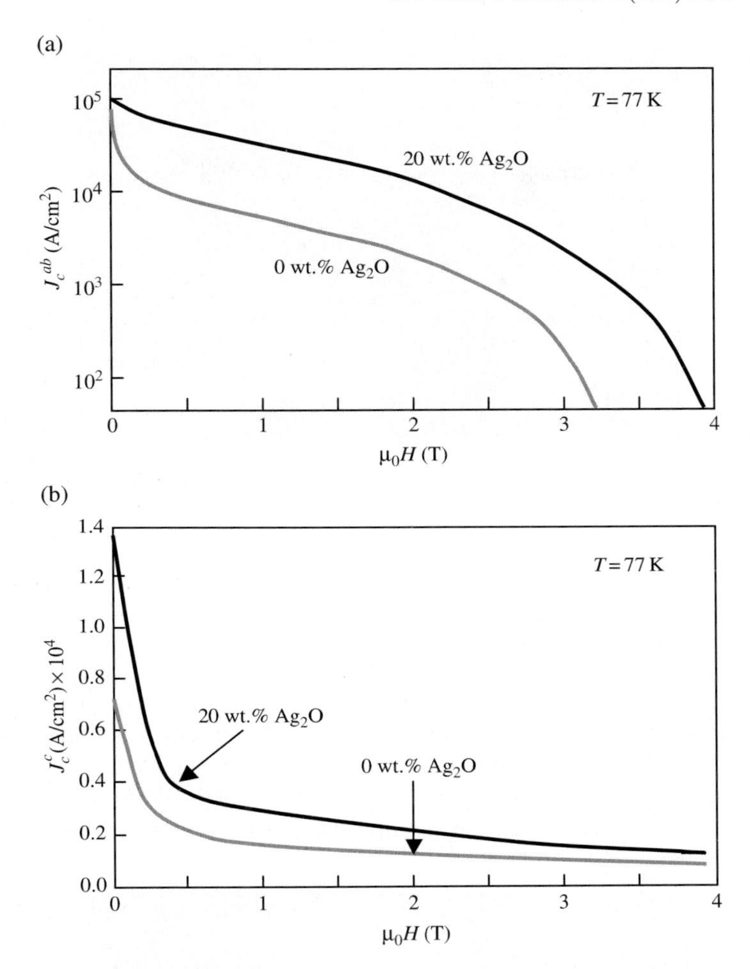

Fig. 3.57. Magnetic field dependencies of the critical current density (at T = 77 K): (**a**) at $H \,||\, c$ and (**b**) at $H \,||\, ab$ for 0 wt% and 20 wt% Ag_2O [694]

A deficiency of the 211 particles can induce tensile residual stresses in the *c* direction in the central *c*-GS and also tensile tangential stress in the *ab*-plane at the sample surface (in the *a*-GS). These stresses can cause *ab*-microcracking and detrimental radial macrocracking [197]. Other reasons for macrocracking are the great thermal gradients at sample cooling [910] and magneto-elasticity effects connected with magnetic flux pinning, causing residual stresses during the sample magnetization [879]. As a result, the cracks grow along the (100)/(010) plane and decrease a trapped magnetic field [709]. At increasing oxidization duration of the melt-textured YBCO samples, an aging of mechanical and superconducting properties [918] may be observed. It is accompanied by increasing dislocation density, formation of great chippings at

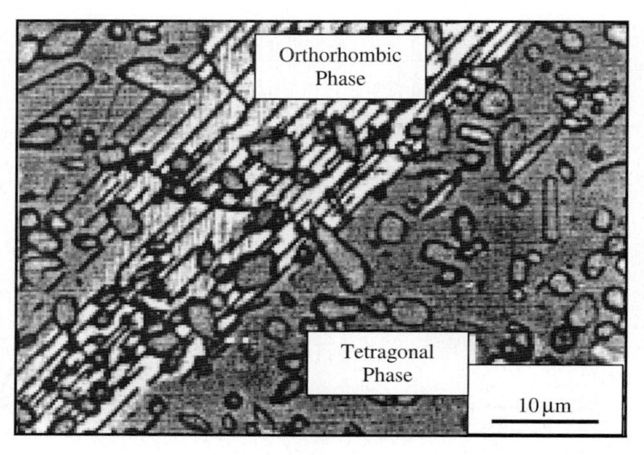

Fig. 3.58. Dense *ab*-microcracks in the orthorhombic Nd-123 phase and no *ab*-microcracking in the tetragonal Nd-123 phase (small 422 particles) [201]

rather low-angle grain boundaries and around 211 inclusions and also by intensive degradation of Y-123 matrix near microcracks and the 211 particles. The studies of textured YBCO ceramics at helium temperatures and high magnetic fields show that the latter can cause the dynamics of twinning dislocations, leading to cleavage stresses, that can fracture the sample along the planes parallel to the basic *ab*-plane [240]. 211 particles deflect crack, increasing its path. In this case, the crack branching at the 211 inclusions [614] and de-lamination of 211/123 interface [197] may be observed. As the 211 particle has a higher Young's modulus compared to the 123 matrix, they can also

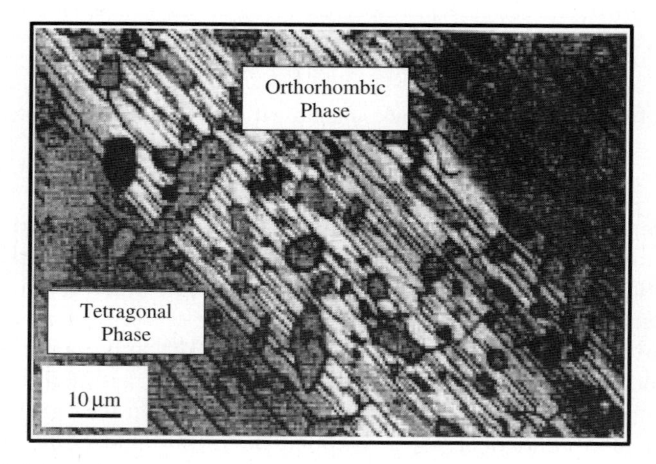

Fig. 3.59. Higher *ab*-microcrack density in orthorhombic than in tetragonal Nd-123 phase (large 422 particles) [201]

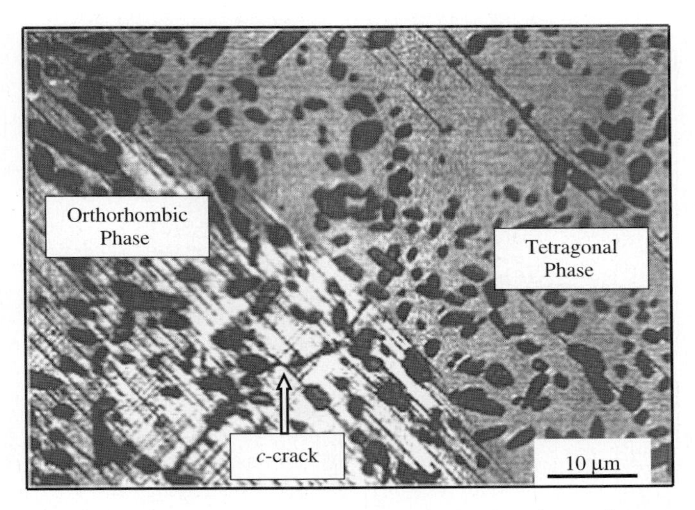

Fig. 3.60. *c*-crack stop at the boundary between orthorhombic and tetragonal phase [198]

act as grains-bridges, pinning down the crack opening and decreasing tensile stresses in the crack tip [338].

Silver dispersion (Ag_2O, Au/Ag) decreases considerably cracking and porosity both in the *ab*-plane and along *c*-axis, increasing J_c significantly [376, 694, 998]. Figure. 3.62 shows the magnetic field distributions in YBCO bulk (0 wt% Ag) at 9.0 and 8.5 T in the decreasing field process from 10 T. Magnetic

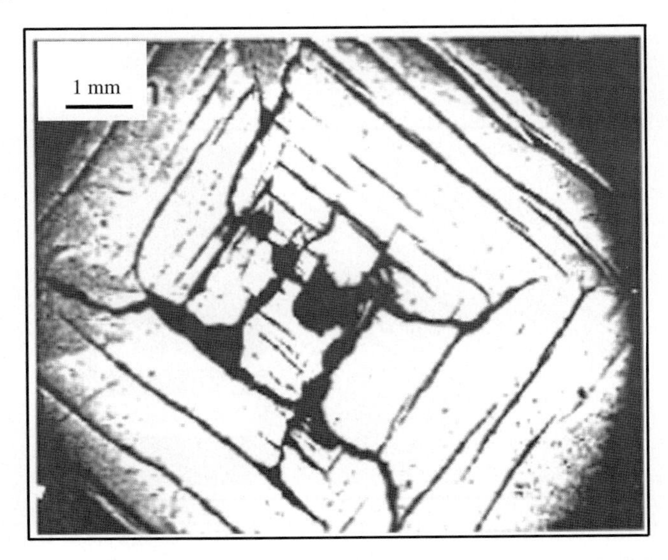

Fig. 3.61. Polarized-light photomicrograph, revealing the higher density of 211 particles (*gray diagonal* regions) in *a*-GSs along 90°-grain boundaries and also bulging of *c*-GSs and macrocracks parallel to *ab*-plane [197]

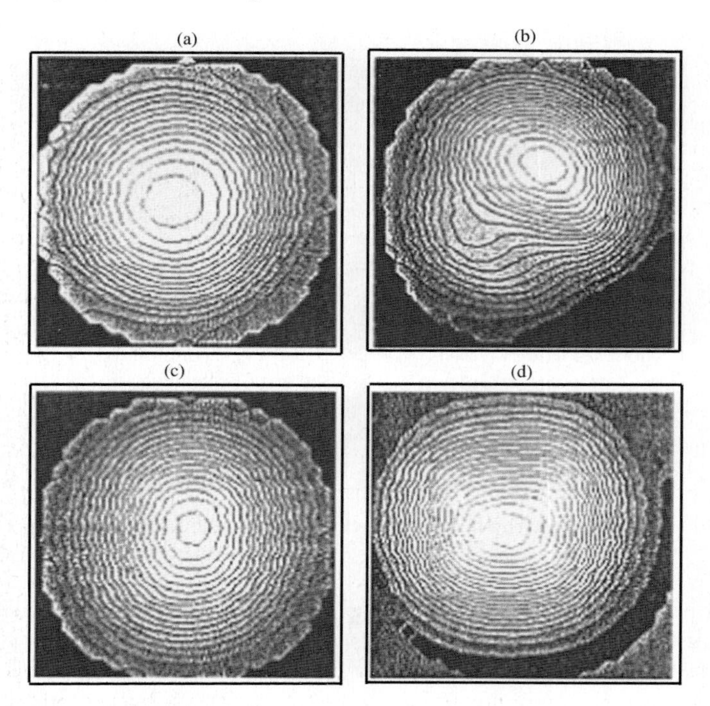

Fig. 3.62. Magnetic field distribution in YBCO samples during decreasing process from 10 T down to: (**a**) 9.0 T, 0 wt% Ag; (**b**) 8.5 T, 0 wt% Ag; (**c**) 0 T, 10 wt% Ag; (**d**) 0 T, 20 wt% Ag [709]

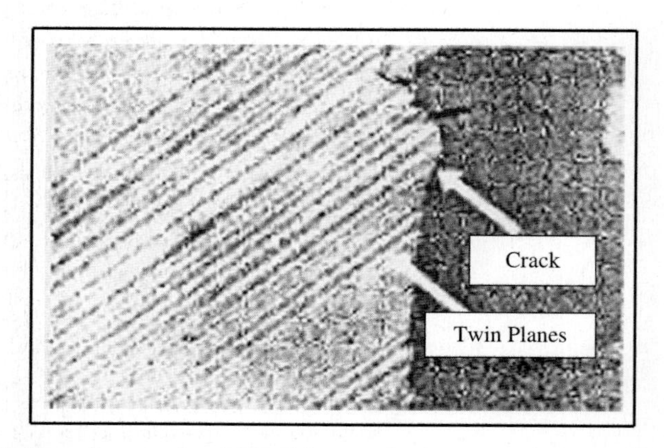

Fig. 3.63. Light micrograph of YBCO sample (without Ag) near the crack edge with higher magnification observed under polarized light. Twin planes can clearly be seen on the sample surface. Note that the crack crosses twin planes at 45°, showing that the cleavage surface is (001) or (010) plane [709]

field distributions show clearly the sample fracture during the decreasing field process from 9.0 to 8.5 T. At the same time, the additives of 10 wt% Ag and 20 wt% Ag prevent the sample fracture at the same decreasing field process even after total removal of external field [709]. The observed fracture of the sample into two pieces by macrocrack occurs along (001) or (010) planes, showing that those are the cleavage planes in YBCO. This is followed from the concept that twin planes are {110}, but the fracture plane crosses at 45° from twin planes (Fig. 3.63). Then, it may be concluded that the cleavage occurs at (001) or (010) planes. The Ag addition (Ag_2O) decreases considerably the *ab*-microcracking, introduced in TOPT by microstresses, caused by 211

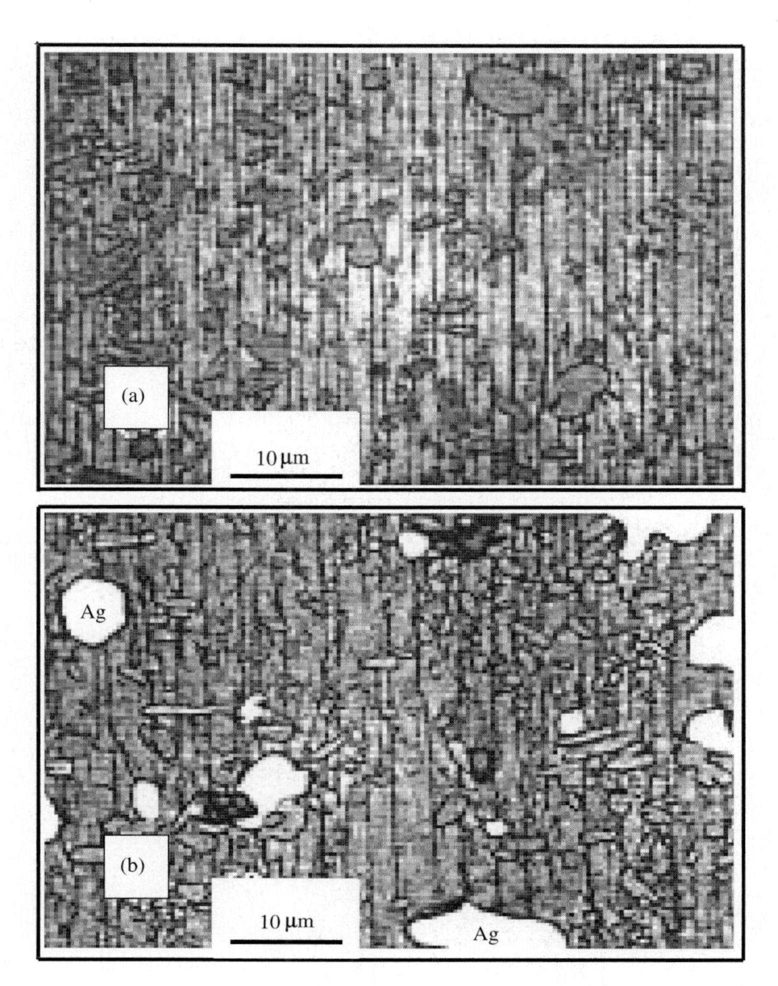

Fig. 3.64. Microcracks in YBCO/Ag_2O samples after etching: (**a**) 0 wt% Ag_2O and (**b**) 12 wt% Ag_2O [198]

Table 3.3. Crack spacing [198]

Sample	$\lambda_{ab} \pm \sigma$(μm)	$\lambda_c \pm \sigma$ (μm)	d_{211}(μm)	V_{211} (wt%)	d_{Ag}(μm)	V_{Ag} (wt%)
0 wt% Ag_2O	1.8 ± 0.1	34.2 ± 17.0	1.95	21.1	–	–
12 wt% Ag_2O	3.7 ± 0.4	133.5 ± 43.2	2.02	20.9	5.2	6.1

λ_{ab} is the *ab*-microcrack spacing; λ_c is the *c*-microcrack spacing; d_{211} is the 211 particle size; d_{Ag} is the Ag particle size; V_{211} is the 211 volume percent; V_{Ag} is the Ag volume percent; σ is the standard deviation

particles (Fig. 3.64). In this case, significant increase of spacing between both *ab*-cracks and *c*-cracks (Table 3.3) [198] is observed.

The crack, initiated due to Vickers micro-indentation, deflects when one reaches the gap between two silver inclusions (Fig. 3.65). After the crack passes through the gap, it continues in the same direction of the initial growth. Thus, a toughening is demonstrated, increasing intrinsic fracture toughness of YBCO sample [652]. Silver increases both maximum size and incline of *R*-curve (Fig. 3.66). This increase of fracture toughness is caused by reinforcing, developing into so-called, *wake zone*, where a plastic strain of Ag inclusions-bridges, pinning the crack surfaces, states most mechanism of toughening and fracture resistance in the absence of material transformations [1171].

An addition of tetragonal ZrO_2 particles can significantly reinforce monolithic YBCO samples due to the martensitic (tetragonal-monoclinic) phase transition [322]. In particular, when 10 mol.% ZrO_2 particles, coated with the Y-211 phase by using a sol–gel process, were added to Y-123, only modest improvement of fracture toughness (K_{Ic}) was observed. However, 20 mol.% ZrO_2 improved the K_{Ic} by nearly 50%. In this case, for the samples with the

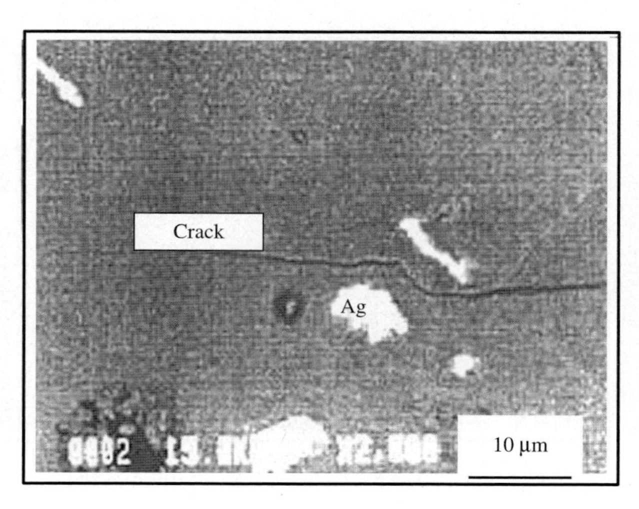

Fig. 3.65. Crack deflection between silver particles, dispersed in YBCO superconducting matrix [652]

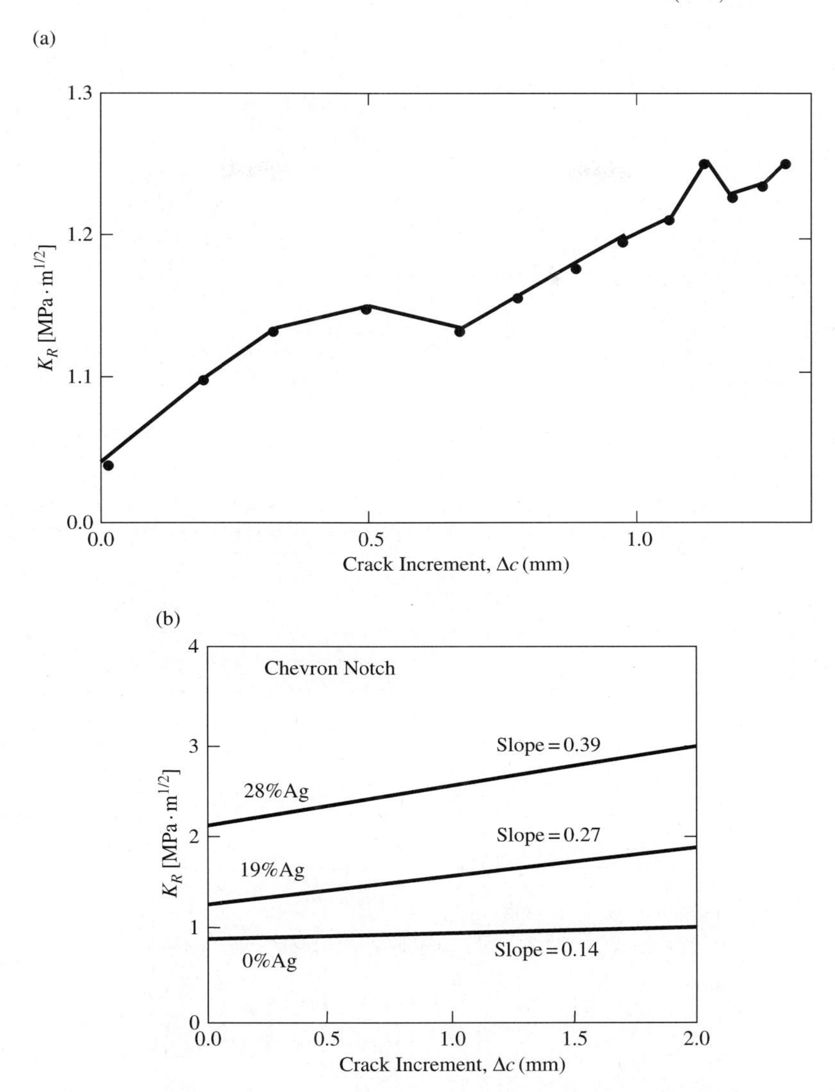

Fig. 3.66. *R*-curves obtained for different YBCO/Ag composites in samples with single edge notch: (**a**) 123 superconductor; (**b**) three composites with various amount of silver. Ag increasing improves both the level and incline of the fracture resistance curves, showing cumulative character of the toughening with the crack growth [1171]

same density, the 10 mol.% ZrO_2 additions also increased K_{Ic} by nearly 50%. At the same time, the average strength of the Y-123 and Y-123/Y-211/ZrO_2 remained nearly constant despite the differences in K_{Ic}. On the contrary, a critical concentration of ZrO_2 additions ($\sim$ 15 wt%) [791] was stated, at which the maximum superconductor strength and plasticity were reached.

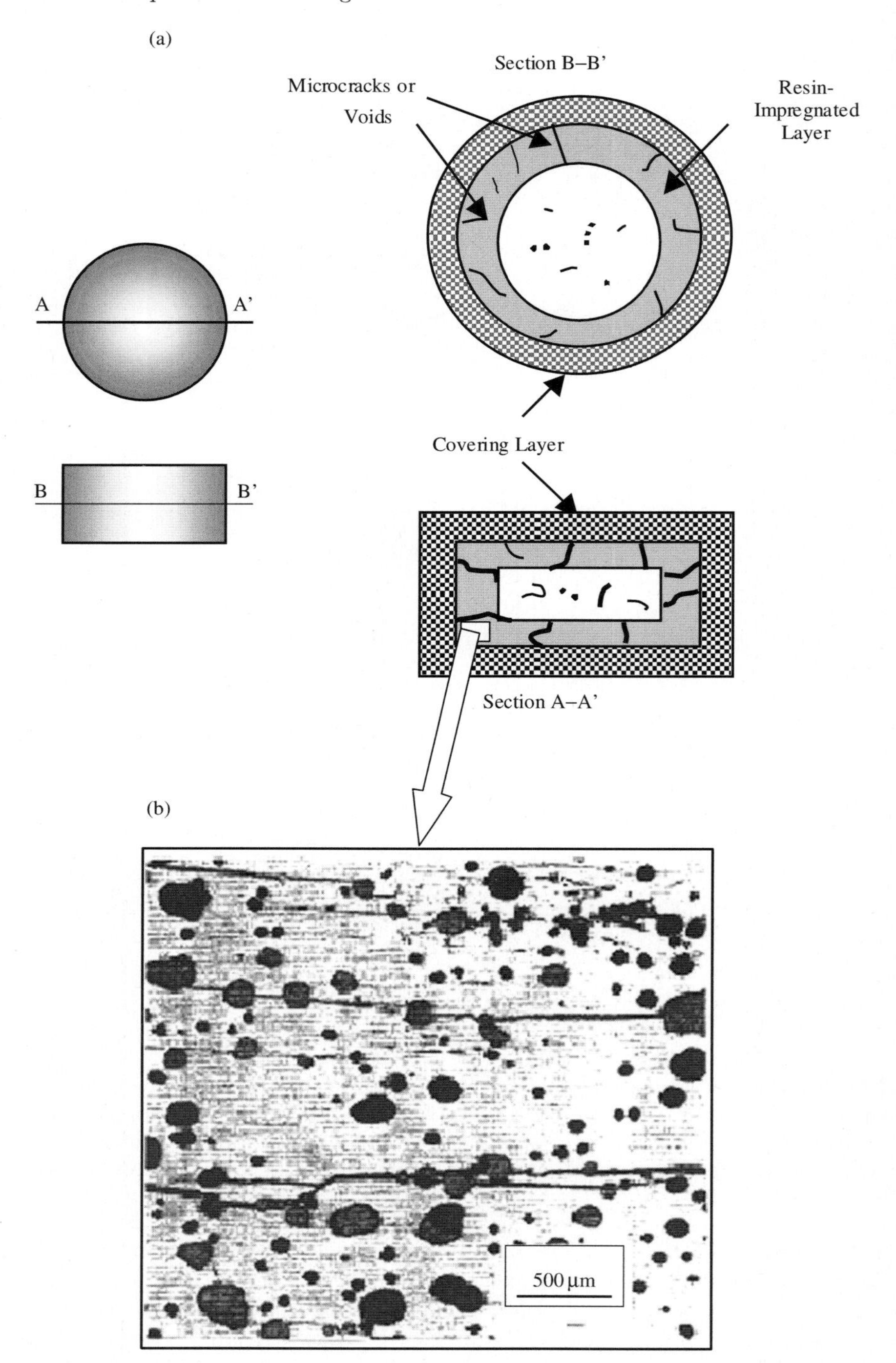

Fig. 3.67. (**a**) Structure of superconducting bulk reinforced by resin impregnation [1070]; (**b**) optical micrograph for cross-section of Sm-123 bulk after the resin impregnation [1073]

(a)

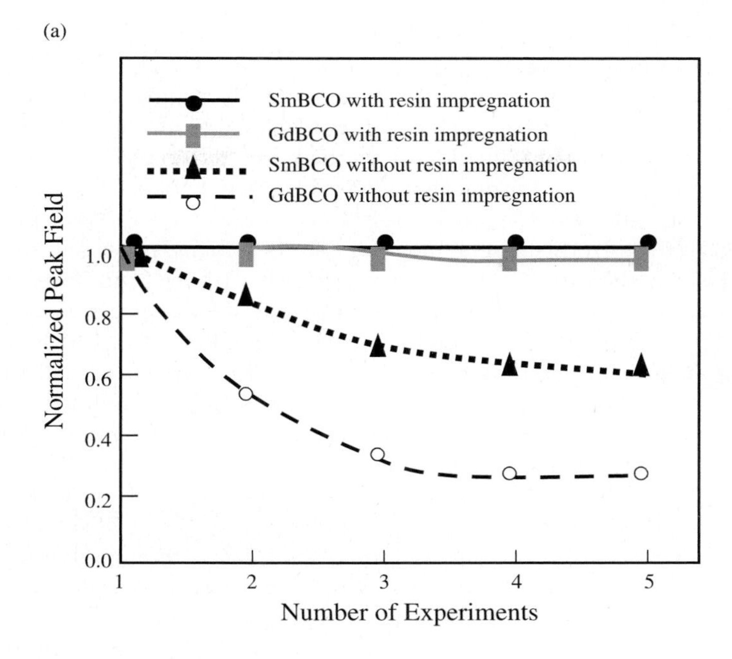

(b)

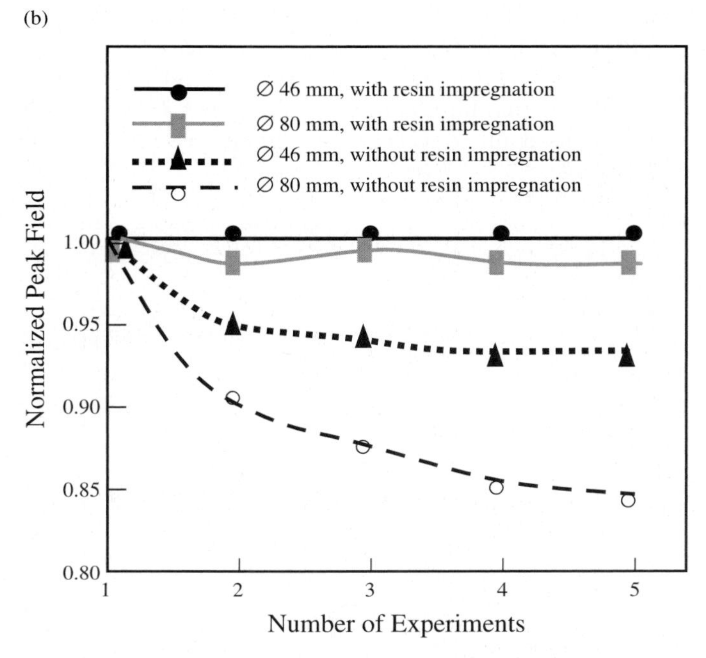

Fig. 3.68. Trapped field values vs the number of thermal cycles (77–293 K): (**a**) SmBCO and GdBCO with and without resin impregnation [1068]; (**b**) the same results for YBCO bulks with different diameter of pellets [1070]

In this case, the plastic deformation occurs in range of high temperatures in mechanism of intercrystallite sliding on nanophase with Zr-contents.

The reinforcement by surrounding the Y(*RE*)BCO bulk with a metal ring is also effective in increasing mechanical properties [727]. In this case, the bulks can be pre-strained in compression, since the thermal expansion coefficient of a metal ring is much smaller than that of the Y(*RE*)BCO bulks. The epoxy resin impregnation into *RE*BCO bulk effectively improves mechanical properties [647, 1069, 1070]. In this case, almost all defects (cracks, pores and scratches) can be filled by the epoxy resin within a distance of about 2 mm from the sample surface (for superconducting pellets of 3 cm in diameter and 2 cm in thickness) (Fig. 3.67) [1069]. These bulks do not show a decreasing of the trapped magnetic field even after intensive thermal cycling (cycles 77–293 K) [1070] and demonstrate an increasing of levitation force [1072]. Compared to usual samples, the characteristics of reinforced specimens are presented in Fig. 3.68. Moreover, a large current of 1000 A could be passed along the *RE*BCO bulk without the transition of the superconductor in normal state [1073]. The epoxy resin could be successfully reinforced by dispersing quartz fillers which also promote a decreasing of thermal expansion difference between polymer and superconductor [1070]. Finally, an additional high pressure at high temperature permits to obtain during short time non-porous practically *RE*BCO bulks with improved mechanical properties at simultaneous preservation or improvement of superconducting properties [868]. The mechanical properties of disc-shaped $YBa_2Cu_3O_7B$ magnets have been reinforced significantly due to an epoxy resin impregnation and introducing of carbon filaments [1071]. Moreover, an aperture has been drilled in the sample center, which has been filled by Bi–Pb–Sn–Cd alloy with aluminum wire. As a result, the static trapped field has been increased up to 17.24 T at magnet cooling with the absence of mechanical damage.

4

Carbon Problem

4.1 YBCO System

During all stages of HTSC preparation from calcining of precursor powder up to fabrication of final product, there is a problem of carbon segregation into sample deteriorating structure-sensitive properties. So, during calcining in preparation of YBCO, the $BaCO_3$, Y_2O_3 and CuO powders form 123 phase reacting according to the following [958]:

$$BaCO_3 \text{ (solid)} + CuO \text{ (solid)} \rightarrow CO_2 \text{ (gas)} + BaCuO_2 \text{ (solid)} \quad (4.1)$$

$$4BaCuO_2 \text{ (solid)} + Y_2O_3 \text{ (solid)} + 2CuO \text{ (solid)} \rightarrow 2YBa_2Cu_3O_x \text{ (solid)} \quad (4.2)$$

When $BaCO_3$ decomposes, according to reaction (4.1), CO_2 is released, and its localized concentration quickly reaches equilibrium value. The localized CO_2 pressure, depending on the temperature and other thermodynamic conditions, can cause other decomposition reactions, forming undesired phases, which in turn contribute to reduce the superconducting properties of the final product. It has been reported [825, 1098] that in the superconductors carbon remains in an amount of 0.1–1.0% or more, even if the powder is fired at high temperatures. In particular, an existence of residual carbon in oxide superconductors, synthesized by a sol–gel method [585, 687, 1095], is possible. However, the residual carbon content of the powder, synthesized by the conventional solid-state reaction method, can be about three times as much as that of the powder, prepared by the sol–gel method for a similar condition of heat (Fig. 4.1).

The powders, synthesized by the sol–gel method, demonstrate superconductivity, when they are heated at temperatures higher than 900°C and the total carbon content is less than 0.4%. Dependencies of the $T_{c,onset}$ property and the magnetic susceptibility of YBCO powders on total carbon contents are shown in Fig. 4.2. Obviously, diamagnetism volumes are increased and $T_{c,onset}$ is improved as the carbon contents are reduced.

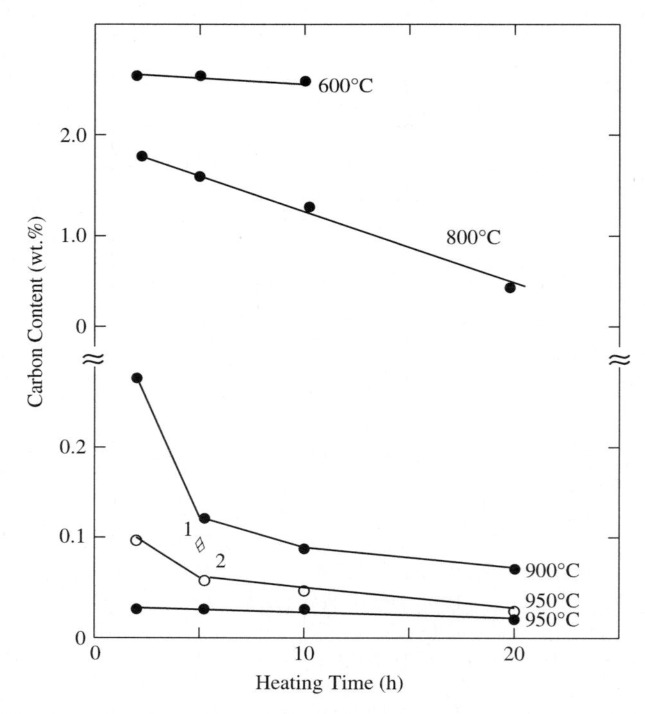

Fig. 4.1. Change of carbon content with heating time for various firing temperatures. (•) in O_2 flow, (o) in air flow, (◊1) and (◊2): samples synthesized by the conventional method [675]

The next hot isostatic pressing (HIP) can lead to the following improvement of the powder's superconducting properties. So, the powder, heated at 500°C with HIP, has a higher $T_{c,onset}$ and a larger volume of diamagnetism than the powder without HIP treatment (Fig. 4.3). In this case, the orthorhombic distortion increases when the HIP treatment is performed at 500°C, but the distortion decreases when the temperatures of the HIP treatment are raised to 800 and 900°C (Fig. 4.4).

It should be noted that the carbon content of the powder, prepared by the sol–gel method, decreases more effectively than that of the powder by the conventional solid-state reaction, in spite of the presence of many more carbonaceous materials in the starting materials used in the sol–gel method. The reason for this behavior may be explained by assuming that $BaCO_3$, introduced in the powder synthesized by the sol–gel method, would be decomposed rapidly on heating at higher temperatures than 900°C due to the very small size (some dozens of nanometers) of $BaCO_3$ particles. It is possible that such fine particles may be highly reactive.

When the HIP treatment is performed at 900°C, a degradation of superconducting properties is observed. It is assumed that the degradation is caused

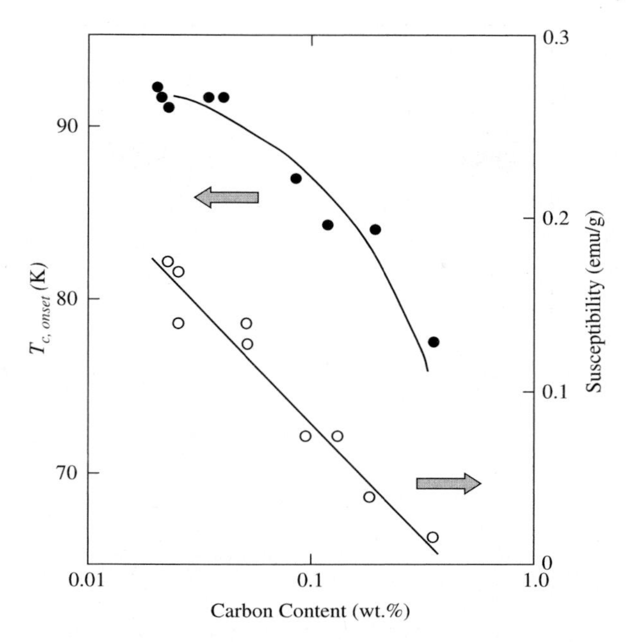

Fig. 4.2. Dependencies of $T_{c,onset}$ and magnetic susceptibility on carbon contents. The susceptibility is measured at 10 K [675]

by an increase of the tetragonal 123 phase over the orthorhombic 123 phase. This agrees with test observation that orthorhombic distortion of the lattice parameters decreases as HIP temperatures increase from 500 to 900°C.

YBCO superconductor reacts during sintering at different temperatures and conditions of gas mixtures. Its reaction with CO_2 may be carried out in two stages. At 815°C, it is given as [307]

$$2YBa_2Cu_3O_{7-x}(\text{solid}) + 4CO_2(\text{gas}) \rightarrow 4BaCO_3(\text{solid}) + Y_2Cu_2O_5(\text{solid}) + 4CuO(\text{solid}) + (0.5 - x)O_2(\text{gas}) \,. \tag{4.3}$$

The complete reaction of YBCO with CO_2 at 950°C is [307]

$$2YBa_2Cu_3O_{7-x}(\text{solid}) + 3CO_2(\text{gas}) \rightarrow 3BaCO_3(\text{solid}) + Y_2BaCuO_5(\text{solid}) + 5CuO(\text{solid}) + (0.5 - x)O_2(\text{gas}) \,. \tag{4.4}$$

For both cases, the decomposition of YBCO is only partially completed and starts at grain boundaries. In the sample annealed at 815°C, the reaction products consist of two phases: $BaCO_3$ and $Y_2Cu_2O_5$. The primary reaction products are all insulators, which essentially coat all grain boundaries. It can make the material lose its superconductivity even if the major phase is still

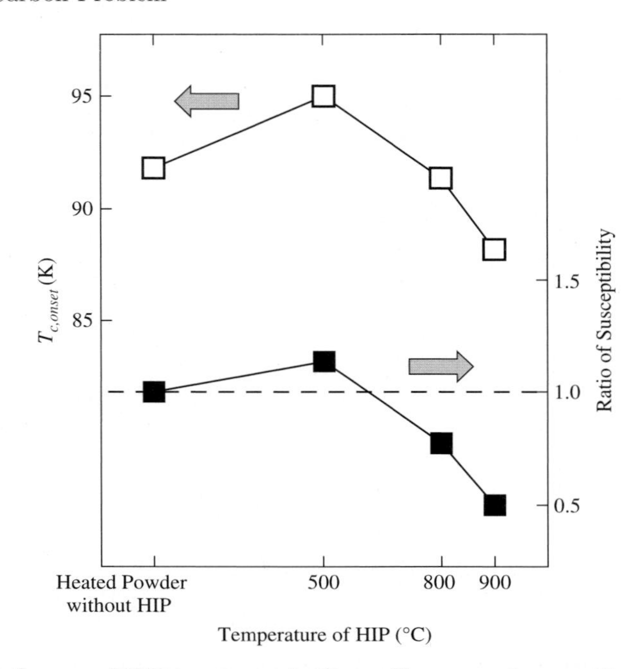

Fig. 4.3. Influence of HIP treatment in O_2 on $T_{c,onset}$ and magnetic susceptibility at 77 K. The HIP was performed with 100 MPa and the duration time 10 h [675]

superconducting. In the case of sintering in a 5% CO_2/O_2 mixture, the reaction products may be nucleated at grain boundaries as separate precipitates rather than as a thin layers, which coat the grain boundaries. In the case of these individual precipitates at grain boundaries, the J_c value may not decrease, as long as most boundaries remain superconducting. On the contrary, since defects, such as twin boundaries and tiny secondary phases, can act as flux pinning centers, the J_c value may in fact increase due to the presence of small precipitates at grain boundaries.

The interaction between YBCO and CO_2 at a total pressure of 1 atm (0.999 atm of O_2 and 0.001 atm of CO_2) leads to two reaction mechanisms [285]:

$$\mathrm{YBa_2Cu_3O_{6+x}(solid) + 2CO_2(gas) \rightarrow 2BaCO_3(solid) + 0.5Y_2O_3(solid) + 3CuO(solid) + (2x-1)/4O_2(gas)}\ , \tag{4.5}$$

$$\mathrm{YBa_2Cu_3O_{6+x}(solid) + 2CO_2(gas) \rightarrow 2BaCO_3(solid) + 0.5Y_2Cu_2O_5(solid) + 2CuO(solid) + (2x-1)/4O_2(gas)}\ . \tag{4.6}$$

Both of these reactions lead to the formation of non-superconducting phases. The partial pressure of CO_2 can affect the partial pressure of O_2, which

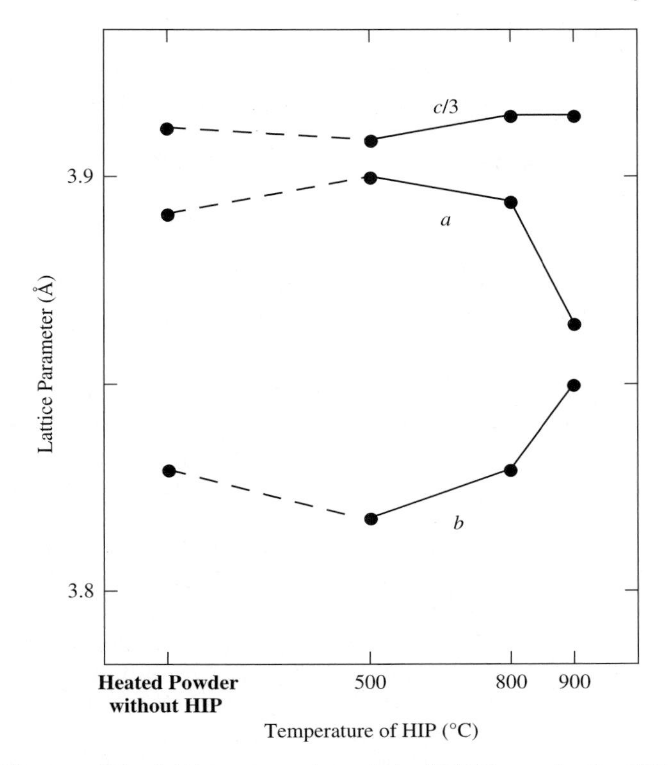

Fig. 4.4. Change of the lattice parameters of the YBCO powder by HIP treatment in O_2 [675]

in turn affects the oxygen content of the YBCO compound formed. It is well known [498, 499, 589] that for $x \approx 1$, $YBa_2Cu_3O_{6+x}$ is in the orthorhombic phase and exhibits superconductivity at 90 K. As the oxygen content, which is dependent on both the temperature and oxygen partial pressure, decreases, the transition temperature of the orthorhombic phase decreases to 60 K, at $0.6 < x < 0.7$. At $x \approx 0.4$, the material becomes tetragonal, and superconductivity is destroyed. The critical current density is affected by both sintering temperature and CO_2 content in the sintering atmosphere (Fig. 4.5). J_c values decrease with decreasing sintering temperature and increasing CO_2 content.

The crystals of the sintered samples in CO_2 atmospheres with 50 or 500 ppm CO_2 can demonstrate twins, whereas those sintered in 5% CO_2 do not show any twinning [958]. The twin structure in YBCO determines its orthorhombic superconducting properties [1031]. Hence, the samples sintered in 5% CO_2 are not orthorhombic, but the extent of the orthorhombic structure decreases with increasing CO_2 in the sintering atmosphere. Moreover, increasing CO_2 contents in the sintering atmosphere increases the electrical resistivity of the sample [958].

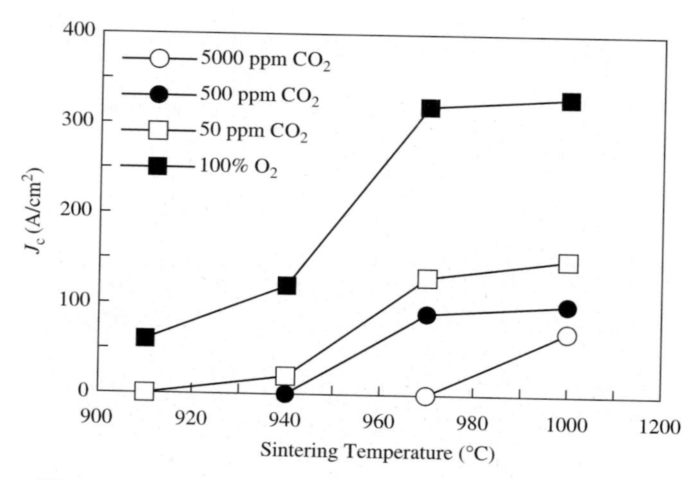

Fig. 4.5. The effect of sintering temperature on J_c [958]

We do not know test evidence of secondary phases at grain boundaries in samples sintered in a 100% O_2 atmosphere. At the same time, two distinct types of grain boundaries can be observed in the samples sintered in 0.5% $CO_2 + 99.5\%O_2$ [958]. Approximately 10% of the grain boundaries are found to be wet by a thin layer of a second phase, as shown in Fig. 4.6, and the remaining boundaries are sharp grain boundaries and free from the presence of secondary phases. The local change of YBCO stoichiometry leads to the eutectic reaction, in accordance with the phase diagram, and formation of $BaCuO_2$ and CuO at the grain boundaries. The decomposition of YBCO phase at the grain boundaries weakens the path for the current, thereby decreasing the overall J_c value. The wetting of the boundaries with non-superconducting phases decreases the effective contact area between the superconducting grains and decreases J_c. From Fig. 4.6, the width of the secondary phase region along the grain boundaries is determined to be approximately 50 nm, which is much larger than the coherence length in this material. Therefore, these secondary phases block the passage of currents.

There are two potential sources of CO_2, when YBCO are processed by solid-state sintering, namely (i) the CO_2, contained in the oxygen gas, used during sintering and/or annealing and (ii) the CO_2, derived from the decomposition of $BaCO_3$ during the calcining step. Each source affects the quality of the final product in different ways and can lead to a drop in the critical current density. An additional potential source is the organic binders that are used in the YBCO processing. The CO_2, formed by the interaction of the organic binders and the oxygen from the gas atmosphere, could have a deleterious effect on the performance capabilities of the final product. By using thermodynamics, it may by shown [958] that for pressure values, $p_{CO_2} < 1$ atm, a sintering temperature must increase as p_{CO_2} increases. Therefore, increasing

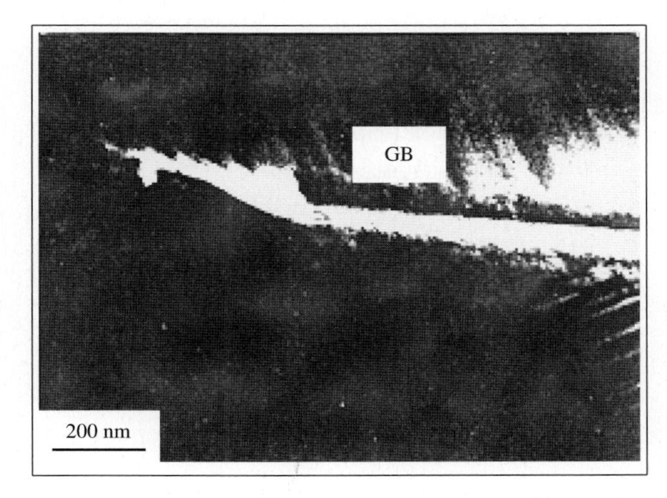

Fig. 4.6. The sample sintered at 970°C in 0.5% CO_2 + 99.5% O_2. TEM micrograph for the grain boundary (GB) which is wet by a layer of $BaCuO_2$ second phase [958]

the temperature results in a higher p_{CO_2}. The sintering temperatures can be lowered if p_{CO_2} can be lowered, and one expedient means of doing this is to remove the CO_2 contained in the O_2 gas.

Carbon also affects negatively the temperature and broadness of superconducting transition [1031]. It is known that YBCO samples, sintered to closed porosity, exhibit broad and suppressed superconducting transitions (Fig. 4.7). The observations suggested that during sintering of the closed-porosity material, "impurities" that suppress the superconducting transition become trapped in the center of the sample, while in the open-porosity sample and near the surface of the closed-porosity material, volatilization of the impurity can occur during sintering.

The highest carbon content of $\approx$1.2 wt% is detected in the grains of the central part from the closed-porosity material. Significantly lower carbon contents are detected in specimens, cut from both the exterior of the closed-porosity material and the central part of the open-porosity material ($\approx$ 0.6 wt%). Carbon removal occurs more readily in the open-porosity material and near the surface of the denser material, consistent with the suggestion that carbon removal is limited by the slow diffusion of carbon out of the YBCO grains.

Under appropriate conditions, carbon has an appreciable solubility in YBCO bulk. The depression of the transition temperature, detected by magnetic susceptibility measurements, correlates well with carbon content. This correlation suggests that incorporation of carbon in the structure is the dominant cause of the lowering of the transition temperature in the high-dense superconductors.

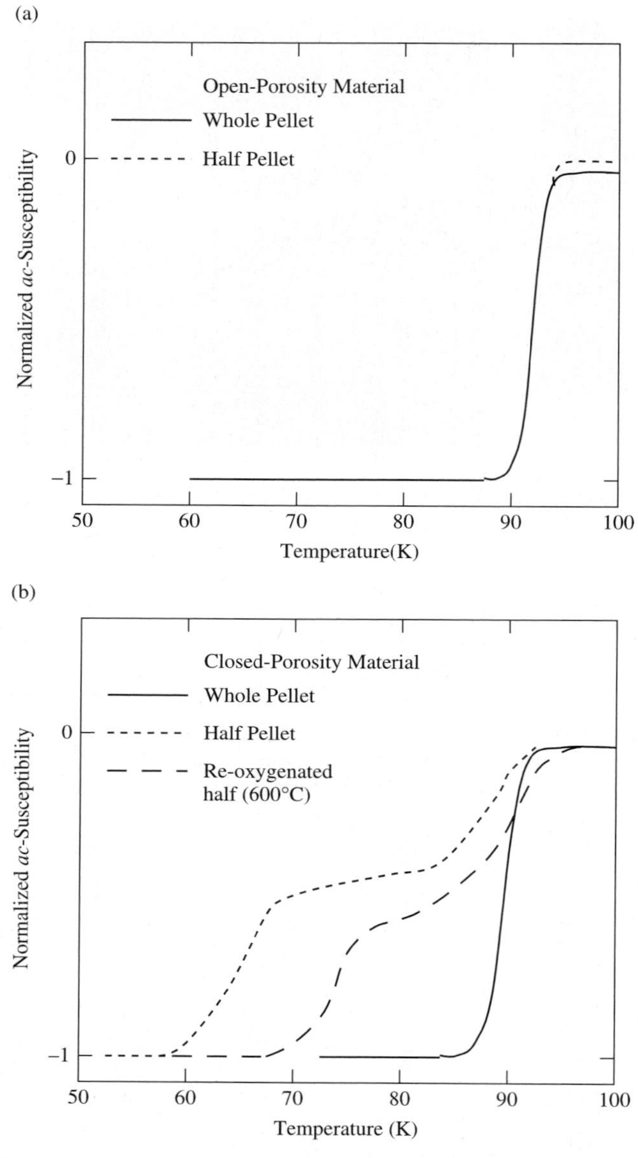

Fig. 4.7. Change in ac-susceptibility with temperature for (**a**) an open-porosity material ($\rho = 87\%$) and (**b**) a closed-porosity material ($\rho = 92\%$). In the open-porosity material both whole and halved samples exhibit sharp transitions, whereas in the closed-porosity material a stepped transition is seen in the halved sample even after an additional oxygenation treatment [1031]

In the interior of the closed-porosity material, CO_2 or CO gas may be trapped in the pores. As a result, a pressure of CO_2 or CO gas (in the range 0.01–0.1 atm) can develop in the pores that is enough to slow down sintering processes. Then, several steps may be taken to lower the amount of carbon, trapped in a dense material, namely:

(1) The external partial pressure of CO_2 should be as low as possible in order to favor decomposition of carbonate in the material.
(2) If the densification process can be delayed (to allow a longer period, in which the sample contains interconnected porosity), more decomposition and gas evolution may occur before the gas becomes trapped.
(3) A successful approach to carbon removal is to heat more slowly to the sintering temperature.

Carbon in the crystal structure of YBCO interacts with Ba and O, reducing the Ba–O coordination [1031]. The carbon connected with Ba ion could be incorporated in the structure either substitutionally or interstitially. In the first case, a direct substitution of oxygen by carbon ions is possible (it is not supported in tests). An alternative possibility is that the carbon lies in an interstitial site adjacent to the barium ion. In the absence of a detailed theoretical understanding of the origin of the superconducting transition in YBCO, the origin of the depression of T_c, produced by the incorporation of carbon, may be speculated. It is known that there is a strong correlation between T_c and the concentration of holes in the material [966]. The presence of the carbon causes localization of some of the holes and thus reduces the mobile holes' concentration in the sample. This suggestion is consistent with the location of the carbon in an interstitial site adjacent to the CuO_2 planes. Because the carbon concentration can approach one atom per unit cell, the location of the carbon in such a site could be expected to cause appreciable local distortions of the CuO_2 planes and thereby lower T_c.

It may be shown [651] that carbon is released from the matrix at temperatures higher than that of the peritectic decomposition. Nevertheless, there is no doubt that even in a melt-processed sample, depression of T_c may still occur. Consider proper technological procedures, leading to decreasing of carbon content in the melt-processed YBCO samples by using as example the results of [1127]. The melt-texturing was performed by a melt-texture growth (MTG) process, using a temperature gradient furnace (temperature gradient about 25 K/cm). As a result, two samples were synthesized, namely, (i) the sample Sr5a, containing large grains (100 μm) with a very dense, closed porosity structure, and (ii) the sample Sr5b, containing small grains (< 10 μm) with open porosity. Their superconducting transition curves are shown in Fig. 4.8. The $T_{c,\text{onset}}$ of sample Sr5a is depressed to 80 K, with a broad transition. Sample Sr5b has $T_{c,\text{onset}}$ at 92 K, which is the optimum value for the Y-123 system. The analysis shows rather a high degree of carbon retention in the two samples: 2600 ppm for sample Sr5a and 1200 ppm for sample Sr5b. Thus,

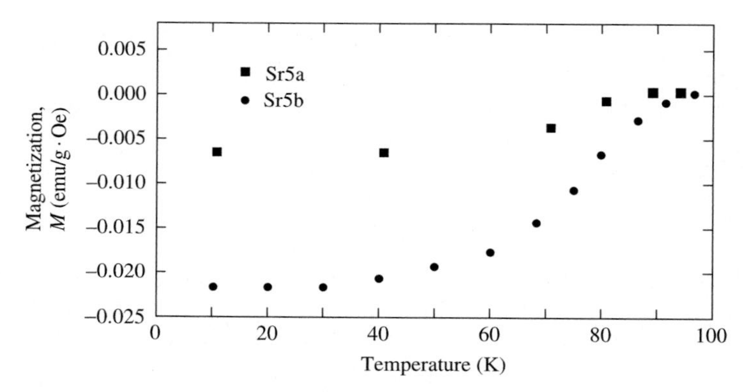

Fig. 4.8. The temperature dependence of magnetization of Y-123 ceramics sintered at different temperatures: Sr5a (sintered at 950°C in oxygen flow) and Sr5b (sintered at 920°C in oxygen flow). Average particle size of precursor (2.5 μm) is the same for both ceramics [1127]

almost all carbon of the initial powder (3000 ppm) is retained in sample Sr5a, and its low $T_{c,onset}$ is in accordance with the correlation stated above.

In sample Sr5a, three experimental features can explain the carbon content in this sample as follows:

(1) The obtained lattice parameters ($a = 3.8288$ Å, $b = 3.889$ Å and $c = 11.667$ Å) showed that the decrease of the c parameter, which was sensitive to the carbon content, was considerably less than that observed for samples prepared under a CO_2 atmosphere ($c = 11.585$ Å) [76].
(2) The characteristic features of the YBCO carbonated phases is the existence of oriented domains, where the c-axes are perpendicular; the formation of these domains is correlated not only to the decrease of the a/b and a/c ratios, but also to the presence of carbonate groups, which are preferentially located at the level of the domain boundaries.
(3) In carbon-rich materials, the existence of distributions of local superstructures, leading to the formation of streaks along a-direction and to short-range aligned areas, is clearly visible on the images.

The above two sintered samples were further processed by MTG. The results presented in Fig. 4.9 show that this procedure allowed to further decrease the carbon content. However, an excessive amount of carbon can still be retained, if the carbon content of the initial precursor is too high, as shown for sample Sr5amt. Therefore, the absolute necessity of controlling the quality of the precursor material before melt texturing is obvious.

Then, in order to examine the influence of the 211 fine-disperse additives and Pt doping on carbon retention, two sets of samples were prepared, one without Pt addition (5avmt) and another with the addition of 0.5 wt% PtO_2 (5avpmt). Both sets had nominal composition $Y_{1.4}Ba_{2.2}Cu_{3.2}O_x$. Addition of

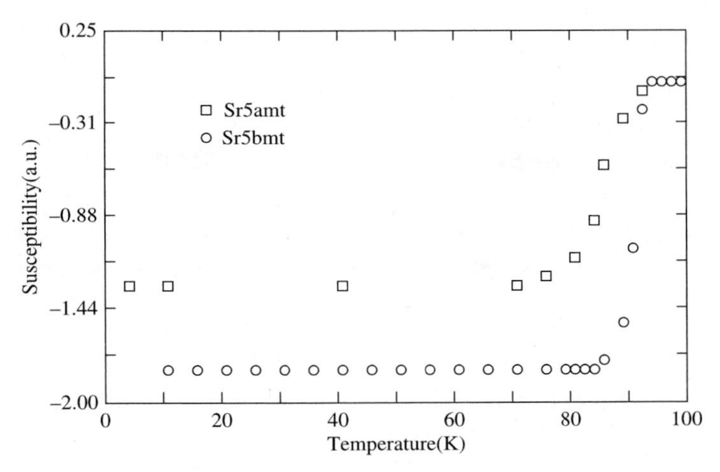

Fig. 4.9. The dependence of magnetic susceptibility on temperature of melt-processed samples. Sr5amt (or Sr5bmt) is an MTG sample, using Sr5a (or Sr5b) as a pre-sintered precursor [1127]

Y-211 does not significantly decrease carbon retention in the Y-123. In contrast, as it follows from Fig. 4.10, for sample 5avpmt with added Pt, $T_{c,onset}$ is enhanced up to the optimum value of 92 K, and a remarkably sharp transition ($\Delta T_c = 2$ K) is obtained. Analysis shows that the carbon content in this sample is as low as 200 ppm. At the same time, the sample 5avmt contains

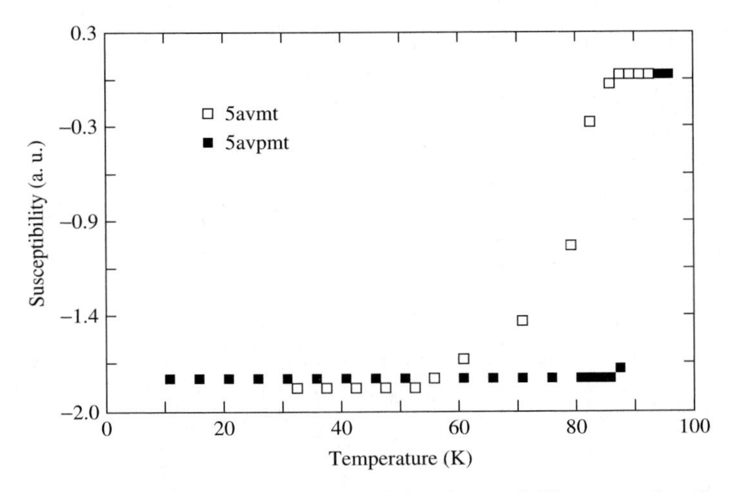

Fig. 4.10. The temperature dependence of the susceptibility curve for the samples processed by an MTG method with starting composition $Y_{1.4}Ba_{2.2}Cu_{3.2}O_x$, with (5avpmt) and without (5avmt) addition of Pt [1127]

about 870 ppm of carbon. Although the exact mechanism by which the platinum acts to decrease the carbon content is unknown, it may be suggested that Pt acts as a catalyst for de-carbonation.

4.2 BSCCO Systems

The behavior of BSCCO samples due to carbon effect somewhat differs from the behavior of YBCO superconductors. The dependencies of sintering temperature and density of Bi-2223 precursor samples on carbon content are shown in Fig. 4.11. Obviously, the higher sintering temperature and increasing density of the Bi-2223 samples cause the lower carbon content.

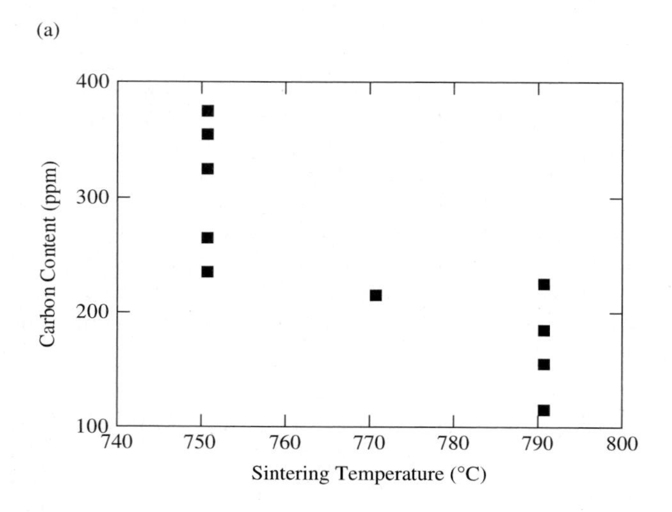

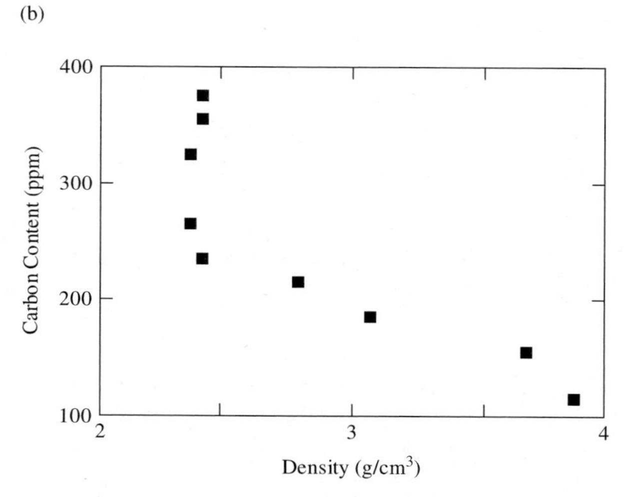

Fig. 4.11. The dependencies of (**a**) sintering temperature and (**b**) density of Bi-2223 precursor samples on carbon content [1164]

Oxide superconducting powder can absorb atmospheric moisture and then react with CO_2, worsening superconducting properties. Therefore, a preparation process, proposed for Bi-2223 superconductor [928], in the first place restricts absorption of moisture in order to suppress the deterioration of the calcined powder. The preparation process includes the processes of dissolving, coprecipitation, filtering, primary calcining ($T = 800°C$), wet pulverizing to obtain a fine powder, drying and secondary calcining ($T = 800–815°C$). Then, in order to fabricate bulk sample the calcined powder is molded by using a coaxial pressing technique. The sample is sintered at 850°C sometimes by using intermediate cold isostatic pressing (CIP). The results obtained for three various samples are presented in Table 4.1. The intermediate pressing leads to increasing of the sample density (Fig. 4.12a) and the corresponding rise of critical current density (Fig. 4.12b). The saturation of the results occurs after third intermediate pressing. The results obtained confirm that superconducting properties of the bulk superconductor are found by content of moisture and carbon in the calcined powder. Carbon and admixture phases segregate at the grain boundaries, forming weak links and decreasing J_c.

The cause of the porosity formation and broken uniformity of Bi-2212/Ag tapes may be as a result of the gas release, such as CO_2, when Bi-2212 phase reforms from the liquid phase [1196]. Carbon exists in the sample in the carbonate form and can transfer into CO_2 gas in proper condition during the heat treatment, such as at high temperature by the reaction:

$$SrCO_3 \rightarrow SrO + CO_2 \,. \tag{4.7}$$

CO_2 gas may be released and causes severe bubble formation and porosity, depending on the carbon content. It may also dissolve into the liquid phase and define its behavior. The carbon content effects on the melting behavior of the Bi-2212 phase [1195] state that increasing the carbon content of the powder can greatly decrease the melting and reformation temperatures of the Bi-2212 phase in the OPIT tapes.

During cooling, when Bi-2212 phase re-forms from liquid, [C] will be released in accordance with the reaction:

$$[C] + O_2 \rightarrow CO_2 \,, \tag{4.8}$$

where [C] is the carbon, which is dissolved in the liquid phase.

Table 4.1. Moisture content and carbon content of calcined powders [928]

Sample	Process	Moisture content (wt %)	Carbon content (wt %)
A	After drying in air	1.01	0.610
	2nd-calcined powder at 800°C	0.13	0.041
B	After drying in vacuum	0.65	0.486
	2nd-calcined powder at 800°C	0.11	0.028
C	After drying in vacuum	0.64	0.482
	2nd-calcined powder at 815°C	0.12	0.025

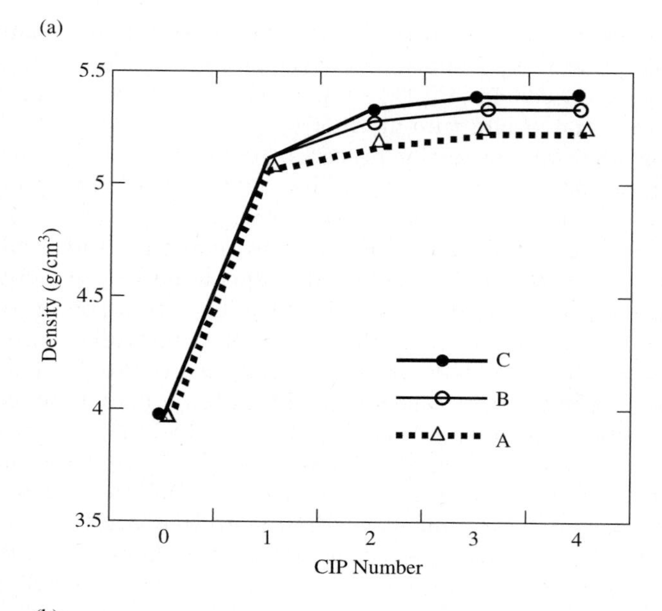

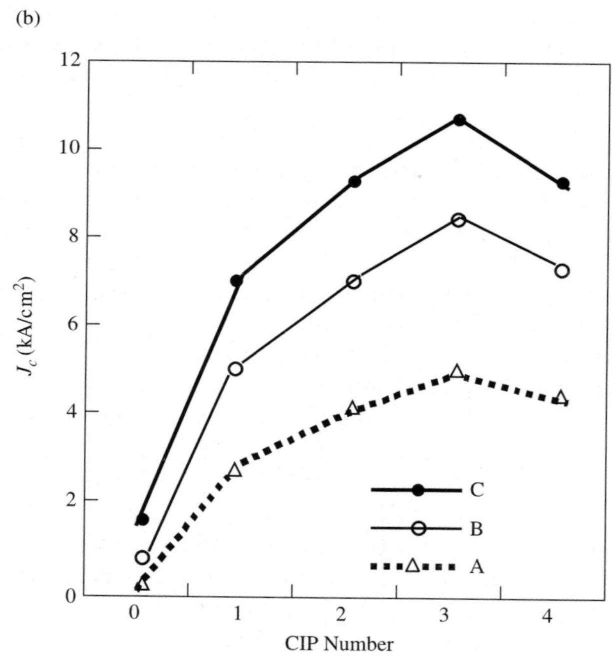

Fig. 4.12. Increase of (**a**) sample density and (**b**) J_c (77 K, 0 T) due to intermediate pressing (CIP) and sintering [928]

The volume change, caused by the formation of CO_2 gas from the solid-state carbon, can be tremendous. A rough calculation indicates that 200 ppm of carbon can cause about 36% porosity in the core, if all carbon form CO_2 at high temperature [1196].

The effect of the introduction of carbon into YBCO and Bi-2212 bulks, obtained by using a partial melting (both samples were heated some above the peritectic temperature), was studied in [962]. The presence of carbon has strong effect on the YBCO melting, but is not essential for Bi-2212. In this case, the exposed carbon excess can cause an increasing of the magnetic flux pinning. At slow cooling of the YBCO and Bi-2212 from the temperature of partial melting, it is necessary to strictly control temperature and gas atmosphere in order to obtain high superconducting properties. Most important factors are the oxygen gas release [143, 408, 742, 961] and the presence of carbon and CO_2 in the sample volume [1101, 1195]. Thus, due to the reaction of YBCO and BSCCO with CO_2 atmosphere, the carbon trapped in the sample can render unfavorable effect on the sample density [1196], intergranular boundaries [307, 958], critical currents [969], and critical current density [1101, 1158]. Besides atmospheric sources, carbon may be introduced due to carbonized precursors and organic binders during sample processing [212] and also by solvents [304]. Finally, the carbon can be precipitated into superconductor due to carbonized gases and liquids. Existing techniques can form carbon particles with size of some nanometers, which can be the centers of magnetic flux pinning into superconducting grains. Then, on one hand, carbon can render useful effect, in particular, using Pb and some rare-earth elements [742], and, on the other hand, it can deteriorate a structure of the intergranular boundaries and structure-sensitive properties of HTSC.

4.3 Carbon Embrittlement and Fracture of YBCO Superconductor

Carbonates, forming in result of the chemical reactions (4.3)–(4.6), are the brittle phase, leading to embrittlement of superconductor. It may lead to delayed carbonate cracking, a subcritical crack growth mechanism with the next formation and fracture carbonate. The carbon-induced embrittlement is a complicated mechanism, which results from the simultaneous operation of several coupled processes, namely (i) carbon diffusion, (ii) carbonate precipitation, (iii) non-mechanical energy flow and (iv) material deformation. In this case, carbon diffusion occurs due to gradients of chemical potential and temperature [189, 974]. Similar to hydride formation in metals [69, 620, 839, 872], it may be proposed that the carbon chemical potential depends on stress, and therefore the carbon diffusion is coupled with material deformationand non-mechanical energy flow. The final carbon dissolution into material depends on the thermal stresses due to carbonate expansion at precipitation defining coupling between carbonate precipitation, material deformation and non-mechanical energy flow. The material deformation depends on all other processes due to material expansion, which is caused by carbon dissolution, carbonate formation and temperature increase. Microscopic models of the HTSC intergranular cracking during carbon segregation, depicting slow, fast and steady-state crack growth, screened by linear dislocation array [812] will be considered in Sect. 4.4. In this section, the governing equations for carbon embrittlement and fracture of HTSC are derived, taking into account thermal stresses in the framework of the thermodynamic theory of irreversible processes [189], based on the thermal diffusion of carbon (Soret's effect). In this case, the delayed fracture of carbonate is modeled by using the de-cohesion model, taking into account de-cohesion energy changed during time due to the time-dependent carbonate precipitation process [809, 810].

4.3.1 Mathematical Model for Carbonate Precipitation and Fracture

The model is developed for a superconductor, forming carbonates. The presence of the carbonates is described by the carbonate volume fraction. Energy flow and diffusion of mass are coupled processes. A temperature gradient leads to flow of matter, and therefore to a concentration gradient. Conversely, a diffusion process gives rise to a small temperature difference. A detailed discussion for thermodynamic treatment of energy flow/diffusion is based on Onsager's principle of microscopic reversibility [189]. In the following, the theory is applied to the process of carbonate diffusion and energy flow, occurring in YBCO interaction with CO_2, forming carbonates and cuprates under stress and temperature gradient.

According to the empirical law of Fourier, heat flux is linearly related to the temperature gradient, which is the thermodynamic force, driving heat flow. In an isothermal system, the flux of a diffusing substance is proportional to the gradient of its chemical potential, that is, chemical potential gradient is the thermodynamic force, driving diffusion under isothermal conditions. When the above-pointed processes operate simultaneously, coupling is taken into account by assuming that the non-mechanical energy and carbon fluxes are linearly related to both thermodynamic forces:

$$J_i^{\mathrm{E}} = L^{\mathrm{E}} X_i^{\mathrm{E}} + L^{\mathrm{EC}} X_i^{\mathrm{C}}, \qquad J_i^{\mathrm{C}} = L^{\mathrm{CE}} X_i^{\mathrm{E}} + L^{\mathrm{C}} X_i^{\mathrm{C}}, \qquad L^{\mathrm{EC}} = L^{\mathrm{CE}} , \quad (4.9)$$

where J_i^{E} and J_i^{C} are the components of the non-mechanical energy flux and the carbon flux, respectively. Note that the energy flux includes conducted heat, described by Fourier's law, and the energy transported by the diffusing carbon; X_i^{E} and X_i^{C} are the thermodynamic forces, driving non-mechanical energy flow and carbon diffusion, respectively; L^{E}, L^{C}, L^{EC} and L^{CE} are phenomenological coefficients. Third relation (4.9) is valid due to Onsager's reciprocity relation.

The thermodynamic forces are related to the gradients of the absolute temperature T and chemical potential of carbon in the carbonate μ^{C} as

$$X_i^{\mathrm{E}} = -\frac{1}{T}\frac{\partial T}{\partial x_i} , \qquad X_i^{\mathrm{C}} = -T\frac{\partial}{\partial x_i}\left(\frac{\mu^{\mathrm{C}}}{T}\right) . \tag{4.10}$$

When carbon and superconductor form carbonate, the carbon flux satisfies the following relation [974]:

$$J_k^{\mathrm{C}} = -\frac{D^{\mathrm{C}}C^{\mathrm{C}}}{RT}\left(\frac{\partial \mu^{\mathrm{C}}}{\partial x_k} + \frac{Q^{\mathrm{C}}}{T}\frac{\partial T}{\partial x_k}\right) , \tag{4.11}$$

where R is the gas constant; C^{C}, D^{C} and Q^{C} are the concentration, diffusion coefficient and heat of transport of carbon in the carbonate, respectively. The concentration of carbon as well as other components or phases is given in moles per unit volume. Relation (4.11) is a special case of (4.9) and (4.10), and therefore satisfies for a HTSC under stress. Because of the absence of carbon diffusion in the cuprate, the total carbon flux in cuprate/carbonate composite is given as

$$J_k^{\mathrm{CT}} = f J_k^{\mathrm{C}} , \tag{4.12}$$

where f is the volume fraction of the carbonate in the material. Mass conservation requires that the rate of total carbon concentration, C^{CT}, inside a volume V, is equal to the rate of carbon flowing through the boundary S:

$$\frac{\mathrm{d}}{\mathrm{d}t}\int_V C^{\mathrm{CT}}\,\mathrm{d}V + \int_S J_k^{\mathrm{CT}} n_k\,\mathrm{d}S = 0 . \tag{4.13}$$

Relation (4.13) is valid for an arbitrary volume. Then, one may derive the respective differential equation by using divergence theorem:

$$\frac{\mathrm{d}C^{\mathrm{CT}}}{\mathrm{d}t} = -\frac{\partial J_k^{\mathrm{CT}}}{\partial x_k} . \tag{4.14}$$

Note that the total carbon concentration, C^{CT}, is only related to the concentration of carbon in the carbonate, C^{C}, because of the absence of carbon in cuprate:

$$C^{\mathrm{CT}} = f C^{\mathrm{C}} . \tag{4.15}$$

This is equal to the carbon terminal solid solubility, C^{TS}, when $f \neq 0$. Value of C^{C} is defined with respect to the volume occupied by the carbonate, that is, fV. Note that in the carbon diffusion model, the effect of carbon trapping in the solid by dislocations and voids is neglected, since the bulk of the material is assumed to behave elastically. However,the effect of traps is more important at low temperatures, where the lattice solubility (i.e., ability to substitute atoms and formation of atom interstices) is relatively small [974]. Corresponding model of carbon trapping by dislocations and crack-like defects can be developed similar to the hydrogen precipitation in metals [426].

Then, the governing equation for non-mechanical energy flow is determined. With this aim, the energy flux is found, that is coefficients L^{E} and L^{EC} are calculated. Comparing (4.9)–(4.11), we have

$$L^{\mathrm{C}} = \frac{D^{\mathrm{C}}C^{\mathrm{C}}}{RT}, \qquad L^{\mathrm{CE}} = L^{\mathrm{EC}} = \frac{D^{\mathrm{C}}C^{\mathrm{C}}}{RT}(Q^{\mathrm{C}} + \mu^{\mathrm{C}}) . \tag{4.16}$$

The remaining coefficient L^{E} can be determined by taking into account Fourier's law for heat conduction:

$$J_i^{\mathrm{E}} = -k\frac{\partial T}{\partial x_i} , \tag{4.17}$$

where k is the thermal conductivity of the superconductor. Note that (4.17) is valid when there is no carbon diffusion. Then, by stating $J_k^{\mathrm{C}} = 0$, we have from (4.11)

$$\frac{\partial \mu^{\mathrm{C}}}{\partial x_i} = -\frac{Q^{\mathrm{C}}}{T}\frac{\partial T}{\partial x_i} . \tag{4.18}$$

Substituting (4.10), (4.17) and (4.18) in (4.9), one may derive L^{E}:

$$L^{\mathrm{E}} = kT + \frac{D^{\mathrm{C}}C^{\mathrm{C}}}{RT}(Q^{\mathrm{C}} + \mu^{\mathrm{C}})^2 . \tag{4.19}$$

Thus, all the coefficients of the phenomenological (4.9) have been determined. Consequently, one may derive the expression for energy flux in a carbonate, substituting (4.16) and (4.19) into (4.9) and taking into account the relation (4.11) for carbon flux:

$$J_i^{\mathrm{E}} = (Q^{\mathrm{C}} + \mu^{\mathrm{C}})J_i^{\mathrm{C}} - k\frac{\partial T}{\partial x_i} . \tag{4.20}$$

The first term of the right-hand side is the energy flux, which is produced by the diffusion of carbon atoms, and the second term is the conducted heat. According to (4.20), the heat of carbon transport is the heat flux per unit flux of carbon in the absence of temperature gradient.

Due to the absence of carbon diffusion in the cuprate, the energy flow in the cuprate is only thanks to heat conduction. It is assumed that the thermal conductivity of the carbonate equals the thermal conductivity of the cuprate; then by using (4.17), the next relation provides the total energy flux in the cuprate/carbonate composite:

$$J_i^{\mathrm{ET}} = fJ_i^{\mathrm{E}} - (1-f)k\frac{\partial T}{\partial x_i} , \tag{4.21}$$

which, when combined with (4.12) and (4.20), leads to

$$J_i^{\mathrm{ET}} = (Q^{\mathrm{C}} + \mu^{\mathrm{C}})J_i^{\mathrm{CT}} - k\frac{\partial T}{\partial x_i} . \tag{4.22}$$

In other to completely describe the process of energy flow, the conservation of energy should be enforced. It requires that the internal energy rate equals the energy input rate due to the external stress power and gradient of the non-mechanical energy flow [661]:

$$\rho \frac{du}{dt} = \sigma_{ij} \frac{d\varepsilon_{ij}}{dt} - \frac{\partial J_k^{\mathrm{ET}}}{\partial x_k} , \tag{4.23}$$

where ρ, u, σ_{ij} and ε_{ij} are the mass density of material, the internal energy per unit mass, the stress tensor and the strain tensor, respectively. The minus sign in (4.23) is due to convention that the energy flux is positive when it leaves the body. According to the discussion of continuum thermodynamics, based on a caloric equation of state, the rate of internal energy is related with the specific entropy, stress–strain state and total carbon concentration rates:

$$\rho \frac{du}{dt} = \rho T \frac{ds}{dt} + \sigma_{ij} \frac{d\varepsilon_{ij}}{dt} + \mu^{\mathrm{C}} \frac{dC^{\mathrm{CT}}}{dt} . \tag{4.24}$$

Equations (4.14) and (4.22–4.24) lead to the following differential equation:

$$\rho T \frac{ds}{dt} = \frac{\partial}{\partial x_i} \left(k \frac{\partial T}{\partial x_i} \right) - Q^{\mathrm{C}} \frac{\partial J_m^{\mathrm{CT}}}{\partial x_m} - J_n^{\mathrm{CT}} \frac{\partial \mu^{\mathrm{C}}}{\partial x_n} . \tag{4.25}$$

Entropy rate is also related to the rates of temperature, carbonate volume fraction and total carbon concentration. This relation is derived taking into account the dependence of entropy on all thermodynamic variables (i.e., temperature, stress as well as carbonate, cuprate and carbon concentrations) [1111]:

$$\rho T \frac{ds}{dt} = \rho c_p \frac{dT}{dt} + \frac{\Delta \overline{H}^{\mathrm{car}}}{\overline{V}^{\mathrm{car}}} \frac{df}{dt} + Q^{\mathrm{C}} \frac{dC^{\mathrm{CT}}}{dt} , \tag{4.26}$$

where c_p is the specific heat of the superconductor at constant pressure; $\Delta \overline{H}^{\mathrm{car}}$ is the enthalpy, associated with the formation of a mole of carbonate and $\overline{V}^{\mathrm{car}}$ is the carbonate molal volume. In the determination of (4.26), the following conditions are taken into account: (i) the total number of HTSC moles in the cuprate and carbonate remains constant; (ii) the partial derivative of entropy with respect to temperature is related to the specific heat of the cuprate/carbonate composite under constant stress, $c_\sigma = T(\partial s / \partial T)_\sigma$ (in the present analysis, $c_\sigma = c_p$ is assumed); (iii) by neglecting thermo-elastic coupling, the partial derivative of entropy with respect to stress is taken equal to zero[1]; (iv) the change of entropy due to carbonate formation is equal to $\Delta \overline{H}^{\mathrm{car}} / T$; (v) the change of entropy due to the addition of a mole of carbon in the carbonate is equal to Q^{C} / T.

[1] Note that, in ceramics, thermo-elastic coupling effects are quite small [74].

Substitution of (4.26) into (4.25), taking into account (4.14), gives the governing equation for the flow of non-mechanical energy:

$$\rho c_p \frac{\mathrm{d}T}{\mathrm{d}t} + \frac{\Delta \overline{H}^{\mathrm{car}}}{\overline{V}^{\mathrm{car}}} \frac{\mathrm{d}f}{\mathrm{d}t} = \frac{\partial}{\partial x_i}\left(k \frac{\partial T}{\partial x_i}\right) - J_n^{\mathrm{C}} \frac{\partial \mu^{\mathrm{C}}}{\partial x_n} . \tag{4.27}$$

Therefore, the variation of the heat content in the cuprate/carbonate composite depends on conducted heat, heat generated during carbon diffusion and heat released during carbonate formation.

According to the above mathematical formulation for carbon diffusion and energy flow, the knowledge of carbon chemical potential and terminal solid solubility is necessary. Both quantities depend on applied stress and are derived below.

The chemical potentials of mobile and immobile components in stressed solids have been derived [620]. The chemical potential of a component B is given as

$$\mu^{\mathrm{B}} = \mu_0^{\mathrm{B}} + \frac{\partial w}{\partial N^{\mathrm{B}}} - W^{\mathrm{B}} , \tag{4.28}$$

where μ_0^{B} is the chemical potential of component B under stress-free conditions, for the same concentration as that under stress; w is the strain energy of the solid and N^{B} is the number of B moles. Therefore, the second term in the right-hand side of (4.28), $\partial w/\partial N^{\mathrm{B}}$, represents the strain energy of the solid per mole of component B; in the determination of $\partial w/\partial N^{\mathrm{B}}$, temperature and stress are held to be constant. Finally, W^{B} is the work performed by applied stresses, σ_{ij}, per mole of addition of component B. For immobile components, since the addition or removal of the component takes place at an external surface or an interface, chemical potential is considered as a surface property. However, this is not the case for mobile components.

Equation (4.28) is applied to barium carbonate under stress. For material particle of volume, V, under uniform stress, we have

$$\begin{aligned}\frac{\partial w}{\partial N^{\mathrm{C}}} &= \frac{\partial}{\partial N^{\mathrm{C}}}\left(\int_0^{\varepsilon_{mn}} V\sigma_{ij}\,\mathrm{d}\varepsilon_{ij}\right) = \frac{\partial}{\partial N^{\mathrm{C}}}\left(\int_0^{\sigma_{mn}} V\sigma_{ij}M_{ijkl}\,\mathrm{d}\sigma_{kl}\right) = \\ &= \left(\int_0^{\sigma_{mn}} \frac{\partial V}{\partial N^{\mathrm{C}}} M_{ijkl}\sigma_{ij} + \frac{\partial M_{rskl}}{\partial \chi^{\mathrm{C}}}\frac{\partial \chi^{\mathrm{C}}}{\partial N^{\mathrm{C}}} V\sigma_{rs}\right)\mathrm{d}\sigma_{kl} ,\end{aligned} \tag{4.29}$$

where M_{ijkl} is the elastic compliance tensor of the superconductor and χ^{C} is the mole fraction of carbon in the carbonate. The work performed by the applied stresses per mole of carbon addition in carbonate is given as

$$W^{\mathrm{C}} = V\sigma_{ij}\frac{\partial \varepsilon_{ij}}{\partial N^{\mathrm{C}}} = V\sigma_{kk}\frac{\overline{V}^{\mathrm{C}}}{3V} = \frac{\sigma_{kk}}{3}\overline{V}^{\mathrm{C}} . \tag{4.30}$$

Substituting (4.29) and (4.30) in (4.28), the final expression is found for the chemical potential of carbon, being in carbonate under stress:

$$\mu^{\mathrm{C}} = \mu_0^{\mathrm{C}} + \overline{V}^{\mathrm{C}} \left(\frac{1}{2} M_{ijkl} \sigma_{ij} \sigma_{kl} - \frac{1}{3} \sigma_{mm} \right) . \tag{4.31}$$

In determination of (4.31), it has been taken into account that the derivative of volume with respect to carbon moles equals the partial molal volume of carbon $\overline{V}^{\mathrm{C}}$. Moreover, it has been assumed in the first approximation that there is no effect of carbon on the elastic modules of the material (this assumption is discussed in detail in Sect. 4.3.2).Consequently, the derivative of the elastic compliance with respect to the mole fraction of carbon is equal to zero. Note that the first term in parenthesis in (4.31) is of the order of σ^2/E, where E is Young's modulus of the superconductor. The second term is of the order of σ and, therefore, is significantly larger than the first term. If the first term in parenthesis of (4.31) is neglected, a relation is derived, which is more often used in the literature.

The obtained relations for the chemical potential of mobile and immobile components are used for derivation of carbon terminal solubility in a superconductor under stress. According to (4.28), carbonate chemical potential in a stressed material is given by

$$\mu^{\mathrm{car}} = \mu_0^{\mathrm{car}} + \frac{\partial w}{\partial N^{\mathrm{car}}} - W^{\mathrm{car}}; \qquad \frac{\partial w}{\partial N^{\mathrm{car}}} = \overline{w}_{\mathrm{acc}} + \overline{w}_{\mathrm{int}} + \overline{w}_{\mathrm{af}} \; ;$$

$$\overline{w}_{\mathrm{acc}} = -\frac{1}{2} \int\limits_{\overline{V}^{\mathrm{car}}} \sigma_{ij}^{\mathrm{I}} \varepsilon_{ij}^{\mathrm{T}} \,\mathrm{d}V \; ; \qquad \overline{w}_{\mathrm{int}} = - \int\limits_{\overline{V}^{\mathrm{car}}} \sigma_{ij} \varepsilon_{ij}^{\mathrm{T}} \,\mathrm{d}V;$$

$$\overline{w}_{\mathrm{af}} = \frac{1}{2} \int\limits_{\overline{V}^{\mathrm{car}}} \sigma_{ij} \varepsilon_{ij} \,\mathrm{d}V, \qquad W^{\mathrm{car}} = \sigma_{\mathrm{n}} \overline{V}^{\mathrm{car}} . \tag{4.32}$$

In determination of relations (4.32) it was taken into account that carbonate formation is accompanied by a deformation, $\varepsilon_{ij}^{\mathrm{T}}$, which is mainly a volume expansion. Under no external loading, the above deformation leads to the development of stresses, σ_{ij}^{I}, in the carbonate, defining strain energy of material per mole of precipitating carbonate $\overline{w}_{\mathrm{acc}}$. Under externally applied stress, σ_{ij}, the interaction energy, $\overline{w}_{\mathrm{int}}$, as well as the strain energy of the applied field, $\overline{w}_{\mathrm{af}}$, should also be taken into account [258]. In the last equation of relations (4.32), σ_n is the normal stress at the location of cuprate/carbonate interface, where the chemical potential is considered.

The chemical potential of the stressed Ba–O component of barium carbonate is defined as

$$\mu^{\mathrm{Ba-O}} = \mu_0^{\mathrm{Ba-O}} + \frac{\partial w}{\partial N^{\mathrm{Ba-O}}} - W^{\mathrm{Ba-O}} \; ; \quad \frac{\partial w}{\partial N^{\mathrm{Ba-O}}} = \frac{1}{2} \int\limits_{(\overline{V}^{\mathrm{car}} - \overline{V}^{\mathrm{C}})} \sigma_{ij} \varepsilon_{ij} \,\mathrm{d}V \; ;$$

$$W^{\mathrm{Ba-O}} = \sigma_{\mathrm{n}} (\overline{V}^{\mathrm{car}} - \overline{V}^{\mathrm{C}}) . \tag{4.33}$$

The carbonate is assumed to be in equilibrium with carbon and cuprate either under stress or under stress-free conditions. Then, taking into account the chemical formula of barium carbonate $BaCO_3$, we have:

$$\mu^{\mathrm{car}} = \mu^{\mathrm{Ba-O}} + \mu^{\mathrm{C}}(C^{\mathrm{TS}}); \qquad \mu_0^{\mathrm{car}} = \mu_0^{\mathrm{Ba-O}} + \mu_0^{\mathrm{C}}(C_0^{\mathrm{TS}}) , \tag{4.34}$$

where C^{TS} and C_0^{TS} are the values of carbon terminal solid solubility under applied stress and stress-free conditions, respectively. By substitution of relations (4.32) and (4.33) in (4.34), one may derive

$$\mu^{\mathrm{C}}(C^{\mathrm{TS}}) - \mu_0^{\mathrm{C}}(C_0^{\mathrm{TS}}) = \overline{w}_{\mathrm{acc}} + \overline{w}_{\mathrm{int}} + \frac{1}{2}\int\limits_{\overline{V}^{\mathrm{C}}} \sigma_{ij}\varepsilon_{ij}\,\mathrm{d}V - \sigma_{\mathrm{n}}\overline{V}^{\mathrm{C}} . \tag{4.35}$$

It has been implied that carbonate and Ba–O equilibrium concentrations do not change significantly with stress. Because of material continuity, the molal volume of the carbonate equals that of the cuprate at the cuprate/carbonate interface; therefore $\overline{V}^{\mathrm{C}} \approx 0$ and the last two terms in the right side of (4.35) can be assumed to be zero. Moreover, in ideal or dilute solutions (Raoult's law), the stress-free carbon chemical potential satisfies the following well-known relation:

$$\mu_0^{\mathrm{C}} = \mu_{\mathrm{RS}}^{\mathrm{C}} + RT\ln(C^{\mathrm{CT}}\overline{V}) , \tag{4.36}$$

where $\mu_{\mathrm{RS}}^{\mathrm{C}}$ is carbon chemical potential in the "standard" (i.e., reference) state and $\overline{V}$ is the molal volume of composite. Then, by invoking (4.15) and (4.36) and substituting (4.31) in (4.35), one may derive the terminal solid solubility of carbon in the composite under stress:

$$C^{\mathrm{TS}} = C_0^{\mathrm{TS}} \exp\left(\frac{\overline{w}_{\mathrm{acc}} + \overline{w}_{\mathrm{int}}}{RT}\right) \exp\left[\frac{\overline{V}^{\mathrm{C}}}{RT}\left(\frac{\sigma_{mm}}{3} - \frac{1}{2}M_{ijkl}\sigma_{ij}\sigma_{kl}\right)\right] . \tag{4.37}$$

The above derivation is based on similar arguments of [620] for the formation of cementite in ferrite. When the compliance term is neglected, (4.37) is similar to the relation for the hydrogen terminal solid solubility [872]. Using (4.37) or its simplified version implies that chemical equilibrium occurs under local thermal stress conditions.

Then, it is assumed that all material phases are elastic, and the elastic properties of carbonate and cuprate are identical in the first approximation and do not depend on carbon concentration. The material strain is coupled with carbon diffusion and energy flow due to the strains, caused by carbon dissolution, carbonate formation and thermal expansion. Similar to the processes of hydride formation in metals [1111] and taking into account carbon absence in cuprate, the following equation may be obtained:

$$\frac{\mathrm{d}\sigma_{ij}}{\mathrm{d}t} = M_{ijkl}^{-1}\left(\frac{\mathrm{d}\varepsilon_{kl}}{\mathrm{d}t} - \frac{\mathrm{d}\varepsilon_{kl}^{\mathrm{C}}}{\mathrm{d}t} - \frac{\mathrm{d}\varepsilon_{kl}^{\mathrm{E}}}{\mathrm{d}t}\right); \quad M_{ijkl}^{-1} = \lambda\delta_{ij}\delta_{kl} + \mu(\delta_{ik}\delta_{jl} + \delta_{il}\delta_{jk}) ;$$

$$\frac{\mathrm{d}\varepsilon_{ij}^{\mathrm{C}}}{\mathrm{d}t} = \frac{1}{3}\delta_{ij}\frac{\mathrm{d}}{\mathrm{d}t}(f\theta^{\mathrm{car}}) ; \quad \frac{\mathrm{d}\varepsilon_{ij}^{\mathrm{E}}}{\mathrm{d}t} = \alpha\delta_{ij}\frac{\mathrm{d}T}{\mathrm{d}t} , \tag{4.38}$$

where λ and μ are Lame constants of the HTSC; $\theta^{car} = \varepsilon_{kk}^{T}$ is carbonate expansion, occurring during its precipitation, and α is the thermal expansion factor of the composite, assumed to be equal to that of the carbonate.

Simulation of fracture by considering cohesive tractions [51, 228] assumes that ahead of a crack tip there is a fracture process zone, where the material deteriorates in a ductile (void growth and coalescence) and/or brittle (carbonate cleavage) mode. According to de-cohesion model, de-cohesion layer with the thickness as fracture process zone is taken off the material, along the crack path. Along the boundaries, created by the cut, the cohesive traction is applied. All information on the damage is contained in the distribution of the cohesive traction, which depends on boundary displacements of the de-cohesion layer. The shape of the traction–displacement function depends on the failure process. In the case of tensile separation, the most important features are the maximum cohesive traction, σ_{max}, and the energy of de-cohesion, ϕ_0:

$$\phi_0 = \int_0^{\delta_c} \sigma_n \, d\delta_n \,, \tag{4.39}$$

where σ_n is the normal cohesive traction and δ_n is the respective normal displacement, which equals the sum of the displacements on both sides of the de-cohesion layer. Moreover, δ_c is the normal displacement, which corresponds to complete fracture and consequently to zero normal cohesive traction. Details of the model for crack growth under plane strain conditions have been presented in [1108, 1109]. The model has been used for the solution of several fracture problems [763, 1089, 1112].

Here, cohesive traction is assumed to vary, according to the next relation:

$$\sigma_n = \begin{cases} E_i \frac{\delta_n}{\delta_0} & \delta_n \le \delta_1 \\ \sigma_{\max} & \delta_1 \le \delta_n \le \delta_f \\ \sigma_{\max} - E_f \frac{\delta_n - \delta_f}{\delta_0} & \delta_f \le \delta_n \le \delta_c \\ 0 & \delta_c \le \delta_n \end{cases} , \tag{4.40}$$

where δ_0 is a constant length of the order of carbonate thickness; E_i and E_f are the de-cohesion modules, which are assumed to be constant; δ_1 is the normal displacement at initiation of damage, at which maximum cohesive traction is reached. Unloading starts when normal displacement exceeds δ_f. Note that δ_f depends on $\sigma_{\max}$ and ϕ_0, according to the following relation:

$$\delta_f = \left[\phi_0 + \frac{1}{2} \sigma_{\max}^2 \delta_0 \left(\frac{1}{E_i} - \frac{1}{E_f} \right) \right] \sigma_{\max}^{-1} \,. \tag{4.41}$$

As shown in [884], the energy of de-cohesion, related to a cohesive zone ahead of a crack tip, in elastic material, is equal to the critical value, J_c, of J-integral, when fracture is imminent:

$$\phi_0 = J_c = \frac{1 - \nu^2}{E} K_{Ic}^2 \,, \tag{4.42}$$

where K_{Ic} is the critical value of stress intensity factor under plane strain conditions; E and ν are, respectively, Young's modulus and Poisson's ratio of the material, which ahead of the crack is a composite made of brittle carbonate and relatively tough cuprate. Therefore, the fracture toughness of the material, expressed by the energy of de-cohesion, depends on carbonate volume fraction, f, along the crack plane. In accordance with a mixture rule, we have

$$\phi_0 = f\phi_0^{\mathrm{car}} + (1-f)\phi_0^{\mathrm{cup}} , \tag{4.43}$$

where ϕ_0^{car} and ϕ_0^{cup} are the de-cohesion energies of the material, which relate with the critical values of the stress intensity factor for carbonate, $K_{\mathrm{Ic}}^{\mathrm{car}}$, and cuprate, $K_{\mathrm{Ic}}^{\mathrm{cup}}$, by using a relationship similar to (4.42). Note that the carbonate is surrounded by cuprate, and its delayed cracking is assumed. The experimental values of the threshold stress intensity factor include the energy required for the generation of the new surface due to crack growth, as well as any plastic dissipation in the cuprate matrix, which surrounds the crack tip carbonate. It is necessary to take these into account by using $K_{\mathrm{Ic}}^{\mathrm{car}}$. The cuprate can sustain different maximum cohesive tractions from that of the carbonate. Therefore, the maximum cohesive traction of the composite material along the crack plane depends on the carbonate volume fraction. Here, the maximum cohesive traction is given as

$$\sigma_{\mathrm{max}} = \sqrt{f\sigma_{\mathrm{car}}^2 + (1-f)\sigma_{\mathrm{cup}}^2} , \tag{4.44}$$

which is derived by assuming that the part of de-cohesion energy during loading satisfies a relation similar to (4.43). In (4.44), $\sigma_{\mathrm{car}}(\sigma_{\mathrm{cup}})$ is the maximum cohesive traction, sustained by the material during crack growth in the presence of the carbonate (cuprate) along the crack plane over a distance from the crack tip $x >> \delta_0$. According to theoretical studies for the crack tip field in elastic plastic materials [225, 460, 885], the maximum hoop stress, along the crack plane, is nearly equal to three times the yield stress of the material. Assuming that σ_{car} is equal to the fracture strength of carbonate, relations (4.43) and (4.44) provide the average properties of the de-cohesion layer over its thickness. Due to carbon diffusion and carbonate formation, carbonate volume fraction changes locally with time, causing a corresponding change of the de-cohesion properties. Maximum cohesive traction is reached when δ_{n} corresponds to a normal traction, satisfying relation (4.44). As time increases, the de-cohesion energy changes, according to (4.43), but unloading starts when relation (4.41) is satisfied. It is assumed that the de-cohesion energy does not change during unloading. The part of de-cohesion energy during unloading is minimized by choosing the largest value of E_{f} for which quasi-static unloading exists [1112].

4.3.2 Discussion of Results

The developed mathematical model of the carbon embrittlement and fracture of YBCO takes into account the coupling of the operating physical processes, namely (i) carbon diffusion, (ii) carbonate precipitation, (iii) non-mechanical energy flow and (iv) cuprate/carbonate composite deformation. Material damage and crack growth are simulated by using the de-cohesion model. Governing equations are obtained for superconductor/carbon system, in which brittle carbonate may precipitate and be accommodated elastically, forming cuprate/carbonate composite. It is assumed that carbon does not affect the elastic modules of the composite. The elastic and thermal properties of the carbonate and composite are taken to be identical in the development of the governing equations. The model can be extended to single or multiphase alloys without any changes, if the additional elements have a very small concentration and do not affect operating processes.

The consideration of elastic behavior in the bulk of the body, causing elastic carbonate accommodation, leads to re-dissolution of the crack tip carbonates, after their fracture, and the reduction of the hydrostatic stresses. In the case of greater plasticity of superconductor, on carbonate precipitation, cuprate matrix should be yielded similar to the hydride behavior in metal matrix [1111]. As a consequence, the crack tip carbonates are more stable and may re-dissolve only partially after fracture. An extension of the present model, which would take into account elastic-plastic carbonate accommodation is useful for simulation of processes beyond crack growth initiation as well as for consideration of all parameters, which play a role in fracture resistance of material. Then, the model of carbon embrittlement and sub-critical fracture of YBCO may be added to earlier developed simulation of toughening mechanisms, acting in HTSC [800, 804, 812, 814, 821, 822]. Another direction for further development of the present model is the consideration of different mechanical and thermal properties of cuprate and carbonate. In this case, the approaches, which are used in composite materials [1175], could be adopted for developing the governing equations. Along this direction is also the consideration of the carbon effect on the elastic modules of carbonate, defining carbon chemical potential in carbonate [620], and therefore on carbonate terminal solid solubility.

Generally, similar to precipitation of hydrogen in metal and intermetallic compounds [616], the carbon effect on the elastic constants of superconductor can be characterized by a factor:

$$r = \frac{(C_{\mathrm{C}} - C_0)/C_0}{C^{\mathrm{C}}/C^{\mathrm{YBCO}}}\,, \qquad (4.45)$$

where C_{C} and C_0 are the elastic moduli of material with and without carbon, respectively; $C^{\mathrm{C}}/C^{\mathrm{YBCO}}$ is the concentration ratio of carbon to superconductor. However, the factor r should be regarded only as a rough approximation of the carbon effect, because the effect is not linear in C^{C}/C^{YBCO} in some cases. The parameter (4.45) should be estimated in experiment and depends on the test temperature and crystallographic properties of HTSC. In particular, the absorption of carbon by a superconductor with free boundaries almost invariably leads to expansion of crystalline lattice and corresponding change of elastic properties.

Similar to the hydrogen effect in intermetallic compounds [616], an existence of the carbon influence on elastic properties (in particular, shear modulus) may be assumed due to an electronic effect. The addition of carbon and the contribution of electrons at the Fermi level moves the Fermi level and results in a corresponding change of temperature dependence of the elastic properties. Carbon contributes electrons in the conduction band and thereby changes the concentration of the conductivity electrons [1030] that also changes electronic contribution to the elastic constants. The sign and magnitude of the effect depend on the electronic band structure and the density of states at the Fermi level. In addition to these long-range effects of the carbon, there are other, more local effects: (i) the direct carbon–superconductor ion potential contributes to the elastic constants, and the interstitial carbon may also affect the superconductor ion–superconductor ion potential [616]; (ii) direct influence on elastic constant of the optical phonons due to the carbon vibrations [311]; and (iii) mechanical relaxation (Snoek effect) of the interstitial carbon in response to strain [655]. However, note that for actual statement of the above carbon effects on elastic properties of HTSC, it is necessary to carry out intensive test investigations.

Then, in the present model simple mixture rules (4.43) and (4.44) have been used for the derivation of the energy of de-cohesion and the maximum cohesive tractions. However, note that the maximum hydrostatic stress, which is expected ahead of the crack tip in an elastic–plastic material before the precipitation of near-tip carbonates, is recovered after their fracture. Consequently, the strong effect of hydrostatic stress distribution on carbon diffusion and carbonate precipitation near the crack tip has been taken into account. At the same time, if perfect cohesion is considered, carbon precipitates at the tip, and no ductile ligament is formed during fracture similar to its occurrence for hydrides in metals [1110]. Therefore, the de-cohesion model improves the performance of the carbon embrittlement model, which is based on elastic behavior of the material. Moreover, the de-cohesion model takes into account the time variation of the de-cohesion energy due to the time-dependent process of carbonate precipitation.

Finally, note that a similar mathematical model and governing equations may be obtained for other HTSC under carbon embrittlement. Finite-element scheme for numerical realization of the governing equations is presented in Appendix B.

4.4 Modeling of Carbon Segregation and Fracture Processes of HTSC

It could be proposed that carbon can segregate not only to grain boundaries, but also to crack surfaces and dislocations, where lattices are distorted. Therefore, two microcracking processes are possible: continuously slow crack growth and discretely rapid crack growth, associated with high amounts of acoustic emissions. Then, the carbon segregation processes can be studied by using the microscopic models of the equilibrium slow and fast crack propagation and also a steady-state crack growth, which are screened by dislocation field [812].

4.4.1 Equilibrium Slow and Fast Crack Growth

Consider an intergranular crack of length, $2a$, in a carbonated HTSC (Fig. 4.13). The crack lies along x-axis in an elastic–plastic isotropic body with shear modulus, G, Poisson ratio, ν, yield strength, σ_{y}, and work hardening factor, n. The body is loaded by a remote stress, σ_{a} parallel to the y-axis at a constant temperature, T. At the x-axis, two linear dislocation arrays with the length, r_{y}, are located at the distance, d, from the crack tips. This model proposes that an intergranular crack tip maintains an atomistic sharpness and a local equilibrium condition in the presence of screening dislocations. It is assumed that all geometrical parameters of the crack tip, presented in Fig. 4.14 (in particular, the size of the arc-shaped crack tips, q, and a crack tip displacement, δ_{c}), remain constant during plastic deformation. The condition

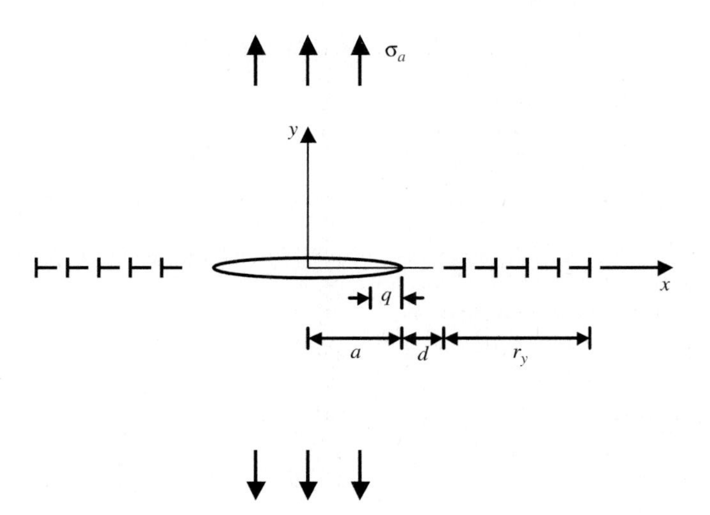

Fig. 4.13. Schematic representation of intergranular crack, screened by linear dislocation array

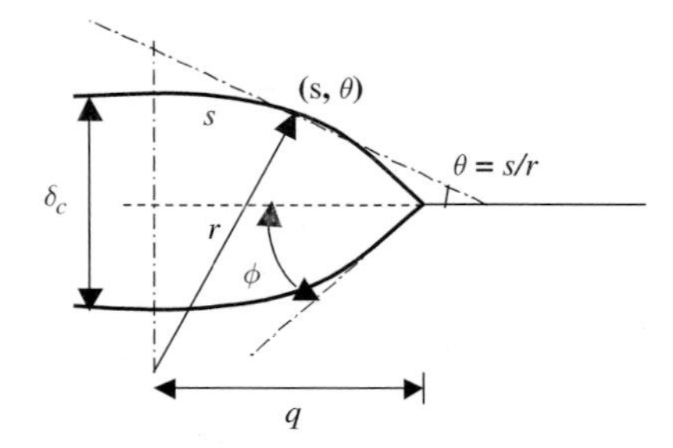

Fig. 4.14. Illustration of crack tip profile, indicating crack tip angle, 2ϕ, and polar coordinates $(s,\ \theta)$

of the local equilibrium at the crack tip is that the crack must be screened by dislocation field and maintains a dislocation-free zone with the size d. The loaded system "crack–dislocation arrays" maintains a local stress, σ_{d}, in the dislocation-free zone and produces the next stress intensity in the screening dislocation zone, given by Hutchinson–Rice–Rosengren model as [884]

$$\sigma_{\mathrm{yy}} = \sigma_{\mathrm{d}} \qquad a < |x| < a + d \tag{4.46}$$

$$\sigma_{\mathrm{yy}} = \beta\sigma_{\mathrm{y}}(K_{\mathrm{a}}/\sigma_{\mathrm{y}})^{2n/(n+1)}(|x| - a)^{-n/(n+1)} \qquad a + d < |x| < a + d + r_{\mathrm{y}} \tag{4.47}$$

where K_{a} is the applied stress intensity and β is the factor, which depends on the elastic and plastic deformation properties (see Fig. 4.15).

The carbon segregation process is found by the crack tip profile and by the stress field ahead of the crack tip. The chemical potentials of carbon and superconductor can be stated, following [507] in various grain boundary and crack surface zones, namely, I is the zone not affected by the stress intensity $(|x| > a+d+r_{\mathrm{y}})$; II is the zone of screening dislocations $(a+d < |x| < a+d+r_{\mathrm{y}})$; III is the dislocation-free zone ahead of the crack tip $(a < |x| < a + d)$; IV is the arc-shaped crack tip zone $(a - q < |x| < a)$; and V is the parallel flat crack surface zone $(|x| < a - q)$.

At equilibrium, the chemical potentials of carbon and superconductor must be the same in all the regions, respectively. So, the equilibrium carbon segregation depends on the binding energies and crack tip conditions. The binding energies of carbon at grain boundaries and crack surfaces $(H_{\mathrm{B}})_{\mathrm{b}}$ and $(H_{\mathrm{B}})_{\mathrm{s}}$, respectively, are found through the standard chemical potentials of C and HTSC as

$$(H_{\mathrm{B}})_{\mathrm{b}} = (\mu_{\mathrm{m0}})_{\mathrm{C}} - (\mu_{\mathrm{b0}})_{\mathrm{C}} - (\mu_{\mathrm{m0}})_{\mathrm{HTSC}} + (\mu_{\mathrm{b0}})_{\mathrm{HTSC}}\ ; \tag{4.48}$$

$$(H_{\mathrm{B}})_{\mathrm{s}} = (\mu_{\mathrm{m0}})_{\mathrm{C}} - (\mu_{\mathrm{s0}})_{\mathrm{C}} - (\mu_{\mathrm{m0}})_{\mathrm{HTSC}} + (\mu_{\mathrm{s0}})_{\mathrm{HTSC}}\ . \tag{4.49}$$

Here and further the subscript m is the matrix, b is the grain boundary and s is the crack surface.

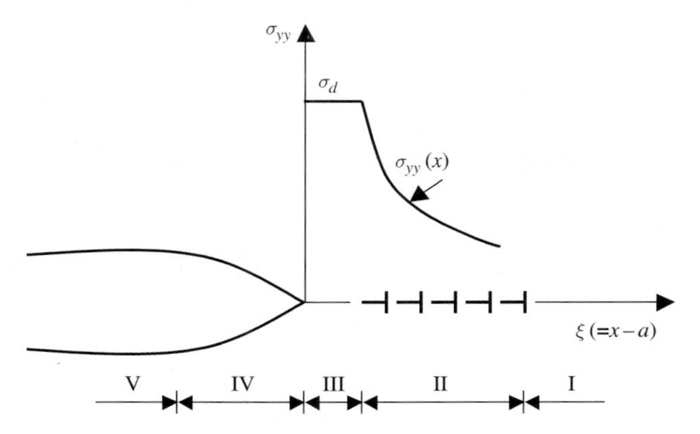

Fig. 4.15. Illustration of local stress and stress intensity ahead of the dislocation-screened crack tip and definition of various regions, where different chemical potentials are given

The basic assumption of the model is that the embrittlement occurs as a reduction of the surface and grain boundary energies due to the carbon segregation. Moreover, it is taken into account that slow fracture occurs when the solute is sufficiently rapid to maintain the same chemical potential of solute between the grain boundary and the crack surface. Fast fracture occurs when the solute concentration remains the same at the grain boundary and the crack surface. Then, from the thermodynamic theory proposed by Seah, Rice and Hirth, the ideal works, expended in slow (γ^{s}) and fast (γ^{f}) fracture, can be obtained as [507]

$$\gamma^{\mathrm{s}} = \gamma_0 - RT(2C_{\mathrm{V}}/\Omega_{\mathrm{s}} - C^{\mathrm{s}}_{\mathrm{III}}/\Omega_{\mathrm{b}}) \; ; \tag{4.50}$$

$$\gamma^{\mathrm{f}} = \gamma_0 - (C^{\mathrm{f}}_{\mathrm{III}}/\Omega_{\mathrm{b}})\Delta\mu \; . \tag{4.51}$$

Here, the equilibrium carbon concentrations in zones III and V have forms:

$$C_{\mathrm{III}} = \frac{C_{\mathrm{m}} \exp\left\{\left[(H_{\mathrm{B}})_{\mathrm{b}} + \sigma_{\mathrm{d}} V_{\mathrm{h}} - \frac{\sigma_{\mathrm{d}}^2 V_{\mathrm{h}}}{4G(1+\nu)}\right]/(RT)\right\}}{1 - C_{\mathrm{m}} + C_{\mathrm{m}} \exp\left\{\left[(H_{\mathrm{B}})_{\mathrm{b}} + \sigma_{\mathrm{d}} V_{\mathrm{h}} - \frac{\sigma_{\mathrm{d}}^2 V_{\mathrm{h}}}{4G(1+\nu)}\right]/(RT)\right\}} \; ; \tag{4.52}$$

$$C_{\mathrm{V}} = \frac{C_{\mathrm{m}} \exp[(H_{\mathrm{B}})_{\mathrm{s}}/(RT)]}{1 - C_{\mathrm{m}} + C_{\mathrm{m}} \exp[(H_{\mathrm{B}})_{\mathrm{s}}/(RT)]} \; , \tag{4.53}$$

where C_{m} is the bulk carbon concentration; V_{h} is the molar volume of carbon; R is the gas constant; $1/\Omega_i$ is the carbon coverage at interfaces; $C^{\mathrm{s}}_{\mathrm{III}}$ and $C^{\mathrm{f}}_{\mathrm{III}}$ are the critical values of carbon concentration in zone III, required for slow and fast fractures, respectively; γ_0 is the ideal work of intergranular fracture in the absence of carbon ($= 2\gamma_{\mathrm{s}0} - \gamma_{\mathrm{b}0}$; $\gamma_{\mathrm{s}0}$ is the surface fracture energy and $\gamma_{\mathrm{b}0}$ is the fracture energy of grain boundary); and $\Delta\mu = RT\,\ln(2C_{\mathrm{V}}/C^{\mathrm{f}}_{\mathrm{III}})$ is the chemical potential difference between the crack surface and the stressed grain boundary.

The equations for constant carbon concentrations can also be found in zones I and IV. In this case, C_{I} coincides with C_{V}, in which $(H_{\mathrm{B}})_{\mathrm{s}}$ is replaced by $(H_{\mathrm{B}})_{\mathrm{b}}$, and for C_{IV} we have

$$\exp\left(\frac{C_{\mathrm{IV}}V_{\mathrm{h}}}{\Omega_{\mathrm{s}}r}\right)+\frac{C_{\mathrm{IV}}(1-C_{\mathrm{m}})}{(1-C_{\mathrm{IV}})C_{\mathrm{m}}}=\exp\{[(H_{\mathrm{B}})_{\mathrm{s}}+\gamma_{\mathrm{s0}}V_{\mathrm{h}}/\Omega_{\mathrm{s}}]/(RT)\}\,, \quad (4.54)$$

where r is the curvature radius of the arc-shaped crack tip (Fig. 4.14). At the same time, due to the variable stress distribution (4.46) and (4.47), the carbon content in zone II is not constant. The relationship between the critical stress intensity, required to propagate the crack (for slow, fast or steady-state fracture) and to change the ideal work due to the carbon segregation, is stated by using the local energy balance condition as [507]

$$-(1-\nu)K_{\mathrm{d}}^2/2G+\gamma^{\mathrm{c}}\leq 0\,, \quad (4.55)$$

where the superscript c corresponds to certain fracture state, K_{d} is the local stress intensity factor connected with the dislocation-free zone size ahead of the crack tip (d) and local stress (σ_{d}) in this zone by the equation, approximately derived from the load balance condition between a crack with a linear stress intensity and that with local stress [507]: $\pi d=(K_{\mathrm{d}}/\sigma_{\mathrm{d}})^2$. Moreover, a relation between $\sigma_{\mathrm{d}}, K_{\mathrm{d}}$ and δ_{c} follows from the condition that the elastic energy release rate is the same as the J-integral, i.e., $\sigma_{\mathrm{d}}=K_{\mathrm{d}}[2(1-\nu^2)/\delta_{\mathrm{c}}]^{1/2}$. Then the threshold apparent stress intensity, $K_{\mathrm{th}}^{\mathrm{c}}$, is given by the relationships (4.47), and (4.55):

$$K_{\mathrm{th}}^{\mathrm{c}}=K_0(\gamma^{\mathrm{c}}/\gamma_0)^{(n+1)/4n}(\delta_{\mathrm{c0}}/\delta_{\mathrm{c}}^{\mathrm{c}})^{(1-n)/4n}\,, \quad (4.56)$$

where K_0 is the fracture toughness, $\delta_{\mathrm{c}}^{\mathrm{c}}$ is the critical crack opening displacement (CCOD), required for various fracture processes (superscript c), and δ_{c0} is the CCOD in the absence of carbon, defined as

$$\delta_{\mathrm{c0}}=\frac{[4G(1+\nu)\gamma_0]^{(n+1)/(1-n)}}{[2\pi(1-\nu^2)]^{2n/(1-n)}\beta^{2(n+1)/(1-n)}\sigma_{\mathrm{y}}^2K_0^{4n/(1-n)}}\,. \quad (4.57)$$

Note that for the crack to maintain the dislocation-free zone during the growth, besides the inequality (4.55), it is necessary to satisfy an additional condition, namely the total energy balance criterion in the form [507]:

$$-(1-\nu)K_{\mathrm{a}}^2/2G+\gamma^{\mathrm{c}}+\gamma_{\mathrm{p}}\leq 0\,, \quad (4.58)$$

where γ_{p} is the plastic work due to the generation and motion of screening dislocations, which could be found numerically, for example, in the case of a linear dislocation array [507, 884].

4.4.2 Steady-State Crack Growth

Assume that the carbon diffusion along stressed boundaries and crack surfaces is the mechanism which controls the intergranular embrittlement and affects the crack growth rate. In this case, the bulk diffusion effects on carbon-induced intergranular cracking (CIIC) are neglected. Under the geometrical and loading conditions of the equilibrium crack growth problem, the steady-state case indicates subcritical intergranular crack growth with constant velocity, ν_c (Fig. 4.16). Taking into account the grain boundary and crack surface zones (II–V), the fluxes of carbon in these regions, J_i^j, can be stated as

$$J_\mathrm{i}^\mathrm{j} = -\frac{D_\mathrm{i} C_\mathrm{i}^\mathrm{j}}{RT} \frac{\mathrm{d}\mu_\mathrm{i}^\mathrm{j}}{\mathrm{d}(x \text{ or } s)} , \tag{4.59}$$

where D_i is the diffusivity of carbon, C_i^j is the carbon concentration, i is the subscript indicating b or s and j is the superscript indicating various interface zones; $\mu_\mathrm{i}^\mathrm{j}$ are the corresponding chemical potentials. The differentiation with respect to s is carried out only in zone IV; in this case s is the variable arc length in the corresponding part of the arc-shaped crack tip (see Fig. 4.14). The continuity equation of fluxes is

$$\frac{\mathrm{d}C_\mathrm{i}^\mathrm{j}}{\mathrm{d}t} + \frac{\mathrm{d}J_\mathrm{i}^\mathrm{j}}{\mathrm{d}(x \text{ or } s)} = 0 , \tag{4.60}$$

where t is the time. Based on (4.46), (4.47), (4.59) and (4.60) and also the relationships between the interface energies, $\gamma_\mathrm{i}^\mathrm{j}$, and the amounts of carbon,

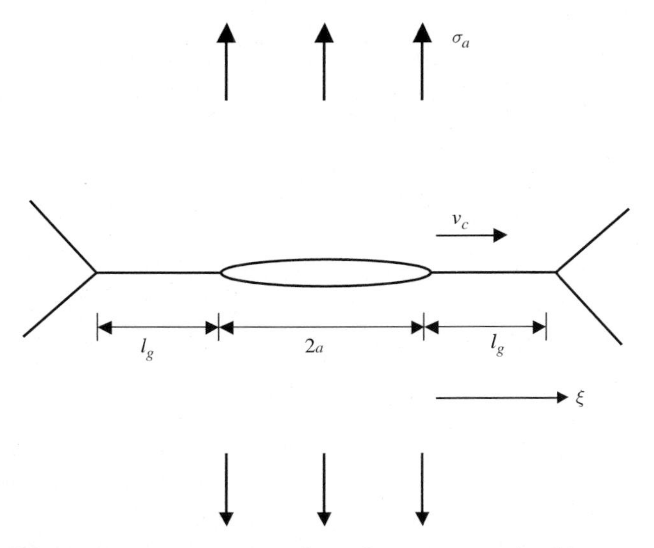

Fig. 4.16. Schematic representation of steady-state growth of intergranular crack and boundary condition

$C_{\rm i}^{\rm j}$, in the various zones, derived from Gibbs theory and dilute solute approximation as $\gamma_{\rm i}^{\rm j} = \gamma_{\rm i0} - (RT/\Omega_{\rm i})C_{\rm i}^{\rm j}$, the second-order differential equations controlling the carbon diffusion in the intergranular cracking regions can be obtained, similar to [508]:

Zone II:

$$J_{\rm b}^{\rm II} = -D_{\rm b}\frac{{\rm d}C_{\rm b}^{\rm II}}{{\rm d}x} - \left(\frac{D_{\rm b}V_{\rm h}}{RT}\right)\frac{n}{(n+1)}\beta\sigma_{\rm y}^{(1-n)/(n+1)} \times K_{\rm a}^{2n/(n+1)}(x-a)^{(-2n-1)/(n+1)}C_{\rm b}^{\rm II}\,; \tag{4.61}$$

$$\frac{{\rm d}^2C_{\rm b}^{\rm II}}{{\rm d}x^2} + \left[\left(\frac{\nu_{\rm c}}{D_{\rm b}}\right) + \left(\frac{V_{\rm h}}{RT}\right)\frac{n}{(n+1)}\beta\sigma_{\rm y}^{(1-n)/(n+1)} \times K_{\rm a}^{2n/(n+1)}(x-a)^{(-2n-1)/(n+1)}\right]\frac{{\rm d}C_{\rm b}^{\rm II}}{{\rm d}x} - - \left(\frac{V_{\rm h}}{RT}\right)\frac{n(2n+1)}{(n+1)^2}\beta\sigma_{\rm y}^{(1-n)/(n+1)} \times K_{\rm a}^{2n/(n+1)}(x-a)^{(-3n-2)/(n+1)}C_{\rm b}^{\rm II} = 0\,. \tag{4.62}$$

Zone III:

$$J_{\rm b}^{\rm III} = -D_{\rm b}\frac{{\rm d}C_{\rm b}^{\rm III}}{{\rm d}x}\,; \tag{4.63}$$

$$\frac{{\rm d}^2C_{\rm b}^{\rm III}}{{\rm d}x^2} + \left(\frac{\nu_{\rm c}}{D_{\rm b}}\right)\frac{{\rm d}C_{\rm b}^{\rm III}}{{\rm d}x} = 0\,. \tag{4.64}$$

Zone IV:

$$J_{\rm s}^{\rm IV} = -D_{\rm s}\frac{{\rm d}C_{\rm s}^{\rm IV}}{{\rm d}s} - \left(\frac{D_{\rm s}V_{\rm h}}{r\Omega_{\rm s}}\right)\frac{{\rm d}C_{\rm s}^{\rm IV}}{{\rm d}s}C_{\rm s}^{\rm IV}\,; \tag{4.65}$$

$$\left[1 + \left(\frac{V_{\rm h}C_{\rm s}^{\rm IV}}{r\Omega_{\rm s}}\right)\right]\frac{{\rm d}^2C_{\rm s}^{\rm IV}}{{\rm d}s^2} + \left(\frac{V_{\rm h}}{r\Omega_{\rm s}}\right)\left(\frac{{\rm d}C_{\rm s}^{\rm IV}}{{\rm d}s}\right)^2 + \sec\phi\left(\frac{\nu_{\rm c}}{D_{\rm s}}\right)\frac{{\rm d}C_{\rm s}^{\rm IV}}{{\rm d}s} = 0\,. \tag{4.66}$$

Zone V:

$$J_{\rm s}^{\rm V} = -D_{\rm s}\frac{{\rm d}C_{\rm s}^{\rm V}}{{\rm d}x}\,; \tag{4.67}$$

$$\frac{{\rm d}^2C_{\rm s}^{\rm V}}{{\rm d}x^2} + \left(\frac{v_{\rm c}}{D_{\rm s}}\right)\frac{{\rm d}C_{\rm s}^{\rm V}}{{\rm d}x} = 0\,. \tag{4.68}$$

The boundary values of carbon concentration, $C_{\rm V}^{\bullet\bullet}$ (at $x = 0$) and $C_{\rm b}^{\bullet\bullet}$ (at $x = a + l_{\rm g}$) used in the solution of the second-order differential equations, have forms:

$$C_{\mathrm{V}}^{\bullet\bullet} = \frac{C_{\mathrm{m}} \exp[(H_{\mathrm{B}})_{\mathrm{s}}/(RT)]}{1 - C_{\mathrm{m}} + C_{\mathrm{m}} \exp[(H_{\mathrm{B}})_{\mathrm{s}}/(RT)]} ; \tag{4.69}$$

$$C_{\mathrm{b}}^{\bullet\bullet} = \frac{C_{\mathrm{m}} \exp\{[(H_{\mathrm{B}})_{\mathrm{b}} + \sigma_{\mathrm{yb}} V_{\mathrm{h}}]/(RT)\}}{1 - C_{\mathrm{m}} + C_{\mathrm{m}} \exp\{[(H_{\mathrm{B}})_{\mathrm{b}} + \sigma_{\mathrm{yb}} V_{\mathrm{h}}]/(RT)\}} , \tag{4.70}$$

where σ_{yb} is the local stress in a triple point of intergranular boundary.

The conditions at the boundaries between zones are given by

Zones II–III:

$$(C_{\mathrm{b}}^{\mathrm{II}})_{x=a+d} = (C_{\mathrm{b}}^{\mathrm{III}})_{x=a+d} ; \tag{4.71}$$

$$\left(\frac{\mathrm{d}C_{\mathrm{b}}^{\mathrm{II}}}{\mathrm{d}x}\right)_{x=a+d} + \left(\frac{V_{\mathrm{h}}}{RT}\right) \frac{n}{(n+1)} \beta \sigma_{\mathrm{y}}^{(1-n)/(n+1)}$$
$$K_{\mathrm{a}}^{2n/(n+1)} d^{(-2n-1)/(n+1)} (C_{\mathrm{b}}^{\mathrm{II}})_{x=a+d} = \left(\frac{\mathrm{d}C_{\mathrm{b}}^{\mathrm{III}}}{\mathrm{d}x}\right)_{x=a+d} . \tag{4.72}$$

Zones III–IV (at the crack tip):

$$\mu_{\mathrm{b}0} + RT\, ln(C_{\mathrm{b}}^{\mathrm{III}})_{x=a} - \sigma_{\mathrm{d}}^{\bullet} V_{\mathrm{h}} + \frac{\sigma_{\mathrm{d}}^{\bullet 2} V_{\mathrm{h}}}{4G(1+\nu)} =$$
$$= \mu_{\mathrm{s}0} + RT \ln(C_{\mathrm{s}}^{\mathrm{IV}})_{s=r\phi} - (V_{\mathrm{h}}/r)\{\gamma_{\mathrm{s}0} - [RT(C_{\mathrm{s}}^{\mathrm{IV}})_{s=r\phi}/\Omega_{\mathrm{s}}]\} ; \tag{4.73}$$

$$D_{\mathrm{b}} \left(\frac{\mathrm{d}C_{\mathrm{b}}^{\mathrm{III}}}{\mathrm{d}x}\right)_{x=a} = 2D_{\mathrm{s}} \left[1 + \left(\frac{V_{\mathrm{h}}(C_{\mathrm{s}}^{\mathrm{IV}})_{s=r\phi}}{r\Omega_{\mathrm{s}}}\right)\right] \left(\frac{\mathrm{d}C_{\mathrm{s}}^{\mathrm{IV}}}{\mathrm{d}s}\right)_{s=r\phi} . \tag{4.74}$$

Zones IV–V:

$$RT \ln(C_{\mathrm{s}}^{\mathrm{IV}})_{s=0} - (V_{\mathrm{h}}/r)\{\gamma_{\mathrm{s}0} - [RT(C_{\mathrm{s}}^{\mathrm{IV}})_{s=0}/\Omega_{\mathrm{s}}]\} = RT \ln(C_{\mathrm{s}}^{\mathrm{V}})_{x=a-q} ; \tag{4.75}$$

$$\left[1 + \left(\frac{V_{\mathrm{h}}(C_{\mathrm{s}}^{\mathrm{IV}})_{s=0}}{r\Omega_{\mathrm{s}}}\right)\right] \left(\frac{\mathrm{d}C_{\mathrm{s}}^{\mathrm{IV}}}{\mathrm{d}s}\right)_{s=0} = \left(\frac{\mathrm{d}C_{\mathrm{s}}^{\mathrm{V}}}{\mathrm{d}x}\right)_{x=a-q} . \tag{4.76}$$

It is assumed that the steady-state crack growth maintains the equilibrium values at the crack center and at the triple point of grain boundaries ahead of the crack. It should be noted that the present boundary condition is a first order approximation because the equilibrium content of carbon at the triple grain junction is difficult to attain, especially at sufficiently high velocity of thecrack. The interface conditions show that the chemical potentials and fluxes of carbon must be the same at each interface in order to maintain the continuity of the carbon flux. So, the boundary-value problem is stated for the solution of which some relationships, defined in the equilibrium crack growth, are to be used, namely: the local equilibrium condition at the crack tip, the geometrical crack tip conditions, and also the crack tip condition, derived from the local energy criterion (4.55). The carbon diffusivity effect is determined by the ideal work of steady-state fracture as

$$\gamma^{\bullet} = \gamma_0 - RT(2C_{\mathrm{V}}^{\bullet}/\Omega_{\mathrm{s}} - C_{\mathrm{III}}^{\bullet}/\Omega_{\mathrm{b}}) + (C_{\mathrm{V}}^{\bullet}/\Omega_{\mathrm{s}} + C_{\mathrm{III}}^{\bullet}/2\Omega_{\mathrm{b}})\Delta\mu^{\bullet} \,, \tag{4.77}$$

where

$$C_{\mathrm{V}}^{\bullet} = (C_{\mathrm{s}}^{\mathrm{V}})_{x=a-q}, \qquad C_{\mathrm{III}}^{\bullet} = (C_{\mathrm{b}}^{\mathrm{III}})_{x=a} \,; \tag{4.78}$$

$$\Delta\mu^{\bullet} = \mu_{\mathrm{b0}} - \mu_{\mathrm{s0}} + RT\,\ln(C_{\mathrm{III}}^{\bullet}/C_{\mathrm{V}}^{\bullet}) - \sigma_{\mathrm{d}}^{\bullet}V_{\mathrm{h}} + \frac{(\sigma_{\mathrm{d}}^{\bullet})^2\,V_{\mathrm{h}}}{4G(1+\nu)} \,. \tag{4.79}$$

The superscript $\bullet$ indicates the steady-state fracture. The boundary-value problem can be solved numerically, for example, by using the Runge-Kutta method. The boundary conditions at the triple points permit to study the effect of grain sizes on the kinetics of CIIC. At the same time, the size effects cannot be estimated in the cases of the equilibrium slow and fast cracks.

4.4.3 Some Numerical Results

Numerical results can be obtained in the case of equilibrium crack growth for different values of the bulk carbon concentration, C_{m}. Equating the right parts of the (4.46), and (4.47) at $|x| = a + d$ and $K_{\mathrm{a}} = K_0$ and considering the relations (4.50), (4.52), (4.53) and (4.55) for slow crack and (4.51), (4.52), (4.53) and (4.55) for fast crack, the problem is reduced to numerical solution of transcendental algebraic equations. These equations state the relationships γ^{c} and the critical tip conditions $(\sigma_{\mathrm{d}}^{\mathrm{c}}, \delta_{\mathrm{c}}^{\mathrm{c}})$ to C_{m}. Then, the values of $K_{\mathrm{th}}^{\mathrm{c}}$ are determined from (4.56), using the calculated values of γ^{c} and $\delta_{\mathrm{c}}^{\mathrm{c}}$. Based on test data, the necessary parameters for numerical calculations are $\gamma_0 = 1\,\mathrm{J/m^2}$, $1/\Omega_{\mathrm{b}} = 1/\Omega_{\mathrm{s}} = 8.1\times10^{-5}\,\mathrm{mol/m^2}$, $(H_{\mathrm{B}})_{\mathrm{s}} = 50\,\mathrm{kJ/mol}$, $(H_{\mathrm{B}})_{\mathrm{b}} = 10\,\mathrm{kJ/mol}, n = 0.1$, $\sigma_{\mathrm{y}} = 10\,\mathrm{MPa}$, $V_{\mathrm{h}} = 2\times10^{-6}\,\mathrm{m^3/mol}$, $R = 8.316\,\mathrm{J/mol\,K}$, $\nu = 0.2$, $K_0 = 1\,\mathrm{MPa\,m^{1/2}}$, $G = 50\,\mathrm{GPa}, T = 1110\,\mathrm{K}$. The numerical results are presented in Tables 4.2 (slow crack) and 4.3 (fast crack) [812].

As shown by the numerical results, all auxiliary parameters ($d^{\mathrm{c}}, \sigma_{\mathrm{d}}^{\mathrm{c}}, K_{\mathrm{d}}^{\mathrm{c}}$, $\gamma^{\mathrm{c}}, \delta_{\mathrm{c}}^{\mathrm{c}}$) change monotonously with C_{m} for both slow and fast fractures. In particular, the normalized parameter, $\gamma^{\mathrm{c}}/\gamma_0$, decreases with an increase of C_{m}. At the same time, the normalized parameter, $\delta_{\mathrm{c0}}/\delta_{\mathrm{c}}^{\mathrm{c}}$, increases along with C_{m}. These alternative contributions to $K_{\mathrm{th}}^{\mathrm{c}}/K_0$ cause its non-monotonous behavior in the dependence on C_{m}. In this case, the strengthening effect on $K_{\mathrm{th}}^{\mathrm{c}}$ (i.e., when $K_{\mathrm{th}}^{\mathrm{c}}/K_0 > 1$) occurs when the segregation of carbon in the crack regions strongly affects the crack tip condition (i.e., reduction of $\delta_{\mathrm{c}}^{\mathrm{c}}$), but does not produce a substantial reduction in γ^{c} (see (4.56)). More evident change of all auxiliary parameters in the case of slow crack compared with the

Table 4.2. Numerical results for slow crack

Parameters	C_m (ppm)					
	50	100	150	200	250	300
d^s (μm)	185	176	167	159	150	142
σ_d^s/σ_y	1.438	1.444	1.451	1.458	1.465	1.472
K_d^s/K_0	0.347	0.340	0.332	0.326	0.318	0.311
γ^s/γ_0	0.963	0.925	0.882	0.850	0.809	0.774
δ_{c0}/δ_c^s	1.050	1.104	1.164	1.222	1.295	1.368
K_{th}^s/K_0	1.007	1.007	0.996	1.004	0.999	0.999

fast crack at the considered range of C_m proposes the greater susceptibility of slow growth on CIIC increase. Then, it is apparent that under the condition of a dislocation-screened crack, carbon segregation induces slow fracture more readily than fast fracture. The weak change of K_{th}^c/K_0 on C_m in both cases of slow and fast cracks is apparently caused by the small range of C_m (while this is a real bulk carbon concentration in HTSC systems). The presented numerical example should be specified with more accurate selection of key parameters for certain HTSC.

For the used local energy criterion, which is controlled by CIIC, the dependence of K_{th}^c on the ideal work of fracture, the crack tip conditions and the plastic deformation properties (see (4.56)) is somewhat similar to that obtained in [1137]. The difference is that the present analysis explicitly includes not only the embrittlement effect of carbon, but also the crack tip conditions, affected by the carbon segregation. It should be noted that the presence of a dislocation-screened crack and the microscopic behavior of plastic deformation, associated with CIIC, have not yet been experimentally verified in HTSC compositions. However, as has been known, an intergranular crack remains atomistically sharp when an energy barrier for the nucleation of a dislocation loop at a crack tip is present. This barrier is produced due to a low level of stress intensity at the crack tip in the presence of screening dislocations, stated by the dislocation sources (e.g., such as intergranular boundaries,

Table 4.3. Numerical results for fast crack

Parameters	C_m (ppm)					
	50	100	150	200	250	300
d^f (μm)	188	182	176	169	163	157
σ_d^f/σ_y	1.436	1.440	1.444	1.450	1.454	1.459
K_d^f/K_0	0.349	0.344	0.340	0.334	0.329	0.324
γ^f/γ_0	0.974	0.947	0.925	0.892	0.866	0.840
δ_{c0}/δ_c^f	1.034	1.067	1.104	1.150	1.192	1.238
K_{th}^f/K_0	1.002	0.997	1.007	1.000	1.000	1.000

particles, defects, etc.) [634]. Thus, the relative strength of energy barriers for dislocation nucleation, produced by the crack tip or/and the other dislocation sources, states whether a crack tip maintains an atomistic sharpness or emits dislocation loops. Obviously, the carbon segregation in HTSC causes a complex effect on the dislocation behavior. Then, the occurrence of the dislocation-screened crack tips is possible in the presence of carbon, depending on how carbon affects the generation of dislocation at the crack tips and the other dislocation sources. Besides the crack tip conditions, the strength of carbon binding at crack surface and intergranular boundaries also controls the carbon embrittlement of the grain boundaries. It is known that the value of $(H_{\mathrm{B}})_{\mathrm{s}}$ is much higher than that of $(H_{\mathrm{B}})_{\mathrm{b}}$, and they are found by the conditions at interfaces such as solute or carbon coverage, structure and roughness. Moreover, a high degree of lattice incoherence (e.g., at the interphase boundary [634]) can also state a higher susceptibility to carbon embrittlement of HTSC. Obviously, there are also other factors controlling the strength of carbon binding at interfaces. So, the values of carbon binding need an accurate experimental foundation for the considered HTSC systems.

The detailed analysis of numerical results for steady-state crack outruns the book. Note, only the parameters corresponding to the obtained ones for slow and fast cracks are found to be dependent on the crack velocity and the carbon diffusivities at grain boundaries and crack surfaces. In particular, the numerical results, obtained for steady-state crack, indicate that the crack growth rate is higher for smaller grain size, that is as the grain size decreases the susceptibility to CIIC increases. In total, the above solutions and numerical results can be used in the finite element formulations and other numerical codes by which the stress–strain state distributions, kinetics and parameters of intergranular defects during HTSC bulk manufacture can be predicted.

5

General Aspects of HTSC Modeling

The problem of fabrication of the oxide superconductors with high structure-sensitive properties suggests as one from primary tasks a design and creation of property monitoring of the HTSC ceramics and composites.[1] This includes (i) observation and modeling of microstructure transformations, causing formation (during processing) and change (during loading) of the material fracture resistance that renders defining influence on superconducting properties; (ii) estimation of change of the microstructure, strength and conducting properties under different loading; and (iii) property prognosis of final product, depending on the sample composition, parameters and features of processing technique. An approximate scheme of numerical monitoring is presented in Fig. 5.1 and includes specific stages of the superconductor preparation and loading, accompanied by microstructure transformations; the modeling stages of strength, superconducting and other structure-sensitive properties; and proper processes, determining the final sample parameters. Note that Fig. 5.1d–f shows modeling in square lattice. However, corresponding realization of the computational scheme in other lattices (i.e., triangular, hexagonal, overlapping elements, etc.) does not present principal difficulties. Table 5.1 shows possibilities of the monitoring and spectrum of superconductor properties, which may be defined at the monitoring realization.

As it has been pointed in the previous chapters, the formation of microstructure defects and weak links during material compaction and sintering renders a significant effect on structure-sensitive properties of HTSC. In this chapter, two important problems are considered, which relate to the technique optimization of HTSC systems, namely the yield criteria are stated, which can describe both displacement into press-powder bulk and its consolidation during compaction [805, 806], and the void formation and transformation due to diffusion processes during sintering are studied [801, 805]. The proposed yield criteria are based on the associated and non-associated flow rules. The

[1] For non-cubic Al_2O_3 ceramic, a numerical scheme of the monitoring has been realized in [803].

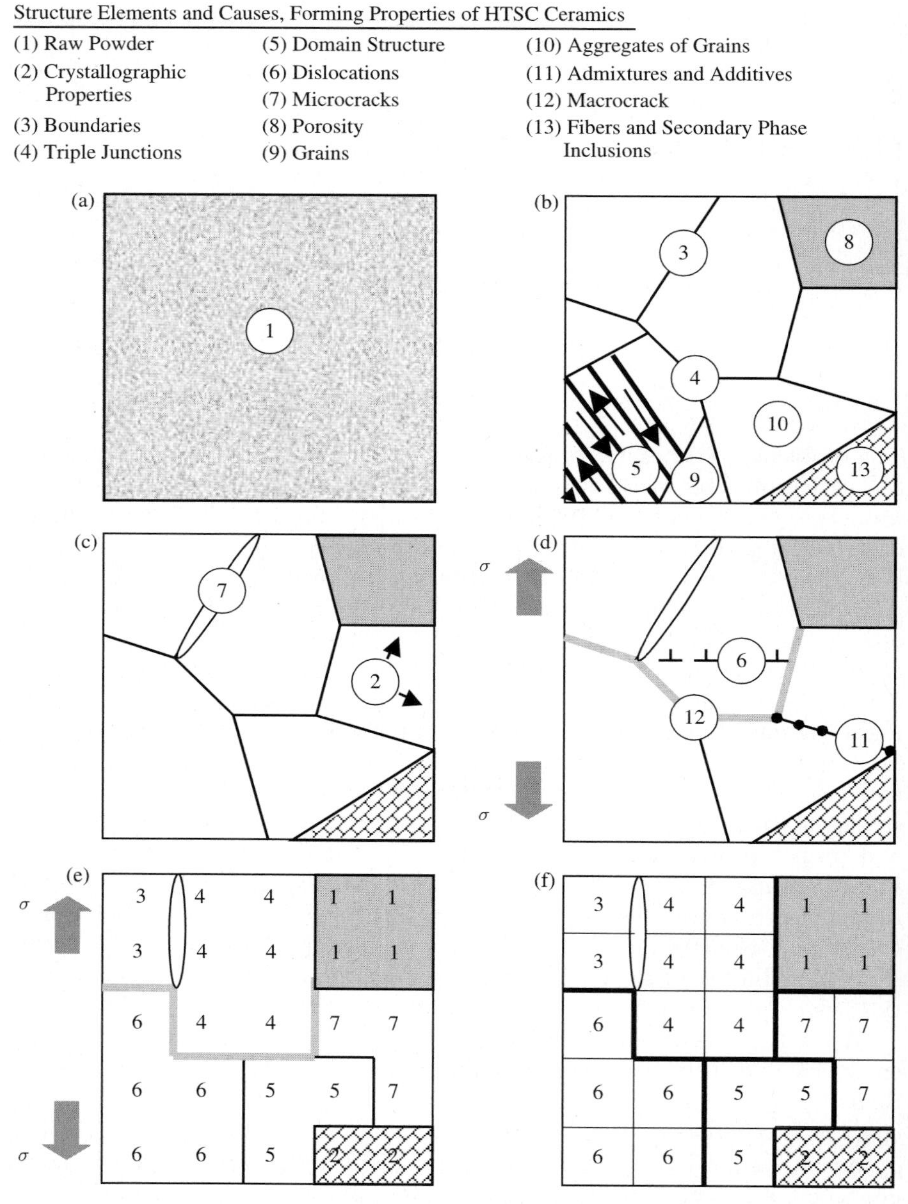

Fig. 5.1. General scheme of computational monitoring of HTSC structure-sensitive properties: (**a**) initial powder, (**b**) sintering, (**c**) cooling, (**d**) macrocrack propagation, (**e**) presentation of structure fragment in PC, (**f**) model structure for study of percolation properties

Table 5.1. Possibilities of computer monitoring and spectrum of estimated properties

Investigated processes and effects	Estimated and accounted properties and characteristics
(1) Thermal and mechanical loading	(1) Elastic anisotropy
(2) Solidification	(2) Thermal expansion anisotropy
(3) Shrinkage	(3) Deformation and temperature mismatch of phases and components
(4) Secondary re-crystallization (grain growth)	(4) Residual (internal) stresses
(5) Phase transitions	(5) External physical and mechanical loading
(6) Superconducting transition	(6) Fracture conditions and types
(7) Non-superconducting fibers, inclusions and additives into superconducting matrix	(7) Microcracking
(8) Superconducting inclusions into non-superconducting matrix	(8) Phase transformations
(9) Defect structures, in particular carbon effect	(9) Twinning and domain re-orientation
(10) Dislocation structures	(10) Crack deflection and twisting
(11) Percolation	(11) Crack branching
(12) Structure effect on magnetic flux pinning	(12) Crack interaction with structure heterogeneity (pores, dislocations, microcracks, inclusions, etc.)
	(13) Crack bridging
	(14) R- (or T-) curve behavior
	(15) Pushing of grains (or fibers) by crack surfaces
	(16) Crack coalescence
	(17) Crack healing
	(18) Crack cohesion
	(19) Scaling factor
	(20) Effective superconducting properties
	(21) Percolation properties
	(22) Magnetic flux pinning

microstructure transformations of porosity are investigated in the framework of phenomenological models of the shrinkage, coarsening and coalescence of intergranular voids, and also their separation from grain boundaries and displacement to interior of grains is possible. Second part of the chapter is devoted to modeling of micro- and macrostructure processes during processing and fracture of HTSC-ceramic, taking into account heating, shrinkage and cooling of material, grain growth and sample microcracking.

5.1 Yield Criteria and Flow Rules for HTSC Powders Compaction

Obviously, the optimization of the process of the HTSC powders compaction plays important role in the preparation of superconductors with optimum structure-sensitive properties. As it has been demonstrated in the previous chapters, the numerous techniques can be used to prepare oxide HTSC materials and propose an application of the complex thermal and mechanical treatments of powdered precursors (e.g., cold one-axis pressing [19, 851, 1043], cold isostatic pressing [282, 479, 851, 1157, 1160], hot isostatic pressing [131], hot forging [1043], etc.). Final aim of the pointed techniques consists of the preparation of strong coupled and aligned structures of grains in the final product. Nevertheless, a formation of microstructure defects and weak links is unavoidable due to intrinsic brittleness of oxide superconductors. In this case, it could be noted that selection of thermal and mechanical loading as rule is badly caused. Therefore, in order to solve the above problem, understanding of densification mechanisms and features of HTSC powder deformation is required. The critical mechanical behavior during HTSC powder compaction may be described by using yield criterion and flow rule, taking into account microstructure properties, material parameters, applied loading, etc.

Below, consider the problem of improved compaction technique for powdered HTSC precursors, coupled with selection of optimum yield criterion with the associated (or non-associated) flow rule, which can describe both displacement into bulk and consolidation of the powder during compaction process. Moreover, a validity of the non-associated flow rules for description of HTSC powder compaction is discussed, taking into account the dissipation energy due to particle re-arrangement and friction.

5.1.1 HTSC Compaction and Yield Criterion

Effective application of finite element method and other numerical methods to modeling of microstructure transformations is needed in detail information about material properties, critical mechanical characteristics and applied loading and densification mechanisms. In particular, density distribution, damage initiation and growth depend on the stress–strain state during HTSC powder compaction. In this case, the densification mechanism consists of two processes, namely displacement into powdered volume and plastic deformation of particles. Because the aggregate includes multiple particles, one may be considered as uniform continuum. Then, it is necessary to state a critical condition or a yielding criterion in order to describe the powder strain and densification. Successive selection of the yielding criterion and flow rule are capable of helping in creation of optimum HTSC composition and microstructure. The deformation, caused by sliding of grains, leads to increase in volume during compaction; at the same time it leads to decrease in volume during consolidation of grains. At the so-called critical state (transition point), there is no tendency for volume change. Different yield criteria for free powder filling, porous and granular specimens have been discussed (see, e.g., [461]). Theories of yielding with an increase in density have been developed in [894]. On the other hand, theories of plasticity with a decrease or no change in density have been discussed in [1008] but stated numerous issues. However, very often, the yield criteria do not take into account the density effect on the sample strain. Moreover, due to volume changes during sample compaction, an effect of hydrostatic pressure should be included into yield criterion. At the same time, the limiting envelopes stated by the Mohr–Coulomb's criterion and similar criteria suggest an infinitely large shear stress to cause slip at compressive stresses. This does not apply to the powder compaction. Therefore, the yield criterion for powder aggregate should be in the form of closed curve (e.g., an ellipse) with asymmetric conditions for compressive and tensile stresses, because the powders are not able to sustain significant tensile stresses. Considering the isotropic case of HTSC powder compaction, a 3D yield criterion has the form [140]:

$$f = \alpha(I_1 + s)^2 + J_2 = \beta Y^2 \ , \tag{5.1}$$

where α, β and s are functions of relative density, $I_1 = \sigma_1 + \sigma_2 + \sigma_3$ is the first invariant of the stress tensor, $J_2 = (1/6)[(\sigma_1 - \sigma_2)^2 + (\sigma_2 - \sigma_3)^2 + (\sigma_3 - \sigma_1)^2]$ is the second invariant of the deviatoric stress tensor (σ_k are the principle stresses) and Y is the yield stress of the solid material.

Using the yield function, f, as the plastic potential, the associated flow rule is obtained by imposing the normality condition between the plastic strain rate and the yield surface:

$$\dot{e}^{\mathrm{p}}_{ij} = \lambda \frac{\partial f(\sigma_{ij})}{\partial \sigma_{ij}} = \lambda[2\alpha(\sigma_{kk} + s)\delta_{ij} + s_{ij}] \ , \tag{5.2}$$

where δ_{ij} is Kronecker's delta function, σ_{kk} and s_{ij} are the hydrostatic and deviatoric stress, respectively. A dot above a symbol implies the material time derivative, and upper superscript p signs plastic component.

Considering the principal strain rates, $\dot{e}^{\mathrm{p}}_k$, and determining the volumetric strain rate, $\dot{e}^{\mathrm{p}}_k = \dot{e}^{\mathrm{p}}_1 + \dot{e}^{\mathrm{p}}_2 + \dot{e}^{\mathrm{p}}_3$, we obtain from (5.1) and (5.2) the positive constant λ, defining the strain amount at a given point, as

$$\lambda = \left[(\dot{e}^{\mathrm{p}})^2/(18\alpha) + \dot{d}^{\mathrm{p}}_{ij}\dot{d}^{\mathrm{p}}_{ij}\right]^{1/2} / (\sqrt{2\beta}Y) \ , \tag{5.3}$$

where the deviatoric deformation rate has the form: $\dot{d}^{\mathrm{p}}_{ij} = \dot{e}^{\mathrm{p}}_{ij} - (1/3)\dot{e}^{\mathrm{p}}\delta_{ij}$. Then, (5.1) can be presented by using normal stress, σ, and shear stress, τ:

$$\frac{(\sigma + s/3)^2}{\left(\sqrt{(1+12\alpha)/9\alpha}\sqrt{\beta}Y\right)^2} + \frac{\tau^2}{\left(\sqrt{\beta}Y\right)^2} = 1 \ . \tag{5.4}$$

The material constants α, β and s can be determined by using shear tests [1086]. For this, the yield locus (Fig. 5.2) is described in the $\sigma - \tau$ plane by

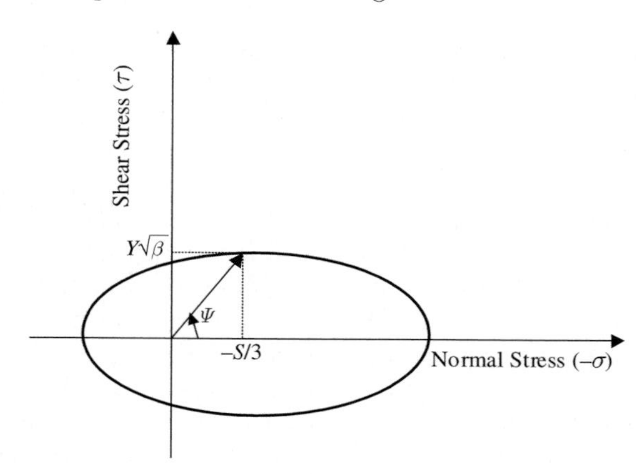

Fig. 5.2. The proposed yield locus for a shear test

an ellipse represented by (5.4). Due to the assumption of an isotropic deformation in a shear test, the transition point almost coincides with the apex of the minor axis of the ellipse. Then, the inclined angle, ψ, of the critical line against the abscissa, σ, and the ratio, R, of the major axis to the minor axis of the ellipse can be considered as material constants, and by using (5.4), they have the forms:

$$tg\psi = \sqrt{\beta}Y/(s/3)\ ; \tag{5.5}$$

$$R = \sqrt{(1+12\alpha)/9\alpha}\ . \tag{5.6}$$

Finally, we select the relation between the density of a powder compact and pressure, needed to achieve that density, as remaining third equation [402]:

$$KP = \ln[(1-\rho_0)/(1-\rho)]\ , \tag{5.7}$$

where P is the applied pressure, ρ_0 and ρ are the average densities of the loose powder and the compressed powder, respectively, and K is the test constant. Then taking into account the relation

$$P = s/3 + R\sqrt{\beta}Y\ , \tag{5.8}$$

the test constants ψ, R and K can be used to define the material constants α, β and s:

$$\alpha = \frac{1}{(9R^2-12)}\ ; \quad s = \frac{3}{K(1+Rtg\psi)}\ln\frac{1-\rho_0}{1-\rho}\ ;$$
$$\beta = \frac{1}{Y^2}\left[\frac{tg\psi}{K(1+Rtg\psi)}\ln\frac{1-\rho_0}{1-\rho}\right]^2 . \tag{5.9}$$

A validity of the proposed yield criterion for description of the HTSC powder compaction can be verified by using, for example, a 3D compaction test (Fig. 5.3a), in which three pairs of anvils compress the powder in three mutually perpendicular directions. However, note that tests of granular materials (e.g., soils) or cemented materials (e.g., concrete or rock), presented in Fig. 5.3b and *c*, have shown that associated flow rule cannot depict test data satisfactorily [1114]. At the same time, satisfactoriness of associated (or non-associated) plasticity has not been stated for compacted HTSC powders. Therefore, it is very important to additionally consider non-associated flow rules for which the plastic strain rate is not orthogonal to the yield surface.

5.1.2 Non-Associated Plasticity of HTSC Powders

Base hypothesis of normality [224], which forms a basis of associated plasticity and is very successful in the description of metallic compositions, can be broken for non-metallic granular materials in free filling, in particular for HTSC powders. The key circumstance is that the associated plasticity cannot validly describe shear dilatancy of granular material (i.e., the change in volume that is associated with shear distortion of an element in the material, consisting of multiple particles as microelements). As has been shown [1114], the material, retaining constant volume under plasticity, reacts otherwise on loading, demonstrating plastic expansion. The differences are related to both the "load–deformation" curve and the value of critical loading. In order to characterize a dilatant material, dilatancy angle Ψ is introduced, presenting the ratio of plastic volume change over plastic shear strain.

An ideal triaxial test should permit independent control of all three principal stresses so that general states of stress could be examined. Typical results in a standard triaxial test of granular material in free filling are shown in Fig. 5.4. In elastic region (I), as usual, the strains are reversible at loading. In hardening regime (II), the strain of granular material becomes more and more inelastic due to particle sliding. Here, non-linear elasticity predicts continuing contraction of the specimen under continued loading in compression. However, such a prediction is disproved by experimental evidence (Fig. 5.5), which shows a dilatant volume increase at subsequent loading [1114]. This phenomenon takes place due to frictional sliding along particles. The elastic strain rate in the hardening regime is almost zero. Moreover, there exists a linear relation between the volume change and the change of the axial strain near the end of the hardening regime (II) and in the softening regime (III) (see Fig. 5.4c). Then, following [1114], the constant dilatancy angle, Ψ, may be introduced as

$$\sin \Psi = \frac{\dot{e}^{\mathrm{p}}}{-2\dot{e}_1^{\mathrm{p}} + \dot{e}^{\mathrm{p}}} \,. \tag{5.10}$$

Equation (5.10) defines constant rate of dilatation, and it is valid under conditions of triaxial compression. For broad row of granular materials, the dilatancy angle, Ψ, approximately equals 0–20°; at the same time, the internal friction angle, Φ = 15–45° [1114]. Because the dilatancy angle can be considerably smaller than internal friction angle, it is necessary to apply for

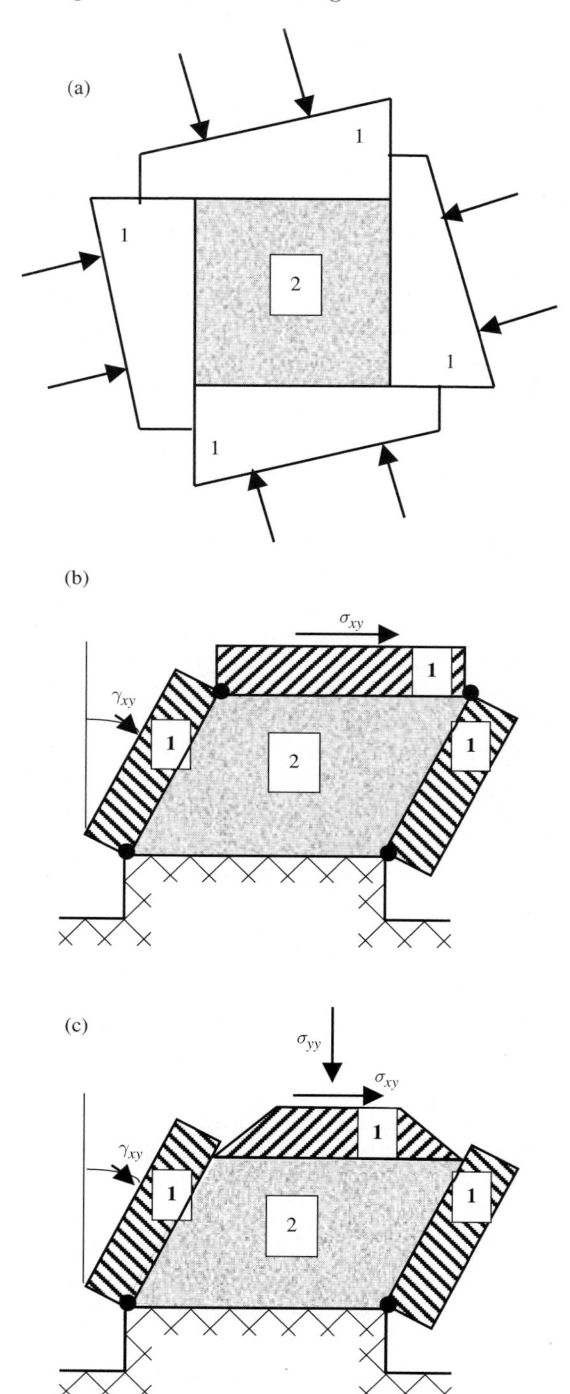

Fig. 5.3. Different schemes of loading for triaxial compression. Third pair of anvils compresses the powder in the direction perpendicular to the figure plane. The numbers sign anvils (1) and powder (2). Arrows show applied loads

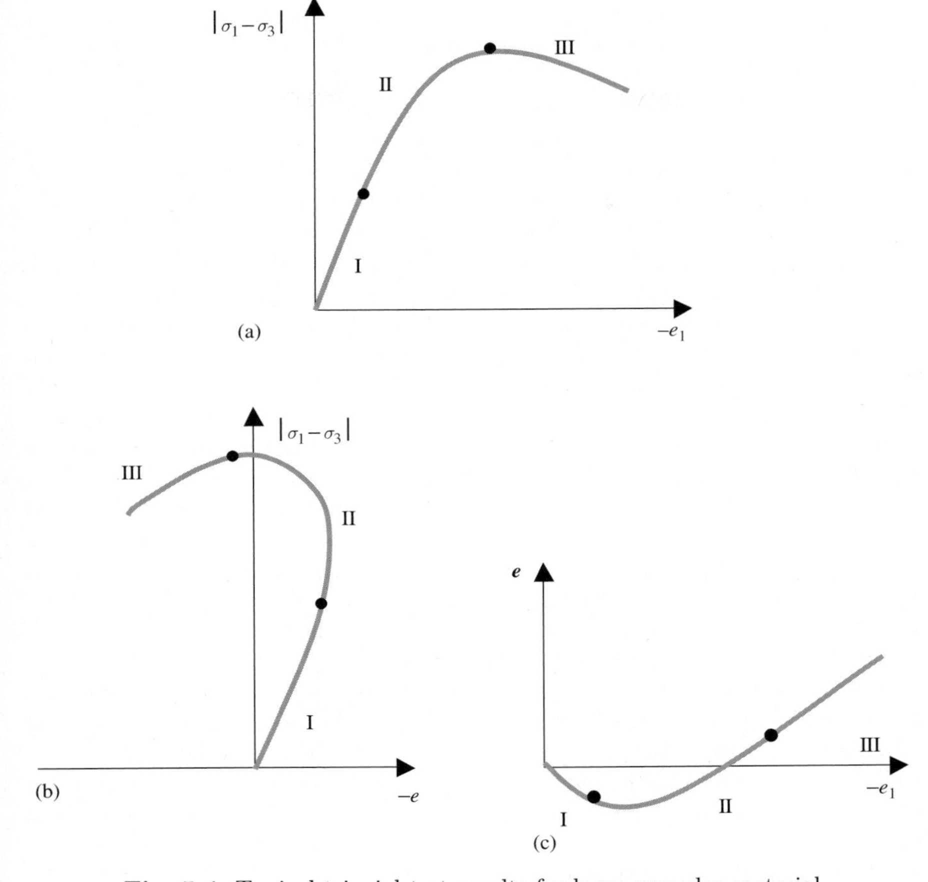

Fig. 5.4. Typical triaxial test results for loose granular material

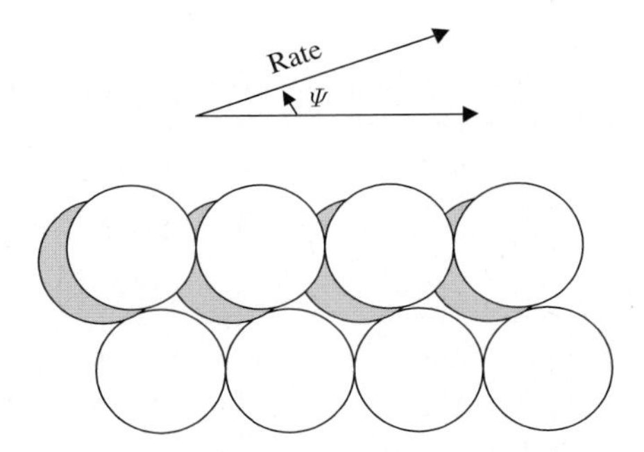

Fig. 5.5. Sliding between groups of particles, leading to dilatation

considered materials the non-associated flow rules. The function of plastic potential, g, used to study plasticity of granular media, coincides nearly with the yielding function, f, applied to separate plastic and elastic states. Difference between these functions consists in that the internal friction angle, Φ, in the equation for f is replaced by the dilatancy angle, Ψ, in relation to g. Thus, in order to state a validity of the associated or non-associated flow rules for HTSC powder compaction, it is necessary to define the dilatancy angle, Ψ, from (5.10) and to compare its value with the friction angle, Φ.

After a pick value in the "stress–strain" curve (see Fig. 5.4a) is reached in the softening regime (III), there is unstable behavior of material, in particular caused by thin shear bands, which separate the specimen in two more or less rigid bodies. For such macroscopic non-uniform deformations, connected with bulging of specimen in triaxial compression test, the strain increment is not measured correctly. At the same time, the strain rate ratio is not so strongly affected by the localization into a shear band. The axial strain–volumetric strain curve of Fig. 5.4c is much more informative, but the dilatancy angle Ψ can be measured with acceptable accuracy.

Violation of the normality law, caused by actual resistance to deformation, comes from two sources: (ii) the dilatation of the bulk material during shear enhances the yield stress under conditions of compression – due to this it is necessary to do a work against the applied pressure and (ii) the frictional dissipation of energy at the contact patches between the granules also enhances the yield stress under applied loading. Below, we will not take into account the work executed during material hardening and softening and also the localization processes at definition of yield surface. Then, assuming that the material is the strain rate independent of small deformation, the strain rate is split up into elastic and plastic components. Each of these strain rates can be also split up into volumetric, $\dot{e}$, and deviatoric, $\dot{d}_{ij}$, components. Consider a typical dilatancy rule for rigid particles caused by the material volume expansion at granule rearrangement [136, 137]:

$$\dot{e}^{\mathrm{p}} = \nu(\dot{d}^{\mathrm{p}}_{ij}\dot{d}^{\mathrm{p}}_{ij})^{1/2} , \tag{5.11}$$

where the constant of proportionality, ν, is a generalization of the dilatancy angle, Ψ. For granules of finite strength, some deformation occurs at the contact patches, so some of this expansion can be alleviated. Hence, it may be assumed that energy is dissipated during this damage at the rate:

$$\dot{D}_1 = l[\nu(\dot{d}^{\mathrm{p}}_{ij}\dot{d}^{\mathrm{p}}_{ij})^{1/2} - \dot{e}^{\mathrm{p}}] , \tag{5.12}$$

where parameter l is proportional to the strength of the granules and the size of the contact patches between granules. Then, the deformation of granular material caused by the granules rolling and sliding defines the energy dissipated due to friction at the contact patches. The rate of this energy can be approximated by

$$\dot{D}_2 = -\mu\sigma(\dot{d}^{\mathrm{p}}_{ij}\dot{d}^{\mathrm{p}}_{ij})^{1/2} , \tag{5.13}$$

where μ is some measure related to the friction factor. Note that both energy dissipation rates are obtained taking into account that (i) there can never be negative dissipation function for all non-zero values of strain increment and (ii) the resistance to deformation is rate independent.

During densification, there are deformation and/or rearrangement of particles depending on the ratio of shear stress to compression. Bearing both mechanisms in mind, a suitable dissipation function might be

$$\dot{D} = \{\mu^2\sigma^2 \dot{d}^{\mathrm{p}}_{ij}\dot{d}^{\mathrm{p}}_{ij} + l^2[\nu(\dot{d}^{\mathrm{p}}_{ij}\dot{d}^{\mathrm{p}}_{ij})^{1/2} - \dot{e}^{\mathrm{p}}]^2\}^{1/2} . \tag{5.14}$$

The rate working per unit volume, $\dot{W}$, by the boundary tractions reduces by Gauss's theorem and the condition of equilibrium to

$$\dot{W} = \sigma_{ij}\dot{e}_{ij} . \tag{5.15}$$

This energy input can be stored as an increase in elastic energy rate, $\dot{U}$, or dissipated, $\dot{D}$. Energy conservation gives $\dot{W} = \dot{U} + \dot{D}$. Splitting the deformation rate into elastic and plastic components, we obtain

$$\sigma_{ij}\dot{e}^{\mathrm{p}}_{ij} - \dot{D} = 0 . \tag{5.16}$$

From Euler's theorem on homogeneous functions, it is clear that

$$\sigma_{ij} = \frac{\partial \dot{D}}{\partial \dot{e}^{\mathrm{p}}_{ij}} . \tag{5.17}$$

Equations (5.14) and (5.17) provide an implicit flow rule, which would satisfy the energy balance:

$$s_{ij} = \frac{\dot{d}^{\mathrm{p}}_{ij}}{\dot{D}}\left\{\mu^2\sigma^2 + \frac{\nu l^2}{(\dot{d}^{\mathrm{p}}_{ij}\dot{d}^{\mathrm{p}}_{ij})^{1/2}}\left[\nu(\dot{d}^{\mathrm{p}}_{ij}\dot{d}^{\mathrm{p}}_{ij})^{1/2} - \dot{e}^{\mathrm{p}}\right]\right\}; \tag{5.18}$$

$$\sigma = \frac{l^2}{\dot{D}}\left[\dot{e}^{\mathrm{p}} - \nu(\dot{d}^{\mathrm{p}}_{ij}\dot{d}^{\mathrm{p}}_{ij})^{1/2}\right] . \tag{5.19}$$

The yield surface can be found in this case by eliminating the strain increments in the energy balance relation (5.16) using (5.18) and (5.19). We obtain

$$\frac{[(s_{ij}s_{ij})^{1/2} + \sigma\nu]^2}{\sigma^2\mu^2} + \frac{\sigma^2}{l^2} - 1 = 0 . \tag{5.20}$$

Again, using (5.18) and (5.19), the flow rule can be derived in terms of the components of stress and the overall rate of energy dissipation:

$$\dot{d}^{\mathrm{p}}_{ij} = \dot{D}s_{ij}\frac{[(s_{ij}s_{ij})^{1/2} + \nu\sigma]}{(s_{ij}s_{ij})^{1/2}\mu^2\sigma^2} ; \tag{5.21}$$

$$\dot{e}^{\mathrm{p}} = \dot{D}\sigma\left\{\frac{1}{l^2} + \frac{\nu[(s_{ij}s_{ij})^{1/2} + \nu\sigma]}{\mu^2\sigma^3}\right\} . \tag{5.22}$$

Equations (5.21) and (5.22) predict that at small σ/l the granules behave in an almost rigid manner, but as $\sigma/l \approx 1$, compaction takes place. This is in accordance with experiments for granular materials [591]. From (5.21), it follows that these flow rules give sensible results as long as

$$(s_{ij}s_{ij})^{1/2} + \nu\sigma > 0 \, . \tag{5.23}$$

Successive development and application of constitutive models for HTSC powder compaction should be based on the three-level hierarchical structure of the models. At first step, the non-hardening perfectly plastic model under consideration provides an excellent introduction to the modeling of the HTSC powders under compression, taking into account the particle cohesion and friction. Second step includes consideration of the isotropic hardening models together with study of isotropic softening and estimation of damage variables, based on ideal plasticity, taking into account non-associated behavior. Finally, in the third step, anisotropic hardening and softening should be studied together with modeling of material instability that leads today to numerous difficulties in the application of finite element codes and other computational methods for numerical solution of corresponding boundary-value problems.

5.2 Void Transformations During Sintering of Sample

The experimental studies of the sintering duration effect on superconducting properties of monocore Bi-2223/Ag tapes [670, 671] have shown that critical current, at first, increases together with annealing time, reaches maximum value and then decreases with reaction time (Fig. 5.6). Monotonous enhancement of weight losses of the sample with simultaneous diminution of Pb content and increase of the Bi-2223 phase homogeneity (Figs. 5.7 and 5.8) at remaining constant average size of superconducting grains about 25 μm has been observed, which has not depended on the annealing duration (in the range of 40–200 h). This behavior of critical current has been related by the authors to acting during calcining competing mechanisms, which initially improve the quality of grain boundaries (or weak links),[2] and then decrease the pinning properties of the superconductor. A decrease in critical current has been explained by lead expelling during annealing and by the corresponding degradation of the pinning strength. At the same time, as rule, during BSCCO/Ag processing, there are significant processes of the void formation and other microstructure transformations due to CO_2 release, in particular

[2] At relatively badly coupled structure of grains, magnetic flux can penetrate intergranular boundaries and trap pores or secondary phases. This non-superconducting volume will be effectively screened in the sample with better intercrystalline properties, and corresponding fraction of superconducting volume will be enhanced.

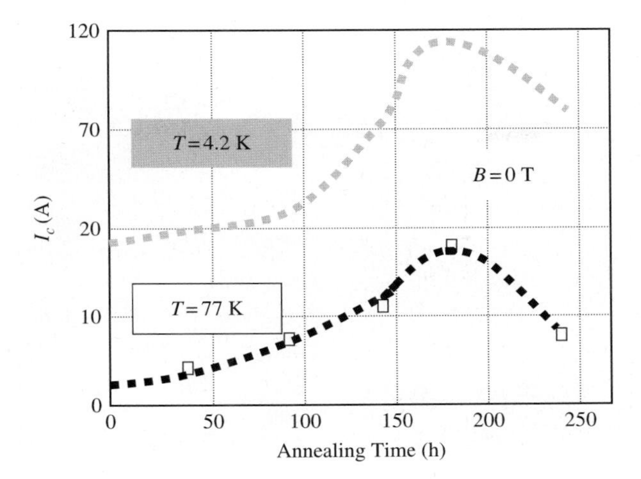

Fig. 5.6. Annealing time dependence of the self-field I_c at 4.2 K and 77 K [670, 671]

leading to bubbles arising and the critical current decrease. As it has been noted in Sect. 4.2, 200 ppm of carbon can cause about 36% porosity in the core of Bi-2212/Ag tapes if all carbon forms CO_2 at high temperature [1196].

Though an average grain size observed in monocore Bi-2223/Ag tapes has remained constant during various duration of annealing, the *grain size distribution* could be altered during reaction. This is supported by considerable

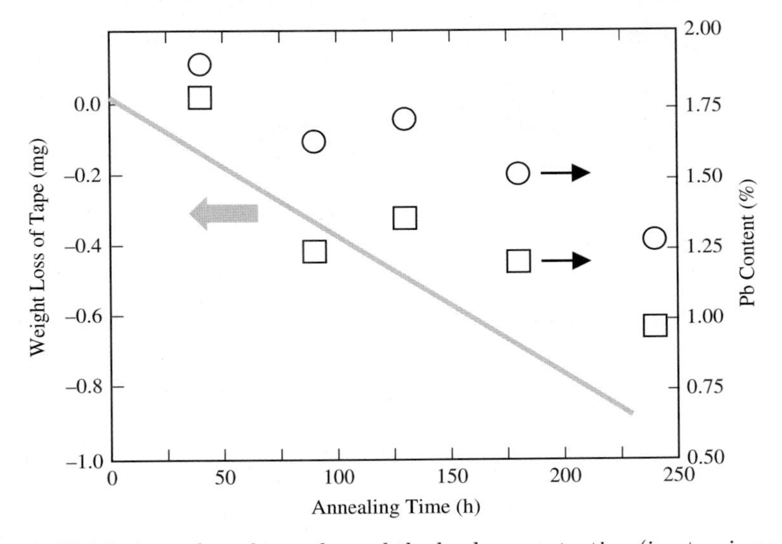

Fig. 5.7. Weight loss of total sample, and the lead concentration (in atomic percent) versus annealing time. The circulars sign values averaged on several single grains and squares show measurements on the large region of about 400 grains [670, 671]

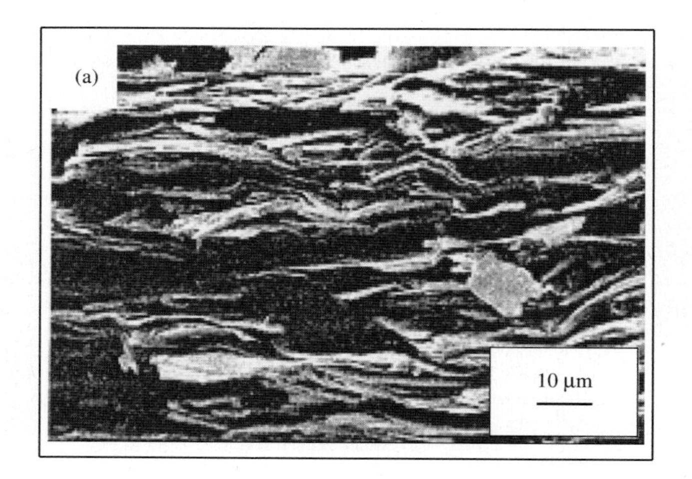

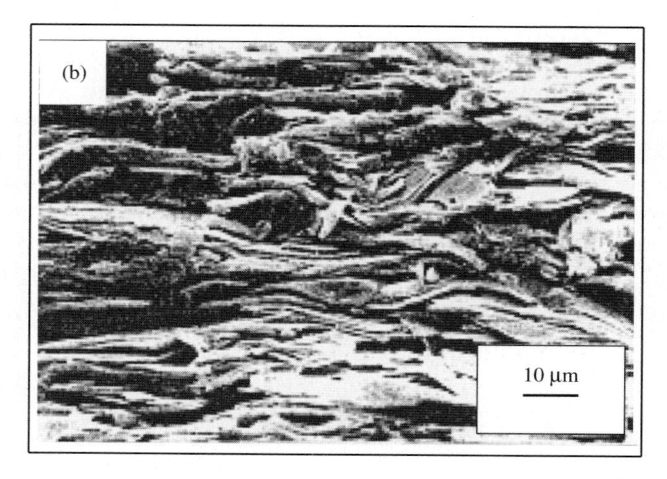

Fig. 5.8. SEM micrographs of a typical transversal fracture surface after (**a**) 40 h and (**b**) 240 h annealing time [671]

lead expelling monotonously increasing with annealing time. But then these additions heterogeneously distributed into the material can inhibit only a local grain growth in the Bi-2223 core. Moreover, as it has been shown in the modeling of fracture processes in YBCO, *the size distributions of microstructure elements* play more important role to compare with average sizes [821]. Hence, it may be assumed that the observed preservation of the *average* grain size in different annealing times cannot assert an absence of grain growth and other microstructure alterations. Therefore, another mechanism of diminution of critical current during prolonged annealing is possible, namely inevitable transformation of pores attached to intergranular boundaries, which can lead to the formation of significant closed porosity. Consider some regimes of change of the void space: their displacement, shrinkage, coarsening, coalescence and possible separation from grain boundaries to the interior of the grains that directly define useful properties of superconductor.

5.2.1 Void Separation from Intergranular Boundary

In discussion of pore breakaway processes, first, it should be noted that pores attached to grain boundaries decrease due to boundary-grain diffusion. However, when a pore separates from grain boundary and locates into the grain, one can shrink only, thanks to much more slow diffusion process of the crystalline lattice. Total separation of pores occurs after their displacement at the boundaries between two grains (Fig. 5.9). Therefore, a preliminary displacement of pores from triple junctions of grain boundaries to interfaces of two grains must precede to total pore breakaway.

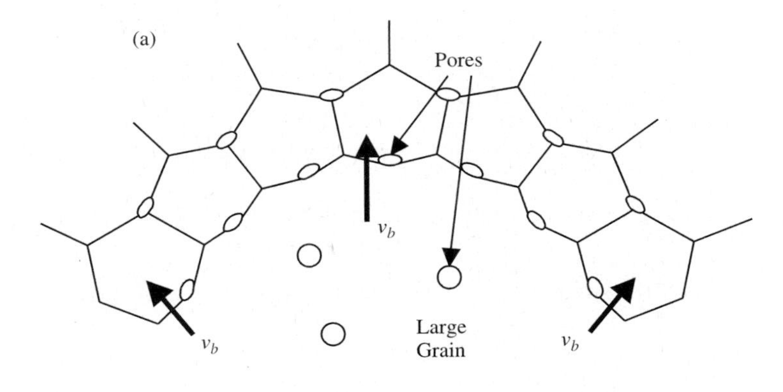

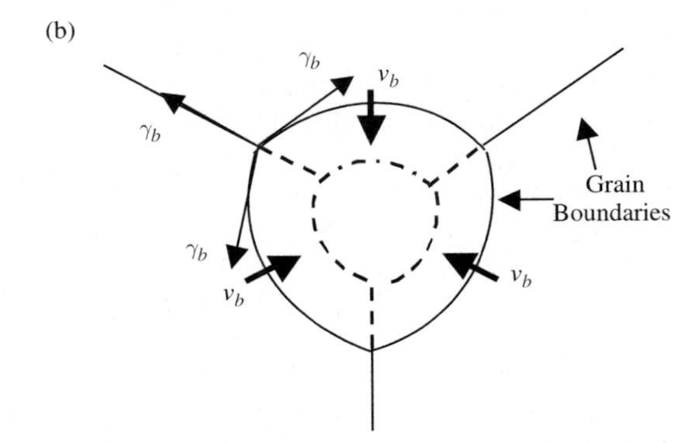

Fig. 5.9. (**a**) A schematic of the grain disappearance process involved in grain growth of large grain at the expense of five-sided small grains located in its periphery; (**b**) the three-sided configuration associated with ultimate grain disappearance

The phenomenological analysis is essentially based on solutions for the interaction of a boundary with a rigid second-phase particle. Specifically, the interaction between a pore and a moving grain boundary assumes a spherical pore moving in isotropic, homogeneous material with a rate determined by the surface diffusion coefficient (into the framework of surface diffusion mechanism, appropriating to small pores). Then, unique pore mobility is derived, retaining the spherical symmetry of the pore (i.e., neglecting the changes in pore shape needed to maintain the atom flux over the pore surface). The corresponding pore mobility is given by [973]

$$M_{\mathrm{p}} = \frac{D_{\mathrm{s}}\delta_{\mathrm{s}}\Omega}{kT\pi a_0^4} , \tag{5.24}$$

where $D_{\mathrm{s}}\delta_{\mathrm{s}}$ is the surface diffusion parameter, a_0 is the pore radius and Ω is the atomic volume. The force F exerted by the grain boundary on the pore (which eventually dictates separation) is derived from the assumption that the contact line between the boundary and the pore can move freely over the pore surface. The force has a maximum value:

$$F_{\mathrm{max}} = \pi a_0 \gamma_b . \tag{5.25}$$

Equations (5.24) and (5.25) yield a peak pore rate as

$$v_{\mathrm{p}}^0 = \frac{\Omega D_{\mathrm{s}}\delta_{\mathrm{s}}\gamma_{\mathrm{b}}}{kTa_0^3} . \tag{5.26}$$

The pore breakaway is considered to occur when the grain boundary rate exceeds this peak pore rate.

Following [445], consider the motion of pores with grain boundaries, taking into account a flux of atoms from the leading to the trailing surface of the pore (see Fig. 5.10). The driving force for the atom flux is caused by the existence of a gradient in the curvature of the pore surface, that is, pore distortion is a necessary consequence of pore motion. The configuration selected is an axi-symmetric pore, moving due to surface diffusion. Initially, steady-state motion (all locations on the surface, moving at the same rate) is considered. The axi-symmetric pore exhibits two curvatures (see Fig. 5.11): an in-plane curvature, k_1, and an axi-symmetric curvature, k_2. These curvatures are related to the coordinates of the problem (x, y) by

$$k_1 = (\mathrm{d}^2y/\mathrm{d}x^2)[1 + (\mathrm{d}y/\mathrm{d}x)^2]^{-3/2} ; \tag{5.27}$$

$$k_2 = (1/x)(\mathrm{d}y/\mathrm{d}x)[1 + (\mathrm{d}y/\mathrm{d}x)^2]^{-1/2} . \tag{5.28}$$

The flux J_{s} of atoms in the presence of these curvatures is given by

$$J_{\mathrm{s}} = -\left(\frac{D_{\mathrm{s}}\delta_{\mathrm{s}}\gamma_{\mathrm{s}}}{kT}\right)\frac{\mathrm{d}(k_1 + k_2)}{\mathrm{d}s} , \tag{5.29}$$

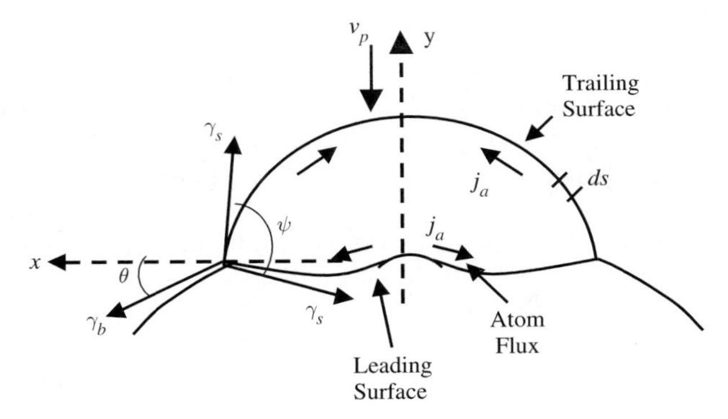

Fig. 5.10. A schematic of a moving pore, indicating the atom flux and the inclination of the grain boundary, θ

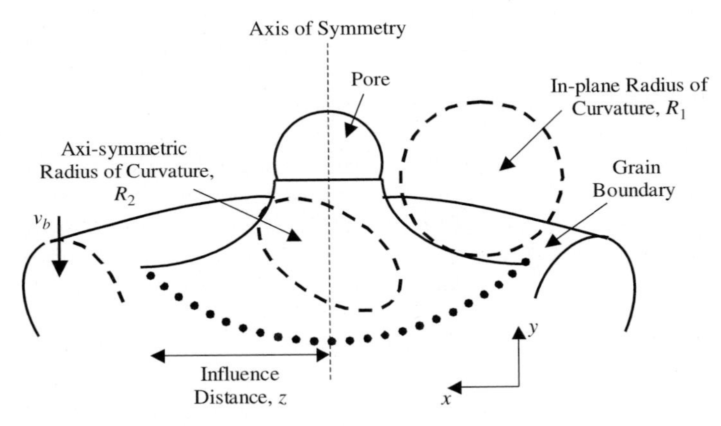

Fig. 5.11. The axi-symmetric configuration associated with pore drag, illustrating the important curvatures, the influence distance and the pore and grain boundary rates

where $\mathrm{d}s$ is an element of pore surface in the flow direction (see Fig. 5.10), γ_s is the surface energy (assumed isotropic). For a pore moving with a rate, v_p, in the y-direction, conservation of matter requires that

$$2\pi x J_\mathrm{a} = \pm \pi x^2 v_\mathrm{p}/\Omega \,, \tag{5.30}$$

where the positive sign refers to the leading surface and the negative sign to the trailing surface.

Using the symbols, p and q ($\equiv \mathrm{d}y/\mathrm{d}x$) for the slopes of the trailing and the leading surfaces, respectively, (5.27)–(5.30) lead to two differential equations of second order. The motion of the pore is subject to the requirement that the total dihedral angle, ψ, between the grain boundary and the pore surfaces (Fig. 5.10) be invariant. Moreover, it is necessary that the chemical potential be continuous at the intersection of the leading and trailing surfaces. Finally, the atom flux and surface slope must be zero at the axis of symmetry. The solution of the differential equations at the pointed conditions is a non-linear problem. A solution is obtained by linearizing a trial solution and then using a finite difference scheme [445].

The inclination, θ, of the grain boundary tangent to the plane of contact between the grain boundary and the pore emerges from the analysis as a unique function of the dihedral angle, ψ, and normalized pore: $\nu_{\mathrm{p}} = (v_{\mathrm{p}}/v_{\mathrm{p}}^{0})(\gamma_{\mathrm{b}}/\gamma_{\mathrm{s}})$. The radius of contact, a, between the boundary and the pore can also be deduced by requiring that the pore volume be independent of pore rate in order to permit a unique comparison between the dimensions of the stationary and moving pores.

Convergent pore shape solutions are found to exist over a limited range of pore rates, implying the existence of a steady-state rate maximum, $v_{\mathrm{p}}^{\mathrm{m}}$. Its existence is associated with an ability to simultaneously satisfy the requirements that (i) the dihedral angle be specified, (ii) the curvature be continuous and finite and (iii) the pore rate be uniform. The consequences of increasing the rate above $v_{\mathrm{p}}^{\mathrm{m}}$, leading to violation of one of these imposed conditions, can be presented when the dihedral angle $\psi \approx \pi$ (Fig. 5.12). Shape changes, which induce a continuous atom flux in the requisite direction for pore motion (i.e., a continuous gradient in surface curvature), cannot be constructed due to an inevitable counter-flux of atoms. Thus, steady-state motion by surface diffusion of a pore with $\psi \sim \pi$ is impossible.[3] The steady-state rate maximum may be required as the equivalent of the peak rate, v_p^0, given by (5.26), and can be expressed as [445]

$$v_{\mathrm{p}}^{\mathrm{m}} = v_{\mathrm{p}}^{0}(17.9 - 6.2\psi)/2\cos(\psi/2)\,. \tag{5.31}$$

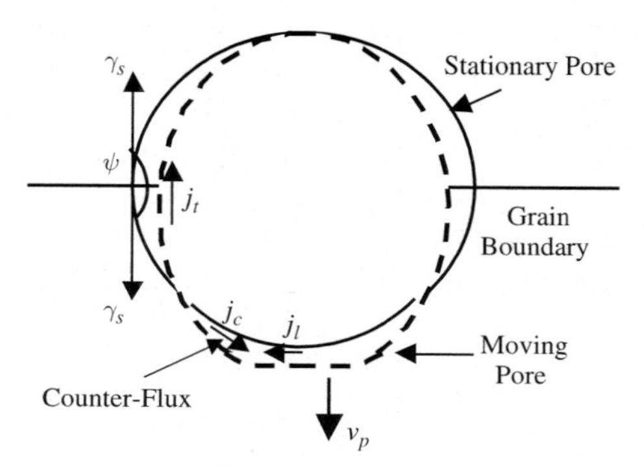

Fig. 5.12. A schematic illustrating the development of a counter flux with $\psi = \pi$ is distorted to achieve a net atom flux and hence pore motion with the grain boundary

[3] The inability of a pore with $\psi \sim \pi$ to exhibit steady-state motion implies that such pores will *always* detach from grain boundaries. This is intuitively clear because when $\psi = \pi$, the grain boundary energy is zero (surface energies are always finite) and there is no preference for pores to locate on grain boundaries.

In this case, for all reasonable choices of the dihedral angle, the peak steady-state rate exceeds the value anticipated by the phenomenological model.

Non-steady-state solutions, in which the rate varies over the pore surface (with a maximum at the axis of the leading surface), can be found for net rates in excess of $v_{\mathrm{p}}^{\mathrm{m}}$. These solutions coincide with a marked change in pore shape and a decreasing in the grain boundary contact radius, a, as shown by the shape depicted in Fig. 5.13. It is presumed that, in this case, the contact radius a will rapidly diminish to zero, causing the grain boundary to converge onto the pore axis and to initiate breakaway. Hence, the upper bound steady-state pore rate can be used as a rate, which, if exceeded, will inevitably result in non-steady-state pore motion and breakaway.

It may be shown that the grain boundary, subjecting to pore drag, exhibits a rate component normal to the axis of symmetry, indicative of a tendency toward instability [445]. Therefore, when the grain boundary rate, v_{b}, exceeds v_p, the axi-symmetric radius of curvature, R_2 (see Fig. 5.11), decreases during the motion of the pore–grain boundary configuration, especially within the immediate vicinity of the pore. The decrease in R_2 exceeds the change in R_1, and therefore, the configuration is intrinsically metastable, but steady state of the complete pore/grain boundary ensemble is impossible. The pore drag

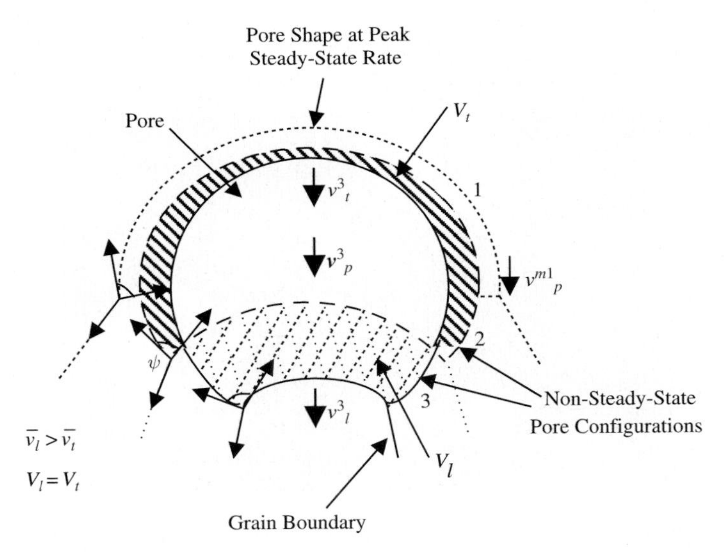

Fig. 5.13. Schematic illustration of the pore shapes under non-steady-state conditions. For the pore volume to remain constant, the matter removed from the leading surface, V_l, must equal the matter deposited on the trailing surface, V_t. Moreover, under non-steady state conditions, the average rate of the leading surface, $\overline{v}_l$, must exceed that for the trailing surface, $\overline{v}_t$. In order to satisfy these requirements, the contact radius a must decrease rapidly with increase in net pore rate

will increasingly distort the boundary and eventually induce separation. It is interesting to note that grain boundary coalescence is likely to result in a dislocation or sub-grain boundary attached to the pore (especially in the presence of some symmetry) [197, 530].

The specific condition that dictates the separation of pores from grain boundaries depends on the grain configuration to which the pore is attached. Consider two configurations that occur during grain growth,[4] namely (i) a pore located on one boundary of a three-sided grain (Fig. 5.9b) (the configuration grain that invariably precedes grain disappearance) and (ii) a pore located on one boundary of a five-sided grain at the perimeter of a large grain subject to exaggerated grain growth (Fig. 5.9a). Pore drag observations (Fig. 5.14) indicate that the pore perturbs grain boundary motion over a certain influence distance, z (see Fig. 5.11). Therefore, it is appropriate to examine the motion of the grain boundary outside this influence distance relative to that of the pore (obviously, it is assumed that the influence distance is less than the grain radius). The rate of the grain boundary outside the influence zone in the direction of pore (Fig. 5.11) is [445]

$$v_{\mathrm{b}} = \sqrt{3}\gamma_{\mathrm{b}} M_{\mathrm{b}} \Omega^{2/3}/R\,. \tag{5.32}$$

By allowing this rate to exceed the peak steady-state pore rate, $v_{\mathrm{p}}^{\mathrm{m}}$, separation should be inevitable, provided a grain boundary displacement

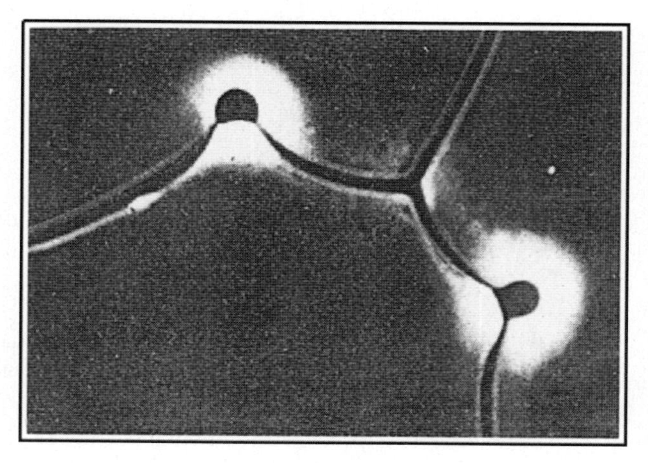

Fig. 5.14. Pores in MgO subjected to motion with the grain boundary [445]

[4] Other configurations, such as several pores on single grain boundary, will provide different separation conditions. However, these separation events usually occur after the first separations have been induced and are, probably, less critical. For example, a modified phenomenological analysis with a pore $\propto R$ gives the two limiting cases [445]: $R \leq 2M_{\mathrm{b}}kT\gamma_{\mathrm{b}}a_0^3/D_{\mathrm{s}}\delta_{\mathrm{s}}\gamma_{\mathrm{s}}\Omega^{1/3}(17.9 - 6.2\psi)$ and $R \geq \delta_{\mathrm{s}}\gamma_{\mathrm{s}}\Omega^{1/3}(17.9 - 6.2\psi)/2\gamma_{\mathrm{b}}a_0$.

(i.e., grain size), sufficient to create boundary convergence at the dragging pore, is available. This condition yields a pore size for separation given by

$$a_0^2 \geq \left(\frac{R}{a_0}\right)\left(\frac{\Omega^{1/3} D_s \delta_s \gamma_s}{kTM_b\gamma_b}\right)\frac{(17.9 - 6.2\psi)}{\sqrt{3}} . \tag{5.33}$$

For the five-sided grain, R remains essentially constant (Fig. 5.15). However, the most stringent condition for breakaway exists when R attains its *smallest* value, which corresponds to the three-sided grain configuration, preceding grain disappearance (Fig. 5.9b). In this case, separation is averted if the pore converges onto the prospective three-grain junction ($R \sim 2a_0$) before (5.33) can be satisfied. Thus, the critical condition becomes

$$\left(a_0^2\right)_c = \left(\frac{\Omega^{1/3} D_s \delta_s \gamma_s}{kTM_b\gamma_b}\right)\frac{2(17.9 - 6.2\psi)}{\sqrt{3}} , \tag{5.34}$$

as shown in Fig. 5.15. This critical pore size represents a lower bound for pore separation at all reasonable values of a_0/R. Obviously, that obtained critical value increases with increasing of surface diffusion or with increasing of grain boundary mobility or dihedral angle.

HTSC materials demonstrate a spectrum of dihedral angles connected, in the first place, with various grain boundaries formed during grain growth. Then, there is a range of the pore critical sizes that the very fine pores can localize at low angle or special boundaries, in the sites of coincidence of the crystalline lattices. Similarly, a distribution of the surface diffusion parameters and grain boundary mobility due to inhomogeneity of admixture distribution should exist. Hence, different spectrum of the pore critical sizes will correspond to various superconducting systems and should be taken into account in consideration with the pore breakaway. The microstructure changes, suppressing the pore separation, should be estimated in comparison with the pore critical size, with the trajectories of alteration of the pore and grain sizes at the final stage of sintering. These processes will be examined in detail in the next section.

5.2.2 Size Trajectories in the Pore/Grain Boundary System During Sintering

The sintering process is generally accompanied by grain growth, pore shrinkage and pore coalescence [161, 1165]. These concomitant processes result in pore/grain size trajectories that typically entail pore size enlargement during the intermediate stage and pore size reduction prior to final densification Moreover, exaggerated grain growth may initiate as a result of the pore separations from grain boundaries [126, 127]. Therefore, it is very important to study the pore/grain size trajectory, especially with reference to the critical size for pore breakaway. In this section, following [1005], concurrent grain growth, pore shrinkage and coalescence are analyzed and compared to the

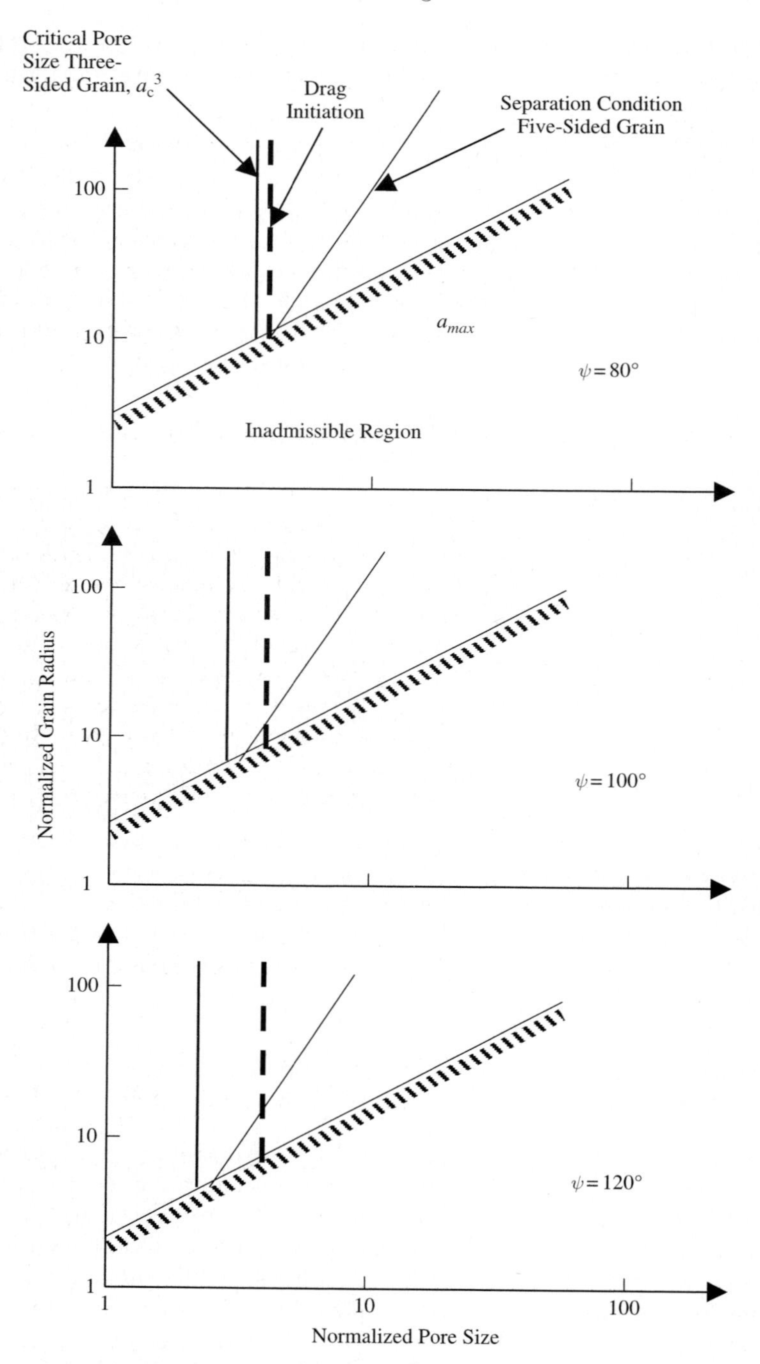

Fig. 5.15. The various separation and admissibility conditions, identified in the analysis of [445], are plotted for three values of the dihedral angle. The abscissa axis presents normalized pore size $(kTM_b\gamma_b/\Omega^{1/3}D_s\delta_s\gamma_s)^{1/2}a_0$; the ordinate axis shows normalized grain size $(kTM_b\gamma_b/\Omega^{1/3}D_s\delta_s\gamma_s)^{1/2}R$

maximum pore size achieved during densification, with the critical pore size at pore–grain boundary separation. Mutual dependencies between the grain and pore sizes, caused by concomitant grain growth, pore coalescence and pore shrinkage, can be used together with expressions for the grain boundary rate and the pore shrinkage rate to establish features of the pore/grain size trajectories during sintering.

It is assumed that pores exist at each three-grain junction (see Fig. 5.16). Subject to this condition, pore coalescence during the disappearance of small grains results in a simple magnification of the microstructure, while concomitant pore shrinkage reduces the size of the pores relative to the grain size. Thus, the relative change in pore volume comprises an increase due to coarsening and a decrease due to shrinkage given directly by

$$\frac{\mathrm{d}V_\mathrm{p}}{V_\mathrm{p}} = \frac{\mathrm{d}V_\mathrm{g}}{V_\mathrm{g}} - \frac{\left|\mathrm{d}V_\mathrm{p}^\mathrm{s}\right|}{V_\mathrm{p}}, \tag{5.35}$$

where $\mathrm{d}V_\mathrm{p}$ is the resultant volume change of the average pore in a small time interval, $\mathrm{d}t$; $\mathrm{d}V_\mathrm{g}$ is the corresponding grain volume change; $\mathrm{d}V_\mathrm{p}^\mathrm{s}$ is the absolute pore volume change due to shrinkage; and V_p and V_g are the average pore and grain volume, respectively. Dividing (5.35) throughout by the time interval $\mathrm{d}t$, and rearranging, we obtain

$$\frac{\mathrm{d}V_\mathrm{p}}{\mathrm{d}V_\mathrm{g}} = \frac{V_\mathrm{p}}{V_\mathrm{g}} - \frac{\left|\dot{V}_\mathrm{p}^\mathrm{s}\right|}{\dot{V}_\mathrm{g}}, \tag{5.36}$$

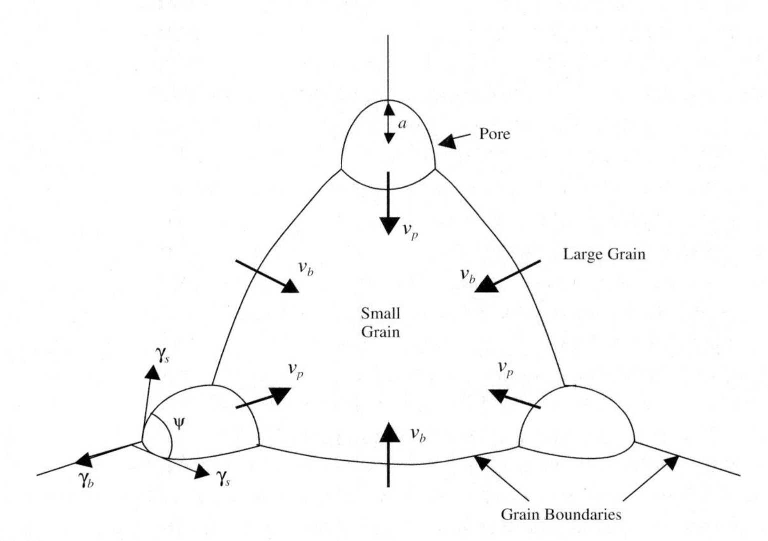

Fig. 5.16. A schematic indicating the grain and pore configurations used to calculate pore drag effects

where a dot above a symbol signs the time derivation. Herein, it will be assumed that the pore coarsening will be balanced by the shrinkage, that is, $\mathrm{d}V_\mathrm{p}/\mathrm{d}V_\mathrm{g} = 0$. In particular, this condition defines the pore volume, V_p^+, at which a transition from pore coarsening to pore shrinkage occurs during normal sintering (i.e., when the average number of pores per grain remains). This is given by

$$V_\mathrm{p}^+ = \frac{V_\mathrm{g}\left|\dot{V}_\mathrm{p}^\mathrm{s}\right|}{\dot{V}_\mathrm{g}} . \tag{5.37}$$

Taking into account the relations:

$$V_\mathrm{g} \propto l^3 ; \quad \dot{V}_\mathrm{g} \propto 3l^2\dot{l} ; \quad V_\mathrm{p} \propto a^2 l ; \quad \dot{V}_\mathrm{p} \propto a^2\dot{l} + 2l\dot{a}a ; \quad \dot{l} = \lambda v_\mathrm{b} , \tag{5.38}$$

where l is the grain facet length, a is the pore size, v_b is the grain boundary rate and λ is a grain shape coefficient $(1/3\sqrt{3} < \lambda < 1/\sqrt{3})$ [421, 1165]. Incorporating (5.38) into (5.36) and (5.37) yields a coarsening pore rate as

$$\mathrm{d}a/\mathrm{d}l = a/l - \left|\dot{a}^\mathrm{s}\right|/\lambda v_\mathrm{b} , \tag{5.39}$$

and a transition pore size as

$$a^+/l = \left|\dot{a}^\mathrm{s}\right|/\lambda v_\mathrm{b} . \tag{5.40}$$

Pore Rate

A moving pore, located at a three-grain junction, must exhibit a distortion with general characteristics (depicted in Fig. 5.17) contingent upon the next requirements. The surface curvature must increase continuously from the leading to the trailing surfaces of the pore in order to establish the net atom flux that constitutes pore motion. Moreover, the equilibrium dihedral angle must be maintained at each grain boundary intersection. Hence, the triangular zone that connects the grain boundary intersections must exhibit an apex angle, 2θ, that diminishes as the pore rate increases. This angular change influences the orientation of the grain boundary tangent at each pore intersection, increasing its curvature and reducing the permissible grain boundary rate (Fig. 5.18).

A linearized analysis of surface diffusion (see Fig. 5.17b) provides an adequate solution as the deviations of the surface tangent from the ordinate are sufficiently small that the following approximations apply:

$$\mathrm{d}s \approx \mathrm{d}x , \qquad 1/r \approx d^2y/dx^2 , \tag{5.41}$$

where $\mathrm{d}s$ is an element of surface and r is the radius of curvature of the surface. Subject to this linearization, the atom flux can be expressed as [154]

$$j_\mathrm{s} = -\frac{D_\mathrm{s}\delta_\mathrm{s}}{kT}\left[\frac{\mathrm{d}}{\mathrm{d}x}(\gamma_\mathrm{s}/r)\right] = -\frac{D_\mathrm{s}\delta_\mathrm{s}\gamma_\mathrm{s}}{kT}\left(\frac{\mathrm{d}^3y}{\mathrm{d}x^3}\right) , \tag{5.42}$$

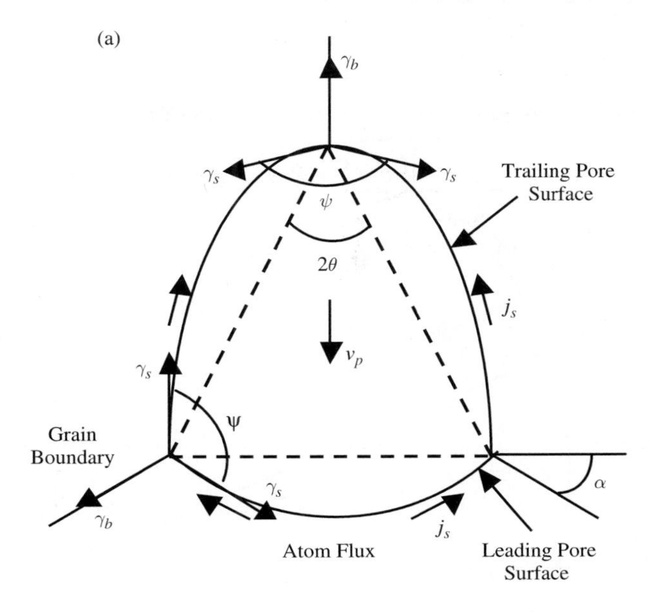

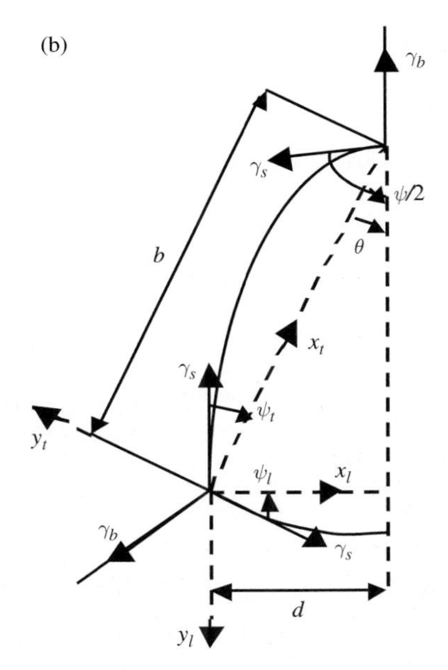

Fig. 5.17. Schematics of the geometry of moving pores at the grain junctions: (**a**) the pore distortion, (**b**) the coordinate system (x, y) used for analysis of the distortion of the leading and trailing surfaces

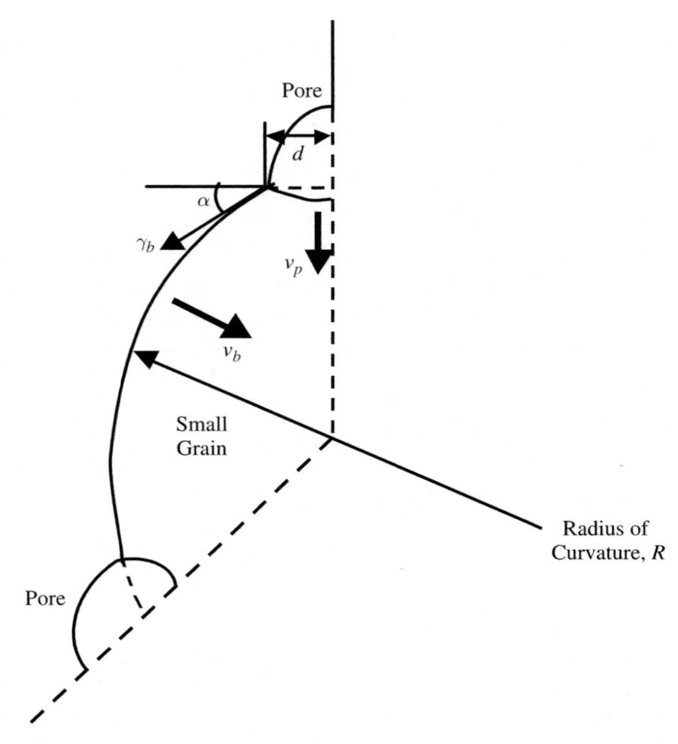

Fig. 5.18. The grain boundary parameters used to determine the grain boundary rate

and the atom flux gradient (based on matter conservation) becomes [154]

$$\frac{\mathrm{d}j_\mathrm{s}}{\mathrm{d}x} = \frac{v_n}{\Omega} , \tag{5.43}$$

where $D_\mathrm{s}\delta_\mathrm{s}$ is the surface diffusivity, γ_s is the surface energy, Ω is the atomic volume and v_n is the component of the pore rate in the y-direction. Combining (5.42) and (5.43), the governing differential equation is given by

$$\frac{\mathrm{d}^4 y}{\mathrm{d}x^4} = -\frac{v_n kT}{D_\mathrm{s}\delta_\mathrm{s}\gamma_\mathrm{s}\Omega} . \tag{5.44}$$

This equation can be solved for the leading and trailing pore surfaces subject to the next boundary conditions (see Fig. 5.17b):
The leading surface: $j_\mathrm{s} = 0$ and $\mathrm{d}y/\mathrm{d}x = 0$, at $x_l = d$, whereupon

$$\frac{y_l}{d} = \frac{\psi_l}{2}\left[1 - \left(\frac{x_1}{d}\right)^2\right] - \frac{\nu_\mathrm{p}(d/a_0)^3}{24}\left[\left(\frac{x_1}{d}\right)^4 - 2\left(\frac{x_l}{d}\right)^2 + 1\right] , \tag{5.45}$$

where ψ_l is the pore surface inclination at the boundary intersection, subscript l refers to the leading surface and $\nu_\mathrm{p} = v_\mathrm{p} kT a_0^3/D_\mathrm{s}\delta_\mathrm{s}\gamma_\mathrm{s}\Omega$ is the dimensionless pore rate identified in Sect. 5.2.1 (a_0 is the initial, zero rate, pore dimension).

The trailing surface: $j_s = 0$ and $dy/dx = \theta - \psi/2$, at $x_t = b$, giving

$$\frac{y_t}{b} = \frac{\nu_p \sin\theta (b/a_0)^3}{24}\left[\left(\frac{x_t}{b}\right)^4 - 4\left(\frac{x_t}{b}\right)^3 + 5\left(\frac{x_t}{b}\right)^2 - 2\left(\frac{x_t}{b}\right)\right] + \left(\frac{\psi}{2} - \theta\right)\left[\left(\frac{x_t}{b}\right) - \left(\frac{x_t}{b}\right)^2\right] . \tag{5.46}$$

Continuity of chemical potential requires that the curvature of the leading and trailing pore surfaces be the same at the boundary intersection. Incorporation of this condition permits the apex angle θ to be related to the pore rate and to the pore dimension, b:

$$\nu_p = \frac{3[2\sin\theta(\psi/2 - \theta) - 2\theta - (\psi - \pi)/2](a_0/b)^3}{\sin^3\theta + (5/4)\sin^2\theta + (1/4)\sin\theta} . \tag{5.47}$$

Finally, requiring that the volume V of the pore be independent of the rate:

$$V/l = \int_0^d y_1\, dx + \int_0^b y_2\, dx + (b^2 \sin\theta\cos\theta)/2 , \tag{5.48}$$

the dimensionless pore rate can be expressed in terms of the apex angle θ and the dihedral angle ψ. Simple functional relations that adequately describe the dependence of θ, α and d on the normalized pore rate in the important application angular range ($\pi/3 < \psi < 2\pi/3$ and $\pi/12 < \theta < \pi/6$) are given by [1005]

$$\theta = (\pi/6)\exp[-\nu_p/(0.07 + 0.4\psi)] ; \quad \alpha \approx \pi/6 + 0.013\nu_p(1 + 3\psi) ;$$
$$d/a_0 \approx 1/2 - 0.013\nu_p(1 + \psi) . \tag{5.49}$$

Grain Boundary Rate

The grain boundary rate, v_b, is found by the grain boundary radii of curvature R_1 and R_2 and also by the boundary mobility, M_b, that is regarded as independent of the driving force:

$$v_b = \gamma_b \Omega^{2/3}\left(\frac{1}{R_1} + \frac{1}{R_2}\right) M_b . \tag{5.50}$$

The dependence of the radii of curvature on the extent of pore distortion, through the pore drag angle, α (Fig. 5.18), establishes a unique link between the boundary rate and the pore rate (with the intermediate determination of α). For a boundary of uniform curvature, the in-plane radius of curvature, R_1, is given by

$$R_1 = \frac{\sqrt{3}l - d}{\sqrt{3}\cos\alpha - \sin\alpha} . \tag{5.51}$$

We have from (5.49) and (5.51):

$$R_1/l \approx \frac{\sqrt{3} + [0.013(1+\psi)\nu_\mathrm{p} - 1/2]a/l}{[2.5 - 0.023\nu_\mathrm{p}(1+3\psi)]} \,. \tag{5.52}$$

Thus, the radius of curvature is dictated by the relative pore size, a/l, the dihedral angle, ψ, and the dimensionless pore rate, ν_p. The out-of-plane curvature is assumed to be of secondary significance because one is independent of the pore movement. Hence, the radius R_2 may be neglected. Then, the grain boundary rate can be obtained from (5.50) and (5.52) as

$$\frac{v_\mathrm{b} l}{M_\mathrm{b}\gamma_\mathrm{b}\Omega^{2/3}} \approx \frac{2.5 - 0.023\nu_\mathrm{p}(1+3\psi)}{\sqrt{3} + [0.013(1+\psi)\nu_\mathrm{p} - 1/2]a_0/l} \,. \tag{5.53}$$

Since the pore and boundary remain attached, it is required that

$$v_\mathrm{b} \cos\alpha = v_\mathrm{p} \,. \tag{5.54}$$

A dimensionless grain boundary rate $\nu_\mathrm{b} = v_\mathrm{b} kTa_0^3/D_\mathrm{s}\delta_\mathrm{s}\gamma_\mathrm{s}\Omega$ can be determined directly from (5.49), (5.53) and (5.54) in terms of a dimensionless parameter, $\xi = M_\mathrm{b} kTa_0^2\gamma_\mathrm{b}/D_\mathrm{s}\delta_\mathrm{s}\gamma_\mathrm{s}\Omega^{1/3}$, that reflects the relative mobility of pores and grain boundaries. The boundary rate can be characterized by two regimes (see Fig. 5.19). For small values of ξ, the motion of the pore/grain boundary ensemble is limited by the motion of the grain boundaries. Within this boundary mobility limited regime, grain boundary motion is not cognizant

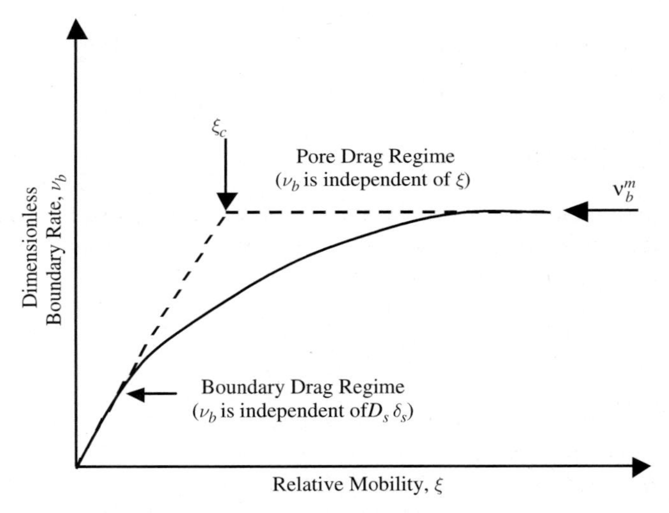

Fig. 5.19. A schematic indicating the pore drag and boundary drag regimes [1005]

of the presence of the pores. Therefore, the boundary rate increases linearly with ξ and has form:

$$\nu_b = M_b \gamma_b \Omega^{2/3} / l \,. \tag{5.55}$$

For large values of ξ, the motion of the ensemble is caused by the mobility of the pores (i.e., by the rate of surface diffusion) and the rate of the boundary becomes independent of ξ (Fig. 5.19), causing $\alpha \to \pi/3$, and the boundary rate approaches the pore drag limit:

$$\nu_b^m = 8/(1 + 3\psi) \,. \tag{5.56}$$

The transition between these two regimes occurs over a range of ξ, associated with a critical value, ξ_c (Fig. 5.19). The grain boundary mobility dominated grain growth is considered to obtain when $\xi < \xi_c$, that is, when

$$a_0^2 (a_0/l) < 8 D_s \delta_s \gamma_s \Omega^{1/3} / M_b k T \gamma_b (1 + 3\psi) \,. \tag{5.57}$$

Pore Shrinkage

The shrinkage of pores at three-grain junctions in a material, consisting of grains of equal size, is given for grain boundary diffusion dominance by [442]

$$\dot{a} a^2 (1 - a/l)^2 = -(8/3) F(\psi) (\Omega D_b \delta_b \gamma_s / k T l) \,, \tag{5.58}$$

where

$$F(\psi) = \sin(\psi/2 - \pi/6) \left\{ 1 + \frac{\sqrt{3}[\psi - \pi/3 - \sin(\psi - \pi/3)]}{2 \sin^2(\psi/2 - \pi/6)} \right\}^{-1} , \tag{5.59}$$

$D_b \delta_b$ is the boundary diffusivity. Evidently, the pore shrinkage rate, $\dot{a}$, increases as the grain facet length, l, diminishes. It might be anticipated that the shrinkage rate in the three-sided grain configuration would increase as the grain boundaries converge. However, the tendency toward enhanced pore shrinkage results in the development of residual tensile stresses. These residual stresses constrain the local shrinkage, such that the shrinkage rate is approximately that of circumventing material, as given by (5.56), in which l is the mean facet length. Similarly, a tendency toward enhanced shrinkage, induced by the change in curvature at the distorted moving pore, is suppressed by the matrix constraint.

Coarsening Trajectories

The pore and grain coarsening that occur during normal sintering may be deduced, inserting (5.58) and (5.59) into (5.39) and (5.40), in the case of the constrained pore shrinkage rate, and inserting (5.55) and (5.56) into (5.39) and

(5.40), in the case of the boundary rate. We obtain in the boundary mobility regime, $\xi < \xi_c$, inserting (5.55), (5.58) and (5.59) into (5.39) and (5.40):

$$\frac{da}{dl} = \frac{a}{l} - \frac{8F(\psi)\omega}{3\lambda(1-a/l)^2a^2} ; \quad (5.60)$$

$$\frac{a^+}{l}\left(1-\frac{a^+}{l}\right)^2 = \frac{8F(\psi)\omega}{3\lambda(a^+)^2} , \quad (5.61)$$

where $\omega = D_b\delta_b\gamma_s\Omega^{1/3}/M_bkT\gamma_b$ is a dimensionless parameter that reflects the ratio of the grain boundary mobility to the pore shrinkage rate by the grain diffusion. The trends in a^+/l indicate that solutions exist over a limited range of ω. For values of ω above a critical value ω_c, the absence of a solution indicates that the *shrinkage rate always exceeds the coarsening rate*, that is, the maximum pore size coincides with initial value, a_0. Therefore, pore coarsening is excluded below a critical pore size, a^*, when [1005]

$$a < a^* \approx (7D_b\delta_b\gamma_s\Omega^{1/3}/M_bkT\gamma_b)^{1/2} . \quad (5.62)$$

Thus, low boundary mobility and large boundary diffusivity are the most desirable conditions for averting pore–boundary separation.

The densification process is accompanied by initial coarsening that causes a decrease in a/l and an increase in pore size. Therefore, if a/l decreases at a sufficiently rapid rate that the transition line $\hat{a}/l$ is attained, further pore coarsening is prohibited and the pore shrinkage commences.

When the grain boundary rate is limited by pore drag, the trajectory is obtained from (5.39), (5.40), (5.56), (5.58) and (5.59) as

$$\frac{da}{dl} = \frac{a}{l}\left[1 - \frac{F(\psi)(1+3\psi)}{3\lambda\Delta(1-a/l)^2}\right] ; \quad (5.63)$$

$$\frac{a^+}{l} = 1 - \left[\frac{F(\psi)(1+3\psi)}{3\lambda\Delta}\right]^{1/2} , \quad (5.64)$$

where $\Delta = D_s\delta_s/D_b\delta_b$. The transition pore size again shows pore coarsening and shrinkage regions. Coarsening is invariably excluded when the diffusivity ratio and dihedral angle satisfy the condition:

$$D_b\delta_b/D_s\delta_s > 3\lambda/F(\psi)(1+3\psi) . \quad (5.65)$$

Otherwise, there is no maximum, and pore coarsening inevitably initiates whenever a/l becomes smaller than a^+/l. Both considered regimes of grain growth can be generalized as it is shown in Fig. 5.20. Recognizing that the transition pore sizes pertinent to the individual grain growth regimes will typically intersect, this intersection may be found by using inequalities (5.62) and (5.65). When both inequalities are closely approached, the intersection occurs within an intermediate range of $a/l(\sim 0.2–0.3)$, and exclusion of pore coarsening is contingent upon less stringent conditions than suggested by either

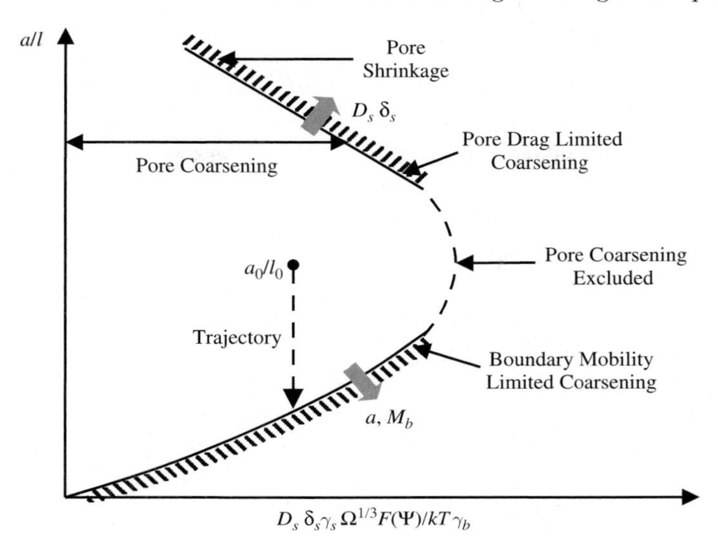

Fig. 5.20. A schematic illustration of the general tendencies in the transition pore size

inequality (5.62) or inequality (5.65) (Fig. 5.20). Otherwise, the coarsening behavior is dominated either by pore drag or by the grain boundary mobility. The intersection is caused by the initial pore size, the grain boundary mobility and the surface diffusivity. In particular, small values of $a_0, D_s\delta_s$ and M_b simultaneously induce intersection behavior, narrowing the region of pore coarsening.

The pore coarsening value outside the exclusion region can be deduced, in principle, by superimposing onto Fig. 5.20 the line that indicates the transition from pore drag to grain boundary mobility limited grain growth and evaluating the coarsening that occurs in each region. An upper bound pore size, when it exists, has a functional form [1005]:

$$\hat{a}/l = F_1(D_s\delta_s/D_b\delta_b)F_2(M_b a_0^2 kT/D_b\delta_b)F_3(\psi^{-1})\,, \tag{5.66}$$

where F_n are increasing functions of the pertinent variables.

5.2.3 Estimation of Pore Separation Effects for HTSC

The coarsening parameter that exerts a major influence on the final microstructure is the peak pore size, $\hat{a}$ (see Fig. 5.21). Its comparison with the critical pore size, a_c, defined by (5.34), states its breakaway from grain boundary. The pore-breaking behavior is caused primarily due to the pores in triple junctions (and pores with this morphology dominate the peak pore size, $\hat{a}$). At the same time, only those pores at two-grain interfaces are amenable to separation from grain boundaries [445] (at pore sizes $> a_c$). Thus, in the comparison

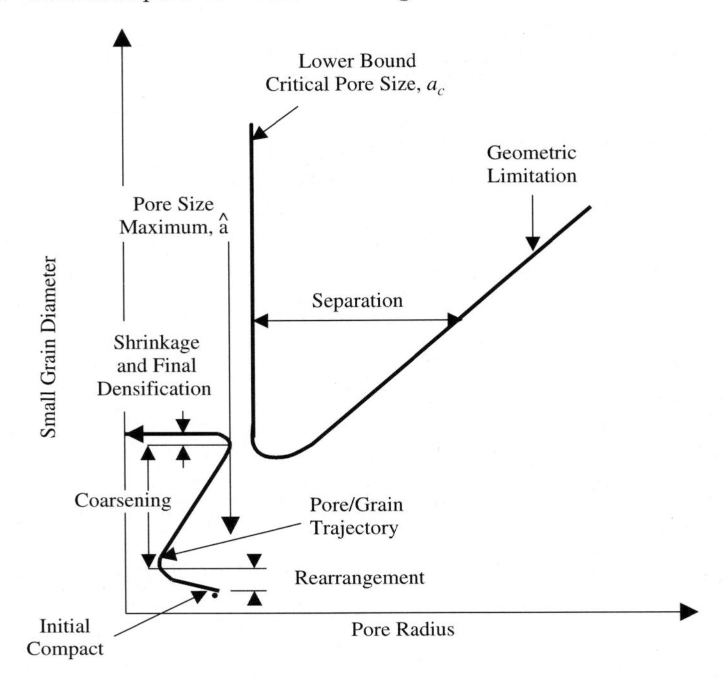

Fig. 5.21. The relation between critical pore size for separation,a_c, and peak pore size, $\hat{a}$

of $\hat{a}$ and a_c, it is necessary to assume the concurrent existence of pores of similar dimensions on both two- and three-grain interfaces. Non-separating pores on two-grain interfaces become inevitably attached to three-grain corners during small grain disappearance [445] that leads to subsequent enlargement by pore coalescence. A part of the coarsened pores may subsequently translate again onto two-grain interfaces causing the process to repeat until either the peak size is reached or the pores on two-grain interfaces become large enough to separate.

Comparing (5.3) and (5.66), the understanding for the possibility of averting pore separation in small microstructures is provided, for which pore shrinkage and pore motion are dominated by grain boundary diffusion and surface diffusion, respectively. Note that requirements of small grain-boundary mobility and large grain-boundary diffusivity are evident. However, these conditions can only be simultaneously satisfied in the presence of appreciable solute (or precipitate) drag throughout the grain disappearance process. Hence, a vital influence of drag-inducing solutes upon the attainment of optimum microstructures is apparent.

The trends connected with the surface diffusion can be stated for two extremes, namely (i) when $a_c >> a_0 >> a^*$ pore breakaway can be avoided by increasing the ratio of the boundary to surface diffusivity, consistent with condition of $a_0 < a_c$, until (5.65) is satisfied. This trend is entirely compatible with that needed to achieve densification during the initial stage of sintering [126, 414]. Conversely, in the case (ii) when $a_0 \sim a^*$, large values of the surface diffusivity are needed to ensure that $a_c > a$ during subsequent pore coarsening [414]. This requirement is distinctly different from the initial stage of densification. Hence, different temperatures are undoubtedly desirable during the initial and final stages of sintering. Also note that the explicit statement of different thermal treatments depends on the values and temperature dependencies of both the surface and boundary diffusivities and also the grain boundary mobility.

The presented analysis of pore transformation as a result of grain growth, based on the surface diffusivity mechanism, is well satisfied for fine pores. Transformations of larger pores, when their motion is caused by the difference of the leading and trailing surface curvatures, can be also studied, taking into account corresponding changes of gas pressure in different points of the pore surface [443]. This case is less interesting for HTSC and therefore is not considered here.

Finally, (5.34) is used for quantitative analysis of possible pore sizes, which break away from grain boundaries during sintering of monocore Bi-2223/Ag tapes. Selecting required parameters as follows [670, 671, 864]: $\Omega = 2.2 \times 10^{-30}\,\mathrm{m}^3$, $D_s\delta_s = 2.5 \times 10^{-21}\,\mathrm{m}^3/c$, $\gamma_c = 2\gamma_b$, $\psi_{max} = \pi/2, T = 1110\,\mathrm{K}, 2R = 25\,\mu\mathrm{m}$, $k = 1.38 \times 10^{-23}\,\mathrm{J/K}$ (here the some parameters is selected for Al_2O_3, because corresponding data is absent for Bi-2223), we obtain a very high value of $M_b \approx 4 \times 10^4\,\mathrm{m/(N \cdot c)}$ even for $a_c = 100\,\mathrm{nm}$. For smaller pores separated from grain boundary, longer grain mobility is required. Obviously, the size of pores, which can be separated from grain boundaries during prolonged annealing, on some orders of magnitude is longer than the coherence length ($\sim 1\,\mathrm{nm}$) in Bi-2223. Therefore, these separated pores cannot serve effective pinning centers and because of percolation features must considerably diminish the critical current. Apparently, in prolonged annealing, this effect is more important than deteriorating pinning strength due to lead expelling [670, 671]. We think that namely numerous pore separations and their movement into grain insides have found the critical current decrease in longer calcining after observed I_c maximum at about 180 hs annealing of monocore Bi-2223/Ag tapes [670, 671]. Thus, the lead expelling causes a decreasing of critical current in long reaction, but rather due to pore transformations, occurred during annealing (because Pb can inhibit only a local grain growth in this case), than thanks to decrease of its pinning efficiency in the grains.

5.3 HTSC Microstructure Formation During Sintering

For investigating the processes of HTSC ceramic preparation, plane sample of superconducting powder compact in gradient furnace is considered. Two-scale modeling, consisting of the macroscopic study of the precursor powder sintering and microstructure formation into the region of the heat front propagation is carried out. For this, the considered rectangular region $[a, b]$ is divided into square lattice with characteristic size of elementary cell, δ, which corresponds to either particle or pore. The sample moves into gradient furnace with constant rate, v (which can be correlated with the temperature change rate of the sample surface). Moreover, it is assumed that temperature distribution T into furnace depends on one coordinate x and consists of sites with constant temperature and linear dependence on this coordinate. In order to solve an initial-boundary problem, the method of summary approximation (MSA) is used (see Appendix C.1).

Microstructure modeling begins from the pore generation in the initial sample by using Monte-Carlo procedure. In this case, it is suggested that the pores are distributed in accordance with the normal distribution, and pore start concentration (i.e., porosity) C_{p}^{0} is given in different variants of the computation. This procedure can be carried out, for example, filling the cells of the initial lattice by arbitrary numbers, using the generator of arbitrary numbers (GAN) and then selecting a number of minimum values that corresponds to the pore number. In order to obtain the statistically reliable results, the computations are accompanied by averaging of the pores and crystallites distributions in the sample microstructure.

The model includes the following main stages, namely (i) a heat front displacement and definition of the material sintering region; where a temperature above the sintering temperature, u_{s}; (ii) a press-powder re-crystallization into the corresponding region; and (iii) a shrinkage of the microstructure formed. The first from pointed steps relates to the macroscopic modeling and the other two to the microscopic modeling.

The computation of effective heat conduction (see Appendix C.2) for non-sintered part of the sample consists of the following stages. First, a coordination number, N_{c}, and the sizes of element with averaged parameters (y_1, y_2) using (C.2.12), (C.2.22) and (C.2.28) is defined. In order to calculate the heat conduction of gas into gaps between particles, δ_{s}, λ_{sr} and λ_{s} from (C.2.37), (C.2.35) and (C.2.34) are computed successively. Then, the heat conduction of frame is computed depending on the porosity of the considered region, using (C.2.43). The porosity of the second-order structure is computed on the basis of (C.2.17) and then, using (C.2.5) and (C.2.6), $c_2 = c$ is calculated. In this case, the heat conduction of gas into pores of the second-order structure, λ_{22}, is defined, using (C.2.20). Finally, the effective heat conduction of the non-sintered part of the sample is calculated using (C.2.13).

In order to study a displacement of the thermal front, the first main problem for quasi-linear equation of heat conduction with a variable $u = T - T_0$ is considered, where T_0 is the environment temperature:

$$\frac{\partial u}{\partial t} = \frac{\partial}{\partial x}\left[k(u, C_{\mathrm{p}})\frac{\partial u}{\partial x}\right] + \frac{\partial}{\partial y}\left[k(u, C_{\mathrm{p}})\frac{\partial u}{\partial y}\right] , \tag{5.67}$$

with initial condition

$$u(0, x, y) = 0 , \tag{5.68}$$

and boundary conditions

$$u(t, 0, y) = u_1(t); \quad u(t, a, y) = u_2(t); \quad u(t, x, 0) = u(t, x, b) = u_3(x, t) . \tag{5.69}$$

Here, $k(u, C_\mathrm{p})$ is the temperature conductivity factor; and C_p is the pore concentration or porosity. The boundary conditions are shown schematically in Fig. 5.22 for different time, t. In this case, the function $u_3(x, t)$ has the form:

$$\text{(a)} \quad u_3(x,t) = \begin{cases} A_0 t^{\frac{(vt-x)}{vt}} & x \le vt \\ 0 & x > vt \end{cases} , \tag{5.70}$$

$$\text{(b)} \quad u_3(x,t) = A_0 t \frac{(a-x)}{a} ; \qquad \text{(c)} \quad u_3(x,t) = u_\mathrm{s} + \frac{(u_{\max} - u_\mathrm{s})(a-x)}{a} , \tag{5.71}$$

where A_0 is the given constant. The calculation is finished in the case of reaching in the whole region the temperature, $u(x, y) \ge u_\mathrm{s}$.

The boundary conditions (5.69)–(5.71) correspond to a scheme of the ceramic gradient sintering [516]. In the case of hot-pressing, these boundary conditions are replaced by the following ones [823]:

$$u(t, 0, y) = u(t, a, y) = u(t, x, 0) = u(t, x, b) = A_0 t . \tag{5.72}$$

In the beginning of the modeling, preliminary, the start porosity, C_p^0, the sintering temperature, $u_\mathrm{s} < u_{\max}$ (where $u_{\max}$ is the maximum temperature into gradient furnace for considered site of sintering) and the sample movement rate, v, are assumed to be known. The temperature conductivity factor at the first stage is found by constant values of heat capacity, c_V, and material density, ρ, and also by heat conduction factor, λ, depending on press-powder porosity. As λ is selected a value of the heat conduction factor, λ_ef, is calculated on the basis of the generalized conductivity principle (Appendix C.2). Hence, $k(u, C_\mathrm{p}) = \lambda_\mathrm{ef}/(c_\mathrm{V}\rho)$.

A solution of the problems (5.67)–(5.71) is obtained on the basis of the method of summary approximation (MSA) by using pure non-evident local one-dimensional scheme (LOS) (Appendix C.1). For this aim, a finite-difference counterpart of the initial-boundary problems (5.67)–(5.71) is written. The finite-difference equation is solved, using the run method. In order to obtain the required temperature field at any step of the considered process, there is a non-linear equation solved by iterations. After definition of temperature distribution into region, where $u \ge u_\mathrm{s}$, a modeling of the material re-crystallization and shrinkage is carried out. Note that the grain formed at any stage can penetrate into earlier "sintered" region if it is allowed by the microstructure porosity. After that, the heat conduction factor is altered

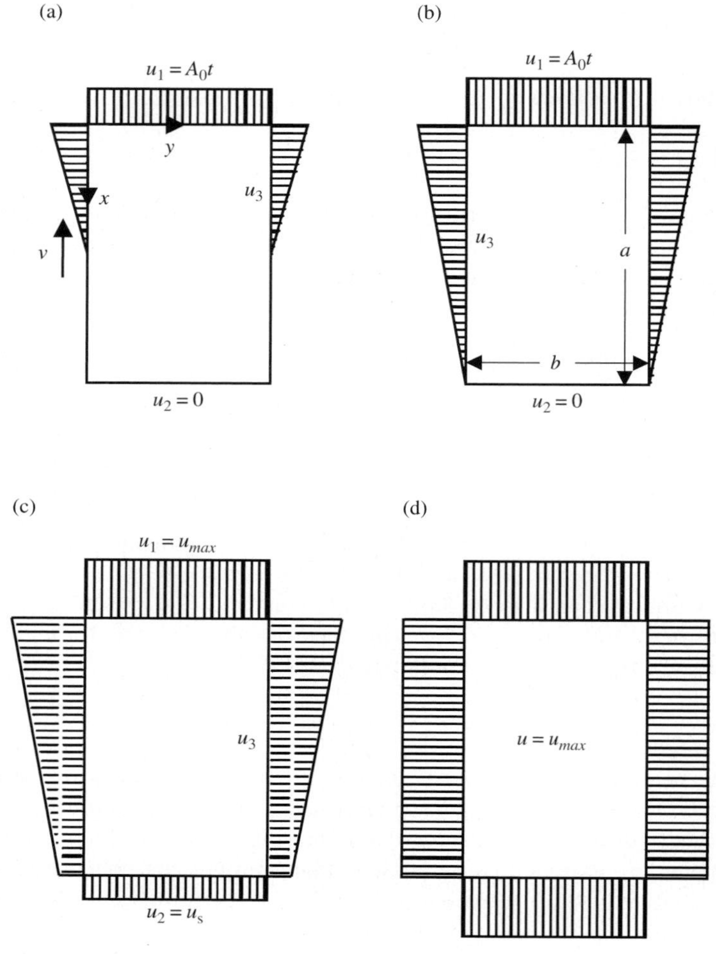

Fig. 5.22. Change of boundary conditions during heating of the sample

and the corresponding value of the temperature conductivity factor is calculated. At the every stage of the microstructure modeling, λ_{ef} will be defined by the heat conduction of sintered and non-sintered regions. The first region possesses heat conduction of fabricated material. At the same time, this parameter is calculated for non-sintered region taking into account an existing porosity by using the generalized conductivity principle (Appendix C.2). Therefore, λ_{ef} in the whole considered region is defined by concentration of both components and by the values of their heat conduction by using the rule of mutual-penetrating components (see (C.2.13)). Then, a transition to the next time interval is carried out. The process finishes after microstructure formation in the whole considered region, which, as a result, consists of pores and grains.

Separately, the microstructure mechanisms of the superconducting press-powder re-crystallization and shrinkage for the formed structure in the temperature region, $u \geq u_{\mathrm{S}}$, are modeled. The process of the crystallite nucleation into precursor press-powder is assumed to be the thermal activated process. Therefore, an arbitrary number corresponds to each remaining (after modeling of porosity) cell of the considered region. This arbitrary number characterizes the initiation time of a single crystallite, and these numbers are obtained by using GAN from the law of exponential distribution [1002]: $P_{ij}(t) = 1 - \exp(-t/\tau_{ij})$, where $\tau_{ij} \sim \exp(U/ku_{ij})$ is the mean expectation time of the crystallite nucleation in the node of i-line and j-column; U is the activation energy of crystallite nucleation; k is Boltzmann constant; u_{ij} is the temperature in the lattice cell with coordinates (i,j). It is suggested that a crystallite with minimum nucleation time, t^*_{ij}, first nucleates, and its nearest neighbors with coordinates (k,l) obtain priority. According to this, the values of t_{kl} are decreased. At the next step, a minimum nucleation time, t^*_{ij}, is again found among of all remaining lattice cells, and a new crystallite nucleates in the corresponding cell. This cell is either the nearest cell to the earlier-nucleated crystallite or far from it. In the first case, there is a grain growth in the re-crystallization process. Generally, the nucleation time of crystallite, t^{c}_{kl}, near with earlier-nucleated one is defined as

$$t^{\mathrm{c}}_{kl} = \left| t_{kl} + \frac{t^*_{ij} - t_{kl}}{S\exp(1-S)} \right| , \tag{5.73}$$

where S is the grain square (which is defined by the corresponding number of cells) near of that it is possible an nucleation of a new crystallite. Note that relation (5.73) causes a decreas of the nucleation time of neighbor cells, until S is sufficiently small, and the time increases together with the grain area due to pushing of secondary phases at the intergranular boundaries during grain growth.

After completion of the crystallite system formation into region of order of the sintering front width, a shrinkage of the sample is modeled. One includes pushing of gaseous component from the sample and decreasing of closed porosity, thanks to grain movement. Computational algorithm of the shrinkage provides successive alternating displacements of grains along two orthogonal directions upon total exhaustion of possibility of their movement. In the displacement process, it is assumed that grains preserve own volume, shape and spatial orientation. This is achieved due to successive definition for each lattice layer of possibility of the grain movement (this considered grain consists of single cells) and its displacement as a single whole. In the calculation, it is assumed that the shrinkage process occurs instantaneously.

The completion of the grain structure formation during the re-crystallization and shrinkage is accompanied by beginning of secondary re-crystallization, that is, abnormal grain growth due to existence of the pores and admixture phases. This grain growth occurs at maximum pressure and a linear character of the temperature change ($0 \leq T \leq T_{\max}$, where $T_{\max}$ is the maximum temperature into furnace) [1034]. The modeling of the abnormal grain growth is carried out on the basis of the Wagner–Zlyosov–Hillert's dynamic growth models [1, 2, 245].

5.4 Microcracking of Intergranular Boundaries at Sample Cooling

A plane sample of sintered ceramic into gradient furnace is considered. After sintering, a uniform temperature, $u = u_{\max}$, is stated in the whole sample. Due to a shrinkage during the sample cooling, the rectangular site of the front with sizes $a_1 \leq a$ and $b_1 \leq b$ is considered, which excludes the layers, including only elements of open porosity. At macroscopic modeling of the cooling, it is assumed that heat conduction factor is found by the thermal conduction of the sintered material and depends only on temperature, u. The modeling of the heat front movement continues down to attainment of room temperature at one of the sample boundaries. Because intergranular boundaries (the main stress concentrators in this case) are subject to cracking at cooling, the required values (i.e., temperatures and stresses) are calculated in the lattice nodes but do not in the lattice cells, as in the case of sintering.

A modeling of microcracking consists of the following stages: (i) a displacement of heat front and definition of temperature field, (ii) a calculation of normal thermal stresses in the lattice nodes, (iii) a division of all intergranular boundaries into separate sections, (iv) a computation of mean normal stress, acting onto given section and (v) a satisfaction to microcracking condition.[5]

Similar to the problem of ceramic sintering, it is assumed that the sample displaces with constant rate, v (which can be also correlated with the temperature change rate of the sample surface), from gradient furnace. The temperature changes in linear law into furnace. In the calculation of temperature field, the initial-boundary problem (5.67)–(5.69) is considered, where initial condition reduces to $u(0, x, y) = u_{\max}$, and boundary conditions are depicted in Fig. 5.23. In this case, a function $u_3(x, t)$ is given as

$$\text{(a)} \quad u_3(x,t) = u_{\max}\,; \tag{5.74}$$

$$\text{(b)} \quad u_3(x,t) = \begin{cases} u_{\max} + \frac{(u_1 - u_{\max})}{vt}(vt - x) & x \leq vt \\ u_{\max} & vt \leq x \leq a_1 \end{cases}, \tag{5.75}$$

As a result of the solution of this thermal conduction problem, a temperature distribution, causing corresponding thermal stresses, is calculated.

[5] Similar numerical algorithm may be applied to modeling of microcracking processes during cooling of oxide superconductor from room temperature down to cryogenic one its application.

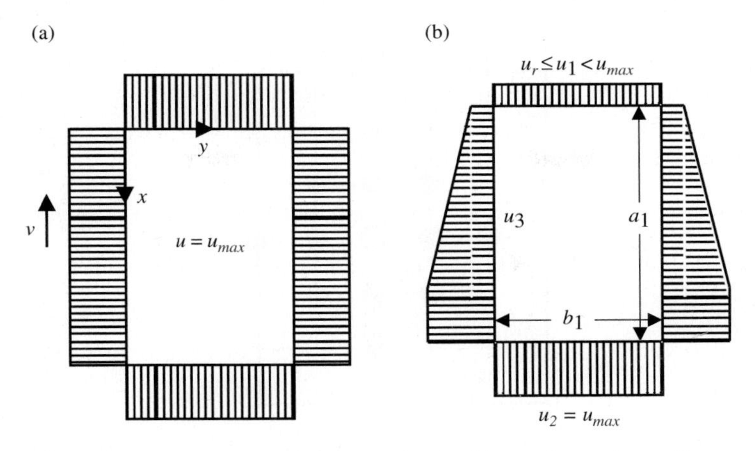

Fig. 5.23. Boundary conditions for problem of sample cooling

In order to define 2D stress state, the method of finite differences [663] is applied. For framework of the thermal stresses problem, stress state is calculated through Airy's function, φ

$$\sigma_x = \frac{\partial^2\varphi}{\partial y^2}; \quad \sigma_y = \frac{\partial^2\varphi}{\partial x^2}; \quad \sigma_{xy} = -\frac{\partial^2\varphi}{\partial x\partial y} . \tag{5.76}$$

In this case, the function φ satisfies the differential equation in partial derivations:

$$\Delta^2\varphi + E\alpha\Delta u = 0 , \tag{5.77}$$

where $\Delta = \partial^2/\partial x^2 + \partial^2/\partial y^2$; $\Delta^2 = \partial^4/\partial x^4 + 2(\partial^4/\partial x^2\partial y^2) + \partial^4/\partial y^4$; $E\alpha =$ const; E is Young's modulus; α is the thermal expansion factor and $u = u(x, y)$ is the temperature, calculated from a some value, in which the thermal stresses are absent.

Replacing the partial derivations in (5.77) by finite differences, we obtain for arbitrary point 0 of the considered region (Fig. 5.24):

$$\begin{aligned} 20\varphi_0 - 8(\varphi_1 + \varphi_2 + \varphi_3 + \varphi_4) + 2(\varphi_6 + \varphi_8 + \varphi_{10} + \varphi_{12}) + \\ + (\varphi_5 + \varphi_7 + \varphi_9 + \varphi_{11}) + E\alpha\delta^2(u_1 + u_2 + u_3 + u_4 - 4u_0) = 0. \end{aligned} \tag{5.78}$$

In the case of simply connected region and absence of external loading, the boundary conditions can be presented as [1062]

$$\varphi = \frac{\partial\varphi}{\partial y} = 0 \quad \text{at } x = 0; x = a_1 ; \tag{5.79}$$

$$\varphi = \frac{\partial\varphi}{\partial x} = 0 \quad \text{at } y = 0; y = b_1 . \tag{5.80}$$

The conditions (5.79) and (5.80) sign a definition of zero values of principal vector and principal moment of the acting forces at the region boundaries.

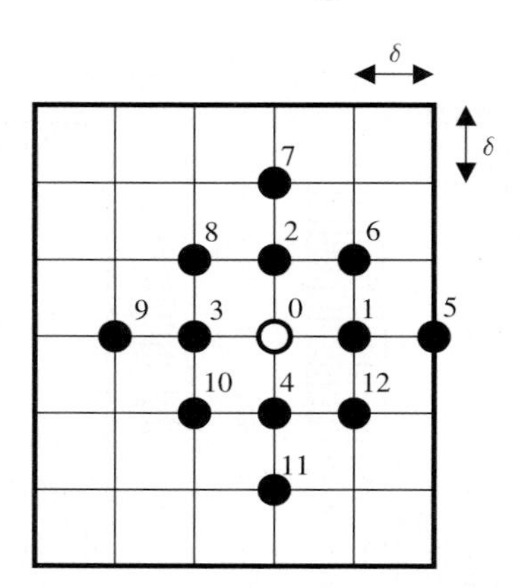

Fig. 5.24. Illustration to the definition of thermal stresses at point 0

Then, by using the function φ, the normal stresses are calculated as

$$(\sigma_x)_0 = \frac{1}{\delta^2}(\varphi_2 - 2\varphi_0 + \varphi_4) \; ; \qquad (\sigma_y)_0 = \frac{1}{\delta^2}(\varphi_1 - 2\varphi_0 + \varphi_3) \; . \tag{5.81}$$

After computation of the thermal stresses, the length of each section at intergranular boundary is estimated. The mean value of normal stress, acting on the given section of intergranular boundary, permits to verify the condition of the boundary microcracking:

$$\overline{\sigma}_n \sqrt{\pi l} \geq K^0_{Ic} \; , \tag{5.82}$$

where $\overline{\sigma}_n$ is the mean value of the normal stress on the intergranular boundary section with length l; K^0_{Ic} is the fracture toughness of ceramic. When for the section, (5.82) is satisfied, it is replaced by a microcrack. Obviously, in the first place, microcracks form on the sufficiently long and stressed boundaries. The microcracking modeling is repeated cyclically together with solution of heat conduction problem and calculation of thermal stresses.

A simpler scheme of the microcracking modeling on intergranular boundaries may be the next. Based on the test data, a critical grain size during spontaneous cracking, D^S_{c}, for considered composition of superconducting ceramic is selected. Assuming, that the grain sizes and facet sizes per grain are subject to the normal distribution, we obtain [176] $D^S_{\mathrm{c}} \approx 2l^S_{\mathrm{c}}$, where l^S_{c} is the critical facet size. Taking into account misorientation of grains, a criterion of microcrack formation at the boundary l is stated [295] as

$$l/l^S_{\mathrm{c}} \geq 2/[1 + \cos(2\Theta_1 - 2\Theta_2)] \; , \tag{5.83}$$

where $\Theta_i (i = 1, 2)$ is the angle between axis of maximum compression in i-grain and the boundary plane of the grain. These angles are calculated by using MonteCarlo procedure [176]. Computational algorithm takes into account that a microcrack grows from triple junctions (i.e., junctions of three or four grains at square lattice) [295, 572] or is initiated by pore and stops at the nearest facet node because the facets are usually subject to internal compressing stresses [295].

5.5 Study of Statistical Properties of the Model Structures

The computational approach, presented in Sects. 5.3 and 5.4, is the good base to investigate the size, quantitative, topological parameters and morphological features and also to define structure-sensitive properties of polycrystalline superconductor. Typical description of ceramic microstructure includes definition of the size, shape and spatial orientation of the pores and grains. There are different classifications of materials in the porosity parameters. Another approach to microstructure description is based on the assumption about its non-regularity. Either microstructure is estimated by own constant statistical properties. The void space is correlated with the solid surfaces and forms by using the grain facets and edges. Architecture of the void space is caused by the shape, size and mutual disposition of crystallites. The description of solid matrix characterizes the material space as united whole, including the void space. Similar approach does not exclude an identification of single spatial structure elements (e.g., pores, grains, etc.) [794].

HTSC microstructure may be described by integral (i.e., global) parameters, defining joint physical properties. An application of integral characteristics permits to abstain from simplified selection of the grain and pore shapes and to characterize the material space by using summary values of volume, surface, curvature, etc., obtained from stereological measurements. Latter estimate only global metrics, that is, volume fraction, surface square, length of linear elements and surface curvature, and also their combined parameters [513].

The method to define the volume structure characteristics by using measurements in the observation plane has been named by the *statistical reconstruction* [145]. It is based on two statistical principles, namely (i) the structure sample should have representative volume and (ii) a statistical correlation of the depicted structure characteristics in the observation plane with actual structure is necessary. Generally, a calculation of necessary number of the measurements to obtain unbiased estimation of any stereological characteristic is carried out [145]:

$$n = (200/y)(\sigma_x/\overline{x}) , \tag{5.84}$$

where y is the accuracy level (%); σ_x is the average quadratic deviation; $\overline{x}$ is the mean value of the stereological characteristic. In the case of given accuracy, the number of measurements depends on the variation factor, $\sigma_x/\overline{x}$, that characterizes the uniformity of analyzed structure element. The procedure for the calculation of necessary measurement number includes the following steps:

(1) the mean value of the stereological characteristic, $\overline{x}$, and variance, σ_x, are estimated for some random sampling;
(2) it is given a necessary accuracy level (y, %) for the mean value of the measured magnitude;
(3) the number of measurements, n, is calculated from (5.84) which ensures the necessary accuracy level.

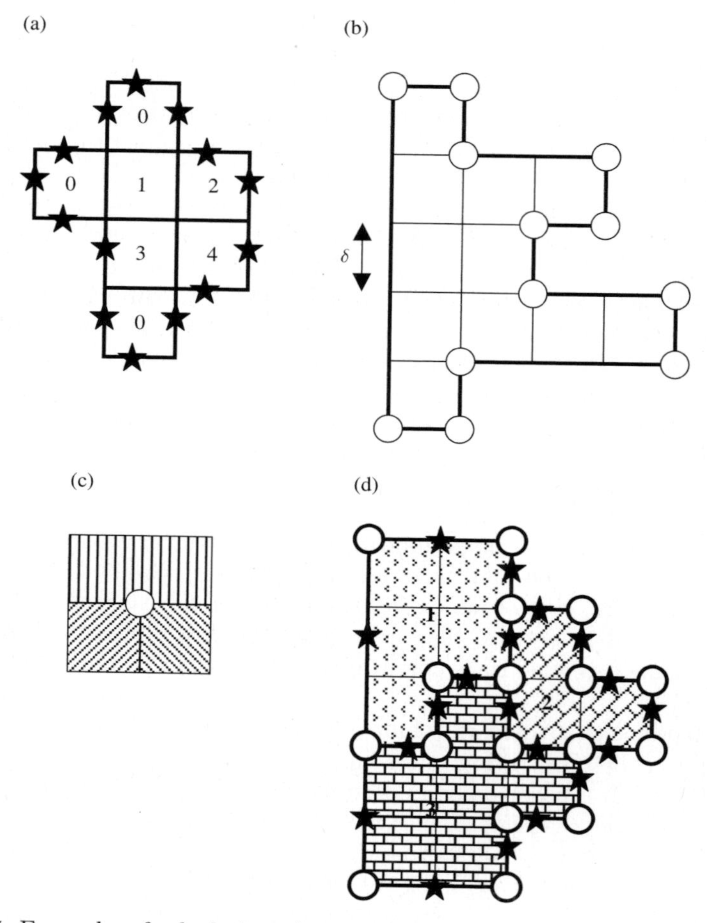

Fig. 5.25. Examples of calculation of some geometrical and topological parameters: (**a**) normalized perimeter of grain, L/δ (5.85), $n = 7$, each from two elements (2, 4) have two neighbors, element (3) has three neighbors and element (1) has four neighbors: $L/\delta = 3 \times 7 - (2 \times 1 + 1 \times 2 + 1 \times 3) = 14$; (**b**) number of grain tip (side) $N_l = 12$; (**c**) triple junction (point of intersection of three grain boundaries in square lattice); (**d**) Euler's relation for polyhedron [145]: $G - E + V = 1$, where G is the grain number (3), E is the side number (19), V is the top number (17)

Note that (5.84) is obtained by using central limit theorem of statistics, according to which average values of characteristic, obtained for representative samplings, are distributed in the normal law relative to the mean value for corresponding universe. This approach to define the required number of measurements is used in the next chapters for study of microstructure and current-carrying parameters of model superconductors.

A square of single grain or pore is usually estimated by using a detailed depiction of microstructure. Here, the realized algorithm defines a disposition of considered grain and computes the total number of its cells. Then, based on these squares, the proper grain sizes (or radii) are calculated. The grain or pore perimeter, L, in 2D case of square lattice may be found (normalized to the elementary cell size, δ) as

$$L/\delta = \begin{cases} 2(n+1) & n \le 2 \\ 3n - \sum\limits_{k=2}^{4} l_k(k-1) & n > 2 \end{cases} , \tag{5.85}$$

where n is the cell number of considered grain (or pore), k is the possible number of nearest neighbors for every cell in 2D case and l_k is the cell number of grain (or pore) with k number of neighbors. An example of (5.85) application is presented in Fig. 5.25a.

5.6 Modeling of Macrocracks

High-temperature superconductors of YBCO or BSCCO are the brittle materials. Cracks, forming in these materials, as rule, are of I Mode, that is, fracture (or critical) loading is perpendicular to the crack plane. The cleavage planes in HTSC usually correspond to one of the direction of [100] type. The algorithm for the definition of arbitrary angles, Θ, formed by the normal to cleavage plane with direction of tensile stresses, based on the method of arbitrary twisting of cube, is presented in monograph [808]. This algorithm models a transgranular cracking of HTSC ceramic, using a critical stress condition. In this case, it is suggested that definite grain fraction is fractured, when the stress normal to cleavage plane $\sigma_n = \sigma \cos^2 \Theta$ attains critical value of σ_n^* (where σ is the tensile stress). At the $\sigma = \sigma_n^*$, the microcrack formation begins in the grains, where the cleavage planes are perpendicular to the tensile axis ($\Theta = 0°$). The following increase of the stress leads to the crack propagation in the grains, for which $0 < \Theta < \Theta^*$, where Θ^* is determined from the relation $\sigma \cos^2 \Theta^* = \sigma_n^*$. When applied stress attains the value of σ_{max}, a grain cracking into cleavage plane occurs. Then, the angle Θ_{max}^* between the normal to the cleavage plane and tensile direction is obtained as

$$\Theta_{max}^* = \arccos(\sigma_n^*/\sigma_{max})^{1/2} . \tag{5.86}$$

Experimentally observed crack growth in oxide superconductor can take place as in grain boundaries (intergranular crack growth), as by cleavage through grain volume (transgranular crack growth). Moreover, there is a mixed character of macrocrack propagation, combining above two fracture types. Moreover, an existence of pores and microcracks into the material structure renders significant influence on crack growth.

Consider a model sample, including layers of rectangular form, $N \times M$ into units of mean size of the elementary cell, δ. As a first mechanism, transgranular fracture through grain volume is considered. It is known that in the case of crack growth arbitrarily oriented to tensile direction, the fracture toughness, K_{Ic}, can be found as [144]

$$K_{Ic} = \frac{K_{Ic}^0}{\cos^2 \Theta}\ , \tag{5.87}$$

where $K_{Ic}^0 = \sqrt{E\gamma_0}$ corresponds to the normal-opening crack, E is Young's modulus, γ_0 is the specific surface fracture energy and Θ is the arbitrary angle between the normal to the crack plane and tension direction.

Assume that there is cleavage crack, crossing crystallites in the planes of {100}-type. Then, the value of $K_{Ic}^{(j)}$ for crack path along j-line of the coordinate lattice with length, h_j, is found as

$$K_{Ic}^{(j)} = \sum_i \frac{K_{Ic}^0}{\cos^2 \Theta_{ij}} \sqrt{\frac{d_{ij}}{h_j}}\ , \tag{5.88}$$

where d_{ij} is the length of i-grain in the j-line and Θ_{ij} is the corresponding arbitrary angle formed by normal to cleavage plane and tension direction. Mean fracture toughness of the considered region may be estimated as

$$K_{Ic} = \frac{1}{N_l} \sum_{j=1}^{N_l} K_{Ic}^{(j)}\ , \tag{5.89}$$

where N_l is the number of lines in the selected rectangular region without the lines, including only elements of open porosity.

Then, consider a macrocrack growth along intergranular boundary without microcracks (Fig. 5.26a) and in the existence of microcracks (Fig. 5.26b) onto the part of the intergranular boundaries. The material fracture toughness, K_{Ic}, as function of the specific surface fracture energy, γ_0, and Young's modulus, E, has the form:

$$K_{Ic} = \sqrt{E\gamma_0}\ . \tag{5.90}$$

In the case of the rectilinear intergranular boundary, $\gamma_0 = \gamma_b$, where γ_b is the specific grain boundary energy. In the actual sample, either crack trajectory is the arbitrary broken line. The value of γ_0 can be found through the ratio of the crack path length, L, to the sample width, h, as

$$\gamma_0 = L\gamma_b/h\ . \tag{5.91}$$

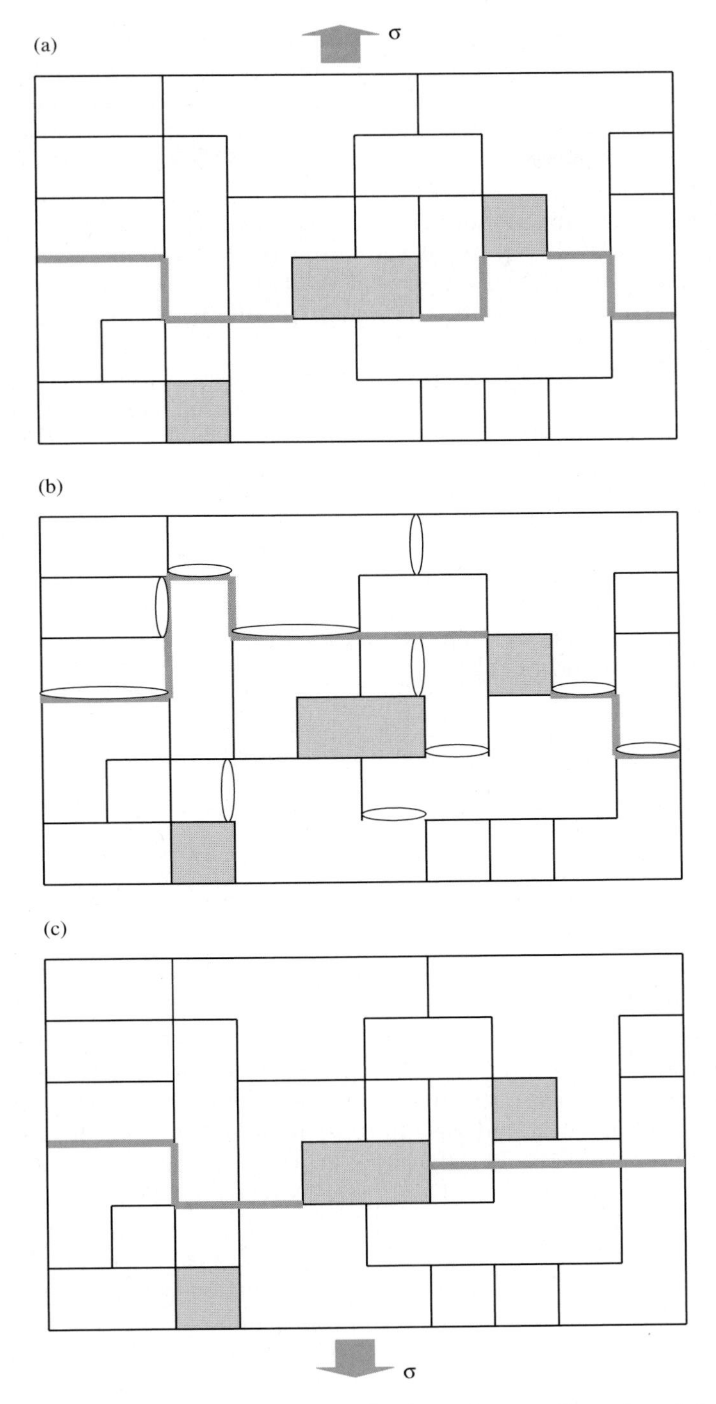

Fig. 5.26. Examples of macrocrack growth, shown by *gray* line, along intergranular boundaries in the case of (**a**) microcracking absence, (**b**) microcracking presence onto part of intergranular boundaries and (**c**) mixed mechanism of macrocrack growth

In order to compute L, we use a presentation about crack path as on the graph branch, joining the points at the opposite sides of the model layer. In this case, the graph branches are constructed, taking into account the intergranular boundary lattice. The beginning of the graph tree coincides with the intersection point of one from intergranular boundaries with the left side of the sample (see Fig. 5.26). This selection of the crack growth beginning corresponds to test data, showing that usually a fracture starts from sample surface. It is clear that a set of the graph tree branches may be selected, connecting a given point at one side of the layer with arbitrary point at the opposite side. From the energy minimum condition, a minimum trajectory corresponds to actual crack path. In order to define this minimum path, the Bellman-Kalaba's algorithm is used [562]. In this case, a problem of minimization of the numerical $(n+1)$-order graph with tips, x_i, is reduced to solution of the next equation set:

$$\begin{cases} V_i = \min(V_j + C_{ij}), \quad i = 0, 1 \ldots, n-1; \quad j = 0, 1 \ldots, n; \quad i \neq j \\ V_n = 0 \end{cases}, \tag{5.92}$$

where V_i is the length of optimum path between points of x_n and x_i; $C_{ij} \geq 0$ is the value corresponding to the graph arc (x_i, x_j). The Bellman-Kalaba's algorithm suggests an iteration method for the solution of the minimization problem (5.92). Supposing $V_i^{(0)} = C_{in}; i = 0, 1 \ldots, n-1; V_n^{(0)} = 0$, successively compute

$$\begin{aligned}
&V_i^{(1)} = \min(V_j^{(0)} + C_{ij}); i = 0, 1 \ldots, n-1; j = 0, 1 \ldots, n; i \neq j \\
&\ldots\ldots\ldots\ldots\ldots\ldots\ldots\ldots\ldots\ldots\ldots\ldots\ldots\ldots\ldots\ldots\ldots\ldots \\
&V_i^{(k)} = \min(V_j^{(k-1)} + C_{ij}); i = 0, 1 \ldots, n-1; j = 0, 1 \ldots, n; i \neq j \\
&\ldots\ldots\ldots\ldots\ldots\ldots\ldots\ldots\ldots\ldots\ldots\ldots\ldots\ldots\ldots\ldots\ldots\ldots \\
&V_n^{(k)} = 0
\end{aligned} \tag{5.93}$$

up to carrying out of the equalities, $V_i^{(k)} = V_i^{(k-1)}; i = 0, 1 \ldots, n-1$. In this case, the values of $V_i^{(k)}$ are the minimum values, which define an optimum branch of the graph tree. $(n-1)$ iterations are sufficient for its determination.

In the considered case, the graph tips coincide with the lattice nodes disposed in intergranular boundaries. In order to accelerate a sorting out of possible graph branches, Viterbi's algorithm is used [72]. Only the graph branches with length, which is not longer than a given inter-nodal distance, l_j (or length of the crack unit jump), are considered. If these paths are to be more than one, a priority is given to the path with the final tip disposed from the initial one at maximum distance along the coordinate perpendicular to tensile direction. The final graph tip, found in this way, corresponds to the crack tip after its jump; at the same time, this tip is assumed to be initial for the next graph tree, and so on. In order to optimize the selection process, it is expedient to limit a number of considered graph tips in accordance with the length of the macrocrack unit jump. The region of possible crack unit jumps is shown in Fig. 5.27a.

In order to define a crack path, taking into account ceramic porosity, it is natural to assume that in the case of the crack hit in a pore, the crack tip blunts, causing a diminution of stress concentration. In our model, the pointed drag is depicted by adding the pore boundary length, crossed by the crack, to the crack length. The normalized boundary length of the pore, L_{p}/δ, is calculated in 2D case using (5.85).

Now, the values of C_{ij} used in (5.93) for calculation of the crack path can be found:

$$C_{ij} = \begin{cases} 0 & i = j \\ L_{\mathrm{b}}/\delta & i, j \text{are two nearest nodes at the grain boundary} \\ L_{\mathrm{p}}/\delta & i, j \text{belong to grain boundary of the same pore} \\ \infty & \text{in other cases} \end{cases} \tag{5.94}$$

where $C_{ij} = C_{ji}$ and $L_{\mathrm{b}} = 2\delta$ is the double length of elementary cell side. Solution of (5.92)–(5.94) defines the whole crack trajectory. The calculation of fracture toughness for the model sample is carried out as

$$K_{I\mathrm{c}} = \frac{1}{N} \sum_{i=1}^{N} K_{I\mathrm{c}}^{(i)} , \tag{5.95}$$

where $K_{I\mathrm{c}}^{(i)} = \sqrt{E\gamma_0^{(i)}}; \gamma_0^{(i)} = (L_i/h_i)\gamma_{\mathrm{b}}$ and N is the number of considered crack paths by which an averaging of the fracture toughness is carried out.

Then, consider a model sample of superconducting ceramic, taking into account microcracks, substituting a part of intergranular boundaries. As it has been demonstrated, these microdefects can nucleate, in particular due to residual thermal stresses, caused by the thermal expansion anisotropy of grains and/or by the deformation mismatch of phases.

It is natural to assume that a microcrack, crossing the macrocrack path, renders a drag effect if the macrocrack crosses the microcrack at a junction of two sections of the microcrack (it is not important, with same direction or disposed at angle). In other cases (i.e., at the macrocrack hit in one of the microcrack tips), an acceleration of the macrocrack growth occurred. In the first case, the length of the microcrack surfaces is added to the macrocrack length. At the same time, in the second case, it is suggested that the macrocrack propagates on the whole microcrack length. Hence, relation (5.94) should be modified as in the following equation:

$$C_{ij} = \begin{cases} 0 & i = j \text{ or at favorable disposition of microcrack} \\ L_{\mathrm{b}}/\delta & i, j \text{ are two nearest nodes at intergranular boundary} \\ L_{\mathrm{p}}/\delta & i, j \text{ belong to the boundary of the same pore} \\ L_{\mathrm{m}}/\delta & \text{at unfavorable disposition of microcrack} \\ \infty & \text{in other case} \end{cases} \tag{5.96}$$

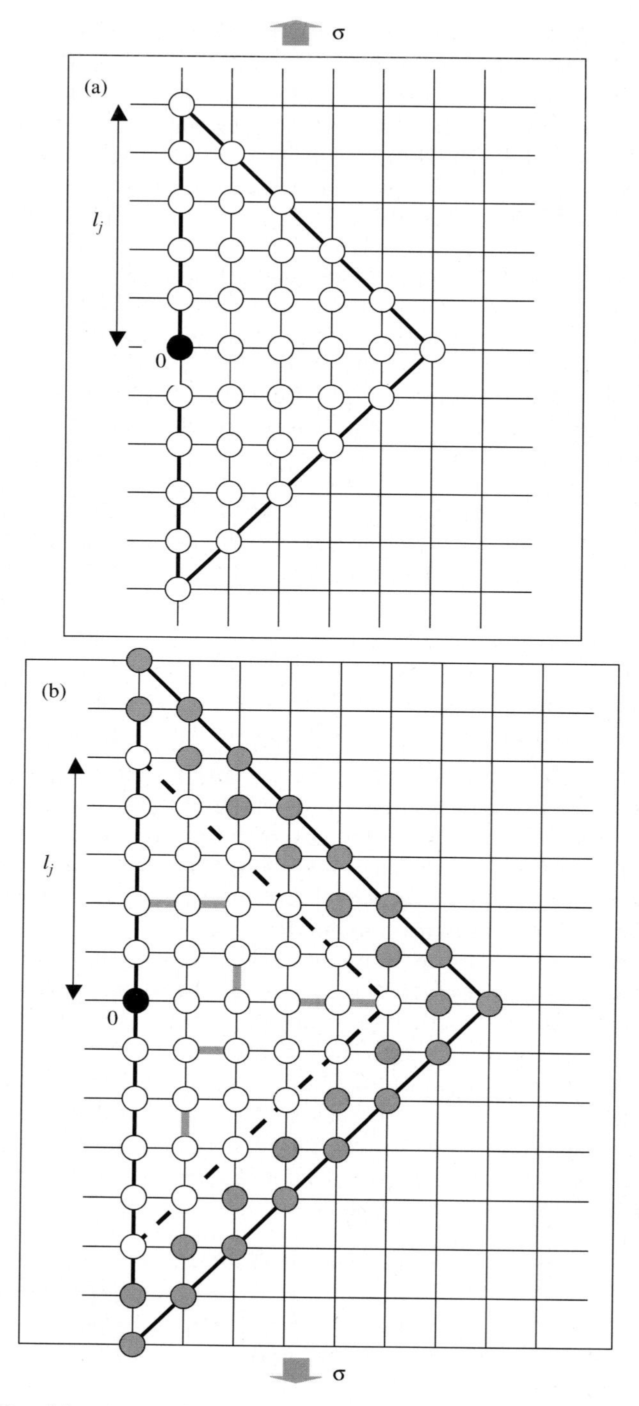

Fig. 5.27. Possible region of macrocrack unit jump during intergranular fracture (where l_j is the length of the crack unit jump): (**a**) boundary microcracking is absent; (**b**) part of intergranular boundaries is replaced by microcracks (*gray*); *circles* show considered nodes; additional nodes, caused by the intergranular nicrocracking, are shown by *gray color*

In this case, $C_{ij} \neq C_{ji}$ and L_m/δ is the normalized length of the microcrack boundary. The microcrack existence leads to a change of the crack unit jump compared to the case of absence of the grain boundary microcracking (see Fig. 5.27b). When there are points for favorable growth of the crack, initially considered region (Fig. 5.27a) should be expanded. The number of additional layers is found by the maximum summary number of rectilinear sections of microcracks disposed in one line (or column) of the initial region (see Fig. 5.27b).

Third mechanism of crack growth is mixed. It implies a possibility of transition from one fracture mechanism to the other at definite conditions (see Fig. 5.26c). Here, at either stage, three possible variants of crack growth are considered (either two intergranular and one transgranular cracks or two transgranular and one intergranular cracks) depending on character of the crack growth at the previous stage. For either from the three competing cracks, the value of $K_{I\mathrm{c}}$ is calculated by using (5.88), (5.89) and (5.95). Then, a minimum value of $K_{I\mathrm{c}}$ states a path of the crack growth at considered stage. Again, a final crack tip becomes an initial one at the next stage of the crack growth. The process is continued up to attainment by the crack of opposite side of the sample or open porosity. The consideration of mixed mechanism is caused by the replacement of the crack growth character at different sections that is supported by existing test results for different ceramics [849]. A fabrication of optimum HTSC microstructures implies an inclusion in the consideration of proper crack shielding (amplification) mechanisms and the material fracture resistance. These mechanisms for YBCO and BSCCO ceramics and composites will be considered in Chaps. 8 and 9.

6

Modeling of BSCCO Systems and Composites

6.1 Transformation of Bi-2212 to Bi-2223 Phase

Understanding the mechanism and kinetics of the Bi-2212-to-Bi-2223 phase transformation is very important to fabricate Bi-2223 superconductors with high structure-sensitive properties. The most widely used method of processing Bi-2223 materials is to mix Bi-2212, $CaPbO_3$ and CuO powders and to anneal the mixture in a sealed tube at about 830°C during 10–100 h [119]. All models proposed for phase transformation of Bi-2212 into Bi-2223 suggest a diffusion-controlled, two-dimensional transformation due to different mechanisms of chemical reactions [33, 352, 358, 635, 650, 726, 1200]. In order to analyze the kinetics of Bi-2223 formation, the Avrami equation is usually used [358, 650, 1200]:

$$\ln\left(\frac{1}{1-C}\right) = K_0\ \exp(-U/RT)t^{\alpha}\ , \tag{6.1}$$

where C is the fraction of Bi-2223 phase transformed at time t, T is the temperature, U is the activation energy, $K_0 = 1.71 \times 10^{-22}$ is the rate constant, R is the universal gas constant and α is the Avrami exponent. The Avrami exponent α obtained from test data can provide some insight into the reaction mechanism. Almost all of the above-mentioned authors conclude that their data support a diffusion-controlled, two-dimensional transformation of Bi-2212 into Bi-2223 phase. The values of the Avrami exponent, which they obtained, varied greatly, ranging from 0.5 [1200] to 1–1.5 [358, 650]. This, seemed to be more consistent with one-dimensional diffusion-controlled transformation mechanism with a varying nucleation rate [453]. At the same time, transmission electron microscope (TEM) investigations, using electron diffraction and lattice imaging, show that during the annealing process, the Bi-2212/ Bi-2223 system consists of fast-growing intercalating Ca/CuO_2 bilayers instead of compact Bi-2223 domains [66]. Unlike the conventional nucleation-and-growth mechanism, where reactant diffuses from grain boundaries into the interior of the bulk material, leading to a compact propagating "front" of the product, the transformation from Bi-2212 to Bi-2223 phase appears to be accomplished via the layer-by-layer intercalation of the extra Ca/CuO_2 planes into the Bi-2212 matrix. In this case, the Ca/CuO_2 bi-layer insertion into the matrix during transformation leads to formation of an edge dislocation [118]. Hence, it becomes obvious that the fast intercalation of individual Ca/CuO_2 plane into the Bi-2212 matrix is more probable as the path for Bi-2212 to Bi-2223 transformation than nucleation and growth of the Bi-2223 compact region.

The growth of the intercalant Ca/CuO_2 planes occurs much faster than the bulk cation diffusion. Then, it may be assumed that a distinct diffusion mechanism acts in the Bi-2212/Bi-2223 system, in which cation diffusion takes place through the cylindrical cavity created by the edge dislocation, which accompanies the insertion of a Ca/CuO_2 plane. These cavities are located at the interfaces between Bi-2212 and Bi-2223 phases, where the reaction occurs. As the transformation progresses, these pores move with the Bi-2212/Bi-2223 interface; thus, the reaction progresses uninhibitedly. Based on the above assumption and the layer-rigidity model [117, 1056] modified to the Bi-2212/Bi-2223 system, we consider the cation diffusion mechanism and calculate the size of pointed void [118, 119].

6.1.1 Edge Dislocations as Channels for Fast Ion Diffusion

The transformation of Bi-2212 phase to Bi-2223 one can be regarded as chemical reaction between the Bi-2212 precursor and the secondary phases such as $CaPbO_3$ and CuO, to provide the required surplus of Ca and Cu. As shown in Fig. 6.1, the only structural difference between Bi-2212 and Bi-2223 is a pair of extra Ca/CuO_2 planes (in the case of Bi-2223) at the corresponding lattice expansion. Therefore, it is convenient to divide the Bi-2212-layered structure into two parts, namely (i) the block layers (called "host layers" below), which consist of BiO, SrO and the Ca/CuO_2 plane in Bi-2212 and (ii) the "gallery layers" into which the extra Ca/CuO_2 planes are inserted in the case of Bi-2223. Thus, we can consider the Bi-2223 as a stacking fault of Bi-2212. Therefore, the interfaces in a transforming system between Bi-2212 and Bi-2223 layers can be considered as an edge dislocation. The approximately cylindrical pore created at the additional half plane of this dislocation is a line of vacancies, which can be an easy path for the additional Ca, Cu and oxygen ions to diffuse from the surface (or grain boundary) into the bulk material, provided the size of the pore is large enough. The size of the pores depends on the rigidity of the block layer as well as the compressibility of the Ca/CuO_2 plane along the c-axis. Obviously, the more rigid the layer the larger the pore. In the limit of infinite layer rigidity, the size of the pore will be infinite.

At the annealing temperature about 830°C, both $CaPbO_3$ and CuO are liquid. The grinding and mixing prior to the annealing make sure that this liquid phase is evenly coated around each Bi-2212 grain. As the Ca/CuO_2 plane nucleates near the surface of the Bi-2212 grain (or grain boundary), the

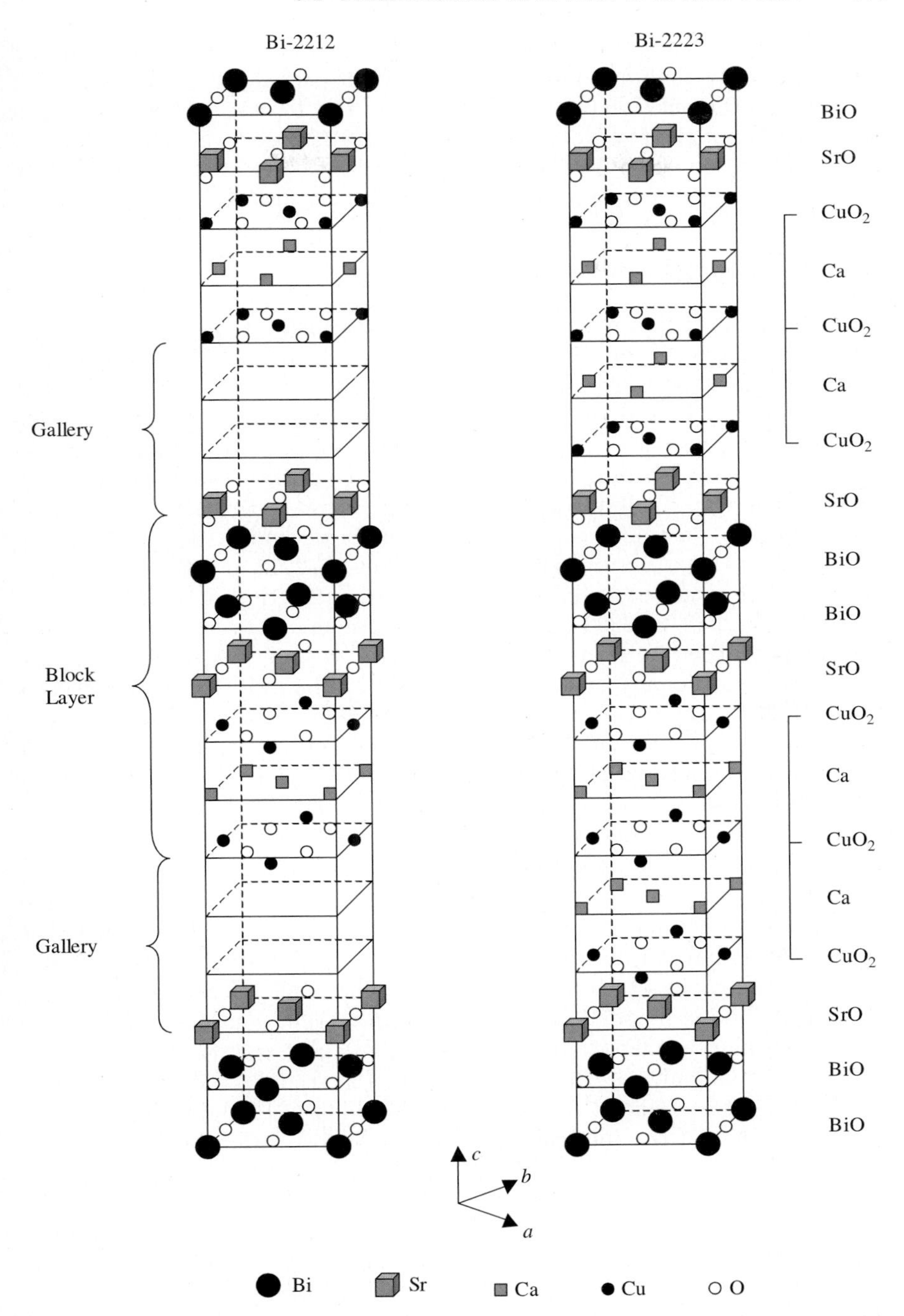

Fig. 6.1. The unit cells of Bi-2212 and Bi-2223. The layered structure of Bi-2212 can be divided into block layers and galleries for intercalation of the additional Ca/CuO_2 plane to form Bi-2223. The "gallery height" in the Bi-2212 structure is shown here expanded from its normal value for clarity of comparison of the two structures [119]

pore created by the partially inserted Ca/CuO_2 plane opens up a channel for the reactant ions to diffuse into the bulk. Thus, it permits the edge dislocation to climb and the reaction to proceed rapidly at this location. Therefore, it can be assumed that the transformation from Bi-2212 phase to Bi-2223 is limited by he nucleation rate of the Ca/CuO_2 plane near the Bi-2212 grain boundary and the diffusion of reactant ions along the moving dislocation lines.

6.1.2 The Layer-Rigidity Model

The volume expansion at the cores of the edge dislocations (caused by additional atomic half-planes) formed during the Bi-2212 to Bi-2223 transformation is connected with the large anisotropy in physical properties of these layered systems. It is reasonable to assume that the major expansion takes place along the direction perpendicular to the layers, denoted as the c-axis. As it has been shown by diffraction tests, the lattice parameters along the a- or b-axis change very small during Bi-2212/Bi-2223 transformation. At the same time, the lattice parameter along the c-axis increases from 30.9 to 37.8 Å [119].

Basing on the structural difference between Bi-2212 and Bi-2223 phases, it is regarded as an insertion of Ca/CuO_2 in the interior of Bi-2212 matrix. Then, the Bi-2212/Bi-2223 system can be considered as a type of intercalation compound in the form of $A_{1-x}B_xL$ with $0 \leq x \leq 1$, where B is the intercalant (additional Ca/CuO_2 layer in Bi-2223 phase), A is a vacant layer in the Bi-2212 phase (see Fig. 6.1), which will be considered as an intercalant of a smaller size, and L denotes the host layer which represents the rest of the structure. The single phases Bi-2212 (AL) and Bi-2223 (BL) present the limiting cases for $x = 0$ and $x = 1$, respectively. The dislocation at the Bi-2212/ Bi-2223 interface can be modeled as an intercalation compound with B, occupying the semi-infinite plane, as shown in Fig. 6.2.

Two different models have been proposed to study the c-axis expansion of intercalation compounds [117, 1056]. In both models, it is assumed that the host layers have finite transverse as well as bending rigidity. At the same time, the intercalants have finite compressibility and different sizes. In the first model (bi-layer model) [1056], it is assumed that compressibility of the host layer is much smaller than that of the intercalant, that is, the correlation between various galleries can be ignored. In the second model (multiplayer model) [117], it is assumed that the compressibility of the host layer is much larger than that of the intercalant, that is, the correlation between various galleries can be mapped into an Ising-type model. It may be suggested that the first type model is more suitable for Bi-2212/Bi-2223 compound. Let the compressibility of the host layer be about 10% of that of the intercalant, because the compressibility of the host layer is inversely proportional to its thickness, but this thickness is much larger than the mean gallery height in the Bi-2212/Bi-2223 compound. Here, a spring model that describes both the layer rigidity and the size and stiffness of the intercalant is considered (see Fig. 6.2).

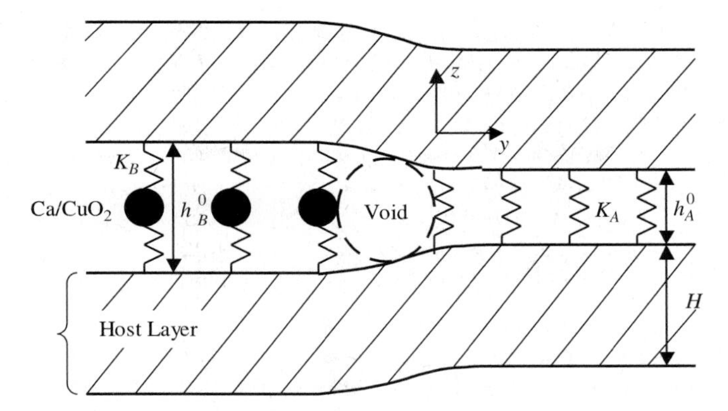

Fig. 6.2. Side view of the intercalants in the transforming Bi-2212/Bi-2223 system. The additional Ca/CuO_2 plane is shown schematically as large intercalant "atoms." The rest of the structural features are shown schematically as the "host layer." The separation between the host layers is given by the local gallery height, h_i [119]

Consider a layered system with composition $A_{1-x}B_xL$, where L represents the host layer of thickness H, which consists of BiO, SrO and Ca/CuO_2 planes in the Bi-2212 phase; A and B are two various types of intercalants, which occupy a set of well-defined lattice sites. The total energy of the system can be presented by a sum of two major contributions: one associated with the interaction between the intercalants and the host atoms and the other between host atoms themselves. It is assumed that the intercalants are "frozen" into the align structure (i.e., the nucleation time of the Bi-2223 phase is much shorter than the time for diffusion of the Ca, Cu and O ions from the grain boundary to the bulk). Thus, the direct interaction energy between the intercalants does not play any role in layer distortion. Because the compressibility of the host layer is much smaller than the intercalant, the interlayer correlation is mostly related to the various thickness of the host layer instead of the correlation between intercalants of various galleries. As the variation of the host layer thickness, ΔH, is very small compared to H, it can be neglected by the interlayer correlation and be considered a single gallery, bounded by two host layers with intercalants, occupying the lattice sites inside the gallery.

Following [119], the total energy of the host layer-intercalant system, presented in Fig. 6.2, can be written as

$$E = \frac{1}{2}\sum_i K_i(h_i - h_i^0)^2 + \frac{1}{2}K_{\mathrm{T}}\sum_{i,\delta}(h_i - h_{i+\delta})^2 + \frac{1}{2}K_{\mathrm{F}}\sum_i\left[\sum_\delta (h_i - h_{i+\delta})\right]^2, \tag{6.2}$$

where h_i is the local gallery height at site i, where an intercalant (either A or B) sits; h_i^0 is the gallery height for the pure system AL or BL($h_A^0 = 2.21$ Å for Bi-2212 (AL) and $h_B^0 = 6.18$ Å for Bi-2223 (BL) from the diffraction measurements); K_i is the spring constant, representing the compressibility of the local intercalant i. The terms, involving the spring constants K_{T} and K_{F}, describe, respectively, transverse and bending rigidity of the host layers [1056]. They are related to the elastic constants of the Bi-2212 compound by the following relations:

$$K_{\mathrm{T}} = \frac{H}{a_0} c_{44} \; ; \tag{6.3}$$

$$K_{\mathrm{F}} = \frac{H^3}{12a_0^3} \left[\frac{(c_{11} + c_{33})c_{33} - 2c_{13}}{c_{33}} \right] , \tag{6.4}$$

where $a_0 = 3.8\,\text{Å}$ is the lattice constant in the ab-plane and $H = 14.35\,\text{Å}$ is the thickness of the host layer. The required elastic constants for Bi-2212 have the next values [119]: $c_{11} = 24.08 \times 10^{11}\,\text{dyn/cm}^2$; $c_{13} = 6.71 \times 10^{11}\,\text{dyn/cm}^2$; $c_{33} = 14.59 \times 10^{11}\,\text{dyn/cm}^2$; $c_{11} = 6.18 \times 10^{11}\,\text{dyn/cm}^2$.

The compressibility $K_A = K_i(A)$ of the intercalant A (the vacancy) can be obtained from the elastic constants as

$$K_A = \frac{a_0}{h_A^0} c_{33} \; . \tag{6.5}$$

The compressibility $K_B = K_i(B)$ of the intercalant B (the Ca/CuO_2 plane) cannot be calculated due to the lack of data for the elastic properties of Bi-2223. However, it can be assumed that $K_A < K_B$. Then, calculate the gallery height for the case of $K_A = K_B$, estimating thus the minimum size of the pore. Minimizing E in (6.2) with respect to h_i, we find

$$\mathbf{Mh} = \mathbf{\Phi} \; , \tag{6.6}$$

where $\mathbf{M}$ is a tridiagonal matrix with elements

$$M_{ii} = K_i + 2K_{\mathrm{T}} + 6K_{\mathrm{F}}; \qquad M_{i,i\pm1} = -K_{\mathrm{T}} - 4K_{\mathrm{F}}; \qquad M_{i,i\pm2} = K_{\mathrm{F}} \; , \tag{6.7}$$

but $\mathbf{h}$ and Φ are two column vectors:

$$\mathbf{h} = \begin{pmatrix} h_1 \\ h_2 \\ \cdots \\ h_N \end{pmatrix} ; \qquad \mathbf{\Phi} = \begin{pmatrix} K_1 h_1^0 \\ K_2 h_2^0 \\ \cdots \\ K_N h_N^0 \end{pmatrix} . \tag{6.8}$$

By diagonalizing matrix $\mathbf{M}$, h_i can be obtained. In the case of $K_A = K_B = K$, the dispersion relation for the algebraic system is given as [119]

$$\lambda_q = K + 2K_{\mathrm{T}}[1 - \cos(qa_0)] + 4K_{\mathrm{F}}[1 - \cos(qa_0)]^2 \; , \tag{6.9}$$

and the expanded form of $(\mathbf{M}^{-1})_{nm}$ can be expressed as [119]

$$(\mathbf{M}^{-1})_{nm} = \frac{1}{N} \sum_q \frac{\exp[\mathrm{i}qa_0(n-m)]}{\lambda_q} \; , \tag{6.10}$$

with $q = 2\pi r/Na_0$; $r = 1, 2 \dots N$ and λ_q being the eigenvalue of the matrix $\mathbf{M}$. Then we have

$$h_n = \frac{1}{N}\sum_m (\mathbf{M}^{-1})_{nm}\Phi_m = \frac{1}{N}\sum_q \frac{K}{\lambda_q}\sum_m \mathrm{e}^{\mathrm{i}qa_0(n-m)}h_m^0 \,. \tag{6.11}$$

Considering a system with half of the yz-plane occupied by intercalant B (i.e., Bi-2223 phase), we assume the dislocation disposes at $y = z = 0$, and the equilibrium height h_n^0 is defined as

$$h_n^0 = \begin{cases} h_B^0, \text{ for } n = -\frac{N}{2}+1, -\frac{N}{2}+2, \dots, -1; \\ h_A^0, \text{ for } n = 1, \dots, \frac{N}{2}-1, \frac{N}{2}. \end{cases} \tag{6.12}$$

Taking the limit of $N \to \infty$ and changing the summation over q by integration, we find that

$$h(y) = \frac{h_A^0 + h_B^0}{2} + \frac{h_A^0 - h_B^0}{4\pi}\int_0^\pi \frac{K\sin[\theta(y/a_0 - 1/2)]}{\lambda_\theta \sin(\theta/2)}\mathrm{d}\theta \,, \tag{6.13}$$

where $\theta = qa_0$ and $\lambda_\theta = K + 2K_\mathrm{T}[1 - \cos(\theta)] + 4K_\mathrm{F}[1 - \cos(\theta)]^2$. The solid line in Fig. 6.3 shows the profile of h_n for $K_A = K_B$ and the values of parameters defined in (6.3)–(6.5). The obtained results can be compared with the model of the stacking fault, presenting an additional CuO plane in the $YBa_2Cu_3O_7$ matrix. For this, the theoretical profile of h_n is calculated, using the elastic constants, obtained from the atomistic simulation of $YBa_2Cu_3O_7$ [44], and is shown by dashed line in Fig. 6.3.

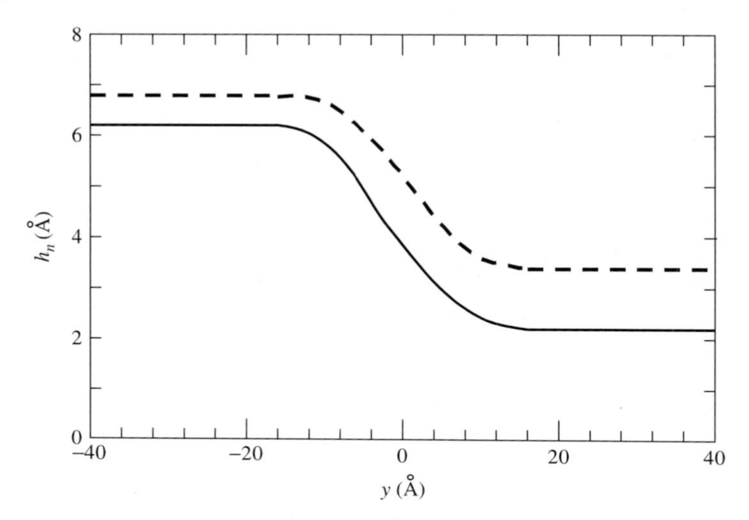

Fig. 6.3. The spatial profile of the gallery h_n. *Solid line* presents Bi-2212/Bi-2223 system, but *dashed line* corresponds to $YBa_2Cu_3O_7/YBa_2Cu_4O_8$ system (at $K_A = K_B = K$)

Of the three types of reactant ions (Ca^{2+}, Cu^{2+} and O^{2-}), the largest is O^{2-}, which has a diameter of about 2.9 Å. As it is shown in Fig. 6.3, the lateral size of the pore is about twice as large as the oxygen ion diameter, if we consider the pore to be the region, where $h_n \geq 3$ Å for $y \geq 0$. Hence, the dislocation line is indeed likely to be an easy path for reactant ions to diffuse from the liquid phase at the Bi-2212 grain boundary into the bulk, causing the edge dislocation (the additional Ca/CuO_2 plane) to rapidly climb into Bi-2212.

6.1.3 Dynamics of Bi-2223 Phase Growth

The Avrami exponent in the Bi-2212-to-Bi-2223 transformation using (6.1) and presented layer-rigidity model is estimated. Because the reactant always contacts with the Bi-2212/Bi-2223 interface, the reaction rate of Bi-2212 to Bi-2223 phase perpendicular to the dislocation line is independent of the volume fraction of Bi-2223 phase, and the growth in this direction is very fast. Therefore, it is necessary to regard only the diffusion along the dislocation line and assume that the reactants are consumed immediately upon reaching the Bi-2212/Bi-2223 interface, because the diffusion is probably a much slower process than any local rearrangement of atoms, which might be required. Then, the growth rate of the product layer along the dislocation line is given as [119]

$$dl/dt = D/l \, , \tag{6.14}$$

where l is the distance from the grain boundary measured along the dislocation line and D is the effective diffusion factor of the reactant ions. Because the diffusion factor and the pore size (see the previous section) are independent of the volume fraction of Bi-2223, dividing variables and integrating (6.14), we obtain

$$l = \sqrt{2Dt} \, . \tag{6.15}$$

The volume of the reactant ions, consumed by the formation of the product phase, is given as

$$R(t) = S\sqrt{2D}t^{1/2} \, , \tag{6.16}$$

where S is the size of the pore. For a Bi-2212 grain with volume V, the number of Ca/CuO_2 planes nucleated at the grain boundary in the time range between t' and $(t' + dt')$ is $I(t')Vdt'$, where $I(t')$ is the nucleation rate per unit area of the grain boundary. Hence, the total volume of reactant ions, consumed at time t (assuming nucleation can occur everywhere at the grain boundary, including the transformed region), is given as

$$V_e^C = A\int_0^t I(t')R(t-t')dt' = S\sqrt{2D}A\int_0^t I(t')(t-t')^{1/2}dt' \, , \tag{6.17}$$

where A is the total area of the grain boundary. Note that V_e^C differ from the actual volume of reactant consumed because it is assumed that nucleation can take place in the transformed region. All regions are considered, including those continuing growth irrespective of other regions. In order to correct this problem, we sign the volume of the transformed region through V^C and state the relation between V^C and V_e^C. Consider a small region, of which a fraction $[1-(V^C/V)]$ remains non-transformed (where V is the total volume of the grain). During the time range, dt, taking into account the total transformed volume, dV_e^C, a fraction $[1-(V^C/V)]$ on the average will be in previously unreacted material and thus contributes to dV^C, while the reminder of the dV_e^C will be in the already transformed region. Then, we obtain

$$\mathrm{d}V^C = \left(1 - \frac{V^C}{V}\right)\mathrm{d}V_e^C \,. \tag{6.18}$$

Dividing variables and integrating, we have

$$V_e^C = -V \ln\left(1 - \frac{V^C}{V}\right) . \tag{6.19}$$

Let $[1-(V^C/V)] = 1 - C$, where C is the volume fraction of Bi-2223. Substituting (6.19) into (6.17), we obtain

$$\ln\left(\frac{1}{1-C}\right) = \frac{S\sqrt{2D}A}{V}\int_0^t I(t')(t-t')^{1/2}\mathrm{d}t' \,. \tag{6.20}$$

The volume fraction of Bi-2223 depends directly on the nucleation rate $I(t)$. It is often assumed that the time dependence of the number of nucleation sites is a classical first-order rate process, that is, [119]

$$\mathrm{d}N(t)/\mathrm{d}t = -fN(t) \,, \tag{6.21}$$

where f is the frequency of an empty gallery in Bi-2212 phase turns into a nucleation site for Bi-2223 phase and $N(t)$ is the number of such nucleation sites at time t. Integrating the above equation yields

$$I(t) = fN(t) = fN_0 \exp(-ft) \,, \tag{6.22}$$

where N_0 is the number of empty galleries in the Bi-2212 grain at $t = 0$. Substituting (6.22) into (6.20) and integrating yields

$$\ln\left(\frac{1}{1-C}\right) = \frac{S\sqrt{2D}N_0A}{V}\left[t^{1/2} + \frac{\mathrm{i}}{2}\sqrt{\frac{\pi}{f}}\exp(-ft)\mathrm{erf}(\mathrm{i}\sqrt{ft})\right] , \tag{6.23}$$

where erf(z) is the error function with complex variable z, which can be expanded as [566]

$$\mathrm{erf}(z) = \frac{2}{\sqrt{\pi}}\int_0^z \mathrm{e}^{-\zeta^2}\mathrm{d}\zeta = \frac{2}{\sqrt{\pi}}\left(z - \frac{z^3}{3} + \frac{1}{2!}\frac{z^5}{5} - \frac{1}{3!}\frac{z^7}{7} + \ldots\right) . \tag{6.24}$$

Now, we can estimate the Avrami exponent α for various values of f.

In the limiting case, when ft is very small, that is, the time to produce a Bi-2223 nuclei is much longer than the time to grow the Ca/CuO_2 plane through the sample, we can expand (5.24) in terms of ft. Taking into account (6.24) and the expansion [566]

$$e^z = 1 + z + \frac{z^2}{2!} + \frac{z^3}{3!} + \cdots , \tag{6.25}$$

we obtain

$$\ln\left(\frac{1}{1-C}\right) = \frac{S\sqrt{2D}N_0A}{V}\left(\frac{2}{3}ft^{3/2} - \frac{4}{15}f^2t^{5/2} + \cdots\right) . \tag{6.26}$$

Comparing (6.26) with (6.1), we have $\alpha \approx 1.5$, which agrees well with the test results for the preheated sample [358, 650]. Note that the assumption of this limiting case is valid only if reactants (Ca, Cu and O) are evenly coated around the Bi-2212 grains and the Bi-2212 grains are small enough. Therefore, the model is more consistent with tests with small precursor powder, which are heat-treated before annealing to ensure the even distribution of the reactants.

On the other hand, if ft very large, that is, the time to produce a Bi-2223 nuclei is much shorter than the time to grow the Ca/CuO_2 planes through the sample, we can consider that all the empty galleries in Bi-2212 have turned into nucleation sites for Bi-2223 before the Ca/CuO_2 planes begin to grow. In this case, we find from (6.23)

$$\ln\left(\frac{1}{1-C}\right) = \frac{S\sqrt{2D}N_0A}{V}t^{1/2} . \tag{6.27}$$

Hence, $\alpha \approx 0.5$. This will be the case if there is local shortage of reactant ions, which will lead to the slow growth of Ca/CuO_2 plane. Tests in [1200] apparently fall into this situation. The pointed study has stated an increase of the Avrami exponent α together with the temperature from 0.5 at 840°C to 0.79 at 870°C. The small value of α in [1200] is due to either the large Bi-2212 grain size as indicated by the high porosity (~40%) or the uneven distribution of the reactant ions. When the temperature rises, the growth of the Ca/CuO_2 plane accelerates due to the increase of ion diffusion constant, D. In this case, the reactant ions distribute more evenly around the Bi-2212 grain, thanks to the decreased viscosity of the ionic liquid of Ca, Cu and O. Thus, the Avrami exponent depends directly on the nucleation rate of the Bi-2223 phase and growth of the additional Ca/CuO_2 plane in the Bi-2212 matrix changing between 0.5 and 1.5.

6.1.4 Formation Energy of Bi-2223 Phase

As it is shown in experiments, less than 30 min is enough to intercalate one Ca/CuO_2 plane into Bi-2212 matrix across a $2-\mu m$ sample [118]. Moreover, due to the presented model, the ion diffusion should be faster in the pore than that in the bulk. Then, we can estimate the formation energy of Bi-2223 from Bi-2212 phase.

The intercalation of the Ca/CuO_2 bi-layer into the Bi-2212 matrix can be presented as a single jog on the otherwise straight dislocation (see Fig. 6.4a). The climb motion of the dislocation can be accomplished by jog nucleation and propagation along the dislocation line. When the reactant ions (Ca^{2+}, Cu^{2+} and O^{2-}) are deposited at the jog, one propagates and the dislocation climbs upward by a distance, Δx, as shown in Fig. 6.4b. Both these processes require

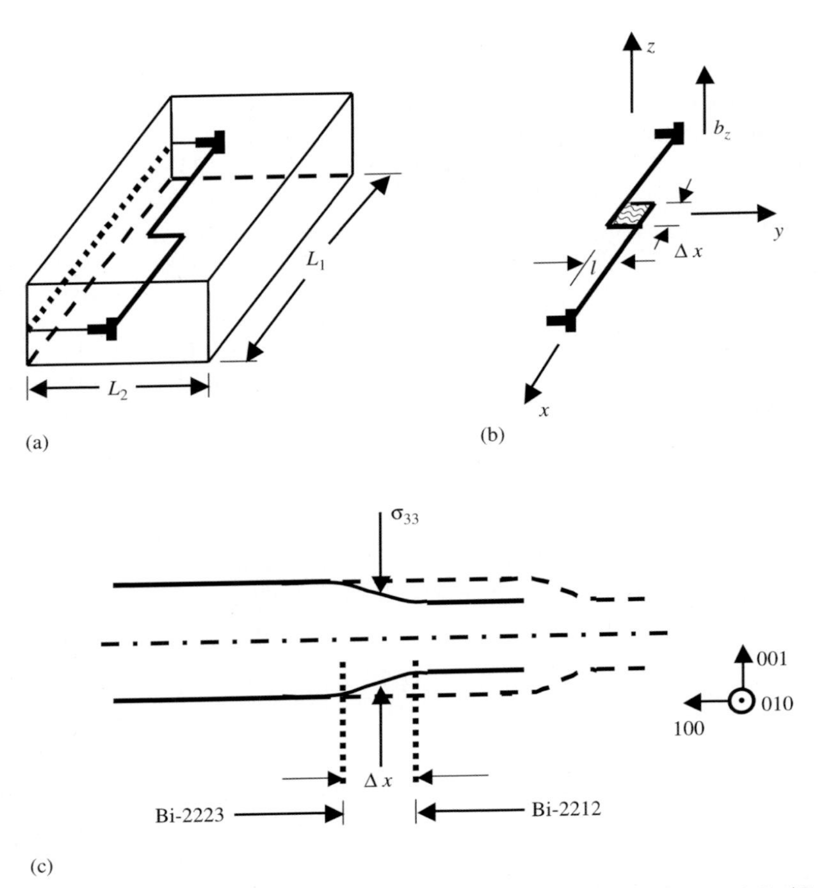

Fig. 6.4. Schematic drawing of the dislocation model for intercalation of Ca/CuO_2 bi-layer into Bi-2212 matrix: (**a**) The insertion of a Ca/CuO_2 bi-layer, described as climb motion of a dislocation in a grain with rectangular platelet shape; (**b**) jog configuration of the dislocation, the shading area shows the transition region; (**c**) core configuration of the dislocation view along the [010] direction, the channel of diffusion for the reactants is shown [118]

net mass transport, whose rate is controlled by the diffusion of the reactant ions. These ions, moving to a preexisting jog, simply translate the jog. This translation does not change core energy and core structure of the dislocation, as shown in Fig. 6.4c.

Considering the reactant ions as particles, suffering random force of their surroundings, the equation of the jog motion can be written as

$$\eta\nu = \mathbf{F}_{\mathrm{c}} + \mathbf{F}_{\mathrm{el}} \;, \tag{6.28}$$

where η is the viscosity of the surrounding matrix, ν is the jog rate, $\mathbf{F}_{\mathrm{c}}$ is the force caused by the reduction of chemical energy at Bi-2212/Bi-2223 transformation and $\mathbf{F}_{\mathrm{el}}$ is the elastic force suffered by the jog.

Assuming the rate of jog motion to be constant, the time t that is necessary to insert a single Ca/CuO_2 bi-layer in Bi-2212 matrix of a $L_1 \times L_2$ rectangular sample is found as

$$t = \frac{L_1 L_2}{\Delta x v} \;, \tag{6.29}$$

and we have from Fig. 6.4c:

$$|\mathbf{F}_{\mathrm{el}}| = \sigma_{33}\varepsilon_{33} b_z l \;, \tag{6.30}$$

where $\sigma_{33} = c_{33}\varepsilon_{33}$ is the stress, ε_{33} is the strain, $c_{33} = 14.59 \times 10^{11}\,\mathrm{dyn/cm^2}$ is the elastic constant of the Bi-2212 matrix and $b_z = 4.4\,\mathring{A}$ is Burgers vector of the dislocation [118]. If we regard the intercalation of a single Ca/CuO_2 bi-layer for each Bi-2212 unit-cell, that is, a local Bi-4435 structure, which is often observed in the early stage of the transformation, then the strain $\varepsilon_{33} = 0.142$ [66]. Moreover, we take into account that $l \approx \Delta x \approx b_z$ [119].

The bulk diffusion factor of O^{2-} in the *ab*-plane of Bi-2212 phase at 850°C is approximately $1.1 \times 10^{-9}\,\mathrm{cm^2/s}$ [118]. The diffusion constant for the reactant ions in the pore depicted in Fig. 6.4c is at least one order of magnitude larger than that. Using the relation between the diffusion constant, D, and the viscosity constant, η,

$$\eta = \frac{k_{\mathrm{B}} T}{D} \;, \tag{6.31}$$

where k_{B} is Boltzmann constant and $T = 850°\mathrm{C}$ is the temperature. Finally, we estimate the formation energy per Ca/CuO_2 bi-layer E_{c} from (6.28) and (6.30) as

$$E_{\mathrm{c}} = -\mathbf{F}_{\mathrm{c}} \cdot \Delta\mathbf{x} \leq 3.5\,\mathrm{eV} \;. \tag{6.32}$$

6.1.5 Effect of Deformation on Bi-2212/Bi-2223 Transformation

Obviously, it may be assumed that deformation processes render direct influence on Bi-2212/Bi-2223 phase transformation. In the case of fabrication of the BSCCO tapes by using oxide-powder-in-tube method, the mechanical deformation effects have been investigated at rolling the wires into tapes with different thickness reduction [625]. Each rolling step produced an approximate 20% thickness reduction. The deformation ratio R is defined as the thickness reduction of the tape:

$$R = (t_0 - t_f)/t_0 , \tag{6.33}$$

where t_0 and t_f are the original and final thicknesses of the superconductor core, respectively. The obtained test dependencies of Bi-2223 phase concentration on deformation and duration of final annealing at $T = 833°C$ are shown in Fig. 6.5. These results demonstrate that the mechanical deformation strongly affects the phase transformation of Bi-2212 to Bi-2223 in the early stages of annealing, implying that different mechanisms may dominate the transformation kinetics above and below definite value of the deformation ratio (here 60%). This unusual behavior, demonstrating minimum value of transformation kinetics at $R = 60\%$, states an existence of two major effects for BSCCO tapes during the mechanical preparation, namely (i) a part of deformation energy is absorbed by the oxide core as fracture energy, splitting the Bi-2212 crystal and producing new interfaces; (ii) a part of the energy is stored in the crystals as the result of lattice distortion of the Bi-2212 crystals. The first effect takes place when the mechanical deformation ratio, R, is relatively low, while the second effect occurs when higher mechanical deformation is applied to the BSCCO tapes [625]. Both effects have different contributions to the Bi-2212/Bi-2223 phase transformation process. Obviously, the area of the fractured surfaces increases as the mechanical deformation ratio enhances, increasing the surface energy. This surface energy causes re-crystallization of the fractured Bi-2212 crystals, reducing the energy of the system during final annealing. At the same time, the lattice distortion increases the internal energy and the phase balance, facilitating phase transformation, forming Bi-2223 instead of the re-crystallized Bi-2212 crystals. Hence, there are two possible types of annealing behavior for the Bi-2212 phase, either the

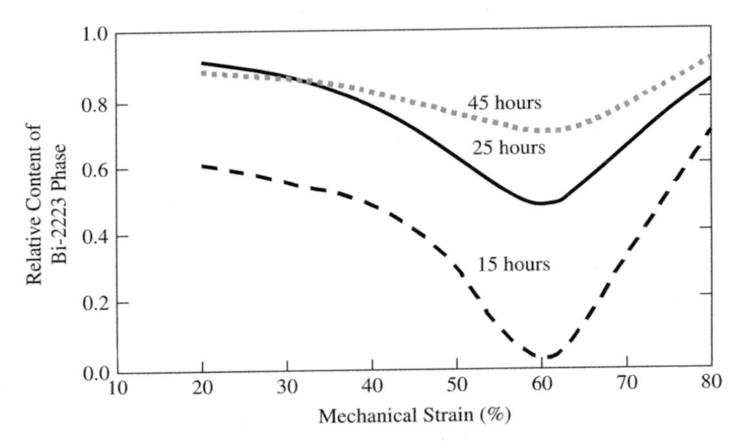

Fig. 6.5. Bi-2223 phase content versus the mechanical deformation ratios [625]

Bi-2212 re-crystallization to minimize the surface area or reaction leading to the phase balance and forming the Bi-2223 phase. Hence, the annealing behavior of the deformed Bi-2212 crystals is determined either by the surface energy or by the chemical potential and internal energy of the Bi-2212 crystals. When the surface energy is higher than the chemical potential and the internal energy, re-crystallization of the Bi-2212 crystals occurs. At the same time, when the chemical potential and internal energy dominate the system, the chemical reaction takes place. Thus, it could be assumed in this case that re-crystallization is probably a favorable process at $R < 60\%$ and the trend increases with the deformation ratio. The surface area rapidly increased with the increasing of R, until the surface area reached a "saturation" value. This slowed down the Bi-2223 formation with increasing mechanical deformation. However, at $R > 60\%$, the lattice distortion of the Bi-2212 crystals increased rapidly, causing an increase of the internal energy with increasing mechanical deformation

Figure 6.6 is a schematic illustration, showing that the driving force for the Bi-2223 phase formation is enhanced by the internal energy. Annealing the sample at temperatures above the equilibrium temperature of the Bi-2212 and Bi-2223 phases, $T_e (T > T_e)$, allows the Bi-2223 phase to form because of the lower free energy. At the same time, the formation of the Bi-2212 phase occurs at lower annealing temperatures ($T < T_e$). As the internal energy of the Bi-2212 phase is increased by the mechanical deformation, the free energy of Bi-2212 phase becomes $G'_{\mathrm{Bi-2212}} = G_{\mathrm{Bi-2212}} + E_{\mathrm{int}}$. One is presented by the curve parallel to the original curve, defining free energy of the Bi-2212 phase ($G_{\mathrm{Bi-2212}}$) without the mechanical deformation, but is shifted

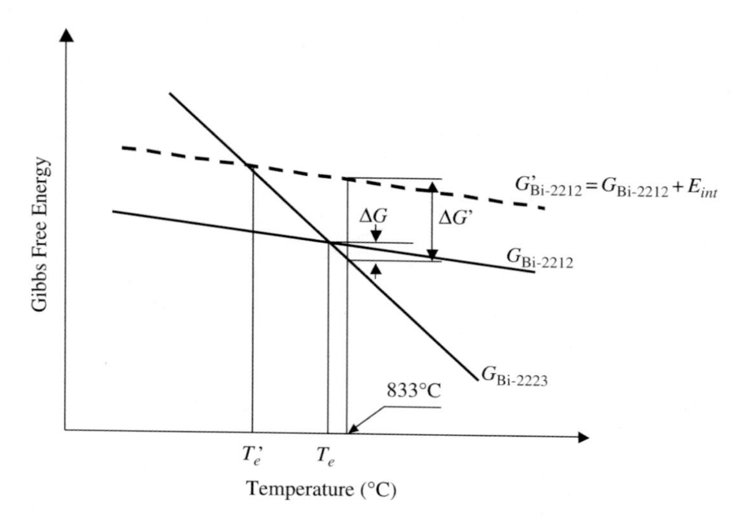

Fig. 6.6. A schematic illustration shows the mechanism, in that the driving force of the Bi-2223 phase formation is enhanced by the internal energy [625]

up to $G'_{\text{Bi}-2212}$ (see Fig. 6.6). Obviously, it results in a decrease in the equilibrium temperature of the Bi-2212 and Bi-2223 phases ($T'_e < T_e$). As a result of the reduced equilibrium temperature, the driving force for the Bi-2223 phase formation from the overheating ($833°\text{C} - T'_e$) increases from ΔG to $\Delta G'$, as shown in Fig. 6.6. In this case, the chemical potential and the internal energy replace the surface energy as the dominant energies of the system. Therefore, after annealing, the content of the Bi-2223 phase increased together with the deformation ratio, when $R > 60\%$ [625]. Moreover, apart from the pointed effects of mechanical deformation, there are other parameters defining the Bi-2212/Bi-2223 phase transformation, namely the effective diffusivity, concentration gradient and grain size. For example, the fine grain size leads to an increase in the grain boundary (interface) area. In this case, the effect of the surface diffusion would be stronger than that of the lattice diffusion during the phase transformation and grain growth processes. This implies that the formation of Bi-2223 phase should increase together with the parameter of R. In addition, the mechanical deformation might also influence the number of nuclei and the nucleation rate. Generally, the nucleation rate is independent of the nucleation sites. The dependence between the number (or volume) of nuclei, N, the nucleation rate, v, the area of the nucleation sites, A, and the annealing time, t, can be expressed as [625]

$$N = \nu A t \,. \tag{6.34}$$

Obviously, for a certain nucleation rate, higher mechanical deformation exposes more surfaces, providing a large amount of Bi-2223 nuclei that facilitate the phase transformation from Bi-2212 to Bi-2223. However, the various mechanical deformation processes may result in different grain boundary areas, defining value of R, corresponding to the minimum formation of Bi-2223 phase. Thus, mechanical deformation causes both the thermodynamics and kinetics of the BSCCO phase formation.

6.2 Modeling of Preparation Processes for BSCCO/Ag Tapes

As it has been noted in the previous chapters, HTSC tapes are subjected to multi-staged and complex thermal, mechanical and magnetic treatments during preparation. Below, we consider several model and computational approaches directed to the process optimization of BSCCO/Ag tape fabrication and to attainment of the improved structure-sensitive and superconducting properties.

6.2.1 Sample Texturing by External Magnetic Field

When superconductor is placed in a magnetic field, the axis of maximum susceptibility for each grain aligns with the magnetic field direction, that is, the grains should align with the c-axis parallel to the external magnetic field. For quantitative estimation of the material texturing degree under magnetic field, consider the rotation of superconductor grains into liquid in the early stages of their growth (i.e., when the particles can rotate without interacting) [275]. We assume in this model that the magnetic field does not influence on the processes of grain nucleation and growth into liquid.

Let an anisotropic grain with volume V be placed in a magnetic field $\boldsymbol{H}$. Then, the change in magnetic energy of the grain with a change in magnetic field can be written as

$$\mathrm{d}E_{\mathrm{m}} = -\mathbf{M}V\,\mathrm{d}\mathbf{H} = -(M_c \cos\theta + M_{ab}\sin\theta)V\,\mathrm{d}H\,, \tag{6.35}$$

where $\mathbf{M}$ is the magnetic moment per unit volume, which can be resolved in the two directions c and ab and θ is the angle between the magnetic field and the c-axis of the grain. For HTSC in their normal state, the magnetic moments M_c and M_{ab} are paramagnetic moments. Then, (6.35) can be reduced to the form:

$$\mathrm{d}E_{\mathrm{m}} = -(\chi_c \cos^2\theta + \chi_{ab}\sin^2\theta)VH\,\mathrm{d}H\,, \tag{6.36}$$

where χ_c and χ_{ab} are the paramagnetic susceptibilities along the c-axis and in the ab-plane, respectively. Integrating (6.36), we obtain the expression for the magnetic energy of a grain:

$$E_{\mathrm{m}} = \int_0^H \mathrm{d}E_{\mathrm{m}} = -(\chi_c\cos^2\theta + \chi_{ab}\sin^2\theta)VH^2/2 \tag{6.37}$$

or

$$E_{\mathrm{m}}(\theta, H) = -(\chi_{ab} + \Delta\chi\cos^2\theta)VH^2/2\,, \tag{6.38}$$

where $\Delta\chi$ is the difference in the volume susceptibilities of the grain.

Subsequent to the nucleation event, nuclei will start their growth under the influence of a magnetic field. In the early stages of growth, the grains are completely surrounded by a liquid phase and thus they can be regarded as small particles, rotating in a free medium without intergrain interactions. Consider the probability, $f(\theta)$, that a grain has an orientation with angle θ under the influence of a magnetic field. It can be expressed, according to a classic Boltzmann statistics, as

$$f(\theta)\mathrm{d}\theta = \frac{\exp[-E(T,H,\theta)/kT]\mathrm{d}\theta}{\int\limits_0^{\pi/2}\exp[-E(T,H,\theta)/kT]\mathrm{d}\theta}\,. \tag{6.39}$$

Then, for total number of grains, n, the mean number of grains with an orientation between θ and $\theta + \mathrm{d}\theta$ can be given as

$$n(\theta)\mathrm{d}\theta = nf(\theta)\mathrm{d}\theta = n\frac{\exp[-E_{\mathrm{m}}(T,H,\theta)/kT]\mathrm{d}\theta}{\int\limits_0^{\pi/2}\exp[-E_{\mathrm{m}}(T,H,\theta)/kT]\mathrm{d}\theta}\,. \tag{6.40}$$

Hence, the distribution, $n(\theta)$, can be related to an alignment parameter, which is used to quantify the degree of texture in melt-processed superconductors under the influence of a magnetic field. This alignment parameter, F, such that $F = 1$ for a completely alignment structure and $F = 0$ for a totally random structure, can be found as

$$F = 1 - \frac{s^2}{s^2_{H=0}} , \tag{6.41}$$

where s^2 is the variance of the grain distribution for a particular processing condition and $s^2_{H=0}$ is the variance of the distribution in the absence of a magnetic field. It is known that the anisotropy of molar magnetic susceptibility $\Delta\chi^{\text{molar}} \approx 22.5 \times 10^{-5}\,\text{cm}^3/\text{mol}$ [53, 1099]. Then, the anisotropy of volume magnetic susceptibility, $\Delta\chi$, used in (6.38), is 1.5×10^{-6}, if we assume a density $5.5\,\text{g/cm}^3$ [862] for the superconductor. Figure 6.7 shows the dependence of the parameter F on external magnetic field for various grain sizes at the temperature, $T = 875°\text{C}$, which is approximately 5°C smaller than the melting temperature of Bi-2212 phase.

As it is followed from Fig. 6.7, when the magnetic field increases, there is a trend for the texture to increase, except in cases where the grain size is small. Therefore, a high degree of alignment can be obtained by increasing the magnetic field and the grain size. In the case of larger grain sizes, the magnetic field tends to saturate and thus increasing the magnetic field has only a negligible effect on the degree of texture.

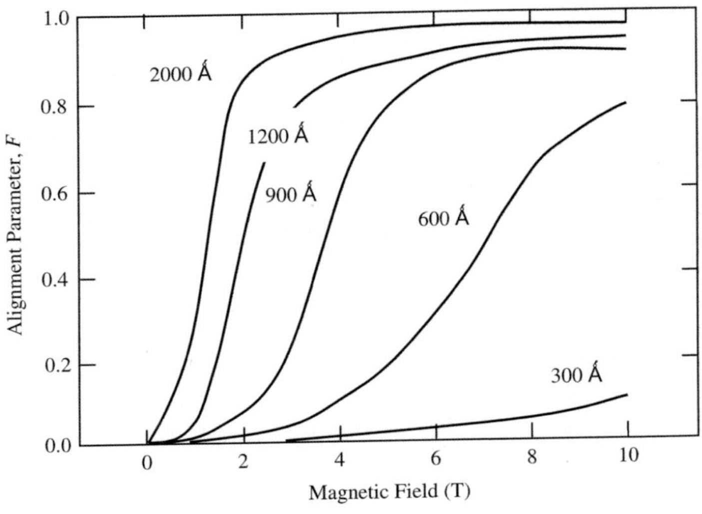

Fig. 6.7. Alignment parameter as a function of magnetic field for various grain sizes. The processing temperature is 875°C [275]

6.2.2 Deformation at Tape Cooling

Deformations caused by compressing stresses, forming into ceramic core during sample cooling are defined. Known estimations of mechanical properties of the silver matrix and oxide superconducting filaments at all considered temperature intervals are used.

Young's modulus. Following [60], we assume a linear dependence of Young's modulus on temperature, $E^{\mathrm{Ag}}(T)$; for BSCCO use data obtained in [321].

Plastic deformation. The linear relation between yielding stress, σ_y^{Ag}, and temperature, suggested in [60], uses for adaptation to the room-temperature results [394] and to value $\sigma_y^{\mathrm{Ag}} = 0$, corresponding to the melting temperature. The yielding strain is found from Hooke's law. The behavior of the pressed multi-phase ceramic core is suggested to be elastic.

Thermal expansion. The behavior of parameters $\alpha^{\mathrm{Ag}}(T)$ and $\alpha^{\mathrm{BSCCO}}(T)$ at low temperatures is stated from results of [781] and is fitted to the value $19 \times 10^{-6}\mathrm{K}^{-1}$ for Ag and to the room-temperature data for BSCCO [321]. At more high temperatures, we suggest a linear dependence up to $22 \times 10^{-6}\mathrm{K}^{-1}$ for Ag and $16.5 \times 10^{-6}\mathrm{K}^{-1}$ for BSCCO (at 1110 K).

At high temperatures, a different thermal compression leads to elastic deformations, both matrix and core. In this case, the deformations at temperature change on the value of ΔT are estimated, using the law "action–counteraction" for two-component tape, consisting of filaments and matrix. From Hooke's law,

$$\Delta\varepsilon^{\mathrm{BSCCO}}(T) = \left[1 + \frac{F E^{\mathrm{BSCCO}}(T)}{(1-F)E^{\mathrm{Ag}}(T)}\right]^{-1} (\alpha^{\mathrm{Ag}} - \alpha^{\mathrm{BSCCO}})\Delta T \; ; \tag{6.42}$$

$$\Delta\varepsilon^{\mathrm{Ag}}(T) = \left[1 + \frac{(1-F)E^{\mathrm{Ag}}(T)}{F E^{\mathrm{BSCCO}}(T)}\right]^{-1} (\alpha^{\mathrm{BSCCO}} - \alpha^{\mathrm{Ag}})\Delta T \; , \tag{6.43}$$

where $F = \%\mathrm{BSCCO}/(\%\mathrm{BSCCO} + \%\mathrm{Ag})$ is the filling factor. The values of $\varepsilon^{\mathrm{BSCCO}}(T)$ and $\varepsilon^{\mathrm{Ag}}(T)$ are defined, integrating (6.42) and (6.43):

$$\varepsilon^{\mathrm{BSCCO}}(T) = \int_{T_\mathrm{m}}^{T} \frac{\Delta\varepsilon^{\mathrm{BSCCO}}}{\Delta T} \mathrm{d}t \; ; \qquad \varepsilon^{\mathrm{Ag}}(T) = \int_{T_\mathrm{m}}^{T} \frac{\Delta\varepsilon^{\mathrm{Ag}}}{\Delta T} \mathrm{d}t \; , \tag{6.44}$$

where T_m is the temperature of the tape annealing. A transition from elastic to plastic regime (i.e., from linear to non-linear dependence) occurs, when $\varepsilon^{\mathrm{Ag}}(T)$ attains the yielding strain, $\varepsilon_y^{\mathrm{Ag}}(T)$. Below corresponding transition temperature, the silver matrix deforms plastically (non-linearly), leading to a loading of the superconductor by constant stress, σ_y^{Ag} [835]. Finally, the required core deformation is found as

$$\varepsilon^{\mathrm{BSCCO}}(T) = \varepsilon_y^{\mathrm{Ag}}(T) \frac{(1-F)E^{\mathrm{Ag}}(T)}{F E^{\mathrm{BSCCO}}(T)} \; . \tag{6.45}$$

The obtained numerical results are presented in Fig. 6.8 [835].

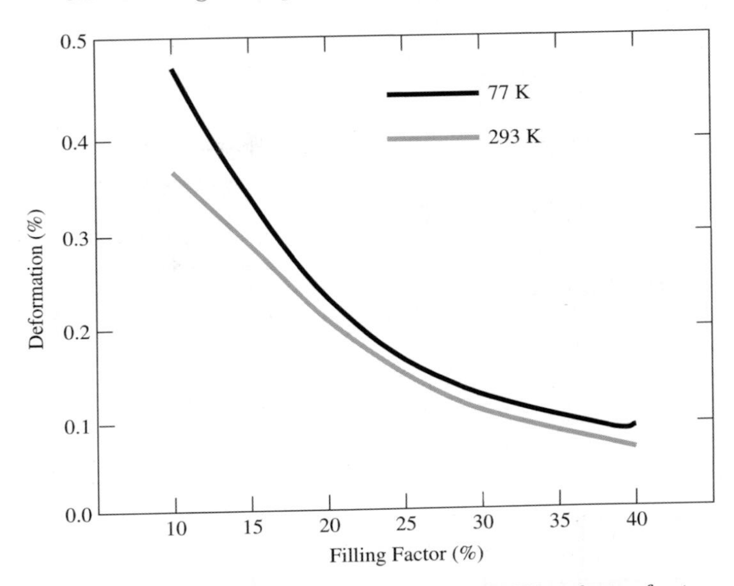

Fig. 6.8. Deformation of BSCCO core as a function of filling factor for tapes, cooled down to room temperature and cryogenic temperature (77 K)

6.2.3 Effects of Mechanical Loading

The introduction of "freedom parameter," Δ_f, estimates qualitatively the influence of different constraint factors on the mass-flow behavior and also presents the models of re-distribution of mass and flux in the powder compact due to plastic deformation of composite [385].

Freedom Parameter

Consider the compression of a two-dimensional slab of an ideal plastic solid between two flat anvils as shown in Fig. 6.9. If there is friction between the slab and the anvils, "barreling" occurs during the process (see Fig. 6.9b). In this case, there is inhomogeneous stress–strain state of material. The pressure at the interface between the anvil and the slab needed to induce plastic deformation is given as [441]

$$P = 2k \exp\left(\frac{2\mu x}{h}\right), \quad 0 \leq x \leq \frac{L}{2} \,. \tag{6.46}$$

The mean pressure is

$$P_{\mathrm{c}} \approx 2k\left(1 + \frac{\mu L}{2h}\right), \tag{6.47}$$

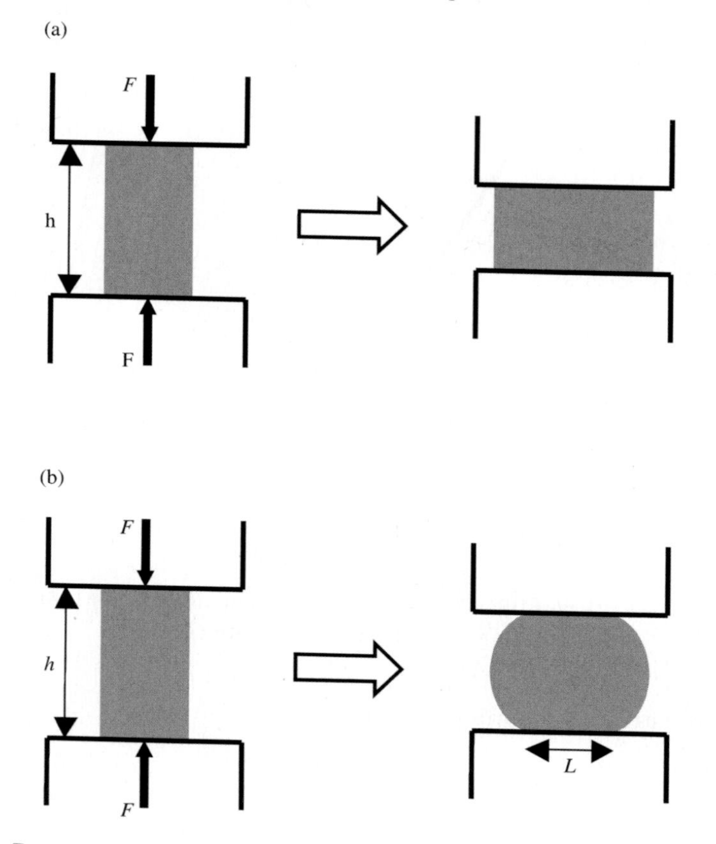

Fig. 6.9. Processes of two-dimensional compression: (**a**) two sides are free, and there is no friction between the slab and the anvils; (**b**) two sides are free, but there is substantial friction between the slab and the anvils

where $2k$ is the flow yield, coinciding with the mean pressure at $\mu = 0$ and/or $L << h$; μ is the sliding friction factor; h is the sample height and L is the contact size. If there is sticking friction at the interface, then the pressure is [441]

$$P = 2k\left(1 + \frac{x}{h}\right), \quad 0 \leq x \leq \frac{L}{2}, \tag{6.48}$$

and the mean pressure is

$$P_c = 2k\left(1 + \frac{L}{4h}\right). \tag{6.49}$$

The increasing pressure, or "friction hill," is caused by friction and reaches a maximum in the center (see Fig. 6.10). The higher the friction, the higher the pressure needed to cause mass flow. Obviously, the pressure strongly depends on the ratio between the height and the contact size, h/L. If the ratio h/L

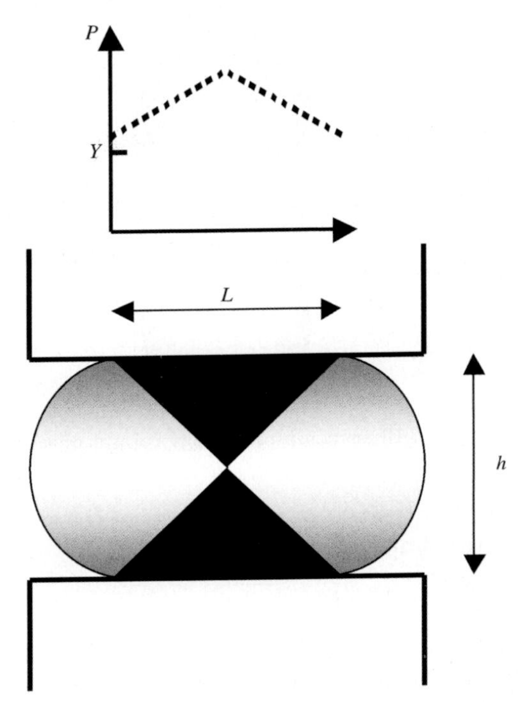

Fig. 6.10. Inhomogeneous deformation under compression (Y is the yield value of the sample). The *shaded regions* have experienced less deformation than the remainder of the slab

is small, then the required pressure to induce plastic flow in the material is high, and vice versa. This means that the material has a large freedom to flow, which is found by the freedom parameter, $\Delta_f = h/L$.

Pressing. The freedom parameter, Δ_f, can be used to describe how freely the material expands. During plane pressing, the larger the Δ_f, the larger the expansion. Therefore, the width increase of a narrow thick tape is larger that that of a wide thin tape under the same pressure. The values of Δ_f will be different if the contact sizes are not the same in the x- and y-directions. For example, if the contact size in the x-direction is shorter than the contact size in the y-direction, then the expansion in the x-direction will be larger than that in the y-direction.

Rolling. Fig. 6.11 presents the flat rolling process. Due to the friction at the interface between the rollers and sample, a friction hill is built up during rolling. Lubrication can reduce the constraint and therefore also reduce the stress in the sample, when other rolling parameters are fixed. If the sample height is much smaller than the roller diameter, (6.49) could be used as a rough estimation with the mean height h:

$$h = (h_0 + h_\mathrm{f})/2\ , \tag{6.50}$$

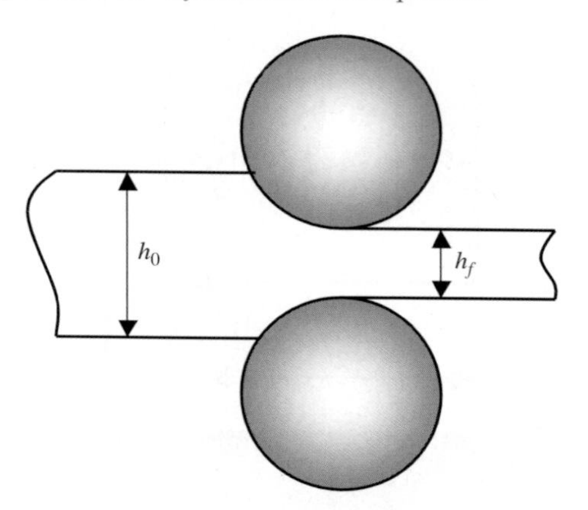

Fig. 6.11. Illustration of the flat rolling process

where h_0 and h_f are the original and final heights of the sample, respectively. The contact size in the sample length direction, L_L, is given as

$$L_\mathrm{L} \approx \sqrt{R(h_0 - h_\mathrm{f})} = \sqrt{R\delta h} \,, \tag{6.51}$$

where R is the roller radius and δh is the height reduction of the sample [441]. If a tensile stress, σ_t, is applied to the sample during rolling, then (6.49) gives

$$P_\mathrm{c} = (2k - \sigma_\mathrm{t}) \left(1 + \frac{L}{4h}\right) . \tag{6.52}$$

This means that the rolling pressure and therefore the stress in the sample could be decreased, applying a tensile stress during rolling.

The freedom parameter in the sample length direction, $\Delta_{\mathrm{f,L}} = h/L_\mathrm{L} = h/\sqrt{R\delta h}$, depends on the roller radius, sample height and the height decreasing. For example, the rolling of a thick tape, using larger rollers, could be similar to that of a thin tape, using small rollers. The freedom parameter in the sample width direction is given by $\Delta_{\mathrm{f,w}} = h/L_\mathrm{w}$, where L_w is the width of the sample. The ratio between the freedom parameters, $\Delta_{\mathrm{f,L}}/\Delta_{\mathrm{f,w}} = L_\mathrm{w}/\sqrt{R\delta h}$, is useful to determine the direction in which the material is most apt to deform. In particular, large rollers, a large decreasing and a small tape width could lead to a significant width increase during rolling and vice versa.

Drawing and extrusion are axis-symmetric deformation processes (see Fig. 6.12), which are in principle similar. There are two main differences between them, namely (i) the compressive stress in the deformation zone during extrusion is larger than that of drawing, especially because the area decreasing during extrusion is often much larger than the decrease in a drawing process[1]; (ii) there is a tensile stress in the exiting wire during drawing, which limits the

[1] This means that the powder density in a BSCCO/Ag composite is larger after extrusion than after drawing.

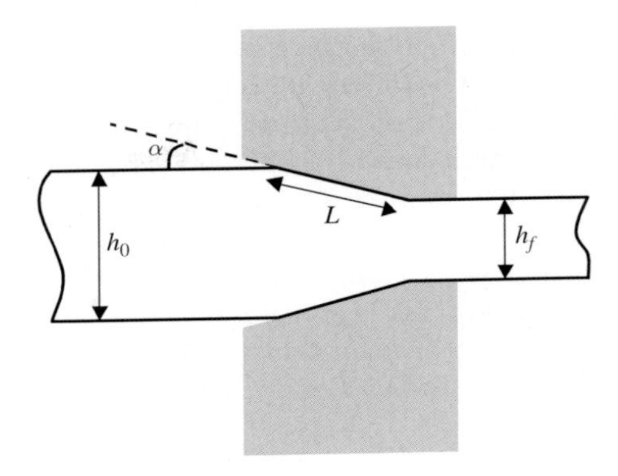

Fig. 6.12. Illustration of the drawing and extrusion process

maximum drawing force and the maximum decrease by the strength of the material. The fact that it is possible to obtain very large decreasing ratios in the extrusion process makes it an efficient process for large-scale production. However, it is rather difficult to obtain homogeneous Bi-2223/Ag composites by extrusion [458].

The freedom parameter, $\Delta_\mathrm{f} = h/L$, could be determined using (6.50) and the contact size, L, as

$$L = \frac{h_0 - h_\mathrm{f}}{2 \sin \alpha} = \frac{\delta h}{2 \sin \alpha}\,, \tag{6.53}$$

where h_0 and h_f is the original and final diameters of the sample, respectively, δh is the diameter decreasing and α is the half-angle of the die.[2]

Thus, the different constraint factors have a strong influence on the deformation process. By introducing the freedom parameter, Δ_f, it is possible to estimate these effects using only one parameter. Table 6.1 gives a summary of the expression of the freedom parameters for various deformation processes.

Table 6.1. Freedom parameters for various deformation processes

Pressing	Rolling	Drawing or extrusion
$\Delta_\mathrm{f} = h/L$	$\Delta_\mathrm{f,L} = h/\sqrt{R\delta h}$ $\Delta_\mathrm{f,w} = h/L_\mathrm{w}$	$\Delta_\mathrm{f} = \frac{2h}{\delta h} \sin \alpha$

[2] It should be noted that the condition of drawing or extrusion is usually far from that of plane pressing. However, the concept of the freedom parameter could be useful for qualitative estimations. For example, the stress in the sample will be large if the die angle is small or the diameter decreasing is large or the diameter of the sample is small.

The presented description of the mechanical deformation process is only for single-element samples. The duplex systems, such as BSCCO/Ag composites, are needed in additional considerations. The difference of the mechanical properties of each component leads to different behavior of the mass flow of each component during deformation, resulting in a "mass re-distribution." As shown in Fig. 6.13 the deformation of a weaker component will be larger than that of a stronger one. A weak component is defined by a lower yield stress or a different flow behavior. This leads to the symmetry change of the component mass distribution for the composites and could be the main reason for the inhomogeneity introduced when BSCCO/Ag composites are mechanically

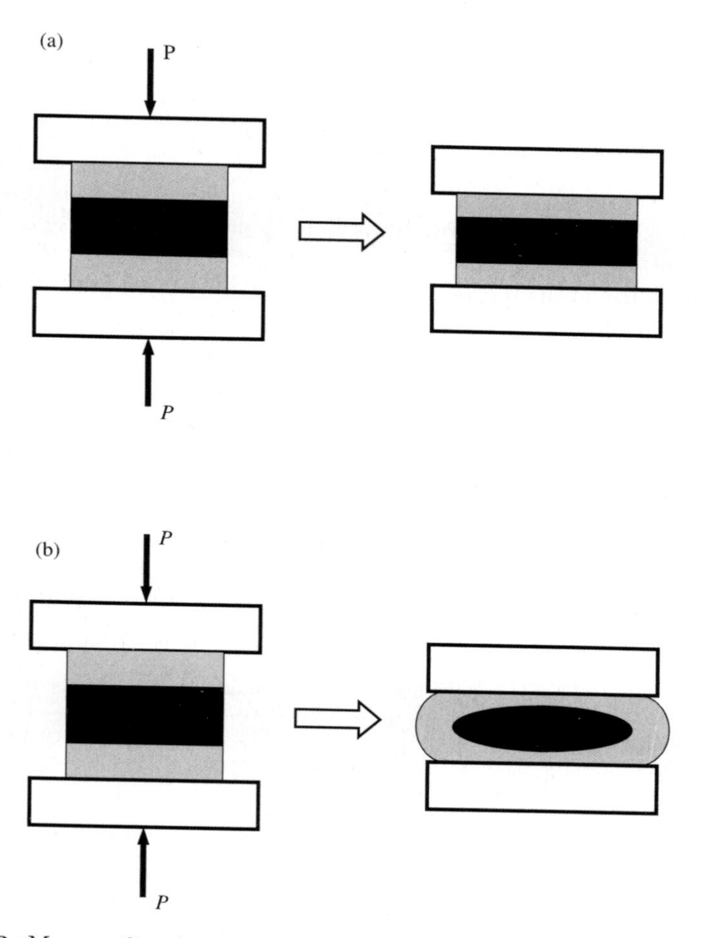

Fig. 6.13. Mass re-distribution during pressing: (**a**) homogeneous deformation at equivalent mechanical properties of both components, and (**b**) inhomogeneous deformation, causing the mass re-distribution (the *dark area* represents a stronger component)

deformed. A lightened deformation of a weak component leads to convey a high pressure to the strong component under certain constraint conditions (see Fig. 6.13b). Due to the friction at the interfaces between the anvils and the sample, between weak and strong components, a high stress could be built up in the center region of the weak parts, especially when the height of the sample is much smaller than the width of the sample, that is, the freedom parameter is small. This leads to transport of pressure through the weak region to the strong part of the sandwich. At the edges of the sandwich, the weak component is less constrained and can transport less direct pressure to the strong part. However, the shear stresses might still be considerable, defining deformation at the edges of the strong part. Obviously, constraint conditions or the freedom parameter have a strong influence on the mass re-distribution process.

If the density of the superconductor is not very high, then the BSCCO/Ag core can be considered as the weaker part of the composite, resulting in a more significant deformation of the core than with the silver sheath. On the other hand, if the core density is very high, then the Ag sheath could be the weaker component unable to convey high enough pressure to the core for further densification of the core. However, by lowering the freedom parameter of the process ($\Delta_{\mathrm{f,L}} = h/\sqrt{R\delta h}$), the mass-flow behavior could be changed so that the Ag sheath could again transport a high enough pressure to increase the density of the superconductor, although the silver part is still the weaker component of the composite. Obviously, when the roller diameter is larger than the wire diameter, the freedom parameter is small, defining larger mass re-distribution.

The freedom parameter also has influence on the sausaging. If the freedom parameter is large, the mass flow of the Ag sheath is large, and the sausaging (at least the sausaging frequency) as well as the density of the core will be decreased. The freedom parameter at rolling ($\Delta_{\mathrm{f,L}} = h/\sqrt{R\delta h}$) assumes that sausaging could be decreased, using small rollers, which has been supported by test results [568]. This also implies that sausaging will be more pronounced when the thickness of the tape becomes small; this has also been demonstrated [568]. It should be pointed that the smaller factor of thickness decreasing causes the larger freedom of silver sheath in the direction of the wire length. Therefore, it is not surprising that the BSCCO core density, obtained in this way, is smaller than the core density obtained, using a larger decreasing factor.

Compatibility line

The stresses developed in a material due to an external loading can be estimated using the so-called Mohr circles [715]. The intrinsic curve, giving the behavior of a material for a given deformation ε, is obtained by drawing the envelope of these Mohr circles. In particular, Fig. 6.14a shows the intrinsic curves for Bi-2223 and Ag at a deformation $\varepsilon = 25\%$. The existence of this intersection (P_{c}) is called the *compatibility point*. Its existence shows

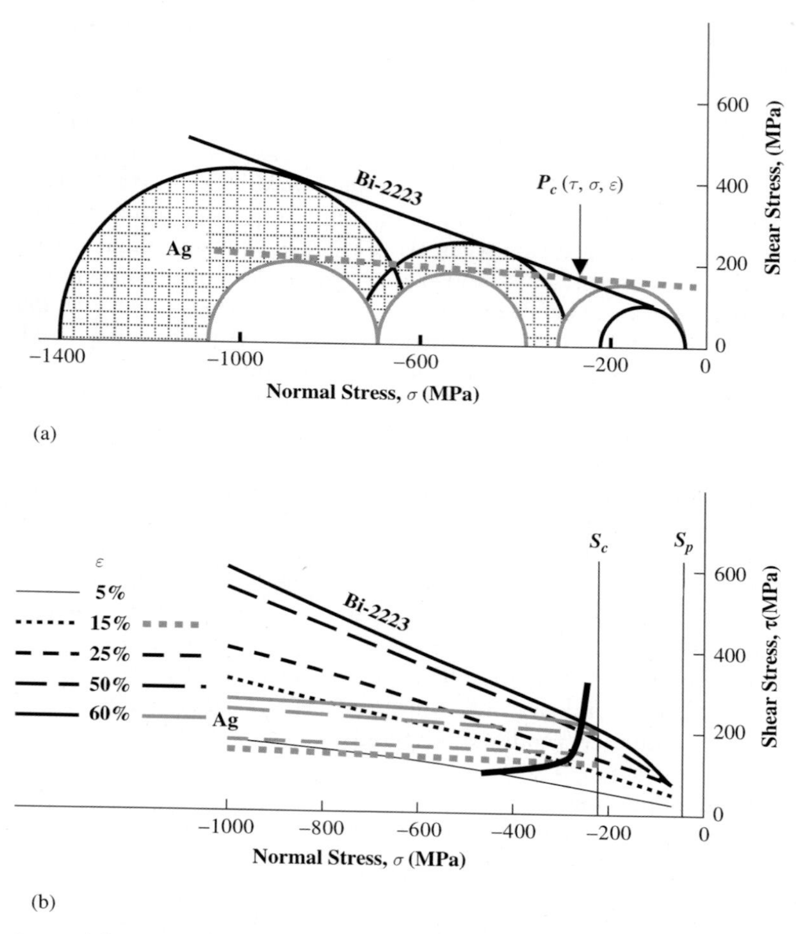

Fig. 6.14. (**a**) Determination of the compatibility point $P_c(\sigma, \tau, \varepsilon = 25\%)$ at the intersection of the intrinsic curves for a Bi-2223 ceramic and metallic Ag; (**b**) the compatibility line between Bi-2223 and Ag: trace of the compatibility points P_c as ε varies from 5 to 60%. S_p corresponds to the plasticity threshold of the Bi-2223 ceramic and S_c to the compatibility threshold before the brittle fracture between the Bi-2223 ceramic and metallic Ag [766]

that in spite of quite different mechanical behavior at atmospheric pressure of the matrix and filaments, there are external mechanical stresses defining the same ductile behaviors of silver and superconductor. By drawing such intrinsic curves for different deformation rates, one can obtain the line along which P_c moves, called the *compatibility line*. One corresponds to conditions where the superconducting powder and the sheath both have the same mechanical behavior, especially in the ductile domain. Then, instead of fabricating wires at atmospheric pressure, that is, out of the plastic domain for the superconducting powder (S_p limit, see Fig. 6.14b) and out of the ductile zone common to the powder and to the sheath (S_c limit, see 6.14b), it is much more expedient to use an analysis of triaxial stress state. This approach helps to minimize Ag/BSCCO interface sausaging, using yielding criterion, defining the deformation compatibility of silver matrix and superconducting core [190].

6.2.4 Finite-Element Modeling of Deformation Processes

Rolling and Pressing of BSCCO/Ag Tapes

The finite element method (FEM) is capable of obtaining qualitative estimations of parameters found by material deformation during BSCCO/Ag tape processing. However, the FEM application to these problems is a complicated task still not widely applied. For instance, the limiting factors for this method are complicated geometry of the composite and great time required to obtain numerical results with necessary accuracy. The application of coarse meshes can decrease the simulation time. However, these meshes could lead to simulation problems and insufficient detailing of results.

In [660, 967, 1048], drawing of monocore wire is modeled by two-dimensional axis-symmetric models. The constitutive equations describing the powder flow are stated using the Drucker-Prager model with an elliptical cap criterion. Process parameters, such as the die angle and degree of reduction in each drawing step, are shown to influence the density of the powder in drawn wire. The distribution of density, being high at the silver/powder interface and lower in the center, is in agreement with test data. FEM modeling of the multi-filament wire drawing has not been found in literature, probably because a full three-dimensional model is necessary.

Numerical simulation of flat rolling also requires a 3D model to describe the deformations in length, width and thickness directions. Simplified 2D simulation can be made, assuming zero deformation in either the width or the length direction. Assuming zero width strain, the pressure distribution along the roll gap is modeled in [238, 633, 939]. In this case, the pressure profile forms either a friction hill or a friction valley, depending on the roll diameter and degree of thickness reduction. These parameters also influence the shear strains in the strain zone. When material flow in the cross-section is analyzed in a 2D model, zero strain in the length direction must be assumed [256]. This technique enables the prediction of filament geometry and density, incorporating the influence of wire geometry and friction.

The input data for numerical properties are essential for the precision of numerical results. Reference [375] presents an extensive investigation of the mechanical properties of BSCCO powder. The powder is evaluated applying a combination of fracture tests and triaxial strain tests. As a result, it is concluded that the Drucker-Prager model is not capable of describing the yield surface of BSCCO powder in detail. Reference [57] shows how the yield surface can be determined by a few relatively simple tests combined with a Drucker-Prager conical cap model. It is demonstrated that an FEM simulation based on this approach gives a rather good prediction of the density in the individual filaments for small reductions of tape thickness in a 2D model.

In more details three FEM approaches for simulation of flat rolling are considered below [257]. The next different approaches are applied, namely (i) 2D pressing in a mesh with 50 × 50 elements, (ii) 3D pressing between non-rotating rolls in a mesh with 17 × 26 × 60 elements (number of elements in width, height and length, respectively) and (iii) full 3D rolling in the same mesh as in the case of the 3D pressing. The superconducting filaments have rectangular shape and lie in a square matrix of pure silver, surrounded by an alloyed silver sheath. Both silver materials are described by Von Mises yield criterion with the flow stress parameters ($\sigma = C\varepsilon^n$) given in Table 6.2. The constitutive plasticity model, describing the powder, is the Drucker-Prager model with a conical cap (see Fig. 6.15). This model is a rough approximation to the real yield surface, but it includes the most important properties of the powder, which are pressure-dependent yield stress, volumetric strain and material hardening. In the $(p, \sqrt{J_2})$ plane, the yield surface consists of two intersecting lines, shown in Fig. 6.15. The Drucker-Prager failure surface is written as

$$F_\mathrm{s} = \sqrt{J_2} - \eta(p + p_\mathrm{t}) = 0 \; , \tag{6.54}$$

where J_2 is the second invariant of the stress tensor, η is the slope of the failure line and p_t is the yield stress in pure, hydrostatic tension. The cap yield surface is a line with a negative slope:

$$F_\mathrm{c} = \sqrt{J_2} + \xi(p - p_\mathrm{c}) = 0 \; , \tag{6.55}$$

Table 6.2. The model parameters for silver, alloy and BSCCO powder

Parameters	Silver	Silver alloy	BSCCO powder
Young's modulus (MPa)	57.32	57.32	13.79
Poisson's ratio	0.38	0.38	0.2
$\sigma = C\varepsilon^n$			
C (MPa)	320	350	–
n	0.3	0.11	–
η	–	–	1.2
ξ	–	–	0.29
p_t (MPa)	–	–	12
$p_\mathrm{c} = a\{\exp[b(\rho - \rho_0)] - 1\}$			
a (MPa)	–	–	27
b	–	–	1
Relative density, ρ_0	–	–	0.6

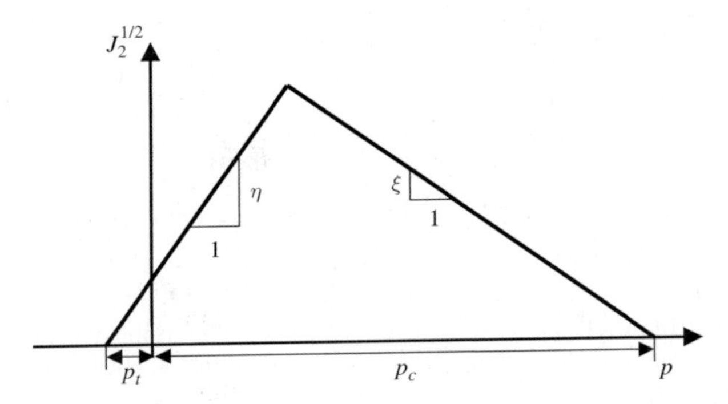

Fig. 6.15. Drucker-Prager model with a conical cap [257]

where ξ is the slope of the cap line and p_c is the hydrostatic compression yield stress. The model parameters are given in Table 6.2 [375].

All three models calculate higher densities in the center than at the edges and overestimate the relative density of filaments. In Fig. 6.16 the relation between tape thickness and tape width is shown for 2D and 3D FEM simulations of rolling and pressing and compared with test data. As it is followed from Fig. 6.16, the 3D rolling simulation fits very well with the test results. At the same time, the FEM simulations of both models overestimate the widening.

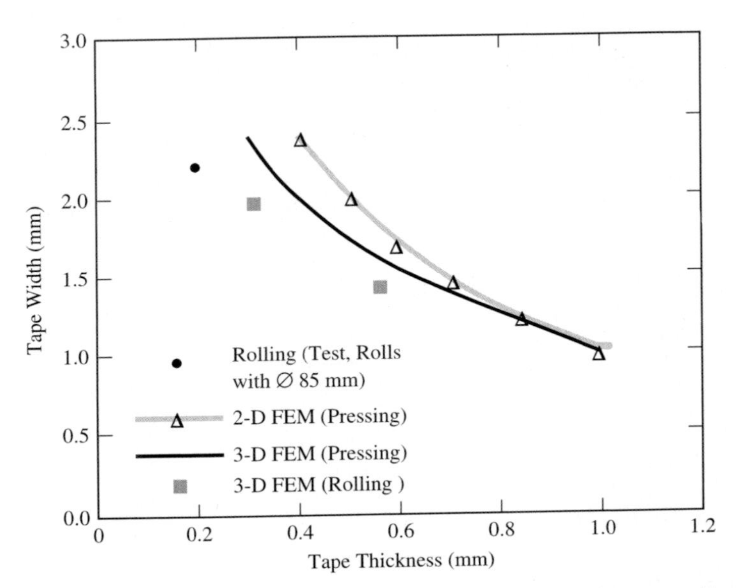

Fig. 6.16. Tape width *vs* tape thickness for experiments with ∅85 mm rolls and 2-D and 3-D FEM simulations [257]

Thermal Cycling of BSCCO/Ag Tapes

FEM is also applied to investigate thermal cycling of BSCCO/Ag tapes. For example, in [776], a simulation of thermal stresses is carried out in monocore tape under the following assumptions:

- The BSCCO/Ag tape is stress-free at the sintering temperature (1113 K).
- The mechanical properties of both components are isotropic.
- Silver behaves elastically and plastically while BSCCO is brittle.
- The Baushinger effect is neglected, that is, the magnitude of the yield stress of silver is assumed to be the same in tension and compression.
- The composite is symmetrical about the x- and y-axes.
- The tape is infinitely long so that plane strain conditions apply.
- The BSCCO core center is assumed to be 65% dense and the outer core is assumed to be 85% dense.

Figure 6.17 [776] shows a section of the mesh used for the thermal cycling analysis. The x-axis is parallel to the width of the tape (3.5 mm), the y-axis is parallel to the height of the tape (200 μm) and the z-axis is parallel to the length, assuming to be infinite. Both the central part of the core tape and external part of the region corresponding to the silver sheath are presented by white color. The gray part defines the external part of the superconductor or interface. The mechanical properties of BSCCO and Ag are shown in Table 6.3,

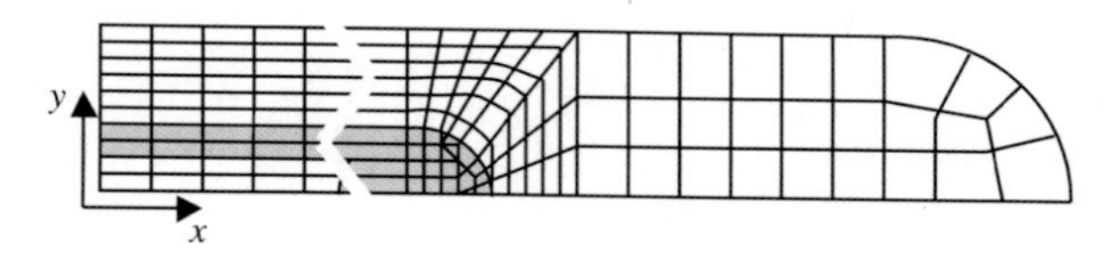

Fig. 6.17. Mesh used for FEM computation. The entire mesh contains 1500 nodes and 250 elements

Table 6.3. Mechanical properties of Ag and BSCCO [776]

Parameters	BSCCO	Ag
$\alpha(K^{-1})$	13.6×10^{-6}	21.9×10^{-6}
E (GPa), 100% dense	127.0	71.0
E (GPa), 85% dense	83.8	–
E (GPa), 65% dense	54.1	–
σ_y (MPa) at 300 K	–	12.6
K_p (GPa) at 300 K	–	0.57
ν	0.14	0.37
σ_y (MPa) at 77 K	–	13.2
K_p (GPa) at 77 K	–	0.7

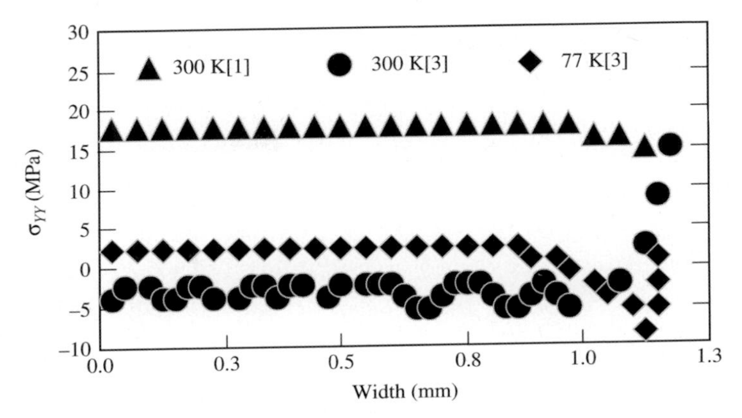

Fig. 6.18. FEM, showing the variation of σ_{yy} with thermal cycling across the width of BSCCO core. Cycles are denoted by the numbers in square brackets [776]

where E is Young's modulus, ν is Poisson's ratio, K_p is the slope of the stress–strain curve after yielding and σ_y is the yield stress. Figure 6.18 shows the variation of σ_yy with thermal cycling across the width of the tape. The stress state is fairly uniform in the center of the core except at the edges, where the curvature of the core can act as a stress concentrator.

7

Modeling of YBCO Oxide Superconductors

7.1 Modeling of 123 Phase Solidification from Liquid

7.1.1 Heterogeneous Mechanism

Initially, for peritectic reaction (basis of melt processing), *heterogeneous mechanism* has been suggested, that is, a formation of 123 phase in direct contact of melt and properitectic ("green") 211 phase followed by the growth of 123 crystallites in account of the component diffusion through product layer [489]. Shortcoming of this mechanism results in the following causes:

(1) Calculated growth rate of 123 phase, limited by diffusion of slowest component ($D_{Y}3+ = 10^{-11}\,cm^2/s$), should be approximately $4\,\mu m/h$ [54], which is some order of magnitude less than test data [54, 890, 891]. In this case, the peritectic reaction, which occurs in diffusion mechanism, should be completed after formation of 123 layer around 211 particles.
(2) It is impossible to describe YBCO platelet structure, because diffusion of components should mostly impede in the direction of dominant growth, that is, in *ab*-plane.
(3) Parallel disposition of platelets is not explained, because arbitrary initial orientation of particles of the 211 properitectic phase into liquid should lead to the absence of anisotropy and to arbitrary orientation of 123 crystallites, presented in Fig. 7.1a.

7.1.2 Models Based on Yttrium Diffusion in Liquid

In realization of melt-processing for preparation of YBCO large-granular samples with improved superconducting properties, it is important a priori to obtain a grain growth in the direction of the gradient to maximize the ratio of "external" thermal gradient (G) to solidification rate (R). From this point of view, it seems better to transfer the sample across a large thermal gradient. However, in this case a large part of the sample stays for a long time at elevated temperature, where Y_2BaCuO_5, $BaCuO_2$ and CuO_2 co-exist. Moreover, an undesirable change of the geometric shape of the sample is possible by liquid flow. Thus, a good compromise between G and R is required, because the G/R ratio governs the stability of the solidification process. The thermodynamic aspects of this problem considered in [155, 475] have led to a model of 211 phase solubility in liquid and *homogeneous formation of nuclei* (see Fig. 7.1b). This model is based on the following assumptions:

(1) Ion concentration of Y^{3+} attains a maximum in liquid near the surface of 211 particle and decreases sharply at the mid-distance of neighbor 211 inclusions. When the distance exceeds this length, the ion concentration of Y^{3+} decreases asymptotically down to a value corresponding to the peritectic melt.
(2) Increasing of yttrium concentration in the liquid enhances a possibility of 123 phase solidification because of metastability of the melt and existence of figurative point in the two-phase region "123 solid phase–melt."
(3) Constitutional undercooling in the system, leading to supersaturation of the melt by yttrium, is found (i) by "external" thermal gradient, (ii) by undercooling in comparison with saturated melt and (iii) by decreasing of solidification temperature for melt, saturated by yttrium relatively to peritectic melt.

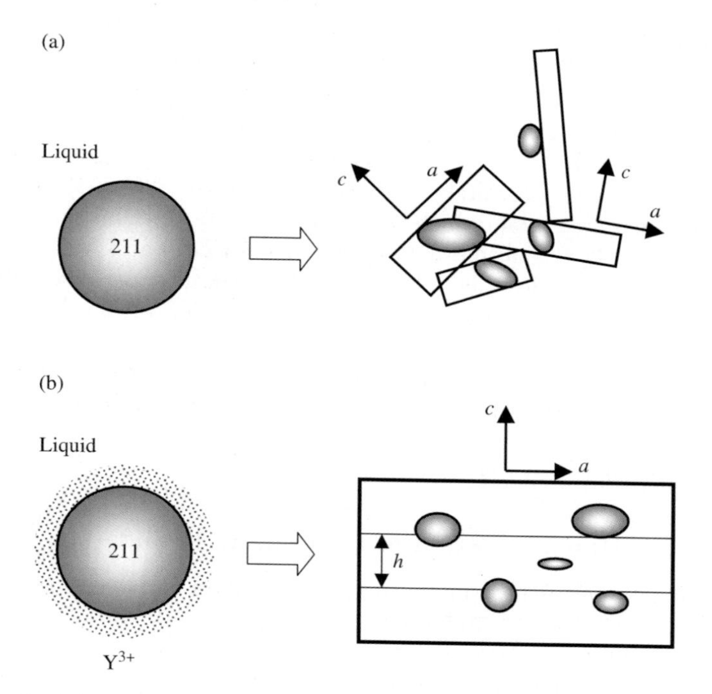

Fig. 7.1. Models of heterogeneous formation of nuclei at 211 particles [489] (**a**) and of solubility of 211 phase into liquid, assisted by homogeneous formation of nuclei (**b**) [155, 475]

The theoretical analysis leads us to introduce a criterion connected to conception of undercooling and to compare the G/R ratio with the relation, $m_L(C_S - C_L)/D_L$ [739], where $C_S(C_L)$ is the ion concentration of Y^{3+} in the solid (liquid) phase at the interface level; D_L is the diffusion coefficient of the ion of Y^{3+} and m_L is the slope of the liquidus curve in the temperature–composition phase diagram. Depending on the various situations presented in Fig. 7.2 two cases are possible:

(1) If G/R ratio is high, that is, G is high, the slope of the curve giving the temperature at the interface level is high, whatever the position of this interface. The comparative behavior of the temperature at the interface level and the temperature of the liquidus is given in Fig. 7.2c. The equation

$$G/R \geq m_L(C_S - C_L)/D_L \tag{7.1}$$

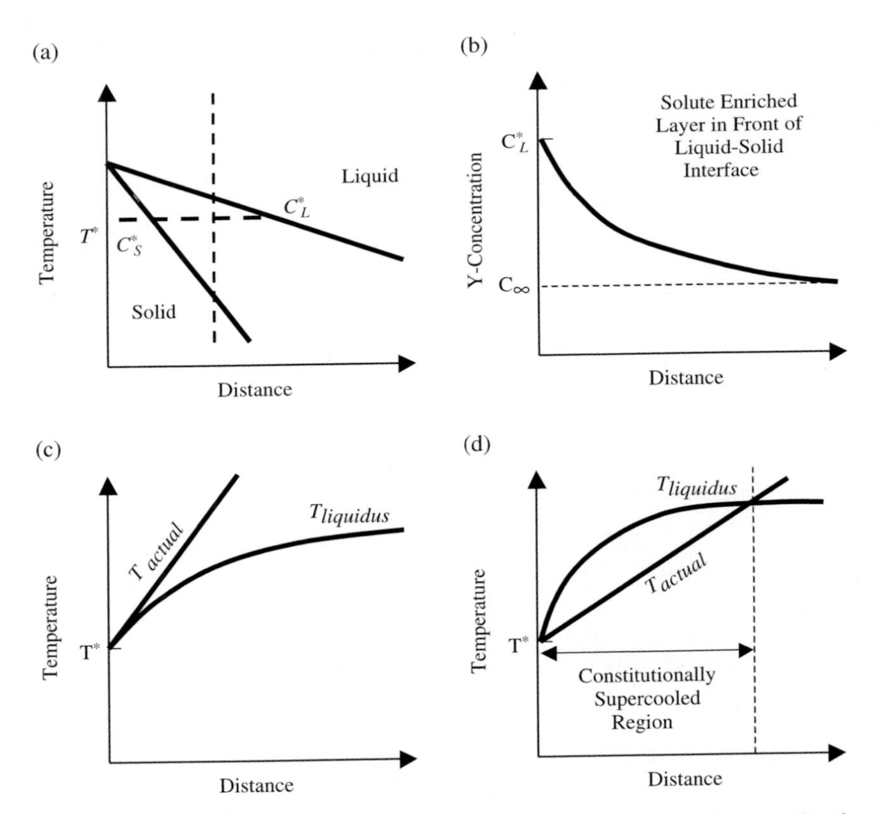

Fig. 7.2. (**a**) Temperature and (**b**) concentration profiles, showing the constitutional undercooling; (**c**) conditions for a plane front; (**d**) unstable case, where imposed temperatures are lower than the temperature of the liquidus [286]. T^* is the initial interface temperature

is continuously satisfied. The interface temperature evolution is stable in time, and the solidification front is flat because the interface temperature follows the temperature of the liquidus.

(2) If the G/R ratio is relatively small, inequality (7.1) is not satisfied, causing instability of the interface temperature (see Fig. 7.2d), because it should be continuously re-adjusted to that of the liquidus. Therefore, the solidification front is not flat, but has cellular (dendritic) shape or consists of equiaxed blocky (see Fig. 7.3), because the interface temperature fluctuates between the two curves, leading alternatively to a liquid or a solid phase.

The above results have been used to explain directed solidification of YBCO after peritectic reaction [890].

The diffusion of yttrium is necessary to re-compose 123 phase in the liquid, that is, for a non-classical peritectic reaction, which is governed by diffusion through melt and additionally demands to attain triple point between primary, secondary and liquid phases [493]. The driving forces inside the diffusion zone, leading to a migration of yttrium, correspond to concentration gradients. Two phenomena, presented in Fig. 7.4, have been established for these concentration gradients [156, 741]: (i) the change of the chemical potential at the "211 phase-liquid" interface caused by the curvature of the 211 particles and (i) constitutional undercooling (ΔT_S), which corresponds to the temperature difference between the actual temperature of the solidification interface and the temperature of the "211 phase-liquid" interface.

The first phenomenon explains a change of 211 particles during melt transformation in 123 crystal. The change of chemical potential, caused by the

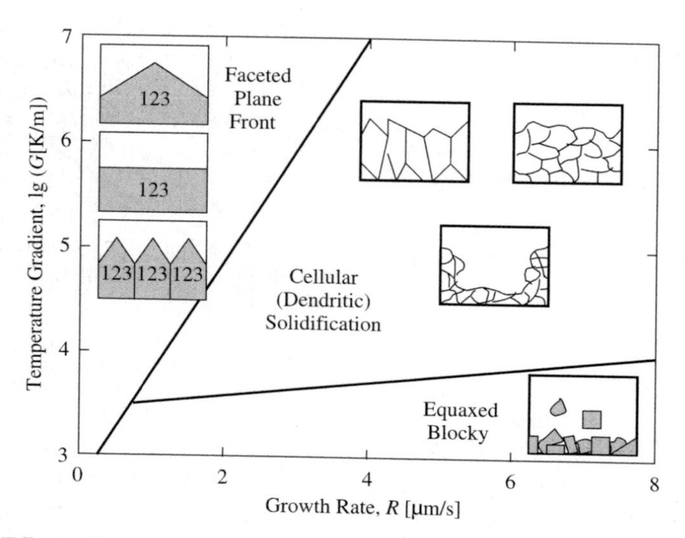

Fig. 7.3. Effects of temperature gradient and growth rate on the morphology of the solidification interface [155]

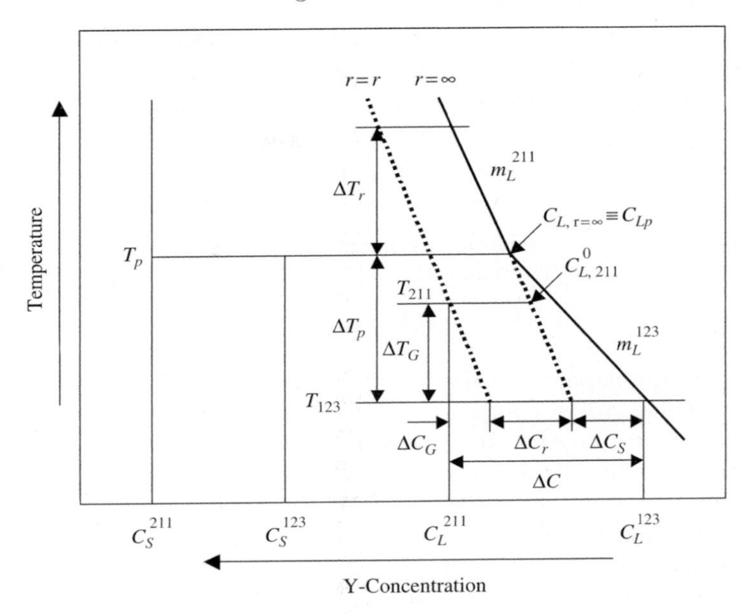

Fig. 7.4. Schematic phase diagram, showing the undercooling and the concentration differences as the driving force for the diffusion of yttrium during the peritectic re-composition of 123 phase [759]

curvature of the 211 particles, relates to the undercooling, ΔT_r, derived from the well-known Gibbs–Thomson relationship:

$$\Delta T_r = 2\Gamma_{211}/r \,, \tag{7.2}$$

where Γ_{211} is Gibbs–Thomson coefficient ($= \sigma/\Delta S$), σ is the interface energy between the 211 particle with radius, r, and the liquid, ΔS is the volumetric entropy of melt. Then, the composition difference of yttrium, ΔC_1, which is considered to be a driving force for diffusion, is expressed using the phase diagram (see Fig. 7.4), as

$$\Delta C_1 = \Delta C_r = C^{211}_{\mathrm{L}(r)} - C^{211}_{\mathrm{L}(r=\infty)} = \frac{2\Gamma_{211}}{r m^{211}_{\mathrm{L}}} \,, \tag{7.3}$$

where $C^{211}_{\mathrm{L}(r)}$ and $C^{211}_{\mathrm{L}(r=\infty)}$ are the yttrium concentrations of liquid at the 211 particle and at the 123 interface, respectively; m^{211}_{L} is the liquidus slope of the 211 phase.

The constitutional undercooling is stated using the second composition change ΔC_2, given by extending towards low temperatures the "211 phase-liquid" liquidus line [476]:

$$\Delta C_2 = \Delta C_{\mathrm{S}} = \left(\frac{1}{m^{123}_{\mathrm{L}}} - \frac{1}{m^{211}_{\mathrm{L}}} \right) \Delta T_{\mathrm{S}} \,. \tag{7.4}$$

Introduction of a third parameter, ΔC_3, taking into account the imposed gradient, completes this model [476]. This term is directly given by the distance, z, between the solidification interface and the considered 211 particle:

$$\Delta C_3 = \Delta C_G = \frac{Gz}{m_{\mathrm{L}}^{123}} , \qquad (7.5)$$

where $\mathbf{m}_{\mathrm{L}}^{123}$ is the liquidus slope of the 123 phase. Thus, total difference of the yttrium concentration, governing the solidification process of the 123 phase is equal to

$$\Delta C = \Delta C_1 + \Delta C_2 + \Delta C_3 . \qquad (7.6)$$

The next development of the model is connected with the assumption of the 211 particles to be spheres and consideration of the concentration difference in yttrium between the 211 phase and the solidification front as the driving force for the diffusion of yttrium in the liquid phase (see Fig. 7.5) [476].

Assume that the distance between 211 particles is equal to 2δ, and the 211 phase remains in final structure. The concentration differences of yttrium near solidification front of 123 phase due to solution of 211 particles are shown in Fig. 7.5b. Moreover, we assume that the 211 particles are in interfacial equilibrium with their surrounding liquid and are large enough that the effect of radius of curvature on melting point is negligible.[1] The phase diagram in the peritectic region, presented in Fig. 7.4, shows that the temperature of the growing 123 front is undercooled by an amount ΔT_{p} below the peritectic temperature, T_{p}, where

$$\Delta T_{\mathrm{p}} = \Delta T_G + \Delta T_{\mathrm{S}} + \Delta T_{\mathrm{C}} , \qquad (7.7)$$

where ΔT_{G} is the depression of the integrated temperature, resulting from the temperature gradient, G; ΔT_{S} is the maximum constitutional supercooling ahead of the interface (i.e., temperature difference between the equilibrium liquidus and the actual temperature at $x = \delta$) and ΔT_{C} is the temperature depression, resulting from the deviation in yttrium concentration at the 211 particle interface from that of the peritectic liquid composition, C_{Lp}. From Fig. 7.5b

$$\Delta T_G = G\delta . \qquad (7.8)$$

The yttrium concentration in equilibrium with the 123 interface (at $x = 0$, $T = T_{123}$) is C_{L}^{123}, and that at $x = \delta$ and $T = T_{123} + \Delta T_G$ in equilibrium with the 211 particles is $C_{\mathrm{L},211}^{0}$. The dependence of the yttrium distribution for $x > \delta$ is given by the equilibrium 211 liquidus. Equating yttrium solvent rejected from the growing 123 front to that diffusing into the liquid gives

[1] If the particles are sufficiently small, this radius of curvature effect could be a significant, even a major, defining driving force of mass transport, leading to 123 phase solidification [477].

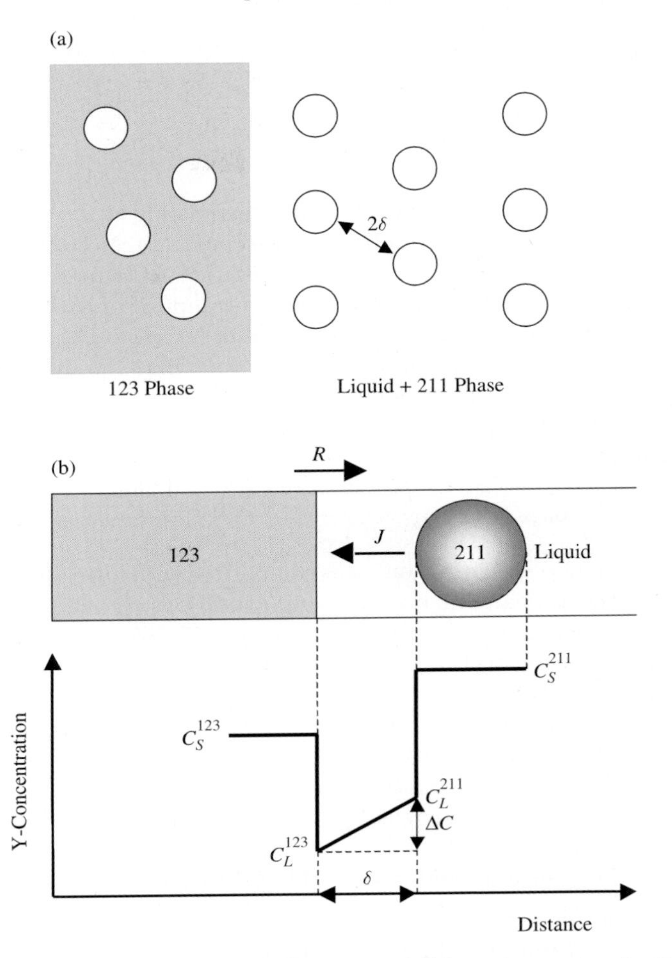

Fig. 7.5. (**a**) Model of peritectic solidification and (**b**) yttrium concentration profile near solidification front of 123 phase due to the dissolution of 211 particles. J is the flux of yttrium ions, which is necessary for growth of the 123 phase

$$-R(C_{\mathrm{S}}^{123} - C_{\mathrm{L}}^{123}) = -D_{\mathrm{L}} \left(\frac{C_{\mathrm{L},211}^{0} - C_{\mathrm{L}}^{123}}{\delta} \right) , \tag{7.9}$$

where R is the growth rate of the solidification front. A relation between interface undercooling (ΔT_{p}) and growth rate, R, can be derived, assuming $C_{\mathrm{S}}^{123} - C_{\mathrm{L}}^{123} \approx C_{\mathrm{S}}^{123} - C_{\mathrm{Lp}}$. Moreover, linear liquidus lines, shown in Fig. 7.4, result in $C_{\mathrm{L}}^{123} = C_{\mathrm{Lp}} - \Delta T_{\mathrm{p}}/m_{\mathrm{L}}^{123}$, and $C_{\mathrm{L},211}^{0} = C_{\mathrm{Lp}} - (\Delta T_{\mathrm{p}} - \Delta T_G)/m_{\mathrm{L}}^{211}$. Substitution of these expressions in (7.9) and making use of (7.8) gives

$$\Delta T_{\mathrm{p}} = \left[\frac{R\delta}{D_{\mathrm{L}}} (C_{\mathrm{S}}^{123} - C_{\mathrm{Lp}}) - \frac{G\delta}{m_{\mathrm{L}}^{211}} \right] \frac{m_{\mathrm{L}}^{123} m_{\mathrm{L}}^{211}}{m_{\mathrm{L}}^{211} - m_{\mathrm{L}}^{123}} . \tag{7.10}$$

Thus, ΔT_p is linearly proportional to R, when G is sufficiently small. Similarly, the next relation can be written between ΔT_S and R:

$$\Delta T_\mathrm{S} = m_\mathrm{L}^{123} \frac{R\delta}{D_\mathrm{L}} (C_\mathrm{S}^{123} - C_\mathrm{Lp}) - G\delta \,. \tag{7.11}$$

The quantity ΔT_S is the constitutional supercooling at $x = \delta$, and it is the maximum constitutional supercooling in the semisolid region. In solidification of this type, there must always be finite constitutional supercooling in front of the growing interface in order to create the compositional driving force for diffusion from the particle surface to the crystal interface. Assuming an existence of maximum undercooling [155] and noting that the distance between the solidification front and considered 211 particle, $z = \delta$, we obtain from (7.11) a relation defining the maximum growth rate, R_max, compatible with a plane solidification front:

$$R_\mathrm{max} = \frac{D_\mathrm{L}}{z(C_\mathrm{S}^{123} - C_\mathrm{Lp})} \left[\frac{(\Delta T_\mathrm{S})_\mathrm{max} + Gz}{m_\mathrm{L}^{123}} \right] \,. \tag{7.12}$$

Thus, the highest solidification rates compatible with a steady-state growth are obtained theoretically at the following conditions: (i) with larger undercooling, (ii) with higher temperature gradient and (iii) with smaller size of 211 particles after peritectic decomposition (z is directly dependent on the particle size). Equation (7.12) shows that the maximum solidification rate is inversely proportional to the distance z between the 211 particles, resulting from the peritectic decomposition of 123 phase. This suggests a decrease in the 211 particle size and the addition of very fine 211 powder to the starting 123 precursor, because an increase of the volume ratio of the 211 phase in the melt for a given particle size results in a reduction of the interparticle spacing z. In fact, such an addition of 211 to 123 offers multiple advantages:

- It allows the geometric shape of the sample to be kept at high temperature owing to the formation of a solid skeleton.
- Diffusion of yttrium favors the supply of this species at solidification.
- It compensates the loss of yttrium in the recombination process, where entrapment of 211 particles takes place.
- Addition of properitectic 211 dispersion refines the grain size of the 211 precipitates, coming from the peritectic decomposition of the 123 phase (properitectic 211 particles act as nucleation centers).
- It improves the mechanical properties of the textured 123 sample.
- The 211 inclusions, in excess trapped in the textured 123 material, act as pinning centers, increasing the critical current.

Compare the terms $(\Delta T_\mathrm{S})_\mathrm{max}$ and Gz, taking into account that in a classical furnace $z \approx 1\,\mu\mathrm{m}$ and $G < 100\,\mathrm{K/cm}$ [190]. Then, it is obvious that $Gz << (\Delta T_\mathrm{S})_\mathrm{max}$ and the solidification rate appears practically insensitive to the temperature gradient and is proportional to the undercooling, ΔT_S. It should be noted that the case of YBCO system, where the slopes of the liquidus curves are quite different for 211 and 123 phases, leading to relatively large driving forces, appears to be favorable for sample texturing.

The directed observations of 123 phase solidification using an IR camera have found a rate of solidification close to 10^{-7} m/s and estimated the diffusion coefficient of yttrium in the liquid by the value 6×10^{-11} m^2/s [310]. From a practical point of view this solidification rate appears to be quite low. From (7.12), it is governed by two parameters: ΔT_{S} and z. During a slow cooling rate, r, in a thermal gradient, G, because of the translation, x, of the "liquid-solid" interface in this gradient the undercooling is given vs time as [190]

$$\Delta T_{\mathrm{S}} = \Delta T_{\mathrm{S0}} + rt - Gx \,. \tag{7.13}$$

This equation shows that solidification at a constant temperature ($r = 0$) is possible but is limited by the thermal gradient. In this case, only a large initial undercooling, ΔT_{S0}, can lead to relatively large superconducting crystals (monodomains).

7.1.3 Models Based on Interface Phenomena

The models [156, 475, 741] considered in the previous section seem in good agreement with test results, but they do not take into account the interface kinetics processes, assuming the crystallization of a pure 123 phase even though some 211 inclusions are always entrapped in the textured 123 material. Moreover, addition of properitectic 211 dispersion to 123 phase can provoke yttrium supersaturation at the solidification interface. Then, the diffusion rate of yttrium is rapid compared with the propagation of the solidification front, so the diffusion of yttrium is no longer the limiting factor for the growth of 123 phase. Therefore, it is followed to introduce in consideration interface kinetics phenomena. The supersaturation can be presented as

$$\sigma = \frac{C_{\mathrm{I}} - C_{\mathrm{L}}^{123}}{C_{\mathrm{L}}^{123}} \,, \tag{7.14}$$

where C_{I} is the yttrium concentration at the "123 phase-liquid" interface and C_{L}^{123} is the yttrium concentration at equilibrium. Three cases can be distinguished (see Fig. 7.6): (i) the growth rate of 123 phase is governed by the yttrium diffusion (curve c); (ii) the growth rate of 123 phase is under mixed control conditions (curve b) and (iii) the growth rate of 123 phase is governed by the interface reaction (curve a).

A model taking into account the interface phenomena [759] introduces two rates of solidification (in the ab-plane and in the c-axis direction): $R_{ab} = k_a \sigma^2$ and $R_c = k_c \sigma$, where k_a and k_c are the kinetic coefficients for the a- (or b-) and c-directions, respectively. The effect of the kinetic coefficient on the growth rate as a function of the undercooling can be taken into account, using either square power law dependence or linear one on the supersaturation [759]. As followed in Fig. 7.7, a linear dependence vs supersaturation leads to

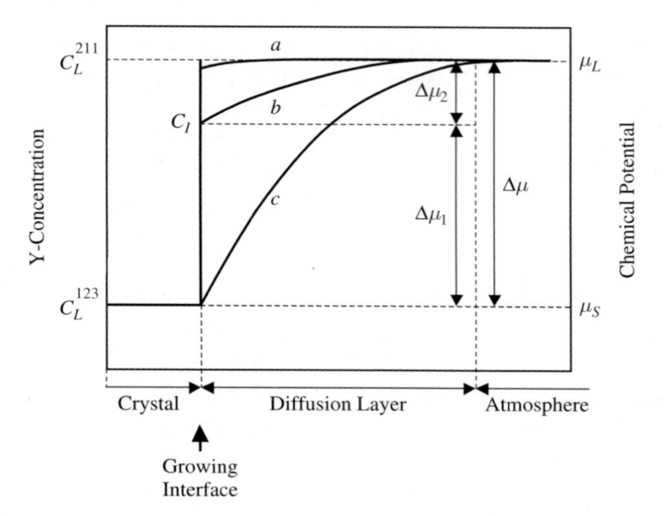

Fig. 7.6. Yttrium concentration profiles in front of the growing interface [759] and differences in chemical potential in the diffusion layer: (**a**) kinetics governed by interface phenomena, (**b**) kinetics governed by both diffusion and interface phenomena and (**c**) kinetics governed by diffusion

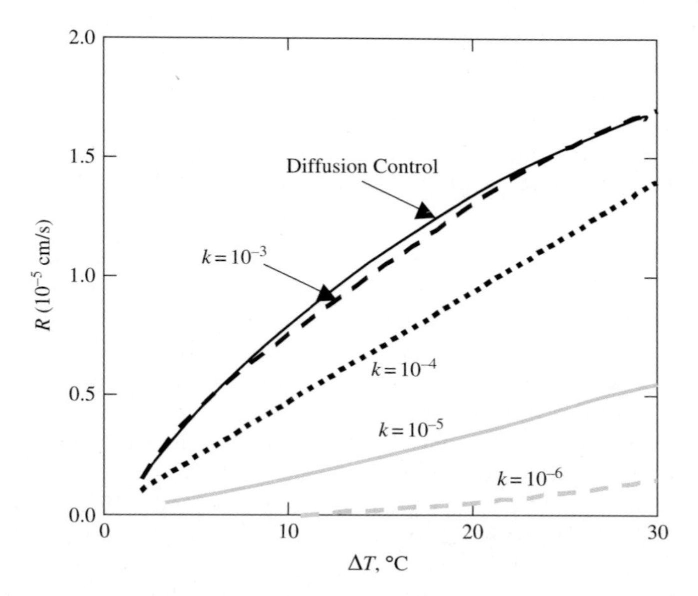

Fig. 7.7. Effect of kinetic coefficient, k, on growth rate, R, as a function of the undercooling in the case of a linear dependence on the supersaturation [759]

a quasi-linear dependence of the growth rate vs supercooling. On the contrary, in the case of a square power law dependence vs supersaturation, two behaviors are possible, depending on the k value: for $k < 10^{-4}$ cm/s the growth rate would show a square power law dependence vs supersaturation, and for $k > 10^{-3}$ cm/s the growth rate would demonstrate a linear dependence vs undercooling, irrespective of the square law dependence of the growth rate vs supersaturation.

In the case of pure 123 phase, a good agreement with experiment [759] is observed for R_{ab} and R_c vs the undercooling. Hence, one can conclude that growth of 123 phase is generally controlled simultaneously by yttrium diffusion and interface kinetics processes. If a 211 properitectic phase is added to 123 phase before the melt process, for a given undercooling the larger amount of 211 phase can supply more yttrium diffusion to the growing front, corresponding to the growth rate of solidification process. At the same time, the change from a quadratic dependence of R_{ab} vs supersaturation to a linear dependence (and vice versa for R_c), observed in [759], is indeed unexpected if one assumes that the yttrium flux reaching the interface is proportional to the 211 phase concentration. Obviously, this means that the growth mechanisms in the considered directions are different for both compositions, in spite of identical growth conditions. Thus, one can modify the growth rate and its anisotropy, changing solidification process and superconductor composition.

The particle motion near an advancing solidification front can be described in terms of an interfacial energy relationship between a particle, solid and liquid in the framework of the pushing/trapping mechanism of the particle by the solidification front (see Fig. 7.8) [1092]:

$$\sigma_{\mathrm{PS}} = \sigma_{\mathrm{PL}} + \sigma_{\mathrm{SL}} , \tag{7.15}$$

where σ_{PS}, σ_{PL} and σ_{SL} are particle/solid, particle/liquid and solid/liquid interfacial free energies, respectively. According to this criterion, a particle present at a growing solid phase is trapped when $\sigma_{\mathrm{PS}} < \sigma_{\mathrm{PL}} + \sigma_{\mathrm{SL}}$. In contrast,

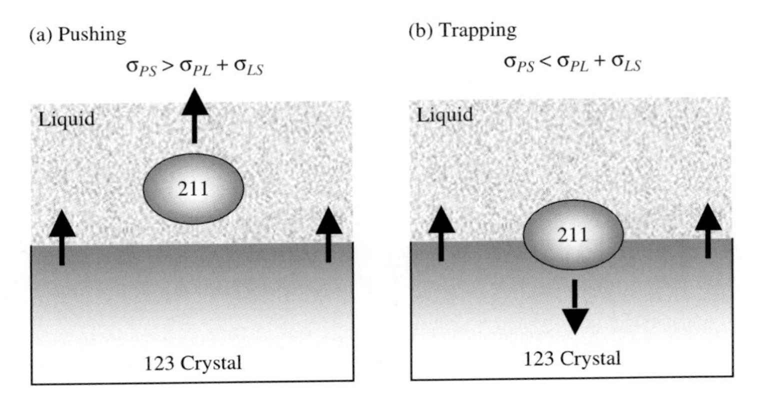

Fig. 7.8. Schematic of interfacial energy criteria on particle pushing and trapping

when $\sigma_{PS} > \sigma_{PL} + \sigma_{SL}$, a particle is pushed out from a growing solid by the fast diffusion of the liquid.

This interfacial energy criterion has been applied to melt-processed YBCO samples in order to explain trapping of 211 particles within 123 crystals [475, 532, 534, 540]. However, this is not enough because it assumes the interaction of the front with only one particle pushed out or trapped by the interface. Moreover, the above criterion should be modified for anisotropically growing crystals of melt-processed YBCO samples. The modified interfacial energy relationship is given as [534]

$$\sigma_{PS(hkl)} = \sigma_{PL} + \sigma_{S(hkl)L} \ , \tag{7.16}$$

where $\sigma_{PS(hkl)}$, σ_{PL} and $\sigma_{S(hkl)L}$ are particle/(hkl) interface, particle/liquid and (hkl) interface/liquid interfacial free energies, respectively. Then, in the isotropically growing system or directionally growing interface, 211 particles will be pushed out or trapped within a 123 domain with a random mode (see Fig. 7.9). In this case, for an anisotropically growing 123 crystal, trapping criteria for each growing plane are different.

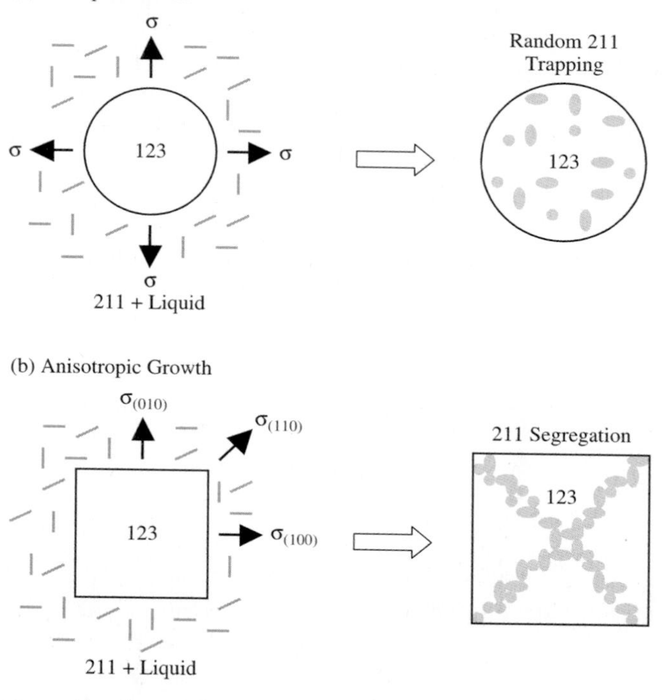

Fig. 7.9. Schematic of particle trapping modes in isotropically (**a**) and anisotropically (**b**) growing systems

The crystal structure of 123 domains at the peritectic temperature has a tetragonal symmetry. Therefore, there are three main growth interfaces of (100), (010) and (001) planes (see Fig. 7.10). In the tetragonal crystal structure, the atomic arrangement of a (100) plane coincides with that of the (010) plane. But the atomic arrangement of a (001) plane is different from that of the (100)/(010) plane. The boundary condition of particle trapping is inferred from the interfacial energy difference among growing (*hkl*) planes. The criteria for 211 particle trapping in a melt-processed YBCO system are given as [530]

$$\Delta\sigma_{(hkl)} = \sigma_{\mathrm{PS}(hkl)} - (\sigma_{\mathrm{PL}} + \sigma_{\mathrm{S}(hkl)\mathrm{L}}); \tag{7.17}$$

$$\Delta\sigma_{(100)} = \Delta\sigma_{(010)} \neq \Delta\sigma_{(001)}; \tag{7.18}$$

$$\Delta\sigma_{(101)} = \Delta\sigma_{(011)} \neq \Delta\sigma_{(110)} \neq \Delta\sigma_{(111)} \,. \tag{7.19}$$

From the interfacial energy model and experimental observations of 211 patterns [481, 532, 534, 1105], it is followed that the crystallographic planes of X-like 211 tracks are the six diagonal planes of 123 domain, which divide the cubic space into three pairs of pyramids (see Fig. 7.11). These diagonal planes may vary with the aspect ratio of grown 123 domains. For example, there may be [110] planes for a cubic domain or [103] planes for an orthorhombic domain [534] (see Fig. 7.12). In the case of planar segregation of 211 phase, a pair of pyramids having (001) growth fronts corresponds to 211-free spaces, while the other two pairs of pyramids having (100) and (010) growth fronts are filled with 211 particles.

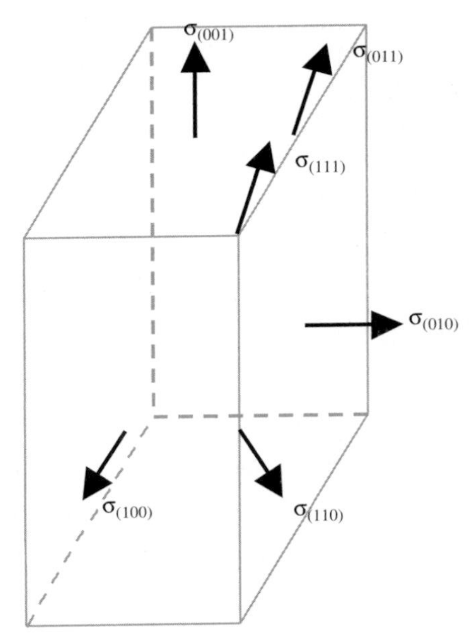

Fig. 7.10. Boundary planes for 211 particle trapping, based on an interfacial energy model

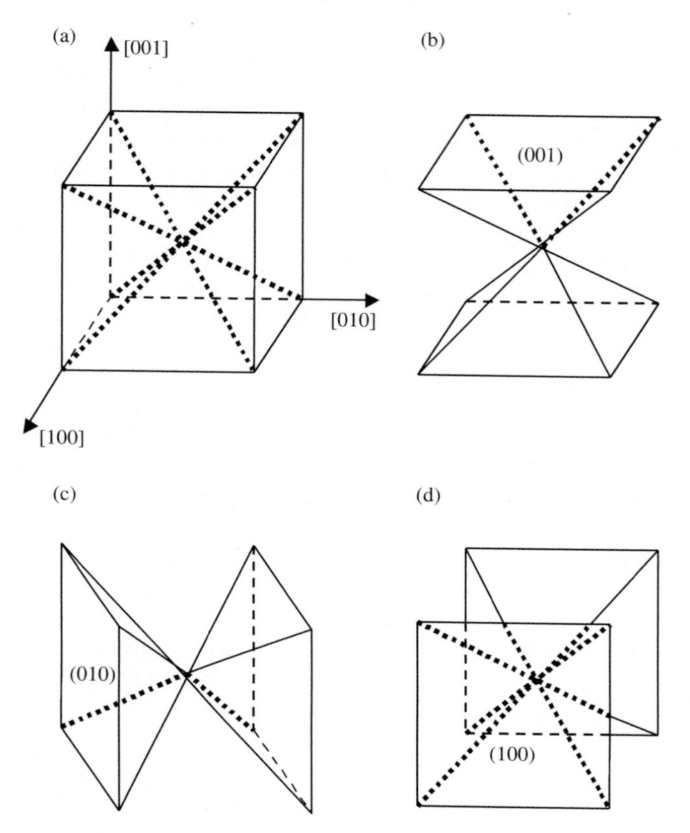

Fig. 7.11. (**a**) Three-dimensional demonstration of the planes, where 211 particles are trapped in a stoichiometric YBCO system, (**b**) the 211-free spaces and (**c, d**) the 211-filled spaces in a 211 phase excess system

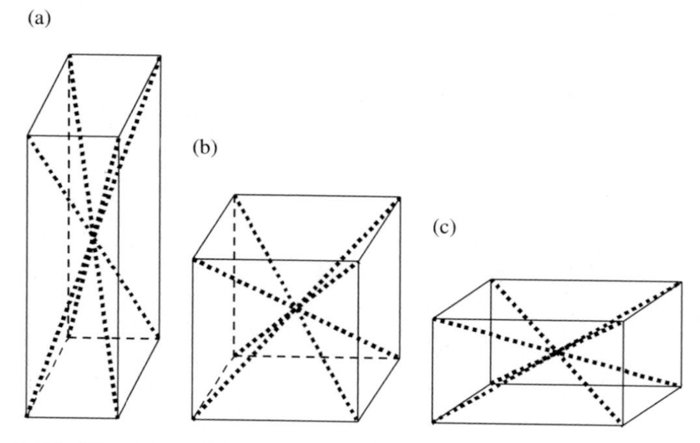

Fig. 7.12. Variation of diagonal planes as a function of grain anisotropy

The crystallographic alignment of 211 particles may also be explained in terms of an interfacial energy relationship between the (*uvw*) plane of a 211 particle, (*hkl*) plane of interface and liquid [530]. The corresponding criterion is given as

$$\sigma_{\mathrm{P}(uvw)\mathrm{S}(hkl)} = \sigma_{\mathrm{P}(uvw)\mathrm{L}} + \sigma_{\mathrm{S}(hkl)\mathrm{L}} \,, \tag{7.20}$$

where $\sigma_{\mathrm{P}(uvw)}$ is an interfacial energy of the (*uvw*) plane of a 211 particle. In this case, these particles have a polygonal shape with faceted surfaces with interfacial anisotropy.

An alternative way to express the interfacial energy criteria is the liquid wetting angle (ϕ) at advancing 123 interfaces. The trapping criterion of the 211 particle can be expressed with the liquid wetting angle and has been proposed for a melt-infiltrated YBCO system [481]. As schematically shown in Fig. 7.13, two different situations can exist for particle trapping. In the case of $\phi = 0°$, a 211 particle should be pushed toward the liquid, because the liquid film is present between the 211 particle and the growth front. In contrast, when $\phi > 0°$ and the 211 particle is not dissolved completely in the liquid, the particle is expected to be easily trapped within the 123 domain.

7.1.4 Models of Platelets-Like Growth of 123 Phase

The model of platelets-like growth of 123 phase, which is more faster in the *ab*-plane compared with the *c*-axis direction in the limit of single-crystalline material and taking into account interaction with 211 particle, is presented in [10]. The 211 particle may contribute to the gap-formation process, when the growth front bipasses this particle, and the 123 material does not envelop the 211 particle completely by growing at nearly identical rates on either side of the particle (see Fig. 7.14d). In this case, the resultant microstructure consists

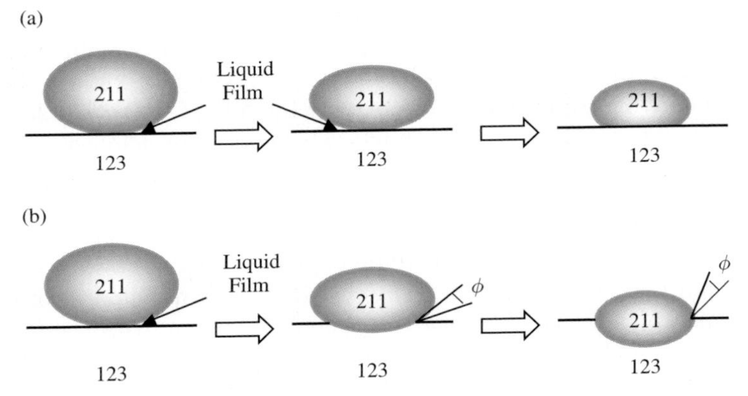

Fig. 7.13. Schematic of two different cases for 211 particle trapping within growing 123 interface [481]. Dihedral angle, ϕ, between 211 and 123 phases is (**a**) $\phi = 0°$ and (**b**) $\phi > 0°$

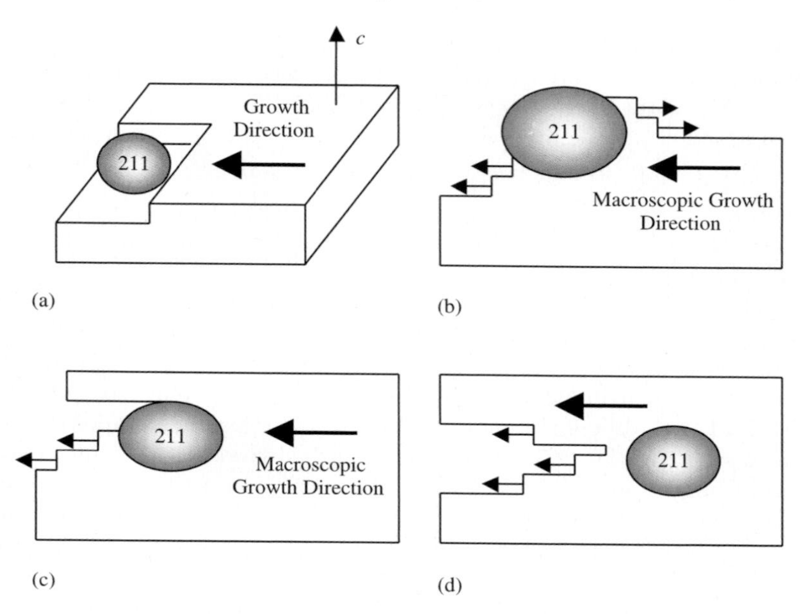

Fig. 7.14. Schematic of possible effect of 211 particles on the growth of melt-processed YBCO crystal: (**a**) 123 platelet abuts a growing platelet; (**b**) heterogeneous nucleation occurs at the "platelet–211 particle–liquid" junction; (**c**) once sufficient growth along the c direction occurs such that the 211 particle can be bypassed, rapid lateral growth will occur; (**d**) this process results in a gap, which may not heal completely [10]

of platelets of identical orientation separated by gaps, which are filled in with the rejected liquid.

The generalization of this model [334] describes the formation of linear and plane gaps between 123 platelets and rejection of liquid at domain boundaries in the case of 211 particle trapping by 123 matrix. As is well known, there is a potential possibility of formation of not only the $YBa_2Cu_3O_{7-x}$ (123) phase, but the $BaCuO_2$ (011) phase from liquid, which does not demonstrate superconducting properties and has solidification temperature about 1015°C that is near to start formation of the 123 phase [34]. Thus, solidification of either 123 phase or 011 phase is possible [317]. The situation, connected with faster solidification of barium cuprate, may be explained on the basis of the phase diagram and taking into account the above-described mechanism of the 123 phase formation [10, 334]. The relatively small (compared with the process of solidification of the 011 phase) rate of establishment of equilibrium between liquid (L) and solid 123 phase can lead to thermodynamic equilibrium in each of these components. The liquid will demonstrate independence because of low concentration of yttrium that is caused by low rate of change between the solid and liquid phases. In this case, a homogeneous formation of 123 nuclei is impossible, and the figurative point, describing melt, displaces into $L + BaCuO_2 + CuO$ region, resulting in the solidification of barium cuprate. There are two causes for breaking of usual equilibrium between liquid and solid 123 phase: (i) a growth of the 211 particles (due to their different sizes and well-known Ostwald ripening), reducing interphase interface with melt, decreasing yttrium flux into liquid, leading to envelopment of these particles by solidification front with rejection of the yttrium-depleted melt at the intergranular boundaries and finally initiating a de-lamination of the system; (ii) a decreasing of the thermal stability of 123 nuclei in the liquid, leading to absence of dominating 123 solidification even at sufficient yttrium concentration [318]. These causes change the stoichiometry of liquid in the case of trapping of large 211 particles [334], and even solidification of pure 011 phase [318].

The development of the model [475] for the case of low G/R ratio is carried out in [936]. The microstructure close to the quenched solid–liquid interface exhibits bridges of 123 material between the solidifying 123 interface and 211 particles. In order to describe their morphologies, a combination of both phenomena, namely a peritectic reaction being mediated by the liquid and a peritectic transformation of the 211 particles, being linked to the solidification front via bridges of 123 phase, is necessary. Note that all the above models neglect peritectic transformation [494]. The principal difference of the model [936] from the models [156, 475, 741] is the account of influence on the process of the 123 phase formation of Lifschitz–Zlyozov boundary effects [291, 629, 997, 1090], and action of the capillary attraction forces between moving front of the 123 phase solidification and 211 particles into liquid.

The entire *local* process of the engulfment of 211 particles into the solidifying interface can be explained, considering four steps: (i) liquid-phase diffusion-controlled growth, following the temperature gradient, if the 211 particle is far away from the phase boundary (see Fig. 7.15a); (ii) bridge formation, when the 123 interface faces an increased Y^{3+} concentration gradient, when being approached by a 211 particle (see Fig. 7.15b); (iii) peritectic surface reaction during the engulfment process (see Fig. 7.15c); and (iv) peritectic transformation (negligible effect compared to previous steps) (see Fig. 7.15d).

Due to the peritectic character of the 123 phase creation, this phase needs a Y-concentration that is not provided by the melt, being in equilibrium with the 211 phase. Therefore, as in classical nucleation theory, a depletion zone arises and the growth of the 123 phase is driven by a concentration gradient in the depletion zone, δ, close to the 123 interface. At the same time, the dissolving 211 particles maintain a medium yttrium concentration in liquid, c_m, corresponding to the Ostwald ripening theory.

Bridge formation starts when the depletion zone ahead of the 123 phase boundary and dissolution region of the 211 particle begin to overlap (see Fig. 7.15b). The increased concentration gradient leads to an accelerated growth of the 123 phase toward the 211 particles, resulting in a bridge.

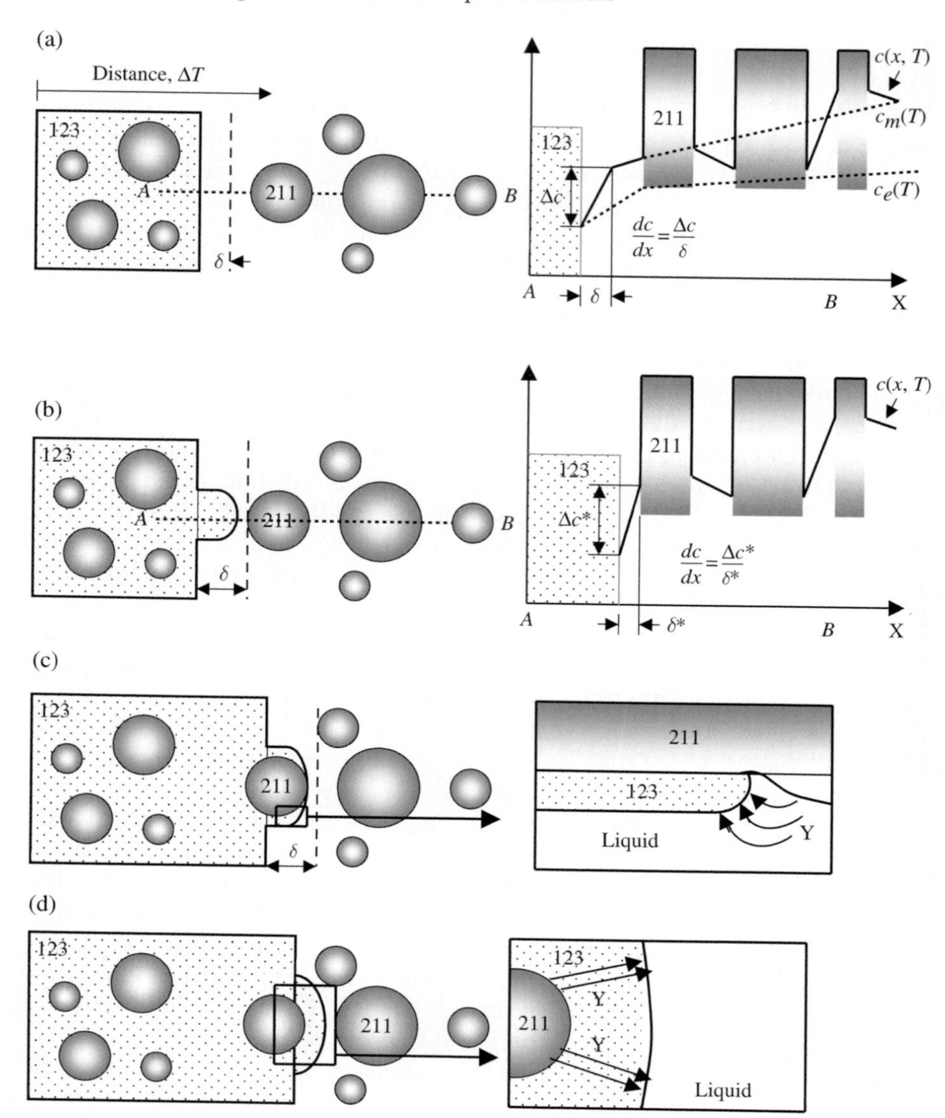

Fig. 7.15. Model of *local* influence of 211 particles on the growth morphology of 123 phase. The parameter $c(x, T)$ corresponds to the *local* yttrium concentration, c_e represents the equilibrium solubility, and c_m the mean concentration, corresponding to the Ostwald ripening theory; $\mathrm{d}c/\mathrm{d}x$ denotes the concentration gradient at the 123 interface. (**a**) Liquid diffusion-controlled growth, (**b**) bridge formation, (**c**) peritectic reaction and (**d**) peritectic transformation [936]

The growing bridge, reaching the 211 interface, defines the start of the peritectic reaction (in its original sense) [422], which subsequently covers the surface of the 211 particle with solid 123 material (see Fig. 7.15c). Once the 211 surface is covered, any further formation of 123 phase is governed by peritectic transformation, that is, strongly limited by diffusion in the solid and negligible (Fig. 7.15d).

As example of the above model, we explain the 1:1 correlation between 211 particle size and thickness of the 123 platelets, observed in [488]. A possible explanation is based on two assumptions: (i) the plate-like growth of the 123 phase is caused by the strong anisotropic growth rates [10], $\nu_{ab} >> \nu_c$, and (ii) the engulfment of the 211 particles is provided by the growing 123 matrix. The anisotropy of the growth rates leads to a preferred growth of the *ab*-planes parallel to the temperature gradient. At the same time, the low growth rate along the *c*-axis direction results in a morphological instability, leading to residual melt enclosed in planar defects between the *ab*-platelets. The platelet A closest to the 211 particle (see Fig. 7.16a) is the first to be influenced by the spherical diffusion region of the 211 particle and therefore will form a bridge as explained above. Next growth in all (a, b and c) directions of this platelet A is governed by the peritectic reaction, occurring along the 211 particle surface. In particular, the *c*-axis growth of platelet A is in competition with the fast *ab*-plane growth of the adjacent platelet B (see Fig. 7.16b). While the growth of platelet B is additionally accelerated, when approaching the 211 particle, the *c*-axis growth of platelet A is more and more decelerated because supply of yttrium is more and more hindered by the growing peritectic reaction layer. Considering the outer platelets C, note there is only competitive growth with one platelet (e.g., platelet B), giving rise to enhanced growth as well as being parallel to the *ab*-planes and the *c*-axis. This leads to an enlarged platelet thickness of the outer platelets after passing the particle (see Fig. 7.16c and *d*). This process automatically results after some iterations in a platelet thickness, corresponding to the mean particle diameter, which is supported by experimental observation presented in Fig. 7.17b. Once the thickness is reached, further growth of the 123 phase will be determined either by two platelets, each passing one side of the 211 particle (see Fig. 7.18a) or one platelet passing the whole particle, respectively (see Fig. 7.18b). Note that even a hypothetic planar interface would change to a cellular morphology, when interacting with 211 particles, also yielding platelet dimensions, according to the 211 particle diameter (see Fig. 7.19). The zipper-like mechanism, acting in this process, results in an oriented growth of multiple connected platelets to a quasi-single crystal. Thus, the considered mechanism of solidification explains both a very high rate of the 123 phase creation and an existence of sufficiently sharp boundary between 211 inclusions and 123 matrix [741]. The models [10, 936] also explain a coincidence of 211 particle sizes with thickness of 123 platelets, observed in [488].

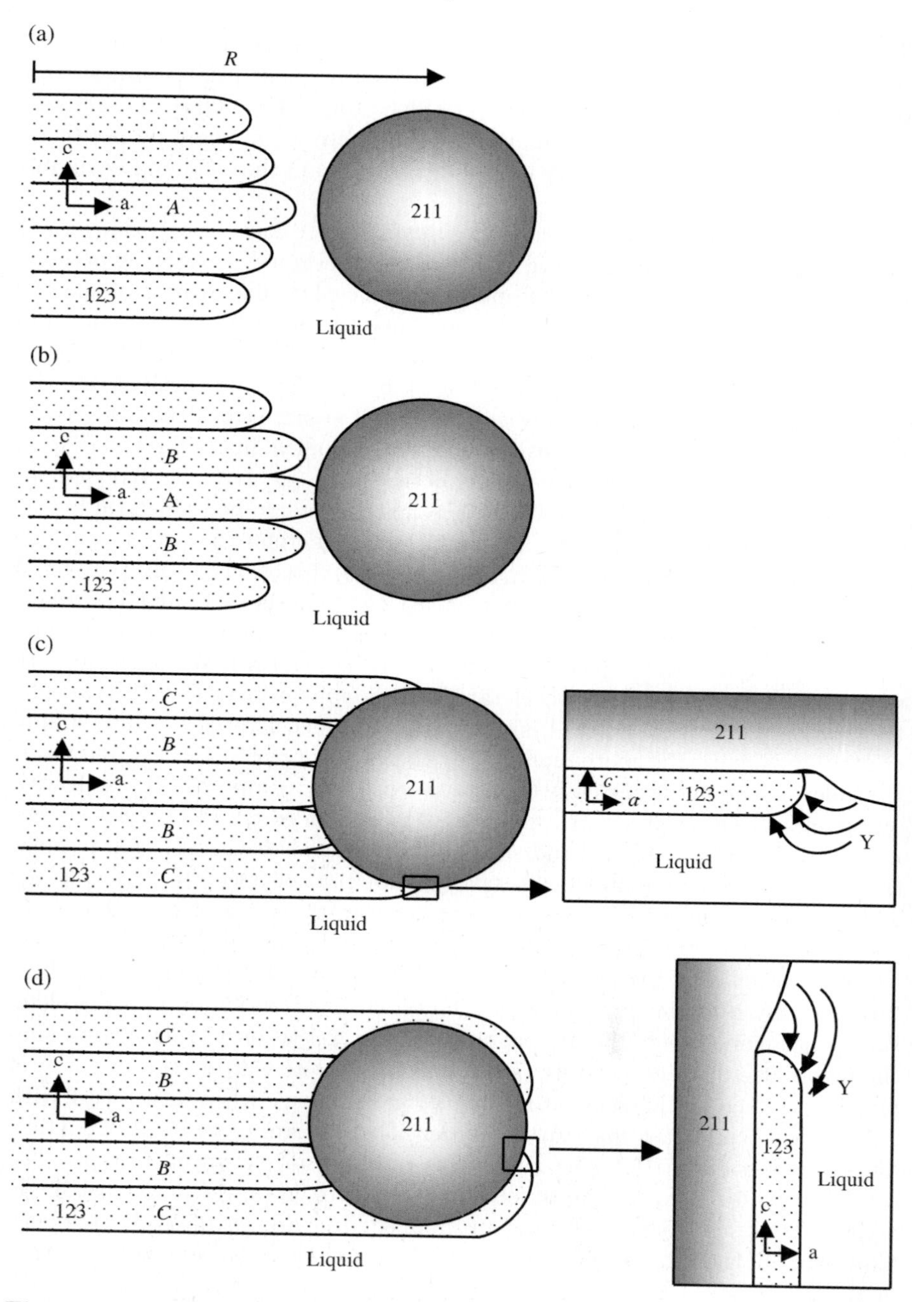

Fig. 7.16. Competing growth of 123 platelets near a 211 particle, leading to quasi-single crystalline material via a zipper-like mechanism. (**a**) The platelet closest to the 211 particle forms a bridge, (**b**) the c-axis growth of platelet A is in competition with the fast ab-plane growth of the adjacent platelet B, (**c, d**) explain enlarged platelet thickness of the outer platelets after passing the particle [936]

(a)

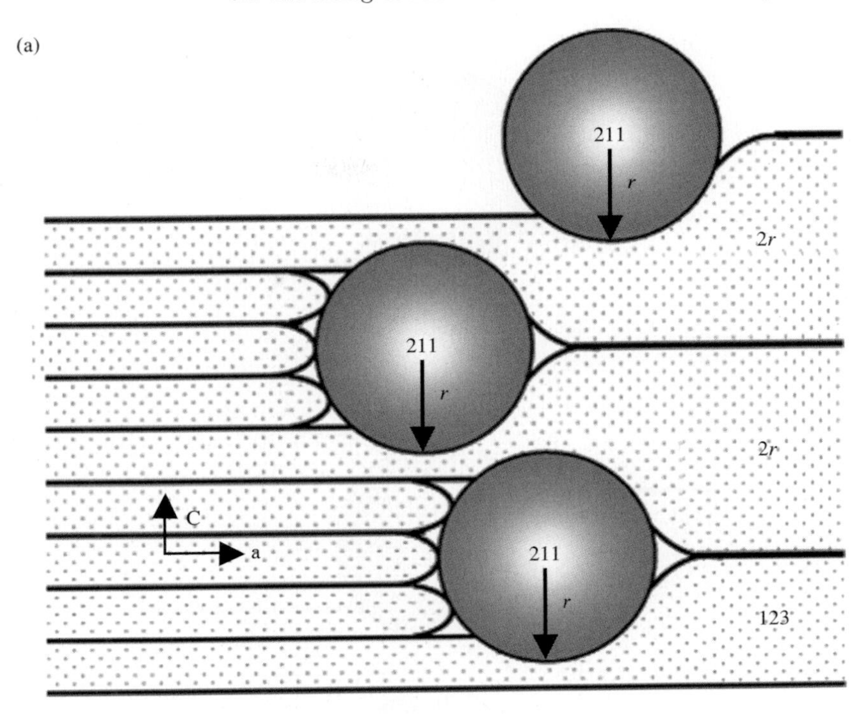

(b)

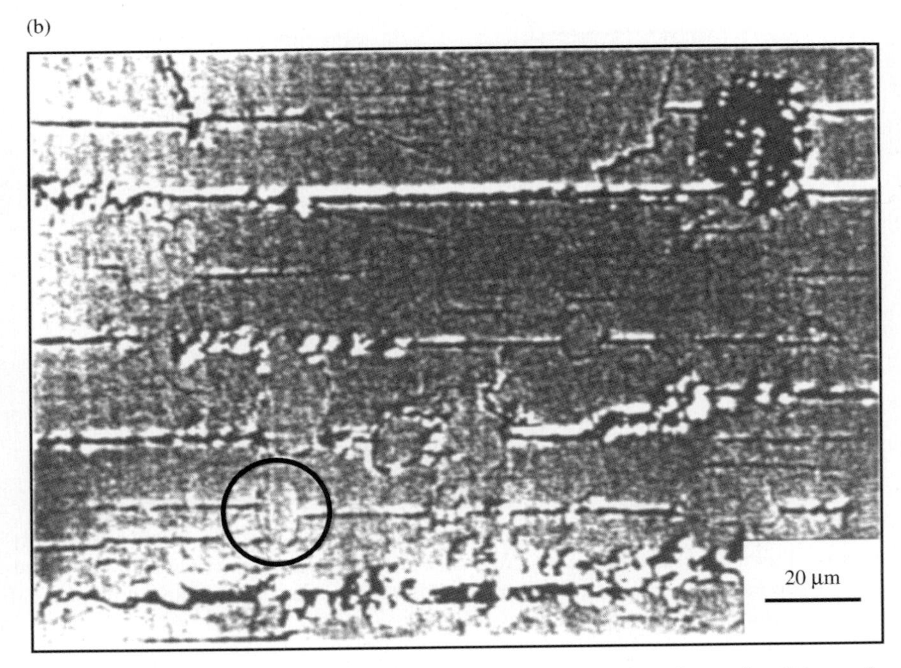

Fig. 7.17. (**a**) A zipper mechanism and (**b**) experimental evidence for this mechanism [936]

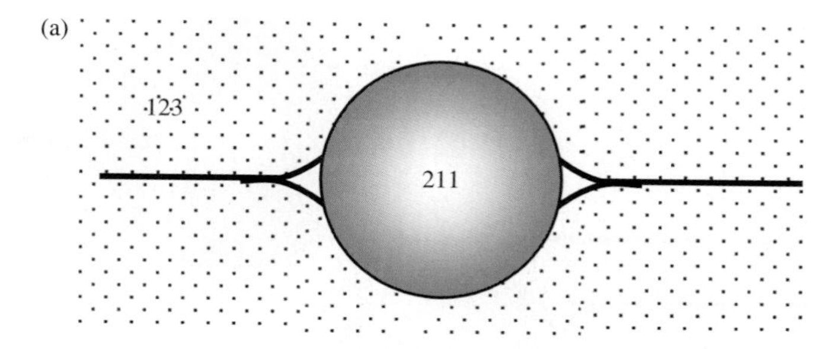

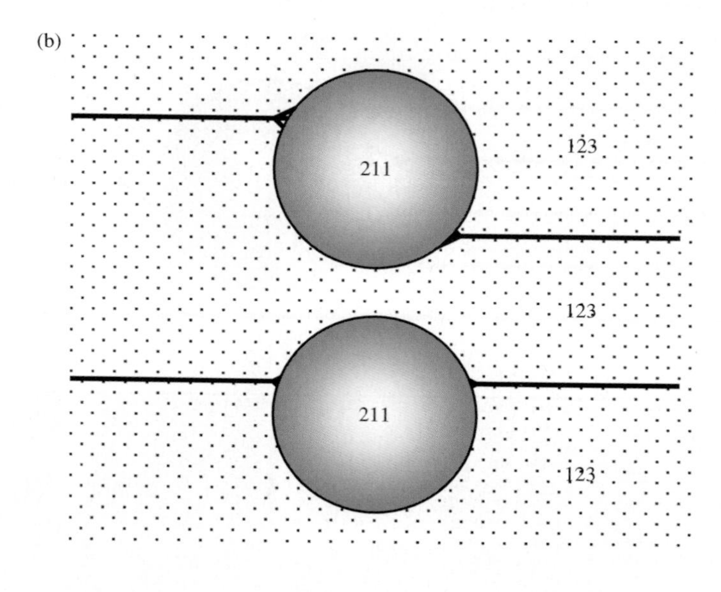

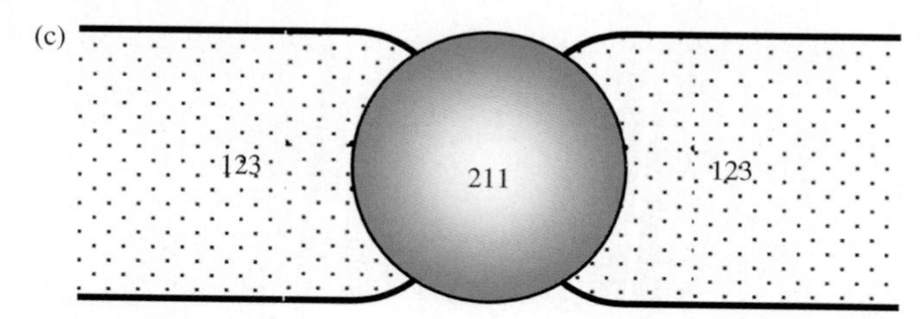

Fig. 7.18. 123 pseudo-grains (platelets), passing a 211 particle of approximately the same dimensions. Predominant growth, when passing the particle is in the paper plane in the case (**a**) and in a plane perpendicular to the paper plane, but parallel to the *ab*-planes, in the cases (**b**) and (**c**) [936]

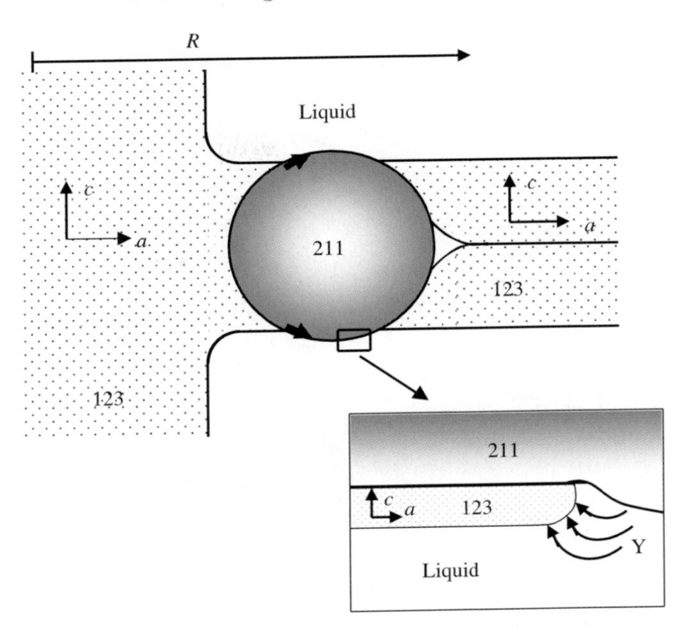

Fig. 7.19. Change in the growth morphology of a planar 123 interface, evoked by a 211 particle [936]

7.1.5 Modeling of Solidification Kinetics

Only thermodynamic and chemical representations cannot totally describe growth processes in the Y(*RE*)BCO compounds. It is also necessary to include in consideration a kinetic process of the 123 phase solidification. As is followed from tests [1104], during the growth processes, the residual phases are pushed and distributed along the growing steps and sometimes are trapped in the 123 matrix, leading to the formation of microcracks. Their shape correlates with the rapid lateral growth that occurs in the *ab*-plane. This can lead to the formation of multi-grain domains as schematically drawn in Fig. 7.20.

Computer simulation of 2D growth kinetics of 123 front near 211 particles in the *ac*- and *ab*-plane is carried out in [1104] on the basis of Eden kinetics model (see Appendix D) for stochastic cell-by-cell growth of compact clusters. On a square lattice, we take a box of size L with periodic boundary conditions in the abscissa direction. The growing crystal is represented by the set of occupied sites on the lattice (see Fig. 7.21). At each step of the growth, the set of empty sites in contact with the growing surface defines the perimeter of the solidification front (the circles show these sites). A site of the perimeter is randomly occupied, following the probability rule described below. This defines a new perimeter configuration, and the same process is repeated.

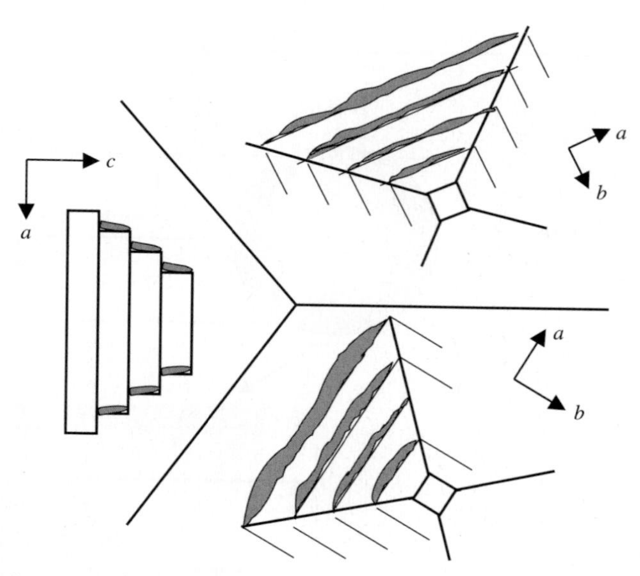

Fig. 7.20. A pyramid-like picture, sketching the formation of a polycrystalline region in the solidification process. Crystallographic orientations are emphasized and residual phase (*gray regions*) along growing steps [1104]

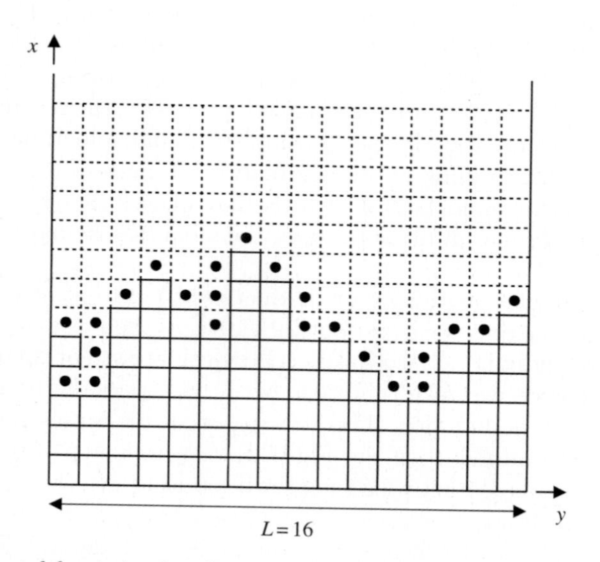

Fig. 7.21. Crystal front simulated using the isotropic Eden model. The *black circles* illustrate the perimeter sites

In such a model, an occupied site can represent a unit cell or a cluster of 123 cells.

It is known that an anisotropic growth rule simulates precipitation processes [689], as well as the growth of cell colonies [52]. In order to simulate an anisotropic growth front in the *ab*- and *ac*-planes, we consider two variants of the Eden model [1104]:
(1) *Model I* simulates the solidification front in the *ac*- or *bc*-plane, considering a square lattice oriented in [100] and [001] directions. At each step of the growth, a growing probability, P, is calculated on each perimeter site, and is given by

$$P \sim \exp(p_a N_a + p_c N_c) \ , \tag{7.21}$$

where N_a and N_c are the number of occupied nearest neighbors (nn) in the a and c directions, respectively; p_a and p_c are the anisotropic growth parameters. Then, the set of these probabilities (defined on the perimeter) is renormalized in the interval [0, 1]. A random number generator chooses the growing site as in a Monte-Carlo simulation. This site is then occupied, defining a new perimeter with the next repetition of the whole process.

Note that the exponential law simulates a curvature effect (or a so-called Gibbs–Thomson effect) [1146]. When $p_a > p_c$, the growth probabilities are more important in empty sites linked to the crystal, following the a-axis direction. It results in a faster growth in the [100] direction than in the [001] direction, simulating an anisotropic solidification process in the *ac*-plane. When $p_a = p_c = 0$, the growth process reduces to a simple Eden's model. When $p_a < p_c$, the growth is trivial: the front remains flat and parallel to the substrate (i.e., to the [001] direction).
(2) *Model II* simulates the solidification front in the *ab*-plane. In order to simulate the [110] as the fast growth direction, the growth probability, P, calculated on each site of a square lattice is given by

$$P \sim \exp(p_{\mathrm{nn}} N_{\mathrm{nn}} + p_{\mathrm{dnn}} N_{\mathrm{dnn}}) \ , \tag{7.22}$$

where N_{nn} and N_{dnn} are the number of occupied nearest neighbor sites (nn) in the [100] or [010] directions and diagonal nearest neighbor sites (dnn) in the [110] directions, respectively; p_{nn} and p_{dnn} are the growth parameters. The interaction with the diagonal nearest neighbors (dnn) is introduced in this model in order to take into account the diagonal fast growth directions in the *ab*-planes. When $p_{\mathrm{nn}} = p_{\mathrm{dnn}} = 0$, the model also reduces to the simple Eden's model.

In both models, the presence of 211 particles can be simulated, avoiding the growth process to be achieved in circle-like regions of the square lattice. Thus, the 211 particles play a passive role, in contrast to, for example, the models [156, 475, 741], in which these particles provide with yttrium solidification front of 123 phase.

The numerical results [1104], obtained on the basis of above models, estimate anisotropic effects of grain growth in the *ab*-plane ($g_{110}/g_{100} \sim 10$)

and in the *ac*-plane ($g_{100}/g_{001} \sim 50$), where g_{hkl} is the growth probability in the crystallographic *hkl* directions. Replacing particles by spins [35], a magnetic field effect can be simulated and the magnetical texture growth processes could be further analyzed [184].

7.1.6 Multi-phase Field Method

In order to minimize the processing time as well as improve texture of YBCO melt-processed bulks, total information about solidification isotherm distribution in time and in space is necessary. This also defines qualitative estimations of a grain growth. The consideration of two-level scaling (in this case the macroscopic study of thermal fields and microscopic grain growth processes) decreases calculation time significantly. The macroscopic simulation of thermal fields in the YBCO samples, sintered into a Bridgman furnace in existence of the thermal insulation layer between the furnace and sample, has been carried out in [956]. In order to simulate YBCO microstructure formation, three different approaches could be used, namely (i) Monte-Carlo technique [800, 819, 822]; (ii) cellular automata models [302]; and (iii) phase field method [113, 1142]. In this section, we consider in more detail the latter method. The phase field conception has been applied to multi-phase system [1016], and also to the microscopic simulation of the 123 platelets growth at the existence of the 211 particles [956].

In the multi-phase field model, describing evolution of interphase boundary, each phase is identified with individual phase field, and phase transformations between any neighbor pairs are considered depending on their own characteristics. The phase field method [142, 1142], based on the Ginzburg-Landau theory of phase transformations [452, 617, 737, 935, 986], was applied to the study of structural phase transformations of the "solid-solid" type in [141]. The theoretical basis was the functional of the local density of free energy, depending on an order parameter of the system and its partial derivatives on the spatial coordinates. The order parameter can be a scalar function, for example, concentration of solid component in the investigation of the phase transformations of the "solid-liquid" type, which changes from 0 (in liquid) to 1 (in solid). [1016] was limited by the parameters, p_i, describing only local phase state of the system. In this case, each phase field is identified either with melt or single 123 grain or 211 particle. Their time evolution is described by the system of non-linear parabolic partial differential equations [1016]:

$$\dot{p}_i = \sum_{k(k \neq i)}^{n} \frac{1}{\tau_{ik}} \left\{ \varepsilon_{ik}^2 (p_k \nabla^2 p_i - p_i \nabla^2 p_k) - \frac{p_i p_k}{2a_{ik}} [p_k - p_i - 2m_{ik}(\Delta T_{ik})] \right\} , \tag{7.23}$$

where the point, as usual, denotes the time derivative; the parameters $\tau_{ik}, \varepsilon_{ik}$, a_{ik} and m_{ik} can be defined through the test values (μ_{ik} is the mobility, σ_{ik} is the surface energy and λ_{ik} is the interface thickness) as [1142]

$$\tau_{ik} = \frac{L_{ik}\lambda_{ik}}{T_{ik}\mu_{ik}}; \tag{7.24}$$

$$\varepsilon_{ik}^2 = \lambda_{ik}\sigma_{ik}; \tag{7.25}$$

$$a_{ik} = \frac{\lambda_{ik}}{72\sigma_{ik}}; \tag{7.26}$$

$$m_{ik} = \frac{6a_{ik}L_{ik}(T_{ik} - T)}{T_{ik}}. \tag{7.27}$$

They calculate through a set of parameters, describing the phase transformation ($i \to k$), namely (i) the equilibrium temperature of the phase transition, T_{ik} (e.g., the melting temperature, T_m, in solidification processes); (ii) the heat release during phase transition, L_{ik} (e.g., the melting heat in solidification problems); (iii) the driving force of the phase transition, m_{ik}, defined by deviation from equilibrium; and (iv) the difference, ΔT_{ik}, between the temperature of local cooling of the phase interface and the temperature, corresponding to a condition for the local equilibrium of i and k phases. These equations are coupled with a diffusion equation, determining the local concentration of yttrium at any position and time. It may by shown that in simulation of the superconducting structure solidification, the solution of constitutive equation for i phase is required to find only near interface boundary with neighbor phase. This set of differential equations is solved by a finite difference technique, using rectangular lattice [1016].

Evolution of single phase, growing by isotropic way (e.g., liquid droplets in gas media), leads during interaction of its structure elements to the formation of 120°-angles in triple points. In the case of multi-phase systems, the triple junctions between different phases can be triple points of the "solid-liquid-gas" type. Then, considerable difference between surface energies for the systems "melt-solid", "solid-gas" and "gas-melt" leads to formation of triple points with angles different from 120°.

Different crystallographic directions cause anisotropic behavior on account of crystallographic structure. The investigation of phase transition from isotropic melt to anisotropic solid leads to corresponding change of equations for phase fields. This is attained by either introduction of kinetic factor or diffusion coefficient for phase field depending on the orientation of phase boundary to the crystallographic directions of neighbor grains or phases. The multi-phase conception may also be used to consider different grains with various spatial orientations, belonging to the same phase. With this aim in [1016], all orientations are divided into 10 orientation classes (the Pott's model), and each class is identified with own order parameter. Two- and three-dimensional computed simulation of grain growth with different crystallographic orientations has shown a dominant grain growth along the direction of the solidification front propagation.

The couple consideration of the phase field equations for thermal and solubility fields allows the process of microstructure formation to be modeled in actual peritectic systems (e.g., YBCO) [604]. The computer simulation results show coupling of the peritectic phase growth with a solubility of the properitectic phase. Obviously, the properitectic particles dissolve in liquid phase, according to their diameters (or surface curvature), increasing local concentration of yttrium in surrounding melt. As has been shown above, a microstructure of forming superconductor does not have not flat front, but creates bridges between solidification front and properitectic particles [936]. When the 211 particles are sufficiently great, the time for their total dissolution before interaction with 123 front is insufficient. As a result, they are trapped by the 123 matrix and can serve as pinning centers (see Fig. 7.22) [956]. Based on the phase field conception [1016], the computer simulation of a single particle in a population of 211 particles ahead of a growing 123 front reveals a "virtual pushing" of the particle, displaying in its initial coarsening before gradual and final dissolution (see Fig. 7.23) [938]. The computer results show that a dissolution/re-precipitation mechanism at least contributes to the displacement of the 211 particles ahead of a growing 123 front.

Another example of computer simulation [937, 956] relates to the geometrically caused grain growth. In this case, the grain selection is found by the conditions of their anisotropic growth. This example may be useful for processing optimization of superconducting tapes. Moreover, the effects of nuclei grains sizes can be investigated in this way. Computer simulation of 211 particles, defining superconducting properties of YBCO and forming X-like precipitation, has been carried out in [937]. This yttrium enrichment of the 123 grain diagonals leads to conclusion observed in experiment that the

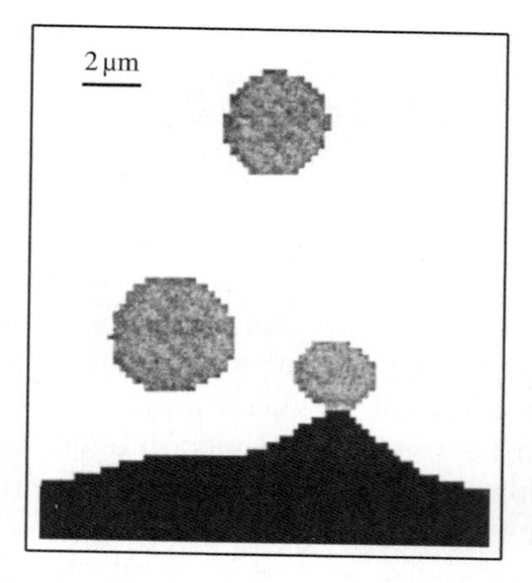

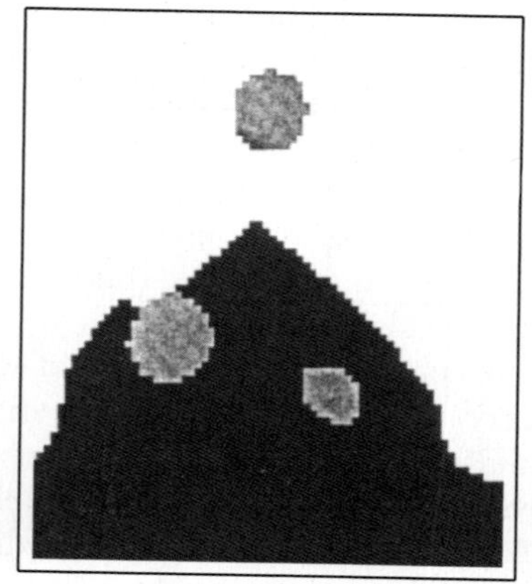

Fig. 7.22. The microscopic simulation results, defining engulfment of 211 particles and bridge formation [956]

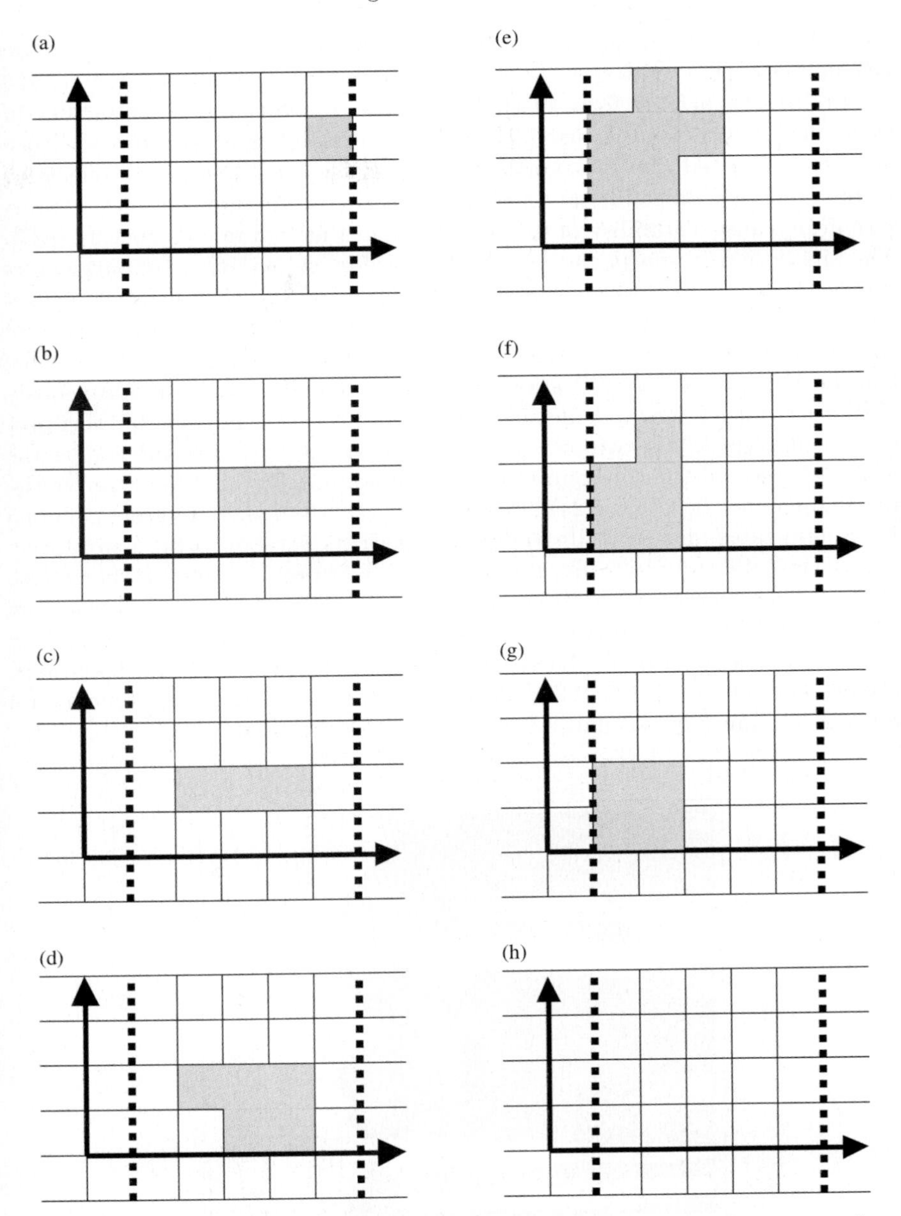

Fig. 7.23. 211 particle ahead of a peritectic solidification front (the front grows from *right to left*). The motion of the particle is shown relative to the grid in combination with initial coarsening and then final dissolution [938]

precipitation of 211 particles is controlled by the 123 phase evolution. Because yttrium concentration is greater at the edges and in the angles of growing 123 crystallite, the driving force of the 211 phase dissolution will be smaller in these sites. Then, the dispersed 211 particles dissolve more slowly and hence are easily trapped along diagonal directions. The neglected small effects of volume diffusion in solid state and corresponding minimum peritectic transformation cause a stability of these particles in the final microstructure [937]. The simulation supports the possibility of effective variation by parameters especially in relation to geometric composition of superconductor. The geometrically caused grain growth does not increase the solidification rate of single 123 grains, but optimizes the processing rate of conductor in whole, considering directed solidification of great number of grains simultaneously with grain nucleation in some sites. The selection of 123 grains for this process occurs, using anisotropic growth conditions. Of particle interest is the grain selection distance, required for the well-oriented grains to overgrow the misaligned ones (see Fig. 7.24) [956]. In this case, the growth of grains, having maximum rates and crystallographic orientations parallel to the main direction of the solidification front propagation, accelerates. In contrast, the growth of grains, perpendicular to the solidification front, slows down. That selection can be used in order to increase processing rate and critical current density in the melt-processed YBCO. Thermodynamic parameters, used in calculations, are presented in Table 7.1 [937]. These parameters are either well-known data for YBCO family or estimations obtained for related materials.

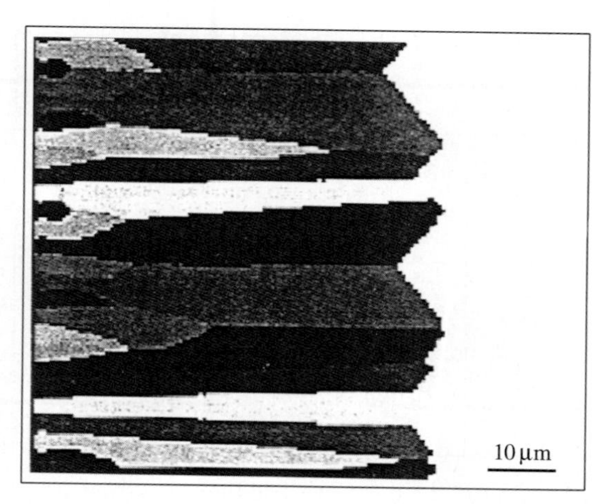

Fig. 7.24. Effect of grain selection investigated by computer simulation method and using phase field approach. Different 123 grains are visualized by different *gray* levels [956]

Table 7.1. Thermodynamic parameters used in modeling

Parameter	Value
Latent heat	$1000\,\mathrm{J/cm^3}$
Diffusion coefficient	$10^{-6}\,\mathrm{cm^2/s}$
Peritectic temperature	1288 K
Cooling temperature	37 K
Yttrium concentration in melt near 211 particle	0.26 mol%
Yttrium concentration in melt near 123 interface	0.14 mol%
Surface energy	$10^{-6}\,\mathrm{J/cm^2}$
Linear kinetic factor	0.186 cm/s K
123 interface thickness	1.018×10^{-4} cm

Thus, the phase field method allows some thermodynamic phases in global non-equilibrium, taking into account local couple interactions of neighbor phases near the point of thermodynamic equilibrium, to be considered. The superposition of these couple interactions causes characteristics of triple point (i.e., multi-phase equilibrium). Then, the solution of equation for multi-phase field defines the phase transition kinetics, which co-relates with limited diffusion during dissolution or with surface energy of interfaces. The equilibrium conditions for phase interface can be selected from existing thermodynamic databases. Numerous phase diagrams for metallic alloys and ceramics define broad potential applications of this method for prediction of HTSC microstructure evolution.

7.2 Stress–Strain State of HTSC in Applied Magnetic Fields

HTSC can trap great magnetic fields (more 10 T) at sufficiently high shielding currents. In this case, there are significant stresses and strains, induced by the Lorentz force between the shielding current and the magnetic field, which is capable of forming and propagating cracks. One-dimensional stress distributions in the totally magnetized superconductor have been considered [492, 879]. Two-dimensional (i.e., axi-symmetric 3D) problem using numerical methods has been investigated in [1039, 1083].

Here, we consider 2D problem for axisymmetric cylinder, presented in Fig. 7.25. In the cylindrical coordinate system, the following equations are obtained for radial and axial displacements, u and ν from the force balance condition for an infinite-small element [1084, 1085]:

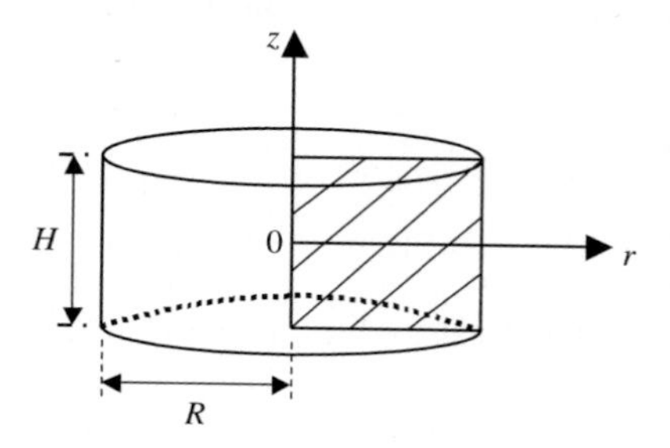

Fig. 7.25. Axisymmetric superconducting cylinder considered in this analysis

$$F_{\mathrm{r}} = \frac{-E}{(1+\nu)(1-2\nu)} \left[(1-\nu)\frac{\partial^2 u}{\partial r^2} + \frac{(1-2\nu)}{2}\frac{\partial^2 u}{\partial z^2} + \frac{(1-\nu)}{r}\frac{\partial u}{\partial r} - (1-\nu)\frac{u}{r^2} + \frac{1}{2}\frac{\partial^2 w}{\partial r \partial z} \right]; \tag{7.28}$$

$$F_{\mathrm{z}} = \frac{-E}{(1+\nu)(1-2\nu)} \left[(1-\nu)\frac{\partial^2 w}{\partial z^2} + \frac{(1-2\nu)}{2}\frac{\partial^2 w}{\partial r^2} + \frac{(1-2\nu)}{2r}\frac{\partial w}{\partial r} + \frac{1}{2}\frac{\partial^2 u}{\partial r \partial z} + \frac{1}{2r}\frac{\partial u}{\partial z} \right], \tag{7.29}$$

where F_{r} and F_{z} are radial and axial Lorentz forces, ν and E are Poisson's ratio and Young's modulus, respectively. Non-zero components of the strain tensor are defined as

$$\varepsilon_{\mathrm{r}} = \frac{\partial u}{\partial r}; \qquad \varepsilon_{\theta} = \frac{u}{r}; \qquad \varepsilon_{\mathrm{z}} = \frac{\partial w}{\partial z}; \qquad \gamma_{\mathrm{rz}} = \frac{\partial u}{\partial z} + \frac{\partial w}{\partial r}, \tag{7.30}$$

where $\varepsilon_{\mathrm{r}}, \varepsilon_{\theta}, \varepsilon_{\mathrm{z}}$ and γ_{rz} are radial, hoop, axial and shear strains, respectively.

Corresponding components of the stress tensor are defined as

$$\sigma_{\mathrm{r}} = \frac{E}{(1+\nu)(1-2\nu)} \left[(1-\nu)\frac{\partial u}{\partial r} + \nu\left(\frac{u}{r} + \frac{\partial w}{\partial z}\right) \right]; \tag{7.31}$$

$$\sigma_{\theta} = \frac{E}{(1+\nu)(1-2\nu)} \left[\nu\frac{\partial u}{\partial r} + (1-\nu)\frac{u}{r} + \nu\frac{\partial w}{\partial z} \right]; \tag{7.32}$$

$$\sigma_{\mathrm{z}} = \frac{E}{(1+\nu)(1-2\nu)} \left[\nu\frac{\partial u}{\partial r} + \nu\frac{u}{r} + (1-\nu)\frac{\partial w}{\partial z} \right]; \tag{7.33}$$

$$\sigma_{\mathrm{rz}} = \frac{E}{2(1+\nu)} \left(\frac{\partial u}{\partial z} + \frac{\partial w}{\partial r} \right). \tag{7.34}$$

Under a condition of $\partial/\partial z = 0$, (7.28)–(7.34) reduce to 1D problem considered in [879].

Macroscopic electromagnetic phenomena in the HTSC are described by the Maxwell equations as

$$\nabla \times \mathbf{E} = -\partial \mathbf{B}/\partial t; \qquad \nabla \times \mathbf{B} = \mu_0 \mathbf{J}; \qquad \mathbf{J} = \mathbf{J}_{\mathrm{SC}} + \mathbf{J}_{\mathrm{ex}} , \tag{7.35}$$

where μ_0, $\mathbf{E}$ and $\mathbf{B}$ are the magnetic permeability in air, the electric and the magnetic fields, respectively. The magnetic field is caused by the shielding current $\mathbf{J}_{\mathrm{SC}}$ and the external current $\mathbf{J}_{\mathrm{ex}}$. A type-II superconductor in a quasi-static field is well described, using the standard critical state model [1024]. Constitutive relationships between the shielding current density $\mathbf{J}_{\mathrm{SC}}$ and the electric field $\mathbf{E}$ are obtained from force balance on a vortex as

$$\mathbf{J}_{\mathrm{SC}} = J_{\mathrm{c}}(|\mathbf{B}|)\mathbf{E}/|\mathbf{E}|, \qquad \text{if} |\mathbf{E}| \neq 0 ; \tag{7.36}$$

$$\partial \mathbf{J}_{\mathrm{SC}}/\partial t = 0, \qquad \text{if} |\mathbf{E}| = 0 . \tag{7.37}$$

When the electric field $\mathbf{E}$ is induced in a local region by change of the magnetic field from (7.35), shielding currents are induced from (7.36). If there is no electric field by the shielding effect, the situation of currents is not changed from (7.37). While the critical current density J_{c} has a strong dependence on

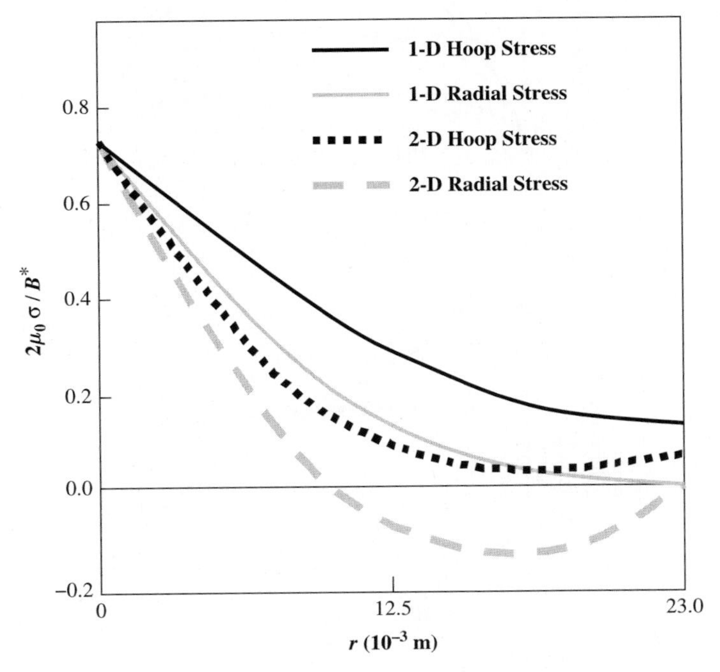

Fig. 7.26. Radial (σ_r) and hoop (σ_θ) stresses vs radial component (at z = 0) in 1D and 2D models [1084]

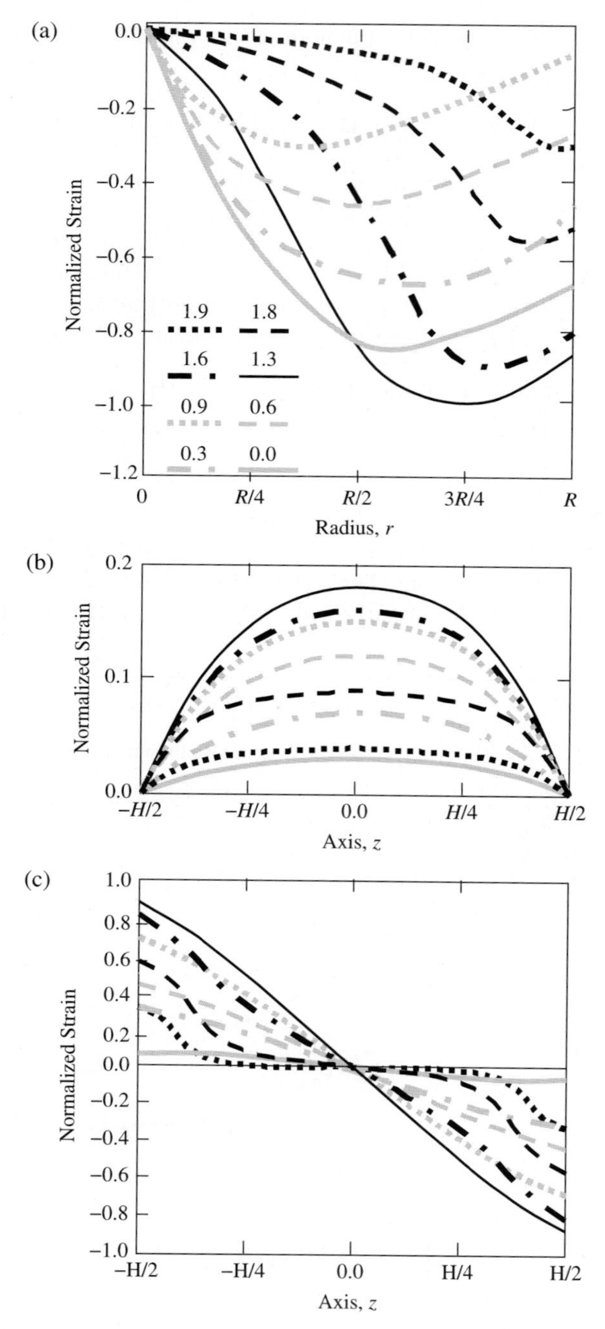

Fig. 7.27. The normalized strains, when the external field is reduced from 2.0 to 0.0 T: (**a**) shear strains (γ_{rz}/γ_{max}) on the upper surface of a bulk HTSC; (**b**)and (**c**) hoop strain ($\varepsilon_{\theta}/\gamma_{max}$) and shear strain ($\gamma_{rz}/\gamma_{max}$) on the side surface of a bulk HTSC [1085]

the magnetic field, the Bean model [737][2] is applied to the present analysis to clarify the basic properties of the stress–strain state. Self-consistent solutions which satisfy the non-linear equations (7.36) and (7.37) can be obtained, using a numerical iterative technique [1024, 1039, 1083] and the following boundary conditions: $\sigma_r = 0$ on the side surface of the cylinder (at $r = R$), and $\sigma_z = 0$ on the upper and lower surfaces of the cylinder (at $z = \pm H/2$). After the Lorentz force calculation from the shielding current distribution at each time step [1039], the finite difference method is applied to solve (7.28) and (7.29). In the iterative calculations, using successive over-relaxation method, the displacements u and w are found in by using turns the boundary conditions until they are converged to definite value. Then, the strain distributions are calculated from (7.30), and the stresses depending on the obtained displacements of u and w are found from (7.31) to (7.34).

Numerical results are obtained for cylinder with the geometrical parameters: $R = 23.0\,\text{mm}$ and $H = 15.0\,\text{mm}$, Young's modulus and Poisson's ratio: $E = 95.9\,\text{GPa}$ and $\nu = 0.14$, respectively. In the Bean model, a standard value of the critical current density is $J_c = 1.0 \times 10^8\,\text{A/m}^2$ [1084, 1085]. Full magnetization with field cooling is obtained when shielding currents are induced in the whole volume of the HTSC (this occurs when the external field is reduced from 2.0 to 0.0 T).

Maximum trapped field B^* in the sample center for 1D and 2D models are 2.9 T and 1.7 T, respectively [1084]). Two-dimensional distributions of the radial (σ_r) and hoop (σ_θ) stresses in the sample center (at $z = 0$) are compared with 1D solutions in Fig. 7.26. The distributions are normalized with corresponding value of B^*. Obviously, largest stresses are obtained in the center of the bulk. Figure 7.27 presents normalized strains on the side, upper and lower surfaces of the cylinder (all strains are normalized with the maximum shear strain, γ_{max}, when the external field is 1.3 T). The distributions change as the external field is reduced from 2.0 to 0.0 T. The large shear strain (γ_{rz}) is obtained by the large Lorentz force between distributed shielding currents and the large external field in the first half of the magnetization. After that, the shear strain distribution reduces as the external field is decreased in the magnetization process.

[2] According to the Bean model [55], for cylindrical body: $\partial B_z/\partial r = \text{const}$, yielding a *linear variation* of B_z with the r-coordinate. In the similar Kim model [545, 546], it is assumed that $B_z \partial B_z/\partial r = \text{const}$, yielding a *parabolic variation* of the function $B_z(r)$.

8

Computer Simulation of HTSC Microstructure and Toughening Mechanisms

8.1 YBCO Ceramic Sintering and Fracture

It is quite obvious that the spectrum of structure-sensitive properties of superconducting ceramics is caused directly by essential inhomogeneous structure, consisting of superconducting grains, secondary phases, pores and microdefects, as a rule disposed on intergranular boundaries. The microstructure formation and fracture occur during sintering, causing internal (residual) stresses, and during the material loading by different thermo-mechanical and electromagnetic fields. Based on computer simulation, a joint study of sintering, cooling and fracture of the structure-heterogeneous material [516, 518, 796] allows a prediction and an optimization of superconductor properties depending on the parameters, including composition, heat rate, initial porosity of material. In Chap. 5 fundamentals of HTSC computer monitoring have been discussed. In this section, an example of $YBa_2Cu_3O_{7-x}$ gradient sintering is considered [61], and proper toughening mechanisms are investigated [807]. The present analysis could be used, in particular, to study heterogeneous mechanism of the YBCO structure formation and to model the cracking processes in *ab*-plane.

8.1.1 Sintering Model of Superconducting Ceramic

During gradient sintering up to microstructure formation, there are the following transformations in the material (Fig. 8.1) [61]:

(1) the primary crystallization, which consists of formation and growth of facetted crystallites of superconducting phase into powder media of the sample ($T \sim 800$–$900°C$);
(2) the formation of non-superconducting phase around crystallites due to active local thermo-diffusion ($T \sim 900$–$920°C$);
(3) the sample shrinkage and formation of isolated (closed) porosity owing to a melting of the non-superconducting phase and its following pushing from

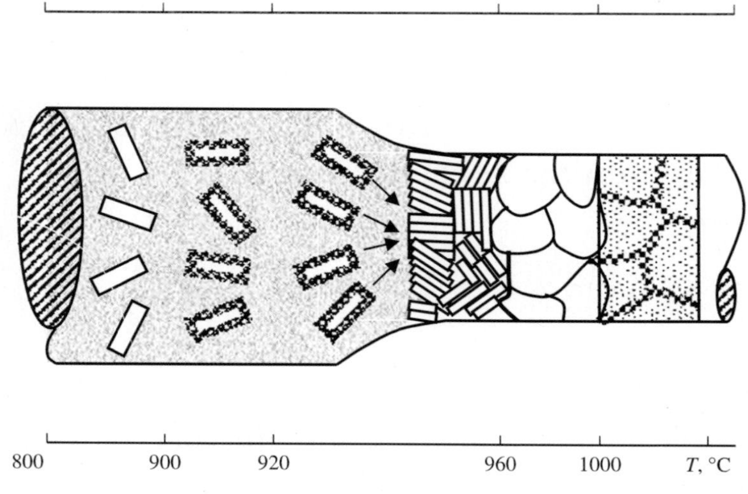

Fig. 8.1. The physical model of microstructure formation of YBCO ceramic during sintering in thermal gradient. The next process zones are shown of primary crystallization (1); non-superconducting phase formation (2); shrinkage (3); abnormal grain growth (4) and structure decomposition (5)

aggregates of superconducting crystallites into pores and at the sample surface ($T \sim 920$–$940°C$);

(4) the secondary re-crystallization, accompanied by breaking of intergranular buffer layers and by creation of big grains with irregular shape ($T \sim 960$–$980°C$);

(5) the structure decomposition ($T > 1000°C$).

The proposed computer model generalizes the research results of gradient sintering of the ferroelectric [516, 518, 796] and YBCO [814] ceramics. It is assumed that temperature in the furnace changes linearly with coordinate. The modeling consists of successive consideration of following stages: (i) the propagation of a heat front in a powder compact with definition of temperature distribution in the sample; (ii) the press-powder crystallization into region of the sintering temperature that is determined by heat front, in conditions of the actual gradient sintering of YBCO [61]; (iii) the sample shrinkage with formation of a closed porosity; (iv) the secondary re-crystallization, which is an abnormal grain growth under conditions of actual inhibition due to existence of a mixed secondary phase. In this consideration, the action regions of the above processes are found by temperature distributions, corresponding to present physical model. The modeling procedure of the forming HTSC microstructure during the sample heat has been written in detail in Sect. 5.3, and the initial parameters, used in computer experiments, are presented in Table 8.1. The finite-difference method is used to solve the first main problem

Table 8.1. Initial parameters used in computer simulation

Parameter	Value	Unit of measurement
Particle properties		
Mean diameter, d	1.5×10^{-6}	m
Thermal conductivity, λ_1	18	W/(m K)
Blackness order, ε	1	–
Density, ρ	3200	kg/m^3
Heat capacity, c_V	0.8	kJ/(kg K)
Air properties		
Atmospheric pressure, H	1010	GPa
Thermal conductivity, λ_g	0.0241	W/(m K)
Adiabatic index, γ	1.4	–
Prandtl criterion, Pr	0.7	–
Accommodation factor, a	0.97	–
Length of molecular free run, Λ_c	0.65×10^{-7}	m

for quasi-linear equation of thermal conduction (see Appendix C.1) for zero initial and corresponding boundary conditions. In this case, the region, considered in ab-plane, is presented by a 2D lattice with 1000 square cells of characteristic size, δ. Every cell is either a grain nucleus of 123 phase or void. The cells with the same numbers form corresponding grain or void (see Fig. 8.2).

The secondary re-crystallization is modeled on the basis of the Wagner–Zlyosov–Hillert's model of abnormal grain growth [1]. In the YBCO ceramic, it is caused by admixture phases at intergranular boundaries. As followed from [61], the material of intergranular buffer layers is $BaCuO_2 - CuO$ system with inclusions of admixtures. In any time t, the abnormal grain growth occurs under condition [1]:

$$|1/R_c - 1/R_j| > I_R/2 , \tag{8.1}$$

where

$$R_c = \sum_{j=1}^{n_c} f_j R_j^2 / \sum_{j=1}^{n_c} f_j R_j ; \qquad f_j = n_j/N_T; \quad \sum_{j=1}^{n_c} f_j = 1; \quad \sum_{j=1}^{n_c} n_j = N_T ; \tag{8.2}$$

n_c is the size-class number of the width ΔR, containing n_j grains with radius R_j in j-class of the discrete space (R, t); $I_R = 6 f_V/(\pi r)$ is the inhibition; f_V is the volumetric fraction of secondary phase particles; r is the mean radius of particles. The inhibition, I_R, depending on the particle radius and on the fraction of secondary phase, governs abnormal grain growth that allows to define the microstructure and strength parameters in dependence on the secondary phase characteristics.

Assume that mass-transfer from one grain to another occurs according to the mechanism of normal crystal growth on non-singular surfaces (i.e., the boundaries with sufficiently high concentration of steps and demonstrating

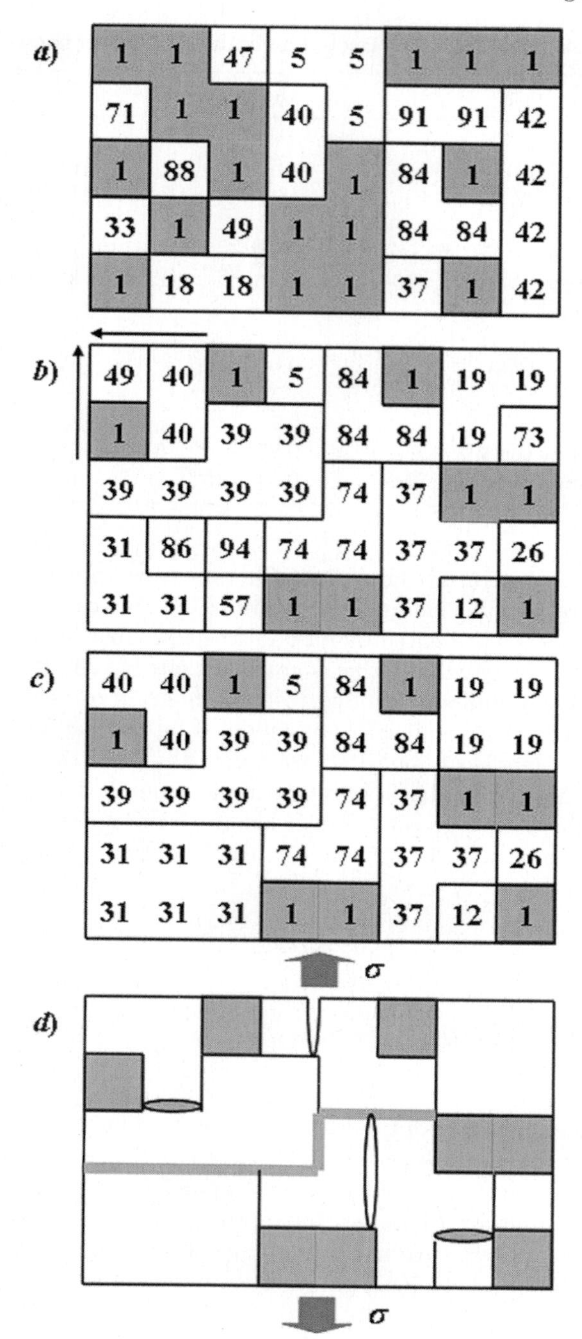

Fig. 8.2. Scheme of 2D fragment of YBCO ceramic structure (*ab*-plane) in PC: (**a**) crystallization (porosity is denoted by 1 and shown by *gray color*); (**b**) shrinkage (*arrows* show its directions); (**c**) abnormal grain growth and (**d**) macrocrack propagation (which is depicted by *gray line*) into model structure (where *white color* shows microcracks formed during the sample cooling, microcracks into process zone have *gray color*). *Numbers* denote a sequence of nucleation of present grain

maximum free surface energy) [1034]. The breaks, covering these facets, causes a possibility of new particles addition practically in any site of the surface. Number of breaks does not limit the re-crystallization rate, and the facets grow self-perpendicular.

Numerical algorithm for modeling of the abnormal grain growth includes the following procedures:

(1) the definition of all nearest neighbors for each grain;
(2) the statement of the nearest neighbors pair (i, j) with $\max_{1 \le i,j \le n_c} |1/R_i - 1/R_j|$;
(3) the growth of larger from grains $(i,\ j)$ at the expense of other;
(4) the verification of the conditions: $|1/R_c - 1/R_j| \le I_R/2$; $j = 1, 2 \ldots, n_c$ and $1/R_{min} - 1/R_{max} \le I_R$;
(5) the end of grain growth in the case of fulfillment, at least of one of the conditions (4) or corresponding change of the parameters: $n_j, n_c, N_T, f_j, R_j, R_c$ with the following repetition of the steps (1)–(5).

An example of model microstructure fragment before and after abnormal grain growth is presented in Fig. 8.2b and c.

8.1.2 Ceramic Cracking During Cooling

The residual stresses form around 211 inclusions into large-grain 123 matrix during cooling due to thermal mismatch in their behavior. Big internal stresses can be created at tetragonal–orthorhombic phase transition in the large-grain matrix (the size of single 123 crystallites can attain some millimeter [431]) that leads to fracture of 123 matrix and to formation of defects. In this case, a density of damage and of defect structures is the function of annealing time at transition from tetragonal to orthorhombic phase.

Thus, the peculiarities of the YBCO ceramic microcracking are connected with the tetragonal–orthorhombic phase transition, causing a spontaneous deformation due to the phase mismatch, and also the thermal expansion anisotropy (TEA) of grains. Due to macroscopic heterogeneity of 211 particles in 123/211 composite, the thermal expansion factors of 123 (α_{123}) and 211 (α_{211}) phases differ significantly (see Table 8.2). As a result, the 123/211

Table 8.2. Some properties of 123/211 composite

Parameter	$YBa_2Cu_3O_7$ (*c*-axis)	$YBa_2Cu_3O_7$ (*ab*-plane)	Y_2BaCuO_5	References
E (GPa)	143	182	213	[910]
ν	0.255	0.255	0.25[a]	[338]
α	3.2×10^{-5}[b]	0.86×10^{-5}[b]	1.24×10^{-5}	[612, 757]
K_{Ic} (MPa m$^{1/2}$)	0.8	0.32	–	[227]

[a] The typical value for oxides.
[b] This includes thermal expansion and oxygen increase.

composite has different average values of thermal expansion factors in various parts of the sample:

$$\alpha_{123/211} = \alpha_{123} V_{123} + \alpha_{211} V_{211} . \quad (8.3)$$

These heterogeneities lead to complicated picture of residual stresses after sample cooling from sintering temperature. Microcracks form in 123/211 polycrystalline structure under the influence of internal stresses between 123 grains during sintering due to anisotropy in their thermal expansion [335]. The 123 grain has a larger compression along the c-axis direction compared to the a- and b-axes during cooling (owing to both thermal expansion and oxygen content increase). This creates residual stresses, leading to tension of the sample along the c-axis.

In contrast to ferroelectric ceramics, where deformation phase mismatch is the main cause of microcracking [796], we consider both effects in the proposed model of HTSC ceramics. The tetragonal–orthorhombic phase transition (occurred for $YBa_2Cu_3O_{7-x}$ at $T \sim 600 - 700°C$ and accompanied by oxygen diffusion [551, 774]) leads to alignment of twinning platelet grains in ab-plane, which is perpendicular to c-axis. However, twinning is practically absent, leading to alignment along the c-axis. The shrinkage along this direction is not compensated by twinning, causing microcracking parallel to (001) [27]. Moreover, a change of elastic stiffness components C_{22}, C_{23}, C_{55} at phase transformation leads to spontaneous deformation of crystalline lattice [345]. Due to this, effect of TEA may be estimated on cooling and spontaneous deformation, caused by the phase transition on the formation of intergranular microcracks. The critical length of cracked boundary, l_c^s, can be estimated as [294]

$$l_c^s = \beta_0 [K_c^b (1 + \nu)/(E\varepsilon)]^2 . \quad (8.4)$$

Here, K_c^b is the fracture toughness of grain boundary; $\beta_0 \approx 3.5$ is the test constant; E is Young's modulus; ν is Poisson's ratio; ε is the deformation of intergranular boundary due to TEA or deformation phase mismatch. In the case of the TEA effect, $\varepsilon = \Delta\alpha\Delta T$, where $\Delta\alpha$ is the thermal expansion factors difference, defined by extreme values of α; ΔT is the temperature difference at cooling. In the case of deformation phase mismatch, ε is the spontaneous deformation caused by change of the elastic stiffness components.

Selecting $\Delta\alpha = 5 \times 10^{-6}\,\mathrm{K}^{-1}$ [294] and $\Delta T = 625\,\mathrm{K}$ (in order to exclude the tetragonal–orthorhombic phase transition) [217], we have for the case of TEA $\varepsilon \sim 3.1 \times 10^{-3}$. Comparing with deformation caused by the phase transition ($\varepsilon \sim 10^{-4}$ [346]), it may be concluded that the main factor defining a microcracking of superconducting ceramic is the TEA. Therefore, we consider below only this case.

The criterion of miqrocrack formation at a boundary with length, l, is selected, taking into account a grain misorientation [295]:

$$l/l_c^s \geq 2/[1 + \cos(2\Theta_1 - 2\Theta_2)] , \quad (8.5)$$

where $\Theta_i (i = 1, 2)$ is the angle between the axis of maximum compression in i-grain and the grain boundary plane. In order to estimate these angles, we use the Monte-Carlo procedure [176]. Consider that the microcracks propagate along the grain boundary after their nucleation at triple junctions. The microcracks are arrested at the neighboring junctions because the adjacent boundary facets are usually subject to internal compression [295]. The computer algorithm of intergranular microcracking during cooling includes the following steps:

(1) the definition of all triple points in model structure;
(2) the definition of sizes of all intergranular boundaries, crossing the triple points;
(3) the re-consideration of the boundaries and their replacement by microcracks in the case of fulfillment of the microcrackimg criterion (8.5) for corresponding boundary.

8.1.3 Formation of Microcracks Around 211 Particles in 123 Matrix

Thermal Stresses on 123/211 Interface

During cooling from the sintering temperature, inside 211 particles and around them in 123 matrix, the thermal stresses form as a result of the thermal expansion factors difference between 123 and 211 phases. Consider the spherical coordinate system $\{r, \Theta, \Phi\}$ with origin coinciding with the 211 particle center. Then, depending on hydrostatic stress P_0 applied to the particle of radius, R_{211}, the radial $\sigma_{123R}(r)$ and tangential stresses $\sigma_{123\Theta}(r) = \sigma_{123\Phi}(r)$ in the matrix (for $R_{211} < r < \infty$) can be defined as [957]

$$-2\sigma_{123\Theta}(r) = -2\sigma_{123\Phi}(r) = \sigma_{123R}(r) = P_0 (R_{211}/r)^3 \; ; \tag{8.6}$$

$$P_0 = \frac{(\alpha_{211} - \alpha_{123})\Delta T}{(1 + \nu_{123})/2E_{123} + 2(1 - 2\nu_{211})/E_{211}} \, . \tag{8.7}$$

For convenience, it is proposed that $\alpha_{211} > \alpha_{123}$; $\Delta T = T_s - T_0$ is the temperature difference at cooling (where T_s and T_0 are the sintering temperature and room temperature, respectively). The stresses on 123/211 interface are independent of the particle size R_{211}. They decrease according to the low of $(R_{211}/r)^3$ and attain 12.5% from maximum value for $r = R_{211}$. Selecting necessary values from Table 8.2 and taking into account the magnitude of $T_s = 925°C$, we obtain stresses on the 123/211 interface, presented in Table 8.3.

The obtained tangential stress in the c-direction (acting on equator of the 211 particle) is sufficiently great and capable to form ab-microcracks around the 211 particles. However, because the 211 particles are under compression, ab-microcracks do not propagate through them, and the 123/211 interface

Table 8.3. Stresses obtained on 123/211 interface

Stress	σ_{123R} (GPa)	$\sigma_{123\Theta}$ (GPa)	$\sigma_{123\Phi}$ (GPa)
c-Axis	−2.0	1.0	–
ab-Plane	0.43	−0.215	−0.215

does not fail unless the size difference between two neighboring particles is too large. Because the radial stress in *ab*-plane is tensile, but both tangential stresses are compressive, then taking into account a higher fracture toughness in the planes, which are parallel to the *c*-axis, it may be expected that microcracks, introduced by the 211 particles will not be parallel to the *c*-direction. General picture of change of the value and sign for stresses σ_{123R} and $\sigma_{123\Theta}$ in the 123 matrix on the 123/211 interface is presented schematically in Fig. 8.3. The complicated picture of the stress state in the 123/211 composite leads to alternative opinions about acting stresses and 123 lattice distortions near 211 particles. In particular, some authors neglect stress state around the 211 particles [748]. Other authors define perturbations of the 123 matrix far from the 211 particles or higher density of crystalline defects near some 123/211 interfaces [37]. The considerable indirect experimental supporting of thermal stresses, existing around 211 particles, is provided with the observed processes of de-twinning or predomination of one twin variant along [100] directions in the case of movement of the twin boundaries due to thermal stresses, caused by 211 (422) particles [199, 200]. The stresses required for the motion of twin boundaries and estimated by the value of 50 MPa [199] are well in agreement with obtained magnitudes in *ab*-plane. The presented analysis of stress state is also supported by observed distribution of microcracks around the 211 particles (see Fig. 8.4). The high radial compressive stresson poles of the 211

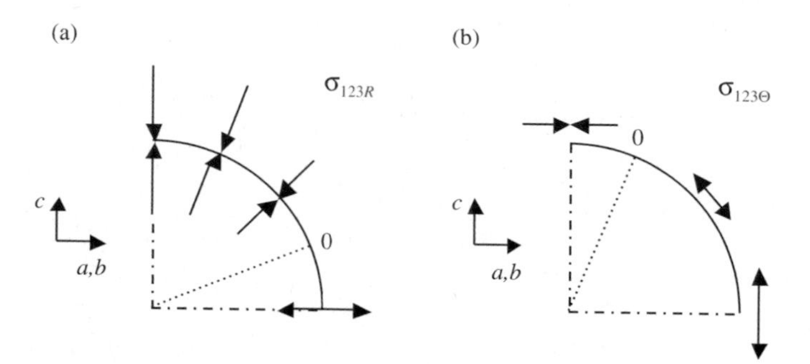

Fig. 8.3. Schematic representations of: (**a**) radial and (**b**) tangential stresses in 123 matrix at 123/211 interface. Sizes and direction of *arrows* demonstrate value and also action regions of tensile and compressing stresses

Fig. 8.4. Microcrack pattern around 211 particles. The microcracks in *ab*-plane avoid 211 particle poles and deflect in radial direction [196]

particle inhibits microcrack growth into *ab*-plane near the poles. Therefore, *ab*-microcracks avoid the 211 particle poles and return to propagation along radial direction in existence of tangential tensile stresses, according to Figs. 8.3 and 8.4.

Microcracking Criterion in *ab*-Plane

Formation of the *ab*-microcracks in the case when the particle demonstrates a smaller thermal expansion compared to matrix demands existence of a microcrack nucleus on the 123/211 interface and supply by energy, which is necessary for the defect growth. This is provided by elastic energies of the particle and surrounding matrix. In isotropic case, total stored energy per unit volume can be defined as [177]

$$U_{\mathrm{T}} = P_0^2 \pi R_{211}^3 \left[\frac{1+\nu_{123}}{E_{123}} + \frac{2(1-2\nu_{211})}{E_{211}} \right] . \tag{8.8}$$

In this case, only part of the stored energy can be used to form *ab*-microcracks. Using the 123 constants for *c*-direction, it may be assumed that the energy approximately equals third part of the total energy, U_{Tc}. Then, the necessary, but insufficient condition of microcrack formation is that the energy of creation of new surface, $U_{\mathrm{s}} = \gamma_{\mathrm{s}} A$ (where γ_{s} is the effective surface energy of matrix and A is the square of the newly formed surface) must not exceed $U_{\mathrm{Tc}}/3$, that is

$$U_{\mathrm{Tc}}/3 \geq U_{\mathrm{s}} . \tag{8.9}$$

Taking into account $\gamma_{\mathrm{s}}^{ab} = (K_{\mathrm{Ic}}^{ab})^2/E^c$ (the upper indexes here and below point the a, b and c directions) and selecting the corresponding data from Table 8.2, we obtain $\gamma_{\mathrm{s}}^{ab} = 7.16 \times 10^{-7}\,\mathrm{MPa \cdot m}$. The square A is found from

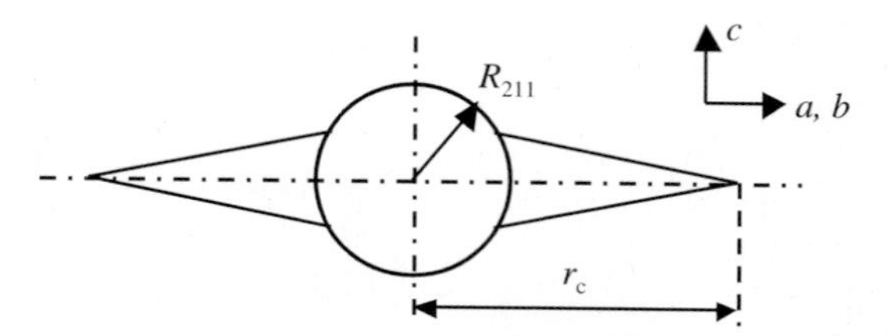

Fig. 8.5. Schematic representation of *ab*-microcrack formation due to release of elastic energy introduced by 211 particle

Fig. 8.5 as $A = 2\pi(r_c^2 - R_{211}^2)$. Considering the usual distance between 211 particle, which is equal to $4R_{211}$ (for 20 mol% of the 211 particles) [196], it is necessary to equate the value of r_c to $2R_{211}$ in order to create the *ab*-microcrack, connecting two adjacent 211 particles. Then we have

$$U_S = 2\pi[(2R_{211})^2 - R_{211}^2]\gamma_s = 6\pi R_{211}^2 \gamma_s \,. \tag{8.10}$$

The existing energy is proportional to R_{211}^3; at the same time the energy absorbed during cracking is proportional to R_{211}^2. Combining (8.8)–(8.10), we define the critical size of the particle, R_{c211}, as

$$R_{c211} \geq \frac{18\gamma_s^{ab}}{(P_0^c)^2[(1+\nu_{123})/E_{123}^c + 2(1-2\nu_{211})/E_{211}]} \,. \tag{8.11}$$

Substituting the obtained values of P_0^c and γ_s^{ab} and also elastic constants from Table 8.2 for the 123/211 composite, we define $R_{c211} = 0.24\,\mu\text{m}$. Another estimation of the critical particle size for microcracking in *ab*-plane (giving the value $< 1\,\mu\text{m}$ [199]) agrees well with the value (R_{c211}) obtained in the present analysis. The strong proof that the *ab*-plane defects are not growth-related defects is that they were not observed in the tetragonal 123 phase [202, 535] as well as in the tetragonal Nd-123 system with small 422 particles (see Fig. 8.6) [201].

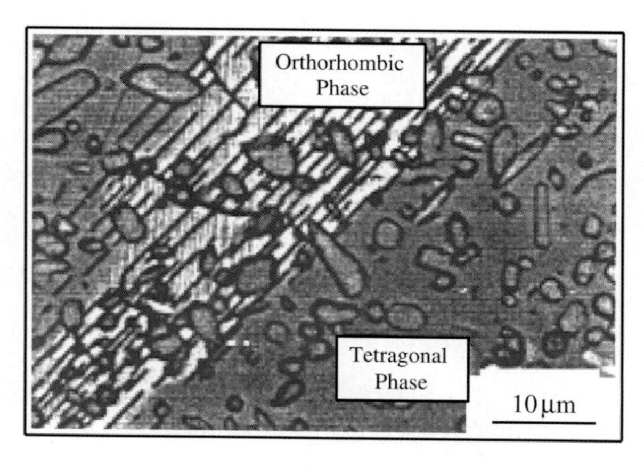

Fig. 8.6. Dense *ab*-micxrocracks in the orthorhombic Nd-123 phase and no *ab*-microcracking in the tetragonal region. The case of small 422 particles [201]

8.1.4 Fracture Features at External Loading

The sufficiently high thermo-mechanical and electromagnetic loading can lead to fracture of material. In order to estimate critical fracture parameters, we use a model of crack growth in anisotropic body [583]. Consider a polycrystalline aggregate of grains with hexagonal shape, containing an annular flaw (Fig. 8.7). The probabilistic condition of misorientation of the adjacent grains leads to initiation of the greatest compression in the central grain, while the surrounding grains will be subject to tension. It is assumed that the central grain is forced into a cavity of diameter, $D = 2R$, surrounded by the annular flaw with length of S. In this case, we shall take into account effects of elastic stress concentration from external tension, σ, and thermal strains on the grain boundaries, $\varepsilon = \Delta\alpha\Delta T$. Then, the summary stress intensity factor (SIF) due to concentration of the thermal stresses and applied external loading is found by

$$K_{\text{tot}} = K_{\text{t}}^{\text{r}} + K_{\text{t}}^{\theta} + K_{\text{a}} \,. \tag{8.12}$$

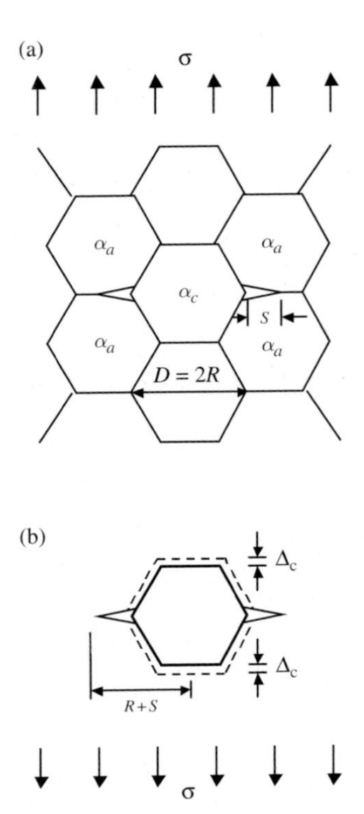

Fig. 8.7. Polycrystalline aggregate, containing annular flaw: (**a**) hexagonal grain under stress; (**b**) crack opening due to TEA (α_a and α_c are the thermal expansion factors along a- and c-axes, respectively)

The first term is determined by the radial component of the thermal stress concentration from the external side of spherical boundary [583, 814] as

$$K_{\mathrm{t}}^{\mathrm{r}} = \frac{2\sigma_{\mathrm{t}} R^{1/2}}{\pi^{1/2}}(1+S/R)^{1/2}\left\{1-\left[1-\frac{1}{(1+S/R)^2}\right]^{1/2}\right\}, \qquad (8.13)$$

and the second term is found by the tangential component [583, 814] as

$$K_{\mathrm{t}}^{\theta} = \frac{2\sigma_{\mathrm{t}} R^{1/2}}{\pi^{1/2}}\left[1-\frac{1}{(1+S/R)^2}\right]^{1/2}\frac{1}{(1+S/R)^{3/2}}. \qquad (8.14)$$

Here,

$$\sigma_{\mathrm{t}} = \frac{2E_{0\mathrm{m}}\Delta\alpha\Delta T}{3(1-\nu_{0\mathrm{m}})}$$

is the thermal stress; $E_{0\mathrm{m}}$, $\nu_{0\mathrm{m}}$ are Young's modulus and Poisson's ratio for porous cracked material, respectively. In order to define elastic constants of porous superconductor, we use the modified cubic model [428]

$$E_{0\mathrm{m}} = E_{\mathrm{m}}(1-P_{\mathrm{V}}^{2/3}); \qquad \nu_{0\mathrm{m}} = \nu_{\mathrm{m}}(1-P_{\mathrm{V}}^{2/3}), \qquad (8.15)$$

where P_{V} is the volumetric fraction of closed porosity; E_{m}, ν_{m} are the elastic constants of cracked no-porous ceramic. The elastic constants E_{m}, ν_{m} can be calculated from the following relationships [892]:

$$\frac{E_{\mathrm{m}}}{E} = 1/\left[1+\frac{16(1-\nu^2)(10-3\nu)}{45(2-\nu)}\beta_{\mathrm{m}}\right]; \qquad (8.16)$$

$$\frac{\nu_{\mathrm{m}}}{\nu} = \frac{1+[(16/45)(1-\nu^2)/(2-\nu)]\beta_{\mathrm{m}}}{1+[(16/45)(1-\nu^2)(10-3\nu)/(2-\nu)]\beta_{\mathrm{m}}}, \qquad (8.17)$$

taking into account arbitrary-oriented microcracks. Here E, ν are the elastic modules of non-defective material; $\beta_{\mathrm{m}} = N_{\mathrm{m}}a_{\mathrm{m}}^3/V_{\mathrm{m}}$ is the microcrack density in the considered volume, V_{m}; N_{m} and a_{m} are the number and characteristic size of microcracks, respectively.

The SIF, caused by applied stress, K_{a}, is defined as [583, 814]

$$K_{\mathrm{a}} = \frac{2\sigma R^{1/2}}{\pi^{1/2}}(1+S/R)^{1/2}\Phi_{\mathrm{e}}, \qquad (8.18)$$

where

$$\Phi_{\mathrm{e}} = \left[1-\frac{1}{(1+S/R)^2}\right]^{1/2}\left\{1+\frac{(4-5\nu_{0\mathrm{m}})}{2(7-5\nu_{0\mathrm{m}})}\frac{1}{(1+S/R)^2}+\frac{9}{2(7-5\nu_{0\mathrm{m}})}\right.$$
$$\left.\times\left[1+\frac{(1+S/R)^2-1}{3}\right]\frac{1}{(1+S/R)^4}\right\}. \qquad (8.19)$$

We define strength of anisotropic body from the critical conditions of a crack initiation: $K_{\mathrm{tot}} = K_{\mathrm{c}}^{\mathrm{b}}$ ($K_{\mathrm{c}}^{\mathrm{b}}$ is the fracture toughness of grain boundary) and $\sigma = \sigma_{\mathrm{f}}$ as

$$\sigma_{\mathrm{f}} = \frac{K_{\mathrm{c}}^{\mathrm{b}}}{\Phi_{\mathrm{e}}} \left[\frac{\pi}{2D(1 + S/R)} \right]^{1/2} - \sigma_{\mathrm{t}} \frac{\Phi_{\mathrm{t}}}{\Phi_{\mathrm{e}}} , \tag{8.20}$$

where σ_{f} is the critical stress;

$$\Phi_{\mathrm{t}} = 1 - \left[1 - \frac{1}{(1 + S/R)^2} \right]^{1/2} \left[1 - \frac{1}{2(1 + S/R)^2} \right] . \tag{8.21}$$

Finally, we write a condition for critical displacement at crack opening (see Fig. 8.7b): $\Delta_{\mathrm{t}} = \Delta_{\mathrm{c}}$ (where Δ_{t} is the linear thermal expansion of the central grain) as [583]

$$D \Delta \alpha \Delta T = \frac{2(1 - \nu_{0\mathrm{m}}^2) K_{\mathrm{c}}^{\mathrm{b}}}{E_{0\mathrm{m}}} \left(\frac{R + S}{\pi} \right)^{1/2} . \tag{8.22}$$

Then, the critical grain size, D_{c}, defining further crack growth, can be obtained, suggesting $D = D_{\mathrm{c}}$ and defining a positive root of quadratic equation (8.22) by [814]:

$$D_{\mathrm{c}} = b + (b^2 + 4Sb)^{1/2}; \qquad b = \frac{(1 - \nu_{0\mathrm{m}}^2)^2 (K_{\mathrm{c}}^{\mathrm{b}})^2}{\pi E_{0\mathrm{m}}^2 (\Delta \alpha \Delta T)^2} . \tag{8.23}$$

Then, we select average grain of model structure as grain size, R, and define value of S (the length of nucleus microcrack) through mean microcrack length, a_{m}, formed during the ceramic cooling. Here and below, an absence of the microcrack interaction is assumed.

At microcrack development in brittle materials, there are processes caused by the material structure properties, which lead to the material toughening. The effects of microcracking, crack branching and bridging and so on render significant, but no simple effect on change of strength properties and fracture resistance. The considerable correction of effects of different toughening mechanisms may occur, taking into account their joint action. A domination of one from the mechanisms is determined by microcrack distribution, and their density causes the toughening value.

8.1.5 Microcracking Process Zone near Macrocrack

In the vicinity of growing macrocrack, a microcracking process zone initiates, which leads to the macrocrack shielding and change of the ceramic fracture toughness [572, 264, 294]. Both phenomena are caused by microcrack distribution. The compliance of the process zone increases the fracture toughness; however, the microcracks, directly neighboring the macrocrack tip, decrease fracture resistance and toughness of material [264, 294]. The critical number of microcracks per volume unit, N_c^*, defining initiation of coalescence in the vicinity of the crack, depends on proper microcrack size, a_m [1125]:

$$N_c^* = 9/(64a_m^3) . \tag{8.24}$$

Define the width of the process zone, $2h_m$, as [107]

$$h_m = a_m I_m^2 , \tag{8.25}$$

where $I_m^2 = \beta_m E_m/(E - E_m)$ is the parameter of elastic interaction of the microcracks; $\beta_m = N_m a_m^3/V_m$ is the microcracking density in the considered sample volume, V_m; N_m is the number of microcracks [892].

The finite-element analysis [572] shows that the stress state of macrocrack to a greater degree promotes growth of microcracks, which are parallel to the macrocrack propagation, and to a lesser degree promotes microcracks of any orientation ahead in the direction of the macrocrack propagation. The stress state of macrocrack also impedes microcracks, which are perpendicular to the macrocrack and disposed at the side from direction of its growth. Moreover, the process zone size in the crack plane is approximately two times smaller than in perpendicular direction. These features are taken into account in computer simulation of microcracking in the process zone and in definition of its extent. The simulation of macrocrack growth along intergranular boundaries (see also a preliminary discussion in Sect. 5.7) is carried out on the basis of the graph theory [514, 515], using the Viterbi's algorithm [72]. During transition from one sequence to other, the process zone width from (8.25) and all triple points in the zone inside are estimated. Then, we model microcracks on proper boundaries with length of l, using the critical facet size: $l_c = 0.4l_c^S$ [294]. The procedure of the microcrack modeling repeats the algorithm used in Sect. 8.1.2 at the stage of spontaneous cracking. In this case, the following are taken into account: the microcracks do not cross the macrocrack and the condition, defining an absence of the microcrack coalescence: $N_m/V_m < N_c^*$.

Obviously, the microcracking density, β_m, increases together with the grain size that guaranties increasing of the process zone size, h_m.[1] As has been shown by calculation of I_m^2 in dependence on the Young's modulus ratio, E_m/E [109], at increasing density of microcracking, the function, I_m^2, attains a maximum value, corresponding to the critical density of microcracks, $\beta_m = \beta_m^c$. This value divides the zones of the material toughening and the crack amplification.

The fracture toughness change due to the alternative trends, caused by the spontaneous cracking and by the process zone of microcracking (Fig. 8.8),

[1] However, in contrast to infinite growth of the process zone, stated in [572], even for grain sizes, which are smaller than the critical one, in reality the process zone must be finite. This contradiction is the consequence of the selected model in [572] for array of hexagonal grains, in which all grain boundaries have the same length, l, and on each of them the triple point (microcrack nucleus) exists. Therefore, for $l \to l_c^S$ the condition, $h_m \to \infty$ is carried out. Our model is free from the above shortcoming.

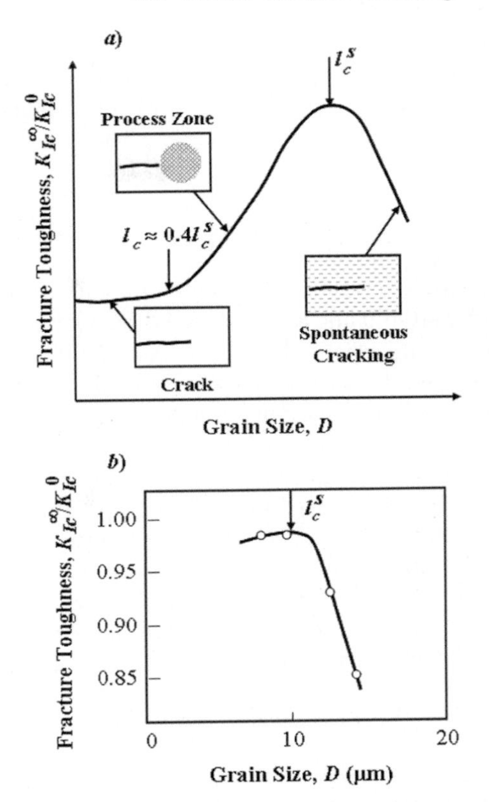

Fig. 8.8. (**a**) Fracture toughness of ceramic vs grain size [294] and (**b**) the simulation results obtained for $PbTiO_3$ ceramic

can be estimated, using Cherepanov–Rice's J-integral [884]. We obtain in the case of stationary macrocrack [294]

$$\frac{K_{\mathrm{Ic}}^{\infty}}{K_{\mathrm{Ic}}^{0}} = (1-\chi_{\mathrm{m}}) \left\{ \frac{1-[(16/45)(1-\nu^2)(10-3\nu)/(2-\nu)]\,\chi_{\infty}}{1-[(16/45)(1-\nu^2)(10-3\nu)/(2-\nu)]\,\chi_{\mathrm{m}}} \right\}^{1/2}, \tag{8.26}$$

where $K_{\mathrm{Ic}}^{\infty}, K_{\mathrm{Ic}}^{0}$ are the ceramic fracture toughness at existence and absence of microcracks, respectively; $\chi_{\mathrm{m}}, \chi_{\infty}$ are the fraction of cracked boundaries in the process zone and in the sintered ceramic. The first multiplier in (8.26) defines a change of local fracture toughness, $K_{\mathrm{Ic}}^{1}/K_{\mathrm{Ic}}^{0}$, in the crack tip. In calculations for χ_{m} and χ_{∞}, the following formulas are used:

$$\chi_{\mathrm{m}} = \frac{N_{\mathrm{m}} S}{2 h_{\mathrm{m}} L N_{\mathrm{g}}}; \qquad \chi_{\infty} = \frac{N_{\infty}}{N_{\mathrm{g}}}, \tag{8.27}$$

where $N_{\infty}, N_{\mathrm{m}}$ is the microcrack number after spontaneous cracking and the boundary number in the sample area, S, respectively; L is the length of the macrocrack path.

The macrocrack shielding caused by microcracking in process zone is estimated, using a model of crack growth under monotonous increasing loading. Assuming that microcracks are the array of isotropic oriented disk-like cracks, it may be obtained [608] that

$$\frac{K_{\mathrm{I}}^{0}}{K_{\mathrm{I}}^{\infty}} = \left[\frac{(1-\nu^2)E_{\mathrm{m}}}{(1-\nu_{\mathrm{m}}^2)E}\right]^{1/2} . \tag{8.28}$$

In this case, Young's modulus, E_{m}, and Poisson's ratio, ν_{m}, for cracked material may be estimated using a model based on averaging of strains on microvolumes, containing misoriented microcracks [892] from (8.16) and (8.17).

The coalescence effect leading to decreasing of the crack growth resistance is estimated, using a coalescence model of half-infinite crack (having tip in the point $x = 0$) with collinear microcrack (having tips in the points: $x = a$ and $x = b, a < b$) that is shown in Fig. 8.9. The collinear microcrack is equivalent to an array of microcracks, distributed in the layer with thickness of $h = N^{-1/3}$, which is measured along normal direction to the fracture plane (N is the mean number of microcracks per volumetric unit). In this case, $a/b = 1 - A$, where $A = N^{2/3}\langle c^2\rangle$ is the fraction of microcracking square in the fracture plane, and $\langle c^2\rangle$ is the mean cracking square, projected on the fracture plane. Then, for the condition of quasi-static crack growth, we obtain [896]:

$$K_{\mathrm{c}}/K_{\mathrm{c}}^{0} = [aE_{\mathrm{m}}/(bE)]^{1/2}C , \tag{8.29}$$

where K_{c} is the SIF, defining the crack coalescence; K_{c}^{0} is the intrinsic fracture toughness of ceramic without microcracks; $C = K(k)/E(k)$; $K(k), E(k)$ are the elliptic integral of 1 and 2 type, respectively; $k = (1 - a/b)^{1/2}$.

Numerical results obtained for the ceramics without voids, possessing the perovskite structure (which HTSC also have): $BaTiO_3(E = 120\,\mathrm{GPa}, \nu = 0.25)$ and $PbTiO_3(E = 80\,\mathrm{GPa}, \nu = 0.3)$ depending on the inhibition parameter of abnormal grain growth (see Table 8.4) lead to the following conclusions. The increasing of mean grain radius, $\overline{R}$, occurs with the decreasing of the inhibition parameter that enhances the fraction of spontaneous-cracked facets and also (that is not obvious) the diminution of microcracking fraction in the process zone. The values of $h_{\mathrm{m}}/l_{\mathrm{c}}^{\mathrm{S}}$ are shown in Table 8.4 for initial stage of the macrocrack growth when the process zone size is stated by structure parameters of sintered ceramic. Taking into account the critical grain size for Al_2O_3 ceramic ($D_{\mathrm{c}}^{\mathrm{S}} = 400\,\mu\mathrm{m}$ [572]), which has strength properties related with the $BaTiO_3$ ferroelectric ceramic, we obtain the process zone size,

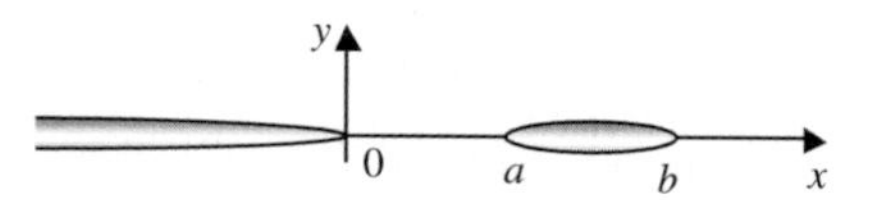

Fig. 8.9. Scheme of interaction of half-infinite crack with collinear microcrack

Table 8.4. Strength properties of $BaTiO_3$ and $PbTiO_3$ ferroelectric ceramics [823]

Ceramic	$I_R\overline{R}_0$	$\overline{R}/\overline{R}_0$	χ_∞	χ_m	h_m/l_c^S	K_{Ic}^1/K_{Ic}^0	K_{Ic}^∞/K_{Ic}^0	K_I^0/K_I^∞
$BaTiO_3$	1.0	1.00	0.121	0.106	1.53	0.894	0.878	0.910
	0.8	1.10	0.161	0.062	1.57	0.938	0.835	0.944
	0.6	1.18	0.183	0.059	1.65	0.941	0.810	0.946
	0.4	1.33	0.187	0.052	1.70	0.948	0.806	0.952
$PbTiO_3$	1.0	1.00	0.121	0. 099	1.56	0.901	0.880	0.912
	0.8	1.10	0.161	0.056	1.59	0.944	0.844	0.947
	0.6	1.18	0.183	0.056	1.67	0.944	0.821	0.947
	0.4	1.33	0.187	0.049	1.73	0.951	0.817	0.951

h_m = 306–340 μm that coincides with analogous test value ($h_m \approx 300$ μm), fixed for aluminum-oxide ceramic [108].

The numerical results for different grain sizes of $PbTiO_3$ ferroelectric ceramic (see Fig. 8.8b) [796] support qualitative trends of characteristic behavior of the fracture toughness depending on microcracking, pointed in Fig. 8.8a [264]. The spontaneous cracking during cooling is the factor determining the fracture toughness change, K_{Ic}^∞/K_{Ic}^0, for grain sizes which are greater than the critical size. Numerical results for ferroelectric ceramics show that in account of the frontal zone of microcracking, the value of the fracture toughness is equal to 0.80–0.88, which is near to 0.9, obtained for Al_2O_3 [264]. The local toughening, K_{Ic}^1/K_{Ic}^0, to the greatest degree defines the fracture process at minimum spontaneous cracking. Finally, the trends in the macrocrack shielding by the process zone coincide with the known results [608]. The decreasing of the stress intensity at the macrocrack tip due to the microcracking increasing enhances the macrocrack shielding. As the numerical results show, the analogous trends are proper to HTSC ceramics.

8.1.6 Crack Branching

The $YBa_2Cu_3O_{7-x}$ ceramic samples heated to the temperature 700°C–930°C have demonstrated strong crack branching at intergranular boundaries [905]. The crack branching due to microcracks in process zone, surrounding the crack tip, enhances the fracture energy. Non-regularity of crack front, causing an increasing of ceramic toughness, often occurs in the case of very high rate of the crack tip. A crack branching in Al_2O_3 ceramic has been observed due to TEA of the grains [1149] that is also proper to superconducting ceramics. The growing crack can absorb microcracks and, on the other hand, the microcracking causes crack branching.

In the modeling of macrocrack propagating into superconducting ceramic structure along intergranular boundaries, the models based on the graph theory can be used, which have been realized in investigation of PZT ferroelectric ceramics [513, 515, 518]. As has been noted above, a process zone of microcracking forms in crack vicinity due to TEA of grains, creating the macrocrack shielding (the process zone size, $2h_{\mathrm{m}}$, is found from (8.25)). The simulation procedure of microcracks into the process zone repeats the algorithm presented in the previous section. Assuming an existence of a branching crack zone with width of $2p$ in the crack tip (see Fig. 8.10b), apparent fracture energy can be found as [303]

$$\frac{\gamma_{\mathrm{gb}}}{\gamma_0} = \left[1 - \frac{\pi\beta_{\mathrm{m}}}{2(3-\pi\beta_{\mathrm{m}})}\right]^{-1} \left\{ [2\cos(\varphi/2)]^{\{\log 2/\log[2\cos(\varphi/2)]-1\}} - \frac{p\beta_{\mathrm{m}}}{a_{\mathrm{m}}(1-\beta_{\mathrm{m}}\pi/2)^{1/2}} \right\}, \tag{8.30}$$

where $\gamma_{\mathrm{gb}}, \gamma_0$ are the fracture energy of grain boundary with and without the macrocrack branching; φ is the branching angle. Note that (8.30) includes a theoretically possible limit case of the microcracking absence ($\beta_{\mathrm{m}} = 0$)taking

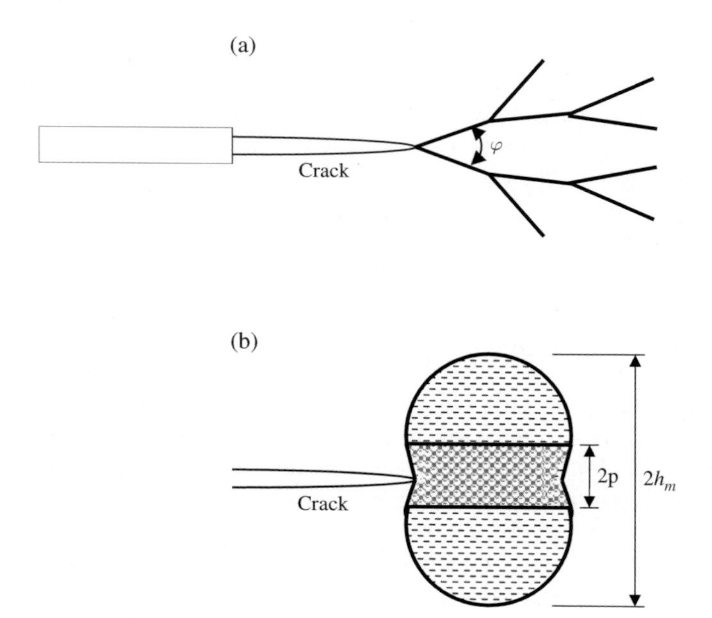

Fig. 8.10. (**a**) Model of crack branching and (**b**) macrocrack tip, showing the zones of crack branching ($2p$) and microcracking near crack tip ($2h_{\mathrm{m}}$)

place, in practice, at creation of fine-grain structure due to corresponding selection of the ceramic processing technique:

$$\frac{\gamma_{gb}}{\gamma_0} = [2\cos(\varphi/2)]^{\{\log 2/\log[2\cos(\varphi/2)]-1\}} . \tag{8.31}$$

Moreover, the fracture toughness change may be estimated in the case of crack branching as

$$K_c/K_c^0 = (\gamma_{gb}/\gamma_0)^{1/2} , \tag{8.32}$$

where K_c is the critical SIF, caused by the crack branching; K_c^0 is the fracture toughness of the ceramic without crack branching. Note that β_m is the summary density of microcracks formed during cooling and in the process zone, $2h_m$. Additionally, in general, the zones of the crack branching and microcracking near the crack tip do not coincide with each other ($h_m \neq p$). Obviously, the actual toughening effects demand fulfillment of the condition $h_m > p$.

8.1.7 Crack Bridging

Other toughening mechanism for ceramics with TEA of grains connects with enhancement of fracture resistance to growing crack due to the formation and fracture of bridges behind the crack front (crack bridging) (see Fig. 8.11). The TEA creates residual compressive stresses, σ_R, keeping the grains that restraint the crack opening in the sites of their disposition. The toughness change as a function of crack length, c (*T-curve*) due to this toughening mechanism may be estimated on the basis of the following assumptions [62]:

(1) The intergranular fracture is only considered.
(2) The grain structure is unimodal and uniform with equal probability of distribution of the compressive and tensile internal stresses (compressed grains form bridges, and tensile grains define a matrix).
(3) The solution is constructed in approximation of a "weak shielding", that is, the effect of compressive stresses on crack is taken into account in the SIF balance, but is neglected in the relationship for displacements of crack surfaces at its opening.
(4) The conception of a "geometric similarity" is used, that is, YBCO ceramic structure changes only in scale without significant alteration of the grain geometry, but parameters, characterizing the geometrical features of the microstructure, are invariant.

Then, T-curve can be constructed as a function of growing crack length, c, as [62]

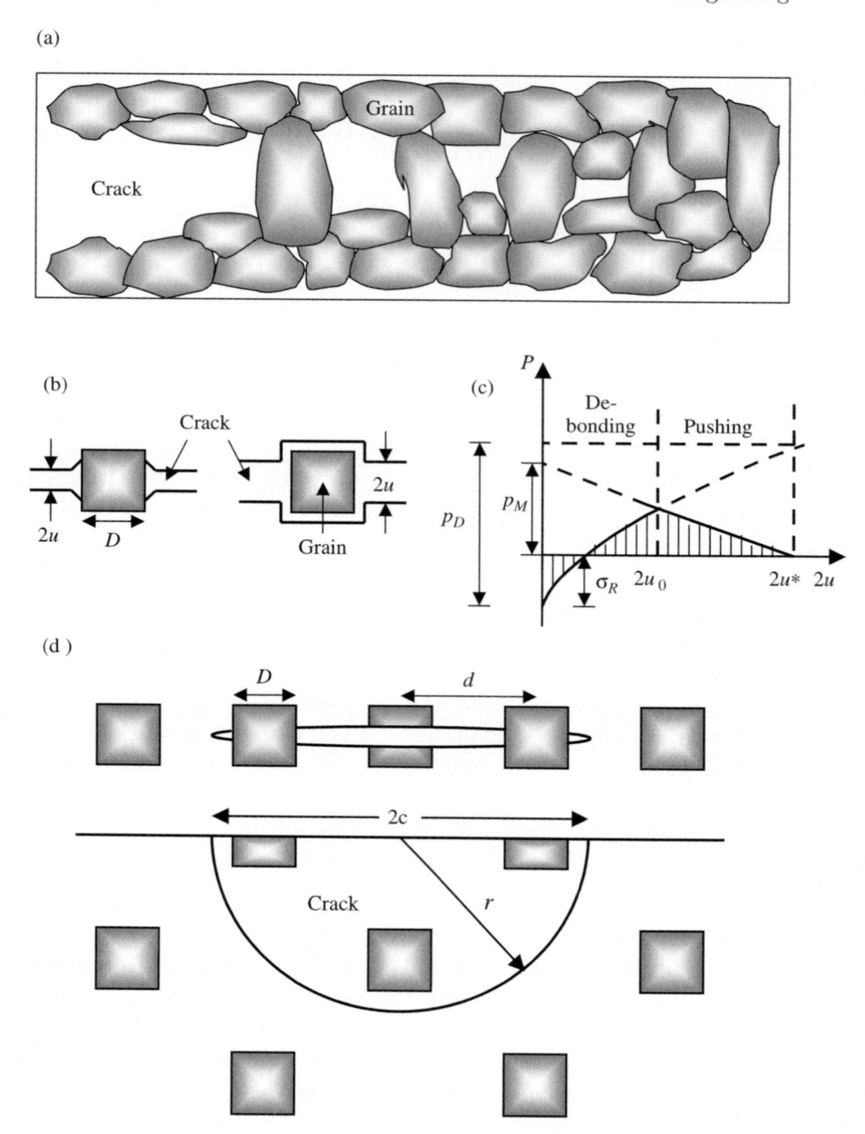

Fig. 8.11. Formation and fracture of bridges behind growing crack front: (**a**) schematic structure fragment; (**b**) two interaction stages of advancing crack with grain-bridge (de-bonding with matrix and pushing of grain by the crack surfaces); (**c**) stress distribution, $p(u)$, inhibiting the crack opening and acting at interfaces; (**d**) penny-shaped crack into region of the crack bridging (sight with side and upper)

$$
\begin{aligned}
T(c) &= T_0 - \psi\sigma_R c^{1/2} \qquad c \le d \\
T(c) &= T_0 - \psi\sigma_R c^{1/2}\left[1 - \frac{2p_D}{3\sigma_R}\left(\frac{c}{c_*}\right)^{1/4}\left(1-\frac{d^2}{c^2}\right)^{3/4}\right] \qquad d \le c \le c_0 \\
T(c) &= T_0 - \psi\sigma_R c^{1/2}\left\{1 - \left(1-\frac{d^2}{c^2}\right)^{1/2} + \left(\frac{c_0}{c}\right)^{1/2}\left(1-\frac{d^2}{c_0^2}\right)^{1/2}\right. \\
&\quad \left.\times\left[1 - \frac{2p_D}{3\sigma_R}\left(\frac{c_0}{c_*}\right)^{1/4}\left(1-\frac{d^2}{c_0^2}\right)^{1/4}\right]\right\} + \\
&\quad + \psi p_M c^{1/2}\left[\left(1-\frac{d^2}{c^2}\right)^{1/2} - \left(\frac{c_0}{c}\right)^{1/2}\left(1-\frac{d^2}{c_0^2}\right)^{1/2}\right] \\
&\quad \times\left\{1 - \frac{1}{2}\left[\left(\frac{c}{c_*}\right)^{1/2}\left(1-\frac{d^2}{c^2}\right)^{1/2}\right.\right. \\
&\quad \left.\left. + \left(\frac{c_0}{c_*}\right)^{1/2}\left(1-\frac{d^2}{c_0^2}\right)^{1/2}\right]\right\} \qquad c_0 \le c \le c_* \\
T(c) &= T_0 + 0.5\psi p_M c_*^{1/2} \qquad c \ge c_*
\end{aligned}
\tag{8.33}
$$

where $p_D = 2(\in_L \mu\sigma_R E_{0m})^{1/2}$ are the shear stresses, forming at the grain de-bonding; c_0 is the crack length, corresponding to a cross of the regions of de-bonding and pushing from matrix of the grains, restraining the crack; $\psi = 1.24$ is the geometrical parameter, corresponding to a penny-shaped surface crack; $\in_L$ is the deformation at the fracture of the links; μ is the factor of the sliding friction. The value of c_0 is calculated from the equation:

$$
c_0^2 - c_*\left(\frac{p_D}{2p_M}\right)^4\left\{\left[1 + \frac{4p_M(p_M+\sigma_R)}{p_D^2}\right]^{1/2} - 1\right\}^4 c_0 - d^2 = 0\,. \tag{8.34}
$$

There are two parameters of the *T-curve* that are need for maximization to attain necessary tolerance of the ceramic to crack growth. These are T_∞/T_0 and c_*/d, where $T_0 = (2\gamma_0 E_{0m})^{1/2}$ is the intrinsic toughness of ceramic, but a toughness at the steady-state crack growth (T_∞) and the crack length in which the far most links from the crack tip begin fracture (c_*) are calculated as

$$
T_\infty = T_0 + 0.5\psi p_M c_*^{1/2}\,; \tag{8.35}
$$

$$
c_* = \left(\frac{\in_L E_{0m} d}{2\psi T_0}\right)^2, \tag{8.36}
$$

where d is the mean spacing between bridging grains, selected as equal to characteristic grain size of the ceramic structure; $p_M = 2 \in_L \mu\sigma_R$ are the stresses caused by the sliding friction; $\gamma_0 = \gamma_s - \gamma_{gb}(1 - f_m)/2$ is the specific surface energy; γ_s is the surface energy in transgranular fracture; γ_{gb} is the fracture energy of grain boundary; f_m is the summary fraction of the cracked facets during the ceramic cooling and in the process zone.

Source of the T-curve is a successive formation and fracture of bridges in material that act as inhibiting links behind the crack tip. Secondary phases at grain boundaries (in our case, the $BaCuO_2$–CuO system with admixture inclusions [61]) play a special role in quantitative interpretation of the T-curve. On one hand, they lead to brittlement of grain boundaries and to decreasing their toughness, but on the other hand they inhibit grain growth, leading to the increasing of ceramic strength. In addition, for related ceramics, there are phenomena which point to a possible stabilization of the crack growth [162]: (i) significant increase of bore applied load after the elastic limit is exceeded, followed by a load drop to non-zero stress (see Fig. 8.12b); (ii) the erratic crack advance is caused by local inhomogeneities in the fracture resistance (see Fig. 8.13); (iii) the discontinuous crack traces are regions of unruptured or frictionally locked material which are the restraining ligaments and which center on large grains (see Fig. 8.14); and (iv) the fracture of the grains-bridges occurs in transgranular mechanism at the primary intergranular failure. These phenomena witness the existence of the processes of the bridges formation behind the fracture front that states an effective toughening mechanism. Due to this, we use results of [162] below and identify three regions of behavior of the penny-shaped crack with the radial coordinate c (see Fig. 8.11), namely (i) the crack motion does not restraint, and $T = T_0$ for short cracks ($c < d$),

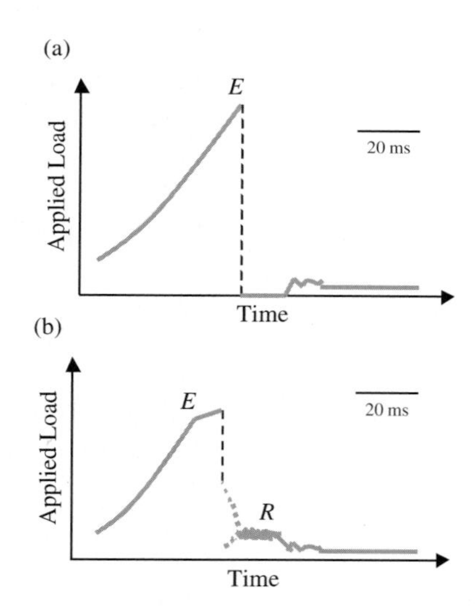

Fig. 8.12. Applied loading as function of time: (**a**) typical response for brittle material, showing spontaneous fall of the loading to zero after exceeding of the elastic threshold (E), and (**b**) response for material at stable crack motion, showing significant increasing bore loading out of elastic limits with following fall of the applied stress down to some non-zero level (R) [162]

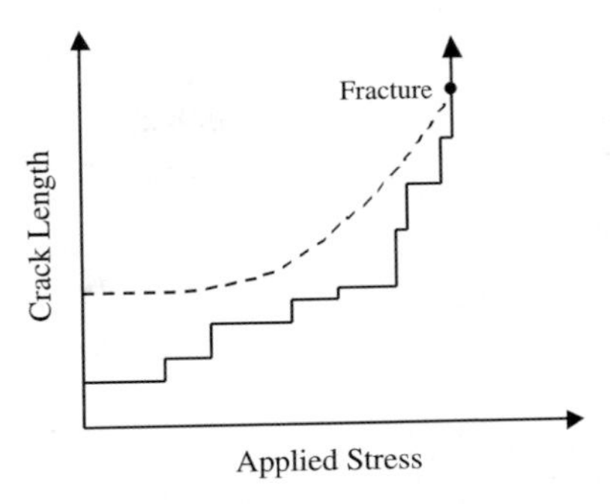

Fig. 8.13. Crack growth vs applied stress (continuous line) for Al_2O_3 sample, demonstrating erratic crack advance, caused by local heterogeneities in fracture resistance. *Dashed curve* corresponds to material with uniform fracture resistance (this line is reduced to the same fracture stress) [162]

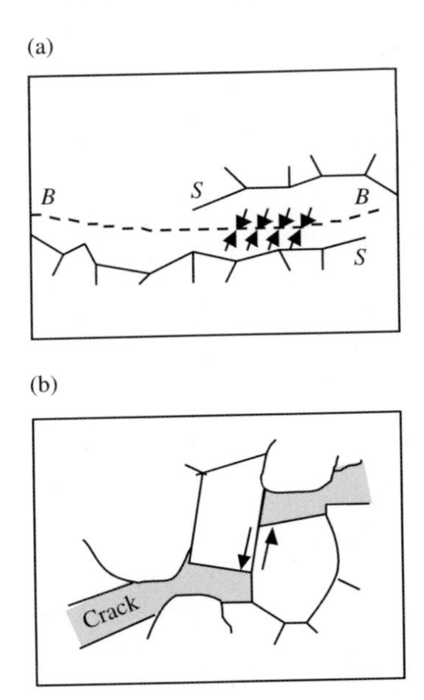

Fig. 8.14. (**a**) Ruptures of crack path, localized at big grains in Al_2O_3 ceramic, which are regions of unfractured material, forming links that limit the crack growth; S the crack surfaces and BB the continuous crack into bulk, depicted by *dashed curve*; the stresses, which close the crack and are caused by bridges at the crack surfaces, are shown by *arrows*; (**b**) grains connected to one another by friction forces; the stresses, closing the crack, and are caused by the crack surface friction are also depicted by *arrows* [162]

where d is the average distance between adjacent grains-bridges; (ii) the zone of restraints is active in the region $d \le c \le c^*$, and $T > T_0$ for intermediate cracks; (iii) the rupture of the restraints occurs at sites distant from the crack tip, leaving a zone of width $c^* - d$ for long cracks ($c > c^*$). The latter zone expands outward together with the crack. In this case, the toughness reaches a maximum value for steady-state crack growth. This toughness corresponds to the crack length, c^*, which is calculated as [800]

$$c^* = \frac{\beta}{2}\left\{1 + \left[1 + 4\left(\frac{d}{\beta}\right)^2\right]^{1/2}\right\}, \tag{8.37}$$

where $\beta = \left(\frac{E_{0m}u^*}{\psi T_0}\right)^2$; $2u^*$ is the opening of the crack with length c^*.

Then, we consider the constitutive equations (the stress-extension dependencies) for inhibiting ligaments of three types [821], namely (i) $\sigma(u) = -\sigma^*(u/u^*)$ for elastic ligaments, where σ^* is the value of the peak restraining stress exerted by the ligament; (ii) $\sigma(u) = -\sigma^*(1 - u/u^*)^m$ for restraints caused by the compressive grains-bridges due to TEA ($m = 1$) and frictional ligaments ($m = 2$). Then, the corresponding microstructure contributions to toughness values are estimated as $T_\mu = 0$ at $c < d$, and $T_\mu = T_\infty - T_0$ at $c > c^*$ (for all three types of the ligaments). In the intermediate zone ($d \le c \le c^*$), the corresponding contributions are estimated for elastic ligaments as

$$T_\mu = (T_\infty - T_0)\left[\frac{c^*(c^2 - d^2)}{c(c^{*2} - d^2)}\right]^{1/2}, \tag{8.38}$$

for ligaments due to TEA ($m = 1$) and frictional ligaments ($m = 2$) as

$$T_\mu = (T_\infty - T_0)\left\langle 1 - \left\{1 - \left[\frac{c^*(c^2 - d^2)}{c(c*^2 - d^2)}\right]^{1/2}\right\}^{m+1}\right\rangle. \tag{8.39}$$

Following [163], we use superposition method in the framework of the force approach and put a local internal stress, caused by microstructure effects of fracture resistance on uniform driving force of crack. Then, total SIF is $K = K_a + K_r$, where $K_a = \psi\sigma_a c^{1/2}$ is the applied SIF from indentation; σ_a is the homogeneous external stress. The local SIF is defined as $K_r = \chi P/c^{3/2}$, where $\chi = 0.004(E_{0m}/H)^{1/2}$ is the constant, depending on the contact geometry and elastic–plastic properties; H is the material hardness; P is the contact loading. We assume a power dependence of toughness on the crack length [163], $T = T_0(c/d)^\tau$, where $0 < \tau < 0.5$ (the toughening is absent for $\tau = 0$, and the catastrophic fracture does not reach for $\tau = 0.5$). Then, from equilibrium condition $K = T$ and the transition condition from stable crack growth to unstable one, $\mathrm{d}K/\mathrm{d}c = \mathrm{d}T/\mathrm{d}c$, we find the critical crack length, c_{cat}, and the applied loading, σ_{cat}, at which the fracture becomes unstable or catastrophic [163, 657]:

$$c_{\mathrm{cat}} = \left[\frac{4\chi P d^{\tau}}{T_0(1-2\tau)}\right]^{2/(2\tau+3)} ; \tag{8.40}$$

$$\sigma_{\mathrm{cat}} = \left[\frac{T_0(2\tau+3)}{4\psi d^{\tau}}\right] c_{\mathrm{cat}}^{(2\tau-1)/2} . \tag{8.41}$$

Then, the inert strength, $\sigma_{\mathrm{m}}(P)$, may be estimated in the kinetic effects absence, considering a condition of unlimited fracture of the ceramic under loading of P due to indentation by using Vickers pyramid [62]. Then, the applied tensile stress, $\sigma_{\mathrm{a}}(c)$, as function of the crack length, c, is calculated, using the toughness magnitude, T, as

$$\sigma_{\mathrm{a}}(c) = \frac{1}{\psi c^{1/2}}\left[T(c) - \frac{\chi P}{c^{3/2}}\right] . \tag{8.42}$$

Generally (for non-cubic, ferroelectric and superconducting ceramics), the function of $\sigma_{\mathrm{a}}(c)$ shows two peaks, which are divided by the value, $c = d$ (see Fig. 8.15b) [803, 811]. The first maximum (for $c < d$) is associated with the residual contact field that dominates. The second maximum (for $c > d$) is associated with the bridging that dominates. The appropriate barrier heights depend on the external load P. When crack overcomes the first barrier (for $c < d$), the crack becomes unstable. However, the crack grows spontaneously up to catastrophic failure if the second barrier (for $c > d$) turns out below the first barrier. Otherwise, the macroscopic fracture is possible only if the load to be sufficient that the crack overcomes the second barrier. Hence, the ceramic strength, σ_{m}, is found by the greater maximum. Note also that the transition point to the horizontal section of the curve $T(c)$ in Fig. 8.15a corresponds to the crack length, $c = c_*$ [814].

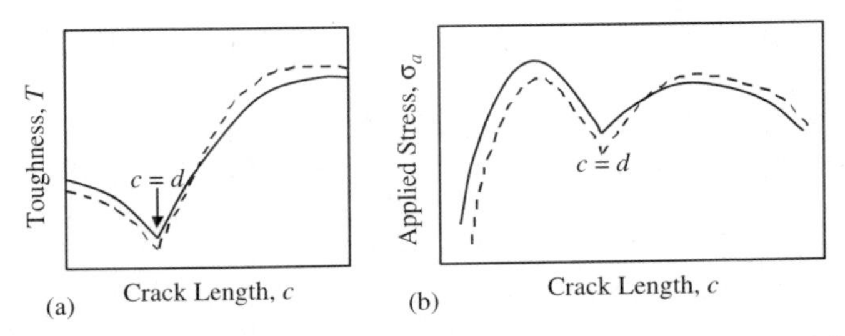

Fig. 8.15. Qualitative trends in the toughness (**a**) and in the applied loading (**b**) as functions of crack length. *Solid* and *dashed curves* demonstrate the dependences before and after grain growth, respectively

8.1.8 Some Numerical Results

In order to obtain numerical results, the known parameters of $YBa_2Cu_3O_{7-x}$ and some data for related ceramics have been used. A necessary number of the computer realization has been established on the basis of stereological approach. The corresponding procedure for definition of unbiased estimation of any stereological characteristic [145] has been presented in Sect. 5.5. We consider the mean grain radius as this stereological characteristic in the computer simulations.

Young's modulus is defined as $E = 3K(1-2\nu)$, using the volume modulus, $K = \gamma c/(\beta V)$, where $\gamma = 1.35$ is Grüneisen's constant; $c/V = 2.72\,\mathrm{MPa/K}$ is the characteristic volume heat; $\beta = 5.86 \times 10^{-5}\,\mathrm{K}^{-1}$ is the mean volume thermal expansion factor; $\nu = 0.21$; $K_\mathrm{c}^\mathrm{b} = 1\,\mathrm{MPa\ m}^{1/2}$ [335]. Selecting $\delta = l_\mathrm{c}^\mathrm{s}$, we estimate an effect of the initial sample porosity before sintering, C_p^0, and the inhibition parameter of grain growth, I_R, on different strength characteristics. The obtained numerical results are shown in Figs. 8.16–8.18 and in Table 8.5 [814].

The calculations show a growth of the mean grain size with the decreasing of C_p^0; at the same time, the insignificant alteration of closed porosity is observed in the considered cases. This causes a formation of longer microcracks at smaller values of C_p^0. However, the effect of grain growth stagnation and existence of a smaller number of triple points (nuclei of microcracks) state the insignificant change of the microcracking density $\beta_\mathrm{m}^{(1)}$ (after cooling) with attainment of its minimum value for $C_\mathrm{p}^0 = 30\%$.

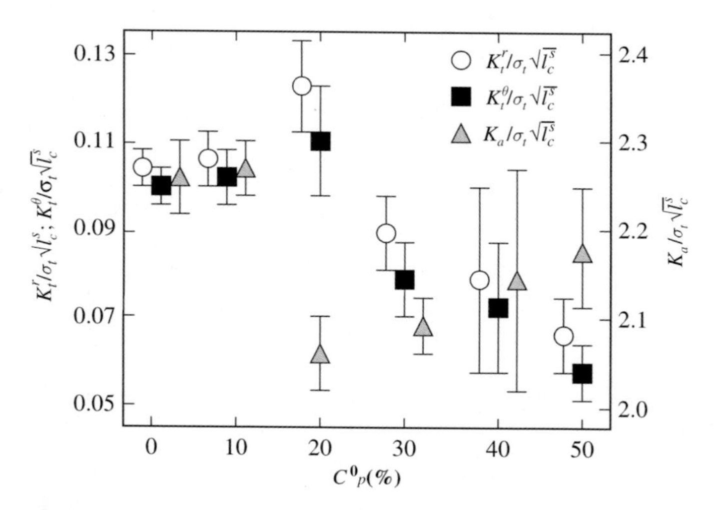

Fig. 8.16. SIF *vs* initial porosity of precursor sample, C_p^0. The deviation of results corresponding to interval change of magnitude of the abnormal grain-growth inhibition parameter ($I_R l_c^\mathrm{s} = 0.4 - 1.0$) is shown

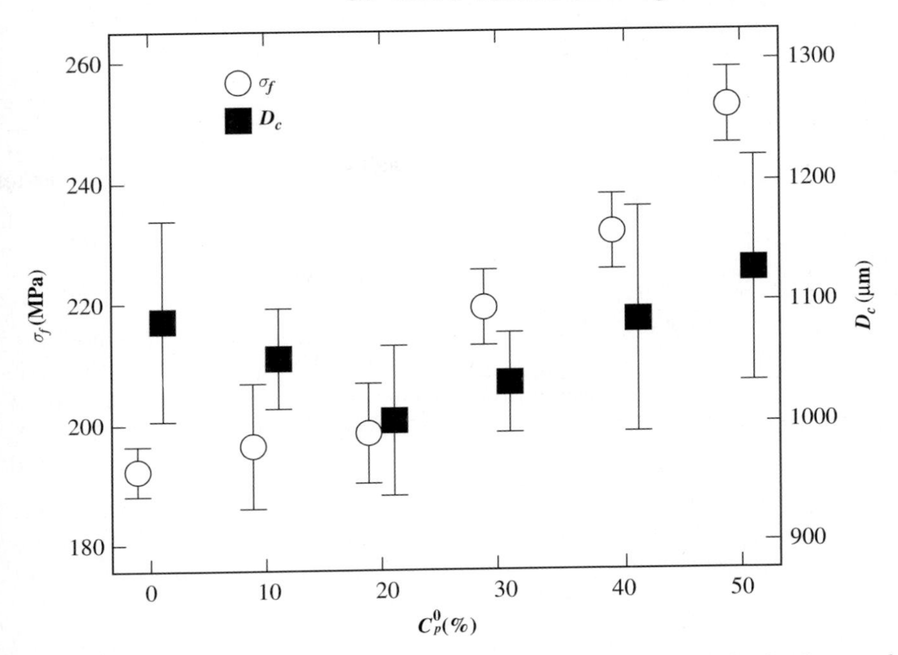

Fig. 8.17. Fracture stress, σ_f, and critical grain size, D_c, causing the further crack growth vs initial porosity of precursor sample, $C_p^0(I_R l_c^s = 0.4 - 1.0)$

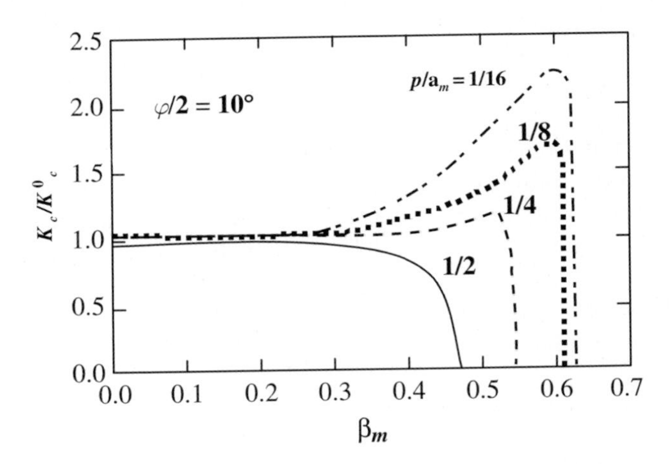

Fig. 8.18. Fracture toughness, K_c/K_c^0, vs microcracking density, β_m, for different sizes of process zone for crack branching (p/a_m) at $I_R D_0 = 2.0$

Table 8.5. Fracture toughness change at crack branching ($I_R D_0 = 2.0$)

Properties	$C_p^0 = 0\%$	$C_p^0 = 10\%$	$C_p^0 = 20\%$	$C_p^0 = 30\%$	$C_p^0 = 40\%$	$C_p^0 = 50\%$
$\beta_m^{(1)}$	0.17	0.17	0.13	0.12	0.15	0.17
$K_c/K_c^0, p/a_m = 1/8$	1.04	1.04	1.03	1.03	1.03	1.04
$K_c/K_c^0, p/a_m = 1/3$	0.99	0.99	1.00	1.00	1.00	0.99
$K_c/K_c^0, p/a_m = 1/2$	0.96	0.96	0.97	0.97	0.97	0.96
$\beta_m^{(2)}$	0.28	0.31	0.32	0.33	0.39	0.44
$K_c/K_c^0, p/a_m = 1/8$	1.08	1.09	1.10	1.10	1.14	1.19
$K_c/K_c^0, p/a_m = 1/3$	0.98	0.98	0.98	0.97	0.96	0.92
$K_c/K_c^0, p/a_m = 1/2$	0.90	0.88	0.87	0.86	0.77	0.62

$\beta_m^{(1)}$ is the microcracking density after cooling; $\beta_m^{(2)}$ is the summary density of microcracks, formed during cooling and in the process zone near macrocrack.

Joint effects of microcracking and grain size on strength properties are found by the ratio, S/R. This parameter owing to above results prevails compared to dependence on the closed porosity of the sample. This explains, at first view, the paradoxical result of the fracture stress, σ_f, increasing with the initial porosity, C_p^0 (see Fig. 8.16). The last result follows from the concept that the fracture stress increases with decreasing of the grain size. In addition, the same trend retains in the dependence of σ_f on S/R (at fixed value of the grain size) [583]. The obtained values of the fracture stress are near the test data observed by the electronic microscopy method for $YBa_2Cu_3O_{7-x}$ ceramic ($\sigma_f = 300\,\text{MPa}$) [567]. The SIF dependencies (K_t^r, K_t^θ, K_a) and the critical grain size for further crack growth (D_c) as function of C_p^0, presented in Figs. 8.16 and 8.17, reach extremes in intermediate points of the interval values of C_p^0. This coincides with the observed trends for non-cubic [583] and ferroelectric [849] ceramics. It should be noted that the SIF, corresponding to the tangential component of thermal stresses (K_t^θ) is smaller than the SIF caused by the radial component (K_t^r). Due to K_t^θ cannot be independently used to define a critical condition of cracking. Similarly, the spontaneous microcracking during cooling to be significant phenomenon of the I mode of fracture in the ceramics with non-cubic symmetry, which demonstrate brittle fracture [609].

The numerical results (Fig. 8.18 and Table 8.5) show that fracture toughness increases together with the crack branching angle and at interaction of macrocrack with microcracks in the short process zone of the crack branching. However, in the case of the process zone increasing, the dominating influence of the crack branching is replaced by the microcracking effect, decreasing the fracture toughness. The transition from toughening to the crack amplification is defined by the process zone size, $p/a_m \approx 1/3$ (for the branching angle, $\varphi/2 = 10°$). The weak dependence of K_c/K_c^0 on C_p^0 at fixed value of p/a_m is caused by the small microcracking density, $\beta_m = 0.12$–0.17 (after the sample cooling). An increase of $\beta_m = 0.28$–0.44 (on account of damage into the microcracking process zone) leads to formation of the fracture toughness dependence on the initial sample porosity. The analysis of the dependence of K_c/K_c^0 on β_m (see Fig. 8.18) shows that an effect of the crack amplification ($K_c/K_c^0 < 1$) takes place for all sizes of the crack branching zone and exists in the case of the microcrack coalescence, at $\beta_m \approx 0.5$. The last value is usually selected as the critical one for initiation of the crack coalescence in ceramics [294].

In order to estimate the effect of the crack bridging, we use the next data [62, 162, 796]: $\in_L = 0.1; \mu = 1.8; \gamma_s \approx 2\gamma_{gb} = 6\,J/m^2; \sigma_R = 100\,MPa; H =$ 19.1 GPa. We take into account in the present case only microcracks formed during cooling. The microcracking due to macrocrack advancing changes quantitative results, but qualitative ones, presented in Fig. 8.15, are retained. In modeling of abnormal grain growth, the characteristic values of $D_0/l_c^s = 2.20$ and 2.84 (the case of no porous ceramic) have been estimated preliminarily for two values of the grain growth inhibition parameter, $I_R D_0 = 2.0$ and 1.2 (D_0 is the grain size before beginning of the grain growth). As is followed from Fig. 8.15a, the effect of the grain size on the change of T-curve is not simple. This is explained by an additional microcracking effect of intergranular boundaries. In the field $c < d$, an enhancement of the value d will lead to corresponding diminishing of left part of the $T(c)$ curve and at some value of parameter d, the $T(c)$ curve crosses the c-axis before fulfillment of the condition: $c = d$. Then, primarily existing microcracks and residual stresses, formed during the ceramic processing, will lead to unstable macrocrack advancing without dependence on applied load. In this case, the spontaneous microcracking prevails over the toughening effects, caused by the crack bridging, and states the ceramic strength. The critical value of the parameter d is estimated from the condition: $T = 0$, at $c = d$. Hence, for the sizes of grains-bridges

$$d \geq (T_0/\psi\sigma_R)^2 \,, \tag{8.43}$$

a positive effect of crack bridging is exceeded by the effects of "deleterious" microcracking.

Other condition for definition of the critical value of the parameter d can be stated from (8.4). Assuming that the sizes of grains and facets per grain are subjected to normal distribution, we obtain [176]: $D_c^s \approx 2l_c^s$. Then, the mean separation between bridging grains, d, coinciding with the characteristic grain size can be estimated as

$$d < 2\beta_0[K_c^b(1+\nu)/(E\varepsilon)]^2 \,. \tag{8.44}$$

The selection of the condition for d as well as for fixing of the grains-bridges into ceramic microstructure should be carried out on the basis of experimental data. Then note that the de-bonding region ($d \leq c \leq c_0$) is very short ($c_0/d \leq 1.01$) and is not observed practically in Fig. 8.15. Thus, pushing of grain by the surfaces of the growing crack mainly contributes to the increasing of fracture resistance on account of existence of the residual thermal stresses. This conclusion is supported by theoretical analysis and numerical results of [62, 1113].

The parameters characterizing a capability of the YBCO ceramics to fracture resistance in the considered cases ($D_0/l_c^s = 2.20$ and 2.84) are, respectively, equal to $T_\infty/T_0 = 2.01$ and 2.17 (from (8.35)); $c_*/d = 15.6$ and 17.1 (from (8.36)). They define a significant effectiveness of present process as the toughening mechanism of superconducting ceramics. Generally, these parameters increase with grain size growth (see Fig. 8.15) for the same initial porosity, C_p^0, that corresponds to experimental observations [1113].

Using the above computational approach, the microstructure parameters required to estimate trans- and intergranular fracture can be calculated [800, 821]. The numerical data, characterizing the YBCO microstructure at different initial porosity of C_p^0, are given in Table 8.6. The computational procedure includes the next stages. First, we define the grain-boundary intercept length, λ, and ratio of the largest grain size in the material to the average one, η. Then, the grains-bridges are selected at the lattice directly (in every case there are 10% of grains from those, that are similar to the experimental data for related compositions [1113]). These are the greatest grains and we select the value of d so that it is equal to the average size of the grains-bridges. Next, the transgranular fracture of restraints and failure of the intergranular boundaries between bridges are simulated (see Fig. 8.19b), using the graph theory for definition of the shortest crack route at the lattice of the intergranular boundaries [72]. After that, the proportion of transgranular failure, A_T, is found. The sections of the transgranular failure are determined using the characteristic size of the bridge, but the sections of the intergranular fracture are defined using the direct crack path along grain boundaries. Further, we choose the parameter u^* for the above ligaments, namely: $u^*/\lambda \sim 0.01$ for elastic and frictional restraints [162] and $u^*/\lambda \sim \eta^2\varepsilon_T$ for compressive traction [1029], where the strain for YBCO due to the TEA is $\varepsilon_T = 3.1 \times 10^{-3}$ (see Sect. 8.1.2). Finally, we determine the values of c^* and $T_\mu/(T_\infty - T_0)$ from (8.37)–(8.39).

The computations demonstrate the value $\lambda \approx 1.83\delta$ without dependence on the initial porosity of C_p^0. Then, the magnitudes of u^*/λ, presented in Table 8.6 for the case of compressive ligaments, coincide well with the cases of

Table 8.6. Some model parameters for YBCO ceramic

Properties	$C_p^0 = 0\%$	$C_p^0 = 20\%$	$C_p^0 = 40\%$	$C_p^0 = 60\%$
η	1.498	1.636	1.758	1.842
A_T	0.312	0.293	0.274	0.442
d/δ	3.423	3.270	3.028	2.646
u^*/λ	0.0070	0.0083	0.0096	0.0105
c^*/d	10.56	11.05	11.92	13.62
	5.32	7.66	10.98	15.00

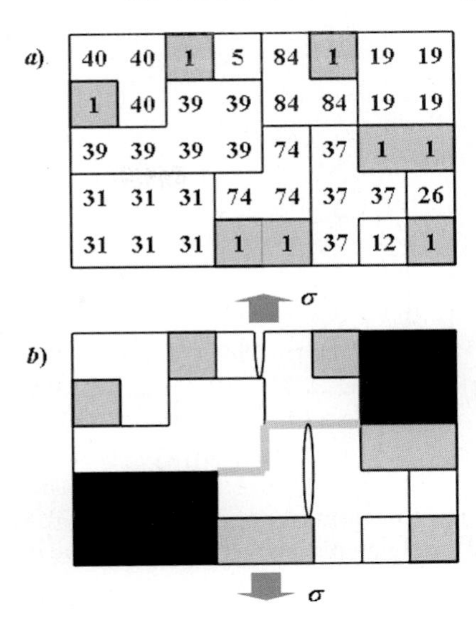

Fig. 8.19. (**a**) Representation of YBCO ceramic structure fragment in PC (voids are denoted by 1 and depicted by *gray color*), and (**b**) macrocrack propagation (*gray line*) into model structure between two grains-bridges depicted by *black color*

the elastic and frictional restraints. The ratio of the largest grain to the average one (η) rises with initial porosity, showing a growth of the heterogeneity of grain sizes. This is due to the outstripping decrease of the average grain size compared with the highest one. Further, there is growth of the grain size and grain-bridge with the initial porosity decrease. Moreover, the numerical data give dependence, $d = k\lambda(1 < k < 2)$, as being similar to alumina ceramics [162]. The intergranular fracture is decreased with the initial porosity growth, because high sinuosity of a crack trajectory corresponds to the coarse-grained structures. However, the fraction of the transgranular fracture (A_{T}) does not change monotonously due to the relative equalizing of the transgranular failures in the various cases of C_{p}^{0} and owing to a rise of the structural parameter (η) with the initial porosity. Thus, the most fine-grained structure with $C_{\mathrm{p}}^{0} = 60\%$, but possessing the greatest value of η, has the increased toughening effects, which correlate with increased area fractions of transgranular failure [162]. On the other hand, this result supports a known thesis about necessary to characterize a polycrystalline structure by a distribution of grain sizes, but no their maximum (or minimum) value.

The parameters which characterize the ability of superconducting ceramics to resist failure are ratios of $T_{\mu}/(T_{\infty} - T_0)$ and c^*/d. The numerical results for the latter (obtained from (8.37)) are shown in Table 8.6 where the upper line corresponds to the elastic and frictional ligaments, but lower line conforms to the compressive traction due to TEA. The observed growth of c^*/d together with the initial porosity for all restraints is caused by appropriate rise of the

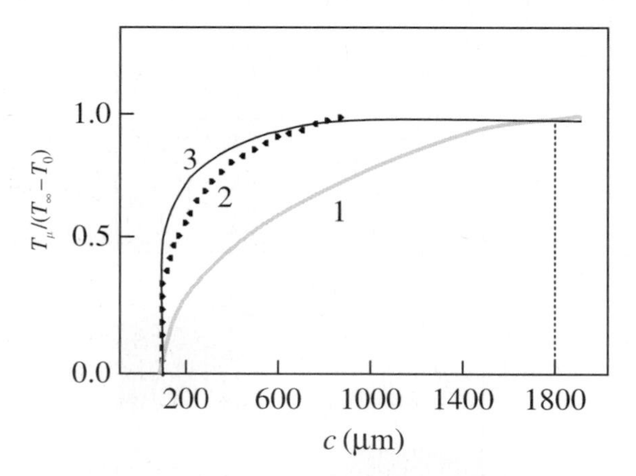

Fig. 8.20. Microstructure contributions of $T_\mu/(T_\infty - T_0)$ to the HTSC ceramic from elastic (1), frictional (2) and compressive (3) ligaments vs crack length, c

structure heterogeneity parameter, η, and coincides with experimental data for different ceramics [62, 162, 1113]. In Fig. 8.20, there are microstructure contributions of $T_\mu/(T_\infty - T_0)$ due to the crack bridging mechanism as function of the crack length c for the above types of ligaments, and $C_\mathrm{p}^0 = 0\%$. Obviously, the main contribution to the toughening is found by the compressive traction due to TEA, and the least one corresponds to the elastic links. These trends are supported by the experiments [62, 1113] as well.

The critical parameters c_cat, and σ_cat (see (8.40) and (8.41)) are shown in Fig. 8.21 depending on the load P and on different toughening powers τ for characteristic grain size, $D_0/\delta = 2.20$. The results show an increasing of the critical crack length, c_cat, and a decreasing of the critical applied stress,

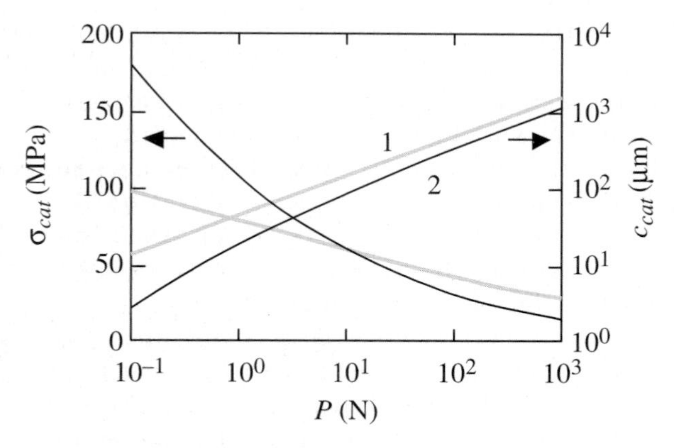

Fig. 8.21. Critical parameters σ_cat and c_cat vs contact load P for two toughening powers: $\tau = 0.25(1)$ and $\tau = 0.10(2)$

σ_{cat}, with growth of the contact load, P, without dependence on features of abnormal grain growth [814]. However, the changes of these parameters from the toughening power have different characters.

Thus, the inhibition parameter of grain growth (I_R) and the mean spacing between bridging grains (d) are parameters which should be taken into account in the calculation and optimization of microstructure and strength parameters of HTSC ceramics. The first parameter is stated by the history of the ceramic processing and by its composition (i.e., thermal treatment, admixture additions in press-powders). The second one is the key parameter of the bridging mechanism, which states a transition from the toughening effects to the amplification crack. At the same time, the account of these parameters does not reject but, on the contrary, proposes to use in calculations other characteristics, in particular, the toughness of intergranular boundaries (T_0), residual stresses (σ_R) and coefficient of the sliding friction at pushing of grains by surfaces of growing crack (μ). Hence, a design of YBCO ceramic microstructure, which is optimum from the view of material strength, is connected with introducing of the grains-bridges on the path of the possible macrocrack propagation at corresponding suppression of "deleterious" microcracking in this zone. This assumes a requirement to form superconducting grains with maximum permissible sizes, which do not exceed the critical value of the spontaneous cracking, and with distribution, demonstrating a maximum possible parameter of the structure heterogeneity, η. The "deleterious" microcracking is formed owing to the grain growth that can be regulated by admixture phases, pulled out during sintering at intergranular boundaries. As a result, it is possible to predict properties of the sintered ceramics even at the earlier stages of the ceramic fabrication in dependence on, for example, the sintering technique, thermal treatment and parameters of secondary phases.

8.2 Twinning Processes in Ferroelastics and Ferroelectrics

It is well known that YBCO ceramics possess ferroelastic properties owing to domain (platelet) structure. Therefore, a sufficiently great attention has been spared to description of microstructure and residual stresses of YBCO polycrystals in the framework of the numerical method for description of twinning processes [27, 270, 840] and also to computer simulation of twins [1123].

Single crystallites in polycrystalline ferroelastics (YBCO) and ferroelectrics ($BaTiO_3$, PZT, etc.) at structure phase transition are subjected to twinning, which is a result of the elastic energy minimization. The twin formation leads to significant decreasing of quasi-uniform stresses in grain volume at simultaneous formation of considerable heterogeneous stresses localized near intergranular and interphase boundaries [27, 261, 840]. A twinning as a sequence of phase transformation, causing a small structure change of unit cell, can be described as mechanical twinning [27, 517] or as martensitic transformation

[192, 262]. The transition accompanied by diffusion, in some cases, also leads to the twinning.[2] The mechanical twinning in ferroelastics can be compared with formation of ferroelectric domains. In the latter case, an energy decrease for electric field is accompanied by the energy diminution for domain walls that leads to minimization of total energy of the system. In the ferroelastics, this summary energy, including elastic energy and energy of domain walls, also becomes minimum. The domain walls displacement can be accompanied by change of the crystallite shape and serves the sources of residual stresses. From the view of influence on fracture toughness, a change of domain structure in vicinity of crack under mechanical loading presents a special interest. This process affects the crack growth. A relationship of domain structures (see Fig. 8.22) and twinning features [27, 840] prompts a possibility of comparative analysis of the ferroelectrics and ferroelastics behavior. An influence of the twinning processes on fracture toughness of these samples and appropriate effects on fracture resistance are estimated below.

8.2.1 Domain Structure and Fracture of Ferroelectric Ceramic

It is well known that the fracture toughness anisotropy of ferroelectric ceramic (FC) is the sequence of structure changes during poling. There are three main causes of formation of the fracture toughness anisotropy [849]: (i) the toughness anisotropy of single crystallites (due to the crack interaction with the boundaries of twins), the greater part of which is oriented along the poling direction; (ii) the anisotropy of internal (residual) stresses into microvolume of piezoelement, leading to the crack advancing to higher values of K_{Ic} in poling plane and to lower values in perpendicular plane; (iii) the plastic deformation of twinning near crack tip. The investigations of (i) and (ii) cases have been carried out using computer simulations in [517, 813]. Here, consider the twinning effects on the strength properties, taking into account the ceramic ferrohardness and initial porosity on example of PZT ceramics,

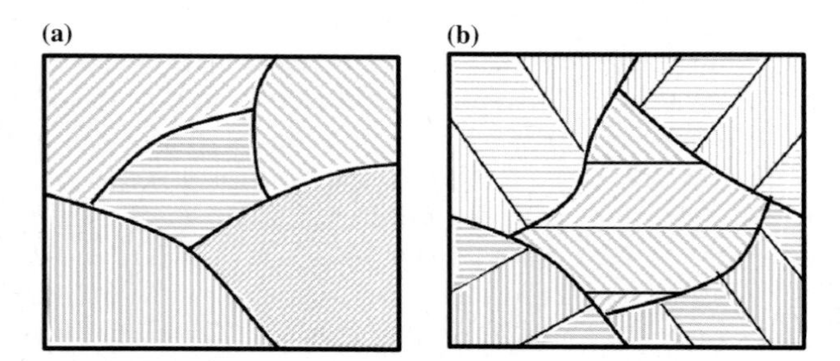

Fig. 8.22. (**a**) Simple platelet twinning in YBCO and (**b**) strip structure of twins in $BaTiO_3$

[2] In particular, the tetragonal–orthorhombic phase transition in YBCO superconductor demands oxygen diffusion into crystalline lattice [277].

sintered under thermal gradient at different initial press-powder porosittes [798]. The FC structure is described by lattice with square cell size, δ. After ceramic fabrication (i.e., sintering, shrinkage, cooling and poling process) a microstructure contains the grains, voids, microcracks at the intergranular boundaries and domain structure. We estimate a microcrcaking by poling due to domain re-orientations which differ from the 180° ones (otherwise the residual stresses do not occur). In this case, the spontaneous strains, ε, for different ferrohardness degrees have been defined by experiments [276] as $\varepsilon_{\mathrm{I}} = 14 \times 10^{-3}, \varepsilon_{\mathrm{II}} = 11 \times 10^{-3}$ and $\varepsilon_{\mathrm{III}} = 8 \times 10^{-3}$ (where indexes I, II and III are concerned with compositions, possessing by ferrohardness, mean ferrohardness and ferrosoftness, respectively). Then, the critical length of the cracked boundary, $l_{\mathrm{c}}^{\mathrm{s}}$, and the facet length subject to cracking, l, for each composition are found by (8.4) and (8.5).

After localization of the microcracks at the intergranular boundaries, we model stress sources at the polydomain grain boundaries by the continuously distributed effective dislocations and obtain an equilibrium domain width as [840]

$$d = \left[\frac{\pi^3 D(1-\nu)\sigma_{\mathrm{d}}}{4.2GS_0^2(2-\nu)}\right]^{1/2} , \tag{8.45}$$

where D is the mean grain size independent of the ferrohardness degree, σ_{d} is the energy of 90° domain boundary, G is the shear modulus, S_0 is the spontaneous shear strain and ν is Poisson's ratio. Note that the critical grain size, D^*, below of which there is no domain twinning, may be found under condition of $D = d$ (a structure of mono-domain grains) as

$$D^* = \frac{\pi^3(1-\nu)\sigma_{\mathrm{d}}}{4.2GS_0^2(2-\nu)} . \tag{8.46}$$

Further, it is known that the FC poling causes a concentration fall of domain boundaries in the poled sample to compare with the non-poled one. The structure transition of the twinning type under stress near crack tip can provoke a reverse process of the domain boundaries density rise, but a twinning process zone, forming around propagating crack, will drag the crack. Obviously, the process zone size depends on the composition ferrohardness and the ceramic piezoelectric properties. Moreover, this size is also defined by the crack propagation direction. It should be noted that the maximum influence is rendered on the crack, advancing along residual poling direction. In contrast, along normal one this structure change is restrained by the stress state near crack tip, that is, the crack propagation is realized without an additional energy loss, and the twinning process zone around the crack does not form.

Around initiated macrocracks favorably oriented to poling direction, there are twinning process zones with thickness h. Then, a toughening due to the twinning may be estimated in the crack shielding terms. The shielding, corresponding to the crack growth nucleation is represented by the solution for stationary crack, but the crack nucleation toughness, K_{c}, has form [192]:

$$K_{\mathrm{c}} = (E_{\mathrm{m}}/H)^{1/2} K_{\mathrm{c}}^0 , \tag{8.47}$$

where $K_c^0 = (2\gamma_0 E_m)^{1/2}$ is the critical SIF, required for the crack growth and reflecting a fracture resistance of twin planes, γ_0 is the fracture energy, H is Vickers hardness, E_m is Young's modulus for cracked ceramic, calculated by (8.16), where β_m is the microcracking density, formed by poling. Neglecting the closed porosity effects, the steady-state crack toughening due to twinning is estimated as [192]

$$\frac{K_{SS}}{K_c^0} = \left\{1 + \left(\frac{E_m}{H} - 1\right)\left[\alpha^2 \ln\left(\frac{h}{d}\right) - \frac{\sigma_0^2 h}{(K_c^0)^2}\right]\right\}^{1/2}, \tag{8.48}$$

where K_{SS} is the steady-state crack toughness, $\alpha^2 = 3/8\pi, \sigma_0$ is the yield stress or the threshold stress, defining the twinning nucleation near crack tip.

The process zone width, h, for different ferrohardness degrees is found, using a calculation of the parameter, $0 < \xi < 1$, reflecting a stability degree of piezoelectric composition to de-poling (i.e., the ferrohardness degree) [572]. These results are based on the energy balance of the modeled process and on application of the finite element method to calculate the zone of feasible microstructure transformations at the crack tip. For PZT ceramic compositions I, II and III, respectively, we obtain the values [572]: $4h/D = 0.2$ (for $\xi = 0.3$), 0.5 (for $\xi = 0.5$) and 1.2 (for $\xi = 0.7$).

On account of a necessary number of computer realizations, defined on the basis of stereological method [145], in order to obtain numerical results, we use known data for PZT ceramics and related materials: $E = 70$ GPa, $\nu = 0.25$ [849], $H = 3$ GPa [848], $G = 20$ GPa, $S_0 = 0.01, \sigma_d = 3 \times 10^{-3}\,\text{J/m}^2$ [840]. The cell size, δ, is selected to be equal to the critical size, l_c^s, for composition with mean ferrohardness. Then, selecting $\gamma_0 = \gamma_s - \gamma_{gb}(1 - f_m)/2$, where γ_s is the surface energy for bulk body ($\gamma_s \approx 2\gamma_{gb} = 6\ \text{J/m}^2$ [796]) and f_m is the fraction of facets cracked in the poling process, we calculate the values of D, β_m and f_m using computer simulation. Finally, the threshold stress, $\sigma_0 = 20\,\text{MPa}$, is used which has been estimated in the experiments on impact loading of $BaTiO_3$ ceramic [578]. The numerical results show that the observed decreasing of the grain size, D, with increasing of the initial porosity, C_p^0, leads to a decreasing of the width of equilibrium domain, d, and the process zone size, h. The latter also decreases at rise of the FC ferrohardness. In the toughening due to twinning, the value of K_c/K_c^0, defining start of the crack growth, changes in the limits: $4.06 \pm 0.15, 4.30 \pm 0.15$ and 4.74 ± 0.08, for I, II and III compositions, respectively, at the initial porosity $C_p^0 = 0{-}60\%$. Similar dependencies of toughening, K_{SS}/K_c^0, on the ferrohardness and the initial porosity in the case of the steady-state crack, are shown in Fig. 8.23 [798]. The results for the case of mean ferrohardness are close to the ratio of the fracture toughness in existence of twinning to its absence in γ-TiAl [192]. In this case, the known growth of the twinning process zone and fracture toughness with increasing of FC ferrosoftness is supported [849]. FC toughening, caused by the twinning process, is a more effective toughening mechanism compared to microcracking near crack tip [518], to crack branching and crack

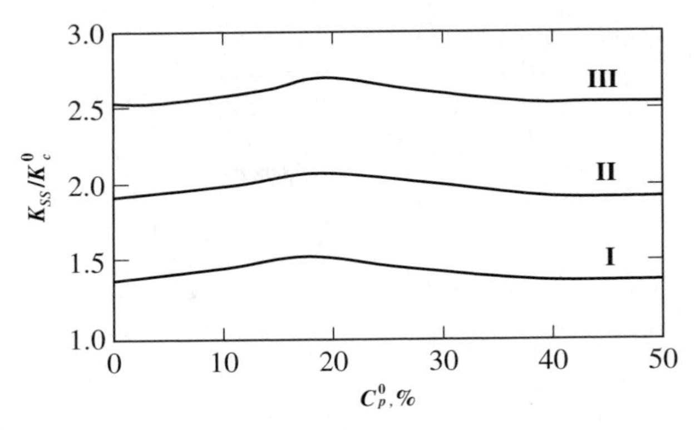

Fig. 8.23. Fracture toughness (K_{SS}/K_c^0) at steady-state crack vs initial porosity (C_p^0) for different ferrohardness of FC

bridging [797], and also for phase transformations due to coexistence of rhombohedral and tetragonal ferroelectric phases in the region of morphotropic transitions. The obtained maximum toughening in the range of change of the initial porosity (at $C_{\mathrm{p}}^0 = 20\%$) could be explained by mutual superposition of complex microstructure effects and is qualitatively supported by similar results for the toughening mechanism of microcracking near crack tip [849].

8.2.2 Fracture Features in Domain Structure of Ferroelectric

In order to consider fracture features in the domain structure of ferroelectric, we study a possibility of sub-critical crack in PZT monocrystal, growing parallel to internal 180° domain or phase boundary in the structure of bi-material [799, 811]. Three mechanisms of the crack braking at 90°-domain boundaries are studied in [512, 517], namely (i) due to change of the crack orientation, (ii) due to the crack interaction with non-coherent boundary and (iii) due to the crack advancing through the boundary with absorbed impurities.

The crystallite with layer domain structure or co-existing phases will be present as a composite system, which consists of homogeneous layer with thickness h and half-infinite substrate (see Fig. 8.24). Let their elastic properties (E, ν) be the same and the homogeneous tensile stresses, acting in the layer, $\sigma_0 = E\varepsilon$, are found by microstrain ε, caused by the thermal and phase properties of the FC material. The stress analysis shows that the most possible crack initiation occurs in substrate and it extends parallel to the internal interface of the composite. We assume a stable (sub-critical) character of the crack growth. When the crack is localized in substrate at the depth λh below the interface, then the layer above the crack is subjected to a bend and some elastic strain energy retains after crack propagation. In order to estimate this energy, we take into account that the contributions of the strain energy due to the compressive force P and bending moment M are additive [1059]. Then,

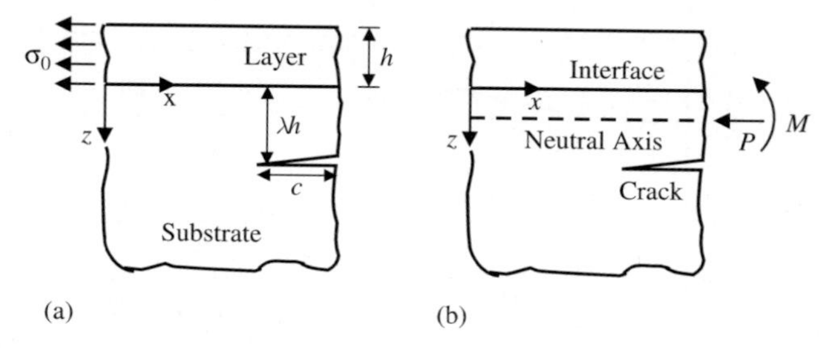

Fig. 8.24. Used composite structure: (**a**) tension σ_0 is caused by microstrain ε in the homogeneous layer; (**b**) equivalent system with applied force P and bending moment M

using the theory of composite beam and based on the approach of [221], we obtain the asymptotic strain energy release rate, G, for steady-state crack with length c as

$$G = \frac{\partial}{\partial c}\left(\frac{P^2 c}{2EA} + \frac{M^2 c}{2EIh^3}\right) , \tag{8.49}$$

where the dimensionless moment of inertia, I, of the beam per unit layer thickness and the effective cross-section, A, are

$$I = (\lambda + 1)^3/12 \; ; \qquad A = h(1 + \lambda) \; ; \qquad \lambda = z/h \; . \tag{8.50}$$

The load P and the bending moment M (per unit thickness) caused by uniform tensile stress σ_0 are defined as

$$P = \sigma_0 h \; ; \qquad M = \sigma_0 h^2 \lambda/2 \; . \tag{8.51}$$

Then, we have from (8.49) to (8.51):

$$G = \frac{\sigma_0^2 h(4\lambda^2 + 2\lambda + 1)}{2E(\lambda + 1)^2} \; . \tag{8.52}$$

Further, the dimensional analysis gives relations for modes I and II stress intensity factors, K_{I} and K_{II}, assuming that each of them has own contributions from the force P and the moment M [221]:

$$K_{\mathrm{I}} = C_1 P h^{-1/2} f(\lambda) + C_2 M h^{-3/2} g(\lambda) \; ; \tag{8.53}$$

$$K_{\mathrm{II}} = C_3 P h^{-1/2} f(\lambda) + C_4 M h^{-3/2} g(\lambda) \; , \tag{8.54}$$

where C_i are unknown constants and $f(\lambda), g(\lambda)$ are unknown functions. An assumption about simplicity of the functions $f(\lambda)$ and $g(\lambda)$ is supported by the computational results of [221]. Comparing (8.49) and (8.52) with the relationship: $GE = K_{\mathrm{I}}^2 + K_{\mathrm{II}}^2$, connecting the strain energy release rate with the SIFs, we obtain:

$$(C_2^2 + C_4^2)g^2 = 1/2I \; ; \qquad (C_1^2 + C_3^2)f^2 = h/2A \; . \tag{8.55}$$

Assume that the factors C_i contribute only to the constant terms of (8.55), while the functions $f(\lambda)$ and $g(\lambda)$ contribute to the variables; then (8.55) are reduced to

$$(C_2^2 + C_4^2) = 1/2\ ; \qquad (C_1^2 + C_3^2) = 1/2\ ; \qquad g = I^{-1/2}\ ; \qquad f = (h/A)^{1/2}\ . \tag{8.56}$$

In order to define finally the constants of C_i, it is necessary to suggest a strict solubility of the problem for the case of either $P = 0$ or $M = 0$. The validity of the last condition in the present formulation of the problem and also the construction and computational solution of the corresponding integral equation for SIF at half-infinite crack propagation near free surface have been represented in [1059]. The unknown constants have the values: $C_1 = 0.434$ and $C_3 = 0.558$. Besides, there is an additional condition for unknown constants: $C_1C_2 + C_3C_4 = 0$. Then, we obtain from (8.56), $C_2 = C_3 = 0.558$; $C_4 = -C_1 = -0.434$; and the relationships for K_{I} and K_{II} are obtained from (8.53) and (8.54), substituting the corresponding parameters, as

$$\frac{K_{\mathrm{I}}}{\sigma_0 h^{1/2}} = \frac{1}{(1+\lambda)^{1/2}}\left(0.434 + 0.966\frac{\lambda}{\lambda+1}\right)\ ; \tag{8.57}$$

$$\frac{K_{\mathrm{II}}}{\sigma_0 h^{1/2}} = \frac{1}{(1+\lambda)^{1/2}}\left(0.558 - 0.752\frac{\lambda}{\lambda+1}\right)\ . \tag{8.58}$$

As is followed from test data for different bimaterial systems and the loading schemes [221, 1059], the crack path, corresponding to the simple Mode I stress intensity demonstrates a surprising stability, and fracture resistance values, obtained under conditions of the Mode I, have a fine reproducibility to compare with the results for mixed mode. Furthermore, as the condition of steady-state crack into ferroelectric domain structure, we select the following one:

$$K_{\mathrm{II}} = 0\ . \tag{8.59}$$

Hence, from (8.58), the depth λ^*, at which the crack has a steady-state trajectory, namely: $\lambda^* = z^*/h = 2.876$ is found. The existence of the asymptotic limit permits to calculate the critical layer thickness, h^*, below which the complete fracture is inhibited. This value is obtained by equating the asymptotic value of K_{I} to the corresponding value of the fracture toughness, K_{Ic}. In the case of the crack growth into substrate along the steady-state trajectory, $K_{\mathrm{II}} = 0$, (8.58) provides the prediction:

$$h^* = 0.755(1+\lambda^*)(K_{\mathrm{Ic}}/\sigma_0)^2\ . \tag{8.60}$$

The above assumption about stable crack growth in FC crystallites demands additional discussions and grounds. First, in [343], it has been shown that a stress concentration near crack tip in the region of the morphotropic boundary at existence of the tetragonal (T) and rhombohedral (R) phases can initiate phase transitions in the ceramic grains. It has been proved that for short cracks ($c < c^*$), at any great stresses, an initiation of phase transformations near crack tip is more advantageously compared to its catastrophic growth. For long cracks ($c > c^*$), the calculation gives a higher critical load compared to Griffith's formula. Secondly, in [105], the mechanism of local phase transition near heterogeneities has been considered, in particular at crack tip. It is shown that an interaction of order parameter with strains leads to the fact that the stresses, concentrating at heterogeneities, can transform the media in other phases [105]. Thus, an occurring of local transformational plasticity in the vicinity of the crack tip is possible, dragging its growth. Thirdly, a non-linear decreasing of the mechanical strength has been observed at a tension with an increasing of a poling field in the poling of the FC on the base of $PbTiO_3$ [350]. This non-linear behavior is explained by non-stationary processes of microcracks nucleation and growth due to their interaction with re-constructing domain structure and defects. In particular, it is assumed that the vacancies, released from connections after disappearance of domain walls during poling, can form "clouds" near tips of the cracks with critical size and restrain their growth. Finally, the cracks in $BaTiO_3$ monocrystals which have been observed near indenter imprint at the (001) facet and grown along the cleavage planes of $\{100\}$, $\{110\}$ and $\{111\}$ can restrain up to total stop at strong fixed 90° domain boundaries which are similar to the twin boundaries [512]. Hence, it is seem to be correct a conclusion about principle possibility of stable, sub-critical crack growth into domain structure of FC.

As has been shown by experiments [588], the strains inside crystallites in paraelectric phase change from 5×10^{-5} to 2×10^{-4} in the morphotropic transition regions in dependence on the sintering temperature of PZT ceramics. Similar strains in grains of the T-phase are independent of temperature within test error. They exceed the microstrains in the paraelectric phase, attaining the values $\sim (6 \pm 3) \times 10^{-4}$. The microstrains in the crystallites of T- and P-phases coincide in the order of magnitude. Therefore, in computer simulations, we consider the microstrain range: $\varepsilon = 5 \times 10^{-5} - 9 \times 10^{-4}$, for definition of the uniform tensile stress of σ_0, leading to the crack advancing.

Select the values of Young's modulus, E = 63–100 GPa [849], and also the fracture toughness of PZT monocrystals, K_{Ic}. In relation to the latter, note that the poling processes lead to the fracture toughness anisotropy as of whole ceramic, as of single crystallites. In the FC ceramics, the value of K_{Ic} attains maximum value along the poling direction and minimum value in the perpendicular direction [813, 849]. However, an anisotropy of the tetragonal ferroelectric monocrystals shows another character, namely: the fracture toughness in a plane perpendicular to the direction of spontaneous poling exceeds significantly the toughness values in other directions. Fracture in this plane is not advantageous energetically, because it leads to a greater surface density of charges with different signs at the opposite surfaces of the crack [512]. This fracture character is experimentally supported by prevailing growth of microcracks through grain bulk in the poling direction [350]. These features of anisotropic behavior can be explained: in the polarized ceramic, a significant part of domains remains oriented in arbitrary direction, not contributing to the anisotropy of K_{Ic}. Moreover, for fine-grain ceramics, an intergranular fracture is proper, which excludes the crack interaction with domain structure [849]. Due to complex behavior of the fracture toughness, we use in calculations the experimental values: $K_{Ic} = 0.4\text{–}1.4\,\text{MPa m}^{1/2}$ [849].

Finally, we define the layer thickness as $h = nh_d$, where $h_d \approx 0.2$ μm is the equilibrium domain width for grain size, $D = 10$ μm, based on the simulation of the stress sources at the boundaries of polydomain grains, using continuously distributed dislocations [840]; n is the domain number in crystallite under uniform tension, σ_0. Obviously, in order for the coordinate of z^* for steady-state crack path to be in grain, it is necessary to select $h_{max} = 3.4$ μm, then $n_{max} = 17$.

Further, (8.57) and (8.60) give $K_I = 0.585\sigma_0 h^{1/2}$, and $h^* = 2.926 (K_{Ic}/\sigma_0)^2$, where K_I is the SIF for steady-state crack growth at the depth of $\lambda^* h$, and h^* is the limit layer thickness, defining a restraint fracture. The selection of the parameters, corresponding to the maximum of K_I and minimum of h^*, gives $\varepsilon = 9 \times 10^{-4}$; $K_{Ic} = 0.4\,\text{MPa m}^{1/2}$; $h = 3.4\,\mu\text{m}$; $E = 100\,\text{GPa}$. As a result, we obtain $K_I = 0.097\ \text{MPa m}^{1/2}$, $h^* = 57.8\,\mu\text{m}$. Obviously, $K_I << K_{Ic}$ and $h^* >> D$ for the considered PZT parameters. Hence, the stable (i.e., subcritical) crack growth is impossible parallel to or along the interfaces in the considered crystallites. The nucleated crack at this boundary will propagate catastrophically until it meets a fixed 90°-domain boundary. Thus, there may be stable crack growth in FC crystallites only in the cases of crack braking by the 90°-domain boundaries [512, 517].

8.2.3 Thermodynamics of Martensitic Transformation in HTSC

Now, investigate the twinning processes in YBCO superconductor [815]. As has been noted above, due to the grain TEA, the cracking of HTSC ceramics occurs during cooling that deteriorates mechanical and strength properties. At the same time, a twinning process zone, forming around an advancing crack, as may be suggested similar to ferroelectric ceramics, should lead to the crack shielding and to the fracture toughness increasing. In this case, a relaxation of internal stresses of second type, nucleated in the HTSC, occurs in two ways [277]: (i) due to the crack nucleation, which is proper for temperatures, $T \leq 500$°C, and (ii) owing to the martensitic process of re-construction of the grain domain structure, that is most intensive at $T \geq 500$°C.

During oxygen annealing of HTSC, two characteristic sites are observed in the curve of oxygen parameter (x) depending on time (t). At the initial "fast" stage of the process, where an accumulation of oxygen by cuprate follows to diffusive regime, the dependence "x–t" is subjected to parabolic law. At the next "slow" stage, the kinetics of reaction are found by relaxation of elastic stresses, and oxidation follows to $\partial x/\partial t \approx$ const regime, that is, to linear law [277, 278]. As has been shown in [277], $YBa_2Cu_3O_{7-x}$ compound possesses the "shape memory effect" that speaks about martensitic mechanism of the stress relaxation at the slow stage of the material oxidation. Therefore, it may be assumed that deletion of the internal stresses of second type initiated near growing macrocrack can occur, according to the martensitic mechanism in account of the energetically advantageous re-construction of domain structure.

We shall assume that the stresses, introduced in the twinning process, to be shear. The normal stresses, defining a boundary microcracking, give a secondary contribution in the energy of plastic deformation and do not take into account here.

Estimate the critical number of twins, η_c, in spherical grain with radius R, corresponding to the martensitic transformation. With this aim, consider alterations in the thermodynamic potential, which accompany formation of the twinned martensite. They include the increment of the mechanical potential, $\Delta\Phi^{m}$, consisting of change of the strain energy and interaction energy, the increment of the surface energy, $\Delta\Phi_{S}$, and also the increment of the chemical potential, ΔF_{c}. The last is independent of geometrical parameters of twins, and for spherical particle with radius R, the increment of the chemical free energy per crystallite is found by [863]

$$\Delta F_{c} = -(4/3)\pi R^3 \Delta F_0 , \tag{8.61}$$

where

$$\Delta F_0 = P_V \delta_{ij} \Delta e_{ij}^{T} = P_V \Delta V , \tag{8.62}$$

where P_V is the stress, leading to the transformation; Δe_{ij}^{T} is the difference of components of the strain tensor; ΔV is the volumetric deformation at the transformation; and Kronecker's delta-function is found as

$$\delta_{ij} = \begin{cases} 1 & i = j \\ 0 & i \neq j \end{cases} \tag{8.63}$$

The surface work is defined by changes occurring at the martensite interface and during formation of the twin boundaries. The total increment of the surface energy, $\Delta\Phi_{S}$, for spherical grain is calculated as [261]

$$\Delta\Phi_{S} = 4\pi R^2 \Delta\Gamma_{i} + (2/3)\pi R^2 (\eta - 1)(\eta + 1)\Gamma_{t}/\eta , \tag{8.64}$$

where $\Delta\Gamma_{i}$ is the difference of the surface energy; Γ_{t} is the energy of twin boundary; $\eta = 2R/d$ is the number of twins per grain; d is the twin extent (see Fig. 8.25).

The difference of the mechanical energy has two terms, namely (i) due to the twinning, $\Delta\Phi_{T}^{m}$, and (ii) owing to macroscopic change of the grain shape, $\Delta\Phi_{m}^{m}$. The component caused by the twinning is calculated using Eshelby's method [259] for study of the tangential stress state, developing in the martensite plates, consisting of twins. Then, for Poisson's ratio ($\nu = 0.2$), we have [261]

$$\Delta\Phi_{T}^{m} = \frac{1.6\pi R^3 G \gamma_{T}^{2}}{3\eta} , \tag{8.65}$$

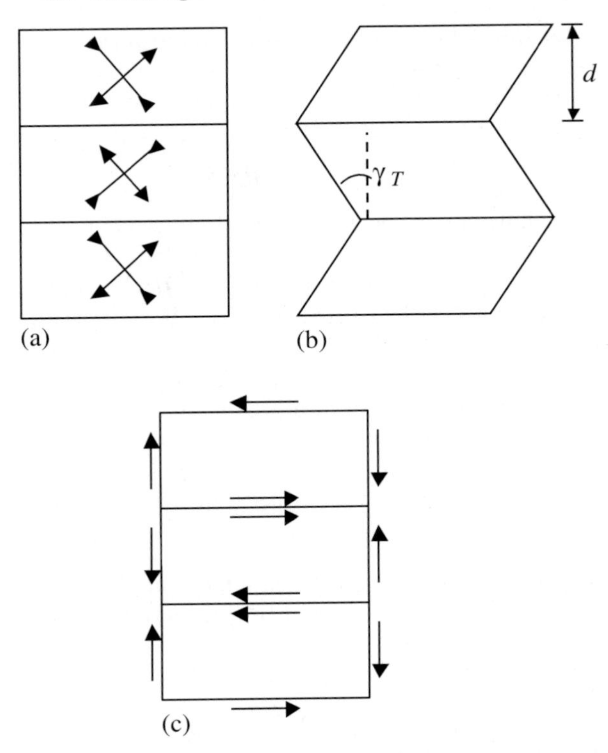

Fig. 8.25. (**a**) No restraint shape of particle before start of twinning; the directions of tension and compression along main axes, characterizing transformation are shown by *arrows*; (**b**) the same particle after twinning (γ_T is the shear angle, d is the twin extent); (**c**) restoration of initial shape, using uniform tangential stresses ($\tau = G\gamma_T/2$), shown by *arrows*

where G is the shear modulus, γ_{T} is the shear deformation at the twinning. This term is determined by the tangential stresses, an action which is restrained by the zone close to the interface "parent phase–twinning product". Moreover, due to neglectfulness by the normal stresses, it may be assumed that (8.65) presents lower boundary for possible values of $\Delta\Phi_{\mathrm{T}}^{\mathrm{m}}$.

Macroscopic component includes the strain energy and the energy of interaction with applied stress p_{ij}^{A}. We are limited by only dilatation component of the applied stress, p^{A} (neglecting deviator component) as while the fraction of transformed particle contains a set of planes, formed by shear, there are altogether alternative planes subjected to shear of opposite sign. Therefore, appreciable macroscopic shear strains apparently are difficult expected even in the case of finite number of twins. Then, $\Delta\Phi_{\mathrm{m}}^{\mathrm{m}}$ is found as [265]

$$\Delta\Phi_{\mathrm{m}}^{\mathrm{m}} = (4/3)\pi R^3 \Delta V (0.14 E \Delta V - p^{\mathrm{A}}) \,, \tag{8.66}$$

where $E = 2G(1+\nu)$ is Young's modulus.

Taking into account (8.61), (8.64)–(8.66), we obtain a change of the total potential due to the transformation as

$$\frac{\Delta\Phi}{(4/3)\pi R^3} = -\Delta F_0 + \frac{6\Delta\Gamma_{\rm i}}{\eta d} + \frac{(\eta^2-1)}{\eta^2 d}\Gamma_{\rm t} + 0.14E\Delta V^2 - p^{\rm A}\Delta V + \frac{0.4}{\eta}G\gamma_{\rm T}^2 \,. \tag{8.67}$$

Thus, the transformation will be defined by four main parameters, namely (i) the chemical free energy (ΔF_0), (ii) the grain radius (R), (iii) the extent of twins (d) and (iv) the applied hydrostatic component of stress ($p^{\rm A}$).

In order to estimate a critical value of $\eta_{\rm c} = 2R_{\rm c}/d$, corresponding to the transformation (where $R_{\rm c}$ is the critical grain radius), we equate the change of the total potential, $\Delta\Phi$, to zero. In this condition, there is minimum value of the critical number domains per grain, $\eta_{\rm c}$, at which the transformation will start. Then, we have from (8.67):

$$-\Delta F_0 + \frac{6\Delta\Gamma_{\rm i}}{\eta_{\rm c} d} + \frac{(\eta_{\rm c}^2-1)}{\eta_{\rm c}^2 d}\Gamma_{\rm t} + 0.14E\Delta V^2 - p^{\rm A}\Delta V + \frac{0.4}{\eta_{\rm c}}G\gamma_{\rm T}^2 = 0 \,. \tag{8.68}$$

Next, estimate the twin extent, d, using a modeling of the stress source at the boundaries of polydomain grains by the continuously distributed dislocations as [840]

$$d = \left(\frac{\pi^3\sigma_{\rm d} D}{12.6GS_0^2}\right)^{1/2} , \tag{8.69}$$

where S_0 is the spontaneous shear strain; $\sigma_{\rm d}$ is the energy of 90°-domain wall. Note that the square root dependence $d \sim D^{1/2}$ has also been predicted for partially stabilized ZrO_2 at $D >> d$ [261]. Moreover, the critical grain size, D^*, lesser than which the domain twinning does not occur, is estimated from the condition, $D = d$. Hence

$$D^* = \frac{\pi^3\sigma_{\rm d}}{12.6GS_0^2} \,. \tag{8.70}$$

This value corresponds to the grain, which has no sufficient elastic energy to render significant influence on the agreement of domain wall, possessing the energy $\sigma_{\rm d}$.

Substituting (8.69) into (8.68) and assuming $p^{\rm A} = 0$ (i.e., an absence of applied stresses), we obtain finally the equation for calculation of the critical value of $\eta_{\rm c}$.

8.2.4 About Toughening of Superconducting Ceramics

Today, a special value has consideration of stable (sub-critical) crack growth at study of fracture processes in the framework of microstructure fracture mechanics. Differing from unstable (catastrophic) fracture, submitting to the classic Griffith–Irwin's theory, it can take into account internal mechanisms of fracture resistance, which are intrinsic for the material structure, and defines essentially true strength properties. In the framework of the force approach by using the SIF for given fracture mode (I, II, III or their combination), the constitutive relationships, defining equilibrium states of crack growth at the pointed stages of its propagation, have the form [163]:

$$\mathrm{d}K/\mathrm{d}c > \mathrm{d}K_\mathrm{c}/\mathrm{d}c \text{ for unstable fracture ;} \tag{8.71}$$

$$\mathrm{d}K/\mathrm{d}c < \mathrm{d}K_\mathrm{c}/\mathrm{d}c \text{ for stable fracture ,} \tag{8.72}$$

where c is the length of the growing crack; K_c is the critical value of SIF, defining the crack start (i.e., the ceramic fracture toughness). If there are internal mechanisms, contributing in the change of the crack growth, then effective driving force, acting in the crack tip (K), can be presented as a sum of applied external load (K_a) and internal residual stresses (K_i):

$$K = K_\mathrm{a} + \sum_i K_i \ . \tag{8.73}$$

The terms of K_i in (8.73) can interpret the shielding and anti-shielding (leading to the crack amplification) influences on transmission of external stresses to the crack tip depending on their signs (i.e., negative or positive, respectively). In particular, the microcracking near macrocrack, crack branching and crack bridging, studied in Sect. 7.1, lead to the shielding effects. At the same time, the microcrack–macrocrack coalescence causes a decreasing of fracture resistance (the anti-shielding effect).

Now, consider directly the toughening of HTSC ceramic, caused by the processes of twinning near growing crack. In this case, the hydrostatic component of stress, required to introduce the transformation, is obtained from the condition $\Delta\Phi \le 0$ and can be written using (8.67) as

$$p^\mathrm{A}\Delta V \ge -\Delta F_0 + \frac{6\Delta\Gamma_\mathrm{i}}{\eta d} + \frac{(\eta^2-1)}{\eta^2 d}\Gamma_\mathrm{t} + 0.14E\Delta V^2 + \frac{0.4}{\eta}G\gamma_\mathrm{T}^2 \ . \tag{8.74}$$

The hydrostatic stress near the crack tip is subjected to the non-equality [261]:

$$p^\mathrm{A} \le \frac{2K(1+\nu)}{3\sqrt{2\pi r}}\cos(\Theta/2) \ , \tag{8.75}$$

where $\{r, \Theta\}$ are the polar coordinates connected with the crack tip. Then, using non-equalities (8.74) and (8.75), we obtain the distance of r_c^* from the crack tip at which single grain will be subjected to the transformation. We have for $\nu = 0.2$:

$$2\pi r_\mathrm{c}^* = \left\{\frac{0.8K\Delta V d\eta^2\eta_\mathrm{c}^2\cos(\Theta/2)}{[2(3\Delta\Gamma_\mathrm{i} + 0.2G\gamma_\mathrm{T}^2 d)\eta\eta_\mathrm{c} - \Gamma_\mathrm{t}(\eta_\mathrm{c}+\eta)](\eta_\mathrm{c}-\eta)}\right\}^2 \ . \tag{8.76}$$

Then, the size of the transformation zone, r_c, can be calculated, using (8.76) at $K = K_\mathrm{c}$ and maximum value of the reached toughness (for $\Theta \approx \pi/3$)

[262, 668]. The latter suggests that each particle is subjected to sufficient shear stress, which introduces the martensitic transformation under condition of thermodynamic limit (8.68) (at $p^{\mathrm{A}} = 0$), superimposed on the particle. Note that the value of the angle $\Theta \approx \pi/3$ divides the zones of the toughening and the crack amplification near the crack tip. The transformed particles at the crack front increase K_{tip} (the stress concentration at the crack tip), while the particles disposing into the region, $|\Theta| > \pi/3$, lead to the opposite effect.

Finally, consider the toughening, increasing fracture resistance of HTSC ceramics due to twinning in the process zone with width $2r_{\mathrm{c}}$, surrounding the advancing crack (see Fig. 8.26). Taking into account the superconductor behavior, caused by the hysteresis curve "σ–ε", we can estimate the toughening for the steady-state crack, caused by the twinning as [102, 192]

$$K_{\mathrm{SS}} = \sqrt{(K_{\mathrm{c}}^{0})^2 + E\Delta J_{\mathrm{SS}}}\ , \tag{8.77}$$

where $\Delta J_{\mathrm{SS}} = 2\int\limits_0^{r_{\mathrm{c}}} U(y)\mathrm{d}y$ is the increment of fracture resistance; U is the square under the curve "σ–ε",calculated at each value of y-coordinate on

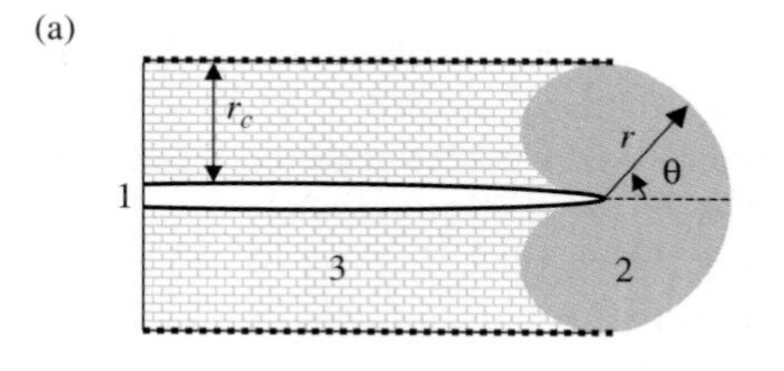

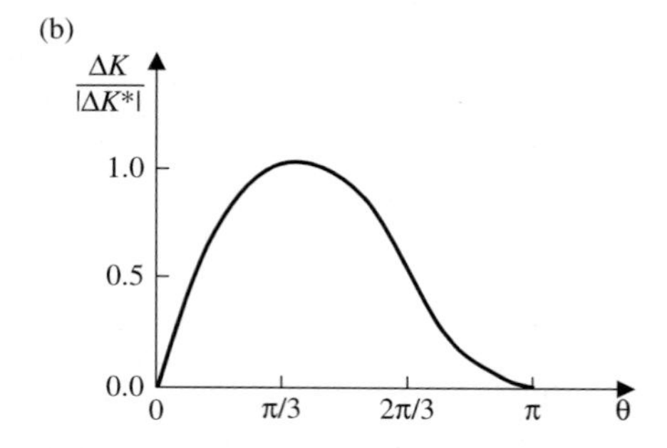

Fig. 8.26. Effect of transformation zone, forming near crack (1), on ceramic toughening: (**a**) frontal (2) and wake (3) zones of toughening (the pointed parameters are discussed in text); (**b**) fracture toughness change, connected with disposition of transformed particle; ΔK is the SIF difference due to total development of the transformation zone, that is, at great values of $\Delta c/r_c$ [262]

the basis of the governing law for the material subject to the twinning. Equation (8.77) is reduced to the following form for one-axis curve "σ–ε" [192]:

$$\frac{K_{\mathrm{SS}}}{K_{\mathrm{c}}^{0}} = \left\{1 + \frac{(E-H)}{H}\left[\frac{3}{8\pi}\ln\left(\frac{r_{\mathrm{c}}}{d}\right) - \left(\frac{\sigma_0}{K_{\mathrm{c}}^{0}}\right)^{2} r_{\mathrm{c}}\right]\right\}^{1/2}, \qquad (8.78)$$

where K_{SS} is the fracture toughness of the superconductor for the steady-state crack; $K_{\mathrm{c}}^{0} = (2\gamma_0 E)^{1/2}$ is the ceramic toughness without the twinning process; γ_0 is the fracture energy; H is the hardness factor of the ceramic at indentation; σ_0 is the threshold stress, stating a transition to non-elastic material behavior due to nucleation of the twinning near the crack tip.

In order to obtain numerical results, we use the known parameters for $YBa_2Cu_3O_{7-x}$, namely: $G = 41\,\mathrm{GPa}, \gamma_{\mathrm{T}} \sim S_0 = 0.018, \mathit{\Gamma}_{\mathrm{t}} \sim \sigma_{\mathrm{d}} = 0.01\,\mathrm{J/m^2}$ [261]. In more detail, we discuss a selection of the values for the volumetric deformation, ΔV, and the threshold stress, $\sigma_0(\sim P_{\mathrm{V}})$. As has been shown in [344], the spontaneous (non-elastic) deformation initiates at temperature $T_{\mathrm{S}} = 920\,\mathrm{K}$, resulting in the structural phase transition from tetragonal phase ($P4/mmm$) to orthorhombic ($Pmmm$). This deformation is defined by the components C_{22}, C_{23} and C_{55} of the elastic stiffness matrix. Below the temperature of T_{S} (into orthorhombic phase), the formed domain structure of HTSC can re-switch from one spontaneous deformed oriented state to the other by external shear load, σ. The shear modulus attains minimum value for $\sigma = \sigma_{\mathrm{c}}$ (where σ_{c} is the coercive stress), because the number of twins is maximum in this case, and the sample is most compliant to mechanical loading. The number of twins decreases at $\sigma > \sigma_{\mathrm{c}}$, leading to greater stiffness of the sample and to increasing of G. Due to this, we select as the threshold value the next one: $\sigma_0 = \sigma_{\mathrm{c}}$. The presented temperature dependencies of σ_{c} [344] in the vicinity of the phase transition give the value of $\sigma_{\mathrm{c}} \approx 3\,\mathrm{MPa}$. This is much less than the corresponding value of the threshold stress, causing the twinning (20 MPa) that is calculated in [578] for $BaTiO_3$ samples, tested on impact load at different rates of the loading, and also than the value for partially stabilized ZrO_2 (4 GPa) [863]. Finally, the spontaneous deformation jump at the phase transition, we suggest, is equal to $\Delta V = 10^{-4}$, according to experimental results of [344]. This is also much less than the corresponding values for $BaTiO_3(54 \times 10^{-4})$ [849] and for $ZrO_2(57 \times 10^{-3})$ [863].

Substitute the necessary values in (8.68) (at $p^{\mathrm{A}} = 0$), calculating the preliminary values of the twin extent, d, from (8.69) for the observed grain sizes of YBCO, $D = 10$–$100\,\mu\mathrm{m}$. The solution (8.68) for the above range of the grain sizes leads to the value of $\eta_{\mathrm{c}} \to 0$. Then, from (8.76) and (8.77) an absence of real toughening for HTSC ceramics, caused by the twinning processes, is found. This is explained by very small magnitudes of ΔV and $\sigma_0(\sim P_{\mathrm{V}})$, defining a spontaneous strain in the YBCO, compared to the corresponding values for partially stabilized ZrO_2 [863] and ferroelectric ceramic $BaTiO_3$ [578, 849], where the twinning processes play the most important rule in material toughening. Thus, the toughening (or crack amplification) at the fracture of $YBa_2Cu_3O_{7-x}$ superconducting ceramic is caused by microcracking due to deformation or/and thermal anisotropy, and by the processes, connected with microcracking (in particular crack branching and crack bridging), but not transformations of the martensitic type.

8.3 Toughening Mechanisms for Large-Grain YBCO

8.3.1 Model Representations

Two main computer simulation approaches, namely Monte-Carlo simulations [814, 821] and phase field method [937, 956, 1016], are usually used to study microstructure transformations, which accompany the YBCO fabrication. The phase field method has been discussed in Sect. 7.1.6. Here, we consider a modified Monte-Carlo scheme. The simulation of the YBCO precursor microstructure will be based on the re-crystallization model for ferroelectric ceramics [516, 823] and on the study of thermal conductivity in mixes and structure-heterogeneous solids [234]. Formation of the final YBCO microstructures, using seeds at existence of dispersed 211 phase, will be found by the Monte-Carlo techniques [22, 1009, 1010].

A precursor microstructure after YBCO pellet modeling is represented by discrete lattice with 2000 square cells of characteristic size δ. Each lattice cell is assigned a number (between 1 and Q), corresponding to the orientation of the grain in which one is embedded. All grains have different orientations, defined during the pellet microstructure simulation, in which lesser numbers correspond to earlier initiated crystallites. This liquidates a neighborhood of the grain of the same orientation. Then, the cells which have neighbors with unlike orientation lie at the grain boundary; in the other case, they are placed in grains. The grain boundary energy is specified, defining an interaction between nearest neighboring lattice sites as [22, 1009, 1010]

$$E = -J \sum_{\text{nn}} (\delta_{S_i S_j} - 1) , \tag{8.79}$$

where S_i is one of the Q orientations on site $i (1 \leq S_i \leq Q), \delta_{ij}$ is the Kronecker's delta. The sum is taken over all nearest neighbor sites (nn), J is a positive constant that sets the energy scale of the simulation. Note that the orientation dependence of the energy of a straight boundary segment exhibits a ratio of the maximum to minimum boundary energy of $2/\sqrt{3}$ for triangular lattice and $\sqrt{2}$ for square one. Both ratios are close to unity, that is, the grain boundary energy is nearly isotropic and only weakly lattice dependent [22].

In simulation of boundary motion kinetics by the Monte-Carlo techniques, a lattice cell is selected at random, and a new trial orientation is randomly changed to one of the other grainorientations. Then, the energy alteration caused by change in orientation (ΔE) is evaluated. The transition probability (P) is given by [22]

$$P = \begin{cases} \exp(-\Delta E/k_\mathrm{B}T) & \Delta E > 0 \\ 1 & \Delta E \leq 0 \, , \end{cases} \tag{8.80}$$

where k_B is Boltzmann constant, T is the temperature. A re-orientation of a site at a grain boundary corresponds to boundary migration. A boundary segment moves with rate, related to the local chemical potential difference (ΔE_i) as [22]

$$v_\mathrm{i} = C[1 - \exp(-\Delta E_\mathrm{i}/k_\mathrm{B}T)] \, , \tag{8.81}$$

where factor C is found by the boundary mobility and symmetry of the lattice. The simulations for the cases of $T \approx 0$ and $T \approx T_\mathrm{m}$, where T_m is the melting point, have shown similar results [1009]; therefore, we use $T \approx 0$, below.

The $(N - N_\mathrm{p})$ attempts of re-orientations, where N is the number of lattice cells and N_p is the total particle number, are arbitrarily used as unit of time and defined as one Monte-Carlo step (MCS) per site. In order to incorporate particles of 211 phase into the model, corresponding cells are assigned by orientation which are distinct from all grain orientations. It is suggested that the particle concentration and sites initially arbitrarily selected are fixed during simulation and are independent from any attempts of cell re-orientation. The insertion of the crystalline seed into microstructure is done by replacing the grains at the center of the microstructure with one large square grain with characteristic size considerably greater than the mean grain radius of the YBCO precursor pellet.

In the computer simulations, we consider three cases of YBCO microstructure evolution after primary re-crystallization (see Fig. 8.27):

(1) particle dispersion of 211 phase into matrix of 123 phase;
(2) insertion of large grain (seed) into matrix of 123 phase;
(3) insertion of large grain (seed) into matrix of 123 phase with dispersed 211 phase.

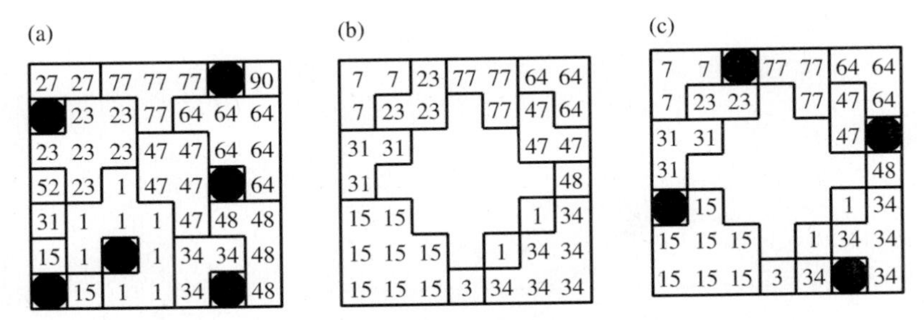

Fig. 8.27. Three models of YBCO microstructure evolution, used in computer simulations: (**a**) dispersion of 211 phase (*black*) in 123 matrix, (**b**) large-grain seed (*white*) in 123 grain matrix and (**c**) 211 dispersion and grain-seed in 123 matrix

As a condition of the computation stop in the first and third cases, an existence of one particle at any intergranular boundary at least (that corresponds to complete pinning of microstructure with dispersed particles [1009, 1010]) is selected. In the second case, the computations are finished when some grain radius attains the seed grain size.

In order to obtain statistically representative results, we use approach [145]. Then, using comparatively small grain aggregates together with stereological method accelerates computations due to the operation with smaller arrays of variables and obtains necessary statistics.

8.3.2 Effect of 211 Particles on YBCO Fracture

As has been shown in the previous chapters, a special effect on the 123 matrix fracture may be exerted by dispersed inclusions of the 211 phase. The platelet structure of 123 superconducting phase covering the 211 single inclusions during solidification from melting via a zipper-like mechanism promotes the formation at the 123/211 phase interface of the defects-faults and increased dislocation concentration. Therefore, the existence of secondary phase inclusions causes the alteration of the strength properties indirectly and directly. In the first case, these inclusions influence formation of certain microstructures during material fabrication with corresponding fracture resistance. In the second case, these inclusions define the acting toughening mechanisms, fracture toughness and strain energy release rate. Therefore, we consider the possible mechanisms of toughening in the case of the composite structure $YBa_2Cu_3O_{7-x}/Y_2BaCuO_5$.

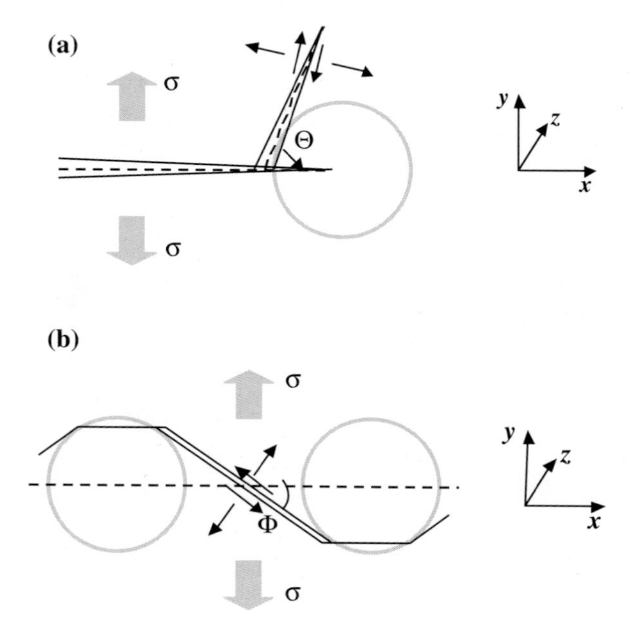

Fig. 8.28. (**a**) Tilt and (**b**) twist of crack at its interaction with inclusion

Crack Deflection

As the 123/211 interface strength is less than YBCO matrix strength, a crack deflection around dispersed particles (see Fig. 8.28) in the form of crack tilt and twist is possible. Then the driving force of the crack, which orients randomly to the main tension vector, is described by the local stress intensity factors k_1, k_2 and k_3, corresponding to the opening, sliding and tearing fracture modes. The crack driving force, governed by the strain energy release rate, G, is found as [268]

$$EG = k_1^2(1-\nu^2) + k_2^2(1-\nu^2) + k_3^2(1+\nu) , \tag{8.82}$$

where E and ν are the elastic modules.

In the case of applied SIF of mode I, K_I, the local SIFs of $k_i^\mathrm{t} (i = 1, 2, 3)$ for tilted crack are given as [268]

$$k_1^\mathrm{t} = F_{11}(\Theta)K_\mathrm{I} ; \qquad k_2^\mathrm{t} = F_{21}(\Theta)K_\mathrm{I} , \tag{8.83}$$

where

$$F_{11}(\Theta) = \cos^3(\Theta/2) ; \qquad F_{21}(\Theta) = \sin(\Theta/2)\cos^2(\Theta/2) . \tag{8.84}$$

For twisted crack the local SIFs of k_i^T are found by [268]

$$k_1^\mathrm{T} = F_{11}(\Phi)k_1^\mathrm{t} + F_{12}(\Phi)k_2^\mathrm{t}; \qquad k_3^\mathrm{T} = F_{31}(\Phi)k_1^\mathrm{t} + F_{32}(\Phi)k_2^\mathrm{t} , \tag{8.85}$$

where

$$\begin{aligned} F_{11}(\Phi) &= \cos^4(\Theta/2)[2\nu\sin^2\Phi + \cos^2(\Theta/2)\cos^2\Phi] ; \\ F_{12}(\Phi) &= \sin^2(\Theta/2)\cos^2(\Theta/2)[2\nu\sin^2\Phi + 3\cos^2(\Theta/2)\cos^2\Phi] ; \\ F_{31}(\Phi) &= \cos^4(\Theta/2)\sin\Phi\cos\Phi[\cos^2(\Theta/2) - 2\nu] ; \\ F_{32}(\Phi) &= \sin^2(\Theta/2)\cos^2(\Theta/2)\sin\Phi\cos\Phi[3\cos^2(\Theta/2) - 2\nu] . \end{aligned} \tag{8.86}$$

The computation of toughening due to crack deflection is the statistical problem, which is connected with averaging of the driving force in all possible tilt and twist angles. The effect of particle volume fraction, V_f, which causes an initial crack path alteration, is equal to [268]

$$G_\mathrm{c}/G_\mathrm{c}^\mathrm{m} = 1 + 0.87V_\mathrm{f} , \tag{8.87}$$

where G_c and G_c^m are the fracture toughness of composite and matrix, respectively. The inclusions with higher characteristic ratio have stronger effects, that is, spherical particles create a lesser toughening compared with disk-shaped and rod-shaped ones. Toughening effect attains saturation at $V_\mathrm{f} \approx 0.2$. An increased dimension of loading generally causes the increase of crack driving force due to the complexity of stress state and decrease of toughening effects.

Crack Pinning by Particles

In the case of small-scale crack bridging in which the bridging zone size (L) is small compared with the crack length, sample sizes and distance from crack tip to the sample boundaries, the crack pinning by elastic particles with size $2a$ (see Fig. 8.29) is possible. Then, the toughening factor, Λ, may be found as [101]

$$\Lambda \equiv \frac{K_c/K_{cm}}{\sqrt{\omega(1-c)}} = \left\{1 + \frac{\pi S^2 ac(1-c^{1/2})}{2K_{cm}^2}\right\}^{1/2} ; \omega = \frac{E(1-\nu_m^2)}{E_m(1-\nu^2)} ; c = (a/b)^2 , \tag{8.88}$$

where K_c and K_{cm} are the critical SIFs for crack growth in composite and in matrix, respectively; S is the particle strength; b is the radius of penny-shaped crack with central pinning particle; E and ν are effective elastic modules of composite (E_m and ν_m are the elastic matrix constants). Thus, the toughening is increased with inclusion size and strength.

Toughening due to Periodically Distributed Inclusions

The internal stress state caused by periodically distributed inclusions is almost sinusoidal at the mean plains between inclusion layers, where a finite crack is placed along the x-axis. Dislocation techniques have been applied to

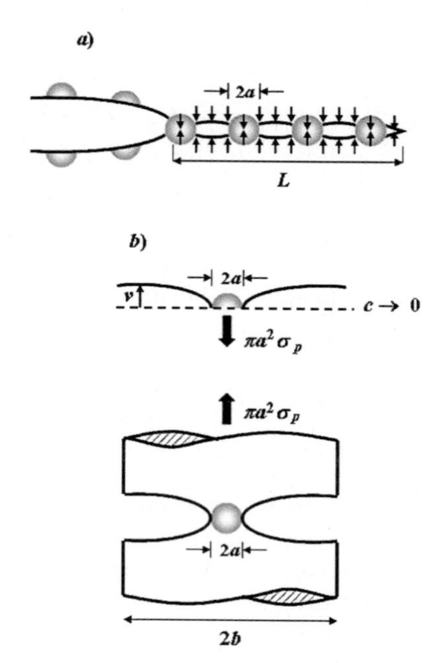

Fig. 8.29. (**a**) Crack pinning by particles and (**b**) crack opening; ν is the crack opening rate; σ_p is the stress, loading the particle

inclusions, modeled by centers of shear, compression, anti-plane and extension (dilatation) (see Fig. 8.30 [621]). These are the following results:

(1) For shear centers:

$$\Delta K_{\mathrm{II}} = 1.6\frac{\mu V_{\mathrm{f}} e_{xy}^{\mathrm{T}} \lambda^{1/2}}{(1-\nu)} \,; \qquad a << \lambda \,; \qquad b_x = h e_{xy}^{\mathrm{T}} \,. \tag{8.89}$$

(2) For horizontal compression centers:

$$\Delta K_{\mathrm{I}} = 1.6\frac{\mu V_{\mathrm{f}} e_{xx}^{\mathrm{T}} \lambda^{1/2}}{(1-\nu)} \,; \qquad a << \lambda \,; \qquad b_x = h e_{xx}^{\mathrm{T}} \,. \tag{8.90}$$

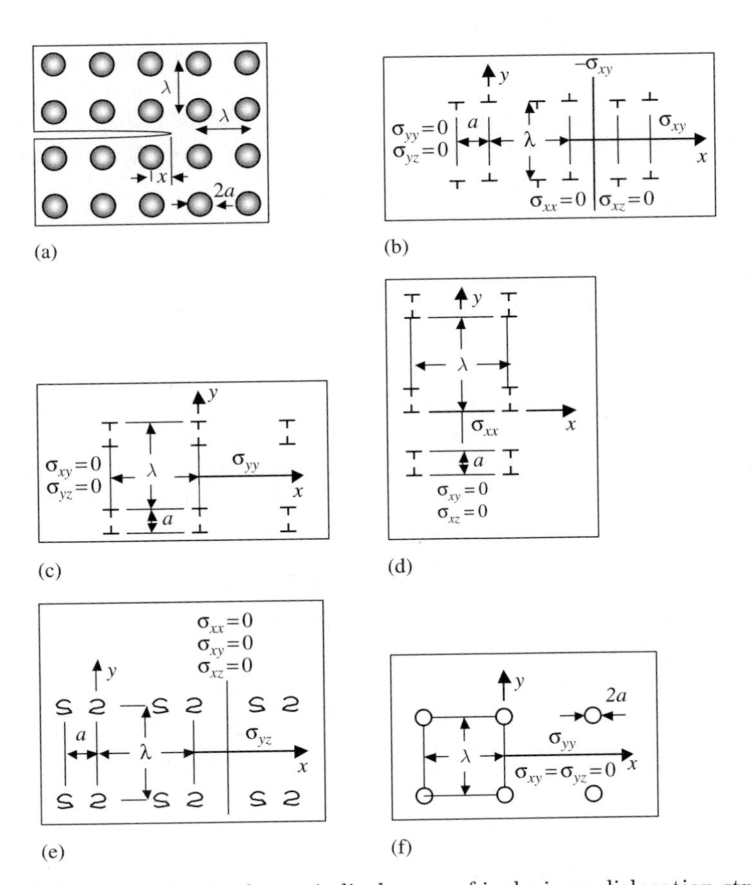

Fig. 8.30. (**a**) Crack toughening by periodical array of inclusions; dislocation structures (with Burgers vector, b), modeling the inclusions: (**b**) shear centers; (**c**) horizontal compression centers; (**d**) vertical compression centers; (**e**) anti-plane shear centers and (**f**) extension (dilatation) centers

(3) For vertical compression centers:

$$\Delta K_{\mathrm{I}} = 2.2\frac{\mu V_{\mathrm{f}} e_{xx}^{\mathrm{T}} \lambda^{1/2}}{(1-\nu)} \; ; \qquad a << \lambda \; ; \qquad b_x = h e_{xx}^{\mathrm{T}} \; . \tag{8.91}$$

(4) For anti-plane shear centers:

$$\Delta K_{\mathrm{III}} = 0.4 \mu V_{\mathrm{f}} e_{yz}^{\mathrm{T}} \lambda^{1/2} \; ; \qquad a << \lambda \; ; \qquad b_x = h e_{yz}^{\mathrm{T}} \; . \tag{8.92}$$

(5) For extension (dilatation) centers:

$$\Delta K_{\mathrm{I}} = 0.8\frac{E V_{\mathrm{f}} e^{\mathrm{T}} \lambda^{1/2}}{(1-2\nu)} \; ; \qquad a << \lambda \; ; \qquad e^{\mathrm{T}} = \Delta V / V \; . \tag{8.93}$$

Thus, the SIF difference of corresponding mode for periodically distributed inclusions is proportional to the elastic modulus, volume fraction of the inclusions, deformation mismatches ($e_{\alpha\beta}^{\mathrm{T}}$) and square root of the inclusion spacing (λ). However, it is independent of the crack length, that is, the dispersion toughening is more effective at the crack initiation to compare with its growth.

8.3.3 Some Numerical Results

The numerical results presented in Table 8.7 [822] have been found for some microstructure properties in all possible variants that are found by the 211 particle concentration ($f = 0.1$ and 0.2) and the seed size ($S_{\mathrm{S}} \approx 2S_0^{\mathrm{m}}$ and $3S_0^{\mathrm{m}}$, where S_{S} is the seed area and S_0^{m} is the maximum grain square after primary re-crystallization before beginning of the 123 grain growth). The final microstructures with the dispersed 211 particles demonstrate dependencies similar to $\bar{S} \approx 4a/f$, where $\bar{S}$ is the mean pinned grain area, a is the particle square. This corresponds to the topological ideas developed in the computer simulations of the grain pinning by dispersed particles [1009]. The behavior of outstripping grain growth around seed coincides well with the computational results of large grain growth in the microstructure of normal grain growth [1010] and the growth morphology of YBCO specimens with seed, solidified at unfavorable temperatures [640]. Moreover, the present computer simulations may be characteristic to the pushing process of the 211 particles due to 123 front advancing [640]. The numerical results show a correlation between the grain area, $\bar{S}$, and superconducting field per grain, $\bar{S}_{\mathrm{SR}}$.

Table 8.7. Numerical results

Parameters	$f = 0.0$, $S_{\mathrm{S}} \approx 2S_0^{\mathrm{m}}$	$f = 0.0$, $S_{\mathrm{S}} \approx 3S_0^{\mathrm{m}}$	$f = 0.1$, $S_{\mathrm{S}} = 0$	$f = 0.1$, $S_{\mathrm{S}} \approx 2S_0^{\mathrm{m}}$	$f = 0.1$, $S_{\mathrm{S}} \approx 3S_0^{\mathrm{m}}$	$f = 0.2$, $S_{\mathrm{S}} = 0$	$f = 0.2$, $S_{\mathrm{S}} \approx 2S_0^{\mathrm{m}}$	$f = 0.2$, $S_{\mathrm{S}} \approx 3S_0^{\mathrm{m}}$
$\bar{S}/\delta^2$	13.89	23.81	45.45	50.03	62.50	25.01	29.41	31.25
$\bar{S}_{\mathrm{SR}}/\delta^2$	13.89	23.81	40.91	45.08	56.25	20.05	23.53	25.00

8.4 Small Cyclic Fatigue of YBCO Ceramics

It is known that the fracture toughness of YBCO is significantly smaller than that of typical ceramics, for example, Al_2O_3, partially stabilized ZrO_2 or SiC [1094]. Therefore, it is very important to consider fracture processes in YBCO, in particular, under cyclic loading. In this section the small cyclic fatigue fracture model for YBCO is stated, based on the microstructure dissimilitude effect [135], using joint considerations of the superconducting ceramic manufacture and fracture for the YBCO samples [819]. Special attention is devoted to the correct definition of specific fracture energy, taking into account porosity, microcracking and cooling features.

8.4.1 Model Representations

Remind that typical techniques for melt-processed YBCO ceramics consist of the precursor powder preparation, formation of the so-called "green" sample, sintering and cooling of the material. The sample cooling and re-crystallization initiate a formation of defects and microcracks due to the deformation mismatches of the 211 and 123 phases, thermal expansion anisotropy of grains and tetragonal–orthorhombic phase transition. Subsequent electromagnetic and thermo-mechanical loading of the ceramic in devices contributes to the damage development (or their initiation) and to corresponding deterioration of the conductive properties of YBCO.

By modeling of the processes which accompany the ceramic fabrication and fracture, the general research scheme, presented in the previous sections, is realized:

(1) The precursor is represented by 2D lattice with 1000 square cells of the characteristic size, δ. Each cell is either a grain nucleus of the 123 phase or a void. Initial disposition of the voids is found by Monte-Carlo procedure, described in Sect. 5.3.
(2) A modeling of the heat front propagation during sintering is carried out. It is assumed that a temperature in gradient furnace changes only along one coordinate with constant rate ν.
(3) By description of the ceramic structure re-crystallization, the MonteCarlo procedure is also used. Here, we neglect the material shrinkage and grain growth, assuming that during formation of closed porosity there are no considerable material densification (see Fig. 8.31).
(4) The microcracks nucleate at triple junctions during sample cooling (in the same conditions for all considered structures) from the melting temperature down to the room temperature. The cooling process consists of a slow decrease (with rate of 1°C/h) of temperature from 1100°C down to 970°C and then quenching, for example, in a liquid nitrogen bath [431].

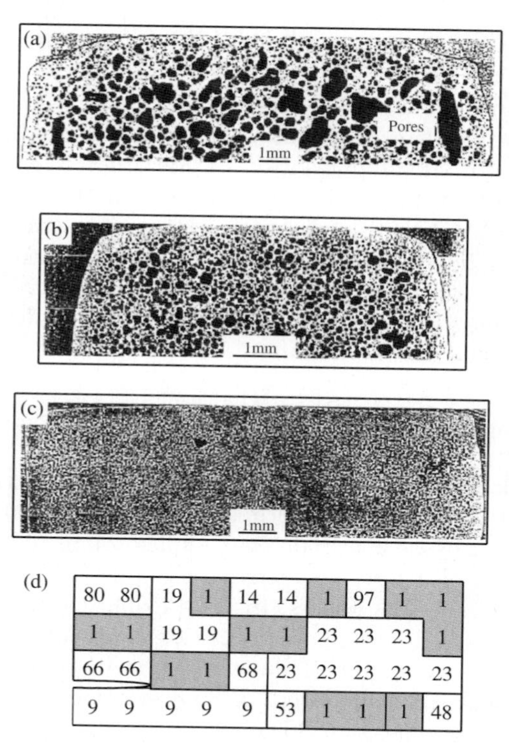

Fig. 8.31. Micrographs of polished cross-sections of YBCO samples, prepared at the heat rates: (**a**) 30°C/h (**b**) 20°C/h and (**c**) 10°C/h [642]; (**d**) fragment of model microstructure; porosity is shown *by gray color*, microcrack is present at intergranular boundary

8.4.2 Microstructure Dissimilitude Effect

It is known that the small fatigue cracks (i.e., cracks that intersect some grains) frequently grow much faster than large ones under other equal conditions, in particular due to the higher stress concentration factors and the decreased closing of the cracks. Microstructure dissimilitude is caused by impossibility to average the strength and fracture properties on the many grains and the existence of the dependence on the crack size. As has been demonstrated, the microstructure dissimilitude can lead to anomalous rapid growth of the small cracks, altering: (i) the intrinsic fracture toughness and cyclic crack growth resistance in the process zone and also (ii) the local crack driving force (ΔK) [135].

Consider a small crack which violates microstructure similitude and uses the modified Barenblatt–Dugdale's (BD) crack tip model that is based on the linear elastic fracture mechanics. The classical BD-crack with process zones (presenting extended cohesive parts at the crack tips) may be replaced by the superposition of two loading configurations (see Fig. 8.32), reflecting a local

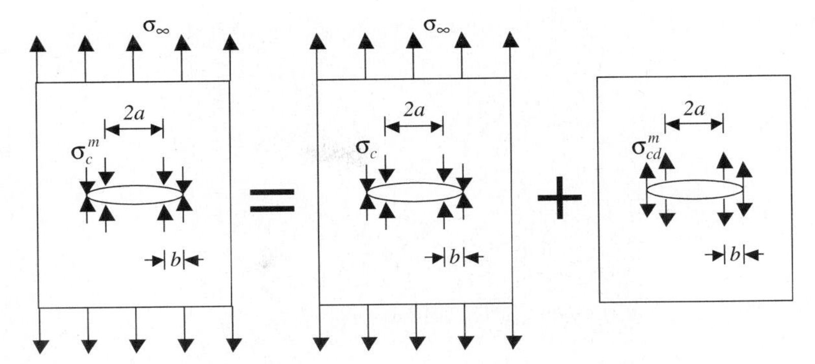

Fig. 8.32. Barenblatt–Dugdale's crack, presenting small crack, which disturbs microstructure similitude. One may be studied in the framework of linear elastic fracture mechanics, introducing additional stresses at the crack surfaces ($\sigma_{\mathrm{cd}}^{\mathrm{m}} = \sigma_{\mathrm{c}} - \sigma_{\mathrm{c}}^{\mathrm{m}}$) that are caused by microstructure dissimilitude

cleavage and general fracture. Then, the fracture toughness increment of large cracks (ΔK_{Ic}) can be divided into two terms for cyclic loading:

$$\Delta K_{\mathrm{Ic}} = \Delta K_{\mathrm{c}} + \Delta K_{\mathrm{cd}} , \tag{8.94}$$

where ΔK_{c} is the fracture toughness increment of small cracks, and ΔK_{cd} represents the local increment of the fracture toughness, induced as a result of microstructure dissimilitude and grain misorientation. Then, ΔK_{cd} for an elastic crack, subjected to a partial tensile load increment on the crack surfaces ($\Delta\sigma_{\mathrm{cd}}^{\mathrm{m}}$), is given as [1033]

$$\Delta K_{\mathrm{cd}} = \frac{2\sqrt{2}}{\pi}\Delta\sigma_{\mathrm{cd}}^{\mathrm{m}}\sqrt{\pi b} , \tag{8.95}$$

where $\Delta\sigma_{\mathrm{cd}}^{\mathrm{m}}$ is taken to be $2\sigma_{\mathrm{cd}}^{\mathrm{m}}$, so that

$$\Delta\sigma_{\mathrm{cd}}^{\mathrm{m}} = 2\sigma_{\mathrm{c}}(1 - \sigma_{\mathrm{c}}^{\mathrm{m}}/\sigma_{\mathrm{c}}) . \tag{8.96}$$

In (8.95) and (8.96), b is the length of the region over which the crack tip cleavage occurs (the process zone size), σ_c is the macroscopic fracture strength of the specimen, $\sigma_{\mathrm{c}}^{\mathrm{m}}$ is the local cleavage strength in a grain, containing the crack tip. Substituting (8.95) and (8.96) in (8.94), we obtain

$$\Delta K_{\mathrm{Ic}} = \Delta K_{\mathrm{c}}[1 + \frac{4\sqrt{2}}{\pi}(\sigma_{\mathrm{c}}/\Delta\sigma_{\infty})\sqrt{b/a}(1 - \sigma_{\mathrm{c}}^{\mathrm{m}}/\sigma_{\mathrm{c}})] , \tag{8.97}$$

where a is the half-length of the crack without process zone, and $\Delta\sigma_{\infty}$ is the increment of the applied remote stress. Similarly, it can be estimated that

$$K_{\mathrm{Ic}} = K_{\mathrm{c}}[1 + \frac{2\sqrt{2}}{\pi}\frac{\sigma_{\mathrm{c}}}{\sigma_{\infty}}\sqrt{b/a}(1 - \sigma_{\mathrm{c}}^{\mathrm{m}}/\sigma_{\mathrm{c}})] . \tag{8.98}$$

Assuming that for small and large cracks there are relationships:

$$K_{\rm c} = \left[\frac{\gamma_{\rm F}^{(1)}}{1-\nu_1^2} E_1\right]^{1/2} ,$$

and

$$K_{\rm Ic} = \left[\frac{\gamma_{\rm F}^{(0)}}{1-\nu_0^2} E_0\right]^{1/2} ,$$

respectively, we obtain from (8.97) and (8.98)

$$\Delta K_{\rm Ic} = \Delta K_{\rm c}[1 + 2(S-1)\sigma_\infty/\Delta\sigma_\infty] , \tag{8.99}$$

where

$$S = \left[\frac{\gamma_{\rm F}^{(0)} E_0(1-\nu_1^2)}{\gamma_{\rm F}^{(1)} E_1(1-\nu_0^2)}\right]^{1/2} . \tag{8.100}$$

Effective elastic modules E_1 and ν_1 for solid with pores and microcracks can be estimated, for example, using the dependencies of strain characteristics for elastic bodies on the concentrations and sizes of spherical pores and disc-like cracks in the self-consistent differential method as [239]

$$E_1 = E_{\rm p}\exp(-\alpha_1\eta_{\rm c}) ; \qquad \nu_1 = \nu_{\rm p}\exp(-\alpha_2\eta_{\rm c}) ; \tag{8.101}$$

$$E_{\rm p} = E_0(1-\eta_{\rm p})^{\beta_1} ; \qquad \nu_{\rm p} = 0.2 + (\nu_0 - 0.2)(1-\eta_{\rm p})^{\beta_2} , \tag{8.102}$$

where $\eta_{\rm c} = (a_{\rm c}/r_{\rm c})^3, \eta_{\rm p} = 1 - \rho/\rho_0$ are the damage parameters, determined through the average size of the disc-like cracks ($a_{\rm c}$) and the average distance between them ($r_{\rm c}$); ρ and ρ_0 are the material and crystalline cell density, respectively; $\alpha_i, \beta_i (i = 1, 2)$ are the given constants; E_0, ν_0 are the elastic modules of the material without defects; $E_{\rm p}, \nu_{\rm p}$ are the elastic modules of sample with voids.

8.4.3 Fracture Energy and Microstructure Features

Obviously, the correct definition of fracture energy, taking into account a microstructure of superconductor, has a significance for understanding of the dynamical crack problem, generally, and the cyclic fractures, particularly. We present the energetic balance condition in the usual form:

$$J = 2\gamma , \tag{8.103}$$

where

$$J = \int_S \left[(U+T)n_1 - \sigma_{jk} n_k \frac{\partial u_j}{\partial x_1}\right] {\rm d}S ; \qquad 2U = \sigma_{jk}\varepsilon_{jk} ; \qquad 2T = \dot{u}_j \dot{u}_j , \tag{8.104}$$

where $\sigma_{jk}, \varepsilon_{jk}, u_j, n_k$ are the components of the stress and strain tensors, vector of the displacement and external normal to surface of S, surrounding a crack tip, respectively; γ is Griffith's parameter. A dot above a symbol implies the material time derivative. There is summation on the repeated indexes from 1 to 3 in (8.104). In the static case, the condition (8.103) reduces to the algebraic relation between physical parameters. The energetic balance condition in the form (8.103) is also applied to the dynamical case. However, the assumption of the J-integral as critical parameter in the criterion of crack advancing [93] reflects only a static nature of the condition (8.103). Therefore, in the dynamical problem of the rectilinear flaw with size, $2L(t)$, it is necessary to take into account that the J-integral depends on the time t and is caused by the selection of the $L(t)$. The J-integral is equal to constant value of 2γ only when a law of the flaw growth for the $L(t)$ satisfies (8.103) in any time. After calculation of the J-integral, this condition reduces to non-linear differential equation, which defines the crack growth during time. Thus, a statement of the dynamical crack problem must include the condition (8.103), which identifies the flaw growth with crack advancing during fracture of solid. Then, specific fracture energy, γ_{F}, to being a limit ratio of the work on separation of atom couples at the fracture surface F to its square at decreasing of this surface to zero, demonstrates a local character. The value of γ_{F} should be significantly distinct from the Griffith's parameter γ, which is a result of the γ_{F} averaging on the fracture surface. Therefore, the physical proper statement demands to replace γ through γ_{F} in condition (8.103). The energetic balance condition as dynamical condition for brittle fracture can be represented as [511]

$$J = \frac{\sigma_\infty^2 \pi (1-\nu^2)\sin^2\Theta_0}{E} L(t) \left\{ F_1^2 \left[\frac{\dot{L}(t)}{c_2}\right] \sin^2\Theta_0 + F_2^2 \left[\frac{\dot{L}(t)}{c_2}\right] \cos^2\Theta_0 \right\} = 2\gamma_{\mathrm{F}} , \tag{8.105}$$

where E and ν are the elastic modules; σ_∞ is the remote tension; Θ_0 is the angle between the load and crack direction at the time t; c_2 is the rate of transversal waves; F_1 and F_2 are the monotonically decreasing functions, determined in the non-stationary problem. It may be shown [511] that the crack growth is irregularly accelerated and gives the relationship:

$$L(t) = L(0) + \frac{a_{\mathrm{c}} t^2}{2} - \frac{L(0)}{3}\left[\frac{a_{\mathrm{c}} t^2}{2L(0)}\right]^2 + O\left\{\left[\frac{a_{\mathrm{c}} t^2}{2L(0)}\right]^3\right\} , \tag{8.106}$$

where the acceleration a_{c} is found as

$$a_{\mathrm{c}} = \frac{\pi(1-\nu)}{8\rho\gamma_{\mathrm{F}}} \frac{\sigma_\infty^2 \sin^2\Theta_0}{\alpha_1(\nu)\sin^2\Theta_0 + \alpha_2(\nu)\cos^2\Theta_0}; \qquad L(0) = \frac{2E\gamma}{\sigma_\infty^2 \pi(1-\nu^2)} . \tag{8.107}$$

The parameters $\alpha_1(\nu)$ and $\alpha_2(\nu)$ are also found, solving a non-stationary problem. Note that (8.105) defines all features of sub-critical crack behavior, in particular, the crack tilt, twist and branching.

The value of γ_F is calculated in the case of fatigue fracture, using both material microstructure and cyclic loading. Estimate this parameter taking into account the structural elements of YBCO (i.e., voids and microcracks), which are small to compare with macroscopic scale of stress–strain state. Because the microcracks and voids, as a rule, are localized at the grain boundaries which are regions of a lesser dense atomic packing compared with grain phase, the curve of interatomic interactions can be approximated by the sinusoid [24]:

$$\sigma(\varepsilon) = \begin{cases} \sigma_a \sin 2\pi\varepsilon & \varepsilon \in [0; 0.5] \\ 0 & \varepsilon \in [0.5; \infty] \end{cases} , \tag{8.108}$$

where $\varepsilon = r/r_0 - 1, r_0$ is the mean equilibrium distance between atoms, $\sigma_a = E_0/2\pi$ is the theoretical strength of atomic link. Then, we obtain for transgranular fracture:

$$\gamma_F^{(0)} = \frac{1}{2}\int_{r_0}^{\infty} \sigma[\varepsilon(r)]\mathrm{d}r = \frac{1}{2}\int_{r_0}^{3r_0/2} \sigma_a \sin 2\pi(r/r_0 - 1)\mathrm{d}r = \frac{E_0 r_0}{4\pi^2} . \tag{8.109}$$

In order to define the mean interatomic spacing at intergranular boundary, r_1, we take into account a proportional dependence between adjacent atoms and the absolute temperature T. Then, $r/r_0^* - 1 = \alpha T$, where α is the thermal expansion factor, r_0^* is the equilibrium spacing between the atoms at the $T = 0\,\mathrm{K}$. Further, in order to define the temperature field in the ceramic sample during cooling from the sintering temperature (T_1) down to the room temperature (T_0), with no destroying of generality, we study the next 1D initial boundary problem:

$$\frac{\partial T}{\partial t} = b_t^2 \frac{\partial^2 T}{\partial x^2} , \qquad b_t^2 = \frac{k}{\rho c} ; \tag{8.110}$$

$$\frac{\partial T}{\partial x} = \pm h(T - T_0) , \quad \text{for } x = \pm l ; \tag{8.111}$$

$$T = T_1 , \quad \text{for } t = 0 , \tag{8.112}$$

where c is the specific heat, k is the thermal conductivity factor, h is the thermal expansion factor, $2l$ is the sample length. The solution of this problem is determined as

$$T(x,t) = T_0 + \sum_{m=1}^{\infty} A_m \exp(-\lambda_m^2 b_t^2 t) \cos \lambda_m x , \tag{8.113}$$

where $A_m = 2 \sin \lambda_m l/(\lambda_m l + \sin 2\lambda_m l)$ and eigenvalues, λ_m, have the following approximate values: $\lambda_m = m\pi(1/l - h/\pi^2)$. Retaining only the first term of the quickly convergent series (8.113), we have finally:

$$\gamma_F^{(1)}/\gamma_F^{(0)} = E_1 r_1/E_0 r_0 = \exp(-\alpha_1 \eta_c)(1-\eta_p)^{\beta_1}[1+\alpha(T_1 - T_0) \times 2 \sin \lambda_1 l/(\lambda_1 l + \sin 2\lambda_1 l)] . \quad (8.114)$$

When $\lambda_1 l$ is small, (8.114) reduces to

$$\gamma_F^{(1)}/\gamma_F^{(0)} = \exp(-\alpha_1 \eta_c)(1-\eta_p)^{\beta_1}[1+2\alpha(T_1 - T_0)/3] . \quad (8.115)$$

8.4.4 Some Numerical Results

The computer simulation results have been obtained for known parameters of YBCO [239, 431]: $T_1 = 970°\mathrm{C}, T_0 = 20°\mathrm{C}, \alpha = 2 \times 10^{-5}\,\mathrm{K}^{-1}, \alpha_1 = 1.65, \alpha_2 = 1.47, \beta_1 = 1.96, \beta_2 = 0.23, \nu_0 = 0.22, E_0 = 64\,\mathrm{GPa}$. Additionally, the values of η_c and η_p are stated by the microstructure modeling, using the formulae: $\eta_c = Na_c^3$ and $\eta_p = C_p$, where N is the crack number per volume unit, C_p is the closed porosity. The numerical results are shown in Fig. 8.33 for different values of the initial porosity of C_p^0 and the heat rate of the sample, ν [819]. The increased rate at the same initial porosity gives a more fine-grain structure, smaller microcracking of grain boundaries and greater value of $\gamma_F^{(1)}/\gamma_F^{(0)}$. Generally, the decreasing heat rate leads to finer sizes of voids and to their homogeneity in sintered samples that improve the YBCO conducting properties (see Fig. 8.31). However, excessive rate decreasing may increase the liquid phase loss. This would be undesirable to a highest degree from stoichiometry

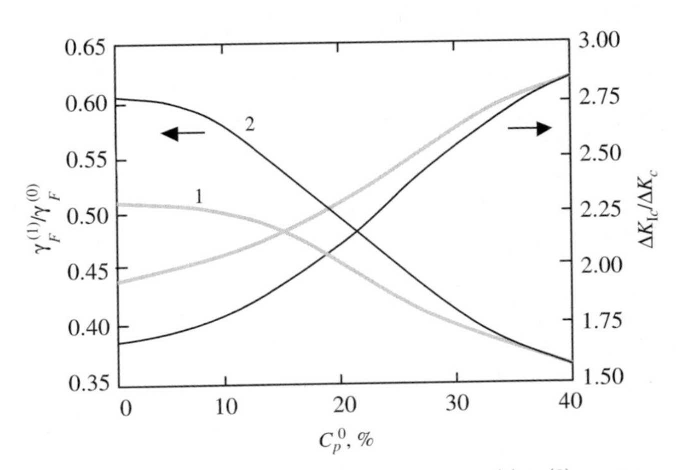

Fig. 8.33. Relative values of specific fracture energies, $\gamma_F^{(1)}/\gamma_F^{(0)}$, and fracture toughness increments, $\Delta K_{Ic}/\Delta K_c$, vs ceramic porosity, C_p^0, at $2\sigma_\infty = \Delta\sigma_\infty$. For two heat rates, $\nu = 10°\mathrm{C}(1)$ and $\nu = 20°\mathrm{C}(2)$

consideration [642]. The increasing initial porosity (at the same heat rate) defines the finer-grain structure with corresponding decreasing of a spontaneous microcracking of intergranular boundaries during cooling. At the same time, this leads to the smaller values of $\gamma_F^{(1)}/\gamma_F^{(0)}$ and proper increasing of $\Delta K_{Ic}/\Delta K_c$. Obviously, the closure of the small cracks under unloading compared with the large crack case is insignificant. Then, the maximum toughening, based on the steady-state crack conditions, can be achieved after a finite crack extension [1012]. This can lead to an overestimation of the toughening in consideration of the cyclic fatigue and, therefore, it should be calculated directly, taking into account the microstructure dissimilitude effects.

8.5 Residual Thermal Stresses in YBCO/Ag Composite

As has been shown in previous chapters, the silver inclusions are widely used to improve both superconducting and mechanical properties of HTSC. Consider effects of residual stresses in this section which form around inclusions in YBCO/Ag composite after cooling from the peritectic temperature down to room temperature and are caused by the great difference in thermal expansion factors of the metal particles and the ceramic matrix. Plastic deformations in ductile silver inclusions should be considered in an estimation of residual stresses around silver particles. The finite element method (FEM) has been applied for evaluating such difficult problem [652]. The using of ANSYS general-purpose software helps to estimate effects of the distribution of the residual stresses on features of macrocrack growth in this composite. For the matrix and inclusion properties given in Table 8.8 and also for the thermal difference ΔT between the peritectic temperature of YBCO/Ag (970°C) and room temperature (20°C), two silver round inclusions with a diameter $D = 10\,\mu\text{m}$ and the distance between them, $l = 5\,\mu\text{m}$, are considered on the isotropic *ab*-plane of the YBCO matrix (see Fig. 8.34a). Figure 8.34b shows the distribution of the residual stresses calculated by the FEM and indicates that expansive (compressive) stresses create after cooling of the composite due to the TEA of the matrix and inclusions.

Figure 8.35 explains the mechanism of the crack deflection between two inclusions. Tensile and compressive stresses remain around the silver particles in the r-axis and θ-axis of the polar coordinate system, respectively. First, the tensile stress along the r-axis assists the crack growth. However, when it

Table 8.8. Material properties used in FEM computations

Material	Thermal expansion factor ($\times 10^{-6}\,\text{K}^{-1}$)	Young's modulus (GPa)	Yield strength (MPa)	Maximum elongation
YBCO (*a*-axis)	13 [524]	182 [338]	–	–
Ag	21 [698]	80 [377]	50 [377]	0.5 [377]

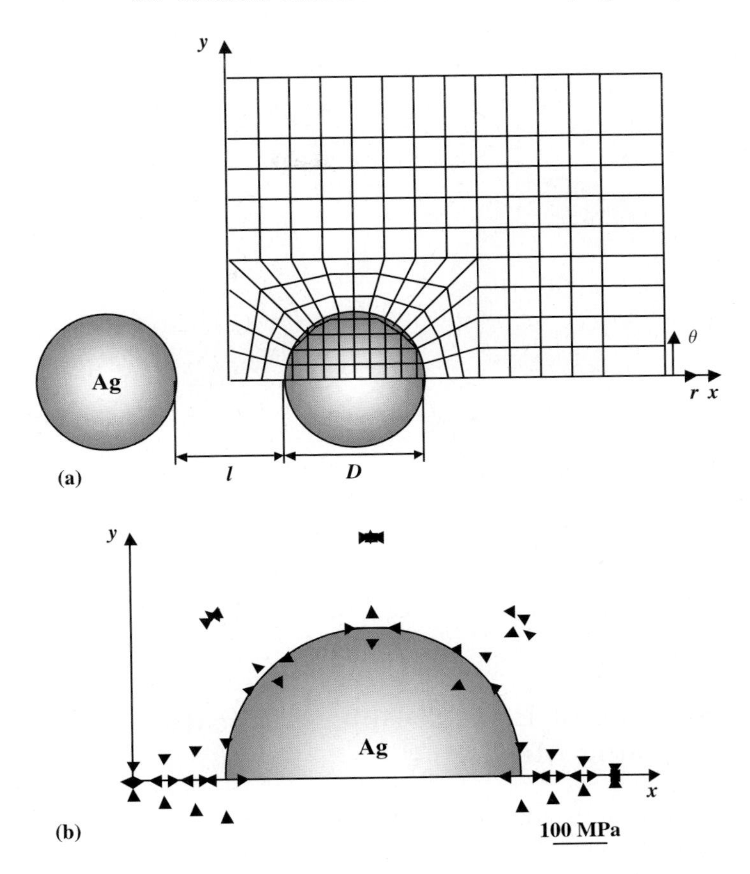

Fig. 8.34. (**a**) Mesh pattern for FEM calculation around silver particles dispersed in YBCO matrix and (**b**) maximum and minimum principal stress distributions around the silver particles in the 123 matrix [652]

reaches a gap between the silver particles, the angle between the vectors of the residual stresses and the crack growth is changed. This means that the tensile (compressive) residual stress promotes (suppresses) the crack growth along the y-axis (x-axis). Therefore, the direction of the crack growth is altered by the residual stresses around the silver particles, according to the energy minimum condition. This also means that the I mode of fracture is changed to the mixing mode (both I and II modes) by this deflection. Such crack deflections ensure the possibility of toughening of the brittle YBCO matrix and increasing its fracture toughness due to the increasing of the deflected crack path.

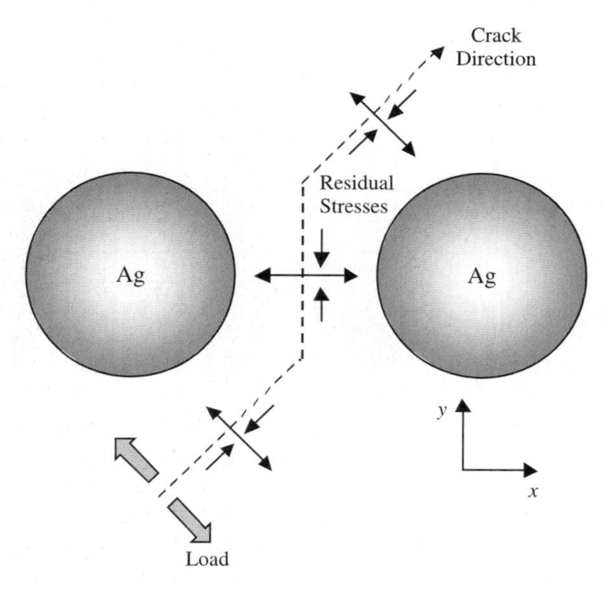

Fig. 8.35. Schematic illustration of the relationship between residual stress and crack propagation [652]

8.6 Toughening of Bi-2223 Bulk, Fabricated by Hot-Pressing Method

Models for processing and fracture of Bi-2223 ceramic which is obtained by hot pressing are discussed in this section. Computer simulation is applied to phenomena occurring during sintering, cooling and following fracture due to the macrocrack growth. The effects of Ag particles dispersed into the Bi-2223 matrix on some strength properties are studied [577].

8.6.1 Microstructure Formation by Processing

The sintering and cooling models for oxide superconductor have been discussed in detail in the previous sections. Therefore, we discuss in some detail the model of abnormal grain growth, caused by the sintering conditions (i.e., pressure and temperature [1034]) and by existence of secondary phases (e.g., $CaCuO_2$ and CuO), defining the Bi-2223 microstructure formation [700]. For different ceramics [796, 814], the Wagner–Zlyosov–Hillert's model [1] can be used. However, this model does not take into account the effects of texture on grain growth, which in this case is controlled by the inhibition parameter, depending on volume fraction and size of the secondary phase particles. At the same time, experiments have shown that the primary re-crystallization, as a rule, initiates a texture [2]. Therefore, in modeling of secondary re-crystallization, we consider the dependence of intergranular boundary energy and its mobility on grain misorientation, using a corresponding texture component for each grain. For simplicity, we limit the texture components to two (A and B), which are quite sufficient to describe considerable number of test data. In particular, we could model the growth processes for considered anisotropic Bi-2223 superconductor along the c-direction and in the ab-plane. Mass transfer between crystallites is based on the following assumptions [245]:

(1) The migration rate of a boundary between two grains i and j is given by

$$\nu_{ij} = m_{ij} p_{ij} = 2\gamma_{ij}(1/R_i - 1/R_j) , \tag{8.116}$$

where p_{ij} is the driving force, m_{ij} and γ_{ij} are the mobility and boundary energy, $(1/R_i - 1/R_j)$ is the average curvature for this grain boundary and $M_{ij} = 2m_{ij}\gamma_{ij}$ is the grain growth diffusivity.

(2) All grains of the same size and orientation experience the same growth rate (homogeneity condition).

(3) The grains, surrounding a given grain, are distributed randomly with respect to size and orientation.

Then, the grain growth rate of the size class i with a texture component A (or B) without the growth stagnation is found by [245]

$$\mathrm{d}R_i^{\mathrm{A(B)}}/\mathrm{d}t = M_*^{\mathrm{A(B)}}[1/R_*^{\mathrm{A(B)}} - 1/R_i^{\mathrm{A(B)}}] , \tag{8.117}$$

where $M_*^{\mathrm{A(B)}}$ and $R_*^{\mathrm{A(B)}}$ are the integrated diffusivity and integrated critical radius, respectively, controlling the grain growth of the component A (or B). Note that the value of $R_*^{\mathrm{A(B)}}$ is the intermediate grain size of the component A (or B) between increasing and decreasing grains, and $M_*^{\mathrm{A(B)}}$ determines the rate of these processes [245]:

$$M_*^{\mathrm{A}} = \frac{F^{\mathrm{A}}\langle R_{\mathrm{A}}^2\rangle M^{\mathrm{AA}} + F^{\mathrm{B}}\langle R_{\mathrm{B}}^2\rangle M^{\mathrm{AB}}}{F^{\mathrm{A}}\langle R_{\mathrm{A}}^2\rangle + F^{\mathrm{B}}\langle R_{\mathrm{B}}^2\rangle}; \quad M_*^{\mathrm{B}} = \frac{F^{\mathrm{A}}\langle R_{\mathrm{A}}^2\rangle M^{\mathrm{BA}} + F^{\mathrm{B}}\langle R_{\mathrm{B}}^2\rangle M^{\mathrm{BB}}}{F^{\mathrm{A}}\langle R_{\mathrm{A}}^2\rangle + F^{\mathrm{B}}\langle R_{\mathrm{B}}^2\rangle}; \tag{8.118}$$

$$R_*^{\mathrm{A}} = \frac{F^{\mathrm{A}}\langle R_{\mathrm{A}}^2\rangle M^{\mathrm{AA}} + F^{\mathrm{B}}\langle R_{\mathrm{B}}^2\rangle M^{\mathrm{AB}}}{F^{\mathrm{A}}M^{\mathrm{AA}}\langle R_{\mathrm{A}}\rangle + F^{\mathrm{B}}M^{\mathrm{AB}}\langle R_{\mathrm{B}}\rangle}; \quad R_*^{\mathrm{B}} = \frac{F^{\mathrm{A}}\langle R_{\mathrm{A}}^2\rangle M^{\mathrm{BA}} + F^{\mathrm{B}}\langle R_{\mathrm{B}}^2\rangle M^{\mathrm{BB}}}{F^{\mathrm{A}}M^{\mathrm{BA}}\langle R_{\mathrm{A}}\rangle + F^{\mathrm{B}}M^{\mathrm{BB}}\langle R_{\mathrm{B}}\rangle} \tag{8.119}$$

where $M^{HK}(H, K = \mathrm{A}$ or $\mathrm{B})$ is the grain-growth diffusivity of a boundary between grains of the orientation classes H and K. The grain fraction of the size class i with the orientation A (or B) is defined as

$$F_i^{\mathrm{A(B)}} = n_i^{\mathrm{A(B)}}/N_{\mathrm{G}}; \quad \sum_{i=1}^{N_{\mathrm{S}}}(n_i^{\mathrm{A}} + n_i^{\mathrm{B}}) = N_{\mathrm{G}}; \quad \sum_{i=1}^{N_{\mathrm{S}}}(F_i^{\mathrm{A}} + F_i^{\mathrm{B}}) = 1; \quad F^{\mathrm{A(B)}} = \sum_{i=1}^{N_{\mathrm{S}}} F_i^{\mathrm{A(B)}} , \tag{8.120}$$

where N_{S} and N_{G} are the total number of size classes and grains, respectively; $n_i^{\mathrm{A(B)}}$ is the number of grains per unit volume of the size class i withorientation

A (or B); $\langle R \rangle$ and $\langle R_{\mathrm{A(B)}} \rangle$ are the mean radius of grains in the whole system and that with the orientation component A (or B), respectively:

$$\langle R \rangle = \sum_{i=1}^{N_{\mathrm{S}}} F_i R_i \; ; \qquad \langle R_{\mathrm{A(B)}} \rangle = \sum_{i=1}^{N_{\mathrm{S}}} F_i^{\mathrm{A(B)}} R_i^{\mathrm{A(B)}} / F^{\mathrm{A(B)}} \; ; \tag{8.121}$$

$$\langle R^2 \rangle = \langle R \rangle^2 + \sigma^2 \; ; \qquad \left\langle R^2_{\mathrm{A(B)}} \right\rangle = \left[\langle R_{\mathrm{A(B)}} \rangle \right]^2 + \left[\sigma^{\mathrm{A(B)}} \right]^2 \; , \tag{8.122}$$

where σ and $\sigma^{\mathrm{A(B)}}$ denote the standard deviations. The size class for a given grain is found by the cell number, contained in this grain. The condition for abnormal grain growth in the grain size class i with the orientation A (or B) in the space (R, t), taking into account (8.117), is found as [245]

$$\left| 1/R_*^{\mathrm{A(B)}} - 1/R_i^{\mathrm{A(B)}} \right| > I_{\mathrm{R}}/2 \; , \tag{8.123}$$

where $I_{\mathrm{R}} = 6 f_\nu / (\pi r)$ is the value of the grain-growth stagnation, f_ν and r are the volume fraction and mean radius of the secondary phase particles. It is assumed that the stagnation parameter is independent of the grain orientation and is calculated as described above. This parameter, as well as the critical radii of the components A and B, governs the abnormal grain growth. This circumstance enables us to define the HTSC ceramic properties and their dependencies on the secondary phase characteristics.

The size parameters, which are necessary for calculations, are found in the simulation of the primary re-crystallization. The orientations A and B are distributed between grains, using the Monte-Carlo procedure, and they do not change during growth. Moreover, it is assumed that mass transfer between the grains of different orientations is absent. As an example, we consider the case of the next diffusion parameters for the orientation classes [2]: $M^{\mathrm{AB}} = M^{\mathrm{BA}} = 2M^{\mathrm{AA}} = 2M^{\mathrm{BB}}$. This sufficiently arbitrary selection is explained by the absence of reliable test data for the Bi-2223 ceramic. The mass transport between grains is simulated in accordance with the grain growth mechanism at the non-singular surfaces [1034].

The computational algorithm for abnormal grain growth consists of the following steps:

(1) The distribution of the orientations H (where H = A or B) between the grains that are formed after primary re-crystallization.
(2) The definition of all neighbors for every grain of both orientation classes.
(3) The determination of adjacent grain couple in each orientation class (i^H, j^H) with $\max\limits_{1 \le i^H, j^H \le N_{\mathrm{S}}} \left| 1/R_i^H - 1/R_j^H \right|$.
(4) The growth of the larger grains from the (i^H, j^H) at the expense of smaller ones.
(5) The checking of the conditions: $\left| 1/R_*^{\mathrm{A(B)}} - 1/R_i^{\mathrm{A(B)}} \right| \le I_{\mathrm{R}}/2$, where $i^H = 1, \ldots, N_{\mathrm{S}}$.

(6) End of the grain growth in the corresponding component H, if the conditions (5) have been satisfied; else a change of the corresponding parameters in (8.118)–(8.122) and fulfillment of the steps: (2)–(6) again.

The simulation of intergranular cracking due to cooling is carried out as before, using (8.4) and (8.5) and the procedure described in detail in Sect. 8.1.2.

8.6.2 Bi-2223 Toughening by Silver Dispersion

As has been noted above, an addition of Ag ductile phase dispersion to the Bi-2223 ceramic causes a considerable increasing in the fracture resistance of the superconductor compared to that of untoughened matrix. It is known that the main mechanism responsible for enhanced toughness of brittle composites with ductile particles appears to be the crack bridging by the ductile phase. Here, we limit ourselves to the most important case for HTSC, when an increasing of the ceramic toughness is independent of the particle size and ductile strength. This corresponds to the state, when the ductile flow has occurred in a considerable zone near the crack tip. Then, the toughness increasing due to the ductile particles (stationary crack case) can be estimated as [988]

$$\sqrt{1-\nu_{\mathrm{ef}}^2}\,\frac{K_{\mathrm{c}}}{K_{\mathrm{c}}^0} = \sqrt{3}\left[1+\frac{10(1-\nu_{\mathrm{ef}}^2)f_{\mathrm{p}}}{(7-5\nu_{\mathrm{ef}})(1-f_{\mathrm{p}})}\right]^{1/2}, \tag{8.124}$$

where K_{c} and K_{c}^0 are the fracture toughness with and without toughening, respectively; f_{p} is the ductile particulate concentration, defined by the fraction of this phase, intercepting the macrocrack path; ν_{ef} is effective Poisson's ratio, defined by the relative concentration of the Bi-2223 and Ag phases, and also by the intergranular microcracks, formed during the composite processing. Under conditions of the modified cubic model, we obtain [428]:

$$\nu_{\mathrm{ef}} = (1-f_{\mathrm{m}}^{2/3})\nu_{\mathrm{c}} + f_{\mathrm{m}}^{2/3}\,\frac{[\nu_{\mathrm{m}} f_{\mathrm{m}}^{1/3}/E_{\mathrm{m}} + \nu_{\mathrm{c}}(1-f_{\mathrm{m}}^{1/3})/E_{\mathrm{c}}]}{[f_{\mathrm{m}}^{1/3}/E_{\mathrm{m}} + (1-f_{\mathrm{m}}^{1/3})/E_{\mathrm{c}}]}, \tag{8.125}$$

where the indexes m and c correspond to the metal inclusions and the ceramic matrix. For the cracked matrix with a microcrack density, β_{cr}, Poisson's ratio, ν_{c}, and Young's modulus, E_{c}, are expressed as [892]

$$\frac{\nu_{\mathrm{c}}}{\nu_0} = \frac{1+[(16/45)(1-\nu_0^2)/(2-\nu_0)]\beta_{\mathrm{cr}}}{1+[(16/45)(1-\nu_0^2)(10-3\nu_0)/(2-\nu_0)]\beta_{\mathrm{cr}}}; \tag{8.126}$$

$$\frac{E_{\mathrm{c}}}{E_0} = 1\Big/\left[1+\frac{16(1-\nu_0^2)(10-3\nu_0)}{45(2-\nu_0)}\beta_{\mathrm{cr}}\right], \tag{8.127}$$

where ν_0 and E_0 are the intrinsic elastic modules.

Since the metallic inclusions have a greater thermal expansion than the ceramic matrix [564], the residual tension (σ_R) occurs in the metal, but the ceramic is compressed. This internal stress state affects the toughening, because the compressive stresses in matrix must be exceeded within the bridging zone before beginning of crack opening. The additive increment of the toughness is estimated as [988]

$$\Delta G_{\mathrm{R}} \approx \alpha f_{\mathrm{p}} \sigma_{\mathrm{R}} u^* , \tag{8.128}$$

where $\alpha(\sim 0.25)$ is a factor that depends on the precise nature of the function $\sigma = \sigma(u)$; u^* is the total crack opening when the ductile material fails (Fig. 8.35a). For cylindrical metal particles in a non-hardening material, the axial residual stress, $\sigma_{\mathrm{R}}^{\mathrm{z}}$, can be obtained as [444, 988]

$$\frac{\sigma_{\mathrm{R}}^{\mathrm{z}}}{E_{\mathrm{m}} \Delta\alpha \Delta T^*} = \frac{3}{3(1-2\nu_{\mathrm{m}}) + 2(1+\nu_{\mathrm{c}})(E_{\mathrm{m}}/E_{\mathrm{c}})} , \tag{8.129}$$

where $\Delta\alpha$ is the thermal expansion factor difference of the phases; ΔT^* is the cooling thermal range, in which the rapid creep provides relaxation.

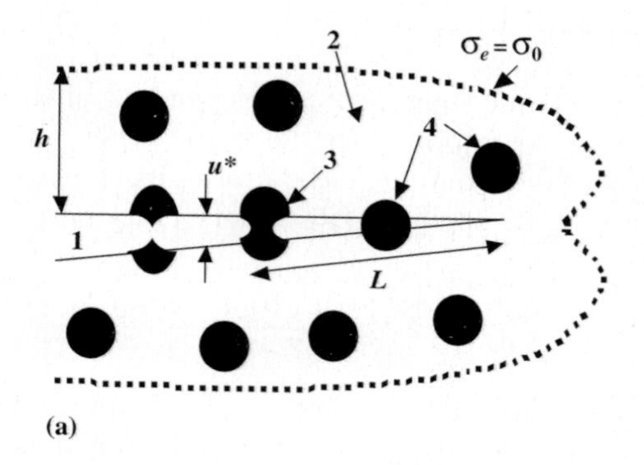

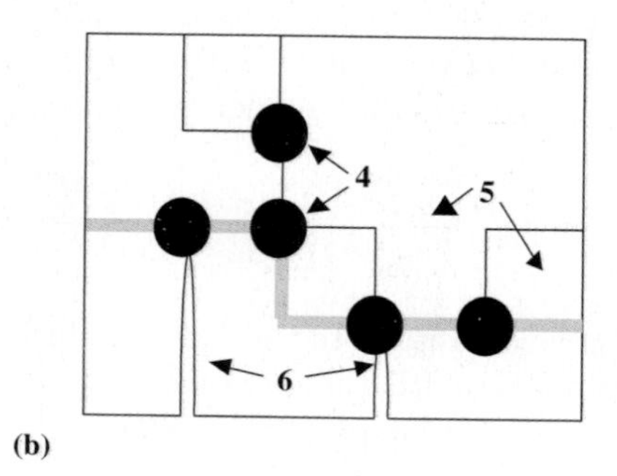

Fig. 8.36. (**a**) Scheme of crack (1) bridging by intercepted ductile silver particles (4); the process (2) and stretch (3) zones of plastically distorted particles are shown (h and L are the corresponding zone sizes; u^* is the residual crack opening at link failure and σ_0 is the yield strength) [988]; (**b**) a model fragment of Bi-2223/Ag composite with the Bi-2223 grains (5) and cooling microcracks (6); macrocrack is denoted by *gray line*

Table 8.9. Numerical results [577]

D/δ	β_{cr}	l_{cr}/b	f_p	K_c/K_c^0
2.40	0.15	1.51	0.12	1.96
2.84	0.22	1.62	0.10	1.92

It has been suggested in computer simulations that the silver inclusions are localized at the triple junctions, where there are usually microdefect sites, healed by the Ag. The necessary parameters for Ag particles are given elsewhere [845]. The optimum Ag volume concentration in the Bi-2223 bulk is assumed to be $f_m = 0.2$ [564]. Finally, we modeled the intergranular macrocrack path (see Fig. 8.36b), using Viterbi's algorithm for graphs [796, 823], taking into account the grain structure and processing microcracks.

Statistically reliable results during computer simulations are obtained again by application of the stereological method [145]. In order to define more accurate estimations of the effects of dispersed Ag particles on the strength properties, we assumed that other toughening mechanisms, considered above, are absent. We obtain numerical results for values of $I_R D_0 = 2.0$ and 1.2 (first and second line in Table 8.9, respectively), where D_0 is the grain size before the grain growth. In the macrocrack simulation which propagates along the intergranular boundaries of the Bi-2223 matrix, the value of f_p is found, taking into account the macrocrack length, the number of triple junctions (or Ag particles) in the crack path and the relationship between the defect size ($2a$) and critical boundary size (l_c^s), namely: $a = 0.1 l_c^s$, which is the same as that for a ceramic with TEA [260]. As it follows from (8.128), the value of f_p is proportional to the toughness difference, ΔG_R. At the same time, a decreasing of the parameter I_R leads to increasing of the matrix grain size and corresponding enhancement of the microcracking density (β_{cr}). Moreover, the number of triple points also decreases with the increasing of the relative macrocrack size l_{cr}/b, where l_{cr} is the macrocrack size, and b is the specimen width due to the longer path, required around the larger grains. The latter causes a decreasing of f_p and corresponding decreasing of ΔG_R. Finally, the smaller values of β_{cr} increase the elastic modules and together with the increased concentration of f_p lead to increasing of K_c/K_c^0 at decreasing of the grain size.

9

Mechanical Destructions of HTSC Josephson Junctions and Composites

As it has been shown in Chap.1, Josephson effects are connected with a behavior of weak links of superconductors. According to the classification of high-temperature superconducting Josephson junctions (HTSC JJs), presented in Sect. 1.7, special interest from view of strength and fracture toughness is excited by the junctions with intrinsic barriers or interfaces, formed by the intergranular boundaries with different crystallographic orientations, and also HTSC JJs with extrinsic interfaces in the fabrication of which, the artificial normal metallic or insulating barriers are used. The strength problems of the JJs, having the extrinsic interfaces (see Fig. 9.1), cause their small effective superconducting area as compared with the geometrical square and can lead to the large parameter spreads. The microstructure destruction can be found by some causes, namely by the deformation (lattice) mismatches and thermal expansion anisotropy [17, 554], by exceeding the critical film thickness [314, 929], by the misorientation effects [314], by the rough or damaged interfaces [929], etc.

The use of buffer layers (e.g., $CeO_2, MgO, YSZ, ZrO_2, SrTiO_3$ [981]) has enabled the increase of superconducting and transport properties of HTSC composite structures due to diminution of TEA and lattice mismatches between HTSC film and substrate and also owing to decrease of a chemical reactivity of the substrate. Nevertheless, the problems of critical mechanical behavior for these laminated structures under conditions of existence of residual stresses and external loads remained in the center of attention of HTSC JJ researchers and engineers.

This chapter presents a set of models for estimation of strength properties for HTSC composites of S-I-S and S-N-S (where S is the superconductor, I is the isolator, N is the normal metal) types, based on the consideration of interfaces, taking into account TEA, geometrical and material parameters, external loads and residual stresses. Moreover, features of the mechanical damage of HTSC composites and proper fracture resistance mechanisms are considered [802, 804, 807].

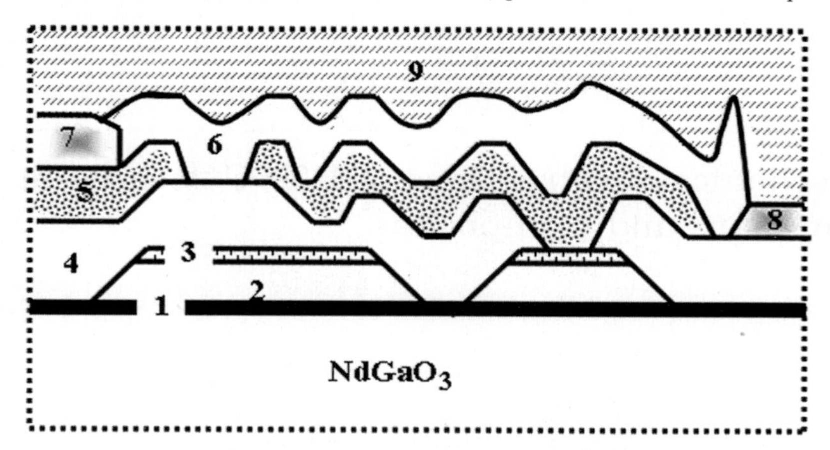

Fig. 9.1. Cross-section of HTSC JJ, consisting of nine layers, on $NdGaO_3$ substrate: 1 – buffer layer (30 nm), 2 – YBCO base layer (150 nm), 3 – transitional superconducting layer (150 nm), 4 – $SrTiO_3$ epitaxial dielectric (150 nm), 5 – YBCO top layer (150–200 nm), 6 – non-epitaxial silicon nitride dielectric (250 nm), 7 – silver contact layer (600 nm), 8 – molybdenum resistor (90 nm) and 9 – silver wiring layer (600 nm) [751]

9.1 Interface Fracture

The interface roughness in HTSC bimaterial and a crack growth near and at the interface introduce inevitably a mixed loading mode. In this case, there are some morphological features of fracture [263, 485, 979], shown in Fig. 9.2, namely (i) an interface fracture; (ii) a crack growth into more brittle component; (iii) an alternative fracture along interface or between the interface and adjacent material; and (iv) a crack deflection from one interface to an other. The experiments on various materials have shown that the crack path is found by both the ratio of interface fracture energy to fracture energy of more brittle component (Γ_i/Γ_2) and the phase angle of loading, Ψ = arctg $(K_{\text{II}}/K_{\text{I}})$, where K_{I} and K_{II} is the SIF of I and II mode, respectively. On the other hand, this angle can be coupled with the ratio of displacements – the shear to the opening (i.e., to the ratio of ν/u) due to the insertion of Dunders parameters (α and β) as [236]

$$\Psi = \text{arctg}(\nu/u) - \in_0 \ln r - \text{arctg}(2 \in_0) \,, \tag{9.1}$$

where

$$\in_0 = \frac{1}{2\pi} \ln\left(\frac{1-\beta}{1+\beta}\right) ; \tag{9.2}$$

$$\alpha = \frac{\mu_1(1-\nu_2) - \mu_2(1-\nu_1)}{\mu_1(1-\nu_2) + \mu_2(1-\nu_1)} ; \quad \beta = \frac{\mu_1(1-2\nu_2) - \mu_2(1-2\nu_1)}{2\left[\mu_1(1-\nu_2) + \mu_2(1-\nu_1)\right]} ; \tag{9.3}$$

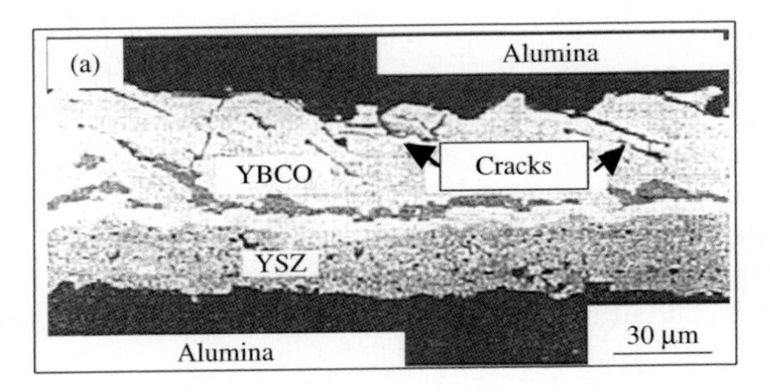

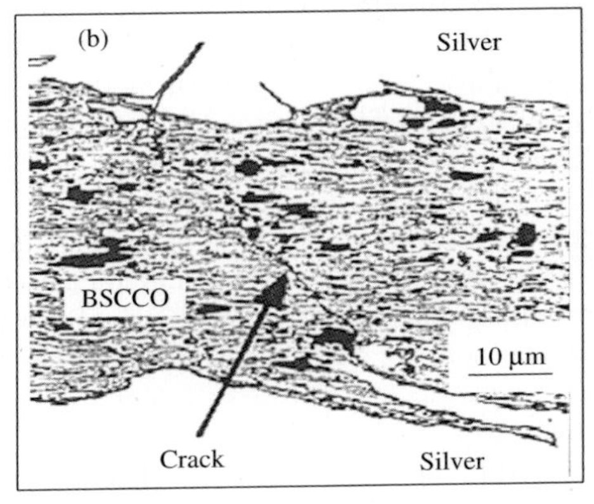

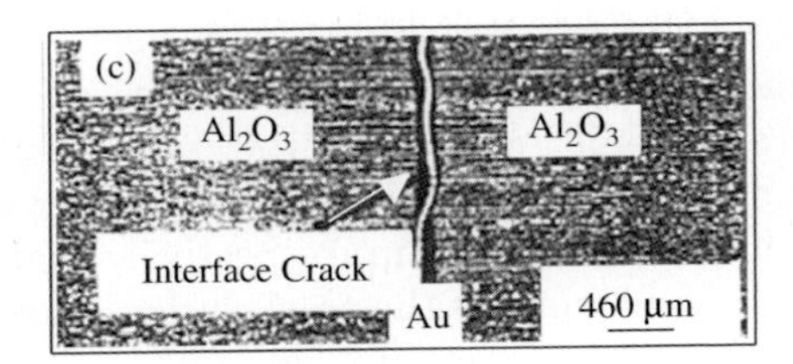

Fig. 9.2. Characteristic morphology of possible fractures at and near interfaces into HTSC composite structures: (**a**) interface fracture and an alternative fracture between the interface and adjacent material (SEM cross-sectional micrograph of YBCO thick film, processed on YSZ barrier layer on alumina [979]), (**b**) crack growth into more brittle component (Back-scattered electron image of transversal cross-section of BSCCO/Ag tape [485]) and (**c**) crack in Al_2O_3 bonded with Au, showing a crack alternating between interfaces [263]

μ_k and $\nu_k (k = 1, 2)$ are the shear modulus and Poisson's ratio, respectively, for k-component and r is the distance measured from the crack tip at the interface. The parameter of α together with the dependence of Γ_i/Γ_2 on Ψ enables to separate the areas of the interface crack and the brittle substrate fracture in the case of an initial crack at the interface. Then, the fracture behavior and the interface fracture energy are very sensible to a sign of the phase angle at the great difference of the fracture energies for both components ($\Gamma_1 >> \Gamma_2$). In the case of the positive value of Ψ, there are both regimes, namely the interface cracking and a crack deflection into more brittle component, depending on the parameter α. The second case (when $\Psi < 0$) is more interesting. Therein, the greater value (i.e., Γ_1) is compared with the interface fracture energy Γ_i. As the condition $\Gamma_1 >> \Gamma_i$ prohibits crack propagation away from the interface, then there are two cases. For the low material strength, a plastic bluntness of the crack at the interface occurs and the failure features are caused by the toughening mechanisms, including an initiation of voids at the interface.[1] In the contrary case, the stress state of the interface crack interacts with microcracks and structural defects which as rule exist in the brittle material and provoke a growth of microcracks in the direction to the interface. This causes a saw-tooth fracture, with chips of the brittle material attached to the interface (see Fig. 9.3).

Further more, because the complete smoothness of the interfaces is impossible (e.g., see Fig. 9.4a), an estimation of the JJs fracture resistance depending on the interface roughness is the actual problem. The crack surfaces, growing along the interface, contact each other either at the roughness or at the facets. In this case, it is possible to obtain different values of the interface fracture resistance, which grow with the phase angle of loading, Ψ. These effects have been observed and estimated for different brittle materials [266, 482]. In particular, a comparative analysis of microstructure properties and fracture parameters, caused by the mixed loading mode, has been fulfilled for the fine-grain ($PbTiO_3$) and coarse-grain ($BaTiO_3$) ferroelectric ceramics [816]. The decrease of the strain energy release rate (or crack shielding) $\Delta G = G - G^{\mathrm{t}}$ (where G and G^{t} are the values of the strain energy release rate connected with applied load and at the crack tip, respectively) can be estimated using two models: (i) the contact zone of the crack surfaces taken into account, but without account of Coulomb's friction [266], and (ii) the inclination angle of the faceted interface, δ, is taken into account [482]. We have for the first model (see Fig 9.4b) [266]:

$$\Delta G/G = [1 - \lambda^{-2}(\alpha)] tg^2\Psi/(1 + tg^2\Psi) \; , \tag{9.4}$$

where $\alpha = (L/l_{\mathrm{m}})/\ln[1/\sin(\pi D_{\mathrm{b}}/2l_{\mathrm{m}})]$. The values of the function, $\lambda(\alpha)$, for various values of α have been tabulated in [101]. The length of the contact zone, L, is found by $L = (\pi/32)[EH/(1 - \nu^2)K_{\mathrm{I}}]^2$. Here l_{m} is the spacing between the facet centers, D_{b} is the facet length and H is the height of the interface step. The numerical approximation is obtained, taking into account the typical geometry of undulating interfaces, assuming that

[1] At the same time, note that those intergranular voids formed, for example, by thermo-mechanical treatment during a multi-stage processing of monocore Bi-2223 tapes, may become the main cause of critical current diminution in the case of prolonged final annealing [801]

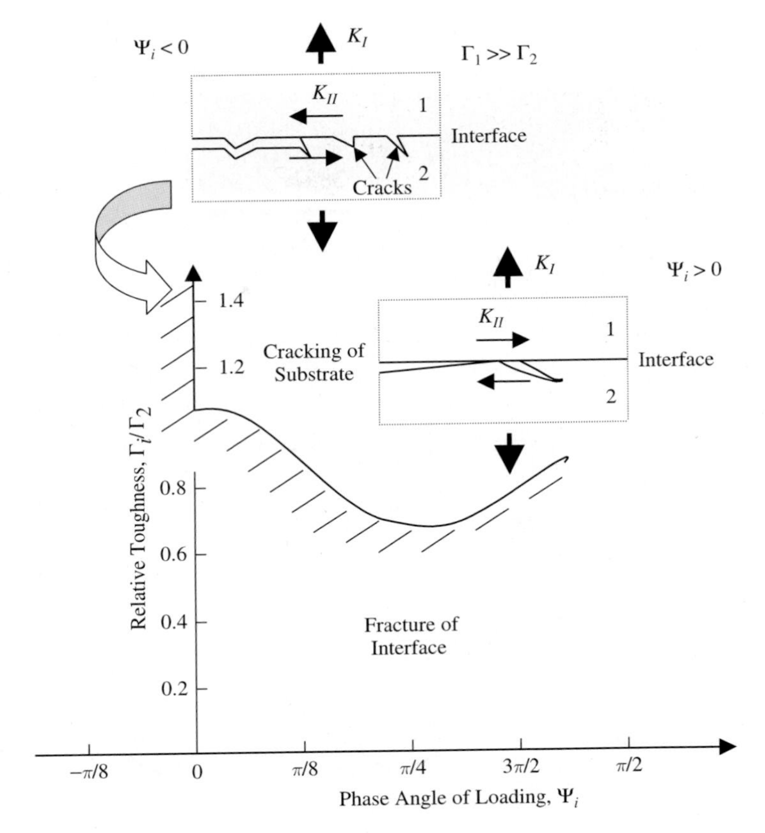

Fig. 9.3. Fracture at interface into bimaterial for $\Gamma_1 >> \Gamma_2$ (result in the case of $\alpha = \beta = 0$). When phase angle, $\Psi < 0$, the interface fracture demonstrates near-boundary segments of material free from cracks [263]

$D_{\mathrm{b}}/l_{\mathrm{m}} \sim 1/2; H/l_m \sim 1/2$; equating K_{I} to the fracture toughness, K_{Ic} (which is approximately equal to $1\,\mathrm{MPa}\cdot\mathrm{m}^{1/2}$ for a broad set of ceramics), and introducing a material parameter, $\chi = E^2H/(1-\nu^2)K_{\mathrm{Ic}}^2$, which defines the length of the contact zone, L. Note that the parameter χ in total causes a fracture behavior in this microstructure consideration. Therefore, one should be measured with a high accuracy in order to obtain acceptable results for the HTSC JJs. For example, the value of $\chi \approx 100$ has been found for the glass–polymer interface [266].

We obtain for the second model (Fig. 9.4c) [482]:

$$\Delta G/G = (\cos\,\delta \sin\,\Psi - \sin\,\delta\ \cos\,\Psi)^2\ , \quad \Psi > \delta; H >> a\ , \qquad (9.5)$$

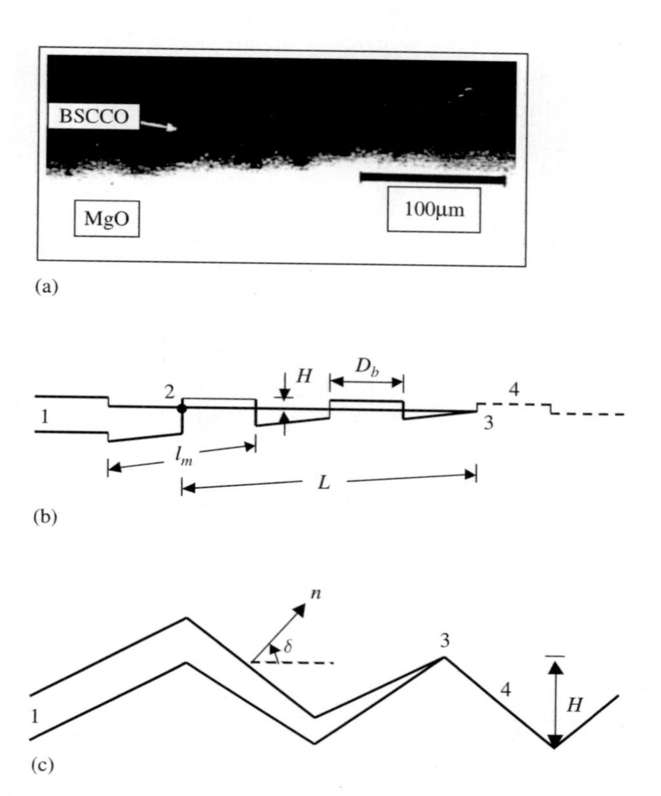

Fig. 9.4. (**a**) Proper interface between melted Bi-2212 thick film and polycrystalline MgO substrate (optical micrograph of cross-section [762]); and two models (**b**) and (**c**) used in the analysis of crack at rough interface [266, 482]. The numbers denote crack (1), last contact point (2), crack tip (3) and interface (4)

where a is the crack size. Then, the condition of facet contact in the second model [482] coincides with the condition of a single crack deflection in the first model [266]. In the case of existence of the facet contacts, the second model as compared with the first one takes into account, additionally, an interaction between various crack deflections. The proper dependencies of the crack shielding ($\Delta G/G$), caused by the interface roughness, on the phase angle of loading (Ψ) and on the different inclinations of the facet sides (δ) are shown in Fig. 9.5. The obtained trends of the crack shielding growth with the phase angle of loading agree well with experimental observations of the interface crack in different ceramics, in which the steady-state crack trajectory is caused by the condition, $K_{\mathrm{II}} = 0$ [263, 266, 482, 816].

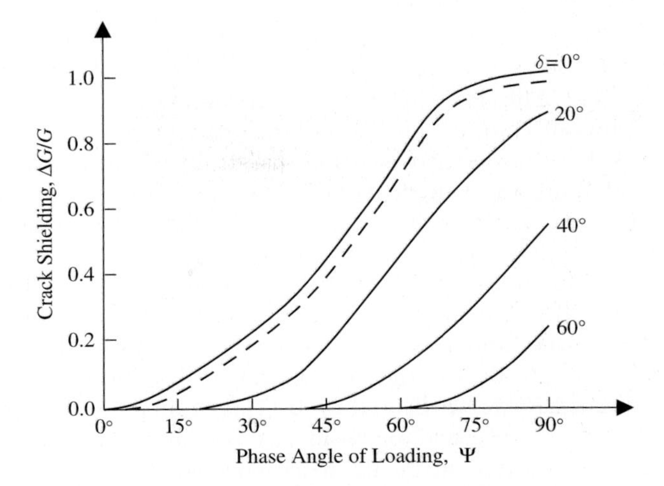

Fig. 9.5. Crack shielding ($\Delta G/G$) vs phase angle of loading (Ψ). The *dashed curve* corresponds to the model of contact zone ($\chi \approx 100$), and the *solid curves* are obtained due to the model for different inclination (δ) of facet sides

9.2 Thin Films on Substrates

During processing of JJ, the residual stresses can initiate in thin films adherent to substrates due to the deformation mismatches or/and thermal expansion anisotropy. The residual stresses in YBCO film structures, caused by different causes (e.g., lattice mismatches, thermal expansion, non-stoichiometry) can reach, in whole, the value of ~100 MPa [333]. A relaxation of these stresses can also lead to the formation of defect structures, in particular the dislocation mismatches form gradually with increase of the film thickness and accumulation of elastic stresses. Due to the high anisotropy of microstructure properties, thin films are subjected to the microcracking and twinning in the specific directions. For example, generally, the cracks form into (001) plane, and spacing between them increases with the film thickness. First, crack occurs in the film at attainment of the critical thickness (H_c), depending on thermo-elastic stress of the lattice disagreement (σ) and fracture toughness (K_c) as $H_c \sim (K_c/\sigma)^2$ [634]. The distinct feature of film from bulk crystals is a cellular domain structure inside single grains with the cell boundaries parallel to the (100) and (010) planes. The cells differed by mutual replacement of directions of the a- and b-axes and coupled coherently in similar to twins.

Together with external electric and magnetic fields, the internal stresses lead to different damage, developing in the film-substrate system, in particular to the film de-cohesion from substrate. The de-cohesion mechanism depends on the residual stress sign (i.e., tension or compression) and existence of the stress gradient. In the film tension, de-laminations initiate at the sample edges and propagate into brittle substrate parallel to the interface [221, 447]. Moreover, an across-film cracking which nucleates at the free surface is possible [446, 447], following its de-lamination and buckling [75]. Under conditions of compression, the film de-lamination and buckling are possible, too [75]. The investigations of mechanical and electromagnetic properties of HTSC films are complicated due to (i) structure defects, (ii) high anisotropy of elastic modules and (iii) existence of structure phase transformations [634].

In order to estimate strength parameters and study fracture features, the composite beam theory is applied, assuming an existence of the initial microdefect crack at the sample edge. Consider the model sample, presented in Fig. 9.6a, which is equivalent to the case of uniform distribution of the thermal stress, $\sigma_0 = E_\mathrm{f}\Delta\alpha\Delta T$, where E_f and ν_f are elastic modules for the film, $\Delta\alpha$ is the difference of the thermal expansion factors between the film and substrate and, ΔT is the temperature difference, covered at cooling. Note that the thermal strain, $\varepsilon = \Delta\alpha\Delta T$, must be replaced in the case of the deformation

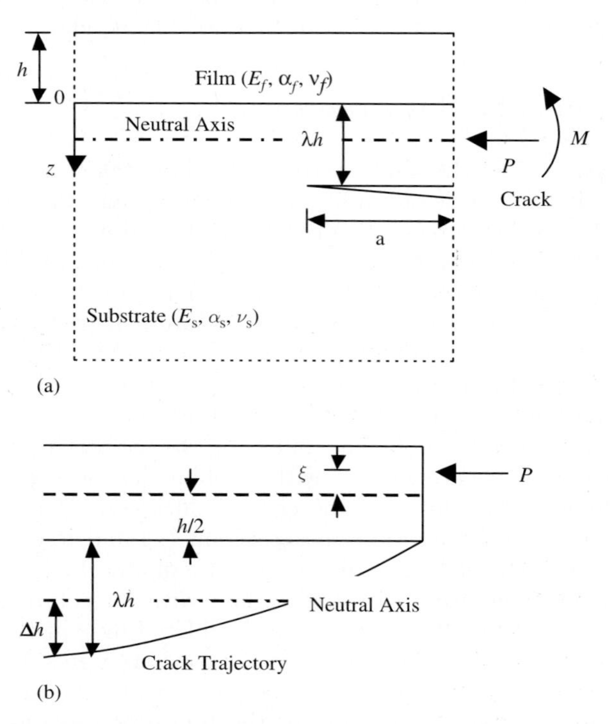

Fig. 9.6. Models of steady-state crack into substrate, taking into account (**a**) thermal expansion difference between film and substrate and (**b**) existence of stress gradient in the film

mismatches by the strain, depending on the crystallographic properties of the system. Then, the corresponding SIFs can be found by [221]

$$K_{\mathrm{I}} = 0.434Ph^{-1/2}(\Sigma + \lambda)^{-1/2} + 0.558M(Ih^3)^{-1/2} ; \tag{9.6}$$

$$K_{\mathrm{II}} = 0.558Ph^{-1/2}(\Sigma + \lambda)^{-1/2} - 0.434M(Ih^3)^{-1/2} , \tag{9.7}$$

where h is the film thickness, $\Sigma = E_{\mathrm{f}}/E_{\mathrm{s}}$ (in the plane stress case) and $\Sigma = E_{\mathrm{f}}(1-\nu_{\mathrm{s}}^2)/E_{\mathrm{s}}(1-\nu_{\mathrm{f}}^2)$ (in the plane strain case); E_{s} and ν_{s} are the elastic modules for the substrate; $I = \{\Sigma[3(\Delta-\lambda)^2 - 3(\Delta-\lambda)+1] + 3\Delta\lambda(\Delta-\lambda) + \lambda^3\}/3$ is the dimensionless moment of inertia; $\Delta = (\lambda^2 + 2\Sigma\lambda + \Sigma)/2(\lambda+\Sigma)$; $\lambda = z/h$; $P = \sigma_0 h$ is the load; and $M = \sigma_0 h^2\lambda(\lambda+1)/2(\Sigma+\lambda)$ is the bending moment (per unit thickness). The steady-state crack path into brittle substrate parallel to the interface is found by the condition $K_{\mathrm{II}} = 0$. Then, equating the SIF of I Mode to the fracture toughness of the substrate ($K_{\mathrm{I}} = K_{\mathrm{Is}}$), we obtain the critical layer thickness (h_{s}), of which complete fracture is inhibited, as

$$h_{\mathrm{s}} = 0.755(\Sigma + \lambda_{\mathrm{s}})(K_{\mathrm{Is}}/\sigma_0)^2 , \tag{9.8}$$

where λ_{s} is the relative depth, governing the steady-state crack path, for which $K_{\mathrm{II}} = 0$.

Consider also the case of the stress gradient, existing on the film thickness (see Fig. 9.6b). In this case, σ_0 is the mean stress in the film and $M = P[(\lambda + 0.5 - \Delta)h + \xi]$, where ξ is the distance from the film center to the force action line in (9.7). Similar to the previous model the parameters h_{s} and λ_{s} are found depending on the value of ξ, that is, on the given stress gradient.

Further more, the simple equations can be obtained for toughness parameters in the case of small-scale yielding of the substrate under condition of steady-state transversal cracking of the film. In this case, the strain energy release rate (G_{ss}) and corresponding SIF (K_{ss}) are found as [446]

$$G_{\mathrm{ss}}E_{\mathrm{f}}/\sigma^2 h = \pi F(\Sigma) ; \qquad K_{\mathrm{ss}}/\sigma\sqrt{h} = \sqrt{\pi F(\Sigma)} , \tag{9.9}$$

where σ is the film tension (residual or/and applied); thefunction of $F(\Sigma)$ is given in Table 9.1 [446] for various values of $\Sigma = \Sigma_{\mathrm{f}}/\Sigma_{\mathrm{s}}$. Then, the typical de-laminations, occurring from the notch orthogonal to the tension direction, are classified in Fig. 9.7a and, b. In the case (a), there is a de-lamination of open type, which grows together with the longitudinal cracking of the film. In the case (b), the longitudinal cracks are absent. The de-lamination remains

Table 9.1. Function $F(\Sigma)$ dependent on ratio of elastic modules, Σ

Σ	4	3	2	1	1/2	1/3
$F(\Sigma)$	0.79	0.75	0.70	0.62	0.57	0.54

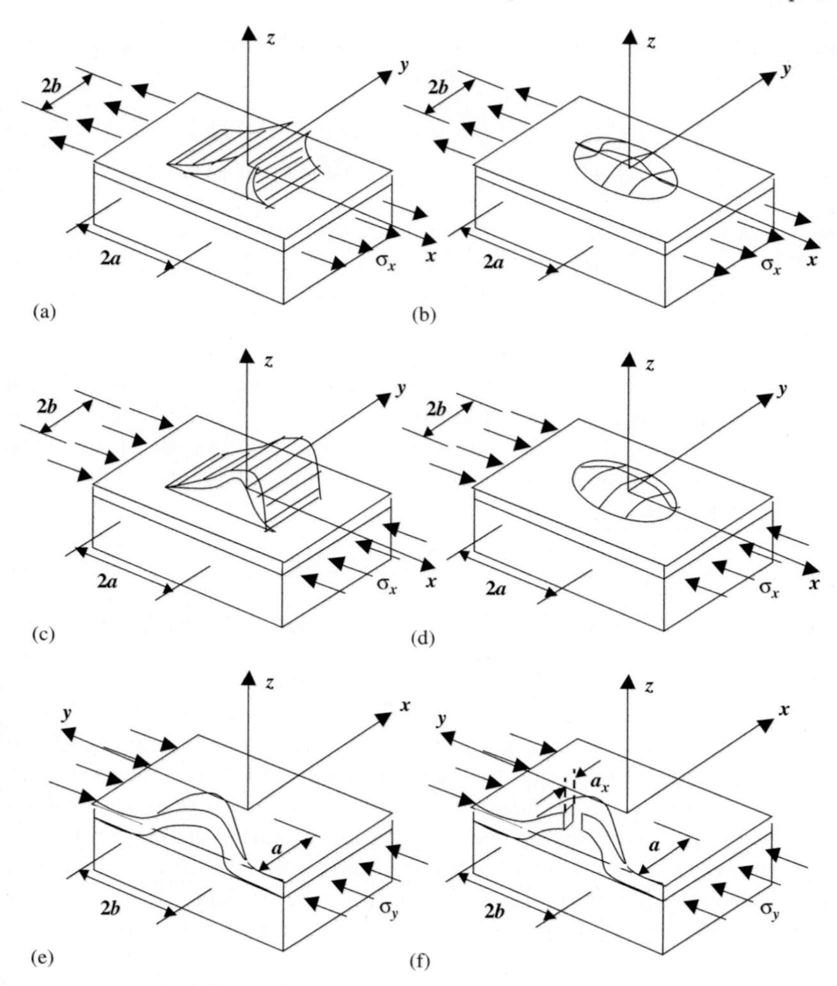

Fig. 9.7. Typical de-laminations in film-substrate structure (see explanation in the text) [75]

to be closed and "pocket-like," and the buckling takes place due to Poisson's effect under sufficiently-high applied stress. The form of this de-lamination can be approximated by ellipse with semi-axes a and b.

Generally, the film de-lamination from the substrate and its buckling in compression as well as under tension can be studied in the framework of the stability theory in fracture mechanics with consideration of different stages for crack-shaped defects [75]. The typical de-laminations under compression are shown in Fig. 9.7c–f. By sufficiently-high compression, the de-lamination opens and is accompanied by a buckling, which is similar to the cylindrical beam-like bending (case (c)). Case (d) represents a close de-lamination in the form of ellipse with semi-axes a and b. The so-called edge de-laminations are presented by cases (e) and (f) and can be approximated by half-ellipse. In case (f), the secondary crack can be observed, which grows across the de-lamination as a result of the de-lamination bending.

In the case of the beam approximation, the critical buckling strain, ϵ_c, can be estimated for acting compressive stresses in the film and substrate (σ^{f}_{xx} and σ^{s}_{xx}, respectively) by [978]

$$\epsilon_c = \frac{\Omega_c}{12}\left[3 + \frac{(1-2\nu_s)}{(1-\nu_s)}\Omega_c + \frac{2}{3}\Omega_c^2 + \frac{(3-4\nu_s)}{12(1-\nu_s)^2}\Omega_c^3\right]\left(1+\frac{\Omega_c^2}{6}\right)^{-1}, \quad (9.10)$$

where

$$\Omega_c^3 = \frac{12(1-\nu_s)^2}{\Sigma(3-4\nu_s)}; \qquad \Sigma = \frac{\sigma^{f}_{xx}}{\sigma^{s}_{xx}} = \frac{E_f(1-\nu_s^2)}{E_s(1-\nu_f^2)}. \quad (9.11)$$

9.3 Step-Edge Junctions

The strength problems, related to above ones arise, considering an inclined interface crack in the step-edge JJs and in the S-N-S edge junctions with ground planes. It is known that the tapered edges of the base electrode with edge angles, being lesser than 45°, are important to avoid a grain boundary formation in the counter-electrode [454]. On the other hand, the shallower edge angles have revealed problems with a void formation on the edges below 15° [455]. So, in order to optimize the strength and damage behavior of these HTS JJs, it is useful to consider a crack problem, when the crack surfaces contact at the kinks. In this case, the SIFs of I and II modes at the crack tip (i.e., K^{t}_{I} and K^{t}_{II}) differ from corresponding applied SIFs by the values, which depend on the kink angle (β), the kink amplitude and Coulomb friction factor (μ) (see Fig. 9.8a). This geometry and a replacement of the inclined frictional force,

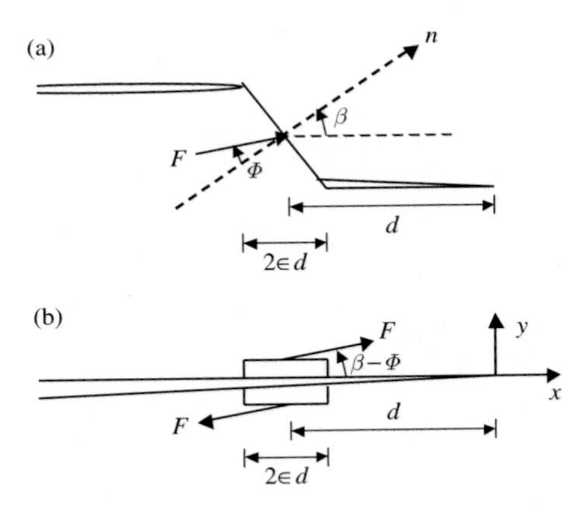

Fig. 9.8. Model of interface crack [266] in step-edge HTSC JJ used in order to analyze effects of the crack shielding

F, acting at the crack surfaces, by the homogeneously distributed tractions, applied to the segment, $2 \in d$ (see Fig. 9.8b), enable to estimate a contact effect through normal and shear forces. In turn, it permits to determine the contributions of this effect on the SIFs at the crack tip and in the corresponding displacements of the crack surfaces. Further more, it is possible to consider the conditions of sliding and locking of the crack surfaces via terms of the phase angle of loading (Ψ), the friction angle ($\Phi = \arctan \mu$) and the kink angle (β). An analysis, which has been carried out for $K_{\rm I} > 0$ [266], enables to distinguish the main types of the crack behavior at the crack asperity (see Fig. 9.9). Then, based on the conditions of a contact and a frictional locking, the crack shielding (ΔG) may be estimated in the case of the sliding contact due to the facet contact as [266]

$$\frac{\Delta G}{G} = 2h(\in)\frac{(\sin\beta + \cos\beta tg\Psi)\left[\sin(\beta - \Phi) + \cos(\beta - \Phi)tg\Psi\right]}{\cos\,\Phi(1 + tg^2\Psi)} - \frac{h^2(\in)(\sin\beta + \cos\beta tg\Psi)^2}{\cos^2\Phi(1 + tg^2\Psi)}\,, \tag{9.12}$$

where $0 < \beta < \pi/2$ (for $\pi/2 < \beta < \pi$, Φ must be replaced by $(-\Phi)$), and due to the crack locking as

$$\Delta G/G = 1 - [1 - h(\in)]^2\,, \tag{9.13}$$

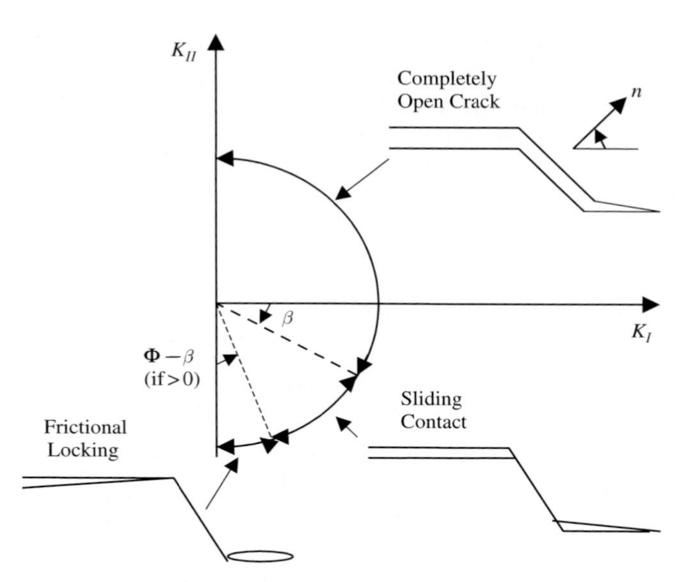

Fig. 9.9. Chart of possible crack behavior types, connected with crack asperity, taking into account loading modes [266]

where $h(\in) = f(\in)/g(\in)$, but the functions $f(\in)$ and $g(\in)$ are found by

$$f(\in) = (\sqrt{1+\in} - \sqrt{1-\in})/\in ; \tag{9.14}$$

$$g(\in) = \frac{1}{2\in}\left[\sqrt{1+\in} - \sqrt{1-\in} + \left(1+\frac{\in}{2}\right)\ln\left(\frac{\sqrt{1+\in}+1}{\sqrt{1+\in}-1}\right)\right.$$
$$\left. - \left(1-\frac{\in}{2}\right)\ln\left(\frac{1+\sqrt{1-\in}}{1-\sqrt{1-\in}}\right)\right] . \tag{9.15}$$

The comparison of $h(\in)$ with the function, defined from exact solution of the problem on a microcrack ahead of a macrocrack [895], shows that the proposed model slightly overestimates a crack shielding, caused by the facet contact. The dependencies of the crack shielding ($\Delta G/G$) on the various parameters for above two mechanisms are presented in Figs. 9.10 and 9.11. the crack behavior due to the facet contact is more interesting. In the both examples of this case, namely without and with a friction (i.e., for $\Phi = 0°$ and 45°, respectively), the crack shielding ($\Delta G/G$) shows non-monotonous dependencies on the phase angle of loading (Ψ). Initially, the value of $\Delta G/G$ grows and then one diminishes with increase of Ψ. A friction presence displaces the maximums of the curves in the direction of the increase of Ψ, restoring usual tendency of a growth of the crack shielding with the phase angle of loading.

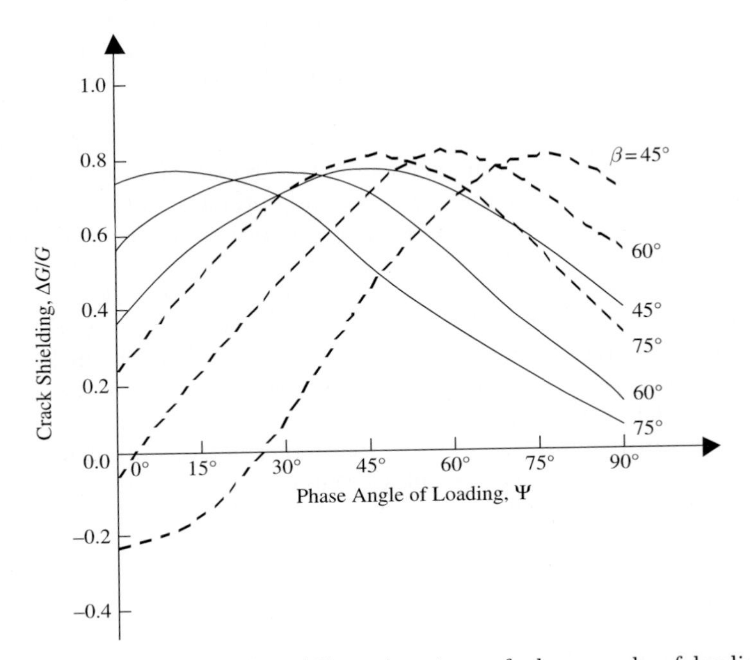

Fig. 9.10. Crack shielding ($\Delta G/G$) as functions of phase angle of loading (Ψ), kink angle (β) and friction angle (Φ). *Solid curves* correspond to the case of friction absence ($\Phi = 0°$), and *dashed curves* find the case of $\Phi = 45°$

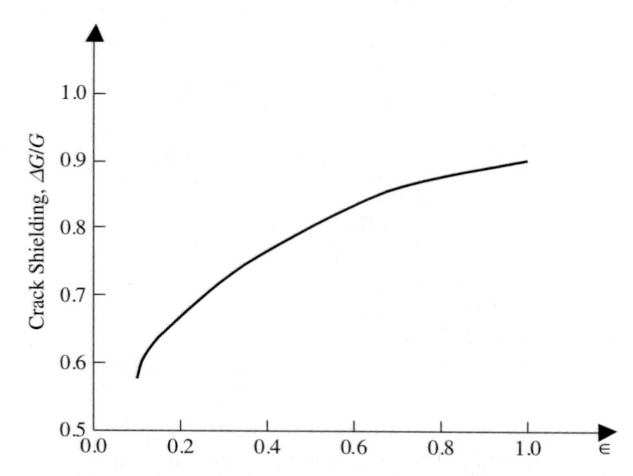

Fig. 9.11. Crack shielding ($\Delta G/G$) vs parameter $\in$ in sliding contact (effect of the crack locking)

At the same time, a displacement of the maximums in the direction of the decrease of Ψ occurs, increasing the kink angle β. The latter corresponds to diminution of the edge angle (α) of the step-edge junction, since $\alpha = 90° - \beta$. Further more, a non-zero friction may introduce the contrary mechanism to the crack shielding, namely the crack amplification ($\Delta G < 0$). This is impossible in the case of $90° < \beta < 180°$, when $\Delta G/G$ grows monotonously with Ψ (into interval of its alteration from 0° up to 90°) [266]. Thus, the crack, which climbs on a step, is more safe as compared with the crack that goes down from the step. Therefore, the maximum values of the crack shielding for the present type of HTSC JJs may be found, thanks to an optimum and simultaneous selection of the parameters β, Φ and Ψ in accordance with the trends, depicted in Fig. 9.10. In the case of the crack locking, a monotonous growth of the crack shielding takes place with the parameter $\in$ (see Fig. 9.11), which finds the step size at given values of β and d. The greater edge size corresponds to the greater microcrack ahead of a macrocrack in the case of the inclined interface, at $\beta =$ const and at the invariable crack shielding. Then, even the small structural defects (with a high probability initiated at the interfaces during HTSC JJ processing, for example, due to the thermal anisotropy or/and crystallographic mismatches) may be sufficient in order to achieve an actual crack shielding, at the small edge size.

9.4 Transversal Fracture

During HTSC JJ processing and a device work under high electromagnetic fields, a nucleation and a propagation of microcrack-like defects in the direction perpendicular to the layers of the HTSC composite is possible (e.g., see

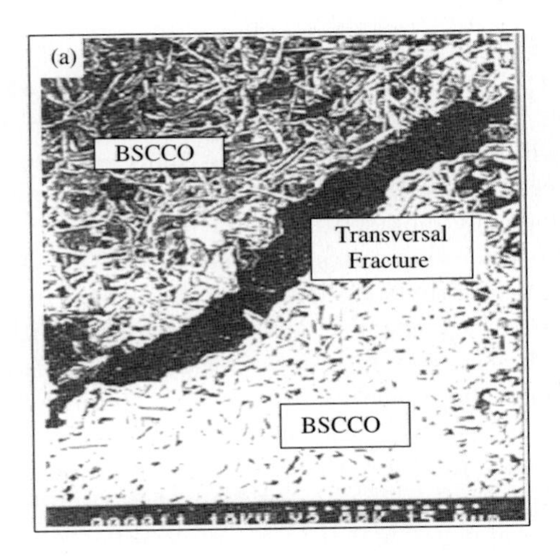

Fig. 9.12. (**a**) SEM micrograph of transversal fracture of BSCCO/Ag rod [775] and (**b**) SEM image of cross-section of Bi-2212/Ag film with transversal crack [275]

Fig. 9.12). On the whole, fracture picture and fracture resistance are not only caused by the material properties and by the geometrical parameters of thick (matrix) layers and thin (buffer) layers, but even rather by the failure features at the interfaces. The de-cohesion (or de-bonding) processes at the material interfaces may be caused by the residual stresses, formed during the HTSC JJ processing. In this case, the internal stresses may be increased by the supplementary stresses due to the thermal change and/or Lorentz forces. The mechanical damage, nucleated by de-cohesion processes, causes an immediate diminution of the functional properties of the junction, decreasing steadily with time, that limits reliability and longevity of the device. At the same time, a partial de-cohesion can play a positive role in the case of a transversal fracture of the HTSC layer composite, increasing its fracture resistance. Hence, it is necessary to estimate different strength parameters and fracture toughness, which should be obtained in the framework of fracture mechanics, applied to the brittle matrix–fiber composites, when a matrix crack is bridged by the one-axially aligned reinforcing fibers [267, 667, 1058]. The use of buffer layers in the HTSC JJs is connected, in particular with attempts to significantly decrease the thermal and lattice mismatches. Therefore, an absence of residual stresses will be assumed below. In this case, the damage processes are caused by external tensile stresses (σ_∞) remotely applied and parallel to the fibers (or analogously modeled buffer layers).

First, note some general features connected with the above fracture mechanism. The main idea of the toughening of the brittle matrix (including, in the HTSC composites) by the brittle fibers is linked with the processes of de-bonding and sliding at the interfaces (see Fig. 9.13). A de-bonding at the matrix–fiber interface should be more favorable compared with a fiber fracture at the matrix crack front in the case of the crack, inhibited by the fibers. The de-bonding is more probable than the fiber fracture, if the interface fracture energy is sufficiently small as compared with the fiber fracture energy. Then, the HTSC composite, which to be under the de-bonding conditions, demonstrates pushing effect of the broken fibers by the crack surfaces. At the same time, an alternative fracture mechanism causes a growth of the matrix crack through the fibers without a de-bonding. So, a non-catastrophic fracture mode could be obtained in the composites with weak interfaces and high-strength fibers (or buffer layers). For this sub-critical fracture mode, the numerous matrix damages prior to a failure of the fibers are proper. Then, the complete strength of the composite is caused by the fracture of the fibers and by the following processes of the broken fiber pushing. The catastrophic mode is found by the fiber fracture in the wake of the main matrix crack during its

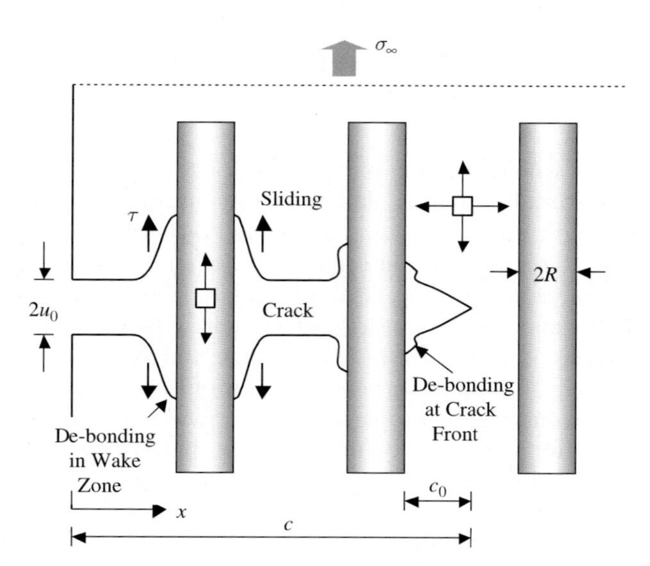

Fig. 9.13. Schematic representation of transversal matrix crack in layer HTSC composite

advancing. In this case, the complete strength is limited by the single dominant crack and is found by a fracture resistance curve (i.e., dependence of toughness on crack size) [267]. Then, a sliding of the broken fibers or thin buffer layers inevitably introduces a fiber pushing by the crack surfaces. This process demands the study of fiber strength statistics, which is usually found by Weibull's distribution with the shape and scaling parameters m and S_0, respectively [1058]. A decrease of m corresponds to the more broad distribution of the fiber strength, that is, to the fracture of greater number of the fibers far from the matrix crack front. This defines an increase of the fiber pushing zone size. High median fiber strength is another useful property with a view of the fracture resistance growth. It is found by the large values of S_0, and causes the sub-critical fracture mode.

For different applications of HTSC JJs, it is more interesting to consider a short crack, that is, the case, when the entire crack contributes in a stress concentration, and a stress required for the crack growth is sensitive to its size. Moreover, it is important to estimate critical parameters at the transition to the steady-state cracking (i.e., to long crack), when the stresses at the crack tip grow with an applied tension but are independent on the whole crack size. The effects of buffer layers on fracture of the HTSC composite can be estimated, introducing as the stresses, which close the crack surfaces, the stresses in the buffer layers, playing role of the bridges between the crack surfaces [667]. The corresponding decrease of SIF at the crack tip is calculated from these surface attractions, using standard Green's function. The crack growth criterion is found, equating the SIF at the crack tip to intrinsic toughness of the matrix without reinforcements ($K_{\mathrm{c}}^{\mathrm{M}}$). Then an analysis of the buffer layer pushing from the matrix, based on the results of [667], enables to find a relationship between the closure pressure (p) and the crack opening (u) as

$$p = [u\tau V_{\mathrm{b}}^2 E_{\mathrm{b}}(1+\eta)/R]^{1/2} , \tag{9.16}$$

where $\eta = E_{\mathrm{b}}V_{\mathrm{b}}/E_{\mathrm{m}}V_{\mathrm{m}}$; $2R$ is the buffer layer thickness; τ is the sliding frictional stress at the interface; E_{m} and E_{b} correspond to Young's modules of the matrix and buffer layer; and $V_{\mathrm{m}} = 1 - V_{\mathrm{b}}$ is the volume fraction of the matrix.

An approximate analytical solution for the short crack can be obtained, assuming that the crack profile at small crack sizes (c) does not differ greatly from that of the crack, subjected to uniform pressure. Then the crack opening is found as [667]

$$u(x) = 2(1-\nu_{\mathrm{c}}^2)K^{\mathrm{L}}c^{1/2}(1-x^2/c^2)^{1/2}/E_{\mathrm{c}}\pi^{1/2} , \tag{9.17}$$

with the limiting displacement ($2u_0$) and the equilibrium stress (σ_∞) defined depending on the crack size, respectively, as

$$u_0 = \sigma_\infty^2 R/\tau V_{\mathrm{b}}^2 E_{\mathrm{b}}(1+\eta) ; \tag{9.18}$$

$$\sigma_\infty/\sigma_{\mathrm{m}} = (1/3)(c/c_{\mathrm{m}})^{-1/2} + (2/3)(c/c_{\mathrm{m}})^{1/4} , \quad c \le c_0 , \tag{9.19}$$

where $E_c = E_m V_m + E_b V_b$; $\nu_c = \nu_m V_m + \nu_b V_b$; and $K^L = K^M E_c/E_m$, K^L and K^M, referring to the SIF of the composite and of the matrix, respectively. The crack size ($c = c_0$), being transitional to the steady-state crack, and the corresponding equilibrium stress ($\sigma_\infty = \sigma_0$), which is independent on the crack size, can be obtained as

$$c_m = 2[K_c^M E_m V_m^2 (1+\eta) R / I^2 \tau V_b^2 E_b (1-\nu_c^2)]^{2/3} ; \tag{9.20}$$

$$\sigma_m = (6.7/\Omega)[I^2(1-\nu_c^2)(K_c^M)^2 \tau E_b V_b^2 V_m (1+\eta)^2 / E_m R]^{1/3} ; \tag{9.21}$$

for straight crack:

$$\sigma_0/\sigma_m = 1.02 ; \quad c_0/c_m = 1.88 ; \quad \Omega = \pi^{1/2} ; \quad I = 1.20 ; \tag{9.22}$$

for penny-shaped crack:

$$\sigma_0 = \sigma_m ; \quad c_0 = c_m ; \quad \Omega = 2/\pi^{1/2} ; \quad I = 2/3 . \tag{9.23}$$

Then, the steady-state toughness increment is found as [897]

$$\Delta G_c = 2 \int_0^{u_0} p(u) \mathrm{d}u , \tag{9.24}$$

and we have, taking into account (9.16),

$$\Delta G_c = (4/3)\alpha_0 u_0^{3/2} , \tag{9.25}$$

where $\alpha_0 = [\tau V_b^2 E_b (1+\eta)/R]^{1/2}$. Limiting by the case of the penny-shaped crack and substituting $\sigma_\infty = \sigma_m$ into (9.18), we have finally:

$$\Delta G_c = 4\sigma_m^3 / 3\alpha_0^2 . \tag{9.26}$$

On the other hand, inserting the buffer layer strength (S) into (9.16) and (9.24), found may be another equation for ΔG_c [267]. Application of the specific dependence of ΔG_c (on the parameter u_0 or S) is very important in order to experimentally determine the value of ΔG_c, namely to select the strength test type, the shape and sizes of experimental sample. Then note that (9.20), (9.21) and (9.26) include considerable number of material parameters; therefore, it is necessary to solve the problem of multi-parametric optimization for their optimum selection with account of the specific loading conditions and possible damages.

9.5 HTSC Systems of S-N-S Type

HTSC JJs of S-N-S type include laminated compositions of the ceramic–metal–ceramic kind. First, consider a pair of the edge cracks, growing symmetrically and parallel to the metallic buffer layer, dividing two half-infinite brittle

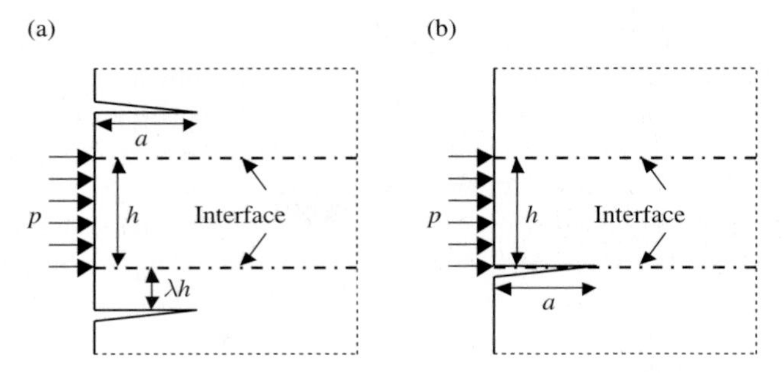

Fig. 9.14. Model representations of damage in composites of the ceramic–metal–ceramic type: (**a**) pair of steady-state cracks into brittle substrates and (**b**) interface crack

half-planes (see Fig. 9.14a). Then, the SIFs determining the crack path may be estimated using uniform pressure in the metal, $p = E_{\rm m}(\Delta\alpha_{\rm m} - \Delta\alpha_{\rm c}/2)\Delta T/(1 - \nu_{\rm m})$, where $\Delta\alpha_{\rm c} = \alpha_{\rm c_1} - \alpha_{\rm c_2}$ and $\Delta\alpha_{\rm m} = \alpha_{\rm m} - \alpha_{\rm c_1}$. The following are analytical representations for the steady-state cracks and like ceramics [121]:

$$K_{\rm I}/p\sqrt{h} = -0.26(1 + 2\lambda)^{-1/2} \; ; \qquad K_{\rm II}/p\sqrt{h} = 0.43(1 + 2\lambda)^{-1/2} \; , \tag{9.27}$$

where λh is the distance from the cracks to the corresponding interface and h is the thickness of the metal layer. At the greater thermal expansion of the metal, that is, for $\alpha_{\rm m} > \alpha_{\rm c}$ (e.g., Ag), $K_{\rm I} < 0$, and edge cracks cannot grow along interface, and only after coalescence near edge are able to propagate into ceramic away from the interface. At smaller thermal expansion of the metal, that is., for $\alpha_{\rm m} < \alpha_{\rm c}$ (e.g., Mo), $K_{\rm I} > 0$ and a fracture along interface is possible, when ductile layers are subjected to large residual stresses and has a requisite thickness [447]. Therefore, the metals, which possess small thermal expansion, are undesirable for bond integrity in HTSC JJs.

In the case of single interface crack (see Fig. 9.14b), it may be shown for ceramic layers with like properties that the SIFs are maximal near the sample edge. The equations for long crack ($a \geq 3h$) are [121]

$$K_{\rm I}/p\sqrt{h} = \Phi_0(h/a)^{1/2} \; ; \qquad K_{\rm II}/p\sqrt{h} = \omega(h/a)^{1/2} \; , \tag{9.28}$$

where $\Phi_0 \approx \pm 0.47$ and $\omega \approx 0.73$, with negative sign of Φ_0 for $\alpha_m > \alpha_c$. The values of the SIFs for short crack ($a/h << 1$) are given as

$$K_{\rm I}/p\sqrt{h} = -0.99(a/h)^{1/2} \; ; \qquad K_{\rm II}/p\sqrt{h} = 0.63(a/h)^{1/2} \; . \tag{9.29}$$

The comparison of the solutions for single crack and pair of cracks defines three different regions, namely (i) the SIFs for both cases are equal at $a << h$, (ii) the SIFs for single crack have larger values at $a < h$ due to the shielding of the crack pair and (iii) the SIFs for the crack pair exceed those for the single crack at $a > h$ because of full release of the strain energy in the metallic layer by the crack pair.

In the case of unlike ceramics, the additive SIF, caused by the difference in their properties, should be superimposed on the solution for like ceramics. Obviously, this value is the SIF for body subject to the equal but opposite surface forces on either side of the bond plane. The crack growth in the case of small-scale yielding is found by the summary value of the SIFs from residual and external stresses. We obtain $K_{\mathrm{I}} > 0$ for the edge cracks at $\alpha_{\mathrm{m}} < \alpha_{\mathrm{c}}$, and the residual stress state will define the interface strength, S, corresponding to maximum stress for the crack growth as [121]

$$S = E_{\mathrm{m}} G_{\mathrm{c}} / [2\sqrt{\pi}(\Phi_0 + \sqrt{\Phi_0^2 + \omega^2}) ph(1 - \nu_{\mathrm{m}}^2)] , \tag{9.30}$$

where G_{c} is the fracture resistance of the interface governed by the phase angle of loading, $\Psi \approx \pi/6$. Then, we obtain, $K_{\mathrm{I}} < 0$ at $\alpha_{\mathrm{m}} > \alpha_{\mathrm{c}}$ and the edge crack growth is initially inhibited, but fracture nucleates from center. The interface strength in this case can be given as [121]

$$S \approx 0.5\sqrt{\pi E_{\mathrm{m}} G_{\mathrm{c}} / a_0 (1 - \nu_{\mathrm{m}}^2)} - p/10 , \tag{9.31}$$

where G_{c} is the fracture resistance governed by $\Psi = 0$ and a_0 is the radius of the interfacial crack, which controls the fracture.

9.6 Toughening Mechanisms

The toughening mechanisms in the S-N-S layered systems are connected with the interface geometry, inducing the crack deflection and the out-of-plane microcracking that constructs the links, bridges, which increase toughness and fracture resistance at the stage of sub-critical crack growth along the ceramic–metal interfaces. Microcracks–voids, formed during processing (e.g., using photo-lithographic techniques, combined with evaporation and diffusion bonding) and localized at the ceramic–metal interfaces, under loading promote the bridging bulges and stretching of the metal film, which can inhibit the crack growth. In this case, a cracking both along plain side of the patterned region and along the patterned side of the metal film is possible. As has been shown by the experiments conducted for glass–copper interfaces, the toughness increase is nearly comparable for the samples, where the crack is driven along the plain interface, wherein the crack deflection is absent, and where the crack grows at the patterned interface, wherein the toughening results from both the crack deflection and bridging [778]. The R-curve behavior (i.e., fracture resistance vs crack extension) states in the second case that the crack bridging is the dominating mechanism for the crack tip shielding (to compare with the crack deflection) in the interface toughening. The corresponding toughening mechanisms connected with the crack bridging and deflection are shown in Fig. 9.15. The estimation of the toughening value in the steady-state crack, induced by stretching of the metal layer, which remains to be joined with the brittle substrate at one side (see Fig. 9.15a),

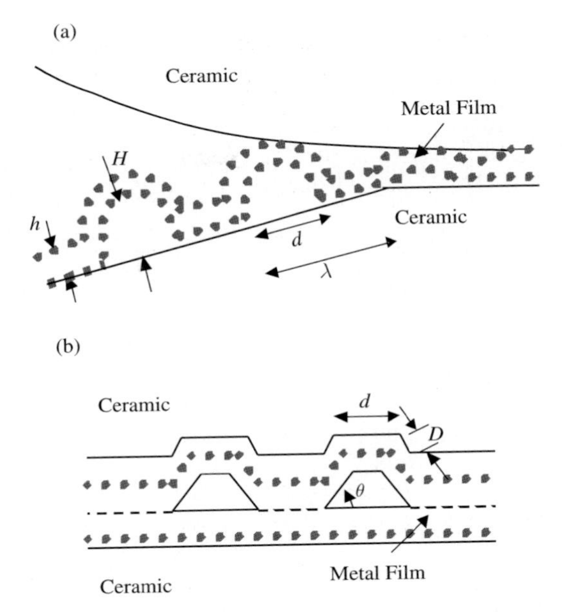

Fig. 9.15. Two models of interfacial toughening due to (**a**) crack bridging under plastic stretching of metal film and (**b**) crack deflection

can be derived, assuming each metallic ligament stretches to the shape of arc with height H, width d and period λ. Then, the toughness increment ($\Delta G_{\mathrm{c}}^{\mathrm{br}}$) is found through thickness h, yielding strength σ_{y} and plastic strain $\in_{\mathrm{p}}$ for metallic layer [778]:

$$\Delta G_{\mathrm{c}}^{\mathrm{br}} = \sigma_{\mathrm{y}} h \in_{\mathrm{p}} = \sigma_{\mathrm{y}} h (8d/3\lambda)(H/d)^2 \,. \tag{9.32}$$

The toughening in the case of the fracture along the patterned interfaces due to the crack deflection can be estimated on the basis of the model for 2D crack with repeated kink segments (see Fig. 9.15b) as [778]

$$\Delta G_{\mathrm{c}}^{\mathrm{def}} = G_0 \left[\left(\frac{D+d}{D \, \cos^2(\theta/2) + d} \right)^2 - 1 \right] , \tag{9.33}$$

where G_0 is the intrinsic toughness, taking into account the inclined and flat interfaces; D and d are the deflected and undeflected crack segment lengths, respectively, and θ is the angle of the crack deflection. Finally, it should be noted that a modest effect of the toughening (compared to the macroscopic crack deflection) occurs from much finer scale surface roughness, causing both the interfacial crack tilts and twists, which inhibit the crack extension and induce greater the plastic stretching of the bridging metal ligaments as the main crack extends [778].

Table 9.2. Material parameters used in calculations

Property	YBCO	$SrTiO_3$	Al_2O_3	Ag	Mo
$\alpha(10^{-6}\,K^{-1})$	16.0	9.4	8.3	18.9	5.3
E(GPa)	64	300	380	75	315
ν	0.22	0.23	0.25	0.37	0.31
σ_y (MPa)	–	–	–	25	570

Table 9.3. Numerical results

Composition	h_s/h	$K_{ss}/\sigma h^{1/2}$	$\in_c$	S(MPa)	$\Delta G_c^{br}/G_0$	$\Delta G_c^{def}/G_0$
YBCO/$SrTiO_3$	4.9	–	1.41	–	–	–
YBCO/Al_2O_3	0.8	–	1.60	–	–	–
YBCO/Ag	–	1.38	–	–	–	–
YBCO/Mo	–	1.28	–	–	–	–
YBCO/Ag/$SrTiO_3$	–	–	–	585	0.02	0.02
YBCO/Mo/$SrTiO_3$	–	–	–	380	0.51	0.02

The known material parameters (see Table 9.2) enable to estimate some numerical results (see Table 9.3) for composite structures, which are proper for HTSC JJs. We use in the computations the following relationships for geometrical parameters: $d/\lambda = 2/3, H/d = 0.1, D/d = 0.15$, and values of $\theta = 35°, G_c = G_0 = 2\,\mathrm{J/m^2}$ [778], $K_{Is} = 1\,\mathrm{MPa\cdot m^{1/2}}, a = a_0 = 3h = 300\,\mathrm{nm}, \Delta T = 800\,\mathrm{K}$.

9.7 Charts of Material Properties and Fracture

The numerical results may also be obtained for different HTSC composites of the S-I-S type using (9.1)–(9.3), (9.20), (9.21) and (9.26). However, even initially, there are great spreads of the values for various mechanical and strength properties, which caused the actual structure of these materials.[2] In particular, the elastic modules for HTSC ceramics can be 1–2 order of magnitude smaller than the modules of the same superconducting crystals. For example, in the case of BSCCO, the fracture toughness (K_c) changes in the range 0.5–3.0MPa $\cdot\, \mathrm{m}^{1/2}$, but Young's modulus lies in the limit 54.1–230 GPa [775]. Therefore, the computations based on (8.20), (8.21) and (8.26) lead to significant differences for the limit values of K_c^M and E_m at other fixed parameters (see Table 9.4). These results are obtained in the simplest case, considering two matrix layers (BSCCO) and one buffer layer (MgO) with the same thickness (i.e., $V_m/V_b = 2$), selecting other parameters as $\nu_m = 0.2$ [648],

[2] The same statement relates also to materials used in HTSC JJs of the S-N-S type, some properties of which are presented in Tables 9.2 and 9.3

Table 9.4. Some numerical results for penny-shaped crack

Properties	$c_{\mathrm{m}}(\tau/R)^{2/3}$	$\sigma_{\mathrm{m}}(R/\tau)^{1/3}$	$\Delta G_{\mathrm{c}}(R/u_0^3\tau)^{1/2}$
$K_{\mathrm{c}} = 0.5\,\mathrm{MPa}\cdot\mathrm{m}^{1/2}$; $E_{\mathrm{m}} = 54.1\,\mathrm{GPa}$	3.516	6.408	14.507
$K_{\mathrm{c}} = 0.5\,\mathrm{MPa}\cdot\mathrm{m}^{1/2}$; $E_{\mathrm{m}} = 230\,\mathrm{GPa}$	5.362	2.298	9.655
$K_{\mathrm{c}} = 3.0\,\mathrm{MPa}\cdot\mathrm{m}^{1/2}$; $E_{\mathrm{m}} = 54.1\,\mathrm{GPa}$	11.608	21.158	14.507
$K_{\mathrm{c}} = 3.0\,\mathrm{MPa}\cdot\mathrm{m}^{1/2}$; $E_{\mathrm{m}} = 230\,\mathrm{GPa}$	17.704	7.588	9.655

$\nu_{\mathrm{b}} = 0.36, E_{\mathrm{b}} = 290\,\mathrm{GPa}$ [47]. The corresponding shear modules are found as $\mu_k = E_k/(\nu_k + 1)$, where $k = 1, 2$. This example is also interesting from the view of the HTSC processing due to the very small chemical reaction between MgO and BSCCO melt [762].

The numerical results lead to qualitative trends in change of some normalized parameters depending on the fracture toughness (K_{c}) and Young's modulus (E_{m}). For example, for the penny-shaped crack into superconducting matrix (see Table 9.4), the crack size, corresponding to its transition to the steady-state crack ($c_0 = c_{\mathrm{m}}$), grows together with the fracture toughness or/and Young's modulus. In this case, the equilibrium stress ($\sigma_0 = \sigma_{\mathrm{m}}$) enhances with K_{c} but diminishes with increase of E_{m}. The fracture toughness increment for the steady-state crack (ΔG_{c}) decreases with increase of E_{m}. The normalizing multiples in the considered dependencies include tangential stress at the interface, taking into account the sliding friction (τ) and the limit displacement at the crack opening ($2u_0$). These parameters state the features of transversal fracture and characterize the material interfaces (in particular, the de-bonding processes), and therefore, preliminary experimental definition of these is necessary.

Similar spreads of data exist for YBCO and other ceramics and metals used in HTSC JJs processing. The main causes, which find these broad ranges of material properties, are the following ones: porosity, microcracks, damages, domain and crystallographic structures, etc. The effects of porosity and microcracking on elastic modules of YBCO and BSCCO ceramics may be taken into account, in particular, using the self-consistent differential method [577, 814]. In any case, mechanical properties every time should be selected with account of specific microstructure, loading and fracture features of materials. So, it is necessary to optimize elastic modules of both material components in the estimation of Dunders parameter α, defined in (9.3).[3] A number of parameters are necessary to be given in the estimation of critical properties of the HTSC-layered composites using (9.20), (9.21) and (9.26).

In order to solve the above selection problem, the material property charts [30] may be useful. Figure 9.16 presents that chart, in which one property is

[3] The parameter α, in particular, is very important in the estimation of the fracture character, because its increase diminishes the fracture of interface cracking to compare with the substrate failure in the plot of dependence of the relative toughness ($\Gamma_i/\Gamma_{\mathrm{s}}$) on the phase angle of loading (Ψ) [263]

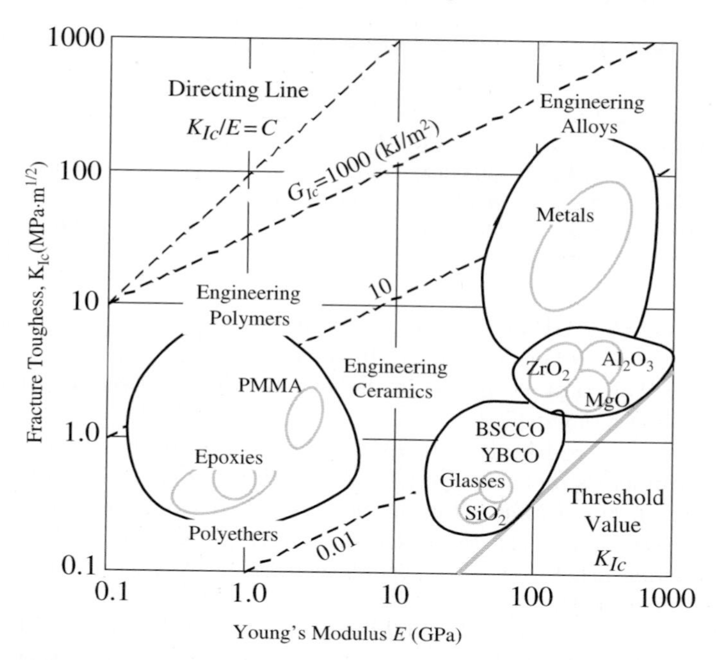

Fig. 9.16. Material property chart, representing K_{Ic} vs E on logarithmic scales. The guide lines K_{Ic}/E = const help in fracture design of materials. Additional property, $G_{\mathrm{Ic}} = K_{\mathrm{Ic}}^2/E$, is shown

plotted against another on logarithmic scales. This displays properties in very clear form and ensures to represent additional fundamental relationships in each chart (in particular, the critical release rate of strain energy, $G_{Ic} = K_{Ic}^2/E$, is shown additionally in Fig. 9.16). These charts help to select range for any property and to state criteria for the estimation of threshold material behavior that are used in HTSC JJs processing, taking into account specific loading conditions, thermal and electromagnetic treatments and so on.

When the crack grows through polycrystalline or composite structure, a significant change of the surrounding material characteristics occurs with increase of the crack size and local stresses at the crack front. Numerous toughening mechanisms are initiated, namely microcracking and twinning near crack, crack branching and bridging, phase transformations and domain re-orientations, interaction of crack with microcracks and mesostructure elements, etc. Due to these, it is complicated to obtain the quantitative estimations of crack growth rate and hence to define a time before fracture or sample durability for given stress (σ) and temperature (T). Thus, it is very difficult to distinguish between action of various mechanisms of fracture resistance and also material failure features. The fracture charts, based on experimental data and plotted for many materials, solve this problem in some way and help to select material properties, taking into account the fracture type. The boundaries in the fracture charts define equal contributions of adjacent regions, which correspond to action of specific fracture mechanism. They displace at alteration of microstructure characteristics of the given material. The fracture charts for b. c. c. metals (e.g., Mo) and f. c. c. metals (e.g., Ag) differ significantly (see Fig. 9.17a and b). F. c. c. metals demonstrate only

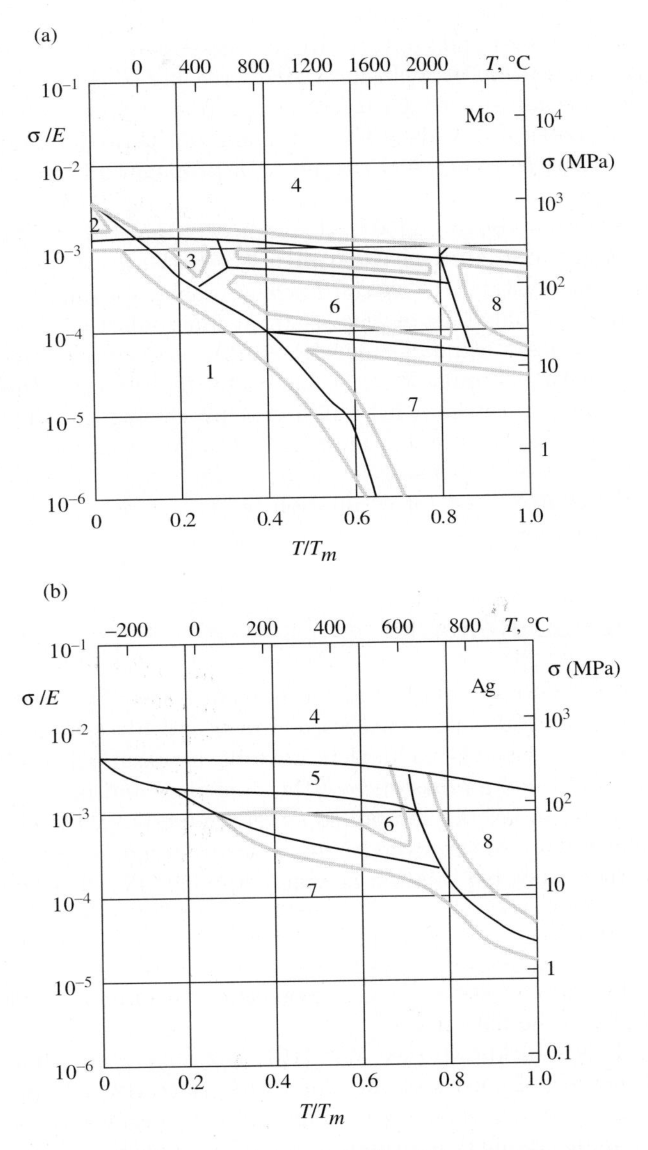

Fig. 9.17. Fracture charts for Mo (**a**) and Ag (**b**). Numbers denote the next fracture types: (1) cleavage I; (2) cleavage II; (3) cleavage III; (4) dynamic fracture; (5) ductile fracture; (6) transgranular and (7) intergranular creep fracture; (8) rupture. *Dashed curves* show limit changes of interfaces

ductile fracture after initial significant plastic deformation. At the same time, b. c. c. metals show brittle fracture at low temperatures.

In the case of silver [31], the material fracture occurs in creep mechanism at low tensile stresses ($\sigma/E < 10^{-3}$). Nucleation of voids and boundary sliding take place in intergranular fracture. Transgranular fracture is accompanied by the void nucleation at inclusions and the void growth together with the creep strain. The strain localizes in weaken cross-section of the sample that leads to growth of local stresses and rupture. High stresses cause ductile fracture also with the void nucleation at inclusions. At increasing of the stresses, fracture occurs in the mechanism, which is proper for polycrystals, when nucleation and growth of cracks and voids at the grain boundaries are facilitated. Finally, the fracture occurs in dynamical regime with propagation of stress waves in the sample bulk.

In the case of molybdenum [301], three regions of brittle fracture by cleavage are added in the fracture chart, namely (i) on the nucleus cracks, preliminary existing in crystal; (II) on the nucleus cracks, forming at coalescence of dislocations in initial stage of plastic strain (microplasticity), and (III) the plastic strain attains a significant level ($\sim$10%), but the fracture takes place owing to the cleavage. Similar fracture charts for high-temperature superconductors can significantly help in design and processing of HTSC JJs with optimum characteristics. Finally, note [487, 1152] as successful examples of development of the system analysis methods, including creation of the material property charts, used for estimation of critical behavior of the HTSC materials and composites.

9.8 Concluding Remarks

The results for different toughening mechanisms, actually acting in HTSC microstructures, and also the features of the fracture resistance alterations due to the growth of crack-like defects have been presented in Chaps.8 and 9. They have confirmed an existence of the numerous and non-simple effects. At the same time, today in the applied superconductivity, the simplified approaches to estimation of the strength and fracture toughness are widespread. For example, the following relation is widely applied [198, 614, 911, etc.]:

$$\sigma_{\mathrm{f}} = Y K_{\mathrm{c}}/c^{1/2} \, , \tag{9.34}$$

where σ_{f} is the fracture stress, Y is a geometric constant, K_{c} is the fracture toughness and c is the flaw size.

However, it is well known (see, e.g. [157, 262, 423, etc.]) that (9.34) has been stated in the framework of the classical Griffith-Irwin's approach [348, 473] for *homogeneous linear-elastic solid under homogeneous loading.* This equation could be applied *only* under conditions of real dominating crack with size considerably greater than microstructure scale and in the absence of internal (residual) stresses. Obviously, an account of these macrocracks is no interesting for HTSC, because the superconducting and transport properties very rapidly degrade even in an insignificant microcracking. Moreover, as has been shown above, the HTSC microstructure is very heterogeneous with heterogeneity scale, being compared with characteristic microcrack size. Finally, the HTSC compositions undergo non-homogeneous loading during working and possess considerable residual stresses, initiated by processing. Note also that another fracture mechanisms act by small fracture stresses. Hence, the use of (9.34) could lead to incorrect conclusions, for example, "K_{Ic} can be improved [in *RE*BCO] by the refinement of the *RE*-211 particle size" (see [911], p. 111), or "K_{c} is independent of flaws in the material (cracks, pores...) and, therefore represents a true property of the Y-123/Y-211 compound" (see [614], p. 2074). Violation of the dependence between σ_{f} and K_{c}, given by (9.34), is evident, for example, from Fig. 8.15. Obviously, (9.34) could be applied only in the absence of acting toughening and crack amplification mechanisms and under conditions of other restrictions described above.

The coming certification of HTSC systems will demand correct definition of the strength properties. The combination of the above considered toughening and crack amplification mechanisms could exert on fracture resistance not only summary effect but also multiplication effect. Therefore, for every actual toughening (or crack amplification) mechanism, ratio $K_{\mathrm{c}}/K_{\mathrm{c}}^0$ or $G_{\mathrm{c}}/G_{\mathrm{c}}^0$ should be defined, where $K_{\mathrm{c}}(G_{\mathrm{c}})$ is the critical SIF (critical strain energy release rate) due to actual toughening (or crack amplification) mechanism and $K_{\mathrm{c}}^0(G_{\mathrm{c}}^0)$ is the corresponding intrinsic parameter without toughening (or crack amplification). Then the total fracture resistance (in the force and energy approach, respectively) should be found as

$$K_{\mathrm{c}}^{\mathrm{tot}}/K_{\mathrm{c}}^0 = \prod_{i=1}^{n} K_{\mathrm{c}}^{(i)}/K_{\mathrm{c}}^0 \,; \tag{9.35a}$$

$$G_{\mathrm{c}}^{\mathrm{tot}}/G_{\mathrm{c}}^0 = \prod_{i=1}^{n} G_{\mathrm{c}}^{(i)}/G_{\mathrm{c}}^0 \,, \tag{9.35b}$$

where n is the common number of the toughening and crack amplification mechanisms.

The research of effectiveness of the HTSC toughening mechanisms should be accompanied by the study of the HTSC conductivity because it is often these effects which are contrary to each other (see, e.g, [577, 822]). The modeling of the current-carrying properties of HTSC system is the main aim of the final chapter of the monograph.

10

Modeling of Electromagnetic and Superconducting Properties of HTSC

10.1 Modeling of Intercrystalline Dislocations

There are two basic approaches for description of intercrystalline structures [43], namely (i) the structural unit approach, focuses on the atomic arrangement at the intergranular boundary and (ii) the intergranular boundary dislocation (IBD) approach, based on the periodic strain field that is observed at many intergranular boundaries. Both models use a *coincidence site lattice* (CSL) description of the intergranular boundary geometry. While these models are equivalent [48], each one has its own advantages in description of the intercrystalline structure. Nevertheless, today studies of intergranular boundaries in HTSCs focus primarily on the IBD description.

The structure of intergranular boundary may be presented, consisting of two different types of parallel conducting channels, disposed along the boundary and defined by dislocation structure (see Fig. 10.1). One of them possesses approximate structure and properties of superconducting crystal, associated with the regions between dislocations. Other channel is normal one or demonstrating a weak superconductivity, and it is compared with dislocation cores or their elastic deformations [406]. When the misorientation angle, θ, increases then the dislocation density increases, the interdislocation spacing diminishes which leads to gradual inhibition of the superconducting channels and to forcing of the weak link behavior. Therefore, an understanding of IBD nature is very important for statement of transition from the strong link behavior to the weak one.

The IBD models are based on the concept that the grain boundary free energy is particularly low for a certain set of special misorientation relationships, θ $[hkl]$. In this case, the boundaries with θ $[hkl]$ values other than special ones relax into a configuration in which sites of a low-energy structure are preserved by a localized rotation of the crystals. The macroscopic θ $[hkl]$ and its deviation from a low-energy misorientation are produced by the introduction of a regular array of dislocations that rotates one crystal relative to the other on a macroscopic scale. On the microscopic scale, these dislocations separate

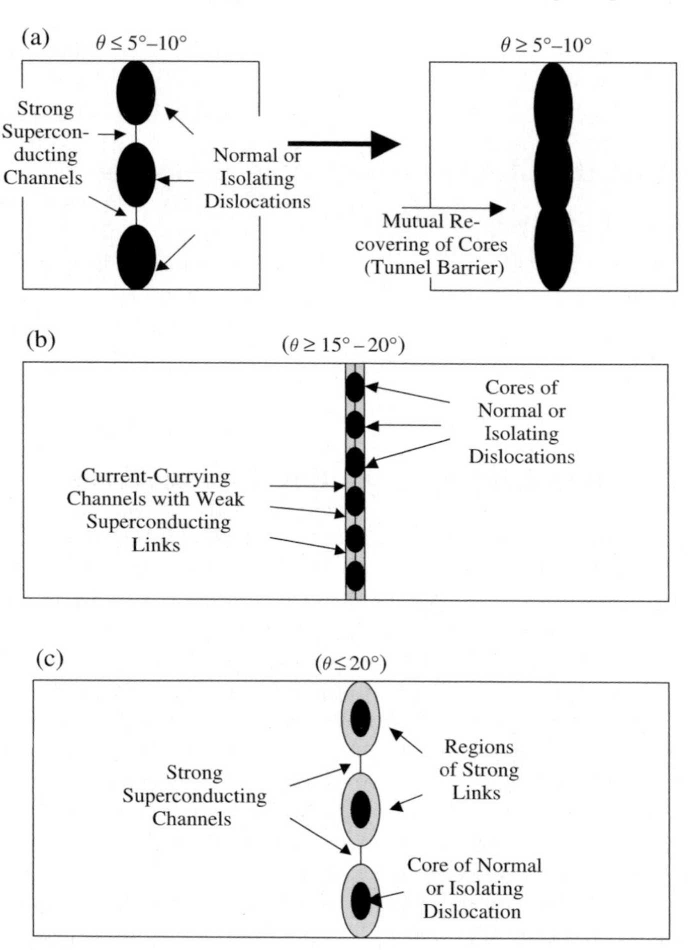

Fig. 10.1. Three models explaining intergranular values of J_c in thin YBCO films with [001] direction vs intergranular boundary structure: (**a**) model of dislocation cores before and after their partial recovering (Dimos model) [149, 205]; (**b**) model of Dayem's bridge [920] and (**c**) model of channel with strong links [280]

the sites of boundary with low-energy structure. Thus, most intergranular boundaries are presented as being divided into regions of "good fit," having low-energy atomic structure, separated by a net of "bad fit" of IBDs. As the difference between the actual misorientation relationship and the low-energy one increases, the spacing of the dislocations decreases and the fraction of bad-fitting region increases. A CSL construction is used to identify the low-energy misorientations and describe the structure periodicity of the boundary. The IBD model suggests that low energy is associated with relatively short-period

CSLs or, equivalently, low-coincidence index, Σ (where Σ is the fraction of lattice sites of one crystal that is coincident with the other [427]).

CSLs can form in orthorhombic and tetragonal crystals only when the squares of the lattice parameters all form rational ratios (i.e., $a^2 : b^2 : c^2$ are rational numbers) [547]. This condition is rarely met in real crystal structures. The approach selected to apply the coincidence concept to these systems is to impose a small strain on the lattice in the boundary vicinity. This strain forces the lattice parameters to meet the rational ratios condition, and the lattice can form CSLs. The CSLs, resulting under strained conditions often are called *constrained CSLs* (CCSLs). Usually, the additional strain is attributed to additional sets of dislocations that are localized in the boundary and become part of the IBD structure. Thus, the IBD structure has two components, one to produce the correct axial ratios and other to accommodate deviations from low-energy misorientations. Because the strain-producing and deviation-accommodating IBDs can have opposite signs, the dislocation density may not increase monotonously with the deviation from the exact CCSL misorientation relationship. As a result, the minimum spacing may not exist at the CCSL misorientation and widely spaced IBD structures may accommodate rather large deviations from lattice coincidence [547]. This leads to a boundary division into seemingly different patches.

There are at least two reasons to explore the (C)CSL/IBD model. First, it is necessary to identify the fundamental origin of the weak-link problem in order to discern whether one is intrinsic to the boundary structure or extrinsic. The second cause is connected with identification of certain types of inergranular boundaries other than low-angle boundaries that have specific properties [43].

Because the structures of boundaries that lie parallel to low-index planes of the CCSL have the shortest wavelength periodicity possible for that misorientation relationship, their energies are expected to be minimum and faceting onto those planes might be expected. In this case, the modeling of intercrystalline structure is simplified in strongly faceting structures, because macroscopic boundaries with a variety of boundary planes can be considered as mixtures of some boundary facets, each of which has a fixed structure. The CSL and CCSL conceptions can be used to optimize intergranular boundary behavior. At the same time, they are not the governing structures due to the following reasons [43]:

(1) Because the unit-cell dimensions are allowed to vary to achieve coincidence, relatively low-Σ misorientations densely populate θ $[hkl]$ space. As a result, for many observed boundary misorientation relationships there are several nearby CCSLs that may be used to the boundary description.
(2) The types of relaxations that occur in these boundaries often assume statement of compromise between the magnitude of the local strain, required to achieve coincidence, the achievement of low Σ values and the minimization of strain energy due to dislocations.

(3) The choice of an appropriate CCSL is complicated by the absence of a suitable estimation for possible deviations from the exact coincidence misorientation that reasonably can be referenced back to a particular CCSL. This criterion must state the compromise between strains, low Σ, the mechanism by which the strain is accommodated and the possible non-monotonic dependence of the dislocation spacing on $\Delta\theta$.

The elastic strains, produced by the IBD arrays in individual facets, can be estimated using the equations derived for isolated pure-edge dislocations in an isotropic medium [1138] and assuming reasonability of this approximation for elastically isotropic material in the *ab*-plane [1079]. Then, the net strain field produced by a finite array of dislocations can be calculated using the superposition method. The numerical results obtained for a boundary containing seven equally spaced edge dislocations show that the material volume within about 20 nm of the facet junction can deform $\sim 0.8\%$ or more. In comparison, the orthorhombic to tetragonal phase transformation is accompanied by an analogous strain $\sim 1\%$ [1079].

Conventional model [427] describes a symmetric low-angle grain boundary (GB) by a chain of edge dislocations with the Burgers vector perpendicular to the GB plane. However, the structure of GBs in HTSC can be more complicated due to existence of partial dislocations and facets at GBs, long-range strain fields and compositional variations near GBs [43, 94]. The IBD structures influence the change of J_c due to an alteration of local composition and hole concentration, additional electron scattering and significant strains near the dislocation cores. The models [241, 714, 920] have been proposed, in which the current channels in GBs are described as an array of parallel point contacts, which demonstrate weak-link behavior if their width becomes smaller than the superconducting coherence length ξ [630]. The theoretical description of the rapid decreasing of $J_c(\theta)$ with increasing θ is usually ascribed by the strain-induced compositional suppression of superconducting order parameter near dislocation cores or in the layer of some thickness near GBs [43, 241, 714, 920]. Due to insufficiency of these phenomenological models, another model is presented in [373], which describes $J_c(\theta)$, taking into account the GB dislocation structure. It is assumed that the strains and excess ion charge of the GB dislocation structure can locally induce a dielectric phase near dislocation cores and cause progressive overall suppression of the superconducting order parameter with θ in a narrow layer of the order of the screening length near GBs. The model provides an intrinsic mechanism for the rapid decreasing of $J_c(\theta)$ with θ, solving the Ginzburg–Landau equation [182, 617], which describes well the practically important temperature range for HTSC ($T > 77\,\mathrm{K}$).

Dislocation model [690] computes a stress state at intergranular boundaries of three types, namely symmetrical and anti-symmetrical tilt ones with periodical system of edge dislocations, and also symmetrical tilt boundaries with arbitrary system of the edge dislocations. The angle dependencies of intercrystalline critical currents are calculated in these computations. Moreover, it is shown that adequate description of superconducting features of the boundaries can be obtained only in the framework of the arbitrary distributed dislocations.

The models of local increasing of the critical temperature (T_c), compared to corresponding values of the bulk critical temperature (T_{c0}) due to structure defects, causing the long-range strains, are presented in [372]. It is shown that the strain-induced T_c variation on defects is markedly enhanced in HTSCs due to small ξ, high T_c and strong anisotropic pressure dependence of the bulk critical temperature [15, 81, 153, 308, 357, 501, 691, 692, 764, 773, 933, 1036, 1136, 1140, 1143]. The letter reflects the characteristic bell-like dependence of T_c on the holes concentration [1041], which changes near defects due to local lattice distortions and electro-neutrality condition. These distortions $\varepsilon(x, y)$ can be quite strong, causing local plastic deformations or structure transformations around dislocation arrays, where the local holes concentration can vary from the critical value, c_s at which $T_c = 0$ up to the optimum concentration, c_m which corresponds to the maximum possible critical temperature, $T_{cm} = \max T_c(\varepsilon)$ in a *deformed* sample. This can give rise to localized superconducting regions coupled by the proximity effect above the bulk critical temperature, T_{c0}. Note that the localized deformations around specific structure defects can be much stronger than those accessible in experiment by applying uniaxial stress which is limited by the overall mechanical strength of the sample. In this case, even the hydrostatic pressure tests show a substantial increase of T_c under pressure from several degrees for optimally doped Bi-2212 single crystals up to $\sim$20 K in Hg-1223 [153, 773]. The effect markedly increases in under-doped HTSCs [15].

The increasing of T_c is estimated in [372] for edge dislocations, low-angle grain boundaries and metastable linear dislocation arrays, taking into account anisotropic strain dependence of T_c in the *ab*-plane and proximity effect, defining the superconducting state at intergranular boundaries. Moreover, the compositional changes are estimated due to the strain states, caused by defects, and effect of the T_c variations on magnetic flux pinning and magnetic granularity.

Being dependent on the sample strain, the T_c changes can be directly affected by applied stress state, which alters the dislocation distribution, GBs, microcracks and so on. In turn, this makes the effect of T_c increasing dependent on the particular deformation pre-history, which can give rise to localized metastable superconducting states on defects above the bulk T_c. For example, maximum ΔT_c is defined by dislocation walls of finite length which can exist in individual crystallites. However, these macrodefects are metastable and can disappear after annealing or redistribution of dislocations in the remnant strain fields.

The strains around defects can also significantly suppress T_c, especially near dislocation walls [372] or high-angle boundaries. Depending on the direction of current, this can manifest itself either as additional pinning or weak links because of the anisotropic flux pinning typical planar crystalline defects [366]. For the perpendicular current, **J**, the defect behaves as a weak link which locally diverts $\mathbf{J}(\mathbf{r})$ as it has been observed in magneto-optical experiments [832]. For parallel current, the dislocation wall enhances flux pinning by producing a deep potential well for vortices [372]. Such a well can trap many vortices, if the length of the dislocation wall, $2L > (\phi_0/H)^{1/2}$, where ϕ_0 is the quantum of magnetic flux and H is the applied field. Moreover, the local non-stoichiometry near GBs or dislocation arrays can significantly increase the strain-induced T_c variations at GBs as compared to the material inside grains.

10.2 Current-Limiting Mechanisms and Grain Boundary Pinning

A dual role of the grain boundaries, which causes increasing of the magnetic flux pinning and, at the same time, effectively block or divert macroscopic supercurrent flow is discussed in [364]. As it has been shown [205], the high-angle GBs, microcracks, stacking faults and other planar defects can lead to significant decreasing of the superconducting order parameter, and thus to diminution of connectivity of the superconducting grains. As a result, HTSCs often demonstrate electromagnetic granularity, which causes significant decrease of current-carrying properties of the superconductor [139, 354, 601, 832, 860]. Moreover, GBs in HTSC are strong current-limiting defects [597] because of the exponential decreasing of the critical current: $J_b(\theta) = J_0 \exp(-\theta/\theta_0)$ across tilt GBs with the misorientation angle of θ between adjacent crystallites, where $\theta_0 = 4$–$5°$ [205].

Low-angle GBs are inhomogeneous on different scales. On a nanoscale, they are formed by edge dislocations spaced by the distance, $d = b/2\sin(\theta/2)$, where b is the Burgers vector (see Fig. 10.2). Besides GBs, in HTSC there are macroscopic facet structures with periods, $D_p \sim 100$–1000 Å, which are much longer than the superconducting coherence length, ξ, along with long-range strains and local non-stoichiometry on the scale of about D_p both across and along GBs [114].

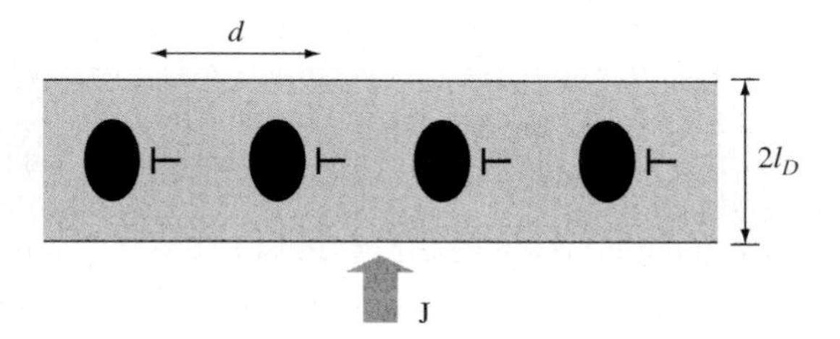

Fig. 10.2. Channel structure of a low-angle GB formed by dielectric dislocation core regions (*black*) and charge space layer (*gray*)

Due to the proximity of HTSC to the antiferromagnetic metal–insulator transition, they demonstrate the strong strain dependence, $T_c = T_{c0} - C_a\varepsilon_a - C_b\varepsilon_b$ (where C_a and C_b define the T_c change under one-axis compression along the a- and b-axis, respectively, for deformations ε_a and ε_b), which increases into dielectric regions of the order of characteristic size b near dislocation cores. These strain-induced dielectric core regions reduce the current-carrying cross-section of GBs, dividing it into a set of current channels. For the anisotropic in-plane dependence $T_c(\varepsilon)$ in YBCO (where $C_a = -C_b$ due to the effect of the CuO_2 chains), these channels have form, very different from channels for the nearly isotropic dependence $T_c(\varepsilon)$ in BSCCO crystals (for which $C_a = C_b = 300\,K$) [373]. Moreover, the long-range deformations, caused by faceting can lead to macroscopic modulations of T_c and J_c along grain boundary on the length of order D_p [114].

Another current-limiting mechanism is due to that the dislocation structure leads to an overall lattice expansion near GB in the layer of thickness of order d, where the ion density N_0 is reduced by $\delta N \sim \zeta < \varepsilon^2 > N_0$, where $\zeta \sim 1$–2 is Grüneisen's number [364]. The screening of the excess GB ion charge by carriers (holes) causes the shift of the electric potential on GB, $\Phi_0(\theta) \propto \theta$, proportional to the GB dislocation density and thus the misorientation angle, θ. This leads to the chemical potential variation at the GB, $\delta\mu(\theta) \propto \theta$, that causes T_c for $\theta \sim 15$–$20°$. In turn, the hole depletion and zone bending near a GB define strong superconductivity suppression in the space-charge layer with thickness of the double Thomas–Fermi screening length, $2l_D \approx \xi$ [373, 418, 419].

As a result, current transport through GBs is determined by the two factors: (i) strain-induced dielectric regions, which impede current flow and (ii) suppression of the superconducting order parameter in the current channels due to charging effects. When the angle θ increases, the chemical potential $\mu(\theta)$ on GB changes toward the under-doped insulating state, leading to progressive suppression of J_b with θ. Calculations of $J_b(\theta)$ are carried out in [373] on the basis of the Ginzburg–Landau equations. They show that above "transistor" model of GB describes well the observed exponential decreasing of $J_b(\theta)$ in HTSC.

The current transport in polycrystals in a magnetic field of B is determined by vortex dynamics and pinning, which are significantly affected by the channel structure of low-angle GBs. The increase in the misorientation angle, θ, directly alters the structure of GB vortices. For very small θ, vortices on a GB are Abrikosov's (A) vortices, localized on the dielectric dislocation core regions. At higher θ, the width of the GB current channels becomes of order ξ, so the normal vortex core disappears turning into a Josephson core, which is a 2π-phase kink of length, $l \approx \xi J_d/J_b$, along GB, where J_d is the decoupling current density [364]. An intermediate Abrikosov vortex with Josephson core (AJ vortex) [362] exists in the region of θ, where the phase core size, l, is smaller than the London penetration depth, λ. This condition ($l < \lambda$) holds in a rather wide region, $J_d/k < J_b(\theta) < J_d$, where $k = \lambda/\xi \sim 100$ is Ginzburg–Landau parameter. For higher θ, $J_{\mathrm{b}}(\theta)$ decreases below of J_{d}/k, and the AJ vortex turns into the Josephson vortex of length $\lambda_{\mathrm{J}} = (c\phi_0/16\pi^2\lambda J_{\mathrm{b}})^{1/2}$ [364]. This continuous A $\rightarrow$ AJ $\rightarrow$ J vortex transition with increase in θ directly affects vortex pinning on GB.

Vortices on grain boundaries are pinned due to both interaction with heterogeneities along GBs (core pinning) and magnetic interaction with GBs and bulk A-vortices. The pinning force, $\mathbf{f}$, is highly anisotropic with respect to the current direction. If $\mathbf{J} \parallel$ GB-direction, bulk A-vortices, spaced by $L < l$ from a GB are strongly pinned due to their magnetic interaction with the GB. This causes the inverse dependence, $J_{\mathrm{c}} \sim \xi J_{\mathrm{d}}/D$, on the grain size, D at low field, B provided that GBs do not block current flow, that is, $J_{\mathrm{c}} < J_{\mathrm{b}}$ [366]. This behavior of $J_{\mathrm{c}}(D)$ alters at higher values of B, either because the global value of $J_{\mathrm{c}}(B)$ is now limited by $J_{\mathrm{b}}(B)$ or pinning becomes collective if the Larkin pinning correlation length $L_{\mathrm{c}}(B) > D$.

When a current crosses GBs, the global value of J_{c} in polycrystals is found by a de-pinning of vortices, displacing along the GBs network. The core pinning of such vortices is caused by their interaction with modulations of the local current density, $J_{\mathrm{b}}(\mathbf{x})$, which are found by current channels or facet structure. For $J_{\mathrm{b}}(\mathbf{x}) = J_{\mathrm{b}0} + J_{\mathrm{a}}\cos(q\mathbf{x})$, the AJ-vortices on a GB become de-pinned, if the component of J perpendicular to the GB exceeds $J_{\mathrm{b}}^{\parallel}$, which is found as [832]

$$J_{\mathrm{b}}^{\parallel} = 2\pi J_{\mathrm{a}} ql \exp(-ql) \ . \tag{10.1}$$

The critical current density, $J_{\mathrm{b}}^{\parallel}(T, B)$, is maximum for the optimum period of the pinning potential, $L_{\mathrm{opt}} = 2\pi l(T, B)$. For fixed q, the optimum pinning occurs in the region of T and B, where $ql(T, B) \approx 1$. Note, that the pinning of GB vortices parallel to the grain boundary dislocations has been observed in [195]. However, because of the large size of the AJ phase core ($l \approx \xi J_{\mathrm{d}}/J_{\mathrm{b}} >> \xi$), the pinning of AJ-vortices is much weaker than that of bulk A-fluxons. As a result, GBs become channels of preferential motion for AJ- or J-vortices.

AJ-vortices, moving along the GB network experience the magnetic potential, $U(\mathbf{x}) = -\phi_0 H(\mathbf{x})/4\pi$, produced by strongly pinned bulk A-vortices. Here, the magnetic field along a GB varies as $H(\mathbf{x}) \approx B + H_{\mathrm{a}}\cos(2\pi\mathbf{x}/a)$, where $a = (\phi_0/H)^{1/2}$ in the inter-vortex spacing and the amplitude, $H_{\mathrm{a}} \approx \phi_0/(2\pi\lambda)^2$ is of the order of the lower critical field, H_{c1}. Thus, the magnetic pinning force, $f^{\parallel} \sim \phi_0 H_{\mathrm{a}} \sin(2\pi x/a)/a \propto H^{1/2}$, increases with H, leading to a non-monotonous field dependence of the global current density, $J(H)$ (so-called "fish-tail" effect) [366] and matching peaks in the dependence, $J_{\mathrm{c}}(H)$ [63]. These matching peaks have also been observed in YBCO bicrystals at fields $H_{\mathrm{n}} = \phi_0 n^2/D^2$, for which the inter-vortex spacing is commensurate with the GB facet structure [114].

10.3 Vortex Structures and Current Lines in HTSC with Defects

The observed influences of defects on flux pattern lead to a classification of the defects into two groups [557]:

(1) The "extended" defects, formed by a weakly superconducting material (typical size between some micrometers up to millimeters), which force the currents to flow along them, thus leading to a facilitation of flux penetration along the defects (these defects form channels for easy flux penetration into the sample).
(2) The small obstacles of non-superconducting material (typical size of micrometers or even less), where the currents can flow around them. There is no enhancement of the flux penetration, but a parabolic discontinuity line of the current is formed.

In Figs. 10.3 and 10.4, schematic current patterns (the critical currents are shown by the arrows) for both defect classes are presented. The characteristic feature of the first type (see Fig. 10.3) is an extended defect line, where flux penetrates the sample preferentially. This leads to the formation of a finger-like flux pattern. Flux density gradients always are pointed away from the defect line and at the end of the flux front the currents are forced to a U-turn. At the beginning of the defect line, the currents form a broad turn towards the defect line as the currents always flow along the sample edges [949] and also along the defect lines.

Figure 10.4 shows the current flow around a small obstacle of the second type, here symbolized as a round object. When the currents come close to this defect, they are forced to a slight turn and now follow the circumference of the obstacle. All the points, where the current lines turn around, lie on

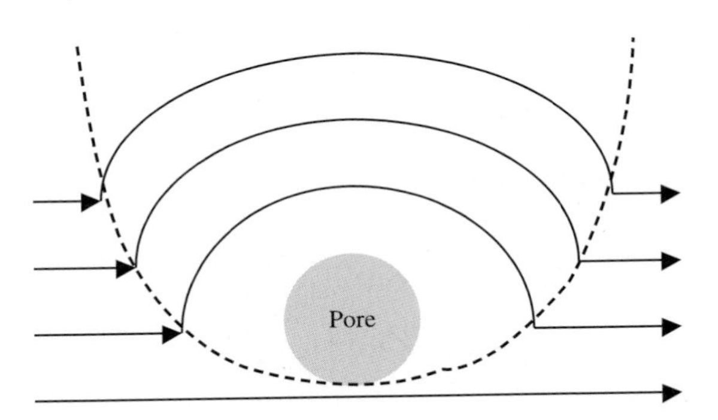

Fig. 10.3. A defect of linear type, symbolized by the dashed curve (current flow around cylindrical pore), leads to increasing flux penetration along it [120]. Therefore, the current flow is found by the flux density gradients. In contrast to the defect, shown in Fig. 10.4, there is no bottleneck between the normal flux front and the flux penetration along the defect line

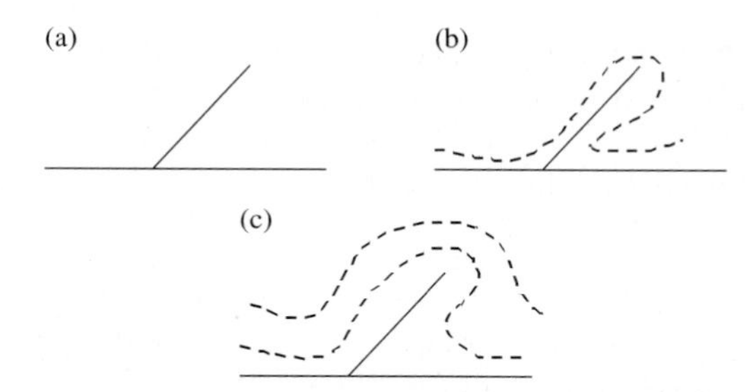

Fig. 10.4. A schematic drawing of the current distribution in the vicinity of a cylindrical cavity near the superconductor edge, located at the bottom of the figure. The current density spatially constant is shown by equidistant current lines (*dashed curves*). When the magnetic flux, penetrating from the sample edge, reaches the cavity, the current lines form sharp bends in order to flow around the hole on arcs of concentric circles. These bending points lie on a parabola, but the penetrating flux to the inside of the parabola always passes the defect, forming a "bottleneck" for the vortices [557]

a parabola, which can be seen in the flux distributions. The defect forms a "bottleneck" for the flux penetration that may trigger flux jumps [949]. If the small obstacle has a rectangular shape, the beginning of the parabola will be linear (forming an angle of 45° [949]), which in some distance from the defect again becomes as in Fig. 10.4.

The sufficiently large density of the defects localizes an area of interaction of two adjacent defects (see Fig. 10.5). At small fields, the current flow is similar to Fig. 10.3. During flux penetration in an increasing external magnetic field the flux fronts meet each other in some moment. This leads to complete

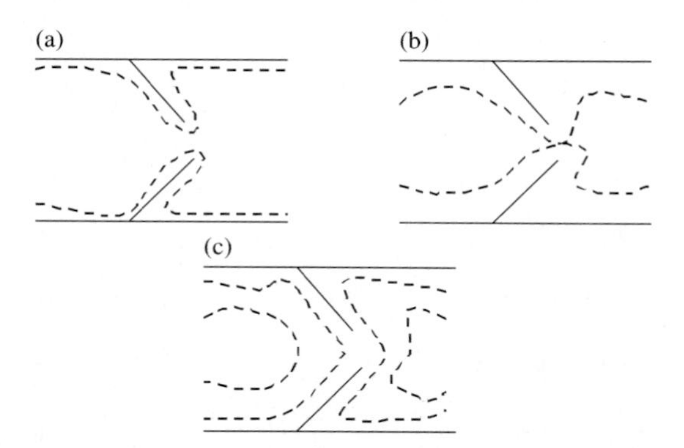

Fig. 10.5. At high defect density, two adjacent defects can be considered in sample (**a**). The flux penetration and current flow are identical to those of Fig. 10.3 as long as the external magnetic field is small (**b**). With rising field, the vortices penetrating the sample from both sides may merge together (**c**). This causes the current flow along both defect lines, leading to formation of a current loop [557]

change of the current pattern, because the currents can now flow along both defect lines and form new smaller current loops. In this case, the length scale of the currents changes and the sample is divided magnetically into smaller pieces. If the external magnetic field is raised further, the parts of the sample will behave completely independent from each other. This leads to magnetically induced *granularity* of the sample, with the defect lines acting as weak links at higher magnetic fields.

Figure 10.6 presents schematically a crystal with many defects. At moderate fields, where Meissner's phase still resides in the sample, the current flow is the same as that shown in Fig. 10.3, and many "flux fingers" are formed along the defect lines. With sufficiently large external magnetic field, many current sub-loops are formed. For this sample, the critical state model, assuming a homogeneous flux distribution is not used. One will severely underestimate the critical current density of such sample, if the critical current is

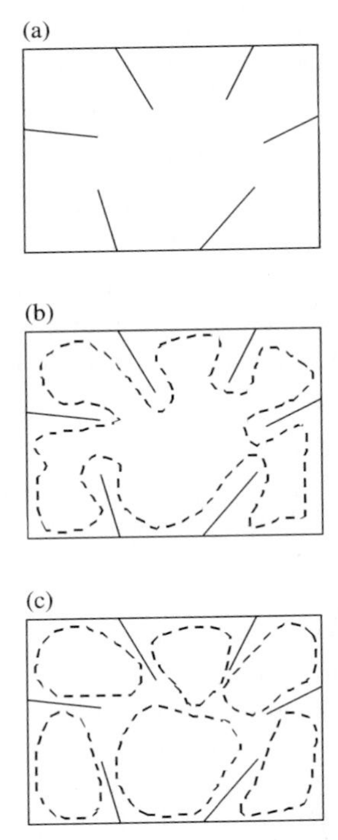

Fig. 10.6. A schematic representation of many defect lines in a rectangular sample (**a**). In moderate external fields (**b**), the flux penetrates easily along the defect lines and only a small flux penetration along the sample edges occurred. The critical currents (shown by *dashed curves*) flow around the defects. If the external magnetic field is sufficiently large (**c**), the flux penetration along the defect lines merges together and current loops are formed, both for the critical currents and for the eventually remaining shielding currents [557]

determined from its magnetic moment (with the tacit assumption of a single current loop, shielding the whole sample). It should be noted that, once a field or current distribution is disturbed, the pattern stays like this even up to high external magnetic fields.

The observed various flux distributions permit to classify the defects in HTSC through defect influence on the flux patterns [557]. In all cases, intrinsic (crystallographic) defects should be distinguished from extrinsic ones, which are caused by external influences (i.e., by cracks, scratches, cutting and etching).

(1) To the extended defects of the first type belong the intrinsic defects, such as twin boundaries [210, 211, 934, 1087, 1088, 1120] and the tilt boundaries found in Bi-2212 single crystals, caused by intergrowths of Bi-2223 phase [560]. Moreover, extrinsic defects, such as cracks in the sample surface, which are typical defects for most Bi-2212 single crystals and of melt-processed YBCO sample [403, 743], and also the defects caused by inhomogeneous film growth on substrates, containing scratches, belong to this type. However, the growth steps, observed in many single crystals, do not influence the current patterns [560]. Grain boundaries also belong to the extended defects, but they act as channels for easy flux penetration even at very small external magnetic fields. This separates a sample magnetically into smaller pieces.
(2) Above-described small obstacles of the second type in HTSC are caused by small flux droplets, observed in thin films and single crystals, by islands of the a-axis-oriented growth in the c-axis-oriented films, by small particles of foreign phases and by extrinsic defects, created owing to irregular cut or etching of the sample.

The presented classification characterizes all defects found in HTSCs. However, this approach is only valid for nearly homogeneous, monocrystalline samples, where a homogeneous flux front will appear in the ideal case. If an internal granularity is presented in the sample, the flux distribution is disturbed by the immediate flux penetration along the grain boundaries. Therefore, the analysis can only be fulfilled for individual grains.

Then, the flux pinning in superconductors with high critical current density is also found by a dense network of planar crystalline defects. In this case, there are two essential features of HTSCs, namely (i) the most effective pinning can be caused by the dense network of planar defects parallel to the flux lines, however (ii) unlike the case of arbitrary distributed point pins, the network of planar defects can block or divert the macroscopic current flow, if the tunneling superconducting current density (j_c) through these defects is smaller than J_c, determined by flux pinning. For $J_c > j_c$, the pinning structure can lead to the magnetic granularity, which manifests itself in a drop of the transport J_c due to the appearance of closed current loops within macroscopic crystalline grains, where densities of circulating magnetization currents become larger than J_c [171, 557].

Note also the importance of geometry of the planar defect network. For example, in twin domains with various orientations of the crystallographic axes, the planar defects can divert the local current flow, limiting the macroscopic value of J_c owing to reduction of the current-carrying cross-section area [288, 593, 622]. On the other hand, in a random network of planar defects with the density of the pinning centers, exceeding the percolation threshold, the supercurrent must cross several defects, which would increase J_c owing to flux pinning. This arbitrary defect network can have a complicated topological structure, in which only a small fraction of the defects belongs to the path of percolation [1013]. Generally, both regimes, in which the planar defects act as pinning centers or block (or divert) the current flow co-exist. Their relative contributions depend on both j_c and the geometry of the pinning network. The theoretical study of flux pinning due to a network of planar high-j_c crystalline defects, which are modeled by the corresponding Josephson contacts is carried out in [366]. A solution of equations of the non-local Josephson electrodynamics [362, 371] for a vortex parallel to the planar defect leads to the calculation of the magnetic field distribution and the transversal pinning force between the vortex and defect. The longitudinal pinning force of vortices along the defect is determined by both their magnetic interaction with pinned transgranular fluxons and local inhomogeneities of the defect. It is shown that the longitudinal component is much smaller than the transversal one, leading to the preferential flux motion along the paths of percolation formed by planar defects.

In highly anisotropic-layered superconductors or artificial superlattices, the interlayer superconducting coupling can be so weak that the coherence length in the direction perpendicular to the layers (ξ_c) becomes smaller than the interlayer spacing (s). In this case, the discreteness of the superconductor at the atomic level directly defines the structure of the vortex core [103, 1121]. There are several models, taking into account the layered structure of superconductor, for example Lawrence–Doniach's model depicting a stack of thin superconducting layers connected by Josephson interaction [607], or the models of S-N-S superlattices consisting of alternating superconducting (S) and normal (N) layers [159, 576, 1121]. An important feature of these models is that, due to weak Josephson interlayer coupling, the maximum supercurrent density (j_c) between the layers is much smaller than the interlayer de-pairing current density (j_d). As a result, the maximum current density that can be locally generated by a vortex parallel to the layers is limited by j_c. Due to the short coherence length ($\xi_c < s$) in layered HTSCs planar defects parallel to the ab-planes can significantly reduce interlayer coupling, limiting the current flow along the c-axis and rising the magnetic field penetration along the defects. In this case, the methods of non-local Josephson electrodynamics can be used to study both static and moving vortex (including non-linear area of Josephson core) at planar defect in layered HTSC and for the description of their structure [365].

The effect of macroscopic defects on localized magnetic flux (fluxons) and critical force of de-pinning is found by the defect size and type, and also by its disposition at grain boundary (or at Josephson junction). The case of the defect width of order of the Josephson penetration depth is considered in [1014]. Using the sine-Gordon model for defective Josephson junction, critical current as a function of magnetic field in the case of asymmetrically disposed defect is calculated. Moreover, the investigation of the interaction between fluxons and defects defines coercive forces of pinning or de-pinning a fluxon from a defect. Consideration, using an analysis of the different mode stability in dependence on the defect disposition at zero magnetic field, finds the bounds of these modes due to two factors, namely (i) the instability at the junction boundaries away from the defect and (ii) the instability due to the fluxon trapping or de-trapping by the defect. When the defect localizes near one of the junction edges, both criteria contribute (in definition) to the instability. In general case, there is coupling between defects and junction edges (surface defects), especially in the case of the numerous defects.

10.4 Non-Linear Current in Superconductors with Obstacles

Following [367], consider features of non-linear current in HTSC systems with existence of obstacles. Macroscopic electrodynamics of type II superconductors in the mixed state is found by the pinning and thermally activated creep of vortex structures, increasing a weakly dissipative critical current, irreversible magnetization and slow current relaxation (flux creep). These phenomena manifest themselves on length scales much longer than the Larkin's pinning correlation length (L_c), on which the critical state is formed [71]. On this macroscopic scale ($L_> > L_c$), the material property, defining the behavior of superconductor in electromagnetic fields, is the next non-linear local relation between the electric field, $\mathbf{E}$, and current density, $\mathbf{J}$:

$$\mathbf{J} = (\mathbf{E}/E)J(E) , \tag{10.2}$$

here $\mathbf{E}(\mathbf{r},t)$ and $\mathbf{J}(\mathbf{r},t)$ correspond to the macroscopic electric field and current density, averaged over all relevant intrinsic scales of pinned vortex structure. Consider an isotropic E–J dependence stated by (10.2), that, in particular, model the nearly two-dimensional current flow in the ab-plane of layered HTSCs. Equation (10.2) combined with the Maxwell equations

$$\partial_t \mathbf{B} = -\nabla \times \mathbf{E} ; \qquad \nabla \times \mathbf{H} = \mathbf{J}(\mathbf{E}) , \tag{10.3}$$

permit to calculate the evolution of heterogeneous distributions $\mathbf{E}(\mathbf{r},t)$ and $\mathbf{B}(\mathbf{r},t)$ and, thus, to describe macroscopic magnetic, transport and relaxation phenomena in superconductors.

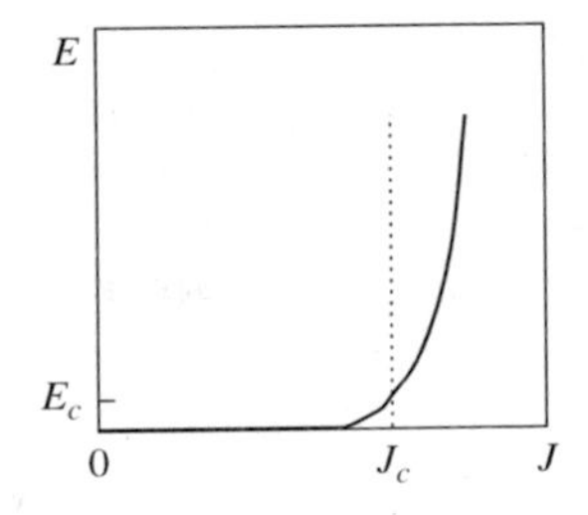

Fig. 10.7. $E(J)$ characteristic of a type II superconductor

A type II superconductor in the mixed state demonstrates a highly non-linear $E(J)$ dependence below a critical current density of J_c, which separates regimes of magnetic flux flow (at $J > J_c$) and its creep (at $J < J_c$) (see Fig. 10.7). The critical behavior of $E(J)$ at $J < J_c$ is determined by thermally activated vortex creep as

$$E(J) = E_c \exp\left[-\frac{U(J,T,B)}{T}\right], \tag{10.4}$$

where $U(J,T,B)$ is the activation barrier, depending on the current density, J, temperature, T, and magnetic induction, B. Here, E_c is the conditional criterion for electric field, which defines corresponding value of J_c through the relation, $U(J_c,T,B) = 0$. For example, the vortex glass/collective creep models [71] determines a dependence, $U = U_c[(J_c/J)^\mu - 1]$, for small values of $J << J_c$, which is observed in transport and magnetization measurements in HTSCs [1173]. Similar, though less singular logarithmic dependence $U = U_c \ln\,(J_c/J)$ corresponds to power-law E–J dependencies, $E = E_c(J/J_c)^n$ with $n(T,B) = U_c/T \sim 3 - 30$, to being a well approximation, especially for layered BSCCO [767, 1103, 1204].

When the distribution $\mathbf{J}(\mathbf{r})$ changes on spatial scales greater than L_c, a superconductor can be considered as a highly non-linear, heterogeneous conductor with a local characteristic of $\mathbf{E}(J,\mathbf{r})$. For example, (10.3) describes non-linear transport current in superconductors with macroscopic obstacles or current percolation in HTSC polycrystals with grain size $>> L_c$ (see Fig. 10.8). This case is important to understand the current-limiting mechanisms of HTSCs, which in addition to grain boundaries, often contain other macrodefects, for example, secondary phases, microcracks and areas of local non-stoichiometry on scales of order 10–1000 μm [597, 598]. These obstacles, blocking current flow, cause macroscopically heterogeneous distributions of transport and magnetization currents [115, 832, 854, 861, 950, 951, 952].

In turn, even relatively weak heterogeneity of local current density, $\mathbf{J}(\mathbf{r})$, can lead to exponentially strong variation of the electric field, $\mathbf{E}(\mathbf{r}) \propto \mathbf{J}$ $\exp[-U(J)/T]$, radically changing global properties of HTSC, $\overline{J}(\overline{E},B,T)$, observed in experiments. Thus, the behavior of $\overline{J}(\overline{E},B,T)$ in superconductor can differ from local characteristics, described by (10.2) and defined by thermally

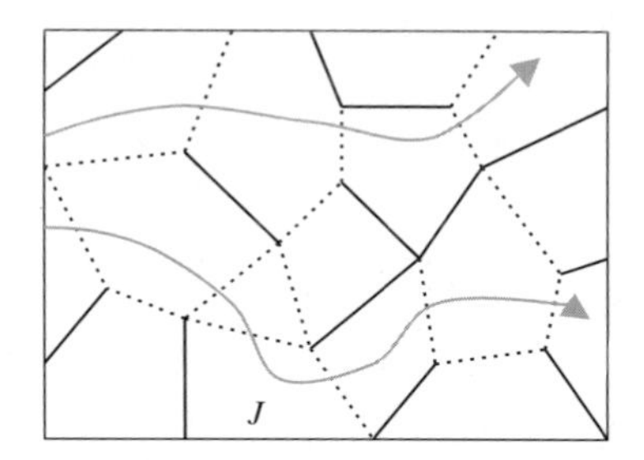

Fig. 10.8. Percolation of current flow in HTSC polycrystals. High-angle grain boundaries, blocking the current flow, are shown by *solid lines*, but low-angle grain boundaries, "transparent" for current flow, are denoted by *dotted lines*

activated vortex dynamics and pinning on mesoscopic scales (at $L < L_c$). This difference increases due to the non-linearity of $E(J)$, makes the effective current-carrying cross-section dependent on T, B and E [339, 360, 361, 369, 374, 398, 424, 611, 1006, 1018].

As examples of two-dimensional current flows for various cases, which could be considered as elements of a more general percolation network, presented in Fig. 10.8, we note proper cases, shown in Fig. 10.9 [367]. These geometries are standard in experimental studies of resistive states in superconductors, thin-film superconducting electronic circuits and HTSC conductors for power applications. In all cases presented in Figs. 10.8 and 10.9, the electric field and current density are heterogeneous on macroscales and demonstrate singularities near the sharp edges and corners. Calculation of the global E–J characteristic demands the solution of highly non-linear (10.2) and (10.3). For large values of n, even weak spatial changes of $\mathbf{J}(\mathbf{r})$ around planar defects in Fig. 10.8 lead to power variation of electric field, $E(r) = E_c[J(r)/J_c]^n$.

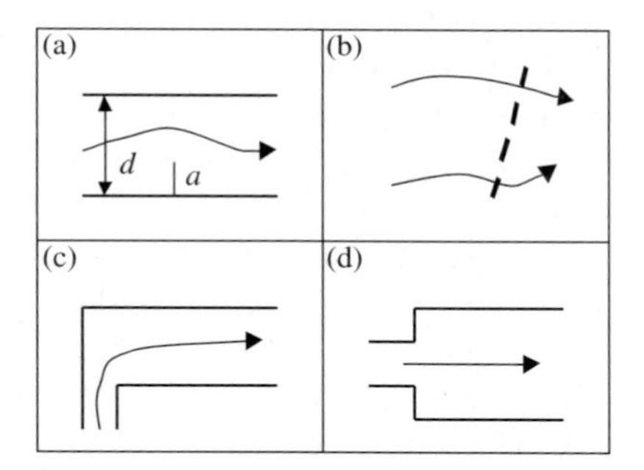

Fig. 10.9. Characteristic cases of two-dimensional non-linear current flow: (**a**) superconducting film with strong current-limiting planar defect (microcrack at high-angle grain boundary), (**b**) current flow through faceted grain boundary with alternating segments of various J_c values, (**c**) current injector (or magnetic flux transformer) and (**d**) microbridge

identical to each other in each group. The corresponding concentrations of the links are denoted by p and $q = 1 - p$.

Divide the present model into two kinds corresponding to (i) weak-field Josephson junction (JJ) and (b) strong-field flux-flow (FF) boundaries.

Binary JJ model. For the JJ case, the weak (strong) links with corresponding concentration, $p(1-p)$, have a critical current, $I_{c1}(I_{c2})$. If $I < I_{ci}$ in such resistor, then $V = 0$; if $I > I_{ci}$, then $V = IR_i$, where $R_i = \text{const}(i = 1, 2)$ and $I_{c1} < I_{c2}$. Note that V is a *discontinuous* function of I for both types of links (see Fig. 10.16).

Binary FF model. For the FF case, the weak (strong) links with corresponding concentration, $p(1-p)$, have a critical current, $I_{c1}(I_{c2})$. If $I < I_{ci}$ in such resistor, then $V = 0$; if $I > I_{ci}$, then $V = (I - I_{ci})R_i$, where $R_i = \text{const}(i = 1, 2)$ and $I_{c1} < I_{c2}$. Here, V is a *continuous* function of I for both types of links (see Fig. 10.16).

Both models differ at microscopic scale. The JJ model represents a situation of low current, and zero applied field. The Josephson junction is assumed to be superconducting for small currents and normal at high currents, when the voltage drop is just ordinary Ohmic loss. There is a discontinuous change between the two regimes. The FF model is intended to simulate weak links because flux pinning in them is weak. In this case, I_c of the link represents the de-pinning current. The losses take place for $I > I_c$ at the movement of flux along the boundary. Experimental measurements on individual GBs show a crossover between the two sorts of behavior. Figure 10.17 shows I–V characteristics for a 10°-GB. At low applied field, the boundary behavior is similar with the JJ model, while at high field it is closer to the FF model.

The considered non-linear network can be an approximation to an actual JJ array in the over-damped limit. The JJ model can represent a set of over-damped junctions with no pinning, in which the McCumber parameter, $\beta_J \equiv \hbar R^2/(2eI_{c0}C)^{1/2} < 1$, where R and C are the resistance and capacitance of the junctions. Then, output I_{tot} represents the *dc*-component of the actual output value of real over-damped junctions. In turn, the FF model is the case, where pinning of Josephson vortices dominates the transport. The result of

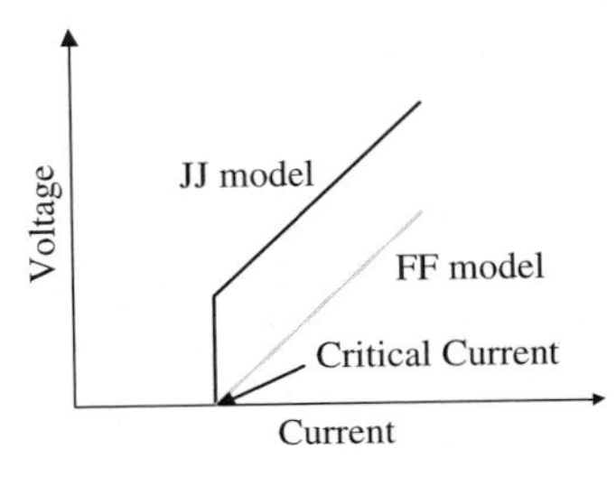

Fig. 10.16. I–V characteristics for Josephson junction (JJ) and flux-flow (FF) models. The first is considered to be appropriate for low fields and the second for high fields [398]

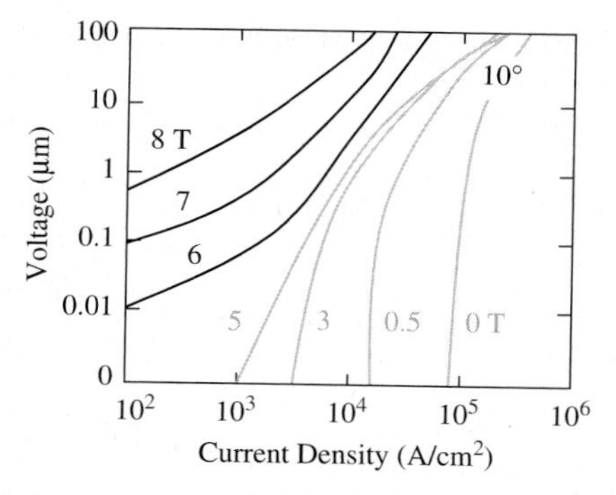

Fig. 10.17. Proper behavior of test I–V characteristics for 10°-GB in YBCO at different applied fields. At low fields, the I–V characteristic is similar with the JJ model. As the applied field increases, there is a crossover at 5.5 T to a behavior that is similar to the FF model [398]

numerical realization of two models is $V_{\text{tot}}(I_{\text{tot}})$ dependence for the system as a whole. Thus, we feed a fixed current at one end and collect it at the other and measure the voltage drop.

Let there be a distribution of non-linear resistors on an infinite mesh, where each resistor have some critical current and occupation probability.[2] This implies that the critical current can be found as

$$I_{\text{c}}^{\text{tot}} = \min_{S} \sum_{l \in S} I_{\text{c}}^{(l)} , \tag{10.13}$$

where the minimum is taken over *all* surfaces that separate the electrodes [883]; l are the wires that pierce S and $I_{\text{c}}^{(l)}$ is the critical current of wire l. Because $I_{\text{c}}^{\text{tot}}$ depends only on $I_{\text{c}}^{(l)}$, (10.13) states the dependence of total critical current of the mesh only upon the critical currents of individual elements and not upon their dissipative behavior at $I > I_{\text{c}}$.

Consider two adjacent infinite surfaces S and S^*, which separate the electrodes. The maximum supercurrent that can be carried through S, considered entirely by itself, is:

$$I_{\text{c}}^{(\infty)} = \sum_{j} p_j I_{\text{c}j} . \tag{10.14}$$

However, it is not clear that the current can penetrate from S to S^*. Indeed, we expect the critical current of the infinite mesh to be less than $I_{\text{c}}^{(\infty)}$, defining an upper bound for $I_{\text{c}}^{\text{tot}}$, independent of model and system dimensionality.

In order to obtain a better approximation for $I_{\text{c}}^{\text{tot}}$, consider a binary model, in which there are only two types of resistors ($I_{\text{c}1}$ and $I_{\text{c}2}$) with occupation

[2] We take a mesh constant of unity to avoid distinguishing between currents and current densities.

probabilities, p and $q = 1 - p$, respectively.[3] Equation (10.14) for this model gives

$$I_{\mathrm{c}}^{\mathrm{tot}} = pI_{\mathrm{c1}} + qI_{\mathrm{c2}} \,. \tag{10.15}$$

Consider $M \times N$ square mesh and suppose there is a current, $I > NI_{\mathrm{c1}}$, transported horizontally across the mesh. Divide the resistors into two groups. Those resistors, which dispose perpendicular to the average current flow direction, are called by "row resistors". These rows are connected by what we will call "column resistors", that is, the resistors dispose parallel to the current direction (see Fig. 10.18). The current flow can be considered as a superposition of two current distributions: NI_{c1}, running directly through the mesh (along the column resistors) and a percolating current, $I - NI_{\mathrm{c1}}$. Considering the percolating current, we can "subtract off" an I_{c1} resistor from each column resistor (a modified dilute resistor network is shown in Fig. 10.18). The columns of this lattice have holes with probability, p, and $I_{\mathrm{c3}} = I_{\mathrm{c2}} - I_{\mathrm{c1}}$ critical current resistors with probability, $q = 1 - p$. This modified mesh has a critical current, $I_{\mathrm{c}M}$. Then, the total critical current is $I_{\mathrm{c}}^{\mathrm{tot}} = I_{\mathrm{c1}} + I_{\mathrm{c}M}$.

The value of $I_{\mathrm{c}M}$ is found by two factors. First, the amount of current that can be transported across any given column of the mesh, and second, whether or not that current can be re-distributed to the I_{c3} resistors in the next column. There are two various regimes of behavior distinguished by whether I_{c3} is less or greater than I_{c1}.

When $I_{\mathrm{c3}} < I_{\mathrm{c1}}$ (or $I_{\mathrm{c2}} < 2I_{\mathrm{c1}}$), it is easy enough to re-distribute the current between columns. Any current flowing through a column resistor can be shunted sideways along a row and re-distributed to the next column as long as another column current does not get in the way. We can approximate the probability that a given column current can be re-distributed along a row without being interfered with in the following fashion. It is equal to the probability that there is an I_{c3} resistor directly across the given column current, plus the sum of probabilities that there is a path at a site s steps along the row, and no paths either coming in or out of the row before that point. By only summing over steps to the right (or left) of the given column current, we avoid the problem of interfering with the paths of other incoming column currents. This re-distribution probability can be written as [398]

$$p_{\mathrm{r}} = q \sum_{s=0}^{\infty} (1 - q)^{2s} = \frac{q}{2q - q^2} \,. \tag{10.16}$$

Taking into account a volume fraction q of I_{c3} row currents, we write down the averaged critical current density as

$$I_{\mathrm{c}}^{\mathrm{tot}} = I_{\mathrm{c1}} + (I_{\mathrm{c2}} - I_{\mathrm{c1}})qp_{\mathrm{r}} = I_{\mathrm{c1}} + (I_{\mathrm{c2}} - I_{\mathrm{c1}}) \times \frac{q^2}{2q - q^2} \,. \tag{10.17}$$

[3] Because, it is considered only the critical current, the exact I–V characteristic of these resistors is unimportant.

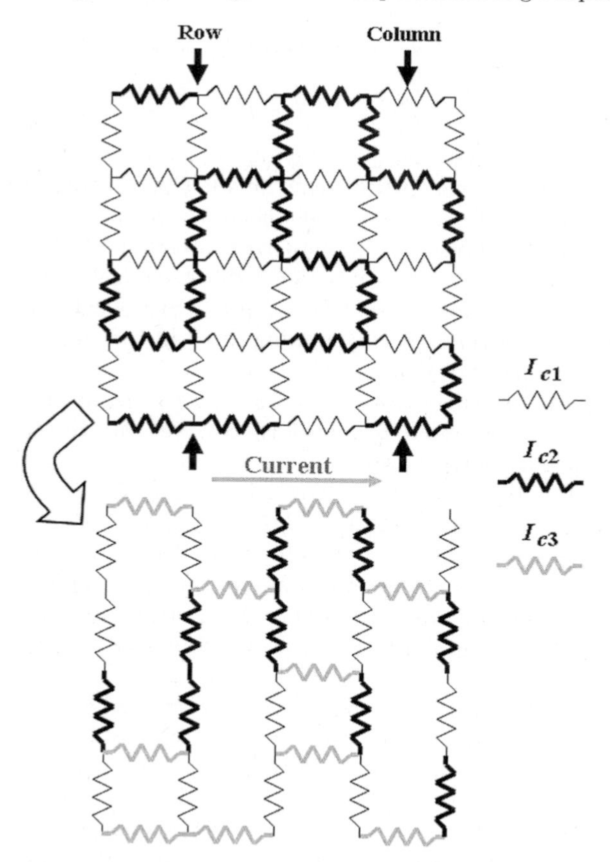

Fig. 10.18. Scheme of two-dimensional network for current flowing to the right. Resistors are divided into those parallel (column) and perpendicular (row) to the current. Then, the actual mesh is reduced to a dilute network, subtracting out a uniform current to the right [398]

The obtained approximate formula (10.17), in particular, neglects the possibility that the percolation current goes backward. However, we expect this path to make a smaller contribution than the ones calculated, because it corresponds to relatively rare configurations of resistors. In particular, approximation (10.17) should be well for the infinite mesh, if $I_{c3} < I_{c1}$ [398].

When $I_{c3} > I_{c1}$, the presented calculation will fail, because it is no longer easy to shunt the current sideways along a row of resistors and hence the redistribution probability will be different. Similar to above, a different critical current is calculated as [398]

$$I_c^{tot} = I_{c1} + q[I_{c1}(p_r - p_r^*) + (I_{c2} - I_{c1})p_r^*] , \qquad (10.18)$$

where

$$p_{\mathrm{r}}^{*} = q \sum_{s=0}^{\infty} (p^2 q)^s = \frac{q}{1 - p^2 q} . \tag{10.19}$$

Then, assume that the distribution of resistive elements is known for a non-linear resistor network. A voltage drop (which will be calculated below for the entire mesh) occurs if a current is imposed across the network. The correct distribution of currents in the mesh can be found by solving Kirchoff's equations, which uniquely define the current distribution in the network. In a network of linear resistors, this reduces a set of linear equations to solution, for which there are many standard methods. We are not aware of any general method for non-linear resistors. The method, developed in [424, 610, 611], can be applied only for the JJ case. The approach for the study of any current densities, which is used below, satisfies current conservation and transport the imposed current across the mesh.

We begin the solution of this problem, choosing a distribution that is thought to be close to the actual one. This choice must conserve current at each mesh node and also transport the imposed current across the lattice. One possibility would be a distribution where the current flows uniformly through the mesh. Obviously, this initial approximation needs to be modified. We do this by superimposing circulation currents on top of the initial distribution as shown in Fig. 10.19. While these circulation currents contributes nothing to the net transport of current they change the current path. Current conservation is obviously satisfied for all values of the circulation currents, which we will consider as free variables. Thus, the search method can study all possible currents.

The requirement to calculate the current distribution, satisfying the voltage Kirchoff's law, can be expressed in terms of a minimization principle. If, for a given resistor, we define the quantity

$$g_j = \int_0^{l_j} V_j(i)\mathrm{d}i , \tag{10.20}$$

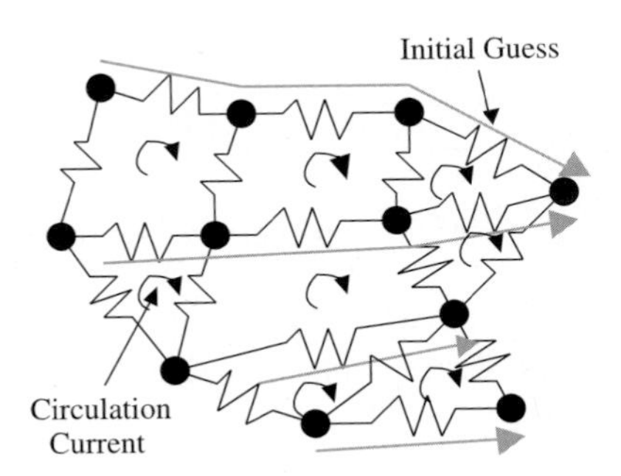

Fig. 10.19. Initial guess for the current distribution and imposition of circulation currents [398]

where $V_j(i)$ is the voltage-current characteristic of the j-resistor (assuming to be known) and I_j is the current flowing across the j-resistor, then the solution of Kirchoff's laws is equivalent to minimizing

$$G = \sum_j g_j \ , \tag{10.21}$$

where the sum is over all resistors in the lattice. We prove this by varying G with respect to the currents in such a way that current conservation is preserved. With this aim, find a variation of G with respect to a circulation current:

$$\frac{\delta G}{\delta C_k} = \sum_j V(I_j) \frac{\delta I_j}{\delta C_k} = 0 \ . \tag{10.22}$$

Because the circulation current, C_k, flows around a specific loop, variations in C_k only affect the currents along that loop. An additional effect to be unit linear in C_k:

$$\frac{\delta I_j}{\delta C_k} = \begin{cases} 1, & \text{for } I_j \text{ on } k \text{ loop} \ ; \\ 0, & \text{otherwise} \ , \end{cases} \tag{10.23}$$

where the sign is positive in the direction of C_k. Then,

$$\frac{\delta G}{\delta C_k} = \sum_{j \in \mathrm{loop}} V_j = 0 \ . \tag{10.24}$$

This equation is valid for any loop in the mesh, that is, the minimization of G with respect to the circulation currents is equivalent to the solution of the Kirchoff's voltage law. Note that, in the case of linear resistors, G reduces to representation

$$G = \frac{1}{2} \sum_j R_j I_j^2 \ .$$

Thus, for the linear case or any network, where all the voltages have the same power-law dependence upon current, the minimization of G is equivalent to minimization of the power loss of the mesh (a familiar result). An equivalent minimization principle for an applied voltage can be formulated, minimizing the sum:

$$\sum_j h_j = \int_0^{V_j} I_j(v) \mathrm{d}v \ , \tag{10.25}$$

where the sum is over the resistors and the free variables are the voltages at each node [1018].

In order to understand what the effect finite size has on the critical current, consider a binary model on a square lattice. Equation (10.14) states the bounds for an infinite mesh as

$$I_{\mathrm{c1}} \leq \frac{I_{\mathrm{c}}^{\mathrm{tot}}}{N} \leq p I_{\mathrm{c1}} + q I_{\mathrm{c2}} \ . \tag{10.26}$$

At the same time, this upper bound decreases for a finite mesh. We have for $M \times N$ lattice of nodes a resistor mesh with $M = N - 1$ columns and N resistors in each column (see Fig. 10.18). Investigate possible fluctuations. The occupation probabilities may be p and $q = 1 - p$. However, obviously, there are great fluctuations from average values in a smaller mesh. Calculate the expectation value for the maximum number of I_{c1} resistors in any of the M columns. This value defines the limiting factor for the critical current. In order to find this expectation value, it is necessary to know the probability $P(a)$ that the maximum number of I_{c1} resistors in any of the M columns is equal to a (out of N).

First, define $f(a)$ to be the probability that a specific column has a resistors of the I_{c1}-type in it:

$$f(a) = \frac{N!}{a!(N-a)!} p^a (1-p)^{N-a} . \tag{10.27}$$

Then, define $P_M(x, a)$ as the probability that x columns (out of M) have a resistors of the I_{c1}-type in them:

$$P_M(x, a) = \frac{M!}{x!(M-x)!} f(a)^x [1 - f(a)]^{M-x} , \tag{10.28}$$

where $0 \leq x \leq M$. Finally, the probability $P(a)$ is given as

$$P(a) = \begin{cases} \sum\limits_{x=1}^{M} [P_M(x,a) \prod\limits_{i=a+1}^{N} P_{M-x}(0,i)], & a \neq N ; \\ \sum\limits_{x=1}^{M} P_M(x,a), & a = N . \end{cases} \tag{10.29}$$

The obtained equation states the probability that x columns in M have a resistors of R_1 multiplied by the probability that there are no columns (in the M–x remaining) that have more than a resistors of R_1 in them, summed over x. Then, we use $P(a)$ to calculate the expectation value of a:

$$\overline{a} = \frac{\sum\limits_{a=0}^{N} aP(a)}{\sum\limits_{a=0}^{N} P(a)} . \tag{10.30}$$

Finally, we get the critical current for the mesh as

$$I_c^{\text{tot}} = \frac{\overline{a} I_{c1} + (N - \overline{a}) I_{c2}}{N} . \tag{10.31}$$

The present analysis is only relevant for small lattices, because for very large meshes $\overline{a}$ will be very close to p. Thus, the critical current is limited by the finite size of the mesh and in that the critical current for the infinite case would be larger.

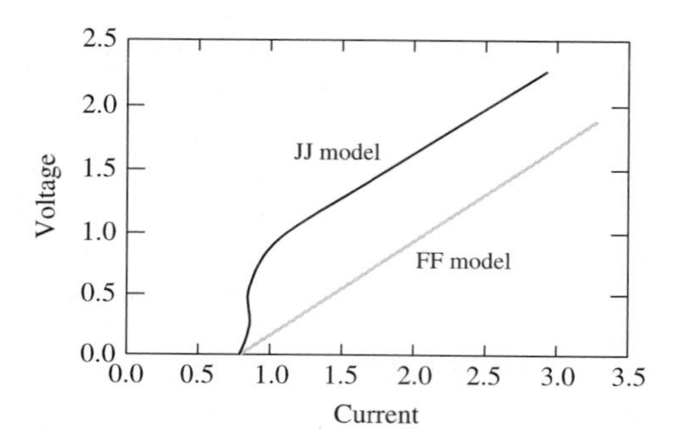

Fig. 10.20. Comparison of I–V characteristics for binary JJ and FF models on 10×10 percolative mesh. Both models possess the same distribution of microscopic critical currents and normal state resistances ($p = 0.5, I_{c1}/I_{c2} = 0.75$ and $R_1/R_2 = 0.75$). The *shape* of the individual circuit elements I–V characteristics is preserved, when they are combined into a mesh

The most important result of the computational realizations is that the overall shape of the individual elements I–V characteristics is preserved when the elements are combined into a mesh. Figure 10.20 compares 10×10 mesh composed of JJ resistors with one of FF ($p = 0.5, I_{c1}/I_{c2} = 0.75$ and $R_1/R_2 = 0.75$ in both cases) [398]. Obviously, the resulting overall characteristics are very different. This in itself is not surprising as the individual components have various properties. However, the fact that the overall characteristics should resemble that of the individual elements so closely is surprising. Statistical averaging, even with strong non-linearity, does not wash out the underlying input.

10.5.2 Simulation of Current Percolation and Magnetic Flux in YBCO Coated Conductors

Despite great progress in texturing of the HTSC layer, grain boundaries remain the current-limiting factor. It has been shown in tests that dissipation processes have been accompanied by viscous vortex flow through low-angle GBs [430]. Magneto-optical studies on coated conductors (CCs) state the mechanism acting in these materials. Based on this, the model [271] uses the simple, piece-wise linear I–V curve of low-angle GBs to calculate the global percolative current transport in CC. Note also that most of the CC simulations are reduced to finding the onset critical currents (I_{c0}) using calculations of "limiting current path" [556, 903, 904, 1007].

A method, which enables to calculate current and flux distribution as well as I–V curves of two-dimensional grain lattices (square or hexagonal) with arbitrary morphology and GB critical current distribution, is presented in [1007]. Two-dimensional consideration of the current flow in a CC is a good approximation, because the thickness of the HTSC layer is much smaller than the other length scales, and the high degree of sample texture ensures that the current flows essentially within the *ab*-planes. The CC is described in this model as a resistor lattice, namely each GB is represented as a pair of resistors, one of them with zero resistance, but finite current capacity (determined by the GB critical current), and the second one possesses resistance proportional to the flux flow resistivity.[4] The GB critical current is found from the GB misalignment angle using an exponential dependence with a plateau at low angles [1188]:

$$I_c(\alpha) = \begin{cases} TL_{gb}J_{c0}\exp[-(\alpha-\alpha_c)/\alpha_0], & \text{for} \quad \alpha \geq \alpha_c\ ; \\ TL_{gb}J_{c0}, & \text{for} \quad \alpha \leq \alpha_c\ , \end{cases} \qquad (10.32)$$

where L_{gb} is the length of the GB; α_c and α_0 are the critical angles.

Analytically, this problem is reduced to a solution of a set of *linear* equations and inequalities for the currents. This set includes the current-limiting equations for the superconducting resistors, the Kirchoff's equations for each grain and each current loop and also equation for the total current in the considered system. This system of equations is under-determined in all non-trivial cases, that is, there are fewer equations than variables. It is solved, simultaneously minimizing the total dissipation in the lattice for a given current. This type of mathematical problem is well known in economic analysis and can be solved by standard methods of linear optimization [393, 752].

The considered method enables to analyze both experimentally measured grain morphologies and model structures and sets of hexagons or squares. The former can be potentially used to predict the performance of a CC from the morphology of its substrate. At the same time, simple model structures are very valuable for fundamental studies of CC properties [903, 904]. As a result of experimental analysis of the grain distributions [1188], there is a possibility of the branching of flux channels and also three scaling regimes of I–V characteristics are identified, namely (i) the onset superconducting current, I_c, is dominated by finite- size scaling; (ii) the intermediate currents are stated by percolation (or power-law) scaling; and (iii) the high currents are described by the linear, Ohmic scaling.

Considerable interest presents a consideration of small model systems with arbitrary distribution of the individual GB misorientation angles or GB critical currents. Figure 10.21 shows an example, where the general grain alignment is determined to be very high (all GB angles $< 3°$), with the existence of two grains (A and B) with misalignment angles of 45°. This configuration leads to a current distribution with the currents flowing around the misaligned grains.

When the current exceeds the critical current, the first flux line enters the conductor along the misaligned grains (see Fig. 10.21b). The path of the first flux line is also the limiting cross-section for the current flow. In an

[4] A similar approach is usually used in lattice analysis, approximating non-linear transport properties [483, 844].

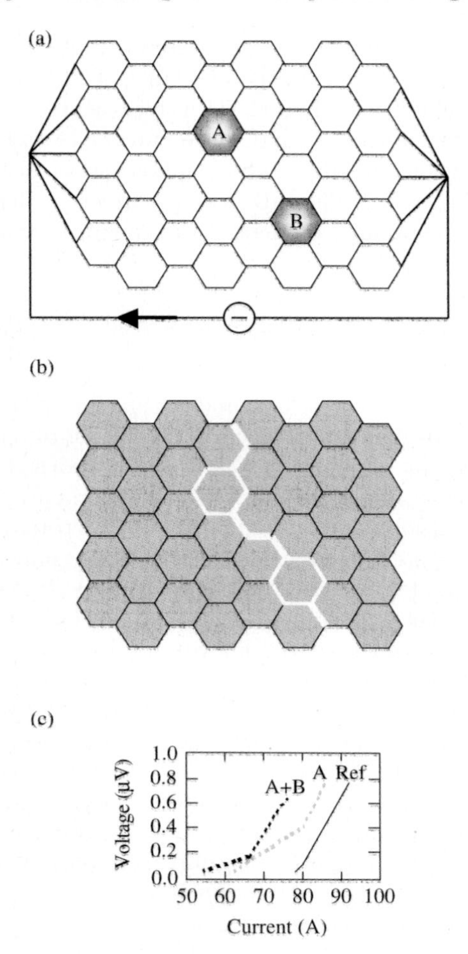

Fig. 10.21. (**a**) Hexagonal model lattice with GB angles between 2 and 3°, except for the grains A and B, having 45° misorientation; (**b**) flux distribution in the same lattice; and (**c**) I–V characteristic for the same lattice, labeled "A + B." Moreover, the curves are shown with only one highly misaligned grain (A) and in the case of no misaligned grains (Ref) [1188]

"undisturbed" lattice of the same topology, the limiting cross-section would be nine GB segments across the conductor. In the case, presented here, the limiting cross-section is reduced to six segments, if one disregards the GB segments of the highly misoriented grains. Accordingly, one expects that the critical current (i.e., the onset current, defined here for non-zero voltage) is decreased by a third.

The I–V curves in Fig. 10.21 show that indeed the critical current diminishes from 78 to 54 A, that is, by 32%. One can also see that the I–V

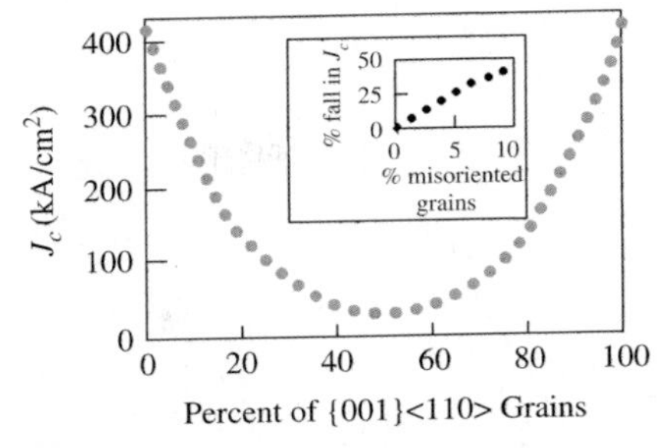

Fig. 10.22. Critical current on lattice of 25 × 50 hexagonal grains in dependence of the percentage of grains with about 45° misalignment

characteristics show pronounced kinking, similar to that found for low-angle GB [430]. This kink indicates the onset of a new flux channel through the conductor or alternatively the branching of an existing channel. The effect of the misaligned grains on the value of I_c is quite drastic: 5% of highly misaligned grains (two grains from 40) lead to 32% reduction of the critical current. The results of analogous simulation for the lattice with 25 × 50 hexagonal grains are presented in Fig. 10.22 [1188]. In this case, the grain angles have been assigned in a statistical manner, using two Gaussian distributions, centered about 0 and 45°, with their full-width half-maximum set at 5°. They lead to two practically separated percolative systems. The insert in Fig. 10.22 shows that in this large system the critical current strongly decreases even at a small percentage of misaligned grains.

Thus, the I–V characteristics in large lattice obey a parameter-independent scaling law typical for percolative processes. In small aggregates, one can find piece-wise linear I–V curves, kinked at a point where new flux channels enter the conductor or existing channels branch out. In this case, highly misoriented grains have a strong detrimental effect on the critical currents.

10.5.3 Modeling of Electromagnetic Properties of BSCCO/Ag Tapes

Based on the actual microstructure of BSCCO/Ag tape, in [833] the superconducting core is modeled as a stack of relatively well-aligned plate-like grains whose *ab*-planes are approximately parallel to the magnetic field, taking into account the limiting of the intergrain current flow by *c*-axis transport through large-area twist boundaries [104], or by a lattice of low-angle "railway switch" connections [150, 411], or by some other factors. The principal consequence of the planar geometry is the current forcing in the direction perpendicular to magnetic field **H** (see Fig. 10.23). In this case, the magnetization currents tend to flow in a plane perpendicular to $\mathbf{H_a}$ (field parallel to the one of the major axis of the slab and plate-like grains), such that the distribution $\mathbf{J}(\mathbf{r})$ is effectively two-dimensional, and $\mathbf{H}(\mathbf{r})$ has only one component $H_z(x, y)$.

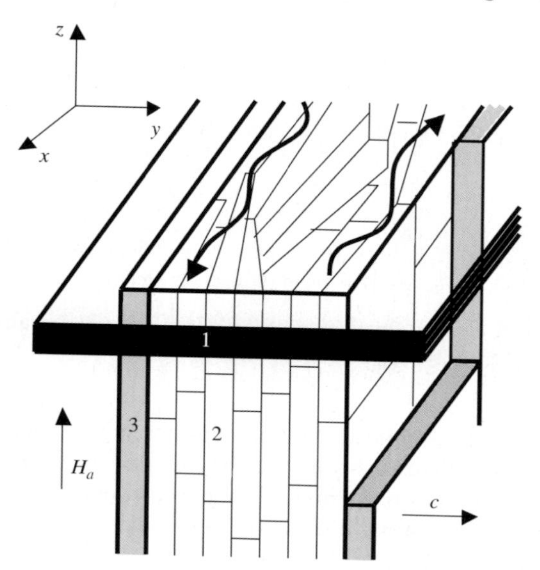

Fig. 10.23. Geometry of sample and magneto-optical indicator film (1). Sketched also are the model grain structure of BSCCO core of the "brick wall" type (2), covered by silver sheath (3). The current streamlines are shown by thick arrows for this planar geometry [833]

Consider the plate-like grain structure, presented in Fig. 10.23, first neglecting the non-superconducting second-phase precipitates and Ag/BSCCO interface irregularities. For the field orientation, shown in Fig. 10.23, the magnetization currents flow macroscopically along the tape, but they must also cross the *ab*-planes at some point in order to permit the current to loop at the end of the tape and to create flow around local barriers. In any case, the intergrain current mostly flows in the plane perpendicular to **H**, either through the large-area GBs normal to the *c*-axis or through small-area low-angle GBs, which separate slightly misoriented *ab*-planes (Fig. 10.23). In this case, the two components $J_x(x,y)$ and $J_y(x,y)$ are related to $H_z(x,y)$ as

$$J_x(x,y) = \frac{\partial H_z}{\partial y}\,; \qquad J_y(x,y) = -\frac{\partial H_z}{\partial x}\,. \tag{10.33}$$

The local $J_c(x,y)$ values are found as $J_c = (J_x^2 + J_y^2)^{1/2}$, which yields

$$J_c(x,y) = \sqrt{\left(\frac{\partial H_z}{\partial x}\right)^2 + \left(\frac{\partial H_z}{\partial y}\right)^2}\,. \tag{10.34}$$

Therefore, for the specific geometry of magneto-optical (MO) measurements, the current distribution is effectively two-dimensional, permitting a reconstruction of $\mathbf{J}(x,y)$ from a real measurement of the $H_z(x,y)$ distribution.

The characteristic spatial scales on which the above two-dimensional scheme can be applied need some discussion. Obviously, second-phase precipitates, intergrowths, cracks, weakly coupled grains and irregularities of the Ag/BSCCO interface can all lead to significant deviations of the local $\mathbf{J}(\mathbf{r})$ from the average current direction (the vector $\mathbf{J}(\mathbf{r})$), generally having both perpendicular ($J_{\perp}$) and parallel ($J_{||}$) components with respect to the local direction of $\mathbf{H}(\mathbf{r})$. As a result, the field $\mathbf{H}(\mathbf{r})$ can acquire a tangential component, becoming dependent on z-coordinate, a feature, which is not taken into account by the two-dimensional model. The characteristic size of L of these inhomogeneities is estimated by the secondary phase and grain sizes. So, these three-dimensional localized field disturbances strongly decay over a distance larger than L. However, if it is necessary to study the current distribution over macroscopic scale (for $l > L$), the MO image can be averaged over the length of L. After such averaging, any information about current loops smaller than L and random tangential fluctuations of $\mathbf{H}(r)$ will be lost. At the same time, the macroscopic inhomogeneities of $\mathbf{J}(r)$ over spatial scales larger than L can be extracted from the MO images, using the two- dimensional (10.33) and (10.34).

Then, it may be supposed that the transport critical current density (J_{c}) in the Bi-2223/Ag tapes is limited by weak links at GBs. As it has been noted, supercurrent usually demonstrates two-dimensional behavior. At a grain boundary, weak link breaks from superconducting to non-superconducting state when current density and/or applied magnetic field are sufficiently high. An electric field is generated locally when all weak links break at a whole cross-section of the column perpendicular to the current flow axis. This behavior can be treated as percolation phenomena based on the stochastic treatment.

For simplicity, consider a single grain as an element in the system in which grains are linked by GBs. Each element has coordination number Z defined by adjacent grains. As it is shown in Fig. 10.24, the system consists of M elements in each column and N elements in each row (here $Z = 4$) [784]. The stochastic computational procedure includes the following stages:

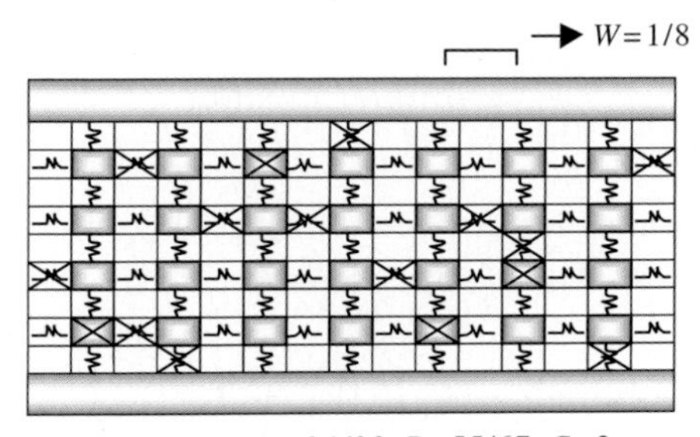

Fig. 10.24. System consisting of M elements in the column and N ones in the row. Each element is connected by weak links with coordination number $Z = 4$

(1) Among the basic elements there select $V\%$ in superconducting state and $(100 - V)\%$ in non-superconducting state. All elements are connected by weak links.
(2) When the applied current to the system increases, the breaking of $P\%$ weak links occurs (the broken links are selected randomly in the system) and they transit in normal state when the current density, crossing weak links, exceeds a critical value, J_{cw}.
(3) Supercurrents cross preferentially the superconducting links. When a sufficiently great number of weak links break, all links, included in a column, are in non-superconducting state. When the current crosses this normal column, an electrical field is generated. The failure of $P\%$ from total weak links leads to transition in normal state of $W\%$ columns from total number.
(4) In order to consider the effects of grain size, connectivity among grains, etc., a parameter C is introduced, defining the minimum size of elements for keeping of superconducting state.
(5) Assuming Weibull's distribution for estimation of strength of the weak links, the probability P, by which the weak links break, is expressed as a function of current density, crossing the respective weak link:

$$P = k(J - J_{\mathrm{cw}}^{\mathrm{min}})^m \,, \tag{10.35}$$

where m is the distribution shape factor. Hence, we have:

$$J = J_{\mathrm{cw}}^{\mathrm{min}} + (P/k)^{1/m} \,. \tag{10.36}$$

For simplicity, it is assumed that $J_{\mathrm{cw}}^{\mathrm{min}} = 0$.
(6) The fraction W of broken columns can be expressed through Weibull's function $f(J_{\mathrm{cw}})$ as

$$W(J) = \int_0^J f(J_{\mathrm{cw}})\mathrm{d}J_{\mathrm{cw}} \,. \tag{10.37}$$

On the other hand, the derivative of electric field is given as

$$\frac{\mathrm{d}E}{\mathrm{d}J} = R_{\mathrm{f}} \int_0^J f(J_{\mathrm{cw}})\mathrm{d}J_{\mathrm{cw}} \,, \tag{10.38}$$

where R_{f} is the resistance for the uniform flux flow. Integrating $W(J)$ with respect to J, the electric field is expressed as

$$E = R_{\mathrm{f}} \int_0^J W(J)\mathrm{d}J \,. \tag{10.39}$$

Hence, a correlation of the parameters (P, W) with the macroscopic quantities (J, E) is followed.

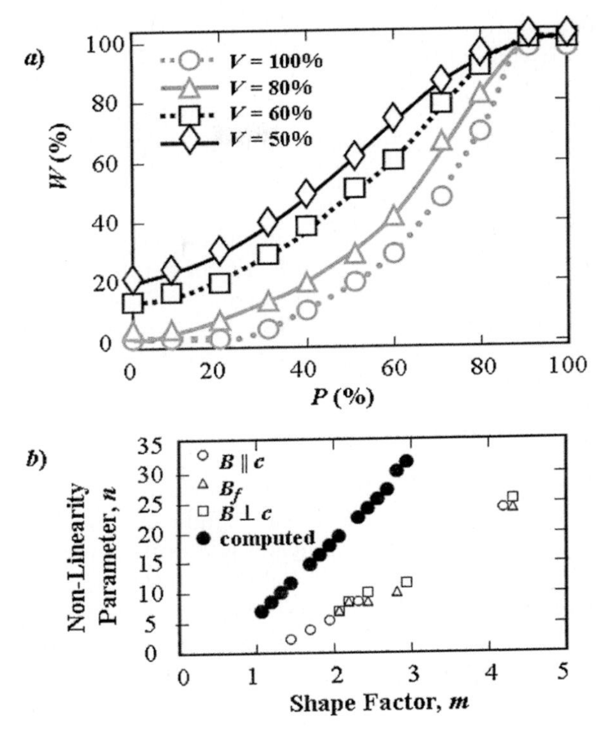

Fig. 10.25. (**a**) Change of degree of broken columns (W) as a function of probability of the broken links (P) for system, including $V\%$ superconducting elements, and (**b**) change of non-linearity parameter, n, as a function of distribution shape factor, m [784]. The computational results (*black circles*) are compared with experimental data (*white figures*) [875]

Figure 10.25a shows W in dependence on P for $M = 400, N = 1000, C = 2$ and variation of V from 50 to 100%. In the case of $V = 100\%$, the fraction W of broken columns starts to increase for $P > 20\%$, that means a generation of an electric field. For $V = 60\%$, W has already finite value even for $P = 0$. After calculation of E–J characteristic, so-called non-linearity parameter, n, is found. As it is shown in Fig. 10.25b, the dependence of n on m, calculated in the considered model, has a similarity with the experimental results [875].

Other approach, based on the statistical distribution of critical current density and used for study of non-linear E–J characteristic, is developed in [1161]. According to this model, the de-pinning probability function (Q) can be described as a function of current density (J) by the next expression:

$$Q(J) = [(J - J_{\mathrm{cm}})/J_0]^m , \qquad (10.40)$$

where J_{cm} is the percolation threshold identical to the minimum value of J_{c}; J_0 is the scale parameter, representing half-width of the statistic J_{c} distribution;

and m is the distribution shape parameter. Then, the E–J characteristic can be obtained integrating $Q(J)$ as

$$E(J) = \rho_{\mathrm{FF}} \int_0^J [(J_\mathrm{c} - J_\mathrm{cm})/J_0]^m \mathrm{d}J_\mathrm{c} , \tag{10.41}$$

where ρ_{FF} is the resistivity for the uniform flux flow [1161]. In particular, this model has permitted to estimate influence of bending strain on transport E–J characteristics for Bi-2223 Ag/AgMg-alloy-sheathed tape [549]. Moreover, this approach permits to separate the components, caused by the grain connectivity and flux pinning, originated from the bending strain.

10.5.4 Aging at Mechanical Loading

In design of energetic HTSC cables, an estimation of electric properties of the current-carrying elements presents the significant interest at mechanical aging of the sample, that is, at long cyclic and static loads. A probabilistic approach is one of the main methods for the solution of this problem [719, 720]. The V–I characteristic can be described, using the Weibull's distribution, and the time behavior of the relevant parameters can be considered during mechanical aging of the cable material, for example, under tensile stress and bend-vibration. In order to model the sample behavior, a probability function, $g(J_c)$, is introduced, which describes the probability that a cross-section of a superconductor has a given value of the critical current density, J_c [67]. It has been shown that $g(J_c)$ could be analytically derived as the second derivative of the E–J characteristic via flux-flow resistivity. This relationship holds, however, for a pure superconductor of type II, that is, neglecting effects of any non-superconducting matrix. Determination of second derivative of E–J characteristic can be found numerically [1135]. A best-fitting distribution must be chosen in order to obtain parameters, which are able to describe the function $g(J_c)$, as well as its integral, that corresponds to the cumulative probability, $G(J_c)$. The probability function considered below is the two-parametric Weibull's distribution, which fits both skewed and unskewed probability density characteristics. The expressions of the two-parametric density and cumulative-probability Weibull's function are found by [527]

$$F(x) = 1 - \exp[-(x/\alpha)^\beta]; \tag{10.42}$$

$$f(x) = (\beta/\alpha)(x/\alpha)^{\beta-1} \exp[-(x/\alpha)^\beta], \tag{10.43}$$

where α and β are scale and shape parameters, whereas the random variable x corresponds to the critical current density J_c.

The Weibull's parameters can be estimated in several ways [112]. In particular, they can be calculated on the base of test data, obtained by the prime derivative of the E–J characteristic, that is, $G(J_\mathrm{c})$, numerically performed and plotted in the coordinate system $\lg\{-\ln[1 - F(x)]\}$ in dependence on $\lg(x)$, which linearizes (10.42), transforming it into a straight line of slope β.

Actually, this probabilistic approach should be only applied to data relevant to the superconductor, while the tested samples have a silver shield, whose contribution is not easily separable from that of the superconductor body. Therefore, the Weibull's function can be used only to the experimentally obtained V–I characteristics with the aim to show how it is correlated with mechanical stress aging.

Examples of V–I characteristics, measured in aged and unaged specimens are shown in Fig. 10.26 for vibrated and tensile-stressed samples. The silver shield V–I characteristic is also presented in Fig. 10.26 for the sake of comparison. As it is followed from this, the mechanical aging process can significantly affect V–I characteristics. Moreover, the V–I curves tend to the silver resistance for values of currents monotonously decreasing as aging time increases.

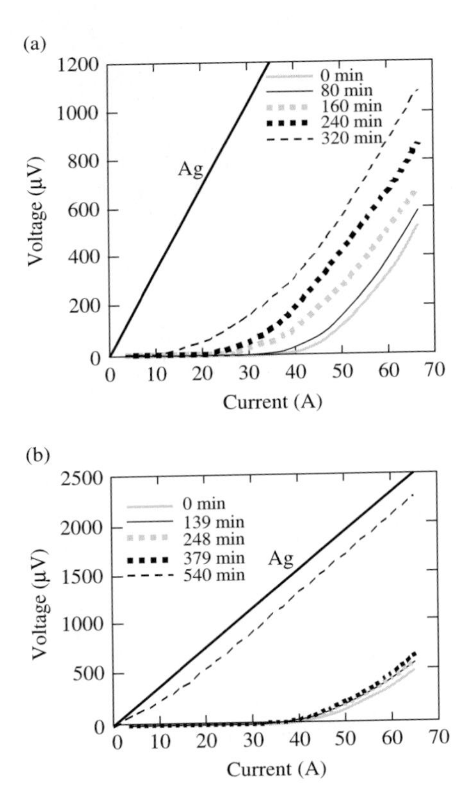

Fig. 10.26. V–I characteristics at various aging time for cases of vibration aging (**a**) and tensile-stress aging (**b**) [720]

Tensile stress and vibration do age specimens in a different way. While the former creates local damage cracks only, the latter leads to overall degradation, when silver shield is no more able to endure the applied stress. This approach permits to obtain the Weibull's functions for cases of cyclic and tensile loads (see Fig. 10.27 [720]) with the normalization factor, defined by the silver characteristics. Moreover, the scale and shape parameters (α, β) of the Weibull's distribution can also be estimated and plotted as a function of aging time (see Fig. 10.28 [720]). Figures 10.26–10.28 permit to estimate effects of the aging level and type on electrical properties of superconductor. So, the

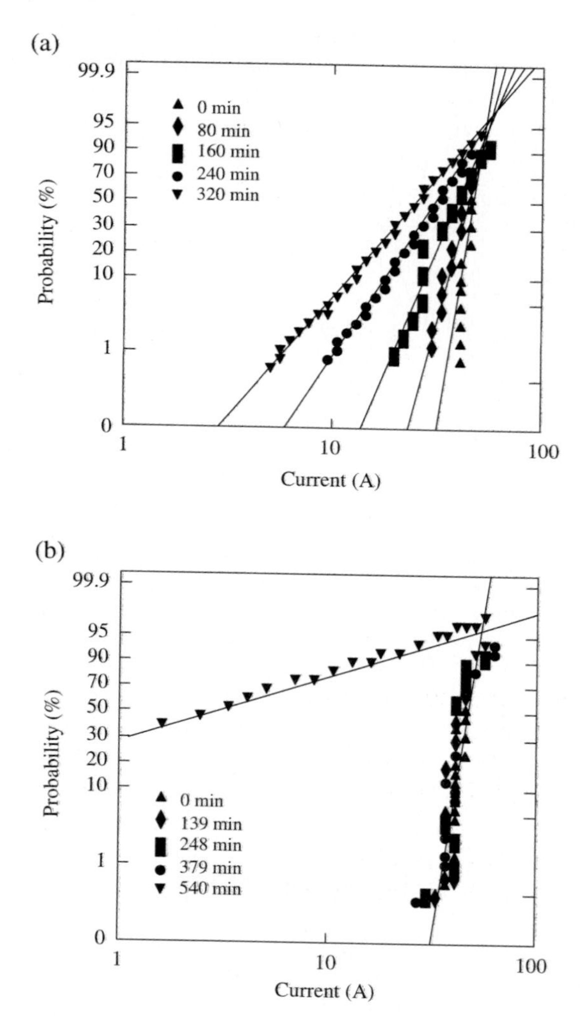

Fig. 10.27. Weibull's plots derived from normalized prime derivative at different aging times for cases of vibration aging (**a**) and tensile-stress aging (**b**)

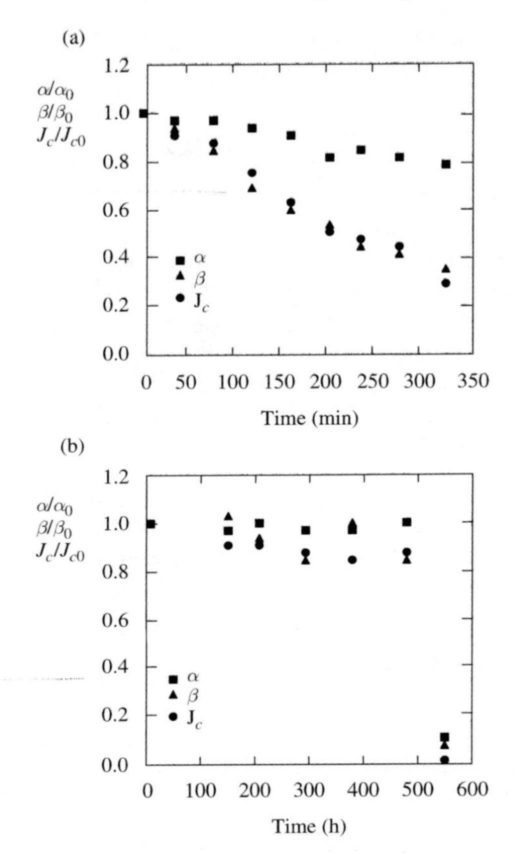

Fig. 10.28. Relative values of scale (α) and shape (β) parameters of the Weibull's distribution and also of critical current density (J_c) vs time (t), for one sample in the case of aging due to vibration (**a**) and tensile stress (**b**)

critical current (I_c) at moderately high and low cumulative probability significantly decreases as aging time increases, sharply in the case of tensile stress, progressively for vibration. Similarly, the value of β significantly diminishes with time, reaching the lowest values for the most aged samples.

10.5.5 Effective Electrical Conductivity of Superconducting Oxide Systems

Finally, based on the percolation theory, we consider current carrying in the model HTSC systems of YBCO and BSCCO, investigated in Chap. 8, and also study corresponding effective characteristics. Comparison of the results of fracture resistance alteration due to action of different toughening mechanisms, with qualitative features of electrical conductivity, stated in this section permits to estimate an effect (positive or negative) of microstructure features on properties of the considered superconductors.

YBCO Ceramic

It is known that electrical conductivity (the inverse value of resistivity) in non-ordered media is proportional to the self-diffusion factor, and hence, the average quadratic deviation of the liquid particles in absence of an external force [332]. Following [800], consider the YBCO model structure as a percolating cluster with cells, occupied by grains, and by free cells, corresponding to voids. Obviously, the percolating (or conducting) properties decrease due to the existence of intergranular microcracks and porosity. However, it is also clear that all model structures considered below, stated in Chap. 8, possess a joining, percolating cluster, because the following inequality always fulfills:

$$C_{\mathrm{p}} + f_{\mathrm{b}} << p_{\mathrm{c}} \,, \tag{10.44}$$

where $C_{\mathrm{p}} = N_{\mathrm{p}}/N$ is the closed porosity of the ceramics (N_{p} is the cell number, occupied by voids and N is the total cell number); $f_{\mathrm{b}} = l_{\mathrm{g}}/l_{\mathrm{l}}$ is the ratio of the cracked facets to the total number of boundaries between cells of the joining cluster (obviously, $f_{\mathrm{b}} < f_{\mathrm{m}} = l_{\mathrm{g}}/l_{\mathrm{i}}$, where l_{i} is the total length of the intergranular boundaries, because $l_{\mathrm{i}} < l_{\mathrm{l}}$); and $p_{\mathrm{c}} = 0.5927$ is the percolation threshold for a square lattice.[5]

In order to estimate effective electrical conductivity of the model structures (considered in Sect. 8.1.1 and 8.1.2), we modify a known algorithm, called *ant into maze*, applied for diffusion description in irregular media [332]. Take into account together with crystallite phase and voids, an existence of grain boundary microcracks, and also GBs, possessing a smaller conductivity compared to intracrystalline space. Consider a chance movement on only occupied cells (on crystallite phase) of the percolating cluster. At any time, the probable numbers, $p_k \in [0, 1]$ (where $k = 1, \ldots, 4$) are generated in all cells, which are adjacent to the main cell. In the case, when the cell under consideration is separated from the main cell by a GB then its probable number is decreased by 0.1 (to design the predominant cluster growth within a grain). If an intergranular boundary has been replaced by a microcrack or the adjacent cell is a void, then the corresponding probable number is found to be zero. The cluster growth results from the occupation of a cell with largest possible number, $p_k \geq p_{\mathrm{c}}$. The cluster growth is impossible when the initial cell is surrounded by voids, microcracks or when all of $p_k < p_{\mathrm{c}}$. Next, all the process is repeated. At each step, including a marking time, the value of t increases by 1. In the time, $t = 0$, some chance cell of the joining cluster begins to

[5] Similar modeling can be fulfilled at other regular lattices having the following values of the percolation threshold, p_{c}, and the shape of unit cell [1075]: 0.6970 (hexagon), 0.5 (triangle), 0.4299 (diamond), 0.3116 (cube), 0.2464 (b.c.c.), and 0.1980 (f.c.c.).

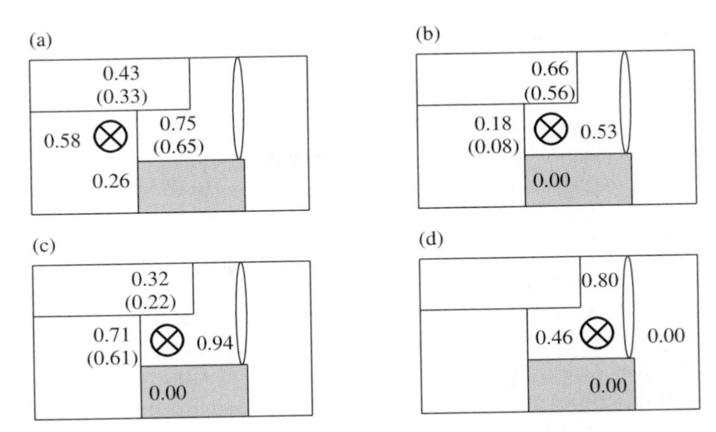

Fig. 10.29. An example of percolating cluster growing in YBCO ceramic. *Numbers* denote probabilities of cell occupation (in brackets there are considered probabilities, taking into account a priority, which are compared with percolation threshold, p_c = 0.5927). A *cross* corresponds to the cluster cell at previous step. Microcracks, formed during cooling, are shown at GBs; porosity is denoted by *gray color*

move. In time t, the square of distance between initial and final sites is calculated. Then, the modeling is repeated several times, and the mean quadratic displacement, R, associated with the conductivity of the modeled structure is calculated. An example of the cluster growth is shown in Fig. 10.29.

Statistically reliable results are obtained by applying the stereological approach. We obtain, again a necessary number of realizations for the statistical process to find unskewed estimation of the considered stereological characteristic as [145]

$$n = (200/y)(\sigma_x/\overline{x}) , \qquad (10.45)$$

where y is the accuracy level (in this case 10%); σ_x is the variance; $\overline{x}$ is the mean value of the stereological characteristic. At the study of the electric conductivity process, first a growth of a joining cluster on the type of occupying percolation [332] in two-dimensional lattice (25 × 40 cells), forming a path from left to right of its boundary, is considered. As $\overline{x}$, we select the lattice width (= 40), and the value of σ_x is calculated by corresponding exceeding of the cluster cell number over lattice width.

The percolation properties of the model microstructures, estimated for t = 100, are presented in Table 10.1. The normalized mean quadratic displacement, R/R_0 (where R_0 is the corresponding value for non-porous material), which is proportional to electrical conductivity decreases with growth of initial porosity as well as the square (cell number) of the joining cluster, S/S_0 (S_0 is the cluster area for nonporous material). These data correlate with increasing GBs, l_i/l_l, at the enhancement of the initial porosity, C_p^0, and decreasing of mean grain size, obtained in [814]. Thus, introducing even small priority for growth of the joining cluster within grain compared to its

Table 10.1. Some model parameters for YBCO ceramics of various structures

Properties	$C_p^0 = 0\%$	$C_p^0 = 20\%$	$C_p^0 = 40\%$	$C_p^0 = 60\%$
l_i/l_l	0.503	0.523	0.557	0.630
S/S_0	1.000	0.956	0.872	0.644
R/R_0	1.000	0.866	0.846	0.785

propagation in adjacent one (in this case, it corresponds to enhancement of probability by 0.1) correlates with experimental data and leads to qualitative confirmation of known results by the results of computer simulation. The presented results point again an important role of GBs and need to obtain coarse-granular YBCO structures. More accuracy estimations can be obtained in the framework of this model under condition of quantitative description of the difference of conductive properties in grain and through GB. Obviously, they will depend on features of domain structure of single crystallites, GBs continuity, secondary phases and so on.

Large-Grain Melt-Processed YBCO

Taking into account the model structures of the melt-textured large-grain YBCO samples, investigated in Sect. 8.3, it is obvious that their transport properties are found by the existence of non-superconducting (211) and superconducting (123) phases, and also by GBs. Then, in the electrical conductivity model, presented in the previous section, only one position is added, namely in the case of the adjacent cell occupation by the 211 particle, the probable number is replaced by zero. An example of the cluster growth is shown in Fig. 10.30, but the obtained numerical results are presented in Table 10.2 [822].

A comparison of the obtained effective characteristics with the model results of Sect. 8.3 (see Table 8.7) shows a correlation between the mean pinned grain area, $\overline{S}$, superconducting field per grain, $\overline{S}_{\mathrm{SR}}$ and ratio of the total intergranular boundary length to the total number of GBs between cells of joining cluster, $l_{\mathrm{i}}/l_{\mathrm{l}}$, on one hand, and change of the seed area, on the other hand.

The percolation properties have been calculated for $t = 100$. The behavior of the mean quadratic displacement, R/R_0, and the joining cluster area, S/S_0 (where R_0 and S_0 are the corresponding values for the case, $f = 0.0$ and

Table 10.2. Numerical results

Property	$f = 0.0$, $S_S \approx 2S_0^m$	$f = 0.0$, $S_S \approx 3S_0^m$	$f = 0.1$, $S_S = 0$	$f = 0.1$, $S_S \approx 2S_0^m$	$f = 0.1$, $S_S \approx 3S_0^m$	$f = 0.2$, $S_S = 0$	$f = 0.2$, $S_S \approx 2S_0^m$	$f = 0.2$, $S_S \approx 3S_0^m$
l_i/l_l	0.325	0.235	0.133	0.116	0.111	0.136	0.099	0.083
R/R_0	0.803	1.000	0.649	0.816	0.865	0.452	0.473	0.544
S/S_0	0.979	1.000	0.929	0.957	0.979	0.893	0.900	0.921

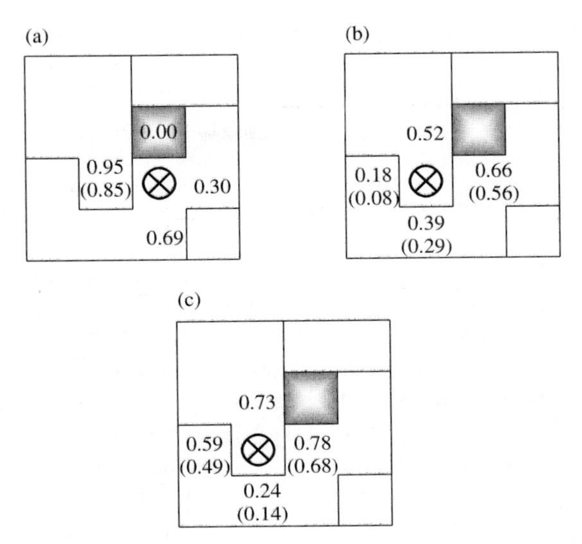

Fig. 10.30. An example of percolating cluster growing in large-grain YBCO sample. *Numbers* denote probabilities of cell occupation (in brackets there are considered probabilities, taking into account a priority). A *cross* corresponds to the cluster cell at previous step. *Gray color* shows 211 particle

$S_S \approx 3S_0^m$), as well as the change in the value of $\overline{S}_{SR}$ coincide with experimental data [318, 640]. We confirm that YBCO conducting properties decrease with increase in the fraction of the 211 normal particles and with decrease in the seed. At the same time, the considered toughening mechanisms intensify (that in turn, influences indirectly an increasing of current-carrying properties) with an increase of the 211 normal particle size and fraction into 123 superconducting matrix. Finally, note that for completeness of investigation of the percolation properties in this case, it is necessary to take into account the GB microcracking during cooling and misorientaion of grain boundaries, which are caused by indirect effects of dispersed phase and also contribute in final superconducting properties.

Effect of Microstructure Dissimilitude

The transport properties of the YBCO model structures, considered in Sect. 8.4, are caused by superconducting granular phase, voids, microcracks and GBs. Again, based on the algorithm *ant into maze* [332], effective characteristics can be calculated, which are presented in Table 10.3 [819]. An example of the cluster growth is shown in Fig. 10.31.

The numerical results show a correlation between the grain area and the ratio of total intergranular boundary length to total number of boundaries between cells of the joining cluster, l_i/l_l. The percolation properties have

Table 10.3. Numerical results

ν, °C/h	l_i/l_l	R/R_0	S/S_0
10	0.51	1.00	1.00
20	0.50	0.91	0.87
30	0.39	0.81	0.74

been calculated for $t = 100$. The behavior of the mean quadratic displacement, R/R_0, and the joining cluster square, S/S_0 (where R_0 and S_0 are the corresponding values for the case $\nu = 10°C/h$), as well as a change of the value of l_i/l_l coincide with test data [642]. Thus, it is confirmed that YBCO superconducting properties diminish with increased heating rate. This corresponds also to the trends in the changes of the specific fracture energy $(\gamma_F^{(1)}/\gamma_F^{(0)})$ and the microstructure dissimilitude $(\Delta K_{Ic}/\Delta K_c)$, following from Fig. 8.33.

BSCCO Bulks

Considering the Bi-2223 model structure as percolating cluster with cells, occupied by grains, note that the percolation properties are found in this case by GBs, microcracks and texture, describing a connectivity of adjacent

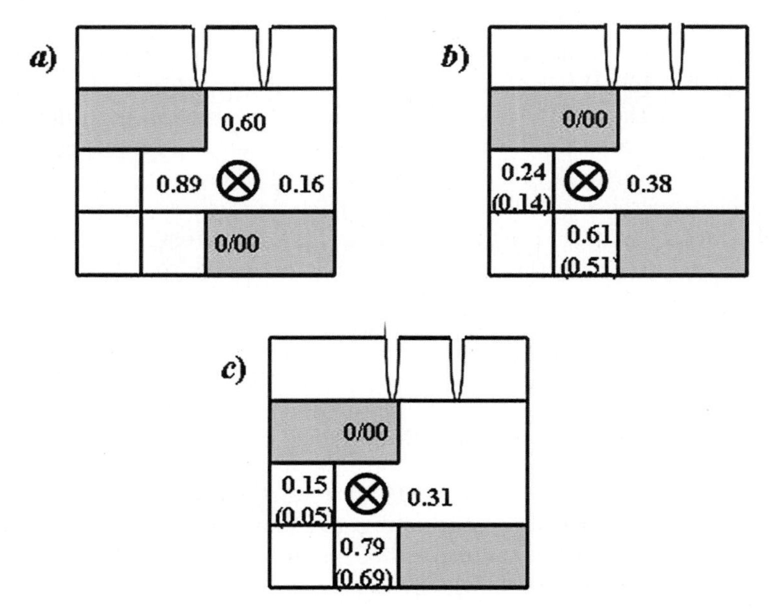

Fig. 10.31. An example of propagation of percolating cluster in the case of microstructure dissimilitude. *Numbers* denote probabilities of cell occupation (in brackets there are considered probabilities). A *cross* corresponds to the cluster cell at previous step. *Gray color* shows voids; microcracks are presented at GBs

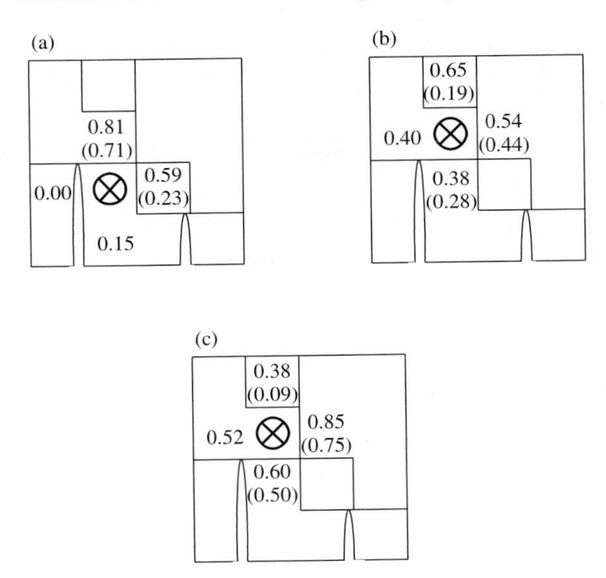

Fig. 10.32. An example of percolating cluster growing in BSCCO bulk. *Numbers* denote probabilities of cell occupation (in brackets there are considered probabilities). A *cross* corresponds to the cluster cell at previous step. Microcracks are presented at GBs

grains. The modification of the *ant into maze* algorithm for this example is the following. When the considered cell is separated from the main one by an intercrystalline boundary and the grains have the same orientation, then its probable number is decreased by 0.1 (for designation of the predominant cluster growth within the grain). If these grains have different orientations, then an additional generated probable number is multiplied by the corresponding value of the considered cell, so taking into account the grain connectivity. An example of the cluster growth is shown in Fig. 10.32 and the obtained results are presented in Table 10.4 [577].

The percolation properties have been calculated for $t = 100$. The computed relative increasing of the mean quadratic displacement, R/R_0, and the joining cluster area, S/S_0 (where R_0 and S_0 are corresponding values for the case $D/\delta = 2.40$), together with a decreasing of total GBs length (l_i/l_i^0) in the case of the larger grain growth coincide with test data (e.g., for large-grain

Table 10.4. Numerical results

D/δ	l_i/l_i^0	R/R_0	S/S_0
2.40	1.00	1.00	1.00
2.84	0.86	1.85	1.52

melt-processed HTSCs, considered in detail in Sect. 3.4). A comparison of the obtained results with estimations of the strength properties, presented in Sect. 8.6, shows that the denser lattice of the intergranular boundaries and, especially, the misorientation of adjacent grains affect the degradation of transport properties in a greater degree in the case of fine-grain ceramic compared to some increased microcracking of larger-grain structures. Thus, the greater toughening of the Bi-2223 bulks by ductile Ag particles (a fine-grain case) and the increased conductivity (a large-grain case) demand optimization of both contributions in the estimation of the HTSC properties.

A

Classification of Superconductors

Some main properties (including superconducting ones) of different chemical elements are shown in color inset. Here, a classification of known superconducting compounds is carried out. One includes 13 types of compounds, namely organic superconductors, *A*-15 compounds, magnetic superconductors, heavy fermions, oxides without copper, pyrochlore oxides, rutheno-cuprates, high-temperature superconductors, rare-earth borocarbides, silicon superconductors, chalcogens, carbon superconductors, MgB_2 and related compounds. The last section is devoted to room-temperature superconductivity.

A.1 Organic Superconductors

The organo-metallic compounds, discovered in 1979, have double bonds and can provide electrons to form Cooper pairs. There are vibration modes so that a phonon-induced mechanism is feasible. Moreover, the electrons may become attractive due to strong correlations, in which a phonon mechanism is not required. Majority of organic superconductors demonstrate co-existence of superconductivity and antiferromagnetism [300]. The critical temperatures, T_c, for BT=BEDT-TTF (BEDT = bis(ethylenedithio), TTF = tetrathiafulvalene) compounds [993] are given in Table A.1. The properties of these superconductors depend on pressure and on annealing procedure. The critical temperatures of most of these compounds are $< 10\,K$.

The mechanical loading leads to change of Fermi surface. The stress parallel to conducting planes of quasi-two-dimensional organic superconductor k-(BEDT-TTF)$_2$Cu(SCN)$_2$, causes an increase of T_c and H_{c2}, but the perpendicular stress leads to their decrease. As it is known, superconductivity can be destroyed by a sufficiently high magnetic field. However, this is not always true. The organic superconductor k-(BEDT-TTF)$_2$Cu(SCN)$_2$ demonstrates Fulde-Ferrell-Larkin–Ovchinnikov (FFLO) effect, namely superconductivity could be observed only in *strong magnetic field.* FFLO superconductivity initiates when an electron with spin "up" interacts with a hole with spin "down."

Table A.1. The properties of several organo-metallic compounds with BT = BEDT − TTF [BEDT = bis(ethylenedithio), TTF = tetrathiafulvalene]

No.	Formula	Phase	Pressure (kbar)	$T_c(K)$	Remarks
1	$(BT)_2ReO_4$	–	4	2.0	
2	$(BT)_2I_3$	α	0	3.3	I_2 doped
3	$(BT)_2I_3$	β_L	0	1.5	
4	$(BT)_2I_3$	β_L	0	2.0	Annealed
5	$(BT)_2I_3$	β_H	0.5	8.1	
6	$(BT)_2I_3$	α_t	0	8.1	
7	$(BT)_2IBr_2$	β	0	2.2–3.0	
8	$(BT)_2AuI_2$	β	0	3.4–5.0	
9	$(BT)_2I_{3-2.5}$	γ	0	2.5	
10	$(BT)_2I_3$	θ	0	3.6	
11	$(BT)_2I_3$	k	0	3.6	
12	$(BT)_4Hg_{2.89}Br_8$	–	0	4.0	
13	$(BT)_4Hg_{2.89}Cl_8$	–	12	1.8	
14	$(BT)_3Cl_2(H_2O)_2$	–	16	2.0	
15	$(BT)_2Cu(NCS)_2$	k	0	10.4	
16	$(BT)_2Cu[N(CN)_2]Br$	k	0	11.5	

In this case, "Cooper pair" of a new type forms, the necessary condition being an external magnetic field directed strictly along BEDT-TTF layers [994].

In other quasi-two-dimensional organic conductor λ-$(BETS)_2FeCl_4$, where BETS = bis(ethylenedithio)tetraselenafulvalene, the superconducting phase is induced by a magnetic field exceeding 18 T [1093]. Crystalline λ-$(BETS)_2$ $FeCl_4$ consists of layers of highly conducting BETS sandwiched between insulating layers of iron chloride. The field is applied parallel to the conducting layers. At the same time this compound, at zero field, is an antiferromagnetic insulator below 8.5 K. By increasing the applied magnetic field value above 41 T, this compound becomes metallic at 0.8 K. The dependence $T_c(B)$ has a bell-like shape with a maximum $T_c \approx 4.2$ K near 33 T. It is interesting to note that a magnetic field of only 0.1 T applied perpendicular to conducting BETS planes destroys the superconducting state. Thus, in this organic conductor, there is a strong superconductivity along the c-axis and weak in the ab-plane (in cuprates, it is the other way round).

A.2 *A*-15 Compounds

Chemical compounds of binary composition $A_{3-x}B_{1+x}$ crystallize into many different structures, depending on the value of x, temperature and pressure. One of the structures existing near $x = 0$, A_3B (where A = Nb, V, Ta, Zr and B = Sn, Ge, Al, Ga, Si) has the structure of beta-tungsten, designated in crystallography by the symbol A-15, and is superconducting. Hardy and Hulm first discovered A-15 superconductor (V_3Si) in 1954. Intermetallic compound Nb_3Ge has a critical temperature $T_c = 23.2\,K$, while Nb_3Ga shows $T_c = 20.7\,K$, Nb_3Al, $T_c = 19.1\,K$ and Nb_3Sn, $T_c = 18.3\,K$. Figure A.1 shows the structure of the binary A_3B compound. The transition temperatures of several $A - 15$ structures are given in Table A.2.

The critical temperature and second critical field increase in Nb_3Al compound, adding Ge and Cu in Nb/Al at the initial stage of the *rapid heating, quenching and transforming* (RQHT) process [463], and they attain for Nb_3Al-Ge($T_c = 19.4\,K, H_{c2} = 39.5\,T$) and for Nb_3Al-Cu($T_c = 18.2\,K, H_{c2} = 28.7\,T$) [462]. Moreover, the superconductors Nb_3Al-Ge, Cu have highest critical current densities among all metallic multifilamentary superconductors at $H > 20\,T$ and $T = 4.2\,K$. Only significantly more expensive HTSC, based on Bi-2223 and Bi-2212, have the near values of supercurrent at $T < 20\,K$.

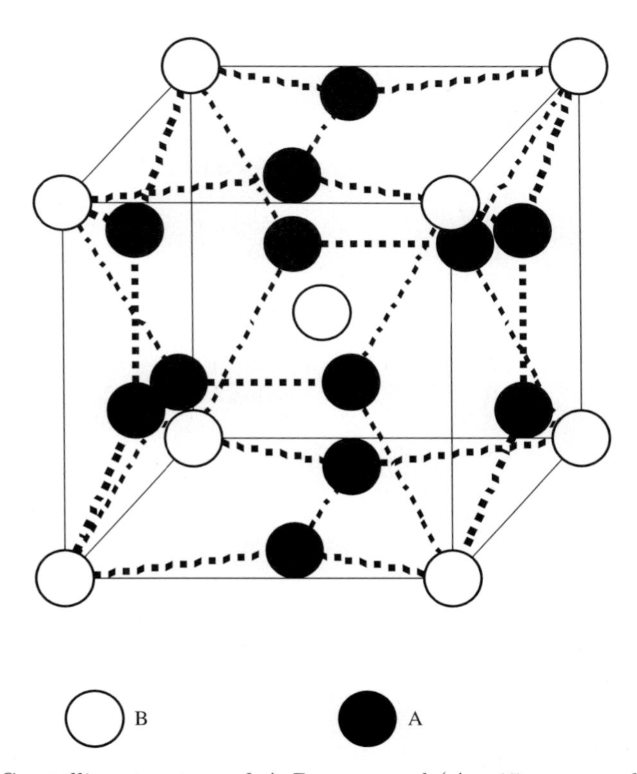

Fig. A.1. Crystalline structure of A_3B compound ($A - 15$ superconductors). The atoms A form one-dimensional chains on each face of the cube. Chains on the opposite faces are parallel, while on the neighboring faces they are orthogonal to each other

Table A.2. The critical temperatures, $T_c(K)$, of A-15 compounds

Ti_3Sb	6.5	Ti_3Ir	4.2
$Zr_{80}Sn_{20}{}^{q}$	0.92	Ti_3Pt	0.5
Zr-Pb	0.76	$V_{29}Re_{71}$	8.4
$Zr_{\sim 3}Bi^{p}$	3.4	$V_{50}Os_{50}$	5.7
$V\text{-}Al^{f}$	11.8	$V_{65}Rh_{35}$	≈ 1
V_3Ga	15.9	$V_{63}Ir_{37}$	1.7
V_3Si	17.0	$V_{\sim 3}Pd$	0.08
$V_{\sim 3}Ge$	6.0–7.5	V_3Pt	3.7
$V_{\sim 3}Ge^{f}$	6.0–11.0	$Nb_{75}Os_{25}$	1.0
$V_{\sim 79}Sn_{\sim 21}$	4.3	$Nb_{75}Rh_{25}$	2.6
$V\text{-}Sn^{q}$	7.0–17.0	$Nb_{72}Ir_{28}$	3.2
$V_{77}As_{23}$	0.2	Nb_3Pt	11.0
$V_{76}Sb_{24}$	0.8	$Ta_{85}Pt_{15}$	0.4
Nb_3Al	19.1	$Cr_{72}Ru_{28}$	3.4
Nb_3Be	10.0	$Cr_{73}Os_{27}$	4.7
Nb_3Ga	20.7	$Cr_{78}Rh_{22}$	0.07
Nb_3Pb	5.6	$Cr_{82}Ir_{18}$	0.75
$Nb_{\sim 3}In^{p}$	8.0–9.2	$Mo_{40}Tc_{60}$	13.4
$Nb_{82}Si_{8}{}^{q}$	4.4	$Mo_{\sim 65}Re_{\sim 35}{}^{f}$	≈ 15 (A-15)
$Nb\text{-}Si^{f}$	9.3	$Mo_{75}Os_{25}$	12.7
$Nb\text{-}Si^{f}$	4.0–8.0	$Mo_{78}Ir_{22}$	8.5
$Nb\text{-}Si^{f}$	11.0–17.0	$Mo_{82}Pt_{18}$	4.6
$Nb\text{-}Ge^{q}$	6.0–17.0	$W_{\sim 60}Re_{\sim 40}{}^{f}$	11.0
$Nb\text{-}Ge^{f}$	23.2	$Ta_{\sim 80}Au_{20}$	0.55
Nb_3Sn	18.3	Zr_3Au	0.9
Nb-Sb	2.0	$V_{76}Au_{24}$	3.0
$Nb_{\sim 3}Bi^{p}$	3.0	$Nb_{\sim 3}Au$	11.5
$Ta_{\sim 3}Ge^{f}$	8.0		
$Ta_{\sim 3}Sn$	8.3		
$Ta_{\sim 3}Sb$	0.7		
Mo_3Al	0.58		
Mo_3Ga	0.76		
$Mo_{77}Si_{23}$	1.7		
$Mo_{77}Ge_{23}$	1.8		

The left column has no transition metals; the right column has transition metals; three alloys of gold are also included. Note: q = quenching, p = pressure, f = film

A.3 Magnetic Superconductors (Chevrel Phases)

In 1971, Chevrel and co-workers discovered a new class of ternary molybdenum sulfides having the general chemical formula $M_x Mo_6 S_8$, where M stands for a large number of metals and rare earths. These superconductors have unusually high values of the upper critical field, B_{c2}, given in Table A.3 [737]. The superconducting compounds $RE Mo_6 X_8$ (where RE = Gd, Tb, Dy, Er and X = S, Se, Te), and $RE Rh_4 B_4$ (where RE = Nd, Sm, Tm) are usually

Table A.3. The critical temperature and the upper critical magnetic field of Chevrel phases

No.	Compound	T_c(K)	B_{c2} (T)
1	$PbMo_6S_8$	15	60
2	$LaMo_6S_8$	7	44.5
3	$SnMo_6S_8$	12	36

related to the Chevrel phases [1028]. Magnetic superconductors demonstrate some novel features not found in conventional type-I superconductors. Upon cooling from the superconducting phase ($T < T_c$), the material becomes normal again at a low temperature and the superconductivity is destroyed. Very often this normal phase is magnetically ordered. Upon cooling from the normal state the system becomes superconducting below T_c, and upon further cooling it becomes magnetically ordered below *Neel temperature*, T_N (where $T_N < T_c$). Thus, the superconducting phase occurs only in a limited range of temperatures, $T_N < T < T_c$. The interaction of conduction electrons with magnetic atoms leads to the formation of a bound state of the electron with the magnetic atom below a certain characteristic temperature called the *Kondo temperature.* It was found that the resistivity of a magnetic alloy, such as Fe impurities in Cu shows a minimum in the resistivity as a function of temperature due to the interaction of conduction electrons of the metal atom with the magnetic moment of the magnetic atom. Such a *Kondo effect* exists in superconductors containing magnetic impurity atoms at low temperatures, because the superconductivity is destroyed. In the case of magnetic superconductors, there is a sub-lattice of magnetic atoms in addition to the lattice of metallic atoms, so that magnetically ordered phase exists at low temperatures below the superconducting phase, $T_N < T_c$.

While ferromagnetic superconductors (e.g., $ErRh_4B_4$ and $HoMo_6S_8$) demonstrate the above behavior, the antiferromagnetic superconductors (e.g., $RE Mo_6S_8$, where RE = Gd, Tb, Dy and Er and also $RE Rh_4B_4$, where RE = Nd, Sm and Tm) present examples of the co-existence of the two phases together with anomalous behavior of the second critical field. In $ErRh_4B_4$, possessing tetragonal structure with $a = 5.299$ Å and $c = 7.588$ Å, a plot of the *ac*-magnetic susceptibility (χ_{ac}) and electric resistance depicts a normal to superconducting transition at $T_{c1} = 8.7$ K, followed by a loss of susceptibility at $T_{c2} = 0.9$ K together with the appearance of the ferromagnetic long-range order. This result takes place at zero magnetic field. At non-zero finite field also, the resistance disappears at some temperature interval within the above domain. The compound $HoMo_6S_8$ becomes superconducting at $T_{c1} = 1.3$ K in zero field. On further cooling, it becomes normal at $T_{c2} = 0.6$ K together with the appearance of ferromagnetic long-range order.

The antiferromagnetic superconductors provide the most striking case of the co-existence of the two kinds of order. These systems (e.g., $RE Mo_6S_8$, where RE = Gd, Tb, Dy, Er and also $RERh_4B_4$, where RE = Nd, Sm, Tm) demonstrate near T_N the antiferromagnetic alignment of rare-earth magnetic moments in the superconducting state of the system. The most important result is the anomalous behavior of the upper critical field as a function of temperature near T_N. In particular, (RE = Gd, Tb and Dy) demonstrate anomalous decreasing of H_{c2} near, but below T_N. The crystal structure of $REMo_6S_8$ is presented in Fig. A.2.

An attempt of T_c increasing in $REMo_6S_8$ owing to change of RE ion radius leads to the structure transition, as a result of which the dielectric gap opens at Fermi level and the superconducting properties disappear together with metallic ones. Based on BCS model, this is explained that the growth of T_c is caused by the approaching Fermi level to the peak of the electron state density, $N(E)$. However, it is disadvantageous energetically for the structure, therefore it suffers phase transition, which suppresses superconductivity.

The superconducting carbosulfide Nb_2SC_x [912] with layered crystalline lattice has critical temperature $T_c = 5\,K$ at $x = 0.8 - 1$. The superconductivity demonstrates volume character, and Nb atom octahedron is the key structure element, in the center of which C atom locates.

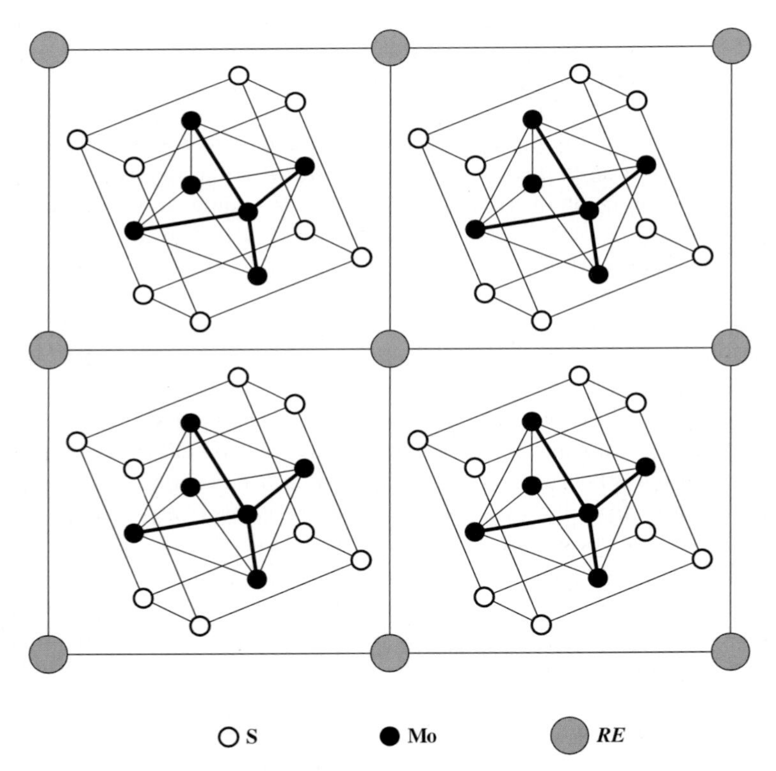

Fig. A.2. Structure of $REMo_6S_8$, which have Mo octahedron inside sulphur cube, which are inside a rare-earth cube [986]

A.4 Heavy Fermion Superconductors

Co-existence of superconductivity and antiferromagnetic order is found in U-based heavy fermions ($UPt_3, U(Pt_{1-x}Pd_x)_3, U_{1-x}Th_xBe_{13}, URu_2Si_2, UNi_2Al_3, UPd_2Al_3, U_6Co$ and U_6Fe) [456, 497, 526, 927] and in the heavy fermions: $RERh_2Si_2$ (where RE = La and Y), $Cr_{1-x}Re_x, CeRu_2, CeIn_3, CePd_2Si_2, CeNi_2Ge_2, CeCoIn_5$ and $Ce_nT_mIn_{3n+2m}$ (where T = Rh or Ir, $n = 1$ or 2; $m = 1$) [674, 842, 874, 1054]. In $CeCoIn_5$ critical temperature $T_c = 2.3$ K [874].

At the same time, the co-existence of superconductivity and ferromagnetism is discovered in the compounds UGe_2 [930], URhGe [25] and $ZrZn_2$ [843]. In $ZrZn_2$, the maximum critical temperature is slightly less than 3 K at ambient pressure and decreases with increasing pressure. The superconductor $YNi_{1.9}B_{1.2}$ demonstrates one of the maximum $T_c \approx 14$ K among known triple compounds with ferromagnetic component [870]. Another superconducting phase $YNiB_3$ has $T_c \sim 4.5$ K (tetragonal lattice with parameters of $a = 0.3782$ nm, $c = 1.1347$ nm, spatial group of symmetry $P4/mmm$) [871].

As it is known, the electronic heat capacity of a solid varies linearly with temperature at low temperatures, $c = \gamma T$, where the constant of proportionality, γ, determines the mass of the electron in the conduction band. It is found that in many compounds of Ce or U, this band mass is very large. Not only the effective mass but also Fermi rate differs by two orders of magnitude from the free electron values found in ordinary metals. For example in $CeCu_2Si_2$, the Fermi rate, $v_F = 8.7 \times 10^8$ m/s and the band mass, $m_{ef} = 220m_0$, where m_0 is the free electron mass. The critical temperature is $T_c \approx 0.5$ K and the London penetration depth approaches a high value of $\lambda_L = 0.2\,\mu$m at zero temperature. The large specific heat is caused by the strong correlations between electrons. In UBe_{13}, a sharp superconducting transition occurs at 0.97 K. The variation in T_c from sample to sample is in the range of 0.88–0.97 K. When Np is added to UBe_{13}, the T_c reduces. For $Np_{0.68}U_{0.32}Be_{13}$, the transition temperature is about 0.8 K. The compound UPt_3 has a superconducting transition temperature of 0.54 K. The extreme low temperature indicates that there are strong magnetic correlations [986].

The superconductor $PuCoGa_5$ with $T_c \sim 18$ K [922] occupies intermediate place between heavy fermion superconductors with $T_c \sim 1$ K and HTSC with $T_c \sim 100$ K. Besides "exotic" superconducting state, the common property for all these materials is the very strong Coulomb correlations between electrons of atomic f-shells (in $CeCoIn_5$ and $PuCoGa_5$) or d-shells (in HTSC), which lead to sharp increase of the effective electron mass or to transition in the state of Mott's dielectric, respectively. Co-existence of the localized and delocalized states, which are close in energy, favors the magnetic mechanism of the electron pairing [169].

Iron is the ferromagnetic, which demonstrates superconductivity at very high pressure. The superconducting transition is registered at $15\,\text{GPa} < P < 30\text{GPa}$ by both resistive and magnetic (Meissner's effect) methods [983].

The dependence $T_c(P)$ is "bell-like" (similar to the dependence of T_c on hole concentration in HTSC) with maximum value of $T_c = 2\,K$ at $P = 21\,GPa$.

Lithium is the metal at normal pressure. Under pressure of $P \approx 30\,GPa$, it becomes a superconductor and T_c attains 20 K at $P = 48\,GPa$. This is the highest T_c for "simple" one-element superconductors [982].

A.5 Oxide Superconductors without Copper

The superconductors of A_xWO_3 type have hexagonal structure of the tungsten bronze, where as A, the large alkaline ions of K, Rb and Cs are used more often (see Fig. A.3a). Many superconducting materials from this family demonstrate $T_c \approx 2$–7 K. The monocrystals of perovskite dielectric WO_3, doped by Na, demonstrate in surface layer of $Na_{0.05}WO_3$ high-temperature superconductivity with $T_c = 91\,K$ [417]. In this case, Na segregates at the grain surface and takes the structure of the WO_3 surface layer.

The superconductor without copper, $Ba_{1-x}K_xBiO_3$ (BKBO) demonstrates maximum critical temperature ($T_c = 31\,K$) among "old" oxide superconductors [737]. The oxide superconductor, $LiTi_2O_4$, without Bi and Cu, also possessing high transition temperature, has the spinel crystalline structure. Two other oxide superconductors: NbO and TiO, showing $T_c \leq 1\,K$, demonstrate clearly the expanded links of the "metal-metal" type as the previous oxide. The oxide superconductor $BaPb_{0.75}Bi_{0.25}O_3$ with the critical temperature $T_c = 12\,K$ has simple perovskite structure, depicted in Fig. A.3b.

Recently, the layered superconductor $Na_xCoO_2 \cdot 1.3H_2O$ with $T_c = 4\,K$ has been fabricated using the chemical oxidation technique [1035]. The conductive layers (CoO_2) in this superconductor alternate with the dielectric buffer layers ($Na_x \cdot 1.3H_2O$), which carry out a function of reservoirs for electric charge. Similar to HTSC, the value of T_c is highest at the optimum level of doping and decreases in overdoped and underdoped samples [931].

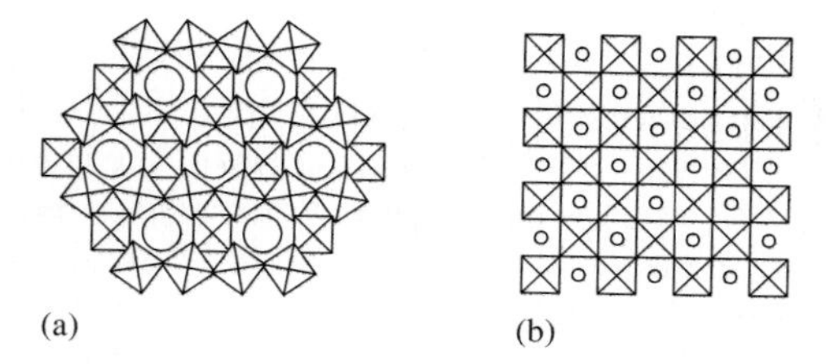

Fig. A.3. (**a**) Hexagonal structure of the tungsten bronze. One elementary layer presented, which is formed by the octahedrons MO_6 with large ions in cavities. (**b**) Perovskite structure [130]

A.6 Pyrochlore Oxides

The superconductors with the structure of pyrochlore oxide has general formula AOs_2O_6, where $A =$ Cs($T_c = 3.3$ K), Rb ($T_c = 6.3$ K) and K ($T_c = 9.6$ K). However, these compounds with the same chemical formula differ sharply in their superconducting properties. If $RbOs_2O_6$ is the BCS superconductor [95, 529], then KOs_2O_6 similar to HTSC relates to oxides of transitional metals, but does not crystallize in the perovskite structure. KOs_2O_6 crystallizes in the pyrochlore oxide structure, based on triangle lattice, which is a classic example of frustration effect in spin system, forming numerous spin structures [1175]. Generally, pyrochlore oxides present a great group of titan-, tantalum- and niobium-containing minerals with cubic crystals and general formula $A_2B_2O_6O'$, where A is the large cation, B is the smaller cation (usually $5d$-transitional metal, i.e., Re, Os or Ir). First, superconductivity in this oxide class has been discovered in $Cd_2Re_2O_7$ compound ($T_c = 1$ K) [386]. The difference between $Cd_2Re_2O_7$ and KOs_2O_6 is in the number of d-electrons in B-cation. It is interesting that KOs_2O_6 compound with fractional degree of oxidation has the critical temperature 10 times higher than $Cd_2Re_2O_7$, possessing even number of $5d$-electrons. In the compounds with the same structure like $RbOs_2O_6$ [1176] and $CsOs_2O_6$ [1174], Os ion has fractional degree of oxidation, +5.5, disposing in intermediate state between $5d^2$ and $5d^3$. In this case, the $5d$-electrons of Os define transport and magnetic properties of these materials, simultaneously, a coupling of which in these oxide compounds is mostly interesting.

The effect of high pressure (up to 10 GPa) on superconductivity in the AOs_2O_6 compounds has been studied for A = Cs, Rb and K [750]. The critical temperature for all three materials increased together with pressure up to a maximum value of $T_c = 7.6$ K (at 6 GPa), $T_c = 8.2$ K (at 2 GPa) and $T_c = 10$ K (at 0.6 GPa) for $A =$ Cs, Rb and K, respectively, after that it diminishes down to total disappearance at 7 and 6 GPa for $A =$ Rb and K, and above 10 for $A =$ Cs.

A.7 Rutheno-Cuprates

Sr_2RuO_4 compound ($T_c \sim 1$ K) is the single example of layered perovskite without cooper, demonstrating superconductivity. This compound relates to the class of "self-doped" conductors due to small ratio, U/W (where U is the energy of Coulomb repulsion and W is the width of Brillouin's zone), that is, there the role of electron correlation is not important compared to cuprates [654]. p-type of pairing (spin-triplet) in Sr_2RuO_4 is realized. This system is also called the system with *ladder structure*. If to seek number of legs of the ladders per cell to infinity, then a transition to two-dimensional structure occurs. In cuprate HTSC, the ladder role could be played by stripes, then, a total analogy between cuprate and ruthenium systems is possible.

Triple perovskite ("hybrid" rutheno-cuprate superconductor) $RuSr_2GdCu_2O_8$ consists of both "superconducting" CuO_2 layers and "ferromagnetic" RuO_2 layers. Herein, superconductivity co-exists with electronic ferromagnetism in microscopic scale (the temperature of ferromagnetic ordering 135 K and T_c = 50 K) [683]. The investigations of the magnetization and magnetic resistance of the $RuSr_2GdCu_2O_8$ have demonstrated influence of the magnetic moments of Ru atoms on the electrons of conductivity.

The intragrain critical temperature, T_c, of HTCS cuprate $RuSr_2(Gd, Ce)_2Cu_2O_{10+d}$ has been studied depending on hole concentration (which is found by oxygen content) [1153]. In this case, T_c changed in very wide limits (17–40 K) with the change of p being only 0.03 holes/CuO_2. Into this range of p, the intragrain superfluid density (which is inversely proportional to square of the magnetic field penetration depth, $1/\lambda^2$) and the value of the diamagnetic jump (found at the sample cooling in magnetic field) have increased more than 10 times. These results contradict to correlations between T_c, p and $1/\lambda^2$, observed in homogeneous HTSC. This is possible due to the phase de-lamination and granularity. Moreover, there is an effect of anomalous increasing of distance between CuO_2 planes during cooling of layered compound $RuSr_2Nd_{0.9}Y_{0.2}Ca_{0.9}Cu_2O_{10}$, doped up to level of the boundary "antiferromagnetism/superconductivity" at the phase diagram [688]. This means negative value of the thermal expansion factor. Difference in volumes of the antiferromagnetic and superconducting phases may be caused by the phase segregation, observed often in weakly doped HTSC.

A.8 High-Temperature Superconductors

Superconducting compound belong to HTSC, if it has one or more CuO_2 planes. The cuprates possess a perovskite, strongly anisotropic crystalline structure, defining majority of their physical and mechanical properties. In conventional superconductors, there are no important structural effects, because the coherence length (ξ) is much longer than the penetration depth (λ). However, it is not the case for cuprates due to $\xi << \lambda$ for them. In general, the high-T_c materials are basically tetragonal (orthorhombic), and superconductivity in cuprates occurs in the copper-oxide planes. These layers are always separated by layers of other atoms such as Bi, O, Y, Ba, La, etc., which provide the charge carriers into CuO_2 planes. In the CuO_2 planes, each copper ion is strongly coupled to four ions, separated by a distance approximately 1.9 Å. At fixed doping level, by increasing the number of CuO_2 planes, T_c first increases, reaching the maximum at $n = 3$ (where n is the number of CuO_2 layers per unit cell), and then decreases. So, nuclear magnetic resonance (NMR) data show that the charge carriers are distributed inhomogeneously between CuO_2 layers (into limits of one elementary cell) in mercurial HTSC at $n > 3$. Their concentration into "internal" layers is smaller than into "external" [569]. Based on the phenomenological approach, it has been shown

Table A.4. Transition temperature and number of CuO_2 planes for different cuprates

No.	Cuprate	CuO_2 planes	T_c (K)	Abbreviation
1	$La_{2-x}Sr_xCuO_4$	1	38	LSCO
2	$Nd_{2-x}Ce_xCuO_4$	1	24	NCCO
3	$YBa_2Cu_3O_{7-x}$	2	93	YBCO
4	$Bi_2Sr_2CuO_6$	1	~ 12	Bi-2201
5	$Bi_2Sr_2CaCu_2O_8$	2	95	Bi-2212
6	$(Bi, Pb)_2Sr_2Ca_2Cu_3O_{10}$	3	110	Bi-2223
7	$Tl_2Ba_2CuO_6$	1	95	Tl-2201
8	$Tl_2Ba_2CaCu_2O_8$	2	105	Tl-2212
9	$Tl_2Ba_2Ca_2Cu_3O_{10}$	3	125	Tl-2223
10	$TlBa_2Ca_2Cu_4O_{11}$	3	128	Tl-1224
11	$HgBa_2CuO_4$	1	98	Hg-1201
12	$HgBa_2CaCu_2O_6$	2	128	Hg-1212
13	$HgBa_2Ca_2Cu_3O_8$	3	135	Hg-1223
14	$HgBa_2Ca_3Cu_4O_{10}$	4	123	Hg-1234
15	$HgBa_2Ca_4Cu_5O_{12}$	5	110	Hg-1245

[133] that due to lower concentration of carriers into "internal" layers of the multilayered elementary cell, a value of pseudogap in these layers is large and superconducting order is suppressed, therefore the following increasing of n does not lead to growth of T_c.

The critical temperature and the number of the CuO_2 planes are presented for different HTSC in Table A.4. In most superconducting cuprates, by changing the doping level at fixed n, the $T_c(p)$ dependence has the bell-like shape, where p is the hole concentration in CuO_2 planes (see Fig. A.4). This shape of $T_c(x)$ dependence is more or less universal in cuprates, where x can be p, n, the lattice constants a, b or c, the buckling angle of CuO_2 layers, etc. Thus, the maximum T_c value can be only be attained when all the required parameters have their optimum values.

In the HTSC also, as in conventional compounds, there is a change in valency due to the change in electron–phonon interaction. The oxygen content defines the transition temperature. As an example, the transition temperature, T_c, as a function of x and δ for the system $Bi_2Sr_2Ca_{1-x}Y_xCu_2O_{8+\delta}$ is presented in Table A.5. Crystal structures of high-temperature superconductors LSCO, YBCO, Bi-2212 and NCCO are shown in Fig. A.5.

Search for superconductors with high values of T_c is continuing. Table A.6 shows a variety of superconducting cuprates, differing in their structural and chemical properties, discovered beginning from 1986. These superconductors are grouped in compositional principle.

The partial substitution of Y by Ca, and Ba by La in $YBa_2Cu_4O_8$ does not have an influence in practice on T_c. At the same time, the substitution of Cu by Zn or Ni leads to rapid degradation of superconductivity [1144].

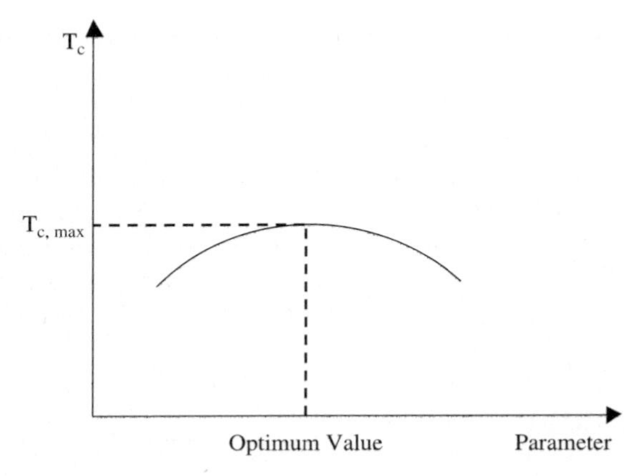

Fig. A.4. Critical temperature as a function of a parameter, which can be the doping level, the number of CuO_2 planes per unit cell, the unit-cell constants, the buckling angle of CuO_2 layers, etc

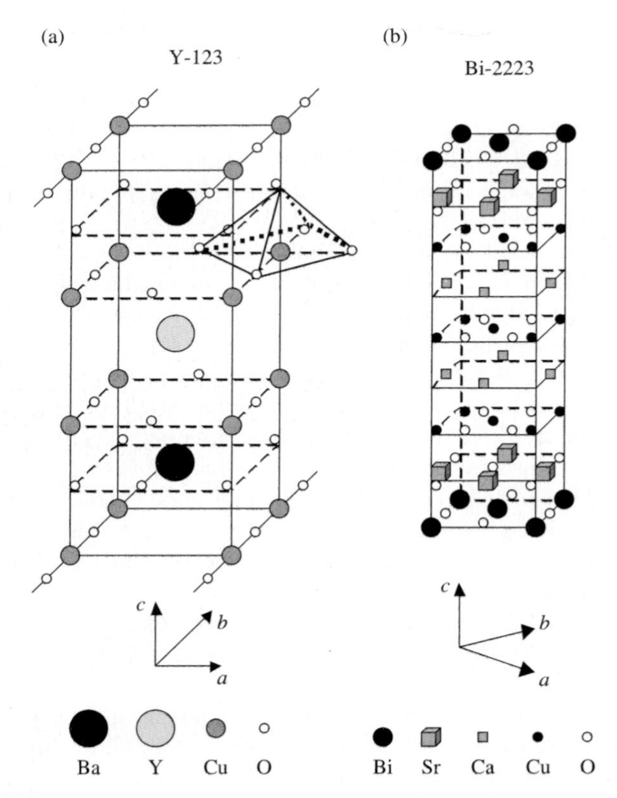

Fig. A.5. Crystal structures of HTSC: (a) YBCO and (b) Bi-2223, (c) LSCO, and (d) NCCO (continued)

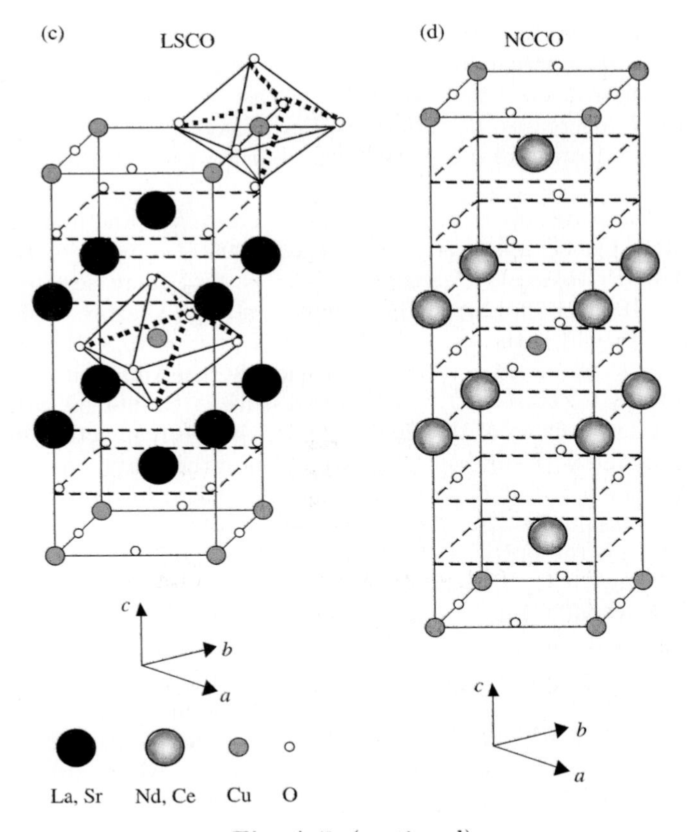

Fig. A.5. (continued)

Table A.5. Composition and oxygen content dependence of critical temperature in the system $Bi_2Sr_2Ca_{1-x}Y_xCu_2O_{8+\delta}$ [986]

No.	x	δ	Cu-valency	T_c (K)
1	0.0	0.23	2.23	70
2	0.1	0.23	2.18	90
3	0.3	0.30	2.15	86
4	0.35	0.30	2.12	74
5	0.4	0.29	2.09	64
6	0.5	0.33	2.08	40
7	0.75	0.41	2.03	< 4
8	1.0	0.51	2.01	< 4

This result agrees totally with two-dimensional picture, in accordance of which high-temperature superconductivity is caused by pairing into CuO_2 layers. If intactness of these layers is violated (as in the case of substitution of the copper atoms), then T_c decreases. If atomic disorder arises outside these layers (as in the case of substitution of the yttrium and barium atoms), then T_c does not change.

High-temperature superconductivity in thin films [70] and in monocrystals [1207] of $PrBa_2Cu_3O_x$ has been discovered in 1996. The value of $T_c = 56.5$ K at $P = 0$ in $PrBa_2Cu_3O_x$ monocrystal ($x = 6.6$) increases up to 105 K at $P = 9.3$ GPa [1170, 1208]. This result contrasts sharply with data for $YBa_2Cu_3O_x$, in which the value of $T_c \sim 90$ K at $x = (6.8–7)$ is almost independent of P up to $P = 10$ GPa. Obviously, the different responses of $PrBa_2Cu_3O_x$ and $YBa_2Cu_3O_x$ on high pressure is connected with various characters of distribution of the charge carriers between structure units of elementary cell, and with different re-distribution of the charge under pressure, respectively. The value of $T_c \approx 90$ K has been attained in the polycrystalline $PrBa_2Cu_3O_{7-d}$ samples [26]. Most probable cause of superconductivity in $PrBa_2Cu_3O_{7-d}$ is the partial substitution of Pr by Ba [760].

HTSC $NdBa_2Cu_3O_x$ (Nd-123) holds set of records among RE-123 ($T_c \sim$ 95 K, high irreversible line, etc.). The main advantages of Nd-123 consists in anomalous peak effect, leading to significant increasing of intragrain currents for account of formation of the effective pinning centers, which begin to work at liquid nitrogen temperature in magnetic field of some Tesla, that are most interesting for HTSC technical applications. Only in Nd-123, it is possible to attain the results, using *chemical methods*, which are comparable in effect to cumbersome, expensive and difficult-accessible methods of *physical* formation of the pinning centers (e.g., due to neutron irradiation or ion bombardment). Additional advantages of the Nd-123 phase are also connected with its higher chemical stability and higher rate of solidification. New pinning centers into Nd-123 are formed during de-lamination of re-saturated solid solution [425, 747]. In this case, the sites of solid solutions form in the basic superconducting matrix. These sites are distributed homogeneously in the matrix and are coupled coherently with it, because they act as effective pinning centers in the external magnetic field. In the case of non-zero magnetic field, superconductivity in them is suppressed sharply, causing the peak effect. Due to these new pinning types, the irreversibility line in the Nd-123 samples displaces magnetic fields above 8 T at 77 K (the record value for RE-123 superconductors). Moreover, Nd and Ba can exchange places in the Nd-system due to favor combination of the ionic radii that forms the defects, serving strong pinning centers. Majority of HTSC have p-type of superconductivity ($YBa_2Cu_3O_7$, $La_{2-x}Sr_xCuO_4$, $Bi_2Sr_2CaCu_2O_8$, etc.). $Nd_{2-x}Ce_xCuO_4$ is the most studied of no numerous HTSC with n-type conductivity. Note that antiferromagnetism is the main alternative to superconductivity as in electronic HTSC also as in hole-doped superconductors [510].

The structure of "infinite-layer" $ACuO_2$ (where A is the alkaline-earth metal) compounds presents a set of parallel CuO_2 layers, separated by the

Table A.6. Superconducting cuprates discovered from 1986

$YBa_2Cu_3O_7 + RE$ equivalents	$Sr_{1-x}CuO_2$
$(Y, Pr)Ba_2Cu_3O_7$	$(Ca, Sr, La, Nd)CuO_2$
$(Y, Ca) Ba_2(Cu, Zn)_3O_7$	$(Ca, Na, Sr, K)_2CuO_2Cl_2$
$(Sm, Nd)Ba_2[Cu_{1-y}(Ni, Zn)_y]_3O_7$	$(Sr, K)_3Cu_2O_4Cl_2$
$TbSr_2(Cu, Mo)_3O_7$	$(Cu, CO_2)(Ba, Sr)CuO_{3+\delta}$
$YBa_2Cu_4O_8 + RE$ equivalents	$(Cu, CO_2)(Ba, Sr)CaCu_2O_{5+\delta}$
$Y_2Ba_4Cu_7O_{15}$	$(Cu, CO_2)(Ba, Sr)Ca_2Cu_3O_{7+\delta}$
$Bi_2(Sr, La)_2CuO_6$	$(Cu, X)Sr_2(Y, Ca)Cu_2O_7$,
$Bi_{2+z}Sr_{2-x-z}La_xCuO_y$	where $X = Pb, CO_2, SO_3, BO_3$,
$Bi_2Sr_2CaCu_2O_8$	Bi, Mo
$Bi_2Sr_2Ca_2Cu_3O_{10}$+(Bi, Pb) equivalents	$GaSr_2(Y, Ca)Cu_2O_7$
$Bi_2Sr_2Ca_2Cu_3O_8F_4$	$NbSr_2(Nd, Ce)_2Cu_2O_{10}$
$Tl_2(Ba, La)_2CuO_6$	(Nd, Pr, Sm) CeCuO
$Tl_2Ba_2CaCu_2O_8$	$(La, Pr, Nd)_{2-x}Ce_xCuO_{4-y}$
$Tl_2Ba_2Ca_2Cu_3O_{10}$	$Sr_2CuO_3F_x$
$Tl(Ba, La)_2CuO_5$	$Nd_2CuO_{4-d}F_\delta$
$TlBa_2CaCu_2O_7$	$(La, Ba, Nd, Ce, Sr)_2CuO_4$
$(Tl_{0.5}, Pb_{0.5})Sr_2(Ca, Y)Cu_2O_7$	$(La, RE)_2CuO_4$, where RE = Y, Lu,
$(Tl_{0.5}Pb_{0.5})(CaSr_{2-x})(La_xCu_2)O_7$	Sm, Eu, Gd, Tb
$(Tl_{0.5}Sn_{0.5})Ba_2(Ca_{0.5}Tm_{0.5})Cu_2O_x$	$(La, Sr)_2CaCu_2O_6$
$TlBa_2Ca_2Cu_3O_9$ + (Tl, Pb)Sr equivalents	$(Eu, Ce)_2(Eu, Sr)_2Cu_3O_9$
$(Cu, Tl)Ba_2Ca_2Cu_3O_y$	$[CaCu_2O_3]_4$
$TlO_x(CO_3)_y(Sr, Ca)_{n+1}Cu_nO_z$	$(Cu_{0,5}Cr_{0,5})Sr_2Ca_{n-1}Cu_nO_{2n+3+d}$
$(X, Cu)(Eu, Ce)_2(Eu, Sr)_2Cu_2O_8$,	$CaBa_2Ca_2Cu_3O_{9-\delta}$
X = Pb, Ga	$CuBa_2Ca_3Cu_4O_{12-d}$
$Hg(Ba, La)_2CuO_{4+\delta}$	$(Sr, Ca)_{14}Cu_{24}O_{41}$
$HgBa_2CaCu_2O_{6+\delta}$	$Pb_2(Sr, La)_2Cu_2O_6$
$HgBa_2Ca_2Cu_3O_{8+\delta}$	$PbSr_2(La, Ca)Cu_3O_8$
$HgBa_2Ca_2Cu_3O_{8+\delta}F_x$	$Ba_2RE(Ru, Cu)O_6$, where RE = Y, Pr
$(Hg, Re)Ba_2Ca_2Cu_3O_{8+\delta}$	$Sr_2RE(Ru, Cu)O_6$, where RE = Y, Pr
$HgBa_2Ca_3Cu_4O_{10+\delta}$	$RuSr_2RECu_2O_8$, where RE = Eu,
$HgBa_2Ca_4Cu_5O_{12+\delta}$	Gd, Y
$Hg_2Ba_2CaCu_2O_{7+\delta}$	$RuSr_2(Nd, Y, Gd, Ca, Ce)_2Cu_2O_{10}$

layers of the A atoms. They are simplest (in structure and chemical composition) copper-oxide superconductors and have no structure blocks, playing the role of the charge reservoirs. Due to this reason, they are ideal candidates for research of basic characteristics of CuO_2 plane, which is the main structure unit of copper-oxide HTSC. The critical temperature, $T_c \sim 60\,K$, has been attained in $CaCuO_2$ monolayers, surrounded from both sides by

$Ba_{0.9}Nd_{0.1}CuO_{2+x}$ layers, which carried out a function of the electric charge reservoirs. In this case, the measured "one-layer" critical current density, $J_c > 10^8\,A/cm^2$ at $T = 4.2\,K$ [46].

The irreversibility line in the infinite-layer copper-oxide electronic HTSC $Sr_{0.9}La_{0.1}CuO_2$ samples without admixtures, synthesized under high pressure, disposed much more higher than in $La_{1.85}Sr_{0.15}CuO_4$ and $Nd_{1.85}Ce_{0.15}CuO_4$ [504]. The critical current density was also much more higher that proposed strong pinning of magnetic flux. Moreover, it has been stated that superconductivity of $Sr_{0.9}La_{0.1}CuO_2$ is three-dimensional in contrast to quasi-two-dimensional superconductivity of all HTSC cuprates [544].

A.9 Rare-Earth Borocarbides

The rare-earth borocarbide $LuNi_2B_2C$ has critical temperature, T_c, approximately 16 K, at the same time, LuNiBC is not superconducting. The superconductor is obtained, adding carbon in Lu plane. On the other hand, LuNiBC is obtained from $LuNi_2B_2C$, adding the other layer of Lu–C. Thus, the superconducting composition is fabricated from isolator by changing corresponding atom ratio. Similarly in $La_{2-x}Sr_xCuO_4$, superconducting phase is formed from isolating phase by changing parameter x [737]. The compounds $Re\mathrm{Ni_2B_2C}$ with Re = Y, Lu, Tm, Er, Ho and Dy are superconducting with moderate high critical temperatures, $T_c \approx 16\,K$. At low temperatures, there is antiferromagnetic phase, and the phase exists in which can co-exist superconductivity and magnetic long-range order. As for the example of $Ho_{1-x}RE_xNi_2B_2C$ (RE = Y, Lu) system with different numbers of Ho(RE)C layers, a clear correlation of T_c with the density of states in Fermi level has been found. The various magnetic structures are derived which co-exist with superconductivity. The highest $T_c = 23\,K$ is attained in YPd_2B_2C.

A.10 Silicon-Based Superconductors

Up to recent time, only one superconducting compound of silicon family, namely $ThSi_2$ with $T_c = 1.56\,K$, has been known. However, recently, it has been stated that semimetal $CaSi_2$ becomes a superconductor with $T_c = 14\,K$ at pressure $P > 12\,GPa$ [919].

In contrary to known crystalline silicon structures, the various cavities can form at slightly distorted angles between links. Clathrate crystals are the compounds consisting of molecules and atoms ("guests") into cavities of frame of the crystalline lattice ("host"). Clathrate structures form on the base of different matters with tetrahedron coordination of atoms. Two types of clathrates are known, namely the silicon compounds M_xSi_{46} and M_xSi_{136} (where M = Na, K, Rb, Cs). If part of the alkaline atoms in this "friable" (but right)

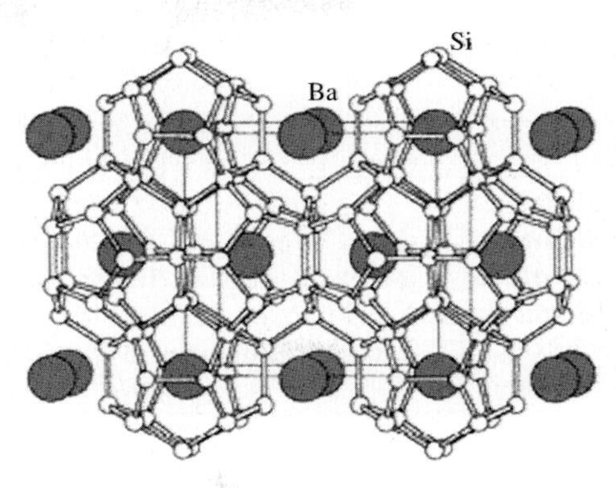

Fig. A.6. Crystal structure of silicon clathrate superconductor Ba_8Si_{46} [1163]

clathrate structure of the silicon compound M_xSi_{46} (where M = Na, K) is substituted by Ba, then this structure becomes superconducting [522]. Clathrates $Na_2Ba_6Si_{46}$ and Ba_8Si_{46} have critical temperatures, $T_c = 4\,K$ [522] and 8 K [1163], respectively. Crystal structure of Ba_8Si_{46} is presented in Fig. A.6.

A.11 Chalcogens

Superconductivity of sulphur with $T_c = 10\,K$ arises under $P = 93\,GPa$, at $P = 200\,GPa$, the value of T_c attains a maximum of 17.3 K, and at further increasing of the pressure again decreases down to 15 K. The similar dependences of $T_c(P)$ are observed in Se and Te. For example, the superconducting transition takes place in Se at $T_c = (4-6)\,K$ in the range of $P = 15-25\,GPa$, while $T_c = 8\,K$ at $P > 150\,GPa$ [342].

The superconducting composites C/S, fabricated from press-powders of graphite and sulphur (23 wt%), which are synthesized in argon, have demonstrated $T_c = 35\,K$ (the superconducting transition has been found by Meissner's effect) [170]. In this case, the magnetization curves demonstrated hysteresis that is proper for type-II superconductors.

The controlled intercalation of copper into di-chalcogen $TiSe_2$ leads to initiation of the waves of charge density (WCD), which are periodical modulations of the density of conductive electrons, and also to arising of superconductivity at the copper content $x = 0.04$ in calculation per formal unit [731]. The maximum $T_c = 4.15\,K$ has been observed at $x = 0.08$. Thus, a correlation of WCD with superconductivity could be studied in the example of Cu_xTiSe_2.

A.12 Carbon Superconductors

One of the forms of carbon has 60 atoms per molecule, C_{60}. A large number of molecules can be made using C_{60}. The compounds A_3C_{60} are superconducting and their properties are given in Table A.7. The doping by electrons reaches

Table A.7. The critical temperatures and the lattice constant of the f. c. c. unit cell [986]

No.	Formula	a_0(Å)	T_c(K)
1	Rb_2CsC_{60}	14.493	31.3
2	Rb_3C_{60}	14.436	29.4
3	Rb_2KC_{60}	14.364	26.4
4	$Rb_{1.5}K_{1.5}C_{60}$	14.341	22.15
5	RbK_2C_{60}	14.299	21.8
6	K_3C_{60}	14.253	19.28

the critical temperature of 33 K in the system $Rb_yCs_xC_{60}$ [1049]. The highest $T_c = 40$ K has been obtained in Cs_xC_{60} [792]. If there are defects in the lattice, $\ln T_c$ varies as the inverse square of the unit cell constant (see Fig. A.7).

Nature of high-T_c of fullerenes is not understood totally. Herein, significant help could be rendered by experimental definition of the power, α, from the isotopic effect $T_c \sim M^{-\alpha}$,where M is the mass of atoms. In [296], ^{12}C has been substituted practically totally (on 99%) by isotope ^{13}C in Rb_3C_{60} monocrystals with $T_c = 31$ K and has been found the value of power $\alpha = 0.21 \pm 0.012$. Based on these calculations, it is concluded that the phonon mechanism of electron pairing with "intermediate" force of electron–phonon interaction is realized in the doped fullerenes.

Then, an existence of Meissner's effect (repulsion of magnetic field from volume of superconductor) in the polycrystalline copper-containing fullerides (fullerene-containing crystallitecompounds) at temperatures up to 110 K has been experimentally stated [869]. The estimation of the experimental data showed that approximately *thousandth part* of the bulk sample has superconducting properties, causing the repulsion force.

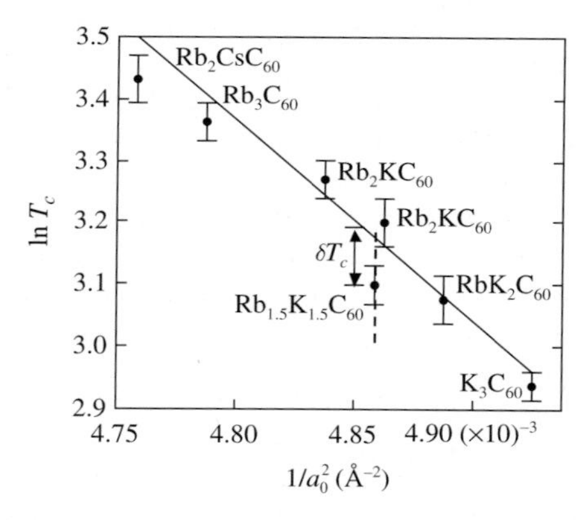

Fig. A.7. The dependence of ln $T_c$vs inverse square of the unit-cell constant [986]

In 2000–2001, J. H. Schön et al. announced that, using an intercalation with $CHCl_3$ and $CHBr_3$ of C_{60} single crystals, directed to expansion of the lattice and formation of high densities of electrons and holes, they reached first the critical temperature, $T_c = 52\,K$ [943], and then $T_c = 117\,K$ in hole-doped $C_{60}/CHBr_3$ [944]. Figure A.8 presents the variation in critical temperature as a function of charge carrier density for electron- and hole-doped C_{60} crystals in different types of intercalation. Obviously, these results rendered doubtful.[1]

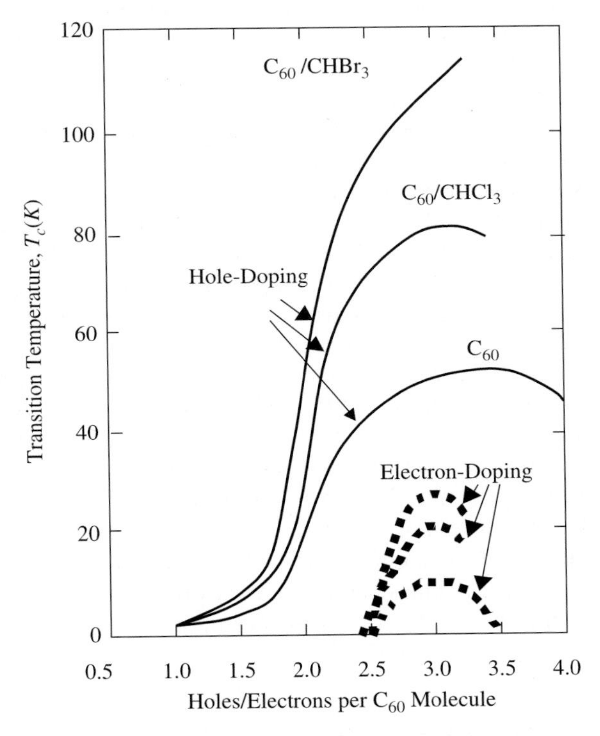

Fig. A.8. Variation in T_c as a function of charge carrier density for electron- and hole-doped C_{60} crystals in incorporation of $CHCl_3$ and $CHBr_3$ [944]

[1] The critical papers [96,964,965] about J. H. Schön et al. results [940–942, 945–947] have been published in 2002 (see also http://www.lucent.com/news_events/researchreview/). The papers [940–942, 945–947] have been devoted to the use of organic molecules, segregated on thin film, as "molecular switches", and also to the use of field transistors to change the charge carrier concentration in different organic matters, providing a regulation to their electric properties (from dielectric to semiconductor, from metal to superconductor and so on). Supposedly, this rendered to induce high-T_c superconductivity of the fullerenes C_{60} by electric field. However, first, total identity of the figures (right up to accidental noise inevitably accompanying any experiment) in [947] and [946] has been found. The same figure has been also presented in [940], devoted to another type of mi-

Superconductivity with $T_c = 7\,K$ has been found in C_{70} monocrystal with size of about 1 mm [319]. Obviously, carbon nanotubes (1D molecular conductors) are ideal candidates for research of 1D superconductivity. Today, maximum critical temperature, $T_c = 15\,K$, for one-wall nanotubes [1047], and $T_{c,\,on} \sim 11.5\,K$ (beginning of resistive transition), $T_c(R = 0) = 7.8\,K$, for multiwall nanotubes [1040] have been reached.

The attempts to switch graphite into superconducting state, using its doping by different chemical elements, led to superconductivity with $T_c < 1\,K$ for cases of K and Na [36, 388]. Introduction of ytterbium and calcium atoms between graphite layers led to the fabrication of superconducting compounds with $T_c = 6.5$ and $11.5\,K$, respectively [1139]. In this case, the role of Yb and Ca atoms is that they supply free charge carriers in the graphite layers.

Then, the oriented (111) boron-doped diamond thin films (with boron concentration 0.53%) have been grown at (001) silicon substrates, using MPCVD technique (modification method of the chemical vapor deposition) [1038]. These films demonstrated $T_{c,\,on} = 7.4\,K$ and $T_c(R = 0) = 4.2\,K$ at $H = 0$. The linear extrapolation gives $H_{c2}(0) = 10.4\,T$ and $H_{irr}(0) = 5.12\,T$, the critical current density $J_c = 200\,A/cm^2$ at $H = 0$. First signs of superconductivity arose at boron concentration of 0.18%. Finally, the materials, combining superconductivity, superhardness and high strength (these materials could be used for research of electric and superconducting properties under pressure) have been synthesized at high static pressures (up to 7.7 GPa) and temperatures (up to 2173 K) in the following systems [226]: (i) diamond/Nb, $T_c = 12.6\,K (\Delta T = 1.5\,K)$, $H_{c2}(4.2\,K) = 1.25\,T$ (diamond matrix and superconducting channels from niobium carbide); (ii) diamond/Mo, $T_c = 9.3\,K, \Delta T = 5\,K$ (diamond matrix and superconducting channels from molybdenum carbide); (iii) composites with matrices from superhard materials (80 wt%) and MgB_2 channels, namely diamond/MgB_2 ($T_c = 37\,K$) and cubic boron nitride/MgB_2 ($T_c = 36.1\,K$), frame of these composites has microhardness in the range of 57–95 GPa.

A.13 MgB_2 and Related Superconductors

The superconducting system MgB_2 discovered in 2001 demonstrates highest volume superconductivity (critical temperature, $T_c = 39\,K$) among non-copper oxide conductors [756].[2] Crystal structure of MgB_2, having spatial

croelectronic devices. In this case, the coincidence has not been in the whole, but separate parts of the plots coincided totally with one another. Then, *eight* cases of the figure coincidence (total or fragmentary) have been found in six papers of J. H. Schön, devoted to different types of devices for various materials and temperatures. After that, 100 *scientific groups* in the world have attempted unsuccessfully to repeat the results of J. H. Schön et al.

[2] WO_3 monocrystals, doped by Na, demonstrate high-temperature superconductivity with $T_c = 91\,K$ in the *surface layer* with content of $Na_{0.05}WO_3$ [417].

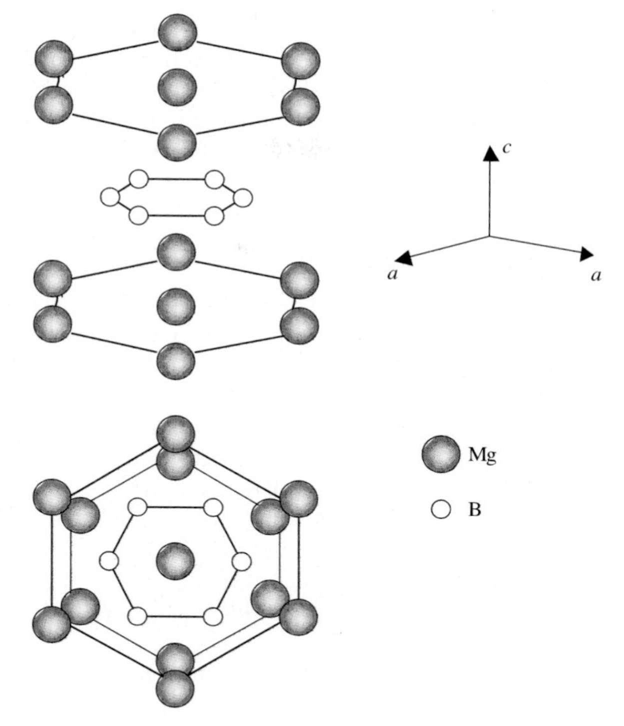

Fig. A.9. Crystal structure of MgB_2 [756]

group of symmetry $P6/mmm$, is shown in Fig. A.9 (where $a = 3.086$ Å, and $c = 3.524$ Å are the parameters of hexagonal unit cell). It is most surprising that simplicity and accessibility of this material, known from the beginning of 1950 [496], combines with complexity of the superconductivity phenomenon.

Even in non-textured multiphase MgB_2 samples, superconducting currents flow in all volume of the sample and are not sensitive to weak magnetic field (which should suppress J_c at the existence of Josephson links between grains). The temperature scaling of the pinning strength practically in all superconducting H–T plane supports the concept that critical current density is found by the magnetic flux pinning [602]. It has been shown in experiments that both the transport (i.e., intergranular) j_c and its dependence on H and T, totally coincide with those for inductive (i.e., intragranular) j_c. The test results prove that grain boundaries in MgB_2 are transparent for superconducting current [194]. It is observed small, but evident anisotropy of critical field, $H_{c2}^{ab}/H_{c2}^{c} = 1.1$ in the bulks [387]. At the same time, it is found that $H_{c2}^{ab}/H_{c2}^{c} = 1.8$–$2.0$ in MgB_2 films with c-axis, oriented perpendicular to the

---

The nature of the observed superconductivity shows that in this case, supercurrents flow on the crystal surface but not into its volume.

film surface [837]. The high-qualitative-oriented MgB_2 films has critical current density, $j_c = 1.6 \times 10^7 A/cm^2$ at $T = 15$ K and $H = 0$ [543]. It is stated that T_c and residual electric resistivity, ρ_0, depend only on thickness, d, of the epitaxial MgB_2 films, but are independent of the film growth rate. An increase in d leads to increase of T_c and decreasing of ρ_0. Now, the values of $T_c = 41.8$ K and $\rho_0 = 0.28 \mu\Omega \cdot$ cm at $d > 300$ nm are maximal and minimal magnitudes, respectively [853]. The record value of $H_{c2} = 52$ T at $T = 4.2$ K has been reached in one sample of thin MgB_2 films in the field parallel to the film surface [274].

The irreversibility field, H_{irr}, in Nb_3Sn, Nb-Ti and MgB is equal to 20, 10 and 14 T at $T = 4.2$ K, respectively [254]. Generally, the characteristics of the superconductors MgB_2, BKBO and $A - 15$ (see Table A.8) [737] show similar values. Therefore, this group of superconductors can be considered as intermediate between the conventional superconductors, subjected to BCS theory, and the superconductors with high critical temperature.

The attained values of critical current density for MgB_2-coated conductors in Fe-sheath [490]: $J_c > 85 kA/cm^2 (T = 4.2$ K and $H = 0)$ and $J_c = 23 kA/cm^2$ ($T = 20$ K and $H = 0$) are nearly sufficient for current-carrying cables.[3] The required components Mg, B and Fe are lightly accessible chemical elements and method of "oxide powder in tube" used in the conductor fabrication be very well tested at preparation both low- and high-temperature large-scale superconductors. Now, this technique can supply high temperatures of industrial processing. All these factors open a tempting perspective of fastest application of MgB_2 systems and MgB_2-based superconducting products.

Several compounds, related to MgB_2, reach the following critical temperatures: BeB_2 ($T_c = 0.79$ K [1182]); Re_3B ($T_c = 4.7$ K [1020]); ZrB_2 ($T_c = 5.5$ K [309]); ReB_2 ($T_c = 6.7$ K [1020]); TaB_2 ($T_c = 9.5$ K [506]); $Nb_{0.76}B_2$ ($T_c = 9.2$ K at $P = 5$ GPa [1162]); CaAlSi ($T_c = 7.8$ K [645]); SrAlSi ($T_c = 5.1$ K [645]); $MgCNi_3$ ($T_c = 8$ K [401]); $MgAlB_4$ ($T_c = 12$ K [623]). The calculations of electronic structure and constant of electron–phonon interaction predict that AgB_2 and AuB_2 should be high-temperature superconductors with $T_c = 59$ and 72 K, respectively [590].

Table A.8. Characteristics of Nb_3Ge (A-15 compound), BKBO ($x \approx 0.4$) and MgB_2

No.	Compound	T_c (K)	$v_F (10^7 cm/s)$	ξ(Å)	$2\Delta/k_B T_c$	B_{c2}(T)
1	Nb_3Ge	23	2.2	35–50	4.2	38
2	BKBO	31	3.0	35–50	4.5	32
3	MgB_2	39	4.8	35–50	4.5	39

T_c is the critical temperature; v_F is Fermi rate; ξ is the coherence length; Δ is the energy gap; k_B is Boltzmann constant; and B_{c2} is the upper critical magnetic field.

[3] MgB_2-coated conductor (or tape), fabricated by laser evaporation on flexible metallic tape has $J_c = 1.1 \times 10^5 A/cm^2$ ($T = 4.2$ K and $H = 10$ T) [563].

A.14 Room-Temperature Superconductivity

There are at least two papers reporting superconductivity above room temperature. In the first paper, superconductivity is observed in a thin surface layer of the complex compound $Ag_{\beta}Pb_6CO_9 (0.7 < \beta < 1)$ at 240–340 K [207]. The second paper claims that superconductivity exists in carbon-based multiwall nanotubes at $T > 400$ K [1197].

B

Finite Element Implementation of Carbon-Induced Embrittlement Model

Consider the finite element implementation of the governing equations (4.14) and (4.27) for carbon diffusion and non-mechanical energy flow, respectively [810]. The next initial and boundary conditions supplement the governing equations:

$$C^{\mathrm{CT}} = C_0^{\mathrm{CT}}, \quad T = T_0, \quad \text{at } t = 0 \; ; \tag{B.1}$$

$$C^{\mathrm{CT}} = C_{\mathrm{b}}^{\mathrm{CT}}, \quad \text{on } S_{\mathrm{b}}; \quad J_k^{\mathrm{C}} n_k = \varphi^{\mathrm{C}}, \text{on } S_\varphi \; ; \tag{B.2}$$

$$T = T_{\mathrm{s}}, \quad \text{on } S_{\mathrm{T}}; \quad -k\frac{\partial T}{\partial x_i} n_i = \varphi^{\mathrm{E}}, \quad \text{on } S_{\mathrm{F}} \; , \tag{B.3}$$

where C_0^{CT} and T_0 are the initial carbon concentration and temperature, which may vary within material volume V. If C_0^{CT} is larger than carbon terminal solid solubility, the initial carbon concentration in carbonate equals the terminal solubility of carbon in carbonate, and the initial carbon volume fraction is calculated according to (4.15). A similar comment is valid for $C_{\mathrm{b}}^{\mathrm{CT}}$, which is prescribed carbon concentration on S_{b}, a part of the bounding surface S. For the calculation of C^{TS}, stress and temperature distributions are taken into account. φ^{C} is the prescribed carbon flux on S_φ; φ^{E} is the prescribed heat flux on S_{F} and T_{s} is the prescribed temperature on S_{T}. Note that $S_{\mathrm{b}} \cup S_\varphi = S_{\mathrm{T}} \cup S_{\mathrm{F}} = S$. The quantities $C_{\mathrm{b}}^{\mathrm{CT}}$, φ^{C}, T and φ^{E} may vary with time.

The finite element equations are derived from variational descriptions of diffusion and energy flow. For this purpose variations of carbon concentration, δC^{C}, and temperature, δT, are considered, which satisfy the boundary conditions. Therefore,

$$\delta C^{\mathrm{C}} = 0 \quad \text{on } S_{\mathrm{b}} \tag{B.4}$$

$$\delta T = 0 \quad \text{on } S_{\mathrm{T}} \tag{B.5}$$

Relation (4.14) is multiplied by a carbon concentration variation satisfying (B.4). Subsequently, it is integrated over the volume V. Then, taking into account (4.11), (4.12) and (4.31), the following expression is derived, which is valid at any time t:

$$\int_V \delta C^{\mathrm{C}} \frac{\mathrm{d}C^{\mathrm{CT}}}{\mathrm{d}t}\mathrm{d}V = -\int_{S_\varphi} \delta C^{\mathrm{C}} \varphi^{\mathrm{C}}\,\mathrm{d}S - \int_V f D^{\mathrm{C}} \frac{\partial C^{\mathrm{C}}}{\partial x_k}\frac{\partial(\delta C^{\mathrm{C}})}{\partial x_k}\mathrm{d}V -$$
$$-\int_V f\left(-\frac{D^{\mathrm{C}}\overline{V}^{\mathrm{C}}}{3RT}\frac{\partial \sigma_{mm}}{\partial x_k} + \frac{D^{\mathrm{C}}Q^{\mathrm{C}}}{RT^2}\frac{\partial T}{\partial x_k}\right) C^{\mathrm{C}} \frac{\partial(\delta C^{\mathrm{C}})}{\partial x_k}\mathrm{d}V\,. \tag{B.6}$$

The above relation is simplified if, within a time increment, Δt, the change of carbonate volume fraction is included in C^{C}. Value of Δt is order of a characteristic time, introduced by diffusion and carbonate size, or smaller. Then

$$\frac{\mathrm{d}C^{\mathrm{CT}}}{\mathrm{d}t} = f_t \frac{\mathrm{d}C^{\mathrm{C}}}{\mathrm{d}t}\,. \tag{B.7}$$

Note that C^{C} is no longer the carbon concentration in carbonate, but the carbon concentration in a part of the material, which at time t is in the form of carbonate and has volume $f_t V$.

Spatial discretization is obtained by introducing the usual finite element interpolation for carbon concentration, carbonate volume fraction, temperature and stress trace. For example, carbon concentration is calculated from nodal values as

$$C^{\mathrm{C}} = a_q C_q^{\mathrm{C}}\,, \tag{B.8}$$

where a_q and C_q^{C} are the interpolation function and the nodal carbon concentration value for q-node, respectively. Substitution of (B.7) into (B.6) and use of spatial discretization lead to the finite element equations for carbon diffusion:

$$C_{pq}\frac{\mathrm{d}C_q^{\mathrm{C}}}{\mathrm{d}t} + \left(D_{pq}^1 + D_{pq}^2\right) C_q^{\mathrm{C}} = F_p\,, \tag{B.9}$$

where

$$C_{pq} = \int_V f_t a_p a_q\,\mathrm{d}V, \quad D_{pq}^1 = \int_V f_t D^{\mathrm{C}} \frac{\partial a_p}{\partial x_k}\frac{\partial a_q}{\partial x_k}\,\mathrm{d}V\,; \tag{B.10}$$

$$D_{pq}^2 = \int_V f_t\left(-\frac{D^{\mathrm{C}}\overline{V}^{\mathrm{C}}}{3RT}\frac{\partial a_r}{\partial x_k}\sigma_{mm}^r + \frac{D^{\mathrm{C}}Q^{\mathrm{C}}}{RT^2}\frac{\partial a_s}{\partial x_k}T_s\right) a_q \frac{\partial a_p}{\partial x_k}\,\mathrm{d}V;$$
$$F_p = -\int_{S_\varphi} a_p \varphi^{\mathrm{C}}\mathrm{d}S\,. \tag{B.11}$$

The time derivative of carbon concentration is approximated by

$$\frac{\mathrm{d}C_q^{\mathrm{C},t+\Delta t}}{\mathrm{d}t} = \frac{1}{\Delta t}\left(C_q^{\mathrm{C},t+\Delta t} - C_q^{\mathrm{C},t}\right). \tag{B.12}$$

Relation (B.9) is taken at time $t+\Delta t$. Equation(B.12) is substituted into (B.9) and the D^2_{pq}-term is transferred to the right hand, leading to the next relation (similar to the one developed in [1003] for hydrogen diffusion due to concentration and stress gradient):

$$\left(\frac{1}{\Delta t}C_{pq}+D^1_{pq}\right)C^{\mathrm{C},t+\Delta t}_q=F_p-D^2_{pq}C^{\mathrm{C},t}_q+\frac{1}{\Delta t}C_{pq}C^{C,t}_q\,. \tag{B.13}$$

The matrices C_{pq}, D^1_{pq} and D^2_{pq} are calculated by using the known nodal values of carbonate volume fraction, temperature and stress from the previous calculation step. The value of carbonate volume fraction corresponds to time t. However, the values of temperature and stress correspond to time $t+\Delta t$, according to the discussion at the end of Appendix B. Vector F_p is calculated from the boundary conditions at time $t+\Delta t$. Note that the solution of (B.13) provides a preliminary carbon concentration value, $C^{\mathrm{C},pr}_{t+\Delta t}$, which according to (B.7) may include carbon in carbonate. This preliminary value can be used for the determination of the total carbon concentration

$$C^{\mathrm{CT}}_{t+\Delta t}=f_t C^{\mathrm{C},pr}_{t+\Delta t}\,, \tag{B.14}$$

as well as the new carbonate volume fraction, $f_{t+\Delta t}$:

$$f^{pr}=\frac{C^{\mathrm{CT}}_{t+\Delta t}-C^{\mathrm{TS}}_{t+\Delta t}}{C^{\mathrm{C}}-C^{\mathrm{TS}}_{t+\Delta t}};\quad f_{t+\Delta t}=\begin{cases}0, & f^{pr}<0;\\ f^{pr}, & 0\le f^{pr}\le 1\,,\end{cases} \tag{B.15}$$

where $C^{\mathrm{TS}}_{t+\Delta t}$ is calculated based on the results for temperature and stress from the previous calculation step.

The new carbon concentration in the carbonate is derived, based on (B.15), as

$$C^{\mathrm{C}}_{t+\Delta t}=\begin{cases}C^{C,pr}_{t+\Delta t}, & f^{pr}<0,\\ C^{\mathrm{TS}}_{t+\Delta t}, & 0\le f^{pr}\le 1\,.\end{cases} \tag{B.16}$$

Then, the finite element implementation of the governing equations for non-mechanical energy flow will be obtained. Relation (4.27) is multiplied by a temperature variation, satisfying (B.5). Subsequently, it is integrated over volume V and the following expression is derived, which is valid at any time t:

$$\int\limits_V \delta T\rho c_p\frac{\mathrm{d}T}{\mathrm{d}t}\mathrm{d}V+\int\limits_V \delta T\frac{\Delta\overline{H}^{\mathrm{car}}}{\overline{V}^{\mathrm{car}}}\frac{\mathrm{d}f}{\mathrm{d}t}\mathrm{d}V$$
$$=-\int\limits_{S_{\mathrm{F}}}\delta T\varphi^{\mathrm{E}}\mathrm{d}S-\int\limits_V k\frac{\partial T}{\partial x_i}\frac{\partial(\delta T)}{\partial x_i}\mathrm{d}V-\int\limits_V \delta T J^{\mathrm{CT}}_n\frac{\partial\mu^{\mathrm{C}}}{\partial x_n}\mathrm{d}V\,. \tag{B.17}$$

As in the case of the carbon diffusion, spatial discretization is introduced and the following finite element equations are derived:

$$H_{qr}\frac{\mathrm{d}T_r}{\mathrm{d}t}+K_{qr}T_r=\Phi^1_q+\Phi^2_q-L_{qs}\frac{\mathrm{d}f_s}{dt}\,, \tag{B.18}$$

where

$$H_{qr} = \int_V \rho c_p a_q a_r \, dV; \quad K_{qr} = \int_V k \frac{\partial a_q}{\partial x_i} \frac{\partial a_r}{\partial x_i} dV; \tag{B.19}$$

$$\Phi_q^1 = -\int_{S_F} a_q \varphi^E dS; \quad \Phi_q^2 = -\int_V a_q J_n^C \frac{\partial \mu^C}{\partial x_n} dV; \quad L_{qs} = \int_V \frac{\Delta \overline{H}^{car}}{\overline{V}^{car}} a_q a_s \, dV \, . \tag{B.20}$$

Assuming that temperature time derivative is given by

$$\frac{dT_r^{t+\Delta t}}{dt} = \frac{1}{\Delta t} \left(T_r^{t+\Delta t} - T_r^t \right) \, , \tag{B.21}$$

and following an approach similar to that for carbon diffusion, one may derive the next algebraic system from (B.18)

$$\left(\frac{H_{qr}}{\Delta t} + K_{qr} \right) T_r^{t+\Delta t} = \Phi_q^1 + \Phi_q^2 - L_{qs} \frac{df_s^t}{dt} + \frac{H_{qr}}{\Delta t} T_r^t \, , \tag{B.22}$$

where Φ_q^1 is calculated from the boundary conditions at time $t+\Delta t$. In order to estimate Φ_q^2, the nodal values of temperature, carbon concentration, carbonate volume fraction and stress from the previous calculation step are used.

A complete calculation cycle is as follows. At time t, all field quantities are known: $u_i^t, \varepsilon_{ij}^t, \sigma_{ij}^t, C_t^C, f_t, T_t, dC_t^C/dt, df_t/dt, dT_t/dt$, where u_i are the components of the displacement vector of a material particle. A time increment Δt is considered and the following calculation steps are performed:

(1) Material deformation problem is solved first. The boundary conditions of the applied traction and/or displacements are defined at time $t+\Delta t$. The isotropic expansion strain rate due to carbon dissolution, carbonate formation and thermal expansion is calculated by using the values and time rates of carbon concentration, carbonate volume fraction and temperature at time t. The parameters of the de-cohesion model are also derived from carbonate volume fraction and temperature distribution at time t. By performing calculation step (1), $u_i^{t+\Delta t}, \varepsilon_{ij}^{t+\Delta t}$ and $\sigma_{ij}^{t+\Delta t}$ are calculated.
(2) The energy flow problem is solved next. The boundary conditions of the applied surface temperature and/or heat flux are defined at time $t+\Delta t$. Φ_q^2-term is calculated, based on the distributions of temperature, carbon concentration and carbonate volume fraction, at time t, as well as on the distribution of stress, $\sigma_{ij}^{t+\Delta t}$, calculated within step (1). The carbonate volume fraction rate, at time t, is used for the calculation of L_{qs}-term. By performing calculation step (2), $T_{t+\Delta t}$ is calculated.
(3) The carbon diffusion problem is solved at the end. The boundary conditions of the applied surface carbon concentration and/or carbon flux are defined at time $t+\Delta t$. In all terms, f_t is used. D_{pq}^2-term is calculated, based on the values $\sigma_{ij}^{t+\Delta t}$ and $T_{t+\Delta t}$, derived using steps (1) and

(2), respectively. By performing calculation step (3), $C^{\mathrm{C}}_{t+\Delta t}$ and $f_{t+\Delta t}$ are calculated.

The material deformation problem is solved, assuming a constant value of Young's modulus. The error in the calculations is further minimized, considering the value of E for the temperature in the crack-tip region, where embrittlement and fracture processes operate. When the variation of temperature, either in space or in time, is significant, its effect on elastic modules should be taken into account (see, e.g., [865]).

Adequate numerical results by using the finite element implementation can be obtained after carrying out preliminary tests, estimating properties of carbon, cuprate, carbonate and YBCO superconductor, which are necessary for calculation.

C

Macrostructure Modeling of Heat Conduction

C.1 Method of Summary Approximation for Quasi-Linear Equation of Heat Conduction

For macrostructure modeling of the heat front propagation during sintering and cooling of HTSC ceramic, the method of summary approximation (MSA) is used as a method for construction of economic schemes for quasi-linear non-stationary equations in the case of arbitrary region and any number of measurements p [916].

The quasi-linear equation of heat conduction without heat sources (heat capacity, c_V, and material density, ρ, are suggested to be constant) has the form:

$$\frac{\partial u}{\partial t} = Lu; \qquad x = (x_1, x_2, \dots x_p) \in G, t > 0 ;$$
$$Lu = \sum_{\alpha=1}^{p} L_\alpha u; \qquad L_\alpha u = \frac{\partial}{\partial x_\alpha}\left[k_\alpha(u, x_\alpha)\frac{\partial u}{\partial x_\alpha}\right] ; \qquad k_\alpha \geq C_\alpha > 0 \quad \text{(C.1)}$$

with boundary and initial conditions:

$$u|_\Gamma = \mu(x,t); \qquad u(x,0) = u_0(x); \quad x \in \overline{G} , \tag{C.2}$$

where u is the temperature; x_α are the spatial coordinates; t is the time and α is the fixed coordinate direction. A boundary Γ of a region G is sufficiently smooth, which is necessary for the existence of smooth solution, $u = u(x,t)$. It is assumed that smooth derivations are required.

As it is made in all economic schemes, the process of approximate solution for multidimensional problem is divided into several stages. Simple problem is solved at every stage. The operator L is presented by the operator sum of more simple structures:

$$L = \sum_{\alpha=1}^{p} L_\alpha . \tag{C.3}$$

We consider multidimensional equation (C.1) and compare to problem (C.1), (C.2) a "chain" of equations:

$$\sum_{\alpha=1}^{p} P_\alpha u = 0; \quad P_\alpha u = \frac{1}{p}\frac{\partial u}{\partial t} - L_\alpha u \,. \tag{C.4}$$

At the interval, $0 \le t \le t_0$, an uniform lattice, $\varpi_\tau = \{t_j = j\tau; j = 0, 1, \ldots, j_0\}$ with a step, $\tau = t_0/j_0$, is introduced. Every interval is divided into p parts, introducing points $t_{j+\alpha/p} = t_j + \alpha\tau/p; \alpha = 1, 2, \ldots, p-1$. Successively (at $\alpha = 1, 2, \ldots, p$),

$$P_\alpha V_{(\alpha)} = 0; \qquad x \in G; \qquad t \in (t_{j+(\alpha-1)/p}; t_{j+\alpha/p}]; \alpha = 1, 2 \ldots, p\,, \tag{C.5}$$

are solved suggesting

$$V_{(1)}(x, 0) = u_0(x); \qquad V_{(\alpha)}(x, t_{j+(\alpha-1)/p}) = V_{(\alpha-1)}(x, t_{j+(\alpha-1)/p})\,. \tag{C.6}$$

Also, it is assumed that the boundary condition of first type is given at Γ. The solution of this problem is named

$$V(x, t_j) = V_{(p)}(x, t_j); \qquad j = 0, 1, \ldots, j_0\,. \tag{C.7}$$

Every of the equations, $P_\alpha V_{(\alpha)} = 0$, it is substituted by the difference scheme (approximating $\frac{\partial u}{\partial t}$ and L_α by corresponding difference relations at uniform lattice ω_h with steps $h_1, h_2, \ldots, h_p$)

$$\Pi_\alpha y_{(\alpha)} = 0; \qquad \alpha = 1, 2, \ldots, p\,. \tag{C.8}$$

The scheme (C.8) approximates the equation $P_\alpha V_{(\alpha)} = 0$ in usual sense, that is

$$\Pi_\alpha u^{j+\alpha/p} - (P_\alpha u)^{j+\alpha/p} \to 0, \quad \text{at } \tau \to 0 \text{ and } h_\alpha \to 0\,. \tag{C.9}$$

The summary approximation of additive scheme (C.8) is attained due to the "chain" of differential equations (C.5), and (C.6) approximates corresponding equations (C.5) in usual sense.

Then, it is assumed that L_α consists of derivations only on variable x_α. Therefore, L_α is the one-dimensional operator, $P_\alpha V_{(\alpha)} = 0$ are the one-dimensional equations and additive scheme (C.8) is the local one-dimensional scheme (LOS). We write LOS, and with this aim the multidimensional equation is replaced by "chain" of one-dimensional equations:

$$\frac{1}{p}\frac{\partial V_{(\alpha)}}{\partial t} = L_\alpha V_{(\alpha)}\,, \tag{C.10}$$

at

$$t_{j+(\alpha-1)/p} < t \le t_{j+\alpha/p}; \; \alpha = 1, 2 \ldots p; \qquad x \in G; \qquad t_{j+\alpha/p} = (j + \alpha/p)\tau\,, \tag{C.11}$$

with conditions

$$V_{(1)}(x,0) = u_0(x); \qquad V_{(\alpha)}(x, t_{j+(\alpha-1)/p}) = V_{(\alpha-1)}(x, t_{j+(\alpha-1)/p}) ;$$
$$V_{(\alpha)} = \mu(x,t), \qquad \text{at } x \in \Gamma_\alpha . \tag{C.12}$$

For difference approximation of the operator L_α in node x_i, a three-point templet is used, consisting of the points: $x_i^{(-1\alpha)}, x_i, x_i^{(+1\alpha)}$ where $x_i^{(\pm 1\alpha)} = [x_1^{(i_1)}, \ldots, x_\alpha^{(i_\alpha)} \pm h_\alpha, \ldots, x_p^{(i_P)}]; x_\alpha^{(i_\alpha)} = h_\alpha i_\alpha; h_\alpha$ is the step of the lattice ω_h in α-direction. A number of internal nodes of the lattice ω_h consists of the points $x = (x_1, x_2, \ldots, x_p) \in G$ of crossing of the hyper-planes $x_\alpha = i_\alpha h_\alpha; i_\alpha = 0, \pm 1, \pm 2, \ldots; \alpha = 1, 2, \ldots, p$, but a number of boundary nodes γ_h consists of the points of crossing of the straight lines C_α, passing through all internal nodes $x \in \omega_h$, with boundary Γ. Also introduced are the next designations: $\gamma_{h,\alpha}$ is the number of boundary nodes in direction x_α; γ_h is the number of all boundary nodes $x \in \Gamma$.

Let $\overline{G} = \{0 \leq x_\alpha \leq l_\alpha\}$ is the parallelepiped, then Γ_α consists of the facets: $x_\alpha = 0$ and $x_\alpha = l_\alpha$. Approximating every heat conduction equation of number α at semi-interval $(t_{j+(\alpha-1)/p}; t_{j+\alpha/p}]$ by a two-layer scheme with weights, a "chain" of p one-dimensional schemes is stated that is called LOS:

$$\frac{y^{j+\alpha/p} - y^{j+(\alpha-1)/p}}{\tau} = \Lambda_\alpha[\sigma_\alpha y^{j+\alpha/p} + (1-\sigma_\alpha) y^{j+(\alpha-1)/p}] , \tag{C.13}$$

where $\alpha = 1, 2, \ldots, p; x \in \omega_h; \sigma_\alpha \in [0,\ 1]$.

In regular nodes, Λ_α has second order of approximation, $\Lambda_\alpha u - L_\alpha u = O(h_\alpha^2)$, and in non-regular nodes, $\Lambda_\alpha u - L_\alpha u = O(1)$. Consider purely implicit LOS ($\sigma_\alpha \equiv 1$):

$$\frac{y^{j+\alpha/p} - y^{j+(\alpha-1)/p}}{\tau} = \Lambda_\alpha y^{j+\alpha/p} , \tag{C.14}$$

and join to this equation the boundary condition:

$$y^{j+\alpha/p} = \mu^{j+\alpha/p}, \quad \text{at } x \in \gamma_{h,\alpha}; j = 0, 1, \ldots, j_0; \alpha = 1, 2, \ldots, p , \tag{C.15}$$

and initial condition:

$$y(x,0) = u_0(x) . \tag{C.16}$$

Let us assume y^j is known. In order to define y^{j+1} at new layer from (C.14) and (C.15), the p equations (C.14) are required to solve together with the boundary condition (C.15), successively suggesting $\alpha = 1, 2, \ldots, p$. In order to define $y^{j+\alpha/p}$, we have the boundary-value problem:

$$A_{i_\alpha} y_{i_\alpha - 1}^{j+\alpha/p} - C_{i_\alpha} y^{j+\alpha/p} + A_{i_\alpha + 1} y_{i_\alpha + 1}^{j+\alpha/p} = 0, \ \text{at } x \in \omega_h; \tag{C.17}$$
$$y^{j+\alpha/p} = \mu^{j+\alpha/p}, \ \text{at } x \in \gamma_{h,\alpha}; \alpha = 1, 2, \ldots, p . \tag{C.18}$$

Here, the lower indexes are pointed only, which change in calculations. The difference equation is written along the section of the straight line, the ends of which coincide with the nodes $\gamma_{h,\alpha}$. Equation (C.17) is solved by the run method at fixed α-direction along corresponding sections. Successively suggesting $\alpha = 1, 2, \ldots, p$ and changing the run directions, $y^{j+1/p}, y^{j+2/p}, \ldots, y^{j+1}$ are calculated, expending $O(1)$ operations per lattice node. Thus, the LOS (C.14)–(C.16) is economic [916]. Moreover, it can be shown that the LOS approximation error tends to zero at $\tau \to 0$ and $h_\alpha \to 0$, and from summary approximation, uniform convergence of the LOS with rate, $O\left(\tau + \max_{1\leq\alpha\leq p} h_\alpha^2\right)$ is followed [916].

C.2 Heat Conduction of Heterogeneous Systems

A study of heat conduction of the heterogeneous systems is a sufficiently complex problem. In the case of HTSC ceramics, it is even more complicated due to significant porosity and different, compared to metal, mechanism of heat conduction. In metal, demonstrating a small porosity and high heat conduction of crystallites, the main mechanism of heat conduction is the convection. In oxide superconductors, possessing relatively greater porosity, together with heat transfer on solid component, the main sources of heat transfer are molecular and radiant components of heat conduction.

C.2.1 Effective Heat Conduction of Mixes and Composites

Existing mixes and composites can be presented by one of the models, depicted in Fig. C.1. A process of ceramic sintering can be described by using structure with mutual-penetrating components (see Fig. C.1b) and the loose granular material (see Fig. C.1e). All components of the structure with mutual-penetrating components are continuous in any direction and geometrically equivalent relative to effective heat conduction:

$$\lambda_{\text{ef}} = f_1(\lambda_1, \lambda_2) = f_2(\lambda_2, \lambda_1), \text{ at } m_1 = m_2 , \tag{C.1}$$

where λ_i and $m_i (i = 1, 2)$ are the heat conduction and concentration of the components.

Granular materials consist of monolithic particles (1) (see Fig. C.1e) and occupy an intermediate state between the structures with impregnation and the structures with mutual-penetrating components. The contacting particles and pores (2), disposed between them, form continuous extent of solid components and cavities in any direction.

Following [234], define heat conduction of granular system. First, consider a structure with mutual-penetrating components. In order to take into account the distortion of the heat flux lines, we use the following formulae for effective

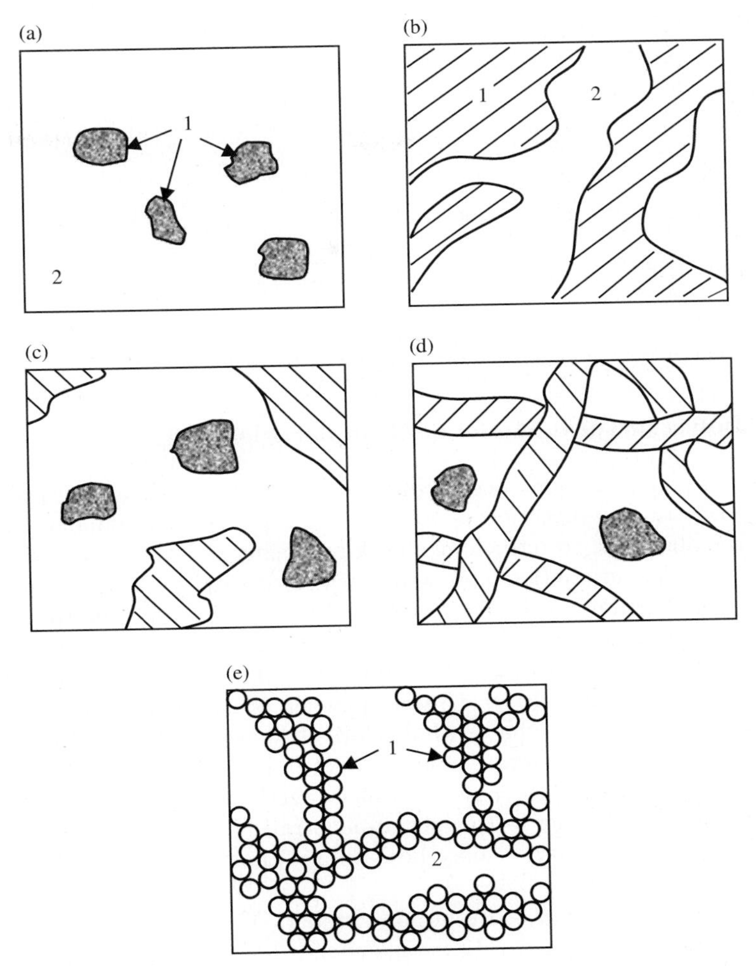

Fig. C.1. Heterogeneous systems with different structure: (**a**) impregnated structure; (**b**) structure with mutual-penetrating components; (**c, d**) combined structures with mutual-penetrating components and impregnation; (**e**) loose granular material

heat conduction in the cases of adiabatic and isothermal division of elementary cell, respectively:

$$\lambda_{\text{ad}} = \lambda_1 \left[c^2 + \nu(1-c)^2 + \frac{2c(1-c)\nu}{1+c(\nu-1)} \right], \quad \nu = \lambda_2/\lambda_1; \tag{C.2}$$

$$\lambda_{\text{is}} = \lambda_1 \left[\frac{1-c}{c^2+\nu(1-c^2)} + \frac{c}{c(2-c)+\nu(1-c)^2} \right]^{-1}. \tag{C.3}$$

here c is the solution of the equation:

$$m_2 = 2c^3 - 3c^2 + 1; m_2 = 1 - m_1 . \tag{C.4}$$

Hence

$$c = 0.5 + A\cos(\varphi/3)\ , \tag{C.5}$$

where

$$A = \begin{cases} -1; \varphi = \arccos\,(1 - 2m_2), \text{ at } 0 \le m_2 \le 0.5\ ; \\ 1; \varphi \ \ = \arccos\,(2m_2 - 1), \text{ at } 0.5 \le m_2 \le 1\ , \end{cases} \tag{C.6}$$

and $3\pi/2 \le \varphi \le 2\pi$. Then, effective heat conduction is defined as

$$\lambda_{\text{ef}} = (\lambda_{\text{ad}} + \lambda_{\text{is}})/2\ . \tag{C.7}$$

The approximate solution (C.7) gives values, which are near to numerical results and preserves property of component invariability.

C.2.2 Polystructural Model of Granular Material

A heat transfer into pores occurs due to molecular collisions and radiation. At the same time, convection is absent, as a rule. The molecular transfer of heat takes place due to interchange by kinetic energy at collisions of moving molecules with one another and with surface of solid or liquid component, limiting pores (grain and liquid surfaces). The heat transfer due to radiation occurs on account of the absorption, emission and dissipation of radiant energy. Both mechanisms can exist together and influence each other mutually.

The granular structure of ceramic powder consists of "frame" (1) (see Fig. C.1e), formed by the chaotic, but relatively dense package of continuously contacting grains (the structure of first order) and spatial lattice of more larger cavities (2), penetrating the powder, which together with the frame form the structure of second order with mutual-penetrating continuous components. Not destroying generality, it may be assumed that the structure of second order occurs at $m_2 \ge 0,4$ [235].

First, a dependence of coordination number, N_{c} (i.e., number of contacts per one particle) on the porosity m_2 is defined. For this, a granular system is considered, consisting of convex rounded particles. In all points of the particle contacts, we depict tangential planes. Then, volume of the system is divided into polyhedrons, circumscribed around any particle. In this case, the facet number of every polyhedron is equal to the number of contacts for given particle (Fig. C.2a).

The porosity of the system is presented through ratio of the difference between volumes of all polyhedrons and particles to volume of the polyhedrons. Determination of required dependence for the parameter N_{c} is carried out, using arbitrary polyhedron, circumscribed around particle with mean radius, r. Then, N_{c} is the mean coordination number for all polyhedrons of the system. For chaotic actual structure, it is assumed that all contacts are uniformly distributed on particle surface. Then, it is suggested that the polyhedron consists of N_{c} identical pyramids with tops at the particle center (Fig. C.2b). In this case, mean porosity of pyramid is equal to mean porosity of the granular system. For simplicity, the pyramid is substituted by cone with the same solid

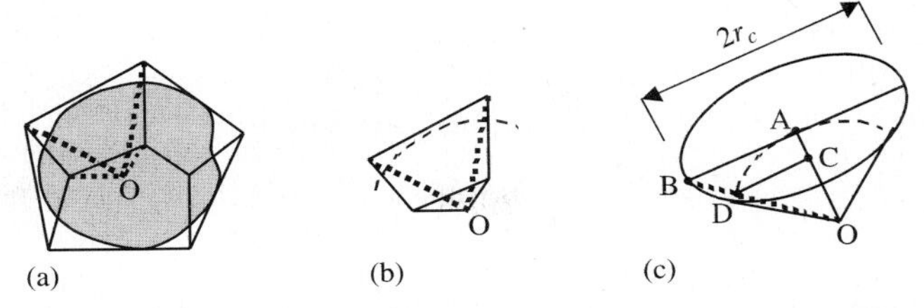

Fig. C.2. Substitution procedure used for definition of coordination number as function of porosity: (**a**) polyhedron formed due to crossing the planes tangential to particle in contact points; (**b**) pyramid is the element of polyhedron; (**c**) straight circular cone

angle and around-contact surfaces are replaced by spherical ones (Fig. C.2c). Then, the porosity of this system is equal to

$$m_2 = 1 - V_{\mathrm{bs}}/V_{\mathrm{c}}; \qquad V_{\mathrm{c}} = 0.33\pi r_{\mathrm{c}}^2 r \,, \tag{C.8}$$

where V_{c} is the cone volume; V_{bs} is the ball sector in the cone; r_{c} is the radius of the cone base: and r is the radius of the spherical surface.

In order to define r_{c}, we use equalities: $F_{\mathrm{b}}/N_{\mathrm{c}} = F_{\mathrm{s}} = 2\pi r h_{\mathrm{s}}$. Hence

$$h_{\mathrm{s}} = 2r/N_{\mathrm{c}} \,, \tag{C.9}$$

where $F_{\mathrm{b}}, F_{\mathrm{s}}$ are the surface squares of the ball and ball segment and h_{s} is the segment height. Because ΔOAB is similar to ΔOCD, then from (C.9), we obtain

$$r_{\mathrm{c}} = 2r(N_{\mathrm{c}} - 1)^{1/2}/(N_{\mathrm{c}} - 2) \,. \tag{C.10}$$

The volume of the ball sector is smaller by N_{c} times than the volume of the ball

$$V_{\mathrm{bs}} = 4\pi r^3/3N_{\mathrm{c}} \,. \tag{C.11}$$

By using the formula for N_{c} and (C.8), (C.10) and (C.11), we obtain finally

$$N_{\mathrm{c}} = [m_{2\mathrm{f}} + 3 + (m_{2\mathrm{f}}^2 - 10m_{2\mathrm{f}} + 9)^{1/2}]/2m_{2\mathrm{f}} \,. \tag{C.12}$$

Then, effective heat conduction of the structure of second order with mutual-penetrating components is calculated by using (C.2), reduced to the form

$$\lambda = \lambda_{\mathrm{f}} \left[c_2^2 + \nu_{\mathrm{f}}(1 - c_2)^2 + \frac{2\nu_{\mathrm{f}} c_2 (1 - c_2)}{\nu_{\mathrm{f}} c_2 + 1 - c_2} \right]; \qquad \nu_{\mathrm{f}} = \lambda_{22}/\lambda_{\mathrm{f}} \,. \tag{C.13}$$

Here, the geometrical parameter c_2 characterizes volume concentration of frame and is connected with the porosity, m_{22} (in the second order structure), through an equation of type (C.4):

$$m_{22} = 2c_2^3 - 3c_2^2 + 1 \,. \tag{C.14}$$

The volume concentration of pores, m_2, in granular system with volume V and pore volume V_2 is equal to $m_2 = V_2/V$, but the porosities of the frame, $m_{2\mathrm{f}}$, and the second order structure, m_{22}, are found as

$$m_{2\mathrm{f}} = V_{2\mathrm{f}}/(V_1 + V_{2\mathrm{f}}); \qquad m_{22} = V_{22}/V \,, \tag{C.15}$$

where $V_1, V_{2\mathrm{f}}$ and V_{22} are the volumes of particles, pores in the frame and pores in the second order structure.

Moreover, there are next relations:

$$V = V_{22} + V_{2\mathrm{f}} + V_1; \qquad V_2 = V_{22} + V_{2\mathrm{f}} \,. \tag{C.16}$$

Then, we obtain from (C.15) and (C.16)

$$m_{22} = (m_2 - m_{2\mathrm{f}})/(1 - m_{2\mathrm{f}}) \,. \tag{C.17}$$

In order to calculate heat conduction, using (C.13), it is necessary to know the heat conduction of component, filling pores of the second order structure, λ_{22}, and heat conduction of the frame, λ_{f}. The value of λ_{22} is the sum of molecular and radiant components. It depends both on the physical properties of gas and on the geometrical and physical parameters of the pores. For structure of granular material, the molecular component is calculated as [234]

$$\lambda_{2\mathrm{m}} = \frac{\lambda_{\mathrm{g}}}{1 + B/(H\delta_{2\mathrm{c}})} \,, \tag{C.18}$$

where $\delta_{2\mathrm{c}} = 3d(1 - c_2)/c_2$ is the mean size of great cavities; $d = 2r$ is the grain diameter;

$$B = \frac{4\gamma(2 - a)\Lambda_{\mathrm{c}}H}{(\gamma + 1)a\mathrm{Pr}} \,;$$

where λ_g is the heat conduction factor of gas in the infinite space at pressure H and temperature T; a is the accommodation factor of gas at uniform walls; Λ_{c} is the mean length of the gas molecule run; $\gamma = c_p/c_V$ is the adiabatic index, being a ratio of isobaric heat capacity to isochoric; $\mathrm{Pr} = \nu/k$ is Prandtl's criterion, being the ratio of kinematic viscosity, ν, to temperature conductivity of gas, k.

Relation for radiant component of the heat conduction factor in pores of the second order structure has the form [234]:

$$\lambda_{2\mathrm{r}} \approx 0.23(T/100)^3 \frac{Yd}{c_2^2(1 - c_2)(2 - \varepsilon)} \,, \tag{C.19}$$

where $Y = f(\tau, \varepsilon)$ is the function, taking into account influence of the optic thickness of sample, $\tau = \beta l_l$, and of the blackness degree, ε, limiting surfaces (walls); for "gray" approximation, $\beta = \alpha_\lambda + \gamma_\lambda$ is the spectrum factor of weakening; α_λ and γ_λ are the volume spectrum factors of absorption and scattering; l_l is the thickness of filling layer. The value $Y \approx 1$ for granular systems with porosity $m_2 < 0.95$ [234]. We obtain from (C.18) and (C.19) relation for the heat conduction of gas component in pores of the second order structure:

$$\lambda_{22} = \lambda_{\mathrm{g}} \left[1 + \frac{Bc_2}{3Hd(1-c_2)} \right]^{-1} + 0.23(T/100)^3 \frac{Yd}{c_2^2(1-c_2)(2-\varepsilon)} \, . \quad \text{(C.20)}$$

C.2.3 Model of Granular System with Chaotic Structure

In order to model the granular system with chaotic structure, consider a system consisting of rounded absolute solid particles with heat conduction factors, that are greater than corresponding parameter of component, occupying pores. Main fraction of heat flux passes through the regions, surrounding point contacts of particles (the sizes of near-contact regions are much smaller than grain diameter). Then, we divide the heat flux into single flux tubes, so that the tube axis in every particle passes successively the near-contact regions at entry and exit of the flux (Fig. C.3a).

Assumption 1. *Heat conduction of any tube is equal to effective heat conduction of all granular system.*

It is assumed that the tube length is much more greater than the cross-section length of particles with non-elongated shape, which fill the granular system volume chaotically. We divide the tube into elements: $i-1, i, i+1, \ldots$. Every element is limited by two planes perpendicular to the heat flow, namely the plane in contact point and the plane δ–δ, dividing particle in half (Fig. C.3). The lateral surface of the tube is formed by adiabatic surface.

Thermal resistance of the tube is equal to the sum of thermal resistance of its elements, which are divided into two types. The elements without (first type) and with (second type) through pores are shown respectively in Fig. C.3b and c. The through pores are present in only those elements of the tube (second type) for which a–a plane (Fig. C.3c) contacts with δ–δ plane within tube.

The averaged element for case of the ordered cubic package of balls is shown in Fig. C.4a. Any ball contacts with six other balls in the points K, L, M, N, O, P. Four contacts (points M, N, O, P) belong to through pores; the cross-section area of the through pores is shaded.

First, consider a heat transfer in the elements of first type (with distorted boundaries of elements). Let the thermal resistance of the distorted element ($A_2A_1A_0B_0B_1B_2$) (see Fig. C.3b) be the resistance of "straightened" element ($A_0A_1ABB_1B_0$) with the lateral adiabatic surfaces parallel to the direction of general heat flux, but cross-section square is the same as initial element square. Then, the distortion of flux lines in the "straightened" element occurs only in the plane of near-contact region.

Assumption 2. *The near-contact regions of particle in any element of flux tube are formed by spherical surfaces with mean radius r.*

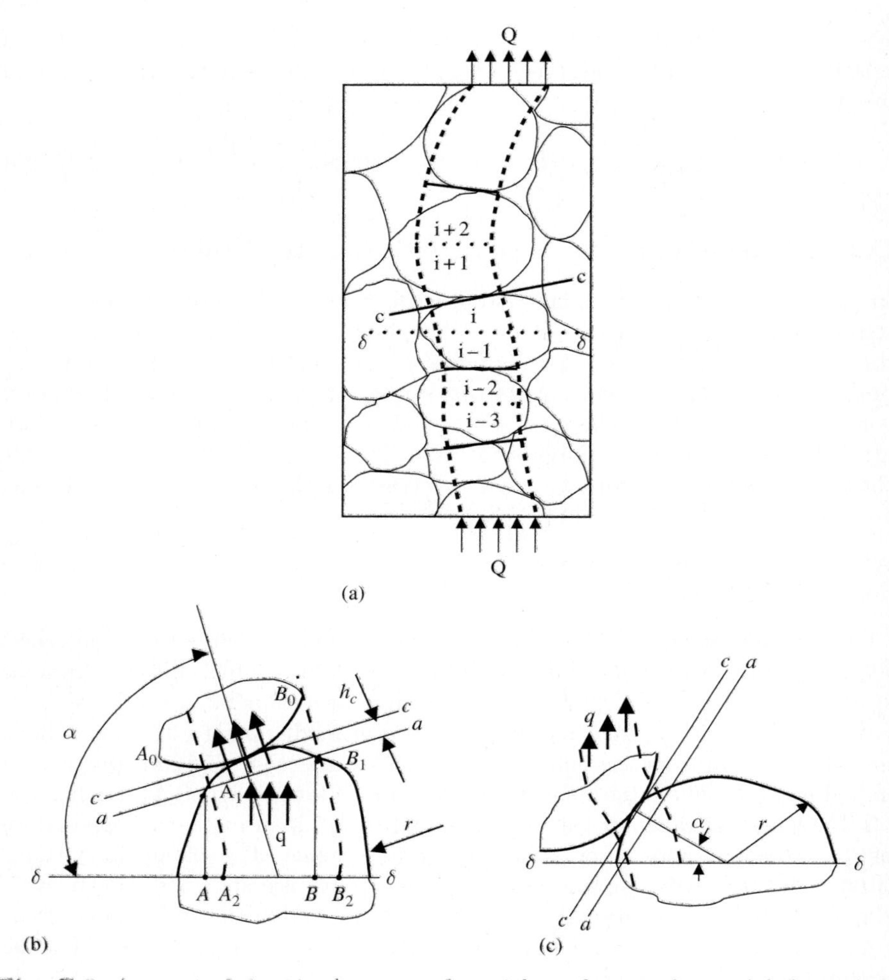

Fig. C.3. Account of chaotic character of particle package in frame: (**a**) flux tubes in the frame; (**b**) element of the flux tube of 1st type; (**c**) element of the flux tube of 2nd type

The square of spherical surface per contact, S_c, is the equal to ratio of total square of the particle surface, $S = 4\pi r^2$, to the coordination number, N_c, that is, $S_c = 4\pi r^2/N_c$. A cross-section of grain part in the flux tube is presented in the form of circle with radius r_1 (Fig. C.4c). Dependency between parameters r_1 and N_c is stated, taking into account the following relations:

$$r_1^2 = h_{bs}(2r - h_{bs}); \quad S_c = 2\pi r h_{bs} , \tag{C.21}$$

where h_{bs} is the height of ball segment. Hence

$$y_1 = r_1/r = 2(N_c - 1)^{1/2}/N_c . \tag{C.22}$$

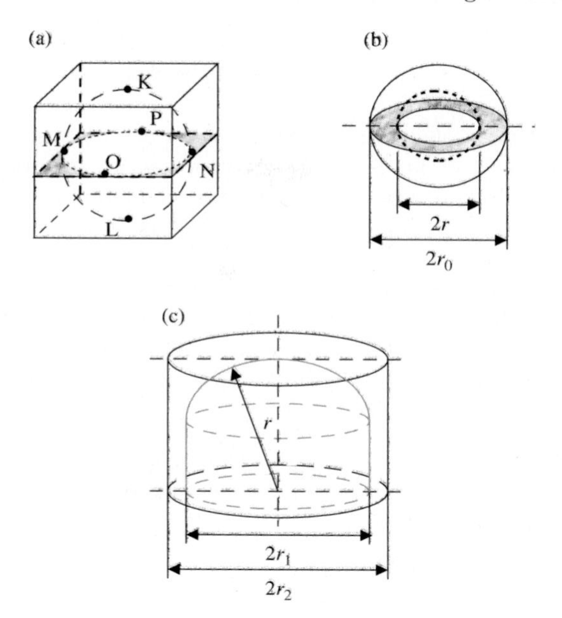

Fig. C.4. Definition of averaged geometrical parameters of elements of the flux tube in the frame: (**a**) disposition of contacts and through pores in cubic package of balls; (**b**) mean cross-section through pores; (**c**) element of chaotic structure with averaged parameters

Using (C.12) and (C.22), the dependence $y_1 = y_1(m_2)$ may be stated. The heat resistance of the first type element is equal to the sum of straight ($A_1A_0B_0B_1$) and truncated (AA_1B_1B) cylinders (Fig. C.3b). Heat resistance of the truncated cylinder is equal to

$$R_{tc} = h_{tc}/(\lambda_1 S_{tc}) , \tag{C.23}$$

where h_{tc} and S_{tc} is the mean height and base square of the truncated cylinder, respectively. Moreover

$$S_{tc} = S_{bs} \sin \alpha , \tag{C.24}$$

where S_{bs} is the base square of the straight cylinder;

$$h_{tc} = (r - h_{bs}) \sin \alpha , \tag{C.25}$$

where r is the ball radius and h_{bs} is the height of the straight cylinder. Then, we have from (C.23) to (C.25)

$$R_{tc} = (r - h_{bs})/(\lambda_1 S_{bs}) . \tag{C.26}$$

Because geometrical parameters of the straight cylinder depend on coordination number only, then the heat resistance of the first type element does not depend on contact location. Therefore, during the study of heat transfer through elements of first type, it may be restricted by only considering the central element.

In the second type elements, contact points are disposed near interface plane and heat transfer depends significantly on fraction of through pores. Due to the existence of first, second and mixed type elements in chaotic structure, it is assumed that effective heat conduction of flux tube is equal to the heat conduction of combined element with averaged parameters, that is, the first type element with mean fraction of the through pores belonging to the second type element. In order to define mean fraction of the through pores, tangential planes, forming spatial polyhedron, are drawn about any particle in the points of contact with neighbors (Fig. C.2a). Average porosity of these polyhedrons and all systems are coincided.

Assumption 3. *Irregular shape of particles (in the form of polyhedron) can be substituted by system of equal volume concentric balls with the same porosity (Fig. C.4b).*

It is followed from Fig. C.4b that

$$r_0/r = (1 - m_2)^{-1/3} = r_2/r \; , \tag{C.27}$$

where $m_2 = (V_0 - V_1)/V_0$; V_0 and V_1 is the volume of external ball and particle, respectively and m_2 is the porosity. Quantitative estimation of the through pores is obtained by using the ratio of the through pore square per particle to the mean cross-section square of the particle.

Assumption 4. *Through pores may be presented in the form of cylinders with annular base, enclosing central element of the tube (Fig. C.4c). In this case, relative square of the cylinder bases is the same as relative square of the through pores:* $\pi(r_0^2 - r^2)/\pi r^2 = \pi(r_2^2 - r_1^2)/\pi r_1^2$.

We have from (C.22) and (C.27)

$$y_2 = r_2/r = y_1(1 - m_2)^{-1/3} \; . \tag{C.28}$$

Equations (C.22), (C.27) and (C.28) define finally the averaged geometrical parameters of the considered system (Fig. C.4). Presented model takes into account the existence of continuous contacts of particles in any direction (the stability condition) and isotropy of the granular system with chaotic structure. One is applied for porosity in the range of $0 \leq m_2 \leq 0.4$.

C.2.4 Heat Flux Through Averaged Element

After estimation of the averaged geometrical parameters of granular system, heat flux through an averaged element is investigated. As this element, we consider an element with bases presenting isothermal planes, and lateral surfaces

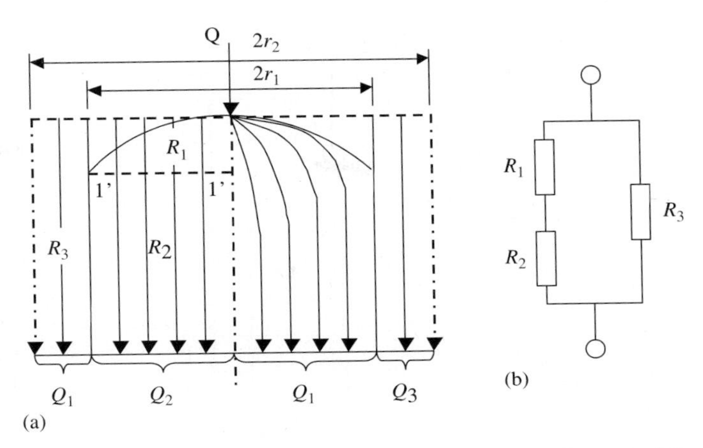

Fig. C.5. Definition of heat resistance of chaotic structure element of the granular system: (**a**) distribution of heat fluxes (actual flux spreading is shown at *right hand*, and idealized picture *at left hand*), (**b**) schematic junction of the heat resistances for single sections

are adiabatic (Fig. C.5). A division by adiabatic surfaces parallel to the heat flux gives overstated value for heat resistance, but a division by isothermal surfaces leads to understated value. A division of mixed type leads to more accurate results with error about 10%.

General heat flux, Q, is presented by the sum: $Q = Q_1 + Q_2$, where fluxes Q_1 and Q_2 pass through particle and through pore in the averaged element, respectively. Then, heat resistances R_1, R_2 and R_3 are found. The resistance R_1 is related to spherical section of solid particle down to intermediate isotherm $1'$–$1'$ (Fig. C.5a). Then, the pointed region is divided into annular adiabatic planes. Conductivity of single layer, $\mathrm{d}\sigma_1$, with thickness, $\mathrm{d}x$, is calculated as (see Fig. C.6) [234]

$$\begin{aligned}\mathrm{d}\sigma_1 &= \left(\frac{l_1^*}{\lambda_s \mathrm{d}S_1} + \frac{l_1^{**}}{\lambda_1 \mathrm{d}S_1}\right)^{-1} \\ &= \mathrm{d}S_1\left(\frac{r-\sqrt{r^2-x^2}}{\lambda_\mathrm{s}} + \frac{\sqrt{r^2-x^2}+\sqrt{r^2-r_1^2}}{\lambda_1}\right)^{-1}, \qquad \text{(C.29)}\end{aligned}$$

where $\mathrm{d}S_1 = x\ \mathrm{d}x\ \mathrm{d}\theta$.

Because the radiant component into gap between spheres is significantly smaller than the molecular one ($\lambda_{\mathrm{sr}} << \lambda_{\mathrm{sm}}$), then in integration of (C.29), we take into account only the molecular heat transfer. Hence, we have from (C.18)

$$\lambda_\mathrm{s} \approx \lambda_\mathrm{sm} = \lambda_\mathrm{g}\left[1 + \frac{B}{2l_1^*(x)H}\right]^{-1}. \qquad \text{(C.30)}$$

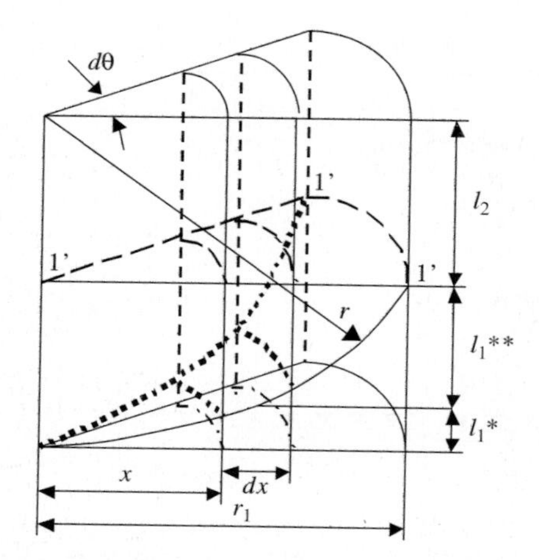

Fig. C.6. Definition of heat resistance of the element's sections with averaged parameters

Total conductivity of the gap between spheres, taking into account only molecular transfer is

$$\sigma_{1\mathrm{m}} = \int_0^{2\pi}\int_0^{r_1} d\sigma_{1\mathrm{m}} = \frac{2\pi\lambda_{\mathrm{g}} r}{1-\nu_{\mathrm{g}}}\left(D - 1 + \omega_{\mathrm{m}} \ln\frac{\omega_{\mathrm{m}} - D}{\omega_{\mathrm{m}} - 1}\right) , \tag{C.31}$$

where

$$D = \sqrt{1 - y_1^2}; \qquad y_1 = r_1/r; \qquad \omega_{\mathrm{m}} = [1 - \nu_{\mathrm{g}} D + B/(Hd)]/(1-\nu_g); \\ \nu_{\mathrm{g}} = \lambda_{\mathrm{g}}/\lambda_1 . \tag{C.32}$$

Now, we take into account a contribution of the radiant component of the heat conduction factor into gap between spheres, summarizing it with the molecular component. We obtain from (C.31) and (C.32)

$$\sigma_1 = \frac{2\pi\lambda_{\mathrm{s}} r}{1-\nu_{\mathrm{s}}}\left(D - 1 + \omega_{\mathrm{s}} \ln\frac{\omega_{\mathrm{s}} - D}{\omega_{\mathrm{s}} - 1}\right) , \tag{C.33}$$

where

$$\omega_{\mathrm{s}} = [1 - \nu_{\mathrm{s}} D + B/(Hd)]/(1-\nu_{\mathrm{s}}); \qquad \nu_{\mathrm{s}} = \lambda_{\mathrm{s}}/\lambda_1; \qquad \lambda_{\mathrm{s}} = \lambda_{\mathrm{sm}} + \lambda_{\mathrm{sr}} . \tag{C.34}$$

The value of λ_{sr} into gap between spheres is estimated approximately, replacing a complex shape of the gap through buffer layer with thickness δ_{s}, which is equal to the gap thickness between spheres, to being average-integral value on square, πr_1^2 (see Fig. C.6), that is,

$$\lambda_{\mathrm{sr}} \approx 4\varepsilon_l \sigma_{\mathrm{S-B}} T^3 \delta_{\mathrm{s}} , \tag{C.35}$$

here

$$\delta_s = \frac{1}{\pi r_1^2} \int_0^{2\pi} \int_0^{r_1} dV_s \, , \tag{C.36}$$

where
$dV_s = l_1^* \, dS_1 = (r - \sqrt{r^2 - x^2}) x \, dx \, d\theta; \varepsilon_l = \varepsilon/(2 - \varepsilon); and\ \sigma_{S-B}$ is Stefan–Boltzmann constant.

At integration of (C.36), it is taken into account that dependence $r_1 = f(N_c)$ is defined by relation (C.22). Then, we have from (C.36)

$$\delta_s = d/N_c \, . \tag{C.37}$$

(a) Finally, the heat resistances are found as

$$R_1 = 1/\sigma_1 \, ; \tag{C.38}$$

(b) the heat resistance, R_2, of the section between intermediate isotherm 1′–1′ and upper base of the element (see Fig. C.5a) is

$$R_2 = \frac{l_2}{\lambda_1 \pi r^2} = \frac{\sqrt{1 - y_1^2}}{\lambda_1 \pi r y_1^2} \, ; \tag{C.39}$$

(c) the heat resistance of the through pore, R_3, is

$$R_3 = \frac{r}{\lambda_{2p} \pi (r_2^2 - r_1^2)} = \frac{1}{\lambda_{2p} \pi r (y_2^2 - y_1^2)}; \qquad y_2 = r_2/r \, , \tag{C.40}$$

where λ_{2p} is the heat conduction of gas into through pores with height equal to the particle radius; one is computed by using (C.20).

By using heat conduction, R_i, of single sections (C.33)–(C.40) and taking into account their schematic junction (see Fig C.5b), effective heat resistance, R, of element with averaged parameters is calculated as

$$R = f(R_i) \, . \tag{C.41}$$

On other hand, we obtain by assuming that all volume of the element is filled by homogeneous matter with effective heat conduction, λ_f, which is equal to effective heat conduction of the granular system frame:

$$R = r/(\lambda_f \pi r_2^2) \, . \tag{C.42}$$

Finally, we find effective heat conduction of the frame, λ_{f}, from (C.41) and (C.42):

$$\lambda_{\mathrm{f}} = \frac{\lambda_1}{y_2^2}\left\{\left[\frac{D}{y_1^2} + \frac{(1-\nu_s)}{2\nu_{\mathrm{s}}}\left(D - 1 + \omega_{\mathrm{s}} \ln\frac{\omega_{\mathrm{s}} - D}{\omega_{\mathrm{s}} - 1}\right)^{-1}\right]^{-1} + \frac{\lambda_{2\mathrm{p}}E}{\lambda_1}\right\}; \tag{C.43}$$

where

$$D = (1 - y_1^2)^{1/2}; \qquad E = y_2^2 - y_1^2; \qquad \omega_{\mathrm{s}} = \frac{1 - \nu_{\mathrm{s}}(1 - y_1^2)^{1/2} + B/(Hd)}{1 - \nu_{\mathrm{s}}}; \tag{C.44}$$

$$\nu_{\mathrm{s}} = \frac{\lambda_{\mathrm{s}}}{\lambda_1}; \qquad \nu_{2\mathrm{p}} = \frac{\lambda_{2\mathrm{p}}}{\lambda_1}; \qquad \nu_{\mathrm{g}} = \frac{\lambda_{\mathrm{g}}}{\lambda_1}; \qquad B = \frac{4\gamma(2-a)\Lambda_{\mathrm{c}}H}{(\gamma+1)a\mathrm{Pr}}. \tag{C.45}$$

This formula could be used to calculate effective heat conduction system at $m_2 < 0.5$.

D

Eden Model

The Eden's model is a sequential model for the stochastic growth of compact clusters [244]. In this model, each new element of a cluster is added at growth site that is chosen with equal probability, from a set of all possible growth sites. These sites are defined each time by a microscopic rule for cluster expansion, which, in general, identifies sites that are on the cluster edge. The set of possible growth sites reflects instantaneous (i.e., in the present time) shape of the cluster (see Fig. D.1).

Eden's clusters are compact in all space dimensions [886], but their surfaces demonstrate self-affine fractal geometry [662, 917]. Computer simulation methods are used extensively to examine the interfacial properties of lattice-based Eden's clusters [502, 503, 852]. In [503], three microscopically distinct versions of the Eden's growth process are considered and it is shown that they all have similar scaling behaviors with a finite size (or roughening) exponent, $\alpha = 0.50 \pm 0.02$, and a dynamic exponent, $\beta = 0.30 \pm 0.03$. The Eden's model describes a large class of irreversible, interfacial growth processes, which includes processes described by continuum equations.

The relaxation processes, which occur concurrently with the growth and disposition of one cell in a group of cells, are substantially unquantified. For example, in the extended Eden's model [52], relaxation is restricted to the synchronous motion of a train of neighboring cells. Computer simulation is performed on a square lattice in 2D strip with size L. Planes at $y = 0$ and $y = L$ are periodic boundaries, and the strip is infinite in the x-direction. Initially, all the lattice sites with $x \leq 0$ are occupied, and all those with $x > 0$ are empty. At each time step, an occupied lattice site $\mathbf{r}$ is chosen randomly. Then, if there are some unoccupied sites, $\mathbf{r_0}$, such that $\mathbf{r} - \mathbf{r_0} = s\mathbf{k}$, where $\mathbf{k}$ is a lattice vector and $s \leq q$ (q is the maximum train length), the site $\mathbf{r}$ becomes, momentarily, doubly occupied. This double occupancy is relaxed by the occupation of a site chosen randomly from the set of unoccupied sites $\mathbf{r_0}$ for which $|\mathbf{r} - \mathbf{r_0}|$ is the minimum. In most cases, this set has only one member. The relaxation is constructed from the simultaneous motions of a train of

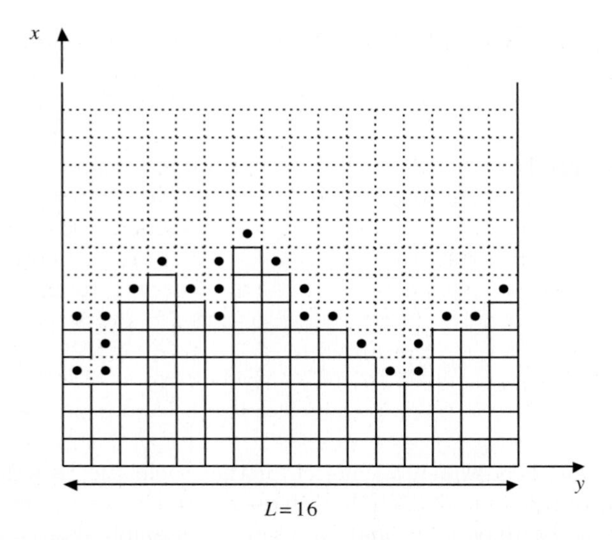

Fig. D.1. Typical cluster obtained, using Eden model. *Circles* denote sites located along the cluster perimeter

neighboring cells in the direction of least resistance with the maximum train length, q. Note that for $q = 1$ this growth process has been modeled in [503].

The growth process leads to a compact cluster with an irregular, rough surface. At any time, the surface of the cluster may be defined by a set of heights, $h(y_i), 1 \leq i \leq L$. The heights, $h(y_i)$, represent the extension of the cluster in the x-direction at a lateral point y_i. For a cluster with constant density, the mean height is found as

$$h_0 = L^{-1} \sum_{i=1}^{L} h(y_i) \ . \tag{D.1}$$

For a particular value of h_0, the standard deviation of the heights

$$\sigma(L, h_0) = \sqrt{L^{-1} \sum_{i=1}^{L} [h(y_i) - h_0(L)]^2} \tag{D.2}$$

represents the surface roughness and has a scaling form [52]

$$\begin{aligned} \sigma(L, h_0) &= L^{\alpha} f(h_0 L^z) \ ; \\ &\propto h_0^{\beta} \quad \text{at } h_0 << L^z \ ; \\ &\propto L^{\alpha} \quad \text{at } h_0 >> L^z \ , \end{aligned} \tag{D.3}$$

where $f(x)$ is a scaling function and $z = \alpha/\beta$. These equations describe a surface with a strip width that initially increases and then, after a time, which depends on the strip width L, reaches a saturated value.

The growth from beneath the surface of an expanding cluster leads to an interface that roughens slowly compared to one that is driven by a simple surface addition process. For the extended model $z \approx 2$, and the temporal spread of height fluctuations is diffusive [52]. This is in contrast to the super-diffusive behavior of pure Eden's model where $z < 2$. Moreover, in the steady state of the extended Eden's model, the fluctuations of the heights are less correlated along the surface than in the case of pure "deposition", that is, in the extended model the surface is less rough. The results are currently insufficient to establish the precise dependence of the exponents on the parameter q, but they establish a *qualitative* change, in an observable property of the cluster morphology, that arises as a result of the subsurface growth.

The averaged Green's function, $g(\mathbf{r} - \mathbf{r_0})$, for the steady-state growth in the extended Eden's model is largely independent of the system size and geometry. It represents, quite generally, the probability that an unoccupied site at $\mathbf{r_0}$ becomes occupied at the same time as a new particle is introduced at site $\mathbf{r}$. The probability of growth increases with the distance beneath the surface because the number of equally weighted growth sites increases with depth through the surface profile. Simulations with $q > \sigma(L, \infty)$ confirm that $g(\mathbf{r}) \to \text{const}$ for $r > \sigma(L, \infty)$ [52]. Thus, the averaged Green's function may be used in a numerical scheme to construct, directly, clusters in different geometry.

References

1. Abbruzzese G., Acta Metall. **33**, 1329 (1985).
2. Abbruzzese G., Lücke K., Acta Metall. **34**, 905 (1986).
3. Abrikosov A. A., Zh. Eksp. Teor. Fiz. **32**, 1442 (1957) [Sov. Phys. JETP **5**, 1174 (1957)].
4. Agarvala P., Srivastava M. P., Dheer P. N., et al., Phys. C **313**, 87 (1999).
5. Aichele T., Görnert P., Uecker R., Mühlberg M., IEEE Trans. Appl. Supercond. **9**(2), 1510 (1999).
6. Akimov I. I., Shikov A. K., Krinitsina T. P., et al., Phys. C **328**, 125 (1999).
7. Aksay I., Han C., Maupin G. D., et al. USA Patent No 5,061,682 (1991).
8. Alamgir A. K. M., Yamada Y., Harada N., et al. IEEE Trans. Appl. Supercond. **9**(2), 1864 (1999).
9. Alarco J. A., Olsson E., Ivanov Z. G., et al., Ultramicroscopy **51**, 239 (1993).
10. Alexander K. B., Goyal A., Kroeger D. M., et al., Phys. Rev. B **45**, 5622 (1992).
11. Alexandrov A. S., cond-mat/9807185 (1998).
12. Alexandrov A. S., *Proc. Int. Conf. on Basic Problem of HTSC* (Moscow, 18–22 October, 2004).
13. Alexandrov A. S., Mott N. F., Rep. Prog. Phys. **57**, 1197 (1994).
14. Alford N. McN., et al., J. Appl. Phys. **66**, 5930 (1989).
15. Almasan C. C., Han S. H., Lee B. W., et al., Phys. Rev. Lett. **69**, 680 (1992).
16. Aloysius R. P., Sobha A., Guruswamy P., Syamaprasad U., Supercond. Sci. Technol. **14**, 85 (2001).
17. Alvarez G. A., Wen J. G., Wang F., et al., IEEE Trans. Appl. Supercoud. **7**, 3017 (1997).
18. Amemiya N., Jiang Z., Yamagishi K., Sasaoka T., IEEE Trans. Appl. Supercond. **12**(1), 1599 (2002).
19. Amm K. M., Sastry P. V. P. S. S., Knoll D. C., Schwartz J., Adv. Cryog. Eng. (Mater.) **44b**, 457 (1998).
20. Amrein T., Seitz M., Uhl D., et al., Appl. Phys. Lett. **63**, 1978 (1993).
21. Anderson J., Cai X. Y., Feldman M., et al., Supercond. Sci. Technol. **12**, 617 (1999).
22. Anderson M. P., Srolovitz D. J., Grest G. S., Sahni P. S., Acta Metall. **32**, 783 (1984).
23. Anderson Ph. W., Science **235**, 1196 (1987).

24. Animalu A. O. *Intermediate Quantum Theory of Crystalline Solids* (New Jersey, 1977).
25. Aoki D., Huxley A., Ressouche E., et al., Nature **413**, 613 (2001).
26. Araujo-Moreira F. M., Filho P. N. L., Zanetti S. M., et al., cond-mat/9909086 (1999).
27. Arlt G., J. Mater. Sci. **25**, 2655 (1990).
28. Armstrong R. A. *Basic Topology* (Springer, Berlin, 1983).
29. Aselage T., Keefer K., J. Mater. Res. **3**, 1279 (1988).
30. Ashby M. F., Acta Metall. **37**, 1273 (1989).
31. Ashby M. F., Gandhi C., Taplin D. M. R., Acta Metall. **27**, 699 (1979).
32. Ashworth S. P., Glowacki B. A., James M. P., IEEE Trans. Appl. Supercond. **5**(2), 1271 (1995).
33. Assmann H., Guenther A., Rigby K., Wohlfart M., Adv. Supercond. **V**, p. 685, Y. Bando, and H. Yamauchi (eds.) (Springer-Verlag, Tokyo, 1993).
34. Aswlage T., Keefer K., J. Mater. Res. **3**, 1279 (1988).
35. Ausloos M., Vandewalle N., Cloots R., Europhys. Lett. **24**, 629 (1993).
36. Avdeev V., et al., Pis'ma JETP **43**, 376 (1986).
37. Ayache J., Odier P., Pellerin N., Supercond. Sci. Technol. **7**, 655 (1994).
38. Babcock S. E., Cai X.Y., Kaiser D. L., Larbalestier D. C., Nature **347**, 167 (1990).
39. Babcock S. E., Cai X.Y., Larbalestier D. C., et al., Phys. C **227**, 183 (1994).
40. Babcock S. E., Larbalestier D. C., Appl. Phys. Lett. **55**, 393 (1989).
41. Babcock S. E., Larbalestier D. C., J. Mater. Res. **5**, 919 (1990).
42. Babcock S. E., Larbalestier D. C., J. Phys. Chem. Solids **55**, 1125 (1994).
43. Babcock S. E., Vargas J. L., Annual Rev. Mater. Sci. **25**, 193 (1995).
44. Baetzold R., Phys. Rev. B **42**, 56 (1990).
45. Balachandran U., Lelovic M., Prorok B. C., et al., IEEE Trans. Appl. Supercond. **9**(2), 2474 (1999)
46. Balestrino G., Medaglia P. G., Orgiani P., et al., Phys. Rev. Lett. **89,** 156402 (2002).
47. Balkevich V. P. *Technical Ceramics* (Stroyizdat, Moscow, 1984).
48. Balluffi R. W., Bristowe P. D., Surf. Sci. **144**, 28 (1984).
49. Balmer B. R., Grovenor C. R., Riddle R., IEEE Trans. Appl. Supercond. **9**(2), 1888 (1999).
50. Bardeen J., Cooper L. N., Schrieffer J. R., Phys. Rev. **108**, 1175 (1957).
51. Barenblatt G. I., Adv. Appl. Mech. **7**, 55 (1962).
52. Barker G. C., Grimson M. J., J. Phys. A: Math. Gen. **27**, 653 (1994).
53. Barret S. E., Durand D. J., Pennington C. H., et al., Phys. Rev. B **41**, 6283 (1990).
54. Bateman C. A., Zhang L., Chan H., et al., J. Am Ceram. Soc. **75**, 1281 (1992).
55. Bean C. P., Phys. Rev. Lett. **8**, 250 (1962).
56. Bean C. P., Rev. Mod. Phys. **36**, 31 (1964).
57. Bech J., Eriksen M., Toussant F., et al., *Proc. Powder Metallurgy World Congress & Exhibition* (2000).
58. Bednorz J.G., Müller K.A., Z. Phys. B **64**, 189 (1986).
59. Beilin V., Goldgirsh A., Yashchin E., et al., Phys. C **309**, 56 (1998).
60. Bell J. F. *The Experimental Foundations of Solid Mechanics. Encyclopedia of Physics*, **1** (Springer, Berlin, 1973).
61. Belyaev A. V., Lebedev V. N., Fisenko E. G., et al. *Problems of HTSC*, part 1, p. 158 (Rostov Sate Univ., 1990).

62. Bennison S. J., Lawn B. R., Acta Metall. **37**, 2659 (1989).
63. Besseling R., Higgebrugge R., Kes P. H., Phys. Rev. Lett. **82**, 3144 (1998).
64. Bever M. B. *Encyclopedia of Materials Science and Engineering*, p. 3354 (Pergamon Press, Oxford, 1986).
65. Bhattacharya R. L., Blaugher R. D. USA. Patent No 5,413,987 (1995).
66. Bian W., Zhu Y., Wang Y. L., Suenaga M., Phys. C **248**, 119 (1995).
67. Biaxeras J., Fcurnet G., J. Phys. Chem. Solids **28**, 1541 (1967).
68. Binning G., Rohrer H., Rev. Mod. Phys. **59**, 615 (1987).
69. Birnbaum H. K., Grosbeck M. L., Amano M., J. Less-Common Metals **49**, 357 (1976).
70. Blackstead H. A., Dow J. D., Chrisey D. B., et al., Phys. Rev. B **54**, 6122 (1996).
71. Blatter G., Feigel'man M. V., Geshkenbein V. B., et al., Rev. Mod. Phys. **66**, 1125 (1994).
72. Bleykhut R. *Fast Algorithms of Digital Signal Processing* (Mir, Moscow, 1989).
73. Bogolyubov N. N., Pis'ma Zh. Eksp. Teor. Fiz. **34**, 58 (1958).
74. Boley B. A., Wiener J. H. *Theory of thermal stresses* (Wiley, New York, 1960).
75. Bolotin V. V. *Stability Problems in Fracture Mechanics* (John Wiley & Sons Inc., New York, 1996).
76. Boullay Ph., Domenges B., Hervieu M., Raveau B., Chem. Mater. **5**, 1683 (1995).
77. Bourdillon A. J., Tan N. X., Ong C. L., J. Mater. Sci. Lett. **15**, 439 (1996).
78. Boutemy S., Kessler J., Schwartz J., IEEE Trans. Appl. Supercond. **7**(2), 1552 (1997).
79. Boyko V. S., Gorbatenko V. M., Krivenko L. F., et al., Fiz. Nizk. Temper. **15**, 988 (1989).
80. Boyko V. S., Krivenko L. F., Demirsky V. V., Natsik V. D., Supercond. Phys. Chem. Tech. **4**, 1815 (1991).
81. Braden M., Hoffels O., Schnelle W., et al., Phys. Rev. B **47**, 12288 (1993).
82. Brandt E. H., Int. J. Mod. Phys.B **5**, 751 (1991).
83. Brandt E. H., Phys.B **169**, 91 (1991).
84. Brandt E. H., Rep. Prog. Phys. **58**, 1465 (1995).
85. Brandt E. H., Phys. Rev. B **54**, 3530 (1996).
86. Brandt E. H., Phys. Rev. B **54**, 4246 (1996).
87. Brandt E. H., Phys. Rev. B **55**, 14513 (1997).
88. Brandt E. H., Phys. Rev. B **58**, 6506 (1998).
89. Brandt E. H., Phys. Rev. B **58**, 6523 (1998).
90. Brandt E. H., Phys. Rev. B **59**, 3369 (1999).
91. Brinker C. J., Sherer G. W. *Sol Gel Science: The Physics and Chemistry of Sol-Gel Processing* (Academic Press, Boston, 1990).
92. Brizhik L. S., Davydov A. S., Fiz. Nizk. Temp. **10**, 358 (1984).
93. Broberg K.B., J. Mech. Phys. Solids **12**, 407 (1971).
94. Browning N., Chisholm M. F., Pennycook S. J., et al., Phys. C **212**, 185 (1993).
95. Brühwiler M., Kazakov S. M., Zhigadlo N. D., et al. Phys. Rev. B **70**, 020503 (2004).
96. Brumfiel G., Nature **417**, 367 (2002).
97. Bruneel E., Persyn F., Hoste S., Supercond. Sci. Technol. **11**, 88 (1998).
98. Bruneel E., Ramirez-Cuesta A. J., van Drische I., Hoste S., Int. Phys. Conf. Ser. **167**(1), 247 (1999).

99. Brusov P. *Mechanisms of High-Temperature Superconductivity* (Rostov State Univ. Press, Rostov-on-Don, 1999).
100. Buckles W., Driscoll D., IEEE Power Eng. Rev. **20**(5), 16 (2000).
101. Budiansky B., Amazigo J. C., Evans A. G., J. Mech. Phys. Solids **36**, 167 (1988).
102. Budiansky B., Hutchinson J. W., Lambropolous J., Int. J. Solids Struct. **19**, 337 (1983).
103. Bulaevskii L. N., Zh. Eksp. Teor. Fiz. **64**, 2241 (1973).
104. Bulaevskii L. N., Clem J. R., Glazman L. I., Malozemoff A. P., Phys. Rev. B **45**, 2545 (1992).
105. Bul'bich A. A. Pis'ma Zh. Tech. Fiz. **12**, 645 (1986).
106. Bunker B.C., et al. In: *High Temperature Superconducting Materials: Preparation, Property and Processing*, W. E. Hatfield, and J. H. Miller, Jr (eds.) (Marcel Dekker Inc., New York, 1988).
107. Buresch F. E., *Proc. Int. Conf. on Residual Stresses Sci. Technol.* (Garmish-Partenkirkhen, 1986), Oberursel, **1**, p. 539 (1987).
108. Buresch F. E., Babilon E., Kleist G., *Proc.* 2nd *Int. Cong. on Residual Stress.* (ICRS2, Nancy, 1988), London, p. 1003 (1988).
109. Buresch F. E., Buresch O., *Proc.* 14th *Int. Conf. Science of Ceramics* – 14 (Canterbury, 1987), Stoke-on-Trent, p.683 (1988).
110. Burns S. J., Phys. C **206**, 97 (1993).
111. Burns S. J., Supercond. Sci. Technol. **7**, 337 (1994).
112. Cacciari M., Mazzanti G., Montanari G. C., IEEE Trans. Dielectr. Electr. Instr. **3**(1), 18 (1996).
113. Caginalp G., Socolovsky E., SIAM J. Sci. Comput. **15**, 106 (1994).
114. Cai X. Y., Gurevich A., Tsu I. - F., et al., Phys. Rev. B **57**, 10951 (1998).
115. Cai X. Y., Polyanskii A. A., Li Q., et al., Nature **392**, 906 (1998).
116. Cai X. Y., Polyanskii A., Li Q., et al., Nature **393**, 909 (1998).
117. Cai Y., Chung J. S., Thorpe M. F., Mahanti S. D., Phys. Rev. B **42**, 8827 (1990).
118. Cai Z. - X., Zhu Y., IEEE Trans. Appl. Supercond. **9**, 2714 (1999).
119. Cai Z. - X., Zhu Y., Welch D. O., Phys. Rev. B **52**, 13035 (1995).
120. Campbell A. M., Evetts J. E., Adv. Phys. **21**, 1191 (1972).
121. Cao H. C., Thouless M. D., Evans A. G., Acta Metall. **36**, 2037 (1988).
122. Cardona A. H., Suzuki H., Yamashita T., et al., Appl. Phys. Lett. **62**, 411 (1993).
123. Cardwell D. A., Kambara M., Hari Babu N., et al., IEEE Trans. Appl. Supercond. **11**(1) (2001).
124. Cardwell D. A., Lo W., Leung H. – T., Chow J. C. L. *Proc* 9th *Int. Symp on Superconductivity* (*ISS'* 96) (Sapporo, Japan, October 21–24, 1996), p. 725 (Springer, Tokio, 1997).
125. Carim A. H., Mitchell T. E., Ultramicroscopy **51**, 228 (1993).
126. Carpay F. M. A. In: *Ceramic Microstructures* '76, R. M. Fulrath, and J. A. Pask (eds.), p. 261 (Westview, Boulder, CO, 1977).
127. Carpay F. M. A., J. Am. Ceram. Soc. **60**, 82 (1977).
128. Carter C. B., Acta Metall. **36**, 2753 (1988).
129. Casalbuoni S., von Sawilski L., Kotzler J., cond-mat/0310565 (2003).
130. Cava R. J., Phys. C **282–287**, 27 (1997).
131. Celotti G., Tampieri A., Rinaldi D., IEEE Trans. Appl. Supercond. **9**(2), 1779 (1999).

132. Chakrapani V., Balkin D., McGinn P., Appl. Supercond. **1**, 71 (1993).
133. Chakravarty S., Kee H. - Y., Völker K., Nature **428**, 53 (2004).
134. Chakraverty B. K., Ranninger J., Feinberg D., Phys. Rev. Lett. **81**, 433 (1998).
135. Chan K. S., Lankford J., Acta Metall. **36**, 193 (1988).
136. Chandler H. W., J. Mech. Phys. Solids. **33**, 215 (1985).
137. Chandler H. W., Solid. Mech. Appl. **39**, 235 (1995)
138. Chaplygin S. A. *On Gas Jets* (Brown University, Providence, RI, USA, 1944).
139. Chaudhari P., Dimos D., Mannhart J. In: *Superconductivity* J. G. Bednortz, K. A. Müller (eds.) (Springer, Heidelberg, 1990).
140. Chen F. - K., Appl. Mech. Eng. **3**, 413 (1998).
141. Chen L. Q., Scripta Metall. Mater. **32**, 115 (1995).
142. Chen L. Q., Khachaturyan A. G., Acta Metall. Mater. **39**, 2533 (1991).
143. Chen N., Shi D., Goretta K. C., J. Appl. Phys. **66**, 2485 (1989).
144. Cherepanov G. P. *Mechanics of Brittle Fracture*, R. de Wit, and W. C. Cooley (eds.) (McGraw Hill, New York, 1979).
145. Chernyavsky K. S. *Stereology in Metallurgical Science* (Metallurgy, Moscow, 1977).
146. Cherradi A., Desgardin G., Mazo L., Raveau B., Supercond. Sci. Technol. **6**, 799 (1993).
147. Chin C. C., Lin R. J., Yu Y. C., et al., IEEE Trans. Appl. Supercond. **7**(2), 1403 (1997).
148. Chiorescu I., Bertet P., Semba K., et al., Nature **431**, 159 (2004).
149. Chisholm M. F., Pennycook S. J., Nature **351**, 47 (1991).
150. Cho J. H., Maley M. P., Willis J. O., et al., Appl. Phys. Lett. **64**, 3030 (1994).
151. Christen D. K., Thompson J. R., Kerchner H. R., et al., In: *AIP Conf. Proc. Superconductivity and its Applications* **273**, 24 (1993).
152. Chu C. W., IEEE Trans. Appl. Supercond. **7**(2), 80 (1997).
153. Chu C. W., Gao L., Chen F., et al., Nature **365**, 323 (1993).
154. Chuang T. - J., Rice J. R., Acta Metall. **21**, 1625 (1973).
155. Cima M. J., Flemings M. C., Figueredo A. M., et al., J. Appl. Phys. **72**, 179 (1992).
156. Cima M. J., Flemings M. C., Figueredo A. M., et al., J. Appl. Phys. **78**, 1868 (1992).
157. Clarke D. R., Faber K. T., J. Phys. Chem. Solids. **48**, 1115 (1987).
158. Clem J. R., Phys. Rev. B **43**, 7837 (1991)
159. Clem J. R., Coffey M. W., Phys. Rev. B **42**, 6209 (1990).
160. Clerk J. P., Giraud G., Laugier J. M., Adv. Phys. **39**, 191 (1990).
161. Coblenz W. S., Dynys J. M., Cannon R. M., Coble R. L., In: *Sintering Processes*, G. C. Kuczynski (ed.), p. 141 (Plenum Press, New York, 1980)
162. Cook R. F., Acta Metall. Mater. **38**, 1083 (1990).
163. Cook R. F., Clarke D. R., Acta Metall. **36**, 555 (1988).
164. Cooper H. T., Gao W., Li S., et al., Supercond. Sci. Technol. **14**, 862 (2001).
165. Crabtree G. W., Kwok W. K., Welp U., et al., *Proc. NATO Advanced Study Institute on the Physics and Materials Science of Vortex States, Flux Pinning and Dynamics* S. Bose, and R. Kossowski (eds.) (Kluwer Academic Publishers, 1999).
166. Crabtree G. W., Nelson D. R., Phys. Today **50**, 38 (1997).
167. Cronstrom C., Noga M. I., cond-mat/9906360 (1999).

168. Cui H. Y., cond-mat/0212059 (2002).
169. Curro N. J., Caldwell T., Bauer E. D., et al., Nature **434**, 622 (2005).
170. da Silva R. R., Torres J. H. S., Kopelevich Y., et al., cond-mat/0105329 (2001).
171. Daemling M., Seuntjens J. M., Larbalestier D. C., Nature **346**, 332 (1990).
172. Dagotto E., Rev. Mod. Phys. **66**, 763 (1994).
173. Daminov R. R., Imayev M. F., Reissner M., et al., Phys. C **408–410**, 46 (2004).
174. Daniels G. A., Gurevich A., Larbalestier D., Appl. Phys. Lett. **77**, 3251 (2000).
175. Danilin B. S. *Reviews on High-Temperature Superconductivity*, **4**(8), 101 (MCNTI, Moscow, 1992).
176. Davidge R. W., Acta Metall. **29**, 1695 (1981).
177. Davidge R. W., Green T. J., J. Mater. Sci. **3**, 629 (1968).
178. Davydov A. S. *Solitons in Molecular Systems* (Naukova Dumka, Kiev, 1988).
179. Davydov A. S. Phys. Rep. **190**, 191 (1990).
180. Davydov A. S. *Solitons in Molecular Systems.* (Kluwer Academic, Dordrecht, 1991)
181. de Arcangelis L., Herrmann H. J., Phys. Rev. B **39**, 2678 (1989).
182. de Gennes P. G. *Superconductivity of Metals and Alloys* (Benjamin, New York, 1966).
183. de Lozanne A. L., Supercond. Sci. Technol. **12**, R43 (1999)
184. de Rango P., Lees M. P., Lejay P., et al., Nature **349**, 770 (1991).
185. Delamare M. P., Hervien M., Wang J., et al., Phys. C **262**, 220 (1996).
186. Delamare M. P., Monot I., Wang J., Desgardin G., J. Electron. Mater. **24**, 1739 (1995).
187. Delamare M. P., Monot I., Wang J., et al., Supercond. Sci. Technol. **9**, 534 (1996).
188. Delamare M. P., Walter H., Bringmann S., et al., Phys. C **323**, 107 (1999).
189. Denbigh K. G. *The Thermodynamics of the Steady State* (Methuen, London, 1951).
190. Desgardin G., Monot I., Raveau B., Supercond. Sci. Technol. **12**, R115 (1999).
191. Deutscher G., Entin-Wohlmann O., Fishman S., Shapira Y., Phys. Rev. B **21**, 5041 (1980).
192. Deve H. E., Evans A. G., Acta Metall. Mater. **39**, 1171 (1991).
193. Dew Hughes D., Cryogenics **36**, 660 (1986).
194. Dhalle M., Toulemonde P., Beneduce C., et al., cond-mat/0104395 (2001).
195. Díaz A., Mechin L., Berghuis P., Evetts J. E., Phys. Rev. Lett. **80**, 3855 (1998).
196. Diko P., Supercond. Sci. Technol. **11**, 68 (1998).
197. Diko P., Supercond. Sci. Technol. **13**, 1202 (2000).
198. Diko P., Fuchs G., Krabbes G., Phys. C **363**, 60 (2001).
199. Diko P., Gawalek W., Habisreuther T., et al., Phys. Rev. B **52**, 13658 (1995).
200. Diko P., Gawalek W., Habisreuther T., et al., J. Microscopy **184**, 46 (1996).
201. Diko P., Kojo H., Murakami M., Phys. C **276**, 188 (1997).
202. Diko P., Pellerin N., Odier P., Phys. C **247**,169 (1995).
203. Diko P., Takebayashi S., Murakami M., Phys. C **297**, 216 (1998).
204. Diko P., Todt V. R., Miller D. J., Goretta K. C., Phys. C **278**, 192 (1997).
205. Dimos D., Chaudhari P., Mannhart J., Phys. Rev. B **41**, 4038 (1990).
206. Dimos D., Chaudhari P., Mannhart J., LeGoues F. K., Phys. Rev. Lett. **61**, 219 (1988).

207. Djurek D., Medunic Z., Tonejc A., Paljevic M., Phys C **351**, 78 (2001).
208. Dolan G. J., Chandrashekhar G. V., Dinger T. R., et al., Phys. Rev. Lett. **62**, 827 (1989).
209. Dorosinskii L. A., Indenbom M. I., Nikitenko V. I., et al., Phys. C **203**, 149 (1992)
210. Dorosinskii L. A., Indenbom M. V., Nikitenko V. I., et al., Phys. C **206**, 360 (1993).
211. Dorosinskii L. A., Nikitenko V. I., Polyanskii A. A., Vlasko-Vlasov V. K., Phys. C **219**, 81 (1994).
212. Dorris S. E., Dusek J. T., Lanagan M. T., et al., Ceram. Bull. **70**, 722 (1991).
213. Dorris S. E., Prorok B. C., Lanagan M. T., et al., Phys. C **212**, 66 (1993).
214. Dou S. X., Horvat J., Wang X. L., et al., IEEE Trans. Appl. Supercond. **7**(2),. 2219 (1997).
215. Dou S. X., Liu H. K., Guo S. J., Supercond. Sci. Technol. **2**, 274 (1989).
216. Dou S. X., Liu H. K., Guo Y. C., Shi D. L., IEEE Trans. Appl. Supercond. **3**(2), 1135 (1993).
217. Doverspike K., Hubbard C.D., Williams R.K., et al., Phys. C **172**, 486 (1991).
218. Dravid V. P., Zhang H., Wang Y. Y., Phys. C **213**, 353 (1993).
219. Driscoll D., et al., IEEE Power Eng. Rev. **20**(5), 39 (2000).
220. Driscoll D., Zhang B., IEEE Trans. Appl. Supercond. **11**(1), 615 (2001).
221. Drory M. D., Thouless M. D., Evans A. G., Acta Metall. **36**, 2019 (1988).
222. Drozdov Y. N., Gaponov S. V., Gusev S. A., et al., Supercond. Sci. Technol. **9**, 1 (1996).
223. Drozdov Y. N., Gaponov S. V., Gusev S. A., et al., IEEE Trans. Appl. Supercond. **7**(2), 1642 (1997).
224. Drucker D. C., Prager W., Quart. Appl. Math. **10**, 157 (1952).
225. Drugan W. J., Rice J. R., Sham T. - L., J. Mech. Phys. Solids **30**, 447 (1982).
226. Dubitzky G. A., Blank V. D., Buga S. G., et al., Pis'ma JETP **81**, 323 (2005)
227. Dubrovina I. N., Zakharov R. G., Kositsyn E. G., et al., Supercond. Phys. Chem. Tech. **3**, 3 (1990).
228. Dugdale D. S., J. Mech. Phys. Solids **8**, 100 (1960).
229. Dul'kin E. A., Supercond. Phys. Chem. Tech. **5**, 104 (1992).
230. Dul'kin E. A., Supercond. Phys. Chem. Tech. **7**, 105 (1994).
231. Dul'kin E. A., J. Supercond., **11**, 275 (1998).
232. Dul'kin E., Beilin V., Yashchin E., et al., Pis'ma Zh. Tech. Fiz. **27**(9) 79 (2001).
233. Dul'kin E., Beilin V., Yashchin E., Roth M., Supercond. Sci. Technol. **15**, 1081 (2002).
234. Dul'nev G. N., Zarichnyak Yu. P. *Thermal Conductivity of Mixtures and Composite Materials* (Energiya, Leningrad, 1974).
235. Dul'nev G. N., Zarichnyak Yu. P., Muratova B. L., Inzh. Fiz. Zh. **16**, 1019 (1969)
236. Dunders J. *Mathematical Theory of Dislocations* (Am. Soc. Mech. Engrs., New York, 1969).
237. Duran L., Kircher F., Regnier P., et al., Supercond. Sci. Technol. **8**, 214 (1995).
238. Dvorkin E. N., Goldschmit M. B., Cavaliere M. A., et al., J. Mater. Process. Technol. **68**, 99 (1997).
239. Dyskin A.V., Ring L.M., Salganik R. L. *Estimation of Damage of* $YBa_2Cu_3O_{7-\delta}$ *Superconducting Ceramic on Data about its Deformability*, Preprint 417 (IPM AN USSR, Moscow, 1989).

240. D'yachenko A. I., Fiz. Tech. Vys. Davl. **6**, 5 (1996).
241. Early E. A., Steiner R. L., Clark A. F., Char K., Phys. Rev. B **50**, 9409 (1994).
242. Eastell C. J., Henry B. M., Morgan C. G., et al., IEEE Trans. Appl. **7**(2),. 2083 (1997).
243. Edelman H. S., Parrell J. A., Larbalestier D. C., J. Appl. Phys. **81**, 2296 (1997)
244. Eden M. *Proc.* 4th *Berkeley Symp. on Mathematics, Statistics and Probability* **4**, 223 (1961).
245. Eichelkraut H., Abbruzzese G., Lücke K., Acta Metall. **36**, 55 (1988).
246. Ekin J. W., Bray S. L., Cheggour N., et al., IEEE Trans. Appl. Supercond. **11**(1), 3389 (2001).
247. Ekin J. W., Finnemore D. K., Li Q., et al., Appl. Phys. Lett. **61**, 858 (1992).
248. Ekin J. W., Hart H. R., Jr., Gaddipati A.R., J. Appl. Phys. **68**, 2285 (1990).
249. Emery V., Kivelson S., Nature. **374**, 434 (1995).
250. Emery V., Kivelson S., Zachar O., Phys. Rev. B **56**, 6120 (1997).
251. Endo A., Chauhan S. H., Egi T., Shiohara Y., J. Mater. Res. **11**, 795 (1996).
252. Endo A., Chauhan S. H., Nakamura Y., Shiohara Y., J. Mater. Res. **11**, 1114 (1996).
253. Endo K., Yamasaki H., Milsawa S., et al., Nature **355**, 327 (1992).
254. Eom C. B., Lee M. K., Choi J. H., et al., Nature **411**, 558 (2001).
255. Eom C. B., Marshall A. F., Suzuki Y., et al., Nature **353**, 544 (1991)
256. Eriksen M., Bech J. I., Bay N., et al., *Proc.* 4th *Europ. Conf on Applied Superconductivity* (*EUCAS'* 99), paper 12–75 (Barselona, 1999).
257. Eriksen M., Bech J. I., Seifi B., Bay N., IEEE Trans. Appl. Supercond. **11**(1), 3756 (2001).
258. Eshelby J. D., Proc. Roy. Soc. **A241**, 376 (1957)
259. Eshelby J. D. *Continuum Theory of Dislocation* (Inostrannaya Literatura, Moscow, 1963).
260. Evans A. G., Acta Metall. **26**, 1845 (1978).
261. Evans A. G., Burlingame N., Drory M., Kriven W. M., Acta Metall. **29**, 447 (1981).
262. Evans A. G., Cannon R. M., Acta Metall. **34**, 761 (1986).
263. Evans A. G., Dalgleish B. J., He M., Hutchinson J. W., Acta Metall. **37**, 3249 (1989).
264. Evans A. G., Fu Y., Acta Metall. **33**, 1525 (1985).
265. Evans A. G., Heuer A. H., J. Am. Ceram. Soc. **63**, 241 (1980).
266. Evans A. G., Hutchinson J. W., Acta Metall. **37**, 909 (1989).
267. Evans A. G., Marshall D. B., Acta Metall. 37, 2567 (1989).
268. Faber K. T., Evans A. G., Acta Metall. **31**, 565 (1983).
269. Fabrega J., Fontcuberta J., Serquis A., Caneiro A., Phys. C **356**, 254 (2001).
270. Favrot D., Déchamps M., J. Mater. Res. **6**, 2256 (1991).
271. Feldmann D. M., Reeves J. L., Polyanskii A. A., et al., Appl. Phys. Lett. **77**, 2906 (2000).
272. Feng Y., Hautanen K. E., High Y. E., et al., Phys. C **192**. 293 (1992).
273. Feng Y., Larbalestier D. C., Interface Sci. **1**, 401 (1994).
274. Ferdeghini, C., Ferrando V., Tarantini C., et al., cond-mat/0411404 (2004).
275. Ferreira P. J., Liu H. B., Vander Sande J. B., IEEE Trans. Appl. Supercond. **9**(2), 2231 (1999).

276. Fesenko E. G., Dantsiger A. Ya., Razumovskaya O. N. *New Piezoceramic Materials* (Rostov State Univ. Press, Rostov-on-Don, 1983).
277. Fetisov A. V., Fotiev A. A., Supercond. Phys. Chem. Tech. **5**, 1071 (1992).
278. Fetisov V. B., Fetisov A. V., Fotiev A. A., Supercond. Phys. Chem. Tech. **3**, 2627 (1990).
279. Field M. B., Cai X. Y., Babcock S. E., Larbalestier D. C., IEEE Trans. Appl. Supercond. **3**(2), 1479 (1993).
280. Field M. B., Pashitski A., Polyanskii A., et al., IEEE Trans. Appl. Supercond. **5**(2), 1631 (1995).
281. Figueras J., Puig T., Obradors X., et al., Nature Phys. **2**, 402 (2006).
282. Fischer B., Kautz S., Leghissa M., et al., IEEE Trans. Appl. Supercond. **9**(2), 2480 (1999).
283. Fischer K. H., Phys. C **178**, 161 (1991).
284. Fischer K. H., Supercond. Rev. **1**, 153 (1995).
285. Fjellvåg H., Karen P., Kjekshus A., et al., Acta Chem. Scand. A. **42**, 178 (1988).
286. Flemings M. C. *Solidification Processing* (McGraw-Hill, New York, 1974).
287. Fleshler S., Fee M., Spreafico S., Malozemoff A. P., *Adv. Supercond. XII*, 12th *Int. Symp. on Superconductivity* (*ISS'*1999), p. 625 (October 17–19, Morioka, Japan, 1999).
288. Fleshler S., Kwok W. - K., Welp U., et al., Phys. Rev. B **47**, 14448 (1993).
289. Flükiger R., Giannini E., Lomello-Tafin M., et al., IEEE Trans. Appl. Supercond. **11**(1), 3393 (2001).
290. Flükiger R., Grasso G., Grivel J. C., et al., Supercond. Sci. Technol. **10**, A68 (1997)
291. Fol'mer M. *Initiation Kinetics of New Phase* (Nauka, Moscow, 1986).
292. Fossheim K., Tuset S. D., Ebbesen T. W., et al., Phys. C **248**, 195 (1995)
293. Friesen M., Gurevich A., *Ext. Abstr. Int. Workshop on Critical Current* (*IWCC*-99), Madison, Wisconsin, USA, July 7–10, 1999, p. 102 (1999).
294. Fu Y., Evans A. G., Acta Metall. **30**, 1619 (1982).
295. Fu Y., Evans A. G., Acta Metall. **33**, 1515 (1985).
296. Fuhrer M. S., Cherrey K., Zettl A., Cohen M. L., Phys. Rev. Lett. **83** 404 (1999).
297. Fujimoto H., Murakami M., Gotoh S., et al., Adv. Supercond. **2**, 285 (1990).
298. Fujimoto H., Murakami M., Koshizuka N., Phys. C **203**, 103 (1992).
299. Fujita S., Godoy S. *Theory of High Temperature Superconductivity* (Kluwer Academic, Dordrecht, 2001).
300. Gabovich A. M., Voytenko A. I., Fiz. Nizk. Temp. **26**, 419 (2000).
301. Gandhi C., Ashby M. F., Acta Metall. **27**, 1565 (1979).
302. Gandin Ch. - A., Rappaz M., West D., Adams B. L., Metallurg. Mater. Trans. **26A**, 1543 (1995).
303. Gao F., Wang T., // J. Mater. Sci. Lett. **9**, 1409 (1990).
304. Gao W., Chen J., Yang C. O., et al., Phys. C **193**, 255 (1992).
305. Gao Y., Bai G., Lam D. J., Merkle K. L., Phys. C **173**, 487 (1991).
306. Gao Y., Merkle K. L., Bai G., et al., Phys. C **174**, 1 (1991).
307. Gao Y., Merkle K. L., Zhang C., et al., J. Mater. Res. **5**, 1363 (1990)
308. Garcia-Munoz J. L., Suaadi M., Foncuberta J., et al., Phys. C **268**, 173 (1996).
309. Gasparov V. A., Sidorov N. S., Zver'kova I. I., Kulakov M. P., Pis'ma JETP **73**, 532 (2001).

310. Gautier-Picard P. *PhD Thesis* (Grenoble, 1998).
311. Geerken B. M., Griessen R., Huisman L. M., Walker E., Phys. Rev. B **26**, 1637 (1982).
312. Giller D., Shaulov A., Prozorov R., et al., Phys. Rev. Lett. **79**, 2542 (1997).
313. Ginzburg V. I., Landau L. D., Zh. Eksp. Teor. Fiz. **20**, 1064 (1950)
314. Gire F., Robbes D., Gonzales C., et al., IEEE Trans. Appl. Supercoud. **7**(2), 3200 (1997).
315. Goldacker W., Quilitz M., Obst B., Eckelmann H., Phys. C **310**, 182 (1998).
316. Golopan R., Roy T., Rajasekharan T. et al., Phys. C **244**, 106 (1995).
317. Goodilin E. A., Kvartalov D. B., Oleynikov N. N., et al., Phys. **235–240**, 449 (1994).
318. Goodilin E. A., Oleynikov N. N., Supercond. Res. Develop. **5/6**, 81 (1995).
319. Goodman S., Nature **414,** 831 (2001).
320. Goretta K., Delaney W., Routbort J., et al., Supercond. Sci. Technol. **9**, 422 (1996)
321. Goretta K. C., Jiang M., Kupperman D. S., et al., IEEE Trans. Appl. Supercond. **7**(2), 1307 (1997).
322. Goretta K. C., Kullberg M. L., Bär D., et al., Supercond. Sci. Technol. **4**, 544 (1991).
323. Goretta K. C., Kupperman D. S., Majumdar S., et al., Supercond. Sci. Technol. **11**, 1409 (1998).
324. Goretta K. C., Loomans M. E., Martin L. J., et al., Supercond. Sci. Technol. **6**, 282 (1993).
325. Gorter C. J., Physica **23**, 45 (1957).
326. Gorter C. J., Casimir H. B. G., Physica **1**, 306 (1934).
327. Gor'kov L. P., Zh. Eksp. Teor. Fiz. **36**, 1918 (1959) [Sov. Phys. JETP **9**, 1364 (1959)].
328. Gor'kov L. P., Zh. Eksp. Teor. Fiz. **37**, 833 (1959).
329. Gor'kov L. P., Zh. Eksp. Teor. Fiz. **37**, 1407 (1959).
330. Gor'kov L. P., Sokol A. V., Pis'ma Zh. Eksp. Teor. Fiz. **46**, 420 (1987).
331. Goto T., Kuji T., Jiang Y. - S., et al., IEEE Trans. Appl. Supercond. **9**(2), 1653 (1999)
332. Gould H., Tobochnik J. *An Introduction to Computer Simulation Methods. Applications to Physical Systems*, part 2 (Addison-Wesley, New York, 1988).
333. Govor L. V., Dobrego V. P., Golubev, V. N., et al., Supercond. Phys. Chem. Tech. **5**, 2142 (1992).
334. Goyal A., Alexander K. B., Kroeger D. M., et al., Phys. C **210**, 197 (1993).
335. Goyal A., Funkenbusch P. D., Kroeger D. M., Burns S., Phys. C **182**, 203 (1991).
336. Goyal A., Norton D. P., Budai J. D., et al., Appl. Phys. Lett. **69**, 1975 (1996).
337. Goyal A., Norton D. P., Kroeger D. M., et al., J. Mater. Res. **12**, 2924 (1997).
338. Goyal A., Oliver W. C., Funkenbusch P. D., et al., Phys. C **183**, 221 (1991).
339. Goyal A., Specht E. D., Kroeger D. M., Manson T. A., Appl. Phys. Lett. **68**, 711 (1996).
340. Grabko D. Z., Boyarskaya Yu. S., Zhitaru R. P., et al., Supercond. Phys. Chem. Tech. **2**, 67 (1989).
341. Grasso G., Jeremie A., Flükiger R., Supercond. Sci. Technol. **8**, 827 (1995).
342. Gregoryanz E., Struzhkin V. V., Hemley R. J., et al., cond-mat/0108267 (2001).

343. Grekov A. A., Egorov N. Ya, Karpinsky D. N. *Proc. Int. Sympos. on Strength of Materials and Construction Elements at Sound and Ultrasound Frequencies of Loading*, p. 187 (Naukova Dumka, Kiev, 1987).
344. Gridnev S. A., Ivanov O. N., Supercond. Phys. Chem. Tech. **5**, 1143 (1992).
345. Gridnev S. A., Ivanov O. N., Dybova O. V., Supercond. Phys. Chem. Tech. **3**, 1449 (1990).
346. Gridnev S. A., Ivanov O. N., Dybova O. V. *Collect. Abstr. Int. Conf. on Transparent Ferroelectric. Ceram.* (TFC'91), p. 77 (Riga, October 2–6, 1991).
347. Griessen R., Phys. Rev. Lett. **60**, 1674 (1990).
348. Griffith A. A., Phil. Trans., R. Soc. London A. **221**, 163 (1920).
349. Griffith M. L., Huffman R. T., Halloran J. W., J. Mater. Res. **9**, 1633 (1994).
350. Grineva L. D., Zatsarinny V. P., Aleshin V. A., et al., Strength Mater. **4**, 34 (1993).
351. Grivel J. - C., Jeremie A., Flükiger R., Supercond. Sci. Technol. **8**, 41 (1995).
352. Grivel J. - C., Jeremie A., Hansel B., Flükiger R., Supercond. Sci. Tech. **6**, 730 (1993).
353. Gross R. In: *High Temperature Superconducting Systems*, S. L. Shindé, and D. Rudman (eds.), p. 176 (Springer-Verlag, New York, 1992).
354. Gross R. In: *Interfaces in High-T_c Superconducting Systems*, S. L. Shinde, and D. A. Rudman (eds.), p. 174 (Springer-Verlag, New York, 1994).
355. Gross R., Alff L. Beck A., et al., IEEE Trans. Appl. Supercond. **7**(2), 2929 (1997).
356. Gubser D. U., Phys. C **341–348**, 2525 (2000).
357. Gugenberger F., Meingast C., Roth G., et al., Phys. Rev. B 49, 13137 (1994).
358. Guo Y. C., Liu H. K., Dou S. X., Phys. C **235–240**, 1231 (1994).
359. Guo Y. C., Liu H. K., Dou S. X., et al. Supercond. Sci. Technol. **11**, 1053 (1998).
360. Gurevich A. Fiz. Tverd. Tela **30**, 1384 (1988).
361. Gurevich A., Phys. Rev. B **42**, R4857 (1990).
362. Gurevich A., Phys. Rev. B **46**, 3187 (1992).
363. Gurevich A., Phys. Rev. B **46**, 3638 (1992).
364. Gurevich A. *Ext. Abstr. Int. Workshop on Critical Current* (*IWCC*-99), p. 2 (Madison, Wisconsin, USA, July 7–10, 1999).
365. Gurevich A., Benkraouda M., Clem J. R., Phys. Rev. B **54**, 13196 (1996).
366. Gurevich A., Cooley L. D., Phys. Rev. B **50**, 13563 (1994).
367. Gurevich A., Friesen M., Phys. Rev. B **62**, 4004 (2000).
368. Gurevich A., Friesen M., Vinokur V., Phys. C **341–348**, 1249 (2000).
369. Gurevich A., Küpfer H., Keller C., Europhys. Lett. **15**, 789 (1991).
370. Gurevich A., McDonald J. Phys. Rev. Lett. **81**, 2546 (1998)
371. Gurevich A., Mints R. G., Rakhmanov A. L. *The Physics of Composite Superconductors* (Begell House, New York, 1997).
372. Gurevich A., Pashitskii E. A., Phys. Rev. B **56**, 6213 (1997).
373. Gurevich A., Pashitskii E. A., Phys. Rev. B **57**, 13878 (1998)
374. Gurevich A., Vinokur V. M., Phys. Rev. Lett. **83**, 3037 (1999).
375. Gurson A. L., Yuan D. W. In: *Net Shape Processing of Powder Materials*, **216** (ASME-AMD, Metals Park, OH, 1995).
376. Gyorgy E. M., Van Dover R. B., Jackson K. A., et al., Appl. Phys. Lett. **55**, 283 (1989)
377. Hamasumi M. *Nonferrous Metals and Alloys*, 3rd edn., p. 210 (Uchidarouho kaku-shinsya, Tokyo. 1984).

378. Hamdan N. M., Sastry P. V. P. S. S., Schwartz J., Phys. C **341–348**, 513 (2000).
379. Hamdan N. M., Sastry P. V. P. S. S., Schwartz J., IEEE Trans. Appl. Supercond. **12**(1), 1132 (2002).
380. Hamdan N. M., Ziq Kh. A., Al-Harthi A. S., Phys. C **314**, 125 (1999).
381. Hamdan N. M., Ziq Kh. A., Shirokoff J., Supercond. Sci. Technol. **7**, 118 (1994).
382. Hammeri G., Bielefeldt H., Goetz B., et al., IEEE Trans. Appl. Supercond. **11**(1), 2830 (2001).
383. Hammeri G., Schmehl A., Schulz R. R., et al., Nature **407**, 162 (2000).
384. Han Z., Freltoft T. Appl. Supercond. **2**, 201 (1994).
385. Han Z., Skov-Hansen P., Freltoft T., Supercond. Sci. Technol. **10**, 371 (1997).
386. Hanawa M., Muraoka Y., Tayama T., Phys. Rev. Lett. **87**, 187001 (2001).
387. Handstein A., Hinz D., Fuchs G., et al., cond-mat/0103408 (2001).
388. Hannay N. B., Geballe T. H., Matthias B. T., et al., Phys. Rev. Lett. **14**, 225 (1965).
389. Harada K., Matsuda T., Bonevich J., et al., Nature **360**, 51 (1992).
390. Harada K., Matsuda T., Kasai H., et al.. Phys. Rev. Lett. **71**, 3371 (1993).
391. Hari Babu N., Kambara M., McCrone J., et al., IEEE Trans. Appl. Supercond. **11**(1), 2838 (2001).
392. Hart H. R., Jr, Luborsky F. E., Arendt R. H., et al., IEEE Trans. Magn. **27**, 1375 (1991).
393. Hartley R. *Linear and Nonlinear Programming* (Wiley, New York, 1985).
394. Harvey D. P., Jolles M. I., Metall. Trans. **21A**, 1719 (1990).
395. Hasegawa K., Yoshica N., Fujinao K., et al., *Proc.* 16th *ICEC/ICMC*, p. 1413. (Elsevier Science, Amsterdam, 1997).
396. Hasegawa T., Hikichi Y., Koizumi T., et al., IEEE Trans. Appl. Supercond. **7**(2), 1703 (1997).
397. Hasegawa T., Kitamura T., Kobayashi H., et al., Phys. C **190**, 81 (1991).
398. Haslinger R., Joynt R., Phys. Rev. B **61**, 4206 (2000).
399. Haugan T., Barnes P. N., Wheeler R., et al., Nature **430**, 867 (2004).
400. Hazen R. M., Finger L. W., Angel R. J., et al., Phys. Rev. Lett. **60**, 1657 (1988).
401. He T., Huang Q., Ramirez A. P., et al., cond-mat/0103296 (2001).
402. Heckel R. W., Trans. Metal. Soc. AIME **221**, 671 (1961).
403. Hedderich R., Schuster T., Kuhn H., et al., Appl. Phys. Lett. **66**, 3215 (1995).
404. Heffner R. H., Mac Laughlin D. E., Sonier J. E., et al., cond-mat/0102137 (2001).
405. Heine K., Tenbrink J., Thöner M., Appl. Phys. Lett. **55**, 2441 (1989)
406. Heinig N. F., Redwing R. D., Tsu I. - F., et al., Appl. Phys. Lett. **69**, 577 (1996).
407. Hellstrom E. E., J. Miner. Met. Mater. Soc. **44**, 44 (1992).
408. Hellstrom E. E., MRS Bull. **17**, 45 (1992).
409. Hellstrom E. E. In: *High-Temperature Superconducting Materials Science and Engineering. New Concepts and Technology*, D. Shi (ed.), p. 383 (Elsevier Science, London, 1995).
410. Hellstrom E. E., Ray R.D. *II. HTS Materials, Bulk Processing and Bulk Applications*, p.354 (World Scientific, Singapore, 1992).
411. Hensel B., Grivel J. - C., Jeremie A., et al., Phys. C **205**, 329 (1993).

412. Hess H. F., Robinson R. B., Dynes R. C., et al., Phys. Rev. Lett. **62**, 214 (1989).
413. Hess H. F., Robinson R. B., Waszczak J. V., Phys. Rev. Lett. **64**, 2711 (1990).
414. Heuer A. H., J. Am. Ceram. Soc. **62**, 317 (1979).
415. Hibino A., Wantabe R., J. Mat. Sci.: Mater. Electronics **1**, 13 (1990).
416. *High Temperature Superconductivity*, V. L. Ginzburg, D. A. Kirzhnitz (eds.) (Consultants Bureau, New York, 1982).
417. High-T_c Update. **13**(9), May 1 (1999).
418. Hilgenkamp H., Mannhart J., Supercond. Sci. Technol. **10**, 880 (1997).
419. Hilgenkamp H., Mannhart J., Appl. Phys. Lett. **73**, 265 (1998).
420. Hilgenkamp H., Mannhart J., Mayer B., Phys. Rev. B **53**, 14586 (1996).
421. Hillert M., Acta Metall. **13**, 227 (1965).
422. Hillert M., In: *Solidification and Casting of Metals*, p. 81 (The Metals Society, New York, 1977).
423. Hills D. A., Kelly P. A., Dai D. N., Korsunsky A. M. *Solution of Crack Problems. The Distributed Dislocation Technique* (Kluwer Academic Publish., Dordrecht, 1996).
424. Hinrichsen E., Roux S., Hansen A., Phys. C **167**, 433 (1990).
425. Hirayama T., Ikuhara Y., Nakamura M., et al., J. Mater. Res. **12**, 293 (1997).
426. Hirth J. P., Carnahan B., Acta Metall. **26**, 1795 (1978)
427. Hirth J. P., Lothe J. *Theory of Dislocations* (Wiley & Sons, New York, 1982).
428. Hisao Banno, Am. Ceram. Soc. Bull. **66**, 1332 (1987).
429. Hjashi N., Diko P., Nagashima K., Murakami M., J. Mater. Sci. Eng. B. **53**, 104 (1998).
430. Hogg M. J., Kahlmann F., Tarte E. J., et al., Appl. Phys. Lett. **78**, 1433 (2000).
431. Hong G. - W., Kim K. - B., Kuk I. - H., et al., IEEE Trans. Appl. Supercond. **7**(2), 1945 (1997).
432. Holstein W. L., Wilker C., Laubacher D. B., et al., J. Appl. Phys. **74**, 1426 (1993).
433. Holtz R. L., IEEE Trans. Appl. Supercond. **11**(1), 3238 (2001).
434. Holtz R. L., Fleshler S., Gubser D. U., Adv. Eng. Mater. **3**(3), 131. (2001).
435. Homes C. C., Dordevic S. V., Strongin M., et al., Nature **430**, 539 (2004).
436. Honma T., Hor P. H.; cond-mat/0602452 (2006).
437. Hopzapfel B., et al., IEEE Trans. Appl. Supercond. **11**(1), 3872 (2001).
438. Hor P. H., Meng R. L., Wang Y. Q., et al., Phys. Rev. Lett. **58**, 1891 (1987).
439. Horn P. M., Keane D. T., Held G. A., et al., Phys. Rev. Lett. **59**, 2772 (1987).
440. Horvat J., Bhasale R., Guo Y. C., et al., Supercond. Sci. Technol. **10**, 409 (1997).
441. Hosford W. F., Caddel R. M. *Metal Forming* (PTR Prentice Hall, Englewood Cliffs, NJ, 1993).
442. Hsueh C. H., Evans A. G., Acta Metall. **29**, 1907 (1981).
443. Hsueh C. H., Evans A. G., Acta Metall. **31**, 189 (1983).
444. Hsueh C. H., Evans A. G., J. Am. Ceram. Soc. **68**, 120 (1985).
445. Hsueh C. H., Evans A. G., Coble R. L., Acta Metall. **30**, 1269 (1982).
446. Hu M. S., Evans A. G., Acta Metall. 37, 917 (1989).
447. Hu M. S., Thouless M. D., Evans A.G. Acta Metall. **36**, 1301 (1988).
448. Hu Q. Y., Liu H. K., Dou S. X., Apperley M., IEEE Trans. on Appl. Supercond. **7**(2), 1849 (1997).

449. Huang Y. B., Grasso G., Marti F., et al., Inst. Phys. Conf. Ser. **158**, 1385 (1997).
450. Huang Y. B., Marti F., Witz G., et al., IEEE Trans. Appl. Supercond. **9**(2), 2722 (1999).
451. Huang Y. K., ten Haken B., ten Kate H. H. J., IEEE Trans. Appl. Supercond. **9**(2) 2702 (1999).
452. Huebener R. P. *Magnetic Flux Structures in Superconductors* (Springer-Verlag, Berlin, Heidelberg, 2001).
453. Hulbert S. F., J. Br. Ceram. Soc. **6**, 11 (1969).
454. Hunt B. D., Foote M. C., Pike W. T., et al., Phys. C **230**, 141 (1994).
455. Hunt B. D., Forrester M. G., Tolvacchio J., et al., IEEE Trans. Appl. Supercond. **7**(2), 2936 (1997).
456. Hunt M., Jordan M., In: *Advances in Solid State Physics*, B. Kramer (ed.), **39**, 351 (Vieweg, Braunschweig/Wiesbaden, 1999).
457. Hušek I., Kováč P., Kopera L., Supercond. Sci. Technol. **9**, 1066 (1996).
458. Hušek I., Kováč P., Pachla W., Supercond. Sci. Technol. **8**, 617 (1995).
459. Hussey N. E., Abdel-Jawad M., Carrington A., et al., Nature **425**, 814 (2003).
460. Hutchinson J. W., J. Mech. Phys. Solids **16**, 337 (1968).
461. Hwang B.B., Kobayashi S., Int. J. Mach. Tools Manufact. **30**, 309 (1990).
462. Iijima Y., Kikuchi A., Inoue K., Cryogenics **40**, 345 (2000).
463. Iijima Y., Kosuge M., Takeuchi T., Inoue K., Adv. Cryog. Eng. Mater. **40**, 899 (1994).
464. Iijima Y., Onabe K., Futaki N., et al., J. Appl. Phys. **74**, 1905 (1993).
465. Ikuta H., Mase A., Hosokawa T., et al., Adv. Supercond. **11**, 657 (1999).
466. Ilyushechkin A. Y., Yamashita T., Williams B., Mackinnon I. D. R., Supercond. Sci. Technol. **12**, 142 (1999).
467. Imayev M. F., Daminov V. A., Popov V. A., Kaibyshev O. A., Inorg. Mater. **41**(5), 1, (2005).
468. Imayev M. F., Daminov V. A., Popov V. A., Kaibyshev O. A., Phys. C **422**, 27 (2005).
469. Indenbom M. I., Kolesnikov N. N., Kulakov M. P., et al., Phys. C **166**, 486 (1990).
470. Indenbom M. V., Nikitenko V. I., Polyanskii A. A., Vlasko-Vlasov V. K., Cryogenics **30**, 747 (1990).
471. Ioffe L. B., Geshkenbein V. B., Feigel'man M. V., et al., Nature **398**, 679 (1999).
472. Ionescu M., Dou S. X., Babic E., et al., *Proc. Int. Workshop on Superconductivity*, Maui, Hawaii, USA, June 18–21, p. 44 (1995).
473. Irwin G. R., J. Appl. Mech. **24**, 361 (1957).
474. Ivanov Z. G., Nilsson P. Å., Winkler S., et al., Appl. Phys. Lett. **59**, 3030 (1991).
475. Izumi T., Nakamura Y., Shiohara Y., J. Mater. Res. **7**, 1621 (1992).
476. Izumi T., Nakamura Y., Shiohara Y., J. Cryst. Growth. **128**, 757 (1993).
477. Izumi T., Nakamura Y., Sung T. H., Shiohara Y., *Proc.* 4th *Int. Symp. on Superconductivity*, T. Ishiguro (ed.) (Tokyo, October 14–17, 1991).
478. Jackson K. A. *Solidification*, p. 121 (American Society for Metals, New York, 1971).
479. Jee Y.A., Hong G. - W., Kim C. - J., Sung T. H., IEEE Trans. Appl. Supercond. **9**(2) 2093 (1999).

480. Jee Y.A., Hong G. - W., Kim C. - J., Sung T. H., IEEE Trans. Appl. Supercond. **9**(2), 2097 (1999).
481. Jee Y, A., Kang S. - J. L., Jung H. - S., J. Mater. Res. **13**, 583 (1998).
482. Jensen H. M., Acta Metall. **38**, 2637 (1990).
483. Jensen P. A., Barnes J. W. *Network Flow Programming* (Wiley, New York, 1980).
484. Jérome D., Mazaud A., Ribault M., Bechgaard K., J. Phys. Lett. (Paris). **41**, L195 (1980).
485. Jiang J., Shields T. C., Abell J. S., et al., IEEE Trans. Appl. Supercond. **9**(2), 1812 (1999).
486. Jiang Y. - S., Kobayashi T., Goto T., IEEE Trans. Appl. Supercond. **9**(2), 1657 (1999).
487. Jimenez-Melendo M., Dominguez-Rodriguez A., Routbort J. L., Scr. Met. Mater. **32**, 621 (1995).
488. Jin S., Kammlott G. W., Tiefel T. H., et al., Phys. C **181**, 57 (1991).
489. Jin S., Kammlott G. W., Tiefel T. H., et al., Phys. C **198**, 333 (1992).
490. Jin S., Mavori H., Bower C., van Dover R. B., Nature **411**, 563 (2001).
491. Jin S., Tiefel T. H., Sherwood R. C., et al., Phys. Rev. B **37**, 7850 (1996).
492. Johansen T. H., Phys. Rev. B **60**, 9690 (1999).
493. John D. H. St., Acta Metall. Mater. **38**, 631 (1990).
494. John D. H. St., Hogan L. M., Acta Metall. **25**, 77 (1977).
495. Johnson S. M., Gusman M. I., Rowcliffe D. J., Adv.Ceram. Mater. **2**, 337 (1987).
496. Jones M., Marsh R., J. Am. Chem. Soc. **76**, 1434 (1954).
497. Jordan M., Hunt M., Adrian H., Nature **398**, 47 (1999).
498. Jorgensen J. D., Beno M. A., Hinks D. G., et al., Phys. Rev. B **36**, 3608 (1987).
499. Jorgensen J. D., Veal B. W., Kwok W. K., et al., Phys. Rev. B **36**, 5731 (1987).
500. Josephson B. D., Phys. Lett. **1**, 251 (1962).
501. Jover D. T., Wijngaarden R. J., Griessen R., et al., Phys. Rev. B **54**, 10175 (1996).
502. Jullien R., Botet R., J. Phys. A: Math. Gen. **18**, 2279 (1985).
503. Jullien R., Botet R., Phys. Rev. Lett. **54**, 2055 (1985).
504. Jung C. U., Kim J. Y., Kim M. - S., et al., Phys. C **366**, 299 (2002).
505. Kabius B., Seo J. W., Amrein T., et al., Phys. C **231**, 123 (1994).
506. Kaczorowski D., Zaleski A. J., Zoga O. J., Klamut J., cond-mat/0103571 (2001).
507. Kameda J., Acta Metall. **34**, 867 (1986).
508. Kameda J., Acta Metall. **34**, 883 (1986).
509. Kammlott G. W., Tiefel T.H., Jin S., Appl. Phys. Lett. **56**, 2459 (1990).
510. Kang H. J., Dai P., Lynn J. W., et al., Nature **423**, 522 (2003).
511. Kaplitsky M. A. *Energetic Method in Dynamical Theory of Brittle Fracture*, Preprint 511-B96 (VINITI, Moscow, 1996).
512. Karpinsky D. N., Kramarov C. O., Orlov A. N., Strength Mater. **1**, 97 (1981).
513. Karpinsky D. N., Parinov I. A., Glass Ceram. **3**, 27 (1991).
514. Karpinsky D. N., Parinov I. A., *Proc. Jt. FEFG/ICF Int. Conf. Fract. of Eng. Mater. and Struct.* (Singapore, 1991), London, p. 327 (1991).
515. Karpinsky D. N., Parinov I. A., Strength Mater. **7**, 34 (1991).

516. Karpinsky D. N., Parinov I. A., J. Appl. Mech. Techn. Phys. **1**, 150 (1992).
517. Karpinsky D. N., Parinov I. A., Ferroelectric Lett. **19**, 151 (1995).
518. Karpinsky D. N., Parinov I. A., Parinova L.V., Ferroelectrics (1992). **133**, 265
519. Karuna M., Parrell J. A., Larbalestier D. C., IEEE Trans. Appl. Supercond. **5**(2). 1279 (1995).
520. Kase J., Morimoto T., Togano K., et al., IEEE Trans. Mag. **27**, 1254 (1991).
521. Kautek W., Roas B., Schultz L. H., Thin Solid Films **191**, 317 (1990).
522. Kawaji H., Horie H., Yamanaka S., Ishikawa M., Phys. Rev. Lett. **74**, 1427 (1995).
523. Kawasaki M., Sarnelli E., Chaudhari P., et al., Appl. Phys. Lett. **62**, 417 (1993).
524. Kawashima J., Yamada Y., Hirabayashi I., Phys. C **306**, 114 (1998).
525. Keil S., Straub R., Gerber R., et al., IEEE Trans. Appl. Supercond. **9**(2), 2961 (1999).
526. Keizer R. J. de Visser A., Menovsky A. A., et al., cond-mat/9903328 (1999).
527. Kendall M.G., Stuart A. *The Advanced Theory of Statistics*, **1–3** (Griffin Publ., London, 1983).
528. Kes P. H., Phys. C **185–189**, 288 (1991).
529. Khasanov R., Eshchenko D. G., Karpinski J., et al., Phys. Rev. Lett. **93**, 157004 (2004).
530. Kim C.-J., Hong G.-W., Supercond. Sci. Technol. **12**, R27 (1999).
531. Kim C.-J., Kim H.-J., Joo J.-H., Hong G.-W., Phys. C **354**, 384 (2001).
532. Kim C.-J., Kim K.-B., Hong G.-W., Lee H.-Y., J. Mater. Res. **10**, 1605 (1995).
533. Kim C.-J., Kim K.-B., Kuk I.-H., Hong G.-W., Phys. C **281**, 244 (1997).
534. Kim C.-J., Kim K.-B., Kuk I.-H., Hong G.-W., J. Mater. Res. **13**, 269 (1998).
535. Kim C.-J., Kim K.-B., Park H.-W., et al., Supercond. Sci. Technol. **9**, 76 (1996).
536. Kim C.-J., Kim K.-B., Won D.-Y., J. Mater. Res. **8**, 699 (1993).
537. Kim C.-J., Kim K.-B., Won D.-Y., et al., J. Mater. Res. **9**, 1952 (1994).
538. Kim C.-J., Kim K.-B., Won D.-Y., Hong G.-W., Phys. C **228**, 351 (1994).
539. Kim C.-J., Kuk I.-H., Hong G.-W., et al., Mater. Res. Lett. **34**, 392 (1998).
540. Kim C.-J., Lai S. H., McGinn P. J., J. Mater. Lett. **19**, 185 (1994).
541. Kim C.-J., Lee Y.-S., Park H.-W., et al., Phys. C **276**, 101 (1997).
542. Kim C.-J., Park H.–W., Kim K.-B., et al., Mater. Lett. **29**, 7 (1996).
543. Kim H.-J., Kang W. N., Choi E.–M., et al., cond-mat/0105363 (2001).
544. Kim M.-S., Jung C. U., Kim J. Y., et al., cond-mat/0102420 (2001).
545. Kim Y. B., Hempstead C. F., Strnad A. R., Phys. Rev. Lett. **9**, 306 (1962).
546. Kim Y. B., Hempstead C. F., Strnad A. R., Phys. Rev. **129**, 528 (1963).
547. King A. H., Singh A., Wang J. Y., Interface Sci. **4**, 347 (1993).
548. Kirtley J. R., Ketchen M. B., Stawiasz K. G., et al. Appl. Phys. Lett. **66**, 1138 (1995).
549. Kiss T., Oda K., Nishimura S., et al., Phys. C **357–360**, 1123 (2001).
550. Kiss T., van Eck H., ten Haken B., ten Kate H. H. J., IEEE Trans. Appl. Supercond. **11**(1), 3888 (2001).
551. Kistenmacher T. J., J. Appl. Phys. **64**, 5067 (1988).
552. Kitaguchi H., Itoh K., Kumakura H., et al., Phys. C. **357–360**, 1193 (2001).
553. Kivelson S., cond-mat/0109151 (2001).
554. Kleinsasser A. W., Delin K. A., IEEE Trans. Appl. Supercoud. **7**(2), 2964 (1997).

555. Klie R. F., Buban J. P., Varela M., et al., Nature **435**, 475 (2005).
556. Knudsen H. A., Hansen A., Phys. Rev. B **61**, 11336 (2000).
557. Koblischka M. R., Supercond. Sci. Technol. **9**, 271 (1996).
558. Koblischka M. R., Johansen T. H., Bratsberg H., Supercond. Sci. Technol. **10**, 693 (1997).
559. Koblischka M. R., Wijngaarden R. J., Supercond. Sci. Technol. **8**, 199 (1995).
560. Koblischka M. R., Wijngaarden R. J., de Groot D. G., et al., Phys. C **249**, 339 (1995).
561. Koch R. H., Foglietti V. Gallagher W. J., et al., Phys. Rev. Lett. **63**, 1511 (1989).
562. Kofman A. *Introduction in Applied Combinatoric* (Nauka, Moscow, 1975).
563. Komori K., Kawagishi K., Takano Y., et al. Appl. Phys. Lett. **81**, 1047 (2002).
564. Kompantseva V. G., Rusanov K. V. Supercond. Res. Develop. **3, 4,** 41 (1994).
565. Kopera L., Kováč P., Hušek I., Supercond. Sci. Technol. **11**,433 (1998).
566. Korn G. A., Korn T. M. Mathematical Handbook for Scientists and Engineers. Definitions, Theorems and Formulas for Reference and Review. (McGraw Hill, New York, 1961).
567. Korshunov A. N., Shevtshionok A. A., Ovtshinnikov V. I., et al., *Collect. Abstr. Int. Conf. Transparent Ferroelectric. Ceram.* (TFC'91), p. 79 (Riga, October 2–6, 1991).
568. Korzekwa D. A., Bingert J. F., Rodtburg E. J., Miles P., Appl. Supercond. **2**, 261 (1994).
569. Kotegawa H., Tokunaga Y., Ishida K., et al., Phys. Rev. B **64**, 064515 (2001).
570. Kotzler J., von Sawilski L., Casalbuoni S., Phys. Rev. Lett. **92**, 067005 (2004).
571. Kourtakis K., Robbins M., Gallagher P.K., J. Sol. State Chem. **84**, 88 (1990).
572. Kovalev S. P., Kuz'menko V. A., Pisarenko G. G. *Computer Simulation of Microstructure Processes in Ceramic Materials* (Preprint AN UkrSSR, IPS, Kiev, 1983).
573. Kovávc P., Hušek I., Cesnak L., Supercond. Sci. Technol. **7**, 583 (1994).
574. Kováč P., Hušek I., Gömöry F., et al., Supercond. Sci. Technol. **13**, 378 (2000).
575. Kováč P., Richens P. E., Bukva P., et al., Supercond. Sci. Technol. **12**, 168 (1999).
576. Koyama T. K., Takezawa N., Tachiki M., Phys. C **172**, 501 (1992).
577. Kozinkina Y. A., Parinov I. A., Adv. Cryog. Eng. (Mater.) **44b**, 449 (1998).
578. Kramarov S. O., Dashko Yu. V., Strength Mater. **10**, 52 (1987).
579. Kresin V., Solid State Comm. **63**, 725 (1987).
580. Kresin V., Wolf S. A., Solid State Comm. **63**, 1141 (1987).
581. Kresin V. Z., Wolf S. A. *Fundamentals of Superconductivity* (Plenum Press, New York, 1990).
582. Kroeger D. M., Choudhury A., Brynestad J., et al., J. Appl. Phys. **64**, 331 (1988).
583. Krstic V.D., J. Mater. Sci. **23**, 259 (1988).
584. Krusin-Elbaum L., Thompson J. R., Wheeler R., et al., Appl. Phys. Lett. (1994). **64**, 3331.
585. Kumagai T., Yokota H., Kawaguchi K., et al., Chem. Lett. **16**, 1645 (1987).
586. Kumakura H., Kitaguchi H., Togano K., et al., IEEE Trans. Appl. Supercond. **9**(2), 1804 (1999).
587. Kumar P., Pillai V., Shah D. O., Appl. Phys. Lett. **62**, 765 (1993).
588. Kupriyanov M. F., Konstantinov G. M., *Proc. Int. Conf. Electr. Ceram. Product. Propert.*, part I, p. 36 (Riga, April 30 – May 2, 1990).

589. Kwok W. K., Crabtree G. W., Umezawa A., et al., Phys. Rev. B. **37**, 106 (1988).
590. Kwon S. K., Youn S. J., Kim K. S., Min B. I., cond-mat/0106483 (2001).
591. Lade P. V., Int. J. Numer. Anal. Methods Geomech. **12**, 351 (1988).
592. Lahtinen M., Paasi J., Sarkaniemi J., et al., Phys. C **244**, 115 (1995).
593. Lairson B. M., Streiffer S. K., Bravman J. C., Phys. Rev. B **42**,10067 (1990).
594. Landau L. D., Lifshitz E. M. *Electrodynamics of Continuous Media*, 2nd edn., (Nauka, Moscow, 1972) [Pergamon Press, Oxford, 1968].
595. Landau L. D., Lifshitz E. M. *Statistical Physics*, 3rd edn., part 1 (Nauka, Moscow, 1976) [Pergamon Press, Oxford, 1980].
596. Landau L. D., Lifshits E. M. *Fluid Mechanics* (Pergamon, Oxford, 1987).
597. Larbalestier D. C., Science **274**, 736 (1996).
598. Larbalestier D. C., IEEE Trans. Appl. Supercond. **7**(2), 90 (1997).
599. Larbalestier D. C., Babcock S. E., Cai X. Y., et al., Phys. C **185–189**, 315 (1991).
600. Larbalestier D. C., Cai X.Y., Feng Y., et al., Phys. C **221**, 299 (1994).
601. Larbalestier D. C., Maley M. P., MRS. Bull. **17**, 50 (1993).
602. Larbalestier D. C., Rikel M. O., Cooley L. D., et al., cond-mat/0102216 (2001).
603. Larkin A. I., Ovchinnikov Yu. N., J. Low Temp.Phys. **34**, 409 (1979).
604. Lauder A., Face D. W., Holstein W. L., et al., Adv. Supercond. **5**, 925 (1993).
605. Laughlin R. B., cond-mat/0209269 (2002).
606. Laval J. Y., Swiatnicki W., Phys. C **221**, 11 (1994).
607. Lawrence W. E., Doniach S. In: *Proc.* 12th *Int. Conf. on Low Temperature Physics*, E. Kanda (ed.), p. 361 (Academic Press, Kyoto, 1971).
608. Laws N., Brockenbrough J. R., Trans. ASME: J. Eng. Mater. Technol. **110**, 101 (1988).
609. Laws N., Lee J.C., J. Mech. Phys. Solids **37**, 603 (1989).
610. Leath P. L., Tang W., Phys. Rev. B **39**, 6485 (1989).
611. Leath P. L., Xia W., Phys. Rev. B **44**, 9619 (1991).
612. Ledbetter H., Lei M., Hermann A., Sheng Z., Phys. C **255**, 397 (1994).
613. Lee H., Siu W., Yu K. W., Phys. Rev. B **52**, 4217 (1995).
614. Leenders A., Ullrich M., Freyhardt H. C., IEEE Trans. Appl. Supercond. **9**(2) 2074 (1999).
615. Leggett A. J., Phys. Rev. Lett. **83**, 392 (1999).
616. Leisure R. G. In: *Encyclopedia of Materials: Science and Technology*, p. 1 (Elsevier, London, 2004).
617. Levy L. - P. *Magnetism and Superconductivity* (Springer-Verlag, Berlin, Heidelberg, 2000).
618. Levy O., Bergman D., J. Phys.: Condens. Mater. **5**, 7095 (1993).
619. Lew D. J., Suzuki Y., Marshall A. F., et al., Appl. Phys. Lett. **65**, 1584 (1994).
620. Li J. C., Oriani R. A., Darken L. S., Zeitschrift für Physikalische Chemie Neue Folge. **49**, 271 (1966).
621. Li J. C., Sanday S. C., Acta Metall. (1986). **34**, 537
622. Li J. N., Menovsky A. A., Franse J. J. M., Phys. Rev. B **48**, 6612 (1993).
623. Li J. Q., Li L., Liu F. M., et al., cond-mat/0104320 (2001).
624. Li S., Bredehöft M., Gao W., et al., Supercond. Sci. Technol. **11**, 1011 (1998).
625. Li S., Gao W., Cooper H., et al., Phys. C **356**, 197 (2001).
626. Li Y., Kilner J. A., Dhallé M., et al., Supercond. Sci. Technol. **8**, 764 (1995).
627. Li Y., Perkins G. K., Caplin A. D., et al., Phys. C **341–348**, 2037 (2000).

628. Li Y., Zhao Z. - X., Phys. C **351**, 1 (2001).
629. Lifshitz I. M., Slyozov V. V., J. Phys. Chem. Solids **19**, 35 (1961).
630. Likharev K. K., Rev. Mod. Phys. **51**, 101 (1979).
631. Likharev K. K., Ulrich B. T. *Systems with Josephson Junctions. The Basics of the Theory* (Moscow State Univ. Press, Moscow, 1978).
632. Lin R. - J., Chen L. - J., Lin L. - J., et al., Jpn. J. Appl. Phys. **35**, 5805 (1996).
633. Lindgren L. E., Edberg J., J. Mater. Process. Technol. **24**, 85 (1990).
634. Lisachenko D. A., Supercond. Phys. Chem. Tech. **6**, 1757 (1993).
635. Liu H. K., Dou S. X., Phys. C **250**, 7 (1995).
636. Liu H. K., Horvat J., Bhasale R., et al., IEEE Trans. Appl. Supercond. **7**(2), 1841 (1997).
637. Liu H. K., Polyanskii A., Chen W. M., et al., IEEE Trans. Appl. Supercond. **11**(1), 3764 (2001).
638. Liu H. K., Wang R. K., Dou S. X., Phys. C **229**, 39 (1994).
639. Lo W., Cardwell D. A., Chow J. C. L., J. Mater. Res. **13**, 1141 (1998).
640. Lo W., Cardwell D. A., Dewhurst C. D., Dung S. - L., J. Mater. Res. **11**, 786 (1996).
641. Lo W., Cardwell D. A., Dung S. - L., Barter R. G., IEEE Trans. Appl. Supercond. **5**(2), 1619 (1995).
642. Lo W., Cardwell D. A., Dung S. - L., Barter R. G., J. Mater. Sci. **30**, 3995 (1995).
643. Lo W., Cardwell D. A., Dung S. - L., Barter R. G., J. Mater. Res. **11**, 39 (1996).
644. London F., London H., Proc. Roy. Soc. **A149**, 71 (1935).
645. Lorenz B., Lenzi J., Cmaidalka J., et al., cond-mat/0208341 (2002).
646. Lorenz M., Hochmuth H., Natusch D., et al., Appl. Phys. Lett. **68**, 3332 (1996).
647. Low I. M., Skala R. D., Mohazzab H. G., J. Mater. Sci. Lett. **13**, 1340 (1994).
648. Lubenets S. V., Natsik V. D., Fomenko L. S., Fiz. Nizk. Temp. **21**, 475 (1995).
649. Lubkin G. B., Phys. Today **49**, 48 (1996).
650. Luo J. S., Merchant N., Maroni V. A. et al., Appl. Supercond. **1**, 101 (1993).
651. Maciejewski M., Baiker A., Conder K., et al., Phys. C **227**, 343 (1994).
652. Maeda J., Nakamura Y., Izumi T., Shiohara Y., Supercond. Sci. Technol. **12**, 563 (1999).
653. Maeda H., Tanaka Y., Fukutomi M., Asano T., Jpn. J. Appl. Phys. **27,** L209 (1988).
654. Maeno Y., Hashimoto H., Yoshida K., et al., Nature **372**, 532 (1994).
655. Magerl A., Berre B., Alefeld G., Phys. Stat. Sol. A **36**, 161 (1976).
656. Maggio-Aprile I., Renner Ch., Erb A., et al., Phys. Rev. Lett. **75**, 2754 (1995).
657. Mai Y. - W. Lawn B. R., Annual Rev. Mater. Sci. **16**, 415 (1986).
658. Majewski P., Bestengen H., Eschner S., Aldinger F., Z. Metall. **86**, 563 (1995).
659. Malachevsky M. T., Esparza D. A., Phys. C **324**, 153 (1999).
660. Malberg M., Bech J., Bay N., et al., IEEE Trans. Appl. Supercond. **9**(2), 2577 (1999).
661. Malvern L. E. *Introduction to the Mechanics of a Continuous Medium* (Prentice-Hall, Englewood Cliffs, NJ, 1969).
662. Mandelbrot B. B., *The Fractal Geometry of Nature* (Freeman, San Francisco, 1982).

663. Manson S. S. *Thermal Stress and Low-Cycle Fatigue* (McGrow-Hill, New York, 1973).
664. Marinel S., Monot I., Provost J., Desgardin G., Supercond. Sci. Technol. **11**, 563 (1998).
665. Marken K. R., Dei W., Cowey L., et al., IEEE Trans. Appl. Supercond. **7**(2), 2211 (1997).
666. Markiewicz R. S., J. Phys. Chem. Solids **58**, 1179 (1997).
667. Marshall D. B., Cox B. N., Evans A. G., Acta Metall. **33**, 2013 (1985).
668. Marshall D. B., Drory M., Evans A. G., Fract. Mech. Ceram. **6**, 289 (1983).
669. Marti F., Daumpling M., Flukiger R., IEEE Trans. Appl. Supercond. **5**, 1884 (1995).
670. Marti F., Dhallé M., Flükiger R., et al., IEEE Trans. Appl. Supercond.**9**(2), 2766 (1999).
671. Marti F., Grasso G., Grivel J. - C., Flükiger R. Supercond. Sci. Technol. **11**, 485 (1998).
672. Marti F., Huang Y. B., Witz G., et al., IEEE Trans. Appl. Supercond. **9**(2), 2521 (1999).
673. Martini L., Supercond. Sci. Technol. **11**, 231 (1998).
674. Martur N. D., Grosche F. M., Julian S. R., et al., Nature **394**, 39 (1998).
675. Masuda Y., Ogawa R., Kawate Y., et al., J. Mater. Res. **8**, 693 (1993).
676. Masur L. J., Kellers J., Li F., et al., IEEE Trans. Appl. Supercond. **12**(1), 1145 (2002).
677. Matsui Y., Maeda H., Tanaka Y., Horiuchi S., Jpn. J. Appl. Phys. **27**, L361 (1988).
678. Matsumoto K., et al., Adv. Supercond. **11**, 773 (1998).
679. Matsunaga K., Nishimura A., Satoh S., Motojima O., Adv. Cryog. Eng. (Mater.) **46**, 691 (2000).
680. Maul M., Schulte B., J. Appl. Phys. **74**, 2942 (1993).
681. Mayer B., Alff L., Träuble T., et al., Appl. Phys. Lett. **63**, 996 (1993).
682. McConnel B. W., IEEE Power Eng. Rev. **20**(6) (2000).
683. McCrone J. E., Cooper J. R., Tallon J. L., cond-mat/9909263 (1999).
684. McGinn P., Black M. A., Valenzuela A., Phys. C **156**, 57 (1988).
685. McGinn P., Chen W., Zhu N., et al., Appl. Phys. Lett. **59**, 120 (1991).
686. McHale J., Myer G. H., Salomon R. E., J. Mater. Res. **10**, 1 (1995).
687. McIntyre P., Cima M. J., Ng M. F., J. Appl. Phys. **68**, 4183 (1990).
688. McLaughlin A. C., Sher F., Attfield J. P., Nature **436**, 829 (2005).
689. Meakin P., Phys. Rep. **235**, 189 (1993).
690. Meilikhov E. Z., Phys. C **271**, 277 (1996).
691. Meingast C., Junod A., Walker E., Phys. C **272**, 106 (1996).
692. Meingast C., Kraut O., Wolf T., et al., Phys. Rev. Lett. **67**, 1634 (1991).
693. Meissner W., Ochsenfeld R., Naturwissenschaften **21**, 787 (1933).
694. Mendoza E., Puig T., Varesi E., et al., Phys. C. **334**, 7 (2000).
695. Meng R. L., et al., J. Supercond. **11**, 181 (1998).
696. Merchant N., Luo J. S., Maroni V. A., et al., Appl. Phys. Lett. **65**, 1039 (1994).
697. Merkle K. L., Ultramicroscopy **37**, 130 (1991).
698. *Metal Data Book.* 3rd edn, J. Maruzen (ed.) (Institute of Metals, Tokyo, Japan, 1993).
699. Miao H., Kitaguchi H., Kumakura H., Togano K., Cryogenics **38**, 257 (1998).
700. Miao H., Lera F., Larrea A., de la Fuente G.F., Navarro R., IEEE Trans. Appl. Supercond. **7**(2), 1833 (1997).

701. Michikami O., Yokosawa A., Wakana H., Kashiwaba Y., Jpn. J. Appl. Phys. **36**, 2646 (1997).
702. Mikhailov B. P., Burkhanov G. S., Leytus G. M., et al. Neorg. Mater. **32**, 1225 (1996).
703. Milne-Thompson L. M. *Theoretical Aerodynamics* (Dover, New York, 1973).
704. Minami N., et al., Jpn. J. Appl. Phys. **31**, 784 (1992).
705. Miroshnichenko I. P., Parinov I. A., Rozhkov E. V., Serkin A. G., Metallurg **7**, 77 (2006).
706. Mishra D. R., Zadeh H. S., Bhattacharya D., Sharma R. G., Phys. C **341–342**, 1931 (2000).
707. Mito T., Chikaraishi H., Hamaguchi S., IEEE Trans. Appl. Supercond. **12**(1), 606 (2002).
708. Miyake K., Schmitt-Rink S., Varma C. M., Phys. Rev. B **34**,6554 (1986).
709. Miyamoto T., Katagari J., Nagashima K., Murakami M., IEEE Trans. Appl. Supercond. **9**(2), 2066 (1999).
710. Miyamoto T., Nagashima K., Sakai N., Murakami M., Supercond. Sci. Technol. **13**, 816 (2000).
711. Miyamoto T., Nagashima K., Sakai N., Murakami M., Phys. C **349**, 69 (2001).
712. Miyase A., Yuan Y. S., Wong M. S., et al., Supercond. Sci. Technol. **8**, 626 (1995).
713. Mkrtychan G. S., Schmidt V. V., Zh. Eksp. Teor. Fiz. **61**, 367 (1971).
714. Moeckly B. H., Lathrop D. K., Buhrman R. A., Phys. Rev. B **47**, 400 (1993).
715. Mohr O., Zeit. Ver. Deut. Ing. **44**, 1524 (1900).
716. Mönch W. *Semiconductor Surfaces and Interfaces* (Springer-Verlag, Berlin, 2001).
717. Monot I., Hihuchi T., Sakai N., Murakami M., Supercond. Sci. Technol. **7**, 783 (1994).
718. Monot I., Wang J., Delamare M. P., et al., Phys. C **267**, 173 (1996).
719. Montanari G. C., Ghinello I., Gherardi L., Caracino P., Supercond. Sci. Technol. **9**, 385 (1996).
720. Montanari G. C., Ghinello I., Gherardi L., Mele R., IEEE Trans. Appl. Supercond. **7**(2), 1303 (1997).
721. Monteverde M., Núñez-Regueiro M., Acha C., et al., Phys. C **408–410**, 23 (2004).
722. Moore G., Kramer S., Kordas G., Mater. Lett. **7**, 415 (1989).
723. Moore J. C., Bisset M. I., Knoll D. C., et al., IEEE Trans. Appl. Supercond. **9**(2), 1692 (1999).
724. Moore J. C., Naylor M. J., Grovenor C. R. M., IEEE Trans. Appl. Supercond. **9**(2), 1787 (1999).
725. Moore J. C., Shiles P., Eastell C. J., et al., Inst. Phys. Conf. Ser. **158**, 901 (1997).
726. Morgan P. E. D., Hously R. M., Porter J. R., Ratto J. J., Phys. C **176**, 279 (1991).
727. Morita M., Nagashima K., Takebayashi S., et al., *Proc. Int. Workshop on Superconductivity*, p. 115 (Okinawa, Japan, 1998).
728. Morita M., Sawamura M., Takebayashi S., et al., Phys. C **235–240**, 209 (1994).
729. Morita M., Takebayashi S., Tanaka M., et al., Adv. Supercond. **3**, 733 (1991).
730. Morita M., Tanaka M., Takebayashi S., Jpn. J. Appl. Phys. **30**, L813 (1991).

731. Morosan E., Zandbergen H. W., Dennis B. S., et al., Nature Phys. **2**, 544 (2006).
732. Mourachkine A., J. Low Temp. Phys. **117**, 401 (1999).
733. Mourachkine A., Europhys. Lett. **55**, 86 (2001).
734. Mourachkine A., Europhys. Lett. **55**, 559 (2001).
735. Mourachkine A., J. Supercond. **14**, 375 (2001).
736. Mourachkine A., Supercond. Sci. Technol. **14**, 329 (2001).
737. Mourachkine A. *High-Temperature Superconductivity in Cuprates. The Nonlinear Mechanism and Tunneling Measurements* (Kluwer Academic, Dordrecht, 2002).
738. Mulet R., Diaz O., Altshuler E., Supercond. Sci. Technol. **10**, 758 (1997).
739. Mulins W. W., Sekerka R. F., J. Appl. Phys. **34**, 323 (1963).
740. Müller K.A., Takashige M., Bednorz J.G., Phys. Rev. Lett. **58**, 1143 (1987).
741. Murakami M. *Melt Processed High-Temperature Superconductors* (World Scientific Publ., Singapore, 1992).
742. Murakami M., Supercond. Sci. Technol. **5**, 185 (1992).
743. Murakami M., Fujimoto H., Gotoh S., et al., Phys. C **185–189**, 321 (1991).
744. Murakami M., Gotoh S., Koshizuka N., et al., Cryogenics **30**, 390 (1990).
745. Murakami M., Morita M., Doi K., Miyamoto K., Jpn. J. Appl. Phys. **28**, 1189 (1989).
746. Murakami M., Sakai N., Higuchi T., Yoo S. I., Supercond. Sci. Technol. **9**, 694 (1996).
747. Murakami M., Sakai N., Higuchi T., Yoo S. I., Supercond. Sci. Technol. **9**, 1015 (1996).
748. Murakami M., Yamaguchi K., Fugimoto H., et al., Cryogenics 32, 930 (1992).
749. Murakami M., Yoo S. I., Higuchi T., et al., Jpn. J. Appl. Phys. **33**, L715 (1994).
750. Muramatsu T., Takeshita N., Terakura C., et al., cond-mat/0502490 (2005).
751. Murduck, J. M. Burch J., Hu R., et al., IEEE Trans. Appl. Supercond. **7**(2), 2940 (1997).
752. Murty K. G. *Linear Programming* (Wiley, New York, 1983).
753. Myers K. E., Face D. W., Kountz D. J., et al., IEEE Trans. Appl. Supercond. **5**(2), 1684 (1995).
754. Myers K. E., Lijie Bao, J. Supercond. **11**, 129 (1998).
755. Nabatame T., Koike S., Hyun O. B., et al., Appl. Phys. Lett. **65**, 776 (1994).
756. Nagamatsu J., Nakagawa N., Muranaka T., et al., Nature **410**, 63 (2001).
757. Nakahara S., Fisanick G. J., Yan M. F., et al., J. Crystal Growth **85**, 639 (1987).
758. Nakamura N., Nakai Y., Kanai Y., et. al., Rev. Sci. Instrum. **75**, 3034 (2004).
759. Nakamura Y., Shiohara Y., J. Mater. Res. **11**, 2450 (1996).
760. Narozhnyi V. N., Eckert D., Nenkov K. A., et al., cond-mat/9909107 (1999).
761. Nature **414**, 6861 (2001).
762. Naylor M. J., Grovenor C. R. M., IEEE Trans. Appl. Supercond. **9**(2), 1860 (1999).
763. Needleman A. A., J. Appl. Mech. **54**, 525 (1987).
764. Neumeier J. J., Zimmermann H. A., Phys. Rev. B **47**, 8385 (1993).
765. Newson M. S., Ryan D. T., Wilson M. N., Jones H., IEEE Trans. Appl. Supercond. **12**(1), 725 (2002).
766. Nhien S., Langlois P. L., Desgardin G., *Proc.* 4th *Journées d'Études SEE*, G. Desgardin (ed.), AP18 (Caen, March, 1997).

767. Nideröst M., Suter A., Visani P., et al., Phys. Rev. B **53**, 9286 (1996).
768. Nomura S., Suzuki C., Watanabe N., et al., IEEE Trans. Appl. Supercond. **12**(1), 788 (2002).
769. Norton D. P., Goyal A., Budai J. D., et al., Science **274**, 755 (1996).
770. Noudem J. G., Beille J., Bourgault D., et al., Phys. C **230**, 42 (1994).
771. Noudem J. G., Beille J., Draperi A., et al., Supercond. Sci. Technol. **6**, 795 (1993).
772. Noudem J. G., Reddy E. S., Tarka M., et al., Phys. C **366**, 93 (2002).
773. Nunez-Regueiro M., Tholence J. - L., Antipov E. V., et al., Science **262**, 97 (1993).
774. O'Bryan H. M., Gallagher P. K., Adv. Ceram. Mater. **2**, 640 (1987).
775. Oduleye O. O., Penn S. J., Alford N. McN., Supercond. Sci. Technol. **11**, 858 (1998).
776. Oduleye O. O., Penn S. J., Alford N. McN., et al., IEEE Trans. Appl. Supercond. **9**(2), 2621 (1999).
777. Ogawa N., Hirabayashi I., Tanaka S., Phys. C **177**, 101 (1991).
778. Oh T. S., Rodel J., Cannon R. M., Ritchie R. O., Acta Metall. **36**, 2083 (1988).
779. Oka T., Itoh Y., Yanagi Y., et al., Phys. C **200**, 55 (1992).
780. Okada M., Tanaka K., Wakuda T., et al., IEEE Trans. Appl. Supercond. **9**(2), 1904 (1999).
781. Okaji M., Nara K., Kato H., et al., Cryogenics **34**, 163 (1994).
782. Onnes H. K., Proc. Section of Sciences **XII**, 1107 (1911).
783. Osamura K., Nanaka S., Matsui M., Phys. C **257**, 79 (1996).
784. Osamura K., Ogawa K., Horita T., et al., *Ext. Abstr. Int. Workshop on Critical Current* (*IWCC*-99), p. 206 (Madison, Wisconsin, USA, July 7–10, 1999).
785. Osamura K., Ogawa K., Thamizavel T., Sakai A., Phys. C **335**, 65 (2000).
786. Osamura K., Sugano M., Phys. C **357–360**,1128 (2001).
787. Osamura K., Sugano M., Wada T., Ochiai S., Adv. Cryog. Eng. (Mater.) **46**, 639 (2000).
788. Ostrovsky I. V., Selivonov I. N., Fiz. Nizk. Temper. **24**, 67 (1998).
789. Pachla W., Kováč P., Marciniak H., et al., Phys. C **248**, 29 (1995).
790. Pachla W., Marciniak H., Szulc A., et al., IEEE Trans. Appl. Supercond. **7**(2), 2090 (1997).
791. Pak S. V., Slipenyuk A. N., Varyukhin V. N., Samelyuk A. V., Metal. Phys. Surf. Technol. **19**(3), 46 (1997).
792. Palstra T. T. M., Zhou O., Iwasa Y., et al., Solid State Commun. **93**, 327 (1995).
793. Pan V. M. *Study of High Temperature Superconductors*, A. Narlikar (ed.), p. 319 (Nova, New York, 1990).
794. Panteleev V. G., Ramm K. S., Agroskin L. S., Solovieva L. V., Glass Ceram. **2**, 24 (1989).
795. Parikh A. S., Meyer B., Salama K., Supercond. Sci. Technol. **7**, 455 (1994).
796. Parinov I. A., Ferroelectrics **131**, 131 (1992).
797. Parinov I. A., Ferroelectric Lett. 19, 157 (1995).
798. Parinov I. A., Ferroelectrics **172**, 253 (1995).
799. Parinov I. A., J. Surface Invest. X-Ray, Synch. Neutron Techn. **11**, 39 (1996).
800. Parinov I. A., Supercond. Res. Develop. **9, 10,** 16 (1998).
801. Parinov I. A., *Ext. Abstr. Int. Workshop on Critical Current* (*IWCC*-99), p. 166 (Madison, Wisconsin, USA, July 7–10, 1999).

802. Parinov. I. A., IEEE Trans Appl. Supercond. **9**(2), 4304 (1999).
803. Parinov I. A., Strength Mater. **4**, 92 (1999).
804. Parinov I. A., Compos. Mech. Des. **6**, 445 (2000).
805. Parinov I. A., *Proc.* 10th *Int. Congress of Fracture*, ICF100470PR (CD-ROM) (Honolulu, Hawaii, USA, December 2–6, 2001).
806. Parinov I. A., Compos. Mech. Des. **8**, 172 (2002).
807. Parinov I. A., *Microstructure Aspects of Strength and Fracture of High-Temperature Superconductors* (*Overview*) http://www.math.rsu.ru/niimpm/strl/p62/p62.htm (2003).
808. Parinov I. A., *Microstructure and Properties of High-Temperature Superconductors* (Rostov State Univ. Press, Rostov-on-Don, 2004).
809. Parinov I. A., Compos. Mech. Des. **11**, 242 (2005).
810. Parinov I. A., Compos. Mech. Des. **11**, 357 (2005).
811. Parinov I. A., Parinova L. I., Ferroelectrics **211**, 41 (1998).
812. Parinov I. A., Parinova L. I., Rozhkov E. V., Phys. C **377**, 114 (2002).
813. Parinov I. A., Parinova L. V., *Proc. Int. Conf. on Struct. and Proper. Brittle and Quasiplast. Mater.* (SPM′94), p. 98 (Latv. Acad. Sci., Riga, 1994).
814. Parinov I. A., Parinova L. V., Supercond. Phys. Chem. Tech. **7**, 79 (1994).
815. Parinov I. A., Parinova L. V., Supercond. Phys. Chem. Tech. **7**, 1382 (1994).
816. Parinov I. A., Parinova L. V., Strength Mater. **1**, 113 (1997).
817. Parinov I. A., Popov A. V., Rozhkov E. V., Prygunov A. G. Russ. J. Non-Destruct. Control **1**, 66 (2000).
818. Parinov I. A., Prygunov A. G., Rozhkov E. V., et al., *Russian Patent* No 2,169,348 (2001).
819. Parinov I. A., Rozhkov E. V., IEEE Trans. Appl. Supercond. **9**(2), 2058 (1999).
820. Parinov I. A., Rozhkov E. V., Compos. Mech. Des. **10**, 355 (2004).
821. Parinov I. A., Rozhkov E. V., Vassil'chenko C. E., IEEE Trans. Appl. Supercond. **7**(2), 1941 (1997).
822. Parinov I. A., Rozhkov E. V., Vassil'chenko C. E., Adv. Cryog. Eng. (Mater.) **44b**, 639 (1998).
823. Parinov I. A., Vasil'eva Yu. S., Strength Mater. **8**, 77 (1994).
824. Park H. - W., Kim K. - B., Lee K. - W., Supercond. Sci. Technol. **9**, 694 (1996).
825. Parmigiani F., Chiarello G., Ripamonti N., Phys. Rev. B **36**, 7148 (1987).
826. Parrell J. A., Dorris S. E., Larbalestier D. C., Adv. Cryogenic Eng. **40**, 193 (1994).
827. Parrell J. A., Dorris S. E., Larbalestier D. C., Phys. C **231**, 137 (1994).
828. Parrell J. A., Feng Y., Dorris S. E., Larbalestier D. C., J. Mater. Res. **11**, 555 (1996).
829. Parrell J. A., Larbalestier D. C., Dorris S. E., IEEE Trans. Appl. Supercond. **5**(2), 1275 (1995).
830. Parrell J. A., Larbalestier D. C., Riley G. N., Jr., et al., J. Mater. Res. **12**, 2997 (1997).
831. Parrell J. A., Polyanskii A. A., Pashitski A. E., Larbalestier D. C., Supercond. Sci. Technol. **9**, 393 (1996).
832. Pashitskii E. A., Gurevich A., Polyanskii A. A., et al., Science **275**, 367 (1997).
833. Pashitskii E. A., Polyanskii A. A., Gurevich A., et al., Phys. C **246**, 133 (1995).

834. Pashkin Yu. A., Yamamoto T., Astafiev O., et al., Nature **421**, 823 (2003).
835. Passerini R., Dhallé M., Witz G., et al., IEEE Trans. Appl. Supercond. **11**(1), 3018 (2001).
836. Patel S., Chen S., Haugan T., et al., Cryogenics **35**, 257 (1996).
837. Patnaik S., Cooley L. D., Gurevich A., et al., cond-mat/0104562 (2001).
838. Pechini M., USA Patent No 3,330,697 (1967).
839. Peisl H. In: *Hydrogen in Metals. Basic Pproperties*, G. Alefeld, and J. Volkl (eds.),**1**, 53 (Springer, New York, 1978).
840. Pertsev N. A., Arlt G., Fiz. Tverdogo Tela **33**, 3077 (1991).
841. Peterson R. L., Ekin J.W., Phys. Rev. B **37**, 9848 (1988).
842. Petrovic C., Movshovich R., Jaime M., et al., cond-mat/0012261 (2000).
843. Pflelderer C., Uhlarz M., Hyden S. M., et al., Nature **412**, 58 (2001).
844. Philips D. T., Garcia-Diaz A. *Fundamentals of Network Analysis* (Prentice-Hall, Englewood Cliffs, NJ, 1981).
845. *Physical Values. Handbook*, I. S. Grigoriev, and E. Z. Meilikhov (eds.), pp. 47, 223 (Energoatomizdat, Moscow, 1991).
846. Phys. Today, p. 11 (July, 1994).
847. Pippard A. B., Proc. Roy. Soc. **A216**, 547 (1953).
848. Pisarenko G. G. *Fracture Resistance of Piezoelectric Ceramics* (Preprint AN UkrSSR, IPS, Kiev, 1984).
849. Pisarenko G. G. *Strength of Piezoceramics* (Naukova Dumka, Kiev, 1987).
850. Pitel J., Kováč P., Hense K., Kirchmayr H., IEEE Trans. Appl. Supercond. **12**(1), 1475 (2002).
851. Plecháček V., IEEE Trans. Appl. Supercond. **9**(2), 2078 (1999).
852. Plischke M., Racz Z., Phys. Rev. Lett. **53**, 415 (1984).
853. Pogrebnyakov A. V., Redwing J. M., Jones J. E., et al., cond-mat/0304164 (2003).
854. Polak M., Parrell J. A., Polyanskii A. A., et al., Appl. Phys. Lett. **70**, 1034 (1997).
855. Polak M., Polyanskii A. A., Zhang W., et al., Adv. Cryog. Eng. (Mater.) **46**, 793 (2000).
856. Polyanskii A. A. Private communication (2004).
857. Polyanskii A., Beilin V., Yashchin E., et al., IEEE Trans. Appl. Supercond. **11**(1), 3736 (2001).
858. Polyanskii A. A., Cai X. Y., Feldman D. M., Larbalestier. In: *NATO Science Series* 3. *High Technology*, I. Nedkov, and M. Ausloos (eds.), **72**, 353 (Kluwer Academic, Netherlands 1999).
859. Polyanskii A. A., Feldmann D. M., Larbalestier D. C. In: *Handbook of Superconducting Materials*, D. Cardwell, and D Ginley (eds.), p. 236 (Institute of Physics Publishing, Bristol, 1999).
860. Polyanskii A. A., Gurevich A., Pashitskii E. A., et al., Phys. Rev. B **53**, 8687 (1996).
861. Polyanskii A. A., Pashitski A. E., Gurevich A., et al., Adv. Supercond. 9, 469 (1997).
862. Poole C. P., Jr., Datta T., Farach H. A., et al. *Copper Oxide Superconductors* (John Wiley & Sons, New York, 1988).
863. Porter D. L., Evans A. G., Heuer A. H., Acta Metall. **27**, 1649 (1979).
864. Porter J. R., Blumenthal W., Evans A. G., Acta Metall. **29**, 1899 (1981).
865. Povirk G. L., Needleman A., Nutt S. R., Mater. Sci. Eng. **A125**, 129 (1990).

866. Presland M. R., et al., Phys. C **176**, 95 (1991).
867. Prester M., Phys. Rev. B **54**, 606 (1996).
868. Prikhna T. A., Gawalek W., Moshchil V., et al., Phys. C **354**, 415 (2001).
869. Prikhod'ko A. V., Kon'kov O. I., Semiconductors **35**, 687 (2001).
870. Prokhorov A. M., Lyakishev N. P., Burkhanov G. S., et. al., Inorg. Mater. **32**, 854 (1996).
871. Prokhorov A. M., Lyakishev N. P., Burkhanov G. S., et. al., Dokl. Phys. **363**, 771 (1998).
872. Puls M. P., Acta Metall. **29**, 1961 (1981).
873. Rábra M., Sekimura N., Kitagushi H., et al., Supercond. Sci. Technol. **12**, 1129 (1999).
874. Radovan H. A., Fortune N. A., Murphy T. P., et al., Nature **425**, 51 (2003).
875. Ramsbottom H. D., Ito H., Horita T., Osamura K., Appl. Supercond. **5**, 157 (1997).
876. Ravinder Reddy R., Muralidhar M., Hari Babu V., Venugopal Reddy P., Supercond. Sci. Technol. **8**, 101 (1995).
877. Read R. T. *Dislocations in Crystals* (McGraw-Hill, New York, 1953).
878. Reddy E. S., Schmitz G. J., Supercond. Sci. Technol. **15,** L21 (2002).
879. Ren Y., Weinstein R., Liu J., et al., Phys. C **251**, 15 (1995).
880. Ren Y., Weinstein R., Sawh R., Liu J., Phys. C **282–287**, 2301 (1997).
881. Ren Z. F., Lao J. Y., Gao L. P., et al., J. Supercond. **11**, 159 (1998).
882. Renner Ch., Revaz B., Kadowaki K. et al., Phys. Rev. Lett. **80**, 3606 (1998).
883. Rhyner J., Blatter G., Phys. Rev. B **40**, 829 (1989).
884. Rice J., In: *Fracture. An Advanced Treatise Mathematical Fundamentals*, **2**, p. 204, H. Liebowitz (ed.) (Academic Press, New York, 1968).
885. Rice J. R., Johnson M. A. In: *Inelastic Behavior of Solids*, M. F. Kanninen, W. F. Adler, A. R. Rosenfield, and R. I. Jaffee (eds.), p. 641 (McGraw-Hill, New York, 1970).
886. Richardson D., Proc. Cambr. Phil. Soc. **74**, 515 (1973).
887. Rikel M. O., Reeves J. L., Scarbrough N. A., Hellstrom E. E., Phys. C **341–348**, 2573 (2000).
888. Rikel M. O., Williams R.K., Cai X.Y., et al., IEEE Trans. Appl. Supercond. **11**(1), 3026 (2001).
889. Ring T. *Fundamentals of Ceramic Powder Processing and Synthesis* (Academic Press, New York, 1996).
890. Rodriquez M. A., Chen B. J., Snyder R. L., Phys. C **195**, 185 (1992).
891. Rodriquez M. A., Snyder R. L., Chen B. J., et al., Phys. C **206**, 43 (1993).
892. Romalis N. B., Tamuzh V. P. *Fracture of Structure-Heterogeneous Solids* (Zinatne, Riga, 1989).
893. Ronnung F., Danerud M., Lindgren M., Winkler D., IEEE Trans. Appl. Supercond. **7**, 2599 (1997).
894. Roscoe K. H., Geotechnique **20**, 129 (1970).
895. Rose L. R. F., Int. J. Fract. **31**, 233 (1986).
896. Rose L. R. F., J. Amer. Ceram. Soc. **69**, 212 (1986).
897. Rose L. R. F., J. Mech. Phys. Solids **35**, 383 (1987).
898. Rosner C. H., IEEE Trans. Appl. Supercond. **11**(1), 39 (2001).
899. Rossat-Mignod J., Regnault L. P., Vettier C., et al., Phys. C **185–189**, 86 (1991).
900. Rouessac V., Poullain G., Desgardin G., Raveau B., Supercond. Sci. Technol. **11**, 1160 (1998).

901. Rouessac V., Wang J., Provost J., Desgardin G., Phys. C **268**, 225 (1996).
902. Routbort J. L., Goretta K. C., Miller D. J., et al., J. Mater. Res. **7**, 2360 (1992).
903. Rutter N. A., Glowacki B. A., IEEE Trans. Appl. Supercond. **11**(1), 2730 (2001).
904. Rutter N. A., Glowacki B. A., Evetts J. E., Supercond. Sci. Technol. **13**, L25 (2000).
905. Sagaradze V. V., Zeldovich V. I., Pushin V. G., et al., Supercond. Phys. Chem. Tech. **3**, 1309 (1990).
906. Sahimi M. *Applications of Percolation Theory* (Taylor & Frances, London, 1994).
907. Saint James D., de Gennes P. G., Phys. Rev. Lett. **7**, 306 (1963).
908. Sakai N., Mase A., Ikuta H., et al., Supercond. Sci. Technol. **13**, 770 (2000).
909. Sakai N., Ogasawara K., Inoue K., et al., IEEE Trans. Appl. Supercond. **11**(1), 3509 (2001).
910. Sakai N., Seo S. - J., Inoue K., et al., Adv. Supercond. **11**, 685 (1999).
911. Sakai N., Seo S. - J., Inoue K., et al., Phys. C **335**, 107 (2000).
912. Sakamaki K., Wada H., Nozaki H., et al., Solid State Commun. **112,** 323 (1999).
913. Salama K., Selvamanickam V., Gao L., Sun K., Appl. Phys. Lett. **54**, 2352 (1989).
914. Salamati H., Babaei-Brojeny A. A., Safa M., Supercond. Sci. Technol. **14**, 816 (2001).
915. Salib S., Mironova M., Vipulanandan C., Salama K., Supercond. Sci. Technol. **9**, 1071 (1996).
916. Samarsky A. A. *Theory of Difference Schemes* (Nauka, Moscow, 1983).
917. Sander L. M. *Solids Far from Equilibrium*, C. Godreche (ed.) (Cambridge University Press, Cambridge, 1992).
918. Sandiumenge F., Vilalta N., Pinol S., et al., Phys. Rev. B. **51**, 6645 (1995).
919. Sanfilippo S., Elsinger H., Núñez-Regueiro M., et al., Phys. Rev. B **61**, 3800 (2000).
920. Sarnelli E., Chaudhari P., Lacy J., Appl. Phys. Lett. **62**, 777 (1993).
921. Sarnelli E., Chaudhari P., Lee W. Y., Esposito E., Appl. Phys. Lett. **65**, 362 (1994).
922. Sarrao J. L., Morales L. A., Thompson J. D., et al., Nature **420**, 297 (2002).
923. Sastry P. V. P. S. S., Schwartz J., IEEE Trans. Appl. Supercond. **9**(2), 1684 (1999).
924. Sata T., Sakai K., Tashiro S., J. Am. Ceram. Soc. **75**, 805 (1992).
925. Sato K., Hikata T., Mukai H., et al., IEEE Trans. Magn. **27**(2), 1231 (1991).
926. Sato N., Kawachi M., Noto K., et al., Phys. C **357–360**, 1019 (2001).
927. Sato S., Aso N., Miyake K., et al., Nature **410**, 340 (2001).
928. Satoh M., Haseyama S., Kojima M., et al. Adv. Cryog. Eng. (Mater.) **44b**, 405 (1998).
929. Satoh T., Hidaka M., Tahara S., IEEE Trans. Appl. Supercoud. **7**(2), 3001 (1997).
930. Saxena S. S., Agarwal P., Ahilan K., et al., Nature **406**, 587 (2000).
931. Schaak R. E., Klimczuk T., Foo M. L., Cava R. J. cond-mat/0305450 (2003).
932. Schätzle P., Krabbes G., Stöver G., et al., Supercond. Sci. Technol. **12**, 69 (1999).

933. Schilling J. S., Klots S. In: *Physical Properties of High Temperature Superconductors*, D. M. Ginsberg (ed.) **3**, 59 (World Scientific, Singapore, 1992).
934. Schlesinger Z., et al., Phys. C **235–240**, 49 (1994).
935. Schmidt V. V. *The Physics of Superconductors*, P. Müller, and A. V. Ustinov (eds.) (Springer-Verlag, Berlin, Heidelberg, 1997).
936. Schmitz G. J., Laakmann J., Wolters Ch., et al., J. Mater. Res. **8**, 2774 (1993).
937. Schmitz G. J., Nestler B., Seeβellberg M., J. Low Temp. Phys. **105**, 1451 (1996).
938. Schmitz G. J., Seeβelberg M., Nestler B., Terborg R., Phys. C **282–287**, 519 (1997).
939. Schoenfeld S. E., Asaro R. J., Ahzi S., Phil. Mag. A **73**, 1591 (1996).
940. Schön J. H., Berg S., Kloc Ch., Batlogg B., Science **287**, 1022 (2000).
941. Schön J. H., Dodabalapur A., Kloc Ch., Batlogg B., Science **290**, 963 (2000).
942. Schön J. H., Kloc Ch., Batlogg B., Appl. Phys. Lett. **77**, 3776 (2000).
943. Schön J. H., Kloc Ch., Batlogg B., Nature **408**, 549 (2000).
944. Schön J. H., Kloc Ch., Batlogg B., Science **293**, 2432 (2001).
945. Schön J. H., Kloc Ch., Haddon R. C., Batlogg B., Science **288**, 656 (2000).
946. Schön J. H., Meng H., Bao Z., Nature **413**, 713 (2001).
947. Schön J. H., Meng H., Bao Z., Science **294**, 2138 (2001).
948. Schubnikow L. W., Nakhutin I. E., Nature **139**, 589 (1937).
949. Schuster Th., Indenbom M. V., Koblischka M. R., et al., Phys. Rev. B **49**, 3443 (1994).
950. Schuster Th., Kuhn T., Brandt E. H., et al., Phys. Rev. B **52**, 10375 (1995).
951. Schuster Th., Kuhn T., Brandt E. H., et al., Phys. Rev. B **54**, 3514 (1996).
952. Schuster Th., Kuhn T., Brandt E. H., et al., Phys. Rev. B **56**, 3413 (1997).
953. Schuster Th., Kuhn H., Indenbom M. V., Phys. Rev. B **52**, 15621 (1995).
954. Schwartz J., Heuer J. K., Goretta K. C, et al., Appl. Supercond. **2**, 271 (1994).
955. Sedov L. I. Two-Dimensional Problems in Hydrodynamics and Aerodynamics (Interscience, New York, 1965).
956. Seeβellberg M., Schmitz G. J., Nestler M., Steinbach I., IEEE Trans. Appl. Supercond. **7**(2), 1739 (1997).
957. Selsing J., Am Ceram. Soc. **44**, 419 (1961).
958. Selvaduray G., Zhang C., Balachandran U., et al., J. Mater. Res. **7**, 283 (1992).
959. Send P., Bandyopadhyay S. K., Barat P., et al., Phys. C **255**, 306 (1995).
960. Sengupta S., JOM, 19 (October, 1998).
961. Sengupta S., Shi D., Wang Z., et al., Phys. C **199**, 43 (1992).
962. Sengupta S., Todt V. R., Goretta K. C., et al., IEEE Trans. Appl. Supercond. **7**(2), 1727 (1997).
963. Serdobol'skaya O. Yu., Morozova G. P., Fiz. Tverd. Tela**31**(8), 280 (1989).
964. Service R. F., Science **296**, 1376 (2002).
965. Service R. F., Science **296**, 1584 (2002).
966. Shafer M. W., Penny T., Olsen B., Phys. Rev. B **36**, 4047 (1987).
967. Shah R., Tangrila S., Rachakonda S., Thirukkonda M., J. Electron. Mater. **241**, 1781 (1995).
968. Shapiro S., Phys. Rev. Lett. **11**, 80 (1963).
969. Shaw T. M., Dimos D., Batson P. E., et al., J. Mater. Res. **5**, 1176 (1990).
970. Shekar S., Venugopal Reddy P., Mod. Phys. Lett. B **7**, 935 (1993).
971. Sheng Z. Z., Hermann A. M., Nature **332**, 55 (1988).
972. Sheng Z. Z., Hermann A. M., El Ali A., et al., Phys.Rev. Lett. **60**, 937 (1988).

973. Shewmon P. G., Trans. Am. Inst. Min. Engrs. **230**, 1134 (1964).
974. Shewmon P. G. *Diffusion in Solids* (The Minerals, Metals & Materials Society, Pennsylvania, Warendale, 1989).
975. Shi D., Appl. Supercond. **1**, 61 (1993).
976. Shi D., Sengupta S., Lou J. S., et al., Phys. C **213**, 179 (1993).
977. Shibutani K., Egi T., Hayashi S. et al., IEEE Trans. Appl. Supercond. **3**, 935 (1993).
978. Shield T. W., Kim K. S., Shield T. T., Trans. ASME. J. Appl. Mech. **61**, 231 (1994).
979. Shields T. C., Langhorn J. B., Watcham S. C., et al., IEEE Trans. Appl. Supercond. **7**(2), 1478 (1997).
980. Shilling A., Cautoni M., Gao J. D., Ott H. R., Nature **363**, 56 (1993).
981. Shimakage H., Kawakami A., Wang Z., IEEE Trans. Appl. Supercond. **9**(2), 1645 (1999).
982. Shimizu K., Ishikawa H., Takao D., Nature **419**, 597 (2002).
983. Shimizu K., Kimura T., Furomoto S., et al., Nature **412**, 316 (2001).
984. Shimoyama J., Kase J., Morimoto T., et al., Jpn. J. Appl. Phys. **31**, L1167 (1992).
985. Shiohara Y., Hobara N., Phys. C **341–348**, 2521 (2000).
986. Shrivastava K. N. *Superconductivity. Elementary Topics* (World Scientific Publ. Corp., Singapore, 2000).
987. Shubnikov L. V., de Haas W. J., Commun. Phys. Lab. Univ. Leiden. **207, 210** (1930).
988. Sigl L. S., Mataga P. A., Dalglish B. J., et al., Acta Metall. **36**, 945 (1988).
989. Silsbee F. B., J. Wash. Acad. Sci. **6**, 597 (1916).
990. Singh H. K., Saxena A. K., Srivastava O. N., Phys. C **273**, 181 (1997).
991. Singh J. P., Joo J., Vasanthamohan N., Poeppel R. B., J. Mater. Res. **8**, 2458 (1993).
992. Singh J. P., Leu H. J., Poeppel R. B., et al., J. Appl. Phys. **66**, 3154 (1989).
993. Singleton J., Rep. Prog. Phys. **63**, 1111 (2000).
994. Singleton J., J. Solid. State Chem. **168**, 675 (2002).
995. Skov-Hansen P., Han Z., Flükiger R., et al., Inst. Phys. Conf. Ser. **167**, 623 (1999).
996. Skov-Hansen P., Koblischka M. R., Vase P., IEEE Trans. Appl. Supercond. **11**(1), 3740 (2001)
997. Skripov V. P., Koverda V. P., *Spontaneous Solidification of Re-Cooled Liquids* (Nauka, Moscow, 1984).
998. Smith P. J., Cardwell D.A., Hari Babu N., Shi Y., *Applied Superconductivity*, X. Obrados, F. Sandiumenge, and J. Fontcuberta (eds.) **2**, 75 (2000).
999. Sobha A., Aloysius R. P., Guruswamy P., et al., Phys. C **307**, 277 (1998).
1000. Sobha A., Aloysius R. P., Guruswamy P., et al., Phys. C **316**, 63 (1999).
1001. Sobha A., Aloysius R. P., Guruswamy P., Syamaprasad U., Supercond. Sci. Technol. **13**, 1487 (2000).
1002. Sobol I. M. *Numerical Monte-Carlo Methods* (Nauka, Moscow, 1973).
1003. Sofronis P., McMeeking R. M., J. Mech. Phys. Solids **37**, 317 (1989).
1004. Solymar L. *Superconductive Tunneling and Applications* (Chapman and Hall, London, 1972).
1005. Spears M. A., Evans A. G., Acta Metall. **30**, 1281 (1982).
1006. Specht E. D., Goyal A., Kroeger D. M., Phys. Rev. B **53**, 3585 (1996).

1007. Specht E. D., Goyal A., Kroeger D. M., Supercond. Sci. Technol. **13**, 592 (2000).
1008. Spencer A. J. M. In: *Mechanics of Solids. The Rodney Hill* 60th *Anniversary Volume*, H. G. Hopkins, M. J. Sewell (eds.), p. 607 (Pergamon Press, Oxford, 1982).
1009. Srolovitz D. J., Anderson M. P., Grest G. S., Sahni P. S., Acta Metall. **32**, 1429 (1984).
1010. Srolovitz D. J., Grest G. S., Anderson M. P., Acta Metall. **33**, 2233 (1985).
1011. Staines M. P., Flower N. E., Supercond. Sci. Technol. **4**, 232 (1991).
1012. Stamp D. M., Budiansky B., Int. J. Solids Struct. **25**, 635 (1989).
1013. Stauffer D. *Introduction to the Percolation Theory* (Taylor and Frances, London, 1985).
1014. Stefanakis N., Flytzanis N., Supercond. Sci. Technol. **14**, 16 (2001).
1015. Stefanescu D. M., Dhindaw B. K., Kacar S. A., Moitra A., Metall. Trans. A **19**, 2847 (1988).
1016. Steinbach I., Pezzolla F., Nestler M., et al., Phys. D **94**, 135 (1996).
1017. Stoev P. P., Papirov I. I., Finkel V. A., Fiz. Nizk. Temper. **23**, 1019 (1997).
1018. Straley J., Kenkel S., Phys. Rev. B **29**, 6299 (1984).
1019. Strobel P., Toledano J. C., Morin D., et al., Phys. C **201**, 27 (1992).
1020. Strukova G. K., Degtyareva V. F., Shovkun D. V., et al., cond-mat/0105293 (2001).
1021. Subramanian M. A. et al., Science **239**, 1015 (1988).
1022. Sudhakar Reddy E., Rajasekharm T., Phys. Rev. B **55**, 14160 (1997).
1023. Sugano M., Osamura K., Ochiai S., IEEE Trans. Appl. Supercond. **11**(1), 3022 (2001).
1024. Sugiura T., Hashizume H., Miya K., Int. J. Appl. Electromagn. Mater. **2**, 183 (1991).
1025. Sugiyama N., et al., J. Jpn. Inst. Metals. **61**, 985 (1997).
1026. Sumida M., Nakamura Y., Shiohara Y., Umeda T., J. Mater. Res. **12**, 1979 (1997).
1027. Sumiyoshi F., Kinoshita R., Miyazono Y., et al., IEEE Trans. Appl. Supercond. **9**(2), 2549 (1999).
1028. *Superconductivity in Ternary Compounds II*, O. Fisher, and M. B. Maple (eds.) (Springer-Verlag, Berlin, 1982).
1029. Swain M. V., J. Mater. Sci. Lett. **5**, 1313 (1986).
1030. Switendick A. C. In: *Hydrogen in Metals I. Topics Applied Physics*, G. Alefeld, and J. Volkl (eds.), **28**, 101 (Springer, Berlin, 1978).
1031. Syono Y., Kikuchi M., Ohishi K., et al., Jpn. J. Appl. Phys. **26**, L498 (1987).
1032. Tachikawa K., Watanabe T., Inoue T., Shirasu K., Jpn. J. Appl. Phys. **30**, 639 (1991).
1033. Tada H., Paris P., Irwin G. *The Stress Analysis of Cracks. Handbook* (Del Research Corp., Hellertown, 1973).
1034. Tairov Yu. M., Tzvetkov V. F. Technology pf Semiconducting and Dielectric Materials (Vysshaya Shkola, Moscow, 1990).
1035. Takada K., Sakurai H., Takayama-Muromachi E., et al., Nature **422**, 53 (2003).
1036. Takahashi H., Mori N. In: *Studies of High Temperature Superconductors*, A.V. Narlikar (ed.), **16**, 1 (Nova Science, New York, 1995).
1037. Takano M., Takada J., Oda K., et al., Jpn. J. Appl. Phys. **27**, L1041 (1988).

1038. Takano Y., Nagao M., Sakaguchi I., et al., Appl. Phys. Lett. **85**, 2851 (2004).
1039. Takashima H., Tsuchimoto M., Onishi T., Jpn. J. Appl. Phys. **40**, 3171 (2001).
1040. Takesue I., Haruyama J., Kobayashi N., et al. Phys. Rev. Lett. **96**, 057001 (2006).
1041. Tallon J. L., Cooper J. R., Adv. Supercond. **5**, 339 (1993).
1042. Talvacchio J., Forrester M. G., Gavaler J. R., Braggins T. T., IEEE Trans. Mag. **27**(2), 978 (1991).
1043. Tampieri A., Celotti G., Calestani G., IEEE Trans. Appl. Supercond. **9**(2), 2010 (1999).
1044. Tamura H., Mito T., Iwamoto A., et al., *IEEE Trans. Appl. Supercond.* 12(1), 1319 (2002).
1045. Tanaka S. *Superconductivity Research Laboratory Publication* (Tokyo, Japan, 1998).
1046. Tang W. H., Supercond. Sci. Technol. **13**, 580 (2000).
1047. Tang Z. K., Zhang L., Wang N., et al., Science **292**, 2462 (2001).
1048. Tangrila S., Shah R., Rachakonda S. In: *Net Shape Processing of Powder Materials*, **216** (ASME-AMD, New York, 1995).
1049. Tanigaki K., Ebbesen T. W., Saito S., et al., Nature **352**, 222 (1991).
1050. ten Haken B., Beuink A., ten Kate H. H. J., IEEE Trans. Appl. Supercond. **7**(2), 2034 (1997).
1051. ten Haken B., ten Kate H. H. J., Tenbrink J., IEEE Trans. Appl. Supercond. **5**(2), 1298 (1995).
1052. Terashima K., Matsui H., Hashimoto D., et al., Nature Phys. **2**, 27 (2006).
1053. Thomas J., Verges P., Schatzle P., et al., Phys. C **251**, 315 (1995).
1054. Thompson J. D., Movshovich R., Fisk Z., et al., cond-mat/0012260 (2000).
1055. Thompson J. R., Sun Y. R., Kerchner H. R., et al., Appl. Phys. Lett. **60**, 2306 (1990).
1056. Thorpe M. F., Jin W., Mahanti S. D., Phys. Rev. B **40**, 10294 (1989).
1057. Thouless M. D., J. Am. Ceram. Soc. **73**, 2144 (1990).
1058. Thouless M. D., Evans A. G., Acta Metall. **36**, 517 (1988).
1059. Thouless M. D., Evans A. G., Ashby M. F., Hutchinson J. W., Acta Metall. **35**, 1333 (1987).
1060. Thouless M. D., Olsson E., Gupta A., Acta Metal. Mater. **40**, 1287 (1992).
1061. Thurston T. R., Wildgruber U., Jisrawi N., et al., J. Appl. Phys. **79**, 3122 (1996).
1062. Timoshenko S. P., Goodier J. N. *Theory of Elasticity*, 3-d edn. (McGraw Hill, New York, 1970).
1063. Ting S. M., Marken K. R., Cowey L., et al., IEEE Trans. Appl. Supercond. **7**(2), 700 (1997).
1064. Tinkham M. *Introduction to Superconductivity* (McGraw-Hill, New York, 1996).
1065. Tkaczyk J. E., DeLuca J. A., Karas P. L., et al., Appl. Phys. Lett. **62**, 3031 (1993).
1066. Todt V. R., Sengupta S., Shi D., et al., J. Electron. Mater. **23**, 1127 (1994).
1067. Togano K., Kumakura H., Kadowaki K., et al., Adv. Cryog. Eng. (Mater.) **38**, 1081 (1992).
1068. Tomita M., Murakami M., Phys. C **341–348**, 2443 (2000).
1069. Tomita M., Murakami M., Supercond. Sci. Technol. **13**, 722 (2000).
1070. Tomita M., Murakami M., Phys. C **354**, 358 (2001).

1071. Tomita M., Murakami M., Nature **421**, 517 (2003).
1072. Tomita M., Murakami M., Sawa K., Tachi Y., Phys. C **357–360**, 690 (2001).
1073. Tomita M., Nagashima K., Murakami M., Herai T., Phys. C **357–360**, 832 (2001).
1074. Tonomura A., Kasai H., Kamimura O., et al., Nature **397**, 308 (1999).
1075. Torquato S. *Random Heterogeneous Materials. Microstructure and Macroscopic Properties* (Springer-Verlag, New York, 2002).
1076. Tranquada J. M., Sternlieb B. J., Axe J. D., et al., Nature **375**, 561 (1995).
1077. Træholt C., Wen J. G., Zandbergen H. W., et al., Phys. C **230**, 425 (1994).
1078. Tsabba Y., Reich S., Phys. C **254**, 21 (1995).
1079. Tsu I. - F., Babcock S. E., Kaiser D. L., J. Mater. Res. **11**, 1383 (1996).
1080. Tsu I. - F., Kaiser D. L., Babcock S. E. *Proc. MSA Annual Meeting*, p. 338 (1996).
1081. Tsu I. - F., Wang J. - L., Babcock S. E., et al., Phys. C **349**, 8 (2001).
1082. Tsu I. - F., Wang J. - L., Kaiser D. L., Babcock S. E., Phys. C **306**, 163 (1998).
1083. Tsuchimoto M., Takashima H., Jpn. J. Appl. Phys. **39**, 5816 (2000).
1084. Tsuchimoto M., Takashima H., Phys. C **341–348**, 2467 (2000).
1085. Tsuchimoto M., Takashima H., Onishi T., Phys. C **357–360**, 759 (2001).
1086. Tsunakawa H., Aoki R., Powder Tech. **33**, 249 (1982).
1087. Turchinskaya M., Kaiser D. L., Gayle F. W., et al., Phys. C **216**, 205 (1993).
1088. Turchinskaya M., Kaiser D. L., Gayle F. W., et al., Phys. C **221**, 62 (1994).
1089. Tvergaard V., Hutchinson J. W., J. Mech. Phys. Solids **40**, 1377 (1992).
1090. Ubbelode A. *Melting and Crystalline Structure* (Mir, Moscow, 1969).
1091. Ueyama M., Hikata T., Kato T., Sato K., Jpn. J. Appl. Phys. **30**, L1384 (1991).
1092. Uhlmann D. R., Chalmers B., Jackson K. A., J. Appl. Phys. **35**, 2986 (1964).
1093. Uji S., Shinagawa H., Terashima T., et al., Nature **410**, 908 (2001).
1094. Ullrich M., Leenders A., Freyhardt H. C., IEEE Trans. on Appl. Supercond. **7**(2), 1813 (1997).
1095. Umeda T., Kozuka H., Sakka S., Adv. Ceram. Mater. **3**, 520 (1988).
1096. Umezawa A., Feng Y., Edelman H. S., et al., Phys. C **198**, 261 (1992).
1097. Umezawa A., Feng Y., Edelman H. S., et al., Phys. C **219**, 378 (1993).
1098. Uno N., Enomoto N., Tanaka Y., Takami H., Jpn. J. Appl. Phys. **27**, L1003 (1988).
1099. Vaknin D., Sinha S. K., Stassis C., et al., Phys. Rev. B **41**, 1926 (1990).
1100. Valkin D., Sinha S. K., Stassis C., et al., Phys. Rev. B **41**, 1926 (1990).
1101. Vallet-Regi M., Ramirez J., Ragel C. V., Gonzales-Calbet J. M., Phys. C **230**, 407 (1994).
1102. van der Beek C. J., Kes P. H., Phys. Rev. B **43**, 13032 (1991).
1103. van der Beek C. J., Kes P. H., Maley M. R., et al., Phys. C **195**, 307 (1992).
1104. Vandewalle N., Cloots R., Ausloos M., J. Mater. Res. **10**, 268 (1995).
1105. Varanasi C., Black M. A., McGinn P. J., J. Mater. Res. **11**, 565 (1996).
1106. Varanasi C., McGinn P. J., Phys. C **207**, 79 (1993).
1107. Vargas J. L., Zhang N., Kaiser D. L., Babcock S. E., Phys. C **292**, 1 (1997).
1108. Varias A. G., Comput. Mech. **21,** 316 (1998).
1109. Varias A. G., Massih A. R., Eng. Fract. Mech. **65**, 29 (2000).
1110. Varias A. G., Massih A. R., J. Nucl. Mater. **279**, 273 (2000).
1111. Varias A. G., Massih A. R., J. Mech. Phys. Solids **50**, 1469 (2002).

1112. Varias A. G., O'Dowd N. P., Asaro R. J., Shih C. F., Mater. Sci. Eng. **A126**, 65 (1990).
1113. Vekinis G., Ashby M. F., Beaumont P. W., Acta Metall. Mater. **38**, 1151 (1990).
1114. Vermeer P. A. In: *Physics of Dry Granular Media* H. J. Herrmann, J. - P. Hovi, and S. Luding (eds.), p. 163 (Kluwer Academic Publishers, Dordrecht, 1998).
1115. Vilatla N., Sandiumenge F., Pinol S., Obrados X., J. Mater. Res. **12**, 38 (1997).
1116. Vinnikov L. Ya., Gurevich L. A., Emel'chenko G. A., Osip'yan Yu. A., Pisma Zh. Eksp. Teor. Fiz. **47,** 109 (1988).
1117. Viouchkov Y., Schwartz J., IEEE Trans. Appl. Supercond. **11**(1) (2001).
1118. Viouchkov Y., Weijers H. W., Hu Q. Y., et al., Adv. Cryog. Eng. (Mater.) **46**, 647 (2000).
1119. Vlasko-Vlasov V. K., Crabtree G. W., Welp U., Nikitenko V. I. In: *Physycs and Materials Science of Vortex States, Flux Pinning and Dynamics*, R. Kossowsky et al. (eds.) NATO ASI, Ser. E **356**, 205 (Kluwer, Dordrecht, 1999).
1120. Vlasko-Vlasov V. K., Goncharov V. N., Nikitenko V. I., et al., Phys. C **222**, 367 (1994).
1121. Volkov A. F., Phys. C **183**, 177 (1991).
1122. Vuchic B. V., Merkle K. L., Dean K. A., et al., Appl. Phys. Lett. **67**, 1013 (1995).
1123. Vul D. A., Salje E. K. H., Phys. C **253**, 231 (1995).
1124. Wallraff A., Schuster D. I., Blais A., et al., Nature **431**, 162 (2004).
1125. Wang Chongmin, Zhe Xiaoli, Zhang Hongtu, J. Mater. Sci. Lett. **7**, 621 (1988).
1126. Wang J., Monot I., Desgardin G., J. Mater. Res. **11**, 2703 (1996).
1127. Wang J., Monot I., Hervieu M., et al., Supercond. Sci. Technol. **9**, 69 (1996).
1128. Wang J. L., Cai X. Y., Kelley R. J., et al., Phys. C **230**, 189 (1994).
1129. Wang J. L., Tsu I. - F., Cai X. Y., et al., J. Mater. Res. **11**, 868 (1996).
1130. Wang R. K., Horvat J., Liu H. K., Dou S. X., Supercond. Sci. Technol. **9**, 875 (1996).
1131. Wang R. K., Liu H. K., Dou S. X., Supercond. Sci. Technol. **8**, 168 (1995).
1132. Wang W. G., Han Z., Skov-Hansen P., et al., IEEE Trans. Appl. Supercond. **9**(2), 2613 (1999).
1133. Wang W. G., Liu H. K., Guo Y. C., et al., Appl. Supercond. **3**, 599 (1995).
1134. Wang Z. L., Goyal A., Kroeger D. M., Phys. Rev. B **47**, 5373 (1993).
1135. Warnes W. H., Larbalestier D. C., Cryogenics **26**, 643 (1986).
1136. Watanabe N., Fukamachi K., Ueda Y., et al., Phys. C **235–240**, 1309 (1994).
1137. Weertman J., Acta Metall. **26**, 1731 (1978).
1138. Weertman J. J., Weertman J. R. *Elementary Dislocation Theory* (Oxford Univ. Press, New York, 1992).
1139. Weller T. E., Ellerby M., Saxena S. S., et al., Nature Phys. **1**, 39 (2005).
1140. Welp U., Grimsditch M., Fleshler S., et al., J. Supercond. **7**, 159 (1994).
1141. Welp U., Gunter D. O., Crabtree G. W., et al., Nature **376**, 44 (1995).
1142. Wheeler A. A., Boettinger W. J., Mc Fadden G. B., Phys. Rev. A **45**, 7424 (1992).
1143. Wijngaarden R. J., Griessen R. In: *Studies of High Temperature Superconductors*, A. V. Narlikar (ed.), **2**, 29.(Nova Science, New York, 1989).

1144. Williams G. V. M., Tallon J. L., Phys. Rev. B **57**, 10984 (1998).
1145. Williams R. K., Alexander K. B., Brynestad J., et al., J. Appl. Phys. **70**, 906 (1991).
1146. Witten T. A., Sander L. M., Phys. Rev. B **27**, 5686 (1983).
1147. Wong M. S., Miyase A., Yuan Y. S., Wang S. S., J. Am. Ceram. Soc. **77**, 2833 (1994).
1148. Wördenweber R., Rep. Prog. Phys. **62**, 187 (1999).
1149. Wu C. Cm., Freiman S. M., Rice R. W., Mecholsky J. J., J. Mater. Sci. **13**, 2659 (1978).
1150. Wu M. K., et al., Phys. Rev. Lett. **58**, 908 (1987).
1151. Wu X. D., Foltyn S. R., Arendt P., et al., Appl. Phys. Lett. **65**, 1961 (1994).
1152. Xu J. - A., Supercond. Sci. Technol. **7**, 1 (1994).
1153. Xue Y. Y., Lorenz B., Baikalov A., et al., cond-mat/0205487 (2002).
1154. Yamada K., et al., Phys. Rev. B **57**, 6165 (1998).
1155. Yamada Y., et al., Phys. C **217**, 182 (1993).
1156. Yamada Y., Hishinuma Y., Yamashita F., et al., IEEE Trans. Appl. Supercond. **7**(2), 1715 (1997).
1157. Yamada Y., Itoh K., Wada K., Tachikawa K., IEEE Trans. Appl. Supercond. 9(2), 1868 (1999).
1158. Yamada Y., Obst B., Flukiger R., Supercond. Sci. Technol. **4**,165 (1991).
1159. Yamada Y., Satou M., Murase S., et al., *Proc.* 5th *Int. Symp. on Superconductivity* (ISS′92), p. 717.(Kobe, Japan, 1992).
1160. Yamada Y., Yamashita F., Wada K., Tachikawa K., Adv. Cryog. Eng. (Mater.) **44b**, 547 (1998).
1161. Yamafuji K., Kiss T., Phys. C **258**, 197 (1996).
1162. Yamamoto A., Takao C., Masui T., et al., cond-mat/0208331 (2002).
1163. Yamanaka S., Enishi E., Fukuoka H., Yasukawa M., Inorg. Chem. **39**, 56 (2000).
1164. Yamashita T., Alarco J. A., Ilyushechkin A. J., et al., IEEE Trans. Appl. Supercond. **9**(2), 2581 (1999).
1165. Yan M. F., Cannon R. M., Bowen H. K. In: *Ceramic Microstructures* '76 R. M. Fulrath, and J. A. Pask (eds.), p. 276 (Westview, Boulder, CO, 1977).
1166. Yang C. S., Hui P. M., Phys. Rev. B **44**, 12559 (1991).
1167. Yanson I. K., Svistunov V. M., Dmitrenko I. M., Zh. Eksp. Teor. Fiz. **48**, 976 (1965) [Sov. Phys. JETP **21**, 650 (1965)].
1168. Yao X., Furuya K., Nakamura Y., et al., J. Mater. Res. **10**, 3003 (1995).
1169. Yao X., Shiohara Y., Supercond. Sci. Technol. **10**, 249 (1997).
1170. Ye J., Zou Z., Matsushita A., et al., Phys. Rev. B **58**, 619 (1998).
1171. Yeh F., White K. W., J. Appl. Phys. **70**, 4989 (1991).
1172. Yeshurin Y., Malozemoff A. P., Phys. Rev. Lett. **60**, 2202 (1988).
1173. Yeshurin Y., Malozemoff A. P., Shaulov A., Rev. Mod. Phys. **68**, 911 (1996).
1174. Yonezawa S., Muraoka Y., Hiroi Z., J. Phys. Soc. Jpn. **73**, 1655 (2004).
1175. Yonezawa S., Muraoka Y., Matsushita Y., Hiroi Z., J. Phys.: Condens. Matter **16**, L9 (2004).
1176. Yonezawa S., Muraoka Y., Matsushita Y., Hiroi Z., J. Phys. Soc. Jpn. **73**, 819 (2004).
1177. Yoo J., Chung H., Ko J., Kim H., IEEE Trans. Appl. Supercond.**7**(2), 1837 (1997).
1178. Yoo S. H., Wong V. W., Xin Y., Phys. C **273**, 189 (1997).

1179. Yoo S. I., Sakai N., Segawa K., et al., Adv. Supercond. **7**, 705 (1995).
1180. Yoshida M., Tsuzuk A., Hirabayashi I., *Proc Int. Workshop on Superconductivity*, p. 270 (Honolulu, HI, 1992).
1181. Yoshino H., Yamazaki M., Fuke H., et al., Adv. Supercond. **6**, 759 (1994).
1182. Young D. P., Adams P. W., Chan J. Y., Fronczek F. R., cond-mat/0104063 (2001).
1183. Yuan C. W., Zheng Z., de Lozanne A. L., et al., J. Vac.Sci. Technol. B **14**, 1210 (1996).
1184. Yuan Y. S., Wong M. S., Wang S. S., Phys. C **250**, 247 (1995).
1185. Yuan Y. S., Wong M. S., Wang S. S., J. Mater. Res. **11**, 8 (1996).
1186. Yuan Y. S., Wong M. S., Wang S. S., J. Mater. Res. **11**, 18 (1996).
1187. Yuan Y. S., Wong M. S., Wang S. S., J. Mater. Res. **11**, 1645 (1996).
1188. Zeimetz B., Rutter N. A., Glowacki B. A., Evetts J. E., Supercond. Sci. Technol. **14**, 672 (2001).
1189. Zeldov E., Amer N. M., Koren G., Gupta A., Appl. Phys. Lett. **56**, 1700 (1990).
1190. Zeldov E., Larkin A. I., Geshkenbein V. B., et al., Phys. Rev. Lett. **73**, 1428 (1994).
1191. Zeldov E., Majer D., Konczykowsky M., et al., Nature **375**,.373 (1995).
1192. Zhai J., Xing Z., Dong S., et al., Appl. Phys. Lett. **88**, 062510 (2006).
1193. Zhang P. X., Maeda H., Zhou L., et al., IEEE Trans. Appl. Supercond. **9**(2), 2770 (1999).
1194. Zhang P. X., Zhou L., Ji P., et al., Supercond. Sci. Technol. **8**, 15 (1995).
1195. Zhang W., Hellstrom E. E., Phys. C **234**, 137 (1994).
1196. Zhang W., Polak M., Polyanskii A., et al., Adv. Cryog. Eng. (Mater.) **44b**, 509 (1998).
1197. Zhao G. - M., Wang Y. S., cond-mat/0111268 (2001).
1198. Zhengping Xi, Zhou Lian, Supercond. Sci. Technol. **7**, 908 (1994).
1199. Zhou L., Zhang P., Ji P., et al., Supercond. Sci. Technol. **3**, 490 (1990).
1200. Zhu W., Nicholson P. S., J. Mater. Res. **7**, 38 (1992).
1201. Zhu Y., Corcoran M., Suenaga M., Interface Sci. **1**, 361 (1993).
1202. Zhu Y., Wang Z. L, Suenaga M., Philos. Mag. A **67**, 11 (1993).
1203. Zhu Y., Zuo J. M., Moodenbaugh A. R., Suenaga M., Philos. Mag. A **70**, 969 (1994).
1204. Zhukov A. A., Küpfer H., Ryabchuk V. A., et al., Phys. C **219**, 99 (1994).
1205. Ziese M., Phys. Rev. B **55**, 8106 (1996).
1206. Ziman J. *Models of Disorder* (Cambridge University Press, Cambridge, 1979).
1207. Zou Z., Oka K., Ito T., Nishihara Y., Jpn. J. Appl. Phys. **36**, L18 (1997).
1208. Zou Z., Ye J., Oka K., Nishihara Y., Phys. Rev. Lett. **80**, 1074 (1998).

Index

Periodic Table of the Chemical Elements

IA	IIA	IIB	IVB	VB	VIB	VIIB	VIII	VIII	VIII	IB	IIB	IIIA	IVA	VA	VIA	VIIA	VIIIA
1 - HEX $a = 3.750$ $c = 6.120$ $1s^1$ - H																	2 - BCC $a = 4.110$ $1s^2$ - He
3 20* HEX $a = 3.111$ $c = 5.039$ $2s^1$ - Li	4 0.026 HEX $a = 2.286$ $c = 3.584$ $2s^2$ - Be											5 - TET $a = 8.800$ $c = 5.050$ $2s^2p^1$ - B	6 - DIA $a = 3.567$ $2s^2p^2$ - C	7 - BCC $a = 5.644$ $2s^2p^3$ - N	8 0.6* MCL $a = 5.403$ $b = 3.429$ $c = 5.086$ $\beta = 132.53°$ $2s^2p^4$ - O	9 - MCL $a = 5.500$ $b = 3.280$ $c = 10.010$ $\beta = 134.66°$ $2s^2p^5$ - F	10 - FCC $a = 4.462$ $2s^2p^6$ - Ne
11 - BCC $a = 4.282$ $3s^1$ - Na	12 - HEX $a = 3.202$ $c = 5.299$ $3s^2$ - Mg											13 1.176 FCC $a = 4.050$ $3s^2p^1$ 105 Al	14 - DIA $a = 5.430$ $3s^2p^2$ - Si	15 - CUB $a = 7.180$ $3s^2p^3$ - P	16 17.3* ORC $a = 10.465$ $b = 12.866$ $c = 24.486$ $3s^2p^4$ - S	17 - ORC $a = 4.500$ $b = 6.290$ $c = 8.210$ $3s^2p^5$ - Cl	18 - FCC $a = 5.300$ $3s^2p^6$ - Ar
19 - BCC $a = 5.247$ $4s^1$ - K	20 - FCC $a = 5.560$ $4s^2$ - Ca	21 - HEX $a = 3.308$ $c = 5.265$ $3d^14s^2$ - Sc	22 0.4 HEX $a = 2.951$ $c = 4.679$ $3d^24s^2$ 60 Ti	23 5.35 BCC $a = 3.028$ $3d^34s^2$ 1400 V	24 - BCC $a = 2.884$ $3d^54s^1$ - Cr	25 - CUB $a = 8.912$ $3d^54s^2$ - Mn	26 2* BCC $a = 2.866$ $3d^64s^2$ - Fe	27 - HEX $a = 2.510$ $c = 4.090$ $3d^74s^2$ - Co	28 - FCC $a = 3.524$ $3d^84s^2$ - Ni	29 - FCC $a = 3.615$ $3d^{10}4s^1$ - Cu	30 0.9 HEX $a = 2.664$ $c = 4.946$ $3d^{10}4s^2$ 55 Zn	31 1.083 ORC $a = 4.526$ $b = 4.519$ $c = 7.657$ $4s^2p^1$ 59.3 Ga	32 - DIA $a = 5.658$ $4s^2p^2$ - Ge	33 - ROM $a = 4.129$ $\alpha = 54.1°$ $4s^2p^3$ - As	34 8.0* HEX $a = 4.363$ $c = 4.959$ $4s^2p^4$ - Se	35 - ORC $a = 4.480$ $b = 6.670$ $c = 8.720$ $4s^2p^5$ - Br	36 - FCC $a = 5.706$ $4s^2p^6$ - Kr
37 - BCC $a = 5.700$ $5s^1$ - Rb	38 - FCC $a = 6.085$ $5s^2$ - Sr	39 - HEX $a = 3.647$ $c = 5.731$ $4d^15s^2$ - Y	40 0.5 HEX $a = 3.223$ $c = 5.147$ $4d^25s^2$ 54 Zr	41 9.25 BCC $a = 3.300$ $4d^45s^1$ 1970 Nb	42 0.92 BCC $a = 3.147$ $4d^55s^1$ 90 Mo	43 7.8 HEX $a = 2.735$ $c = 4.391$ $4d^65s^1$ 1410 Tc	44 0.5 HEX $a = 2.706$ $c = 4.281$ $4d^75s^1$ 69 Ru	45 0.0003 FCC $a = 3.790$ $4d^85s^1$ - Rh	46 - FCC $a = 3.882$ $4d^{10}$ - Pd	47 - FCC $a = 4.086$ $4d^{10}5s^1$ - Ag	48 0.56 HEX $a = 2.960$ $c = 5.630$ $4d^{10}5s^2$ 29.6 Cd	49 3.4 TET $a = 4.583$ $c = 4.936$ $5s^2p^1$ 281.5 In	50 3.75 DIA $a = 6.504$ $5s^2p^2$ 305 Sn	51 - ROM $a = 6.085$ $\alpha = 57.1°$ $5s^2p^3$ - Sb	52 - HEX $a = 4.457$ $c = 5.929$ $5s^2p^4$ - Te	53 - ORC $a = 4.774$ $b = 7.250$ $c = 9.772$ $5s^2p^5$ - I	54 - FCC $a = 6.250$ $5s^2p^6$ - Xe
55 - BCC $a = 6.141$ $6s^1$ - Cs	56 - BCC $a = 5.019$ $6s^2$ - Ba	57 4.88 HEX $a = 3.770$ $c = 12.159$ $5d^16s^2$ 808 La	72 0.13 HEX $a = 3.195$ $c = 5.051$ $5d^26s^2$ 12.1 Hf	73 4.4 BCC $a = 3.307$ $5d^36s^2$ 831 Ta	74 0.015 BCC $a = 3.165$ $5d^46s^2$ 1.15 W	75 1.7 HEX $a = 2.757$ $c = 4.463$ $5d^56s^2$ 200 Re	76 0.65 HEX $a = 2.750$ $c = 4.319$ $5d^66s^2$ 70 Os	77 0.14 FCC $a = 3.831$ $5d^76s^2$ 19 Ir	78 0.05 FCC $a = 3.200$ $5d^96s^1$ 0.07 Pt	79 - FCC $a = 4.070$ $5d^{10}6s^1$ - Au	80 4.16 $\alpha = 57.28°$ $a = 3.463$ $c = 6.740$ $5d^{10}6s^2$ 411 Hg	81 2.4 HEX $a = 3.450$ $c = 5.514$ $6s^2p^1$ 181 Tl	82 7.23 FCC $a = 4.950$ $6s^2p^2$ 803 Pb	83 - ROM $a = 3.475$ TET $6s^2p^3$ - Bi	84 - CUB $a = 3.359$ $6s^2p^4$ - Po	85 - $6s^2p^5$ - At	86 - FCC $6s^2p^6$ - Rn
87 - BCC $7s^1$ - Fr	88 - BCC $7s^2$ - Ra	89 - FCC $a = 5.311$ $6d^17s^2$ - Ac	104 - - $6d^27s^2$ - Db														

La are elements used in superconducting compounds

P are elements not used in superconducting compounds

Key (for element 57):

ATOMIC NUMBER: 57	CRITICAL TEMPERATURE (K)*: 4.88
STRUCTURE	HEX
LATTICE PARAMETERS (Å)	$a = 3.770$, $c = 12.159$
ATOMIC CONFIGURATION	$5d^16s^2$
CRITICAL MAGNETIC FIELD (Oe): 808	SYMBOL: La

* signs "under pressure"

Lanthanides	57 4.88 HEX $a = 3.770$ $c = 12.159$ $5d^16s^2$ 808 La	58 - FCC $a = 4.85$ $4f^15d^16s^2$ - Ce	59 - HEX $a = 3.664$ $c = 11.867$ $4f^36s^2$ - Pr	60 - HEX $a = 3.658$ $c = 11.799$ $4f^46s^2$ - Nd	61 - HEX $a = 3.650$ $c = 11.650$ $4f^56s^2$ - Pm	62 - ROM $a = 3.626$ $\alpha = 26.18°$ $4f^66s^2$ - Sm	63 - BCC $a = 4.572$ $4f^76s^2$ - Eu	64 - HEX $a = 3.636$ $c = 5.783$ $4f^75d^16s^2$ - Gd	65 - HEX $a = 3.604$ $c = 5.698$ $4f^96s^2$ - Tb	66 - HEX. $a = 3.592$ $c = 5.655$ $4f^{10}6s^2$ - Dy	67 - HEX $a = 3.577$ $c = 5.619$ $4f^{11}6s^2$ - Ho	68 - HEX $a = 3.560$ $c = 5.595$ $4f^{12}6s^2$ - Er	69 - HEX $a = 3.537$ $c = 5.558$ $4f^{12}5d^16s^2$ - Tm	70 - FCC $a = 5.483$ $4f^{14}6s^2$ - Yb	71 0.1 HEX $a = 3.505$ $c = 5.553$ $4f^{14}5d^16s^2$ 400 Lu
Actinides	89 - FCC $a = 5.311$ $6d^17s^2$ - Ac	90 1.37 FCC $a = 5.086$ $6d^27s^2$ - Th	91 1.3 TET $a = 3.631$ $c = 3.236$ $5f^26d^17s^2$ - Pa	92 1.1 ORC $a = 2.844$ $b = 5.869$ $c = 4.960$ $5f^36d^17s^2$ - U	93 0.075 ORC $a = 4.721$ $b = 4.888$ $c = 3.670$ $5f^46d^17s^2$ - Np	94 - MCL $a = 6.183$ $b = 4.822$ $c = 10.963$ $\beta = 101.79°$ $5f^67s^2$ - Pu	95 - HEX $a = 3.462$ $c = 11.223$ $5f^77s^2$ - Am	96 - HEX $a = 3.496$ $c = 11.331$ $5f^76d^17s^2$ - Cm	97 - $5f^86d^17s^2$ - Bk	98 - $5f^{10}7s^2$ - Cf	99 - $5f^{11}7s^2$ - Es	100 - $5f^{12}7s^2$ - Fm	101 - $5f^{13}7s^2$ - Md	102 - $5f^{14}7s^2$ - No	103 - $5f^{14}6d^17s^2$ - Lr